Meteorology

For Scientists and Engineers

Second Edition

A Technical Companion Book with Ahrens' *Meteorology Today*

ROLAND STULL
The University of British Columbia

Australia • Brazil • Japan • Korea • Mexico • Singapore • Spain • United Kingdom • United States

Meteorology For Scientists and Engineers, Second Edition
Roland Stull

Publisher: Gary Garlson
Sponsoring Editor: Nina Horne
Marketing Representative: Rachel Alvelais
Marketing Assistant: Kelly Fielding
Editorial Assistant: John-Paul Ramin
Cover Design: Christine Garrigan
Print Buyer: Micky Lawler
Production Editor: Scott Brearton

ISBN-13: 978-0-534-37214-9

ISBN-10: 0-534-37214-7

Brooks/Cole
10 Davis Drive
Belmont, CA 94002-3098
USA

Cengage Learning is a leading provider of customized learning solutions with office locations around the globe, including Singapore, the United Kingdom, Australia, Mexico, Brazil, and Japan. Locate your local office at: **www.cengage.com/international**

Cengage Learning products are represented in Canada by Nelson Education, Ltd.

To learn more about Brooks/Cole, visit **www.cengage.com/brookscole**

Purchase any of our products at your local college store or at our preferred online store **www.ichapters.com**

Printed in the United States of America
9 10 11 12 13 13 12 11 10 09

CONTENTS

CONTENTS SUMMARY

CHAPTER 1 • The Atmosphere 1

CHAPTER 2 • Radiation 23

CHAPTER 3 • Heat 43

CHAPTER 4 • Boundary Layers 65

CHAPTER 5 • Moisture 95

CHAPTER 6 • Stability 119

CHAPTER 7 • Cloud Formation 143

CHAPTER 8 • Precipitation 159

CHAPTER 9 • Dynamics 179

CHAPTER 10 • Local Winds 205

CHAPTER 11 • Global Circulation 223

CHAPTER 12 • Air Masses & Fronts 251

CHAPTER 13 • Cyclones 281

CHAPTER 14 • Numerical Weather Prediction 311

CHAPTER 15 • Thunderstorms 337

CHAPTER 16 • Hurricanes 357

CHAPTER 17 • Air Pollution Dispersion 379

CHAPTER 18 • Climate Change 399

CHAPTER 19 • Optics 413

APPENDIX A • Fundamentals of Science 435

APPENDIX B • Constants and Conversion Factors 445

APPENDIX C • Notation 447

APPENDIX D • Additional Reading Material 459

APPENDIX E • Answers to Selected Exercises 463

APPENDIX F • Syllabus 467

APPENDIX G • Chapter Correspondences 469

APPENDIX H • Turbulence Closure 471

INDEX 481

PREFACE

PREFACE TO THE SECOND EDITION

This textbook provides undergraduate students with a physical and quantitative background in general meteorology. It is designed for science and engineering majors in their first or second year in university, and does not require knowledge of calculus or differential equations. I did this by presenting differential equations in finite-difference form, which requires only algebra and trigonometry.

Meteorology for Scientists and Engineers uses equations to focus on the physics of the atmosphere. Important descriptive concepts of meteorology are intentionally excluded to keep this book concise, and because they are well covered in other textbooks such as *Meteorology Today* by C. Donald Ahrens. These two books are designed to be complementary; an introductory course in general meteorology for scientists and engineers should use both.

I reorganized the chapters a bit from the first edition, because of the didactic need to present equations in a logical order. Chapter 2 covers radiation, and Chapter 3 covers heat. The optics chapter is moved to the end of the book (Chapter 19). Chapter 4 covers atmospheric boundary layers, which collects some of the material previously scattered in other chapters, and adds much new material. Details of turbulence closure are in Appendix H. Chapter 5 on moisture remains in place, but chapters 6 and 7 are switched. Now, Chapter 7 on clouds immediately precedes Chapter 8 on precipitation. The other chapters have not been moved, but most have been enhanced. Appendix G lists the correspondences between chapters in this book, and those in Ahrens' *Meteorology Today*, and *Essentials of Meteorology*.

Many new features are added. Each chapter summary now includes a "Threads" subsection, to help tie all the chapters together. Also, at the end of each chapter are new exercises on "Understanding & Critical Evaluation", "Web-Enhanced Questions", and "Synthesis Questions", in addition to the "Numerical Problems" from before. Motivations and methods for these new exercise subsections are given at the end of Chapter 1. Answers to selected Numerical Problems are in Appendix E as a study aid for students.

Within all chapters are special feature boxes. "Focus" boxes on current issues or historical anecdotes appear as before. In addition, I've added "On Doing Science" boxes, which present some of the broader philosophical issues of science in addition to the problem-solving tools of the trade. A table listing these "On Doing Science" boxes is at the end of Appendix A (Fundamentals of Science). I also added "Science Graffito" boxes here and there with insightful quotes to provide interludes for the reader.

Although calculus is not required for this book, numerous instructors requested some optional calculus, to interest those students with sufficient background. New "Beyond Algebra" boxes have been added where appropriate. These are not meant to be inclusive, but give only a taste of how atmospheric physics are described using differential equations. These boxes may be skipped with no loss in understanding of the other chapter material.

There are roughly 100 more pages and a larger page format than before. The book has more material than can be comfortably covered in one term. Instructors should be selective in their reading and exercise assignments. Key equations are now identified with bullets (•).

New topics include heat index, tornado scales, hurricane scale, Beaufort scale, saturation as an equilibrium condition, boiling temperature, isentropic cross sections, heat and moisture fluxes in convective conditions, cyclostrophic wind, Coriolis-force explanation, more examples of real weather maps, Bernoulli equation, bora winds, thermal wind, omega equation, thickness maps, jet-stream cross sections, advection analysis, parameterization and modeling methods, lightning detection, origin of tornadic rotation, cold vs. warm core cyclones, and hurricane power.

Thanks to Dave Whiteman, Zbigniew Sorbjan, Phil Austin, Josh Hacker, Bob Bornstein, Susan Allick and Anders Persson for corrections and additions since the first edition. Scott Shipley, David Finley, George Taylor, two anonymous referees, and publisher Nina Horne are gratefully acknowledged for their suggestions. A large effort in proofreading, problem solving, and indexing was undertaken by Larry Berg, Scott Krayenhoff, Phoebe Jackson, Maria Furberg, and Stephanie Meyn. Thanks again to my wife Linda, who was understanding and supportive while I wrote this new edition.

Roland Stull, 2000

WEB-BASED LABORATORY

A web-based lab to supplement this book will be available at the Brooks/Cole/ITP web site (try http:// www.brookscole.com/geo). The labs are not designed to duplicate or reinforce material from the book. Instead, they cover additional topics such as METAR weather observation codes, satellite and radar image interpretation, operational numerical weather forecast models, thunderstorm and tornado behavior, hurricane characteristics, and El Niño.

These web labs give questions, and provide links to other sites around the world for the content needed to answer the questions. Beware that web sites are short lived, so some of the links might fail.

PREFACE TO THE FIRST EDITION

[The first edition was entitled *Meteorology Today for Scientists and Engineers*, and was published in 1995 by West Publ. Co.]

About This Book

A wide mix of students often enroll in college-level meteorology survey courses. Most students find descriptive textbooks, such as *Meteorology Today* by Ahrens, to be quite appealing, because the books are thorough, up-to-date, accurate, stimulating, and attractive.

However, some students have the desire and math ability to go beyond descriptive aspects. It is for this group of students that I wrote *Meteorology Today for Scientists and Engineers*.

In a nutshell, my book has the equations that are omitted from *Meteorology Today*. It is **not** a lab manual nor a workbook. There is no calculus in my book — only algebra, geometry and trigonometry. Finite difference equations are substituted in place of differential equations. As a result, and also by design, there are virtually no derivations. Instead, my focus is to quantify the concepts that are already well described in Ahrens.

This book is intended to be a concise **companion** to Ahrens' *Meteorology Today*. Each of my chapters corresponds to his. (If you use Ahrens' *Essentials of Meteorology*, see Appendix G.) For students learning meteorology for the first time, it is important to use my book with Ahrens' book, because I have not duplicated his explanations. His book provides the glue that binds together and places into context the quantitative topics that I cover.

The contents of my chapters do not perfectly match Ahrens', because of the didactic need to present equations in a logical order. For example, I could not wait until chapter 9 for the ideal gas law – you will find it in my chapter 1. Also, I intentionally did not write a Chapter 19, because the quantitative aspects of climate were covered in previous chapters.

To demonstrate how to use the equations, I include numerous solved examples. Homework exercises are presented at the end of each chapter, with answers to selected questions in Appendix E. Tables are given with constants needed for the various equations. Appendix D lists more-advanced secondary references for further reading.

Instructors and other experts in meteorology will notice a number of heuristic models in my chapters that have not previously appeared in the refereed meteorological literature. These are analytical models that I have fit to atmospheric observations, or which are simplifications to more complex theories. I contrived these models for pedagogical reasons, to demonstrate with simple mathematics the dominant characteristics of certain atmospheric phenomena and processes.

My book takes advantage of spreadsheet programs, such as are available for most personal computers. By solving equations on spreadsheets (both in my examples and in the homework exercises), whole curves can be plotted. This often gives more insight than the few numbers that could be laboriously obtained with a hand calculator. Exercises requiring the use of spreadsheets are flagged with the symbol (§).

Every book has a unique personality inherited from its author. Readers will discover enhanced sections on boundary layers and turbulence, reflecting my personal bias. I hope readers will find these sections useful, because we all live in the boundary layer and our agriculture, commerce, and transportation are affected by it.

Meteorology Today for Scientists and Engineers might be appealing to a broader audience of students than those in survey courses. It could serve as an introductory textbook for students majoring in meteorology, atmospheric science, air-quality, environmental science, geography, climate, and earth-system science.

These students could use this book early in their university studies, before completing calculus. By then proceeding to more advanced courses earlier, they could graduate in four years with an enhanced meteorology background.

Those students already knowledgeable in meteorology might find the book useful as a reference or handbook. It might also serve as a supplement in engineering and physics courses.

Students are the future architects of our society. They will decide the balance between the our environment and the quality of life. These decisions will be implemented both by the legislation they endorse and by the products they buy. The aim of this book is to enhance their understanding of the atmospheric environment so they can make sound decisions.

Highlights

I've tried to include in most chapters some unique or provocative sections. Examples are:

- box-counting techniques to determine the fractal geometry of clouds (Chapter 6)
- astronomical formulae for sunrise and sunset (Chapter 3)
- demonstrations of nonunique ray angles in rainbows and sundogs (Chapter 4)
- gentle spreadsheet derivations (Chapters 2 & 5) of the individual isopleths in a thermodynamic diagram, leading to a full spreadsheet thermo diagram (Chapter 7)
- a series of "Focus" boxes building the history of scientific thought during the scientific revolution, including some failures by famous scientists (scientists are human)
- modern concepts of nonlocal stability and turbulence (Chapters 2 & 7).
- break from historical meteorological conventions toward modern scientific standards, including kiloPascals for pressure in Chapter 1 (instead of hectoPascals to mimic millibars), kilometers for geopotential height instead of decameters (dam), and liquid-water potential temperature instead of equivalent potential temperature as labels for the moist adiabats (Chapters 5 & 7).
- alternative or new formulae: for wind chill (Chapter 3), and raindrop terminal velocity (Chapter 8).
- unique ways to approach certain subjects (such as budgets of heat and momentum).
- analytical models for hurricane state (Chapter 16), deterministic pollutant dispersion in convective mixed layers (Chapter 17), and thunderstorm gust fronts (Chapter 15).
- chaos and predictability, ensemble weather forecasts, and the Lorenz strange attractor (Chapt. 14)
- Daisyworld (Chapter 18)

For Students

Appendix A is a review of science fundamentals, to help those of you with weaker technical backgrounds. An important "Focus" on Problem Solving is included in Appendix A, which is used as a model for the solved examples throughout the book. Appendix B contains useful constants and conversion factors, and Appendix C lists the notation. Appendices D and E give additional reading material, and answers to selected exercises.

For Instructors

There is too much material combined in Ahrens' and my books to teach during a one-semester survey course. Instead, you can assign bits and pieces of my book as appropriate to supplement material covered in lecture or in other readings. See Appendix G if you use Ahrens' *Essentials of Meteorology*.

Those of you wanting to routinely integrate more equations into your freshman-level survey course might find Appendix F interesting. In it is a syllabus that I developed during the past several years of teaching survey courses. This approach respects those students who hate math, and gives all the students options to allow them to gain the type of weather knowledge they seek.

Finally, before assigning exercises as homework or using them on tests, be aware that answers to some of the exercises appear in Appendix E.

Acknowledgements

I wrote [the first edition of] this book while working at the University of Wisconsin in Madison, and I am grateful for the support of the faculty, staff, and students there. Ed Hopkins, Paul Menzel, Pao Wang, Greg Tripoli, Jon Foley, Dave Houghton, Kit Hayden, and Jon Martin from the Department of Atmospheric & Oceanic Sciences contributed ideas, data, figures, and proof-read portions of the book.

John Cassano and Larry Berg checked the solved examples and computed answers for the homework exercises. Cassano also compiled the notation list in Appendix C. He also produced the first-draft computer drawings of synoptic charts from original analyses by Jon Martin.

Figures and data from the European Centre for Medium-Range Weather Forecasts (ECMWF) are reproduced with permission of the director. Anders Persson at ECMWF is gratefully acknowledged for his effort producing weather maps, for his suggestions for the synoptics chapters, and for providing a detailed and very interesting history of numerical weather prediction.

Horst Böttger at ECMWF and Jim Hoke at the National Meteorological Center of the US National Weather Service provided information about forecast skill. Additional suggestions for improvement came from Judy Curry, Richard Peterson, Mankin Mak, C.

Dale Elifrits, Robert Sica, Jon Kahl, and Ian Lumb. To all those too numerous to list who influenced this book, I thank you. Any errors that remain in the book are my own responsibility.

The StormForce Company provided the computer hardware and software with which this book was written. The books *On Writing Well* and *Writing to Learn* by William Zinsser motivated my attempts to write clearly, simply, and concisely.

Finally, I am grateful to my wife Linda, who was understanding and supportive while I wrote this book.

Roland Stull [1995]

THE ATMOSPHERE

CONTENTS

1 **Meteorology** is classical Newtonian physics applied to the atmosphere. Motions obey Newton's second law. Heat satisfies the laws of thermodynamics. Air mass and moisture are conserved. When applied to a fluid such as air, these physical processes describe fluid mechanics. Meteorology is fluid mechanics applied to the atmosphere.

The **atmosphere** is a complex fluid system — a system that generates the chaotic motions we call weather. This complexity is caused by a myriad of interactions between many physical processes acting at different locations. For example, temperature differences create pressure differences that drive winds. Winds move moisture about. Water vapor condenses and releases heat, altering the temperature differences. Such feedbacks are nonlinear, and contribute to the complexity.

But the result of this chaos and complexity is a fascinating array of **weather** phenomena — phenomena that are as inspiring in their beauty and power as they are a challenge to describe. Thunderstorms, hurricanes, tornadoes, cyclones, turbulence, fronts, snow flakes, jet streams, rainbows. Such phenomena touch our lives by affecting how we dress, how we travel, what we can grow, where we live, and sometimes even how we feel.

In spite of the complexity, much is known about atmospheric behavior. Only a small fraction of this understanding has come from solutions of the full governing equations. Most of the equations are either too complex or imprecise to be solved, so empirical relationships are used instead. This book presents some of what we know about the atmosphere for use by scientists and engineers.

INTRODUCTION

In this book are six major components of meteorology: thermodynamics, physical meteorology, dynamics, synoptics and mesoscale forecasting, applied meteorology, and climate change. The thermodynamics of heat, moisture, and radiation are in Chapters 1 to 3, 5, and 6. Physical meteorology including cloud physics, turbulence, and optics is in Chapters 4, 7, 8, and 19. Dynamics of winds and the general circulation are in Chapters 9 to 11. Weather forecasting of synoptic and mesoscale storms is in Chapters 12 to 16. Applied meteorology for air pollution is in Chapt 17, and climate change in 18.

Starting into the thermodynamics topic now, the state of the air in the atmosphere is measured by its pressure, density, and temperature. Changes of state associated with weather and climate are just small perturbations compared to the average (standard) atmosphere. These changes are caused by well-defined processes.

Equations and concepts in meteorology are similar to those in physics or engineering, although the jargon and conventions might look different. For a review of basic science, see Appendix A.

METEOROLOGICAL CONVENTIONS

Although the earth is approximately spherical, we need not always use spherical coordinates. For the weather at a point or in a small region such as a town, state, or province, we can use local right-hand **Cartesian** (rectangular) coordinates, as sketched in Fig. 1.1. Usually, this coordinate system is aligned with x pointing east, y pointing north, and z pointing up. Other orientations are sometimes used.

Velocity components U, V, and W correspond to motion in the x, y, and z directions. For example, a positive value of U is a velocity component from west to east, while negative is from east to west. Similarly, V is positive northward, and W is positive upward (Fig. 1.1).

In polar coordinates, horizontal velocities can be expressed as a direction (α), and speed or magnitude (M). Historically, horizontal wind directions are based on the compass, with 0° to the north (the positive y direction), and with degrees increasing in a **clockwise** direction through 360°. Negative angles are not usually used. Unfortunately, this differs from the usual mathematical convention of 0° in the x direction, increasing **counter**-clockwise through 360° (Fig 1.2).

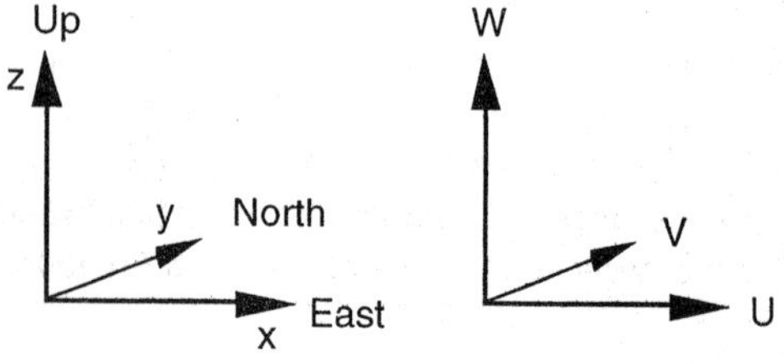

Figure 1.1
Local Cartesian coordinates and velocity components.

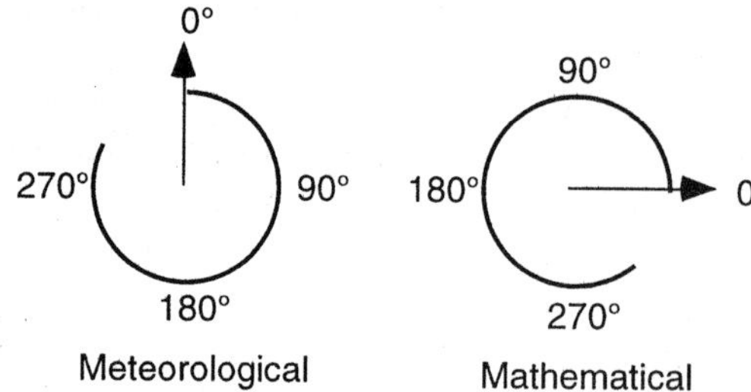

Figure 1.2
Comparison of meteorological and mathematical angle conventions.

Also, historically winds are named by the direction they come **from**, while in mathematics angles give the direction toward which things move. Thus, a **west** wind is a wind from the west; namely, from 270°. It corresponds to a positive value of U, with air moving in the positive x direction.

Because of these differences, the usual trigonometric equations cannot be used to convert between (U, V) and (α, M). Use the following equations instead, where α is the compass direction from which winds come.

Conversion to Speed and Direction:

$$M = \left(U^2 + V^2\right)^{1/2} \qquad \bullet(1.1)$$

$$\alpha = 90° - \frac{360°}{C} \cdot \arctan\left(\frac{V}{U}\right) + \alpha_o \qquad \bullet(1.2a)$$

where α_o = 180° if $U > 0$, but is zero otherwise. C is the angular rotation in a full circle (C = 360° = 2·π radians). Some computer languages and spreadsheets allow a two-argument arctan function:

$$\alpha = \frac{360°}{C} \cdot \text{atan2}(V, U) + 180° \qquad (1.2b)$$

Caution, in the C and C++ languages, switch U & V.

Some calculators, spreadsheets or computer functions use angles in degrees, while others use radians. If you don't know which units are used, compute the arccos(–1) as a test. If the answer is 180, then your units are degrees; otherwise, an answer of 3.14159 indicates radians. Use whichever value of C is appropriate for your units.

Conversion to U and V:

$$U = -M \cdot \sin(\alpha) \qquad \bullet(1.3)$$

$$V = -M \cdot \cos(\alpha) \qquad \bullet(1.4)$$

[NOTE: Bullets • identify key equations that are either fundamental, or are important for understanding later chapters.]

In three dimensions, **cylindrical** coordinates (M, α, W) are sometimes used for velocity instead of Cartesian (U, V, W), where horizontal velocity components are specified by direction and speed, and the vertical component remains as before (see Fig. 1.3).

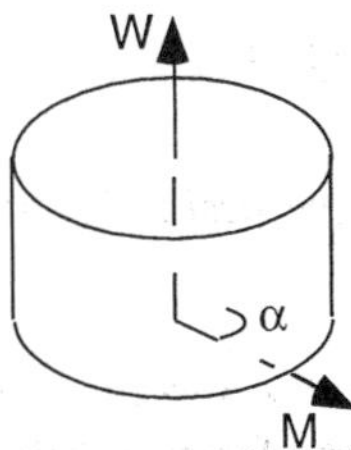

Figure 1.3
Notation used in cylindrical coordinates for velocity.

Solved Example
Find wind speed and direction, given eastward component 3 m/s, and northward 4 m/s.

Solution
(Problem-solving methods are reviewed in Appendix A.)

Given: U = 3 m/s. eastward wind comp.
V = 4 m/s. northward wind comp.
Find: M = ? m/s. wind speed
α = ? degrees. wind direction

Sketch:

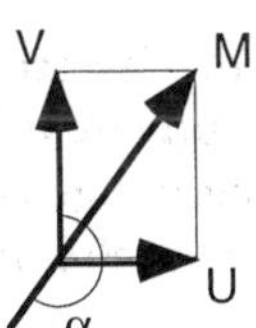

Use eq. (1.1):

$$M = [\, U^2 + V^2 \,]^{1/2}$$
$$= [(3\text{ m/s})^2 + (4\text{ m/s})^2]^{0.5}$$
$$= (9 + 16)^{0.5} \cdot [(\text{m/s})^2]^{0.5}$$
$$= (25)^{0.5}\text{ m/s} = \underline{5\text{ m/s}}.$$

Use eq. (1.2a):

$$\alpha = 90° - (360°/C)\cdot\tan^{-1}(V/U) + 180°$$
$$= 90° - (360/360)\cdot\tan^{-1}[(4\text{ m/s})/(3\text{ m/s})] + 180°$$
$$= 90° - \tan^{-1}(1.333) + 180°$$
$$= 90° - 53.13° + 180° = \underline{216.87°}.$$

Check: Units OK. Sketch OK. Values physical.
Discussion: Thus, the wind is from the south-southwest (SSW) at 5 m/s.

Most meteorological graphs are like graphs in other sciences, with **dependent** variables on the **ordinate** (vertical axis) plotted against an **independent** variable on the **abscissa** (horizontal axis). However, in meteorology the axes are often switched when height (z) is the independent variable. This axis switching makes locations higher in the graph correspond to locations higher in the atmosphere (Fig 1.4).

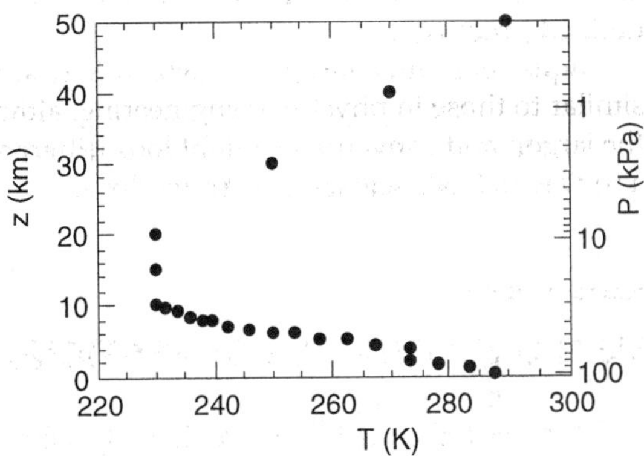

Figure 1.4
Hypothetical temperature T profile in the atmosphere, plotted such that locations higher in the graph correspond to locations higher in the atmosphere. The independent variable can be height z (left axis) or pressure P (right axis).

ON DOING SCIENCE • Descartes and the Scientific Method

From René Descartes we get more than the name "Cartesian". In 1637 he published a book *Discours de la Méthode*, in which he defined the principles of the modern scientific method:

- Accept something as true only if you know it to be true.
- Break difficult problems into small parts, and solve each part in order to solve the whole problem.
- Start from the simple, and work towards the complex. Seek relationships between the variables.
- Do not allow personal biases or judgements to interfere, and be thorough.

This method formed the basis of the scientific renaissance, and marked an important break away from blind belief in philosophers such as Aristotle.

THERMODYNAMIC STATE

The thermodynamic state of air is measured by its pressure, density, and temperature.

Pressure

Pressure P is the force F per unit area A acting perpendicular (normal) to a surface:

$$P = F / A \qquad \bullet(1.5)$$

Solved Example

The picture tube of a TV and the CRT display of a computer are types of vacuum tube. If there is a perfect vacuum inside the tube, what is the net force pushing against the front surface of a big screen 24 inch (61 cm) display that is at sea level?

Solution

Given: Picture tube sizes are quantified by the diagonal length d of the front display surface. Assume the picture tube is square. The length of the side s of the tube is found from: $d^2 = 2\,s^2$. The frontal surface area is

$A = s^2 = 0.5 \cdot d^2 = 0.5 \cdot (61 \text{ cm})^2 = 1860.5 \text{ cm}^2$
$= (1860.5 \text{ cm}^2)\cdot(1 \text{ m}/100 \text{ cm})^2 = 0.186 \text{ m}^2$.

At sea level, atmospheric pressure pushing against the outside of the tube is 101.325 kPa, while from the inside of the tube there is no force pushing back because of the vacuum. Thus, the pressure difference across the tube face is $\Delta P = 101.325 \text{ kPa} = 101.325\text{x}10^3 \text{ N/m}^2$.

Find: ΔF = ? N, the net force across the tube.

Sketch:

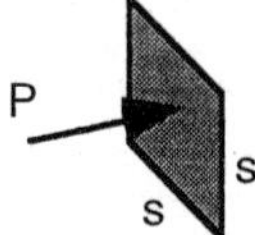

$\Delta F = F_{outside} - F_{inside}$, but $F = P \cdot A$. from eq.(1.5)
$= (P_{outside} - P_{inside})\cdot A$
$= \Delta P \cdot A$
$= (101.325\text{x}10^3 \text{ N/m}^2)\cdot(0.186 \text{ m}^2)$
$= 1.885\text{x}10^4 \text{ N} =$ **18.85 kN**

Check: Units OK. Physically reasonable.
Discussion: This is quite a large force, and explains why picture tubes are made of such thick heavy glass. For comparison, a person who weighs 68 kg (150 pounds) is pulled by gravity with a force of about 667 N (= 0.667 kN). Thus, the picture tube must be able to support the equivalent of 28 people standing on it!

a)

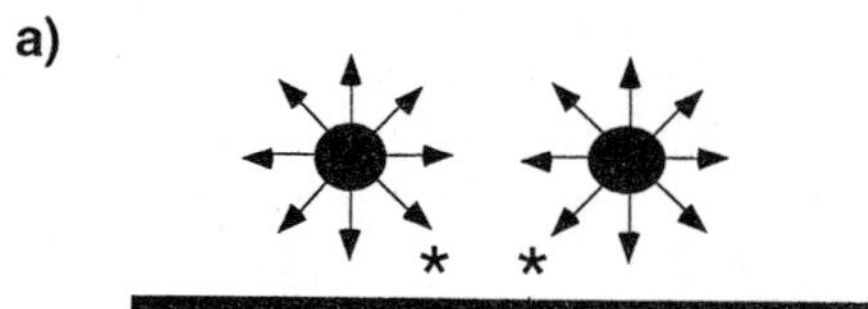

b)

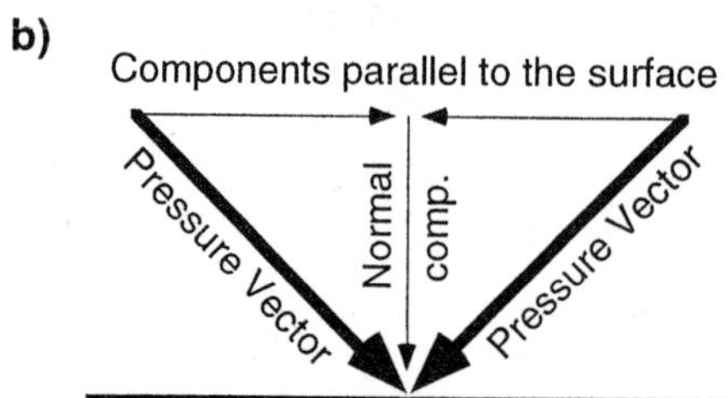

Figure 1.5
*(a) Pressure is isotropic. (b) Dark vectors correspond to those marked with * in (a). Components parallel to the surface cancel, while those normal to the surface contribute to pressure.*

Static pressure (i.e., pressure in calm winds) is caused by randomly moving molecules that bounce off of each other and off of surfaces they hit. In a vacuum the pressure is zero.

In the MKS system, a Newton is the unit for force, and m^2 is the unit for area. Thus, pressure has units of Newtons per square meter, or $N{\cdot}m^{-2}$. One Pascal (Pa) is defined to equal a pressure of 1 $N{\cdot}m^{-2}$. Atmospheric pressures are given in kiloPascals (kPa). The average (standard) pressure at sea level is P = 101.325 kPa. Pressure decreases nearly exponentially with height in the atmosphere, below 105 km.

While kiloPascals will be used in this book, standard sea-level pressure in other units are given in Table 1-1 for reference. Ratios of these units can be formed to allow unit conversion (see Appendix A).

In fluids such as the atmosphere, pressure force is **isotropic**; namely, at any point it pushes with the same force in all directions (see Fig 1.5a). Similarly, any point on a solid surface experiences pressure forces in all directions from the neighboring fluid elements. At such solid surfaces, all forces cancel except the forces normal (perpendicular) to the surface (Fig 1.5b).

Atmospheric pressure that you measure at any altitude is caused by the weight of all the air molecules above you. As you travel higher in the atmosphere there are fewer molecules still above you; hence, pressure decreases with height. Pressure can also compress the air causing higher density (i.e., more molecules in a given space). Compression is greatest where the pressure is greatest, at the bottom of the atmosphere. As a result of more molecules being squeezed into a small space near the bottom than near the top, ambient pressure decreases faster near the ground than at higher altitudes.

Table 1-1. Standard (average) sea-level pressure in various units.

Value	Units
101.325 kPa	kiloPascals
1013.25 hPa	hectoPascals
101,325. Pa	Pascals
101,325. $N \cdot m^{-2}$	Newtons per square meter
101,325 $kg_m \cdot m^{-1} \cdot s^{-2}$	kg-mass per meter per s^2
1.033227 $kg_f \cdot cm^{-2}$	kg-force per square cm
1013.25 mb	millibars
1.01325 bar	bars
14.69595 psi	pounds-force / square inch
2116.22 psf	pounds-force / square foot
1.033227 atm	atmosphere
760 Torr	Torr
Measured as height of fluid in a barometer:	
29.92126 in Hg	inches of mercury
760 mm Hg	millimeters of mercury
33.89854 ft H_2O	feet of water
10.33227 m H_2O	meters of water

Pressure change is approximately exponential with height, z. For example, if the temperature were uniform with height (which it is not), then:

$$P = P_o \cdot e^{-(a/T) \cdot z} \qquad (1.6a)$$

where a = 0.0342 K/m, and where average sea-level pressure on earth is P_o = 101.325 kPa. For more realistic temperatures in the atmosphere, the pressure curve deviates slightly from exponential, as discussed in the next section on atmospheric structure.

[**CAUTION: symbol "a" represents different constants for different equations, in this textbook.**]

Eq. (1.6a) can be rewritten as:

$$P = P_o \cdot e^{-z/H_p} \qquad (1.6b)$$

where H_p = 7.29 km is called the **scale height** for pressure. Mathematically, H_p is the **e-folding distance** for the pressure curve.

In Fig 1.6 is plotted the relationship between P and z on both linear and semi-log graphs, for T = 280 K. In the lowest 3 km of the atmosphere pressure decreases nearly linearly at about 10 kPa per 1 km of altitude.

FOCUS • e-Folding Distance

Some curves never end. In the figure below, curve (a) ends at $x = x_a$. Curve (b) ends at $x = x_b$. But curve (c), the exponentially decreasing curve, asymptotically approaches $y = 0$, never quite reaching it. The area under each of the curves is finite, and in this example are equal to each other.

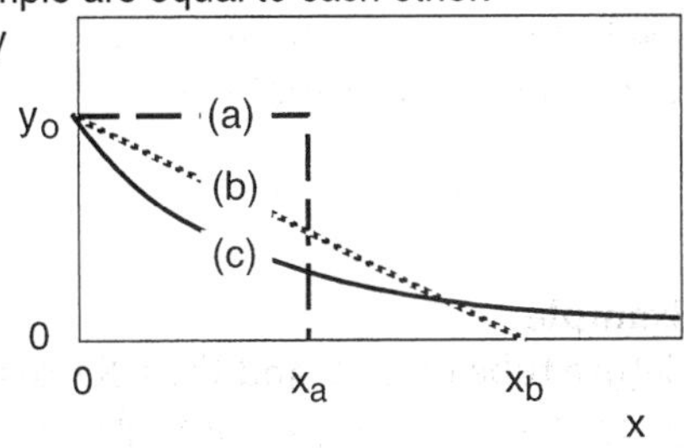

Although the exponential curve never ends, there is another way of quantifying how quickly it decreases with x. That measure is called the **e-folding distance** (or **e-folding time** if the independent variable is t instead of x). This is the distance x at which the curve decreases to 1/e of the starting value of the dependent variable, where e = 2.71828 is the base of natural logarithms.

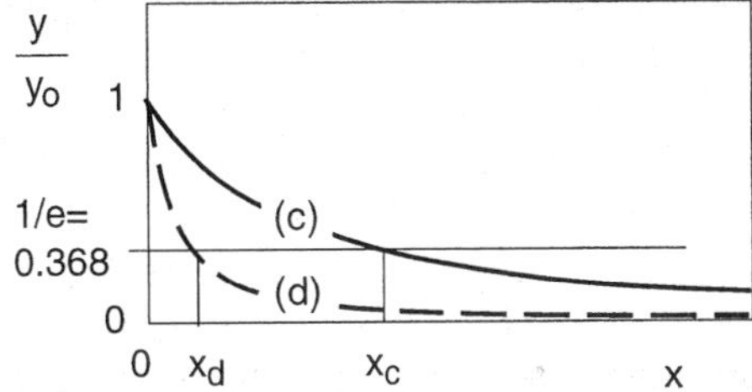

In the example above, both curves (c) and (d) are exponentials, but they drop off at different rates, where x_c and x_d are their respective e-folding distances. Generically, these curves are of the form:

$$y / y_o = e^{-x/x_{e\text{-}fold}} = \exp(-x / x_{e\text{-}fold})$$

Another useful characteristic is that the area A under the exponential curve is $A = y_o \cdot x_{e\text{-}fold}$.

Because of the monotonic decrease of pressure with height, pressure can be used as a surrogate measure of altitude. (**Monotonic** means that it changes only in one direction, even through the rate of change might vary.) Fig 1.4 shows such an example, where a reversed logarithmic scale (greater pressure at the bottom of the axis) is commonly used for P. Aircraft also use pressure to estimate their altitude.

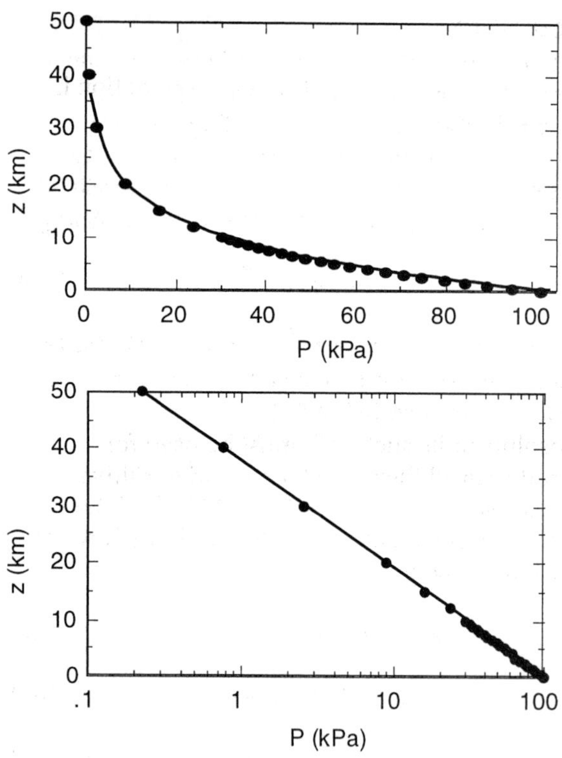

Figure. 1.6
Height z vs. pressure P in the atmosphere, plotted on linear (top) and semi-log (bottom) graphs. (See Appendix A for a review of relationships and graphs.)

> **Solved Example**
> Compare the pressures at 10 km above sea level for average temperatures of 250 and 300 K.
>
> **Solution**
> Given: z = 10 km
> (a) T = 250 K, (b) T = 300 K
> Find: (a) P = ? kPa, (b) P = ? kPa
>
> (a) Use eq. (1.6):
> P=(101.325kPa)·exp[(–0.0342K/m)·(10^4m)/250K]
> P = **<u>25.8 kPa</u>**
> (b)
> P=(101.325kPa)·exp[(–0.0342K/m)·(10^4m)/300K]
> P = **<u>32.4 kPa</u>**
>
> **Check**: Units OK. Physically reasonable.
> **Discussion**: Pressure decreases more slowly with height in warmer air because the molecules are further apart.

Density

Density ρ is defined as mass m per unit volume V.

$$\rho = m / V \qquad \bullet(1.7)$$

Density increases as the number and molecular weight of molecules in a volume increase. Average density at sea level is given in Table 1-2.

Because gases such as air are compressible, air density can vary over a wide range. Density decreases roughly exponentially with height in an atmosphere of uniform temperature.

$$\rho = \rho_o \cdot e^{-(a/T) \cdot z} \qquad (1.8a)$$

or

$$\rho = \rho_o \cdot e^{-z/H_\rho} \qquad (1.8b)$$

Table 1-2. Standard atmospheric density at sea level, for a standard temperature 15°C.

Value	Units
1.2250 kg·m^{-3}.	kilograms per cubic meter
0.076474 lb$_m$/ft^3	pounds-mass per cubic foot
1.2250 g/liter	grams per liter
0.001225 g/cm^3	grams per cubic centimeter

> **Solved Example**
> At sea level, what is the mass of air within a room of size 5 m x 8 m x 2.5 m ?
>
> **Solution**
> Given: L = 8 m length of room
> W = 5 m width of room
> H = 2.5 m height of room
> Find: m = ? kg air mass
>
> Sketch:
>
> 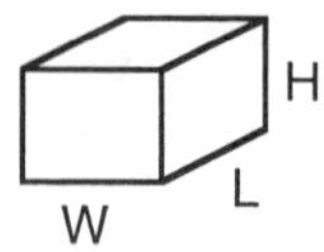
>
>
> The volume of the room is
> $V = W \cdot L \cdot H$ = (5m)·(8m)·(2.5m) = 100 m^3.
> Rearrange eq. (1.7) and solve for the mass:
> $m = \rho \cdot V$
> = (1.225 kg·m^{-3})·(100 m^3) = **<u>122.5 kg</u>**
>
> **Check**: Units OK. Sketch OK. Physics OK.
> **Discussion**: This is roughly twice the mass of a person.

where a = 0.040 K/m, and where average sea-level density is ρ_o = 1.2250 kg·m^{-3}, at a temperature of 15°C = 288 K. The shape of the curve described by eq (1.8) is similar to that for pressure, (see Fig 1.7). The scale height for density is H_ρ = 8.55 km.

Although the air is quite thin at high altitudes, it still can affect many observable phenomena: twilight (scattering of sunlight by air molecules) up to 63 km, meteors (incandescence by friction against air molecules) from 110 to 200 km, and aurora (excitation of air by solar wind) from 360 to 500 km.

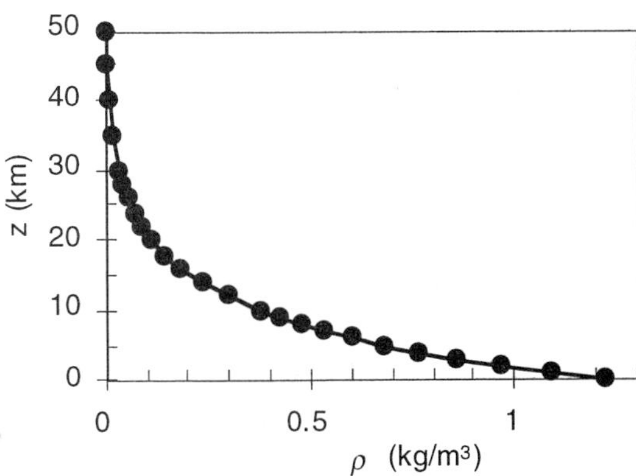

Figure 1.7
Density ρ vs. height z in the atmosphere.

Solved Example
What is the air density at a height of 2 km in an atmosphere of uniform temperature of 15°C?

Solution
Given: z = 2000 m
ρ_o = 1.225 kg/m^3
T = 15°C = 288.15 K
Find: ρ = ? kg/m^3

Use eq. (1.8): ρ =
(1.225 kg/m^3)· exp[(–0.04 K/m)·(2000 m)/288 K]
ρ = **0.928 kg/m^3**

Check: Units OK. Physics reasonable.
Discussion: This means that aircraft wings generate 24% less lift, and engines generate 24% less thrust because of the reduced air density.

Temperature

When a group of molecules (microscopic) move predominantly in the same direction, the motion is called wind (macroscopic). When they move in random directions, the motion is associated with temperature. Higher temperatures T are associated with greater average molecular speeds v, according to:

$$T = a \cdot m_W \cdot v^2 \qquad (1.9)$$

where a = 4.0x10^{-5} K·m^{-2} s^2 is a constant. Molecular weights m_W for the most common gases in the atmosphere are listed in Table 1-3.

Absolute units such as K must be used for temperature in all thermodynamic and radiative laws. At absolute zero (T = 0 K = –273.15°C) the molecules are essentially not moving. Temperature conversion formulae are:

$$T_{°F} = (9/5)\cdot T_{°C} + 32 \qquad \bullet(1.10a)$$

$$T_{°C} = (5/9)\cdot [T_{°F} - 32] \qquad \bullet(1.10b)$$

$$T_K = T_{°C} + 273.15 \qquad \bullet(1.11a)$$

$$T_{°C} = T_K - 273.15 \qquad \bullet(1.11b)$$

Standard (average) sea-level temperature is T = 15.0°C = 288 K = 59°F. Actual temperatures can vary considerably over the course of a day or year. Temperature variation with height is not as simple as the curves for pressure and density, and will be discussed in the Standard Atmosphere section below.

Solved Example
What is the average random velocity of nitrogen molecules at 20°C ?

Solution:
Given: T = 273.15 + 20 = 293.15 K.
Find: v = ? m/s avg mol. velocity

Sketch:

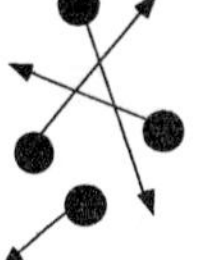

Get m_W from Table 1-3. Solve eq. (1.9) for v:
$v = [T/a\cdot m_W]^{1/2}$
= [(293.15 K) / (4.0x10^{-5} K·m^{-2}·s^2) ·(28.01)]$^{1/2}$
= **511.5 m/s** .

Check: Units OK. Sketch OK. Physics OK.
Discussion: Faster than a speeding bullet.

Table 1-3. Characteristics of gases in the air near the ground. Molecular weights are in g/mole. The volume fraction indicates the relative contribution to air in the earth's lower atmosphere. EPA is the USA Environmental Protection Agency.

Symbol	Name	Molecular Weight	Volume Fraction%
Constant Gases			
N_2	Nitrogen	28.01	78.08
O_2	Oxygen	32.00	20.95
Ar	Argon	39.95	0.93
Ne	Neon	20.18	0.0018
He	Helium	4.00	0.0005
H_2	Hydrogen	2.02	0.00005
Xe	Xenon	131.30	0.000009
Variable Gases			
H_2O	Water vapor	18.02	0 to 4
CO_2	Carbon dioxide	44.01	0.035
CH_4	Methane	16.04	0.00017
N_2O	Nitrous oxide	44.01	0.00003
EPA Air Quality Standards			
CO	Carbon monoxide	28.01	0.0035
SO_2	Sulfur dioxide	64.06	0.000014
O_3	Ozone	48.00	0.000012
NO_2	Nitrogen dioxide	46.01	0.000005
Mean Condition for Air			
	air	28.96	100.0

FOCUS • Escape Velocity

If an air molecule moves fast enough and doesn't hit any other molecules, it can escape into space by trading its kinetic energy (speed) for potential energy (height). High in the atmosphere where the air is thin, there is a better chance for molecules to not hit other molecules. The lowest altitude at which this can happen on earth is at about 550 km above ground, and this marks the base of the **exosphere** (region of escaping gases). This same altitude, when measured from the center of the earth, is about 6920 km, and is called the critical radius, r_c.

The escape velocity, v_e, is given by

$$v_e = \left[\frac{2 \cdot G \cdot m_{planet}}{r_c}\right]^{1/2}$$

where $G = 6.67\text{x}10^{-11}$ $m^3 \cdot s^{-2} \cdot kg^{-1}$ is the gravitational constant, and m_{planet} is the mass of the planet. The mass of the earth is 5.975×10^{24} kg. Thus, the escape velocity from earth is roughly

v_e = 10,732 m/s. *(continued next column)*

FOCUS • Escape Velocity (continued)

We can use this velocity in eq. (1.9) to determine the temperature necessary for average-speed molecules to escape. H_2 needs a temperature of 9,222 K, while He needs a much greater temperature of 18,445 K because it is a heavier molecule. Temperatures in the exosphere (also known as the thermosphere) are not hot enough to allow average-speed hydrogen and helium to escape. However, some atoms are faster than average, and do escape.

Other molecules such as O_2 are so heavy that the necessary escape temperature of 147,562 K is not achieved. Thus, most have stayed in the earth's atmosphere, to the benefit of life.

EQUATION OF STATE – IDEAL GAS LAW

Because pressure is caused by the movement of molecules, we might expect the pressure P to be greater where there are more molecules (i.e., greater density ρ), and where they are moving faster (i.e., greater temperature T). The relationship between pressure, density, and temperature is called the **Equation of State.** Different fluids have different equations of state, depending on their molecular properties.

For dry air the gases in the atmosphere have a simple equation of state known as the **ideal gas law**:

$$P = \rho \cdot \Re_d \cdot T \qquad \bullet(1.12)$$

where $\Re_d = 0.287053 \text{ kPa} \cdot K^{-1} \cdot m^3 \cdot kg^{-1} = 287.053 \text{ J} \cdot K^{-1} \cdot kg^{-1}$

is called the **gas constant** for dry air. Absolute temperatures (K) must be used in the ideal gas law.

For moist air, the gas constant changes because water vapor is less dense than dry air. To simplify things, a **virtual temperature** T_v can be defined to include the effects of water vapor on density:

$$T_v = T \cdot (1 + 0.61 \cdot r) \qquad \bullet(1.13)$$

where r is the water-vapor mixing ratio ($g_{water\ vapor}$ / $g_{dry\ air}$, see Chapt. 5), and all temperatures are in absolute units (K). In a nutshell, moist air of temperature T behaves as dry air with temperature T_v.

If there is both water liquid and vapor in the air, then this virtual temperature must be modified to include the **liquid water loading** (i.e., the weight of the drops):

$$T_v = T \cdot (1 + 0.61 \cdot r - r_L) \qquad \bullet(1.14)$$

where r_L is the liquid-water mixing ratio ($g_{liquid\ water} / g_{dry\ air}$).

With these definitions, a more useful form of the ideal gas law can be written for air of any humidity:

$$P = \rho \cdot \Re_d \cdot T_v \qquad \bullet(1.15)$$

where $\Re_d$ is still the gas constant for *dry* air.

Solved Example

What is the average (standard) surface temperature for dry air, given standard pressure and density?

Solution:

Given: P = 101.325 kPa, ρ = 1.225 kg·m^{-3}

Find: T = ? K

Solving eq. (1.12) for T gives: $T = P / (\rho \cdot \Re_d)$

$$T = \frac{101.325 kPa}{(1.225 kg \cdot m^{-3}) \cdot (0.287 kPa \cdot K^{-1} \cdot m^3 \cdot kg^{-1})}$$

= 288.2 K = **15°C**

Check: Units OK. Physically reasonable.

Discussion: The answer agrees with the standard surface temperature of 15°C discussed earlier, a cool but pleasant temperature.

Solved Example

In an unsaturated tropical environment with temperature of 35°C and mixing ratio of 30 $g_{water\ vapor}/kg_{dry\ air}$, what is the virtual temperature?

Solution:

Given: T = 35°C, r = 30 $g_{water\ vapor}/kg_{dry\ air}$

Find: T_v = ? °C

First, convert T and r to proper units

T = 273.15 + 35 = 308.15 K.

r =(30 g_{water}/kg_{air})·(0.001 kg/g) = 0.03 g_{water}/g_{air}

Next use eq. (1.13):

T_v = (308.15 K)·[1 + 0.61 · 0.03]

= 313.6 K = **40.6°C**.

Check: Units OK. Physically reasonable.

Discussion: Thus, high humidity reduces the density of the air so much that it acts like dry air that is 5°C warmer, for this case.

HYDROSTATIC EQUILIBRIUM

As discussed before, pressure decreases with height. Any thin horizontal slice from a column of air would thus have greater pressure pushing up against the bottom than pushing down from the top (Fig 1.8). This is called a **vertical pressure gradient**, where the term **gradient** means change with distance. The net upward force acting on this slice of air, caused by the pressure gradient, is $F = \Delta P \cdot A$, where A is the horizontal cross section area of the column, and $\Delta P = P_{bottom} - P_{top}$.

Also acting on this slice of air is gravity, which provides a downward force (weight) given by

$$F = m \cdot g \qquad \bullet(1.16)$$

where g = – 9.8 m·s^{-2} is the gravitational acceleration. A negative value for gravity, g, gives a negative (downward) force. (Remember that the unit of force is 1 N = 1 kg·m·s^{-2}). The mass m of air in the slice equals the air density times the slice volume; namely, $m = \rho \cdot (A \cdot \Delta z)$, where Δz is the slice thickness.

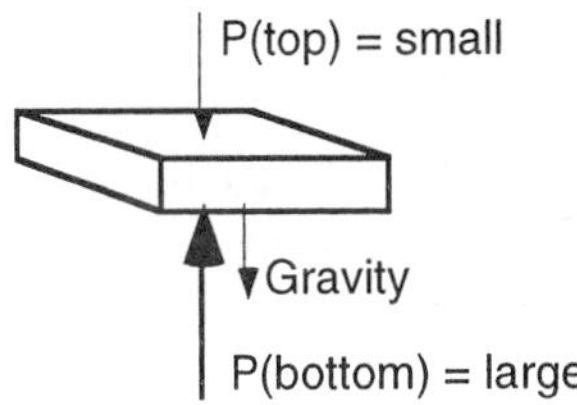

Figure 1.8
Hydrostatic balance of forces on a thin slice of air.

Solved Example

What is the weight (force) of a person of mass 75 kg at the surface of the earth?

Solution

Given: m = 75 kg

Find: F = ? N

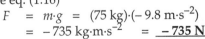

Sketch:

Use eq. (1.16)

$F = m \cdot g$ = (75 kg)·(– 9.8 m·s^{-2})

= – 735 kg·m·s^{-2} = **– 735 N**

Check: Units OK. Sketch OK. Physics OK.

Discussion: The negative sign means that the person is pulled toward the earth instead of repelled away from it.

For situations where pressure gradient force approximately balances gravity force, the air is said to be in a state of **hydrostatic equilibrium.** The corresponding **hydrostatic equation** is:

$$\Delta P = \rho \cdot g \cdot \Delta z \quad (1.17a)$$

or

$$\frac{\Delta P}{\Delta z} = -\rho \cdot |g| \quad \bullet(1.17b)$$

BEYOND ALGEBRA • Physical Interpretation of Equations

What is beyond algebra?

These boxes contain supplementary material that use calculus, differential equations, linear algebra, or other mathematical tools beyond algebra. They are not essential for understanding the rest of the book, and may be skipped. Science and engineering students with calculus backgrounds might be curious about how calculus is used to describe atmospheric physics.

Physical Interpretation of Equations

Equations such as (1.17b) are finite-difference approximations to the original equations that are in differential form:

$$\frac{dP}{dz} = -\rho \cdot |g| \quad (1.17c)$$

The calculus form eq. (1.17c) is useful for derivations, and is the best description of the physics. The algebraic approximation eq. (1.17b) is often used in real life, where one can measure pressure at two different heights [i.e., $\Delta P/\Delta z = (P_2 - P_1)/(z_2 - z_1)$].

The left side of eq. (1.17c) describes the infinitesimal change of pressure P that is associated with an infinitesimal local change of height z. It is the vertical gradient of pressure. On a graph of P vs. z, it would be the slope of the line. The derivative symbol "d" has no units or dimensions, so the dimensions of the left side are kPa/m.

Eq. (1.17b) has a similar physical interpretation. Namely, the left side is the change in pressure associated with a finite change in height. Again, it represents the slope of a line, but in this case, it is a straight line segment of finite length, as an approximation to a smooth curve.

Both eqs. (1.17b & c) state that rate of pressure decrease (because of the negative sign) with height is greater if the density ρ is greater, or if the magnitude of the gravitational acceleration $|g|$ is greater. Namely, if factors ρ or $|g|$ increase, then the whole right hand side (RHS) increases because ρ and $|g|$ are in the numerator. Also, if the RHS increases, then the left hand side (LHS) must increase as well, to preserve the equality of LHS = RHS.

The term **hydrostatic** is used because it describes a stationary (static) balance in a fluid (hydro) between pressure pushing up and gravity pulling down. The negative sign indicates that pressure decreases as height increases. This equilibrium is valid for most weather situations, except for vigorous storms with large vertical velocities.

Solved Example

Near sea level, a height increase of 100 m corresponds to what pressure decrease?

Solution

Given: $\rho = 1.225$ kg·m^{-3} at sea level
$\Delta z = 100$ m

Find: $\Delta P = ?$ kPa

Sketch: P_{top} z = 100 m
P_{bottom}

Use eq. (1.17a):

$\Delta P = \rho \cdot g \cdot \Delta z$
$= (1.225 \text{ kg·m}^{-3}) \cdot (-9.8 \text{ m·s}^{-2}) \cdot (100 \text{ m})$
$= -1200.5 \text{ kg·m}^{-1}\text{·s}^{-2}$
$=$ **−1.20 kPa**

Check: Units OK. Sketch OK. Physics OK.
Discussion: This answer should not be extrapolated to greater heights.

HYPSOMETRIC EQUATION

When the ideal gas law and the hydrostatic equation are combined, the result is an equation called the **hypsometric equation** that allows us to calculate how pressure varies with height in an atmosphere of arbitrary temperature profile:

$$z_2 - z_1 \approx a \cdot \overline{T_v} \cdot \ln\left(\frac{P_1}{P_2}\right) \quad \bullet(1.18)$$

where $\overline{T_v}$ is the average virtual temperature between heights z_1 and z_2. The constant $a = \Re_d / |g| =$ 29.3 m/K. The height difference of a layer bounded above and below by two pressure levels P_1 and P_2 is called the **thickness** of that layer.

To use this equation across large height differences, it is best to break the total distance into a number of thinner intervals, Δz. In each thin layer, if the virtual temperature varies little, then we can approximate $\overline{T_v}$ by T_v. By this method we can sum all of the thicknesses of the thin layers to get the total thickness of the whole layer.

For the special case of a dry atmosphere of uniform temperature with height, eq. (1.18) simplifies to eq. (1.6a). Thus, eq. (1.18) also describes an exponential decrease of pressure with height.

Solved Example (§)

What is the thickness of the 100 to 90 kPa layer, given the temperatures below?

P(kPa)	T(°K)
90	275
100	285

(ATTENTION. The symbol § means that the problem is easier to solve on a computer spreadsheet.)

Solution

Given: the observations at the top and bottom of the layer

Find: $\Delta z = z_2 - z_1$

Assume: temperature varies linearly with height

Also assume the air is dry, so $T = T_v$.

Solve eq. (1.18) on a computer spreadsheet (§) for many thin layers 0.5 kPa thick. Results for the first few thin layers, starting from the bottom, is:

P(kPa)	T_v (K)	$\overline{T_v}$ (K)	Δz(m)
100	285	284.75	41.82
99.5	284.5	284.25	41.96
99.0	284	etc.	etc.

Sum of all Δz = **864.11 m**

Check: Units OK. Physics reasonable.

Discussion: Thus, in an aircraft you must climb 864.11 m to experience a pressure decrease from 100 to 90 kPa, for this particular temperature sounding.

If we had not broken the thick layer into thin ones, the answer would have been Δz = 864.38 m, which is very close to the answer above.

BEYOND ALGEBRA • Hypsometric Eq.

In order to derive eq. (1.18) from the ideal gas law and the hydrostatic equation, one must use calculus. It cannot be done using algebra alone. However, once the equation is derived, the answer is in algebraic form.

The derivation is shown here only to illustrate the need for calculus. Derivations will NOT be given for most of the other equations in this book. Students can take advanced meteorology courses, or read advanced textbooks, to find such derivations.

(continued next column)

BEYOND ALGEBRA • Hypsometric Eq.

(continuation)

Derivation of the hypsometric equation:

Given: the hydrostatic eq:

$$\frac{dP}{dz} = -\rho \cdot |g| \qquad (1.17c)$$

and the ideal gas law:

$$P = \rho \cdot \Re_d \cdot T_v \qquad (1.15)$$

First, rearrange eq. (1.15) to solve for density:

$$\rho = P / (\Re_d \cdot T_v)$$

Then substitute this into (1.17c):

$$\frac{dP}{dz} = -\frac{P \cdot |g|}{\Re_d \cdot T_v}$$

One trick for integrating equations is to separate variables. Namely, for our example, move all the pressure factors to one side, and all height factors to the other. Therefore, multiply both sides of the above equation by dz, and divide both sides by P.

$$\frac{dP}{P} = -\frac{|g|}{\Re_d \cdot T_v} dz$$

Compared to the other variables, g and $\Re_d$ are relatively constant, so we will assume that they are constant and separate them from the other variables. However, usually temperature varies with height: $T(z)$. Thus:

$$\frac{dP}{P} = -\frac{|g|}{\Re_d} \cdot \frac{dz}{T_v(z)}$$

Next, integrate the whole eq. from some lower altitude z_1 where the pressure is P_1, to some higher altitude z_2 where the pressure is P_2:

$$\int_{P_1}^{P_2} \frac{dP}{P} = -\frac{|g|}{\Re_d} \cdot \int_{z_1}^{z_2} \frac{dz}{T_v(z)}$$

where $|g| / \Re_d$ is pulled out of the integral on the RHS because it is constant.

The LHS integrates to become a natural logarithm (consult tables of integrals). However, the RHS is more difficult, because we don't know the functional form for the vertical temperature profile. We could assume a linear profile, but on any given day, the profile has a complex shape that is not conveniently described by an equation that can be integrated. Instead, we will invoke the mean value theorem of calculus to bring T_v out of the integral. The overbar denotes an average (over height, in this context). That leaves only dz on the RHS.

After integrating, we get:

$$\ln(P)\Big|_{P_1}^{P_2} = -\frac{|g|}{\Re_d} \cdot \overline{\left(\frac{1}{T_v}\right)} \cdot z\Big|_{z_1}^{z_2}$$

(continued next column)

BEYOND ALGEBRA • Hypsometric Eq.

(Continuation of hypsometric derivation.)

Plugging in the upper and lower limits gives:

$$\ln(P_2) - \ln(P_1) = -\frac{|g|}{\Re_d} \cdot \overline{\left(\frac{1}{T_v}\right)} \cdot (z_2 - z_1)$$

But the difference between two logarithms can be written as the ln of the ratio of their arguments:

$$\ln\left(\frac{P_2}{P_1}\right) = -\frac{|g|}{\Re_d} \cdot \overline{\left(\frac{1}{T_v}\right)} \cdot (z_2 - z_1)$$

Recalling that $\ln(x) = -\ln(1/x)$, then:

$$\ln\left(\frac{P_1}{P_2}\right) = \frac{|g|}{\Re_d} \cdot \overline{\left(\frac{1}{T_v}\right)} \cdot (z_2 - z_1)$$

Rearranging this eq, and approximating $\overline{T_v} \approx 1/\overline{(1/T_v)}$ (which is NOT an identity), then one finally gets the hypsometric eq:

$$(z_2 - z_1) \approx \frac{\Re_d}{|g|} \cdot \overline{T_v} \cdot \ln\left(\frac{P_1}{P_2}\right) \qquad (1.18)$$

ATMOSPHERIC STRUCTURE

Atmospheric structure refers to the state of the air at different heights. The true vertical structure of the atmosphere varies with time and location due to changing weather conditions and solar activity.

Standard Atmosphere

The "1976 U.S. Standard Atmosphere" (Table 1-4) is an idealized, dry, steady-state approximation of the atmospheric state as a function of height. It has been adopted as an engineering reference. It approximates the average atmospheric conditions, although it was not computed as a true average.

A **geopotential height**, H, is defined to compensate for the decrease of gravitational acceleration g above the earth's surface:

$$H = R_o \cdot z \ / (R_o + z) \qquad \bullet(1.19a)$$

$$z = R_o \cdot H \ / (R_o - H) \qquad \bullet(1.19b)$$

where the average radius of the earth is R_o = 6356.766 km. An air parcel raised to height z would have the same potential energy as if lifted only to height H under constant gravitational acceleration. By using H instead of z, we can use g = 9.8 m/s^2 as a constant in our equations, even though in reality it decreases slightly with altitude.

Table 1-4. Standard atmosphere

H (km)	T (°C)	P (kPa)	ρ (kg/m^3)
-1	21.5	113.920	1.3470
0	15.0	101.325	1.2250
1	8.5	89.874	1.1116
2	2.0	79.495	1.0065
3	-4.5	70.108	0.9091
4	-11.0	61.640	0.8191
5	-17.5	54.019	0.7361
6	-24.0	47.181	0.6597
7	-30.5	41.060	0.5895
8	-37.0	35.599	0.5252
9	-43.5	30.742	0.4664
10	-50.0	26.436	0.4127
11	-56.5	22.632	0.3639
13	-56.5	16.510	0.2655
15	-56.5	12.044	0.1937
17	-56.5	8.787	0.1423
20	-56.5	5.475	0.0880
25	-51.5	2.511	0.0395
30	-46.5	1.172	0.0180
32	-44.5	0.868	0.0132
35	-36.1	0.559	0.0082
40	-22.1	0.278	0.0039
45	-8.1	0.143	0.0019
47	-2.5	0.111	0.0014
50	-2.5	0.076	0.0010
51	-2.5	0.067	0.00086
60	-27.7	0.02031	0.000288
70	-55.7	0.00463	0.000074
71	-58.5	0.00396	0.000064
80	-76.5	0.00089	0.000015
84.9	-86.3	0.00037	0.000007
89.7	-86.3	0.00015	0.000003
100.4	-73.6	0.00002	0.000000
105	-55.5	0.00001	0.000000
110	-9.2	0.00001	0.000000

The difference between geometric z and geopotential height increases from 0 to 16 m as height increases from 0 to 10 km above sea level.

Table 1-4 gives the standard temperature, pressure, and density as a function of geopotential height H above sea level. Temperature variations are linear between key altitudes indicated in boldface.

Below a geopotential altitude of 51 km, eqs. (1.20) and (1.21) can be used to compute standard temperature and pressure. In these equations, be sure to use absolute temperature as defined by T(K) = T(°C) + 273.15 .

(1.20)

$T = 288.15\text{ K} - (6.5\text{ K/km})\cdot H$ for $H \le 11$ km

$T = 216.65$ K $11 \le H \le 20$ km

$T = 216.65\text{ K} + (1\text{ K/km})\cdot(H-20\text{km})$ $20 \le H \le 32$ km

$T = 228.65\text{ K} + (2.8\text{ K/km})\cdot(H-32\text{km})$ $32 \le H \le 47$km

$T = 270.65$ K $47 \le H \le 51$ km

For the pressure equations, the absolute temperature T that appears must be the standard atmosphere temperature from the previous set of equations. In fact, those previous equations can be substituted into the equations below to make them a function of H rather than T.

(1.21)

$P = (101.325\text{kPa})\cdot(288.15\text{K}/T)^{-5.255877}$ $H \le 11$km

$P = (22.632\text{kPa})\cdot\exp[-0.1577\cdot(H-11\text{ km})]$ $11 \le H \le 20$ km

$P = (5.4749\text{kPa})\cdot(216.65\text{K}/T)^{34.16319}$ $20 \le H \le 32$ km

$P = (0.868\text{kPa})\cdot(228.65\text{K}/T)^{12.2011}$ $32 \le H \le 47$ km

$P = (0.1109\text{kPa})\cdot\exp[-0.1262\cdot(H-47\text{ km})]$ $47 \le H \le 51$ km

Solved Example
Is eq. (1.6a) a good fit to std. atmos. pressure?

Solution
Assumption: Use $T = 270$ K in eq. (1.6a) because it minimizes pressure errors in the bottom 10 km.
Method: Compare on a graph.

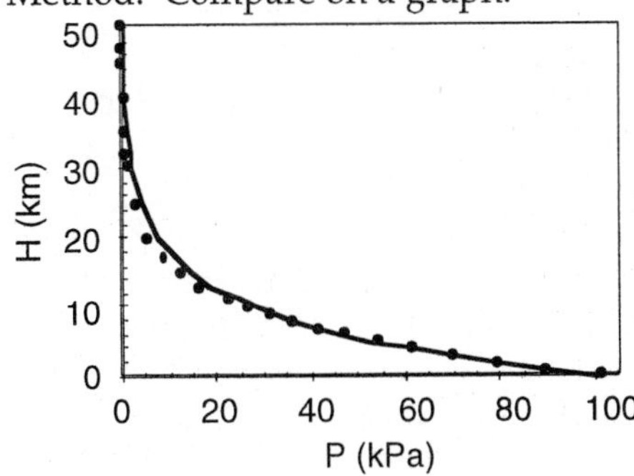

where the solid line is eq. (1.6a) and the data points are from Table 1.4.

Discussion: Over the lower 10 km, the simple eq. (1.6a) is in error by no more than 1.5 kPa. If more accuracy is needed, then use eqs. (1.17) and (1.18).

These equations are a bit better than eqs. (1.6a) and (1.8) because they do not make the unrealistic assumption of uniform temperature with height.

Density is found using the ideal gas law eq. (1.12).

Layers of the Atmosphere

The standard atmospheric temperature is plotted in Fig 1.9. The following layers are defined based on this temperature structure.

Thermosphere	$84.9 \le H$ km
Mesosphere	$47 \le H \le 84.9$ km
Stratosphere	$11 \le H \le 47$ km
Troposphere	$0 \le H \le 11$ km

Almost all clouds and weather occur in the **troposphere**.

The top limits of the bottom three spheres are named:

Mesopause	H = 84.9 km
Stratopause	47 km
Tropopause	11 km

The three relative maxima of temperature are a result of three altitudes where significant amounts of solar radiation are absorbed and converted into heat. Ultraviolet light is absorbed by ozone near the stratopause, visible light is absorbed at the ground, and most other radiation is absorbed in the thermosphere.

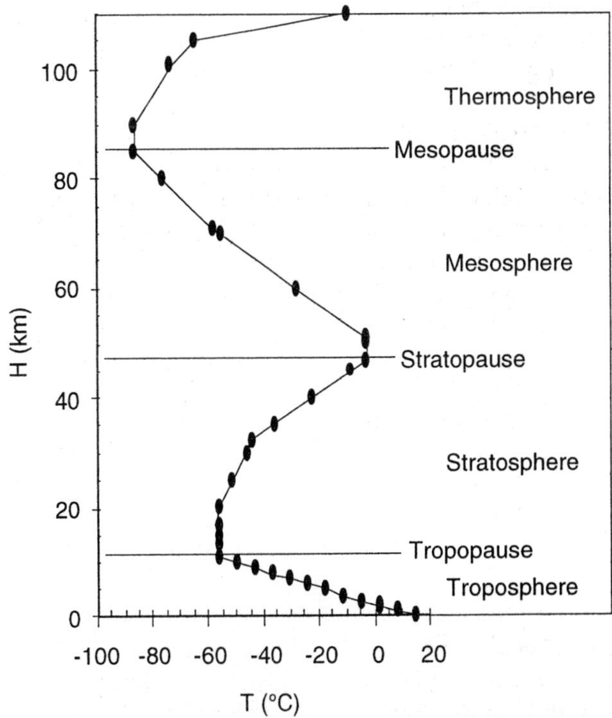

Figure 1.9
Standard temperature T profile vs. geopotential height H.

PROCESS TERMINOLOGY

Processes associated with constant temperature are **isothermal**. For example, eqs. (1.6) and (1.8) apply for an isothermal atmosphere. Those occurring with constant pressure are **isobaric**. A line on a weather map connecting points of equal temperature is called an **isotherm**, while one connecting points of equal pressure is an **isobar**. Table 1-5 summarizes many of the process terms.

Table 1-5. Process names.

Name	Constant or equal
adiabat	entropy (no heat exchange)
contour	height
isallobar	pressure tendency
isallohypse	height tendency
isallotherm	temperature tendency
isanabat	vertical wind speed
isanomal	weather anomaly
isentrope	entropy or potential temp.
isobar	pressure
isobath	water depth
isobathytherm	depth of constant temperature
isoceraunic	thunderstorm activity or freq.
isochrone	time
isodop	radial wind speed
isodrosotherm	dew-point temperature
isogon	wind direction
isohel	sunshine
isohume	humidity
isohyet	precipitation accumulation
isohypse	height (similar to contour)
isoneph	cloudiness
isopleth	(generic, for any quantity)
isopycnic	density
isoshear	wind shear
isotach	speed
isotherm	temperature

Solved Example
Name the process for constant density.

Solution:
From Table 1-5: It is an **isopycnal** process.

Discussion: Isopycnics are used in oceanography, where both temperature and salinity contribute to density.

SUMMARY

Pressure, temperature, and density describe the thermodynamic state of the air. These state variables are related to each other by the ideal gas law. Change one and one or both of the others must change too. Ambient pressure decreases roughly exponentially with height, as given by the hypsometric equation. This vertical pressure gradient is balanced by the pull of gravity, according to the hydrostatic equation.

Density variation is also exponential with height. Temperature, however, exhibits three relative maxima over the depth of the atmosphere, caused by absorption of radiation from the sun. Thermodynamic processes can be classified. The standard atmosphere is an idealized model of atmospheric vertical structure, and is used to define atmospheric layers such as the troposphere and stratosphere.

Threads

Have you ever discovered a loose thread on your clothes, and upon pulling it, found that it leads you to places you didn't expect? With this analogy, a "Threads" section will appear at the end of most chapters, to tie new important concepts to past and future concepts, and to wrap things up.

In this chapter, we introduced the special meteorological coordinate system used in all other chapters to convert wind information into everyday terms that people such as sailors and pilots can understand and use. The standard atmosphere temperature structure determines the ability of storms to grow (Chapts. 15 and 16), and confines many weather circulations to the bottom 11 km of the atmosphere, on the average (Chapts. 4, 11 and 13).

The hydrostatic equation will be used in Chapt. 9 to explain why vertical winds are often weaker than horizontal winds. Virtual temperature will be used in Chapts. 6, 7, and 15 to help define buoyancy and cloud formation. The hypsometric relationship will help explain the jet stream, and will allow a "thermal wind" relationship to be defined in Chapt. 11. The ideal gas law and the hydrostatic equation will be used in Chapt. 2 to show why air parcels become colder as they rise — important for condensation and rain. Process terminology will be used in weather maps and thermodynamic diagrams (Chapts 13 & 6).

EXERCISES

Numerical Problems

These are essentially "plug & chug" exercises. They are designed to ensure that you are comfortable with the equations, units, and physics by getting hands-on experience using them. None of the problems require calculus.

While most of the numerical problems can be solved using a hand calculator, many students find it easier to compose all of their homework answers on a computer spreadsheet. It is easier to correct mistakes using a spreadsheet, and plotting graphs of the answer is trivial. Even if you don't solve all of the exercises on a spreadsheet, the few exercises indicated by symbol § are easiest to do on a spreadsheet.

Most modern spreadsheets also allow you to add objects called text boxes, note boxes or word boxes, to allow you to include word-wrapped paragraphs of text, which is handy for the "Problem" and the "Discussion" parts of the answer.

A spreadsheet example is given below. Normally, to make your printout look neater, you might use the page setup or print option to turn off printing of the row numbers, column letters, and gridlines. Also, the borders around the text boxes can be eliminated, and color could be used if you have access to a color printer. Format all graphs to be clear and attractive, with axes labeled and with units, and with tic marks having pleasing increments.

	A	B	C	D	E
1					
2	N21)	What is air density at height			
3		2 km in an atmosphere of			
4		uniform T of 15 C?			
5					
6		**Given:**	z =	2000	m
7			rho_o=	1.225	kg/m3
8			T =	15	C
9		**Find:**	rho =	?	kg/m3
10					
11		First convert T to Kelvin.			
12			T =	288.15	K
13		Then use eq (1.8), where			
14			a =	0.0342	K/m
15			rho =	**0.966**	kg/m3
16					
17		**Check:** Units OK. Physics OK.			
18		**Discussion:** This means that			
19		aircraft wings generate			
20		0.966/1.225 = 79% of the lift			
21		they would at sea level.			
22					
23	Note to students: the eq used in				
24	cell D15 was: =D7*EXP(-D14*D6/D12)				
25					

Format Guidelines for Your Homework

Good scientists and engineers are not only creative, they are methodical, meticulous, and accurate. To encourage you to develop these good habits, many instructors require your homework to be written in a clear, concise, organized, and consistent format. Such a format is described below, and is illustrated in all the Solved Examples in this book. The format below closely follows steps you typically take in problem solving (Appendix A).

Format:

1. Give the exercise number, & restate the problem
2. Start the solution section by listing the "Given" known variables (WITH THEIR UNITS)
3. List the unknown variables to find, with units.
4. Draw a sketch if it clarifies the scenario.
5. List the equation(s) you will use.
6. Show all your intermediate steps and calculations (to maximize your partial credit), and be sure to ALWAYS INCLUDE UNITS with the numbers when you plug them into eqs.
7. Put a box around your final answer, or underline it, so the grader can find it on your page amongst all the coffee and pizza stains.
8. Always check the value & units of your answer.
9. Briefly discuss the significance of the answer

Example:

Problem : What is air density at height 2 km in an atmosphere of uniform temperature 15°C?

Solution

Given: $z = 2000$ m

$\rho_o = 1.225$ kg/m^3

$T = 15°C = 288.15$ K

Find: $\rho = ?$ kg/m^3

Use eq. (1.8): $\rho =$ (1.225 kg/m^3)· exp[(–0.0342K/m)·(2000m)/288K]

$\rho =$ <u>0.966 kg/m^3</u>

Check: Units OK. Physics reasonable.

Discussion: This means that aircraft wings generate 21% less lift, engines generate 21% less power, and propellers 21% less thrust because of the reduced air density. Thus compounding effect causes aircraft performance to decrease rapidly with increasing altitude, until the service ceiling is reached where it can't climb any higher.

N1. Find the wind direction and speed, given the (U, V) components:

a. (-5, 0) knots
b. (8, -2) m/s
c. (-1, 15) mi/h
d. (6, 6) m/s
e. (8, 0) knots
f. (5, 20) m/s
g. (-2, -10) mi/h
h. (3, -3) m/s

N2. Find the U and V wind components, given wind direction and speed:

a. west at 10 knots
b. north at 5 m/s
c. 225° at 8 mi/h
d. 300° at 15 knots
e. east at 7 knots
f. south at 10 m/s
g. 110° at 8 mi/h
h. 20° at 15 knots

N3. Convert the 7 pressure events marked with arrows in Ahrens' Fig 9.4 into units of kPa.

N4. (i). Suppose that a typical airline window is circular with radius 15 cm, and a typical cargo door is square of side 2 m. If the interior of the aircraft is pressured at 80 kPa, and the ambient outside pressure is given below in kPa, then what are the magnitudes of forces pushing outward on the window and door?

(ii). Your weight in pounds is the force you exert on things you stand on. How many people of your same weight standing on a window or door are needed to equal the forces calculated in part a. Assume the window and door are horizontal, and are near the earth's surface.

a. 30
b. 25
c. 20
d. 15
e. 10
f. 5
g. 0
h. 40

N5. Find the pressure in kPa at the following heights above sea level, assuming an average T = 250K:

a. -100 m (below sea level)
b. 1 km
c. 11 km
d. 25 km
e. 30,000 ft
f. 5 km
g. 2 km
h. 15,000 ft

N6. Use the definition of pressure as a force per unit area, and consider a column of air that is above a horizontal area of 1 square meter. What is the mass of air in that column:

a. above the earth's surface.
b. above a height where the pressure is 50 kPa?
c. between pressure levels of 70 and 50 kPa?
d. above a height where the pressure is 85 kPa?
e. between pressure levels 100 and 20 kPa?
f. above height where the pressure is 30 kPa?
g. between pressure levels 100 and 50 kPa?
h. above a height where the pressure is 10 kPa?

N7. Find the virtual temperature for air of:

	a.	b.	c.	d.	e.	f.	g.
T(°C)	20	10	30	40	50	0	–10
r(g/kg)	10	5	0	40	60	2	1

N8. (i). What are the escape velocities from a planet for each of their main atmospheric components listed in Ahren's Table 1 (Focus on the Atmosphere of Other Planets)? (For simplicity, use the planet radius instead of the "critical" radius at the base of the exosphere.). Relative to earth mass, m_{Venus}=0.82, m_{Mars}=0.11, m_{Jup}=318, m_{Sat}=95, m_{Ur}=15, m_{Ne}=17.

(ii). What are the most likely velocities of those molecules at the surface, given the average surface temperatures given in that table? Comparing these answers to part (i), which of the constituents (if any) are most likely to escape?

a. Venus
b. Earth
c. Mars
d. Jupiter
e. Saturn
f. Uranus
g. Neptune
h. Pluto

N9. Convert the following temperatures:

a. 15°C = ?K
b. 50°F = ?°C
c. 70°F = ?K
d. 15°C = ?°F
e. 303 K = ?°C
f. 250K = ?°F
g. 2000°C = ?K
h. –40°F = ?°C

N10. a. What is the density of air, given P = 80 kPa and T = 0 °C ?
b. What is the temperature of air, given P = 90 kPa and ρ = 1.0 kg·m^{-3} ?
c. What is the pressure of air, given T = 90°F and ρ = 1.2 kg·m^{-3} ?
d. Give 2 combinations of pressure and density that have a temperature of 30°C.
e. Give 2 combinations of pressure and density that have a temperature of 0°C.
f. Give 2 combinations of pressure and density that have a temperature of –20°C.
g. How could you determine air density if you did not have a density meter?
h. What is the density of air, given P = 50 kPa and T = –30 °C ?

N11. Given a sealed rigid box with air in it. If initially the air has T = 10°C and P = 100 kPa, what is the new pressure if the temperature is increased to 50°C?

N12. Using the hypsometric equation, what is the relationship between the 100 to 80 kPa thickness, and the average temperature within that layer.

N13. Name the isopleths would would be drawn on a weather map to indicate regions of equal

a. pressure
b. temperature
c. cloudiness
d. precipitation
e. humidity
f. wind speed
g. dew point
h. pressure tendency

N14. What is the geometric height, given the geopotential height?

a. 10 m b. 100 m c. 1 km d. 11 km

What is the geopotential height, given the geometric height?

e. 500 m f. 2 km g. 5 km h. 20 km

N15. What is the standard atmospheric temperature, pressure, and density at each of the following geopotential heights?

a. 1.5 km b. 12 km c. 50 m d. 8 km
e. 200 m f. 5 km g. 40 km h. 25 km

N16. What are the geometric heights (assuming a standard atmosphere) at the top and bottom of the:

a. troposphere b. stratosphere
c. mesosphere d. thermosphere

N17. Is the inverse of an average of numbers equal to the average of the inverses of those number? (Hint, work out the values for just two numbers: 2 and 4.) This question helps explain where the hypsometric equation given in this chapter is only approximate.

N18(§). Using the standard atmosphere equations, re-create the numbers in Table 1-4.

Understanding & Critical Evaluation

These questions usually require more thought, and are often an extension of the material in the chapter. They might require you to combine two or more equations, or concepts from different parts of the chapter, or from earlier chapters. You might also have to critically evaluate an approach. Some of these questions require a numerical answer, others are "short-answer" essays.

They often require you to make assumptions, because insufficient data is given to solve the problem. Whenever you make assumptions, justify them first. To give you an idea of what is expected for these questions, a solved example is given below.

Solved Example - Critique

What are the limitations of eq. (1.6a), if any? How can those limitations be eliminated?

Solution

Eq. (1.6) for P vs. z relies on an average temperature over the whole depth of the atmosphere. Thus, eq. (1.6a) is accurate only when the <u>actual temperature is constant with height</u>.

(continued next column)

Solved Example - Critique *(continuation)*

As we learned later in the chapter, a typical or "standard" atmosphere temperature is NOT constant with height. In the troposphere, for example, temperature decreases with height. On any given day, the real temperature profile is likely to be even more complicated. Thus, eq. (1.6a) is inaccurate.

A better answer could be found from the hypsometric equation, by taking the exponential of both sides of (1.18). This gives:

$$P_2 = P_1 \cdot \exp\left(-\frac{z_2 - z_1}{a \cdot \overline{T_v}}\right) \qquad \text{with } a = 29.3 \text{ m/K.}$$

By iterating up from the ground over small increments $\Delta z = z_2 - z_1$, one can use any arbitrary temperature profile. Namely, starting from the ground, set $z_1 = 0$ and $P_1 = 101.325$ kPa. Set $z_2 = 1$ km, and use the average virtual temperature value in the hypsometric equation for that 1 km thick layer from $z = 0$ to 1 km. Solve for P_2. Then repeat the process for the layer between $z =$ 1 and 2 km, using the new T_v for that layer.

Because eq. (1.6a) came from eq. (1.18), we find other limitations.

1) Eq. (1.6a) is for <u>dry air</u>, because it uses temperature rather than virtual temperature.

2) The constant "a" in eq. (1.6a) equals $|g|/\Re_d =$ (1/29.3) K/m. Hence, on a different planet with different gravity and different gas constant, "a" would be different. Thus, eq. (1.6a) is limited to <u>earth</u>.

Nonetheless, eq. (1.6a) is a reasonable first-order approximation to the variation of pressure with altitude, as can be seen by using standard-atmos P values from Table 1-4, and plotting them vs z. The result (which was shown in the solved example after Table 1-4) is indeed close to an exponential decrease with altitude.

U1. What are the limitations of the "standard atmosphere"?

U2. For any physical variable that decreases exponentially with distance or time, the **e-folding** scale is defined as the distance or time where the physical variable is reduced to 1/e of its starting value. For the atmosphere the e-folding height for pressure decrease is known as the **scale height**. Given eq. (1.6a), what is the algebraic and numerical value for atmospheric scale height (km)?

U3(§). Invent some arbitrary data, such as 5 data points of wind speed M vs. pressure P. Although P is the independent variable, use a spreadsheet to plot it on the vertical axis (i.e., switch axes on your graph so that pressure can be used as a surrogate measure of height), change that axis to a logarithmic scale, and then reverse the scale so that the largest value is at the bottom, corresponding to the greatest pressure at the bottom of the atmosphere.

Now add to this existing graph a second curve of different data of M vs. P. Learn how to make both curves appear properly on this graph because you will use this skill repeatedly to solve problems in future chapters.

U4. Does hydrostatic equilibrium (eq. 1.17) always apply to the atmosphere? If not, when and why not?

U5. a. Plug eqs. (1.1) and (1.2a) into (1.3), and use trig to show that $U = U$. b. Similar, but for $V = V$.

U6. What percentage of the atmosphere is above a height of : a. 2 km b. 5 km c. 11 km d. 32 km e. 1 km f. 18 km g. 47 km h. 8 km

U7. What is the mass of air inside an airplane with a cabin size of 5 x 5 x 30 m, if the cabin is pressurized to a cabin altitude of sea level? What mass of outside air is displaced by that cabin, if the aircraft is flying at an altitude of 3 km? The difference in those two masses is the load of air that must be carried by the aircraft. How many people cannot be carried because of this excess air that is carried in the cabin?

U8. Given air of initial temperature 20°C and density of 1.0 kg/m^3.

a. What is its initial pressure?

b. If the temperature increases to 30°C in an isobaric process, what is the new density?

c. If the temperature increases to 30°C in an isobaric process, what is the new pressure?

d. For an isothermal process, if the pressure changes to 20 kPa, what is the new density?

e. For an isothermal process, if the pressure changes to 20 kPa, what is the new T?

f. In a large, sealed, glass bottle that is full of air, if you increase the temperature, what if anything would be conserved (P, T, or density)?

g. In a sealed, inflated latex balloon, if you lower it in the atmosphere, what thermodynamic quantities if any, would be conserved?

h. In a mylar (non stretching) balloon, suppose that it is inflated to equal the surrounding atmospheric pressure. If you added more air to it, how would the state change?

U9(§). Starting from sea-level pressure at $z = 0$, use the hypsometric equation to find and plot P vs. z in the troposphere, using the appropriate standard-atmosphere temperature. Step in small increments to higher altitudes (lower pressures) within the troposphere, within each increment. How is your answer affected by the size of the increment? Also solve it using a constant temperature equal to the average surface value. Plot both results on a semi-log graph, and discuss meaning of the difference.

U10. Use the ideal gas law and eq. (1.6) to derive the equation for the change of density with altitude, assuming constant temperature.

U11. What is the standard atmospheric temperature, pressure, and density at each of the following geopotential heights?

a. 65 km b. 55 km c.–0.5 km d. 80 km

U12. The ideal gas law and hypsometric equation are for compressible gases. For liquids (which are incompressible, to first order), density is not a function of pressure. Compare the vertical profile of pressure in a liquid of constant temperature with the profile of a gas of constant temperature.

U13. At standard sea-level pressure and temperature, how does the average molecular speed compare to the speed of sound? Also, does the speed of sound change with altitude? Why?

U14. For a standard atmosphere below $H = 11$ km:

a. Give an equation for pressure as a function of H.

b. Give an equation for density as a function of H.

U15. Use the hypsometric equation to derive an equation for the scale height for pressure, H_p.

Web-Enhanced Questions

These questions allow students to solve problems using current data, such as satellite images, weather maps, and weather observations that can be downloaded through the internet. With current data, exercises can be much more exciting, timely, and relevant. Such questions are necessarily more vague than the other questions, because one is never guaranteed of finding a particular weather phenomenon on any given day.

Many of these questions are worded in a way to encourage students to acquire the weather information for locations near where they live.

However, the instructor might suggest a different location if a better example of a weather event is happening elsewhere. Even if the instructor does not suggest alternative locations, the student should feel free to search the country, the continent, or the globe for examples of weather that are best suited for the exercise.

Web URL (universal resource locator) addresses are very transient. Web sites come and go. Even a persisting site might change its web address. For this reason, the web-enhanced questions do not usually give the URL web site for any particular exercise. Instead, the student is expected to become proficient with internet search engines. Nonetheless, there still might be occasions where the data does not exist any where on the web. The instructor should be aware of such eventualities, and be tolerant of students who are unable to complete the web-enhanced exercise.

In many cases, the student will want to print out the weather map or satellite image to turn in with the homework. Instructors should be tolerant of students who have access to only black and white printers. Students with black and white printouts should take a colored pencil or pen to highlight the particular feature or isopleths of interest, if it is otherwise difficult to discern among all the other black lines on the printout.

Students should always list the URL web address from which you acquired the data or images. This is just like citing books or journals from the library. At the end of each web-enhanced exercise, include a "References" box listing the web addresses used, and any comments you have regarding the quality of the site. Part of the ethic of being a good scientist or engineer is to give proper credit to the sources of ideas and data, and to avoid plagiarism.

Laboratory Exercises.

In addition to these web-enhanced questions at the end of each chapter, there is web-based laboratory home page for "Meteorology for Scientists and Engineers". At the time of writing this book, the web address has not been finalized. However, you can get to it from the publisher's web page: **http://www.brookscole.com/geo/**.

These utilize web-based tutorials on practical subjects beyond those covered in the book. Topics include satellite image interpretation, radar image interpretation, decoding METARs and TAFs, models used in numerical weather prediction, tutorials on thunderstorms and tornadoes, hurricanes, El Niño and climate change, and more.

W1. Download from the web a map of sea-level pressure, drawn as isobars, for your area. Become familiar with the units and symbols used on weather maps.

W2. Download from the web a map of near-surface air temperature, drawn is isotherms, for your area. Also, download a surface skin temperature map valid at the same time, and compare the temperatures.

W3. Download from the web a map of wind speeds at a height near the jet stream level (20 or 30 kPa, for example). This wind map should have isotachs drawn on it. If you can find a map that also has wind direction or streamlines in addition to the isotachs, that is even better.

W4. Download from the web a map of humidities (e.g., relative humidities, or any other type of humidity), preferably drawn is isohumes. These are often found at low altitudes, such as for pressures of 85 or 70 kPa.

W5. Search the web for info on the standard atmosphere. This could be in the form of tables, equations, or descriptive text. Compare this with the standard atmosphere in this textbook, to determine if the standard atmosphere has been revised.

W6. Search the web for the air-pollution regulation authority in your country (such as the EPA in the USA), and find the regulated concentrations of the most common air pollutants (CO, SO_2, O_3, NO_2, volatile organic compounds VOCs, and particulates). Compare with the results in Table 1-3, to see if the regulations have been updated in the USA, or if they are different for your country.

W7. Search the web for surface weather station observations for your area. This could either be a surface weather map with plotted station symbols, or a text table. Use the reported temperature and pressure to calculate the density.

W8. Search the web for updated information on the acceleration due to gravity, and how it varies with location on earth.

W9. Access from the web weather maps showing thickness between two pressure surfaces. One of the most common is the 1000 - 500 mb thickness chart (i.e., the 100 - 50 kPa thickness chart). Comment on how thickness varies with temperature (the most obvious example is the general thickness decrease further away from the equator).

W10. Access from the web an upper air sounding (e.g., Stuve, Skew-T, Tephigram, etc.) that plots temperature vs. height or pressure for a location near you. We will learn details about these charts later, but for now look at only temperature vs. height. If the sounding goes high enough (up to 10 kPa or so), can you identify the troposphere, tropopause, and stratosphere.

W11. Often weather maps have isopleths of temperature (isotherm), pressure (isobar), height (contour), humidity (isohume), potential temperature (adiabat or isentrope), or wind speed (isotach). Search the web for weather maps showing other isopleths. (Hint, look for isopleth maps of precipitation, visibility, snow depth, cloudiness, etc.)

Synthesis Questions

These are "what if" questions. They are often hypothetical ... on the verge of being science fiction. By thinking about "what if" questions you can gain significant insight about the physics of the atmosphere, because often you cannot apply existing paradigms.

"What if" questions are often asked by scientists, engineers, and policy makers. For example, what if the amount of carbon dioxide in the atmosphere doubled, then what would happen to world climate?

For many of these questions, there is not a single right answer. Different students could devise different answers that could be equally insightful, and if they are supported with reasonable arguments, should be worth full credit. Often one answer will have other implications about the physics, and will trigger a train of related ideas and arguments.

A solved example of a synthesis question is presented next. This solution might not be the only correct solution, if it is correct at all.

Solved Example - Synthesis

What if liquid water (raindrops) in the atmosphere caused the virtual temperature to increase [rather than decrease as currently shown by the negative sign in front of r_L in eq. (1.14)]. What would be different about the weather?

Solution

More and larger raindrops would cause warmer virtual temperature. This warmer air would act more buoyant (because warm air rises). This would cause updrafts in rain clouds that might be fast enough to prevent heavy rain from reaching the ground. *(continued next column)*

Solved Example - Synthesis *(continuation)*

But where would all this rain go? Would it accumulate at the top of thunderstorms, at the top of the troposphere? If droplets kept accumulating, they might act sufficiently warm to rise into the stratosphere. Perhaps layers of liquid water would form in the stratosphere, and would block out the sunlight from reaching the surface.

If less rain reached the ground, then there would be more droughts. Eventually all the oceans would evaporate, and life on earth as we know it would die.

But perhaps there would be life forms (insects, birds, fish, people) in this ocean layer aloft. The reasoning is that if liquid water increases virtual temperature, then perhaps other heavy objects (such as automobiles and people) would do the same.

In fact, this begs the question as to why liquid water would be associated with warmer virtual temperature in the first place. We know that liquid water is heavier than air, and that heavy things should sink. One way that heavy things like rain drops would not sink is if gravity worked backwards.

If gravity worked backwards, then all things would be repelled from earth into space. This textbook would be pushed into space, as would your instructor. So you would have never been assigned this exercise in the first place.

Life is full of paradoxes. Just be certain that you don't get a sign wrong in any of your equations ... who knows what might happen as a result.

S1. What if the meteorological angle convention is identical to that shown in Fig 1.2, except for wind directions which are given by where they blow towards rather than where they blow from. Create a new set of conversion equations (1.1 - 1.4) for this convention, and test them with directions and speeds from all compass quadrants.

S2. Find a translation of Aristotle's *Meteorologica* in your library. Discuss one of his erroneous statements, and how the error might have been avoided if he had following the Scientific Method as later contrived by Descartes.

S3. As discussed in a solved example, the glass on the front face of CRT and TV picture tubes is thick in order to withstand the pressure difference across it. Why is the glass not so thick on the other parts of the picture tube, such as the narrow neck near the back of the TV?

S4. Equations (1.6a) and (1.8) show how pressure and density decrease nearly exponentially with height.

a. How high is the top of the atmosphere?

b. Search the library or the web for the effective altitude for the top of the atmosphere as experienced by space vehicles re-entering the atmosphere.

S5. What is "ideal" about the ideal gas law? Are there equations of state that are not ideal?

S6. What if temperature as defined by eq (1.9) was not dependent on the molecular weight of the gas. Speculate on how the composition of the earth's atmosphere might have evolved differently since it was first formed.

S7. When you use a hand pump to inflate a bicycle or car tire, the pump usually gets hot near the outflow hose. Why? If pressure in the ideal gas law was proportional to the inverse of absolute virtual temperature ($P = \rho \cdot \Re_d / T_v$), would the tire-pump temperature be any different?

S8. In the definition of virtual temperature, why do water vapor and liquid water work in opposite directions (i.e., why do they have opposite signs in that equation).

S9. Virtual temperature includes a term for liquid-water. How should the equation (1.14) be modified to also include ice crystals in the air? Should it be modified to include flying birds?

S10. Meteorologists often convert actual station pressures to the equivalent "sea-level pressure" by taking into account the altitude of the weather station. The hypsometric equation can be applied to this job, assuming that the average virtual temperature is known. What virtual temperature should be used below ground to do this? What are the limitations of the result?

S11. Starting with our earth and atmosphere as at present, what if suddenly gravity were to become zero. What would happen to the atmosphere? Why?

S12. Suppose that gravitational attraction between two objects becomes greater, rather than smaller, as the distance between the two objects becomes greater.

a. Would the relationship between geometric altitude and geopotential altitude change? If so, what is the new relationship?

b. How would the vertical pressure gradient in the atmosphere be different, if at all?

c. Would the orbit of the earth around the sun be affected? How?

Science Graffito

Spray painted on the sidewalk of a large university was: "Obey gravity. It's the law."

RADIATION

CONTENTS

2 Solar energy powers the atmosphere. This energy warms the air and drives the motions we feel as winds. The seasonal distribution of this energy depends on the orbital characteristics of the earth around the sun.

The earth's rotation about its axis causes a daily cycle of sunrise, increasing solar radiation until solar noon, then decreasing solar radiation, and finally sunset. Most of this solar radiation is absorbed at the earth's surface, and provides the energy for photosynthesis and life.

Downward **infrared** (IR) radiation from the atmosphere to the earth is usually slightly less than upward IR radiation from the earth, causing net cooling at the earth's surface both day and night. The combination of daytime solar heating and continuous IR cooling yields a **diurnal** (daily) cycle of **net radiation**.

ORBITAL FACTORS

Planetary Orbits

Johannes Kepler, the 17th century astronomer, discovered that planets in the solar system have elliptical orbits around the sun. For most planets in the solar system, the ellipticity is relatively small, meaning the orbits are nearly circular. For circular orbits, he also found that the time period Y of each orbit is related to the distance R of the planet from the sun by:

$$Y = a \cdot R^{3/2} \tag{2.1}$$

Parameter $a \cong 0.1996$ d·(Gm)$^{-3/2}$, where d is earth days, and Gm is gigameters = 10^6 km.

Figs 2.1a & b show the orbital periods vs. distances for the planets in our solar system. These figures show the duration of a year for each planet, which affect the seasons experienced on the planet.

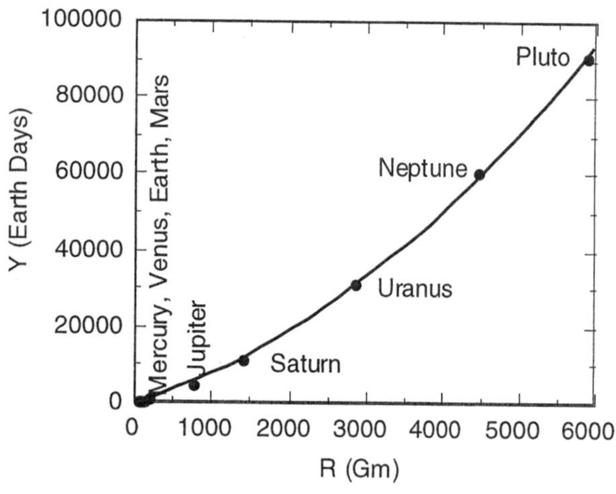

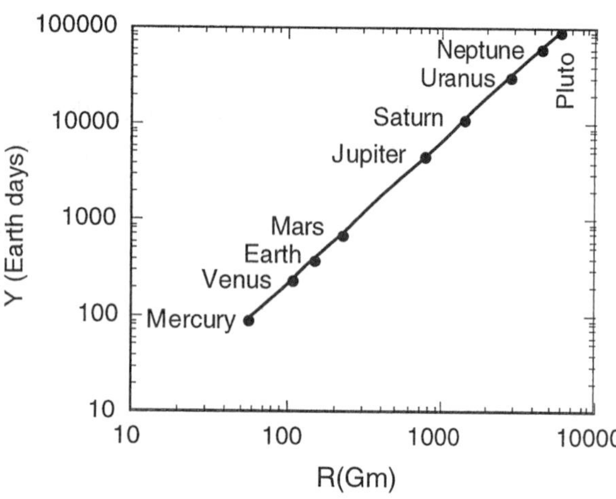

Figure 2.1 a(linear) & b(log-log)
Planetary orbital periods versus distance from sun.

Solved Example
Verify that eq. (2.1) gives the correct earth year.

Solution:
Given: R = 149.6 Gm avg. distance sun to earth.
Find: Y = ? days, the orbital period for earth

Use eq. (2.1):
$Y = (0.1996 \text{ d}\cdot(\text{Gm})^{-3/2}) \cdot [(149.6 \text{ Gm})^{1.5}]$
$=$ **365.2... days**.

Check: Units OK. Sketch OK. Almost 1 yr.
Discussion: In 365.0 days, the earth does not quite finish a complete orbit. After four years this shortfall accumulates to nearly a day, which we correct using a leap year with an extra day.

FOCUS • Satellite Orbits

Artificial satellites such as weather satellites orbiting the earth obey the same orbital mechanics as planets orbiting around the sun. For satellites in near-circular orbits, the pull by the earth's gravity f_G is balanced by centrifugal force f_C:

$$f_G = \frac{G \cdot M \cdot m}{R^2} \quad \bullet(2.2)$$

$$f_C = \left(\frac{2\pi}{t_{orbit}}\right)^2 \cdot m \cdot R \quad \bullet(2.3)$$

where R is the distance between the center of the earth and the satellite, m is the mass of the satellite, M is the mass of the earth ($5.98\text{x}10^{24}$ kg), and G is the gravitational constant ($6.67\text{x}10^{-11}$ N·m^2·kg^{-2}).

Solving for the orbital time period t_{orbit} by setting $f_G = f_C$:

$$t_{orbit} = \frac{2\pi \cdot R^{3/2}}{\sqrt{G \cdot M}} \quad \bullet(2.4)$$

Orbital period does not depend on satellite mass, but increases as satellite altitude increases.

Solved Example
The orbital period of geostationary satellites equals the rotation rate of the earth , 24 hours. At what altitude above the earth's surface must the satellite be parked to achieve this orbital period?

Solution
Given: t_{orbit} = 24 h = 86400 s.
Find: z = ? km

Rearrange eq. (2.4):
$R = (t_{orbit} / 2\pi)^{2/3} \cdot (G \cdot M)^{1/3} = 42{,}251$ km
From this subtract earth radius R_o = 6370 km to get height above the surface:
$z = R - R_o$ = **35,881 km**

Check: Units OK. Physics OK.
Discussion: The advantage of geostationary satellites is that they always see the same portion of the globe, and can take a sequence of pictures that can be looped as a movie to display cloud motions. The disadvantage is that they are so far from the earth that large magnification is needed on the satellite to resolve the smaller clouds.

Orbit of the Earth

The earth and the moon rotate with a period of 27.32 days around their common center of gravity, called the earth-moon **barycenter**. Because the mass of the moon is only 1.23% of the mass of the earth (earth mass is 5.9742x10^{24} kg), the barycenter is much closer to the center of the earth than to the center of the moon. This barycenter is 4671 km from the center of the earth, which is below the earth's surface (earth radius is about 6371 km).

To a first approximation, the earth-moon barycenter orbits around the sun in an **elliptical** orbit (Fig 2.2, thin-line ellipse) with period of P = 365.25463 days. Length of the **semi-major axis** (half of the longest axis) of the ellipse is a = 149.457 Gm, which is the definition of one **astronomical unit** (au).

Semi-minor axis (half the shortest axis) length is b = 149.090 Gm. The center of the sun is at one of the foci of the ellipse, and half the distance between the two foci is c = 2.5 Gm, where $a^2 = b^2 + c^2$. The orbit is close to circular, with an **eccentricity** of only about $e \equiv c/a = 0.0167$ (a circle has zero eccentricity).

The closest distance (**perihelion**) between the earth and sun is $a - c$ = 146.96 Gm and occurs on τ = 3 January. The farthest distance (**aphelion**) is $a + c$ = 151.96 Gm and occurs on 4 July. Because the earth is rotating around the earth-moon barycenter while this barycenter is revolving around the sun, the location of the center of the earth traces a slightly wiggly path as it orbits the sun. This path is exaggerated in Fig 2.2 (thick line).

The angle at the sun between the perihelion and the location of the earth (actually to the earth-moon barycenter) is called the **true anomaly** ν (see Fig 2.2). This angle increases during the year as time t increases from the perihelion date τ = 3 January. According to **Kepler's second law**, the angle increases more slowly when the earth is further from the sun, such that a line connecting the earth and the sun will sweep out equal areas in equal times.

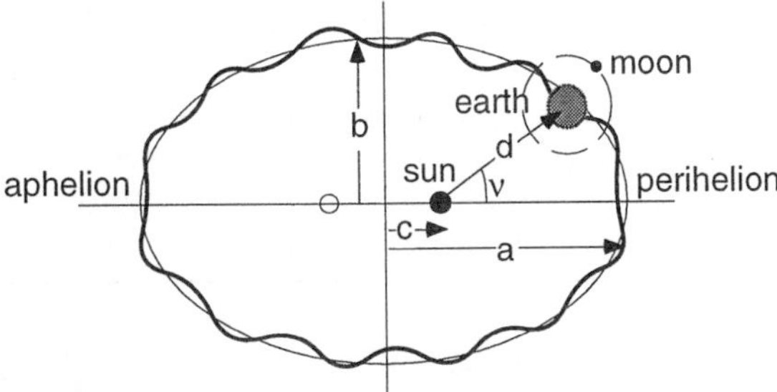

Figure 2.2
Geometry of the earth's orbit. (Not to scale.)

An angle called the **mean anomaly** M is a simple, but good approximation to ν. It is defined by:

$$M = C \cdot \frac{t - \tau}{P} \qquad (2.5)$$

where P is the orbital period and C is the angle of a full circle. ($C = 2 \cdot \pi$ radians = 360°. Use whichever is appropriate for your calculator, spreadsheet, or computer program.) Because the earth's orbit is nearly circular, $\nu \cong M$. An even more exact approximation to the **true anomaly** for the elliptical earth orbit is

$$\begin{aligned} \nu = M \ & +0.0333988 \cdot \sin(M) + 0.0003486 \cdot \sin(2 \cdot M) \\ & +0.0000050 \cdot \sin(3 \cdot M) \end{aligned} \qquad (2.6)$$

The distance R between the sun and earth (actually to the earth-moon barycenter) as a function of time is

$$R = a \cdot \frac{1 - e^2}{1 + e \cdot \cos(\nu)} \qquad (2.7)$$

where e is eccentricity. If the simple approximation of $\nu \cong M$ is used, then angle errors are less than 2° and distance errors are less than 0.06%.

Solved Example (§)

Use a spreadsheet to find the true anomaly and sun-earth distance for several days during the year. Use the simple approximation for ν.

Solution

Given: τ = 3 Jan. P = 365.25 days.
Find: ν = ?° and R = ? Gm.

Sketch: (same as Fig 2.2)
Use eqs. (2.5), (2.7) and $\nu \cong M$

t (date)	M (deg)	R (Gm)
3-Jan	**0.0**	**146.96**
17-Jan	**13.8**	**147.03**
31-Jan	**27.6**	**147.24**
14-Feb	**41.4**	**147.57**
28-Feb	**55.2**	**148.00**
14-Mar	**69.0**	**148.53**
28-Mar	**82.8**	**149.10**
11-Apr	**96.6**	**149.70**
25-Apr	**110.4**	**150.29**

Check: Units OK. Physics OK.
Discussion: As time increases during winter and spring, so does the angle and the distance.

Seasonal Effects

The tilt of the earth's axis relative to the **ecliptic** (i.e., the orbital plane of the earth around the sun) is Φ_r = 23.45° = 0.409 radians. By definition, this angle equals the latitude of the **Tropic of Cancer** in the northern hemisphere (Fig 2.3). Latitudes are defined to be positive in the northern hemisphere. **Tropic of Capricorn** in the southern hemisphere is the same angle, but with a negative sign.

The **solar declination angle** δ_s is defined as the angle between the ecliptic and the plane of the earth's equator (Fig 2.3). Because the direction of tilt of the earth's axis is nearly constant with respect to the fixed stars, the solar declination angle varies from +23.45° on 22 June (**summer solstice** in the northern hemisphere) to –23.45° on 22 December (**winter solstice**).

Define a relative **Julian Day,** d, as the day of the year. For example, 15 January corresponds to d = 15. For 5 February, d = 36 (= 31 days in Jan + 5 days in Feb). For non-leap years, the summer solstice on 22 June corresponds to $d_r \equiv d = 173$. The number of days per year is $d_y = 365$. On a leap year, use d_y = 366.

The solar declination angle for any day of the year is given by

$$\delta_s = \Phi_r \cdot \cos\left[\frac{C\cdot(d-d_r)}{d_y}\right] \qquad \bullet(2.8)$$

This equation is only approximate, because it assumes the earth's orbit is circular rather than elliptical.

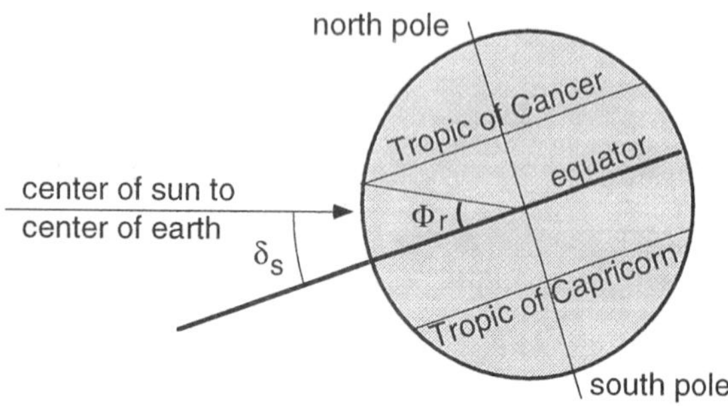

Figure 2.3
Relationship of declination angle δ_S to tilt of the earth's axis.

Solved Example

Find the solar declination angle on 5 March.

Solution

Assume: Not a leap year.
Given: d = 31 Jan + 28 Feb + 5 Mar = 64.
Find: δ_s = ?°.

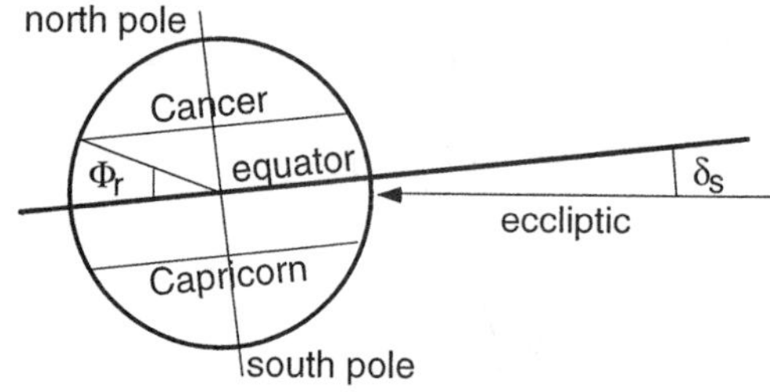

Use eq. (2.8):

$$\delta_S = 23.45°\cdot\cos\left[\frac{360°\cdot(64-173)}{365}\right]$$
$$=23.45°\cdot\cos[-107.51°] = \underline{-7.05°}$$

Check: Units OK. Sketch OK. Physics OK.
Discussion: On the **vernal equinox** (21 March), the angle should be zero. Before that date, it is winter and the declination angle should be negative. In spring and summer, the angle is positive. Because 5 March is near the end of winter, we expect the answer to be a small negative angle.

Daily Effects

As the earth rotates about its axis, the **local elevation angle** Ψ of the sun above the local horizon rises and falls. This angle depends on the latitude ϕ and longitude λ_e of the location:

$$\sin(\Psi) = \sin(\phi)\cdot\sin(\delta_s) - \cos(\phi)\cdot\cos(\delta_s)\cdot\cos\left[\frac{C\cdot t_{UTC}}{t_d} - \lambda_e\right] \qquad \bullet(2.9)$$

Time of day t_{UTC} is **Coordinated Universal Time** (**UTC**, also known as Greenwich Mean Time **GMT** or Zulu **Z** time), C = 2π radians = 360° as before, and the length of the day is t_d. For t_{UTC} in hours, then t_d = 24 h. Latitudes are positive north of the equator, and longitudes are positive west of the Greenwich meridian. The sin(Ψ) relationship is used later in this chapter to calculate the daily cycle of solar energy reaching any point on earth.

The local **azimuth angle** α of the sun relative to north is

$$\cos(\alpha) = \frac{\sin(\delta_s) - \sin(\phi)\cdot\cos(\zeta)}{\cos(\phi)\cdot\sin(\zeta)} \qquad (2.10)$$

where $\zeta = C/4 - \Psi$ is the **zenith angle**. After noon, the azimuth angle might need to be corrected to be $\alpha = C - \alpha$, so that the sun sets in the west instead of the east. For example, Fig 2.4 shows the elevation and azimuth angles for Vancouver (latitude = 49.25°N, longitude = 123.1°W) during the solstices and equinoxes.

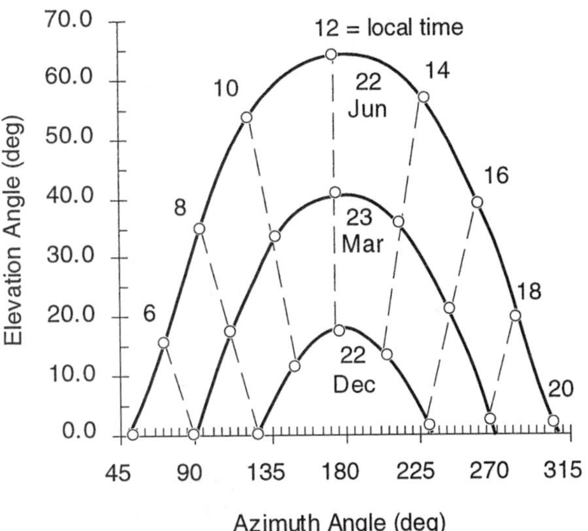

Figure 2.4
Position (solid lines) of the sun for Vancouver, Canada for various seasons. September 21 and March 23 nearly coincide. Isochrones are dashed. All times are Pacific standard time.

Solved Example

Find the local elevation angle of the sun at 3 PM local time on 5 March in Vancouver, Canada.

Solution

Assume: Not a leap year. Also, Vancouver is in the Pacific time zone, where $t_{UTC} = t + 8$ h.

Given: t = 3 PM = 15 h. Thus, t_{UTC} = 23 h.
ϕ = 49.25°N, λ_e = 123.1°W for Vancouver.
δ_s = –7.05° from previous solved example.

Find: Ψ = ?°.

Use eq. (2.9): $\sin(\Psi) =$
$= \sin(49.25°)\cdot\sin(-7.05°)\ \cos(49.25°)\cdot\cos(-7.05°)\cdot\cos[360°\cdot(23h/24h) - 123.1°]$
$= 0.7576\cdot(-0.1227) - 0.6527\cdot 0.9924\cdot\cos(221.9°)$
$= -0.09296 + 0.4821 = 0.3891$
$\Psi = \arcsin(0.3891) =$ **<u>22.90°</u>**

Check: Units OK. Physics OK.
Discussion: The sun is above the local horizon, as expected for mid afternoon. Beware of other situations such as night that give negative elevation angle. Normally, when the sun is below the horizon, set the elevation angle = 0.

Solved Example (§)

Use a spreadsheet to plot elevation angle vs time at Vancouver, for 22 Dec, 23 Mar, and 22 Jun. Plot these three curves on the same graph.

Solution

Given: Same as previous solved example, except d = 355, 82, 173.

Find: Ψ = ?°.

A portion of the tabulated results are shown below, as well as the full graph.

	Ψ (°)		
t (h)	22 Dec	23 Mar	22 Jun
3	0.0	0.0	0.0
4	0.0	0.0	0.0
5	0.0	0.0	6.6
6	0.0	0.0	15.6
7	0.0	7.8	25.1
8	0.0	17.3	34.9
9	5.7	25.9	44.5
10	11.5	33.2	53.4
11	15.5	38.4	60.5
12	17.2	40.8	64.1
13	16.5	39.8	62.6

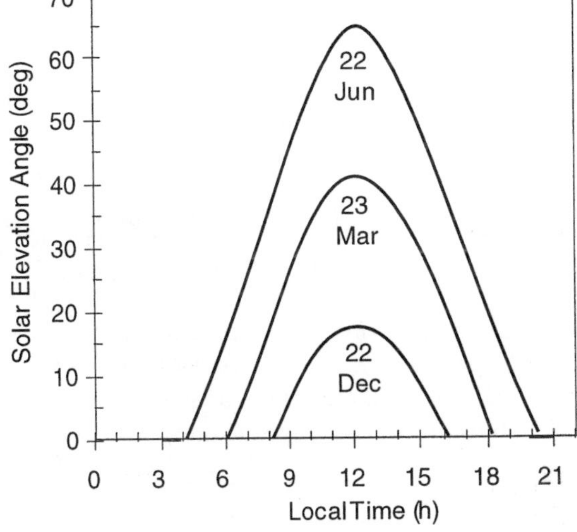

Check: Units OK. Physics OK. Graph OK.
Discussion: Summers are pleasant with long days. The peak elevation does not happen precisely at local noon, because Vancouver is not centered within its time zone.

Sunrise, Sunset & Twilight

Geometric sunrise and sunset occur when the center of the sun has zero elevation angle. **Apparent sunrise/set** are defined as when the *top* of the sun crosses the horizon, as *viewed* by an observer on the surface. The sun has a finite radius corresponding to an angle of 0.267° when viewed from earth. Also, refraction (bending) of light through the atmosphere allows the top of the sun to be seen even when it is really 0.567° below the horizon. Thus, apparent sunrise/set occurs when the center of the sun has an elevation angle of –0.833°.

When the apparent top of the sun is slightly below the horizon, the surface of the earth is not receiving *direct* sunlight. However, the surface can still receive *indirect* light scattered from air molecules higher in the atmosphere that are illuminated by the sun. The interval during which scattered light is present at the surface is called **twilight**. Because twilight gradually fades as the sun is lower below the horizon, there is no precise definition of the start of sunrise twilight or the end of sunset twilight.

Arbitrary definitions have been adopted by different organizations to define twilight. **Civil twilight** occurs whenever the sun center is no lower than –6°, and is based on the ability of humans to see objects on the ground. **Military twilight** occurs for a sun no lower than –12°. **Astronomical twilight** ends when the skylight becomes sufficiently dark to view certain stars, at solar elevation angle –18°.

Table 2-1 summarizes the solar elevation angle Ψ definitions used for sunrise/set and twilight. All of these angles are at or below the horizon.

Time-of-day corresponding to these events can be found by rearranging eq. (2.9):

$$t_{UTC} = \frac{t_d}{C} \cdot \left\{ \lambda_e \pm \arccos\left[\frac{\sin\phi \cdot \sin\delta_s - \sin\Psi}{\cos\phi \cdot \cos\delta_s} \right] \right\} \qquad (2.11)$$

where the appropriate elevation angle is used from Table 2-1. Where the ± sign appears, use + for sunrise and – for sunset. If any of the answers are negative, add 24 h to the result. Don't forget to convert from UTC to the time in your local time zone.

Table 2-1. Elevation angles for diurnal events.

Event	Ψ (°)	Ψ (rad)
Sunrise/sunset:		
Geometric	0	0
Apparent	–0.833	–0.01454
Twilight:		
Civil	–6	–0.10472
Military	–12	–0.20944
Astronomical	–18	–0.31416

Solved Example (§)

Use a spreadsheet to find the Pacific standard time for all the events of Table 2-1, for Vancouver, Canada during 22 Dec, 23 Mar, and 22 Jun.

Solution

Given: Julian dates 355, 82, & 173.
Find: t = ? h (local standard time)
Assume: Pacific time zone: $t_{UTC} = t + 8$ h.

Use eq. (2.11) and Table 2-1.:

	22Dec	23Mar	22Jun
Morning: (Pacific standard time in hours)			
geometric sunrise	8.22	6.20	4.19
apparent sunrise	8.11	6.11	4.09
civil twilight starts	7.49	5.58	3.36
military twilight starts	6.80	4.96	2.32
astron. twilight starts	6.16	4.31	n/a
Evening:			
geometric sunset	16.19	18.21	20.22
apparent sunset	16.30	18.30	20.33
civil twilight ends	16.93	18.83	21.05
military twilight ends	17.61	19.45	22.09
astron. twilight ends	18.26	20.10	n/a

Check: Units OK. Physics OK.
Discussion: During the summer solstice (22 June), the sun never gets below –18°. Hence, it is astronomical twilight all night in Vancouver in mid summer.

FLUX

A **flux density**, $\Im$, called the **flux** in this book, is the rate of transfer of a quantity per unit area per unit time. The area is taken perpendicular (normal) to the direction of flux movement. Examples with metric (SI) units are mass flux (kg· $m^{-2} \cdot s^{-1}$) and heat flux, ($J \cdot m^{-2} \cdot s^{-1}$). Using the definition of a watt (1 W = 1 $J \cdot s^{-1}$), the heat flux can also be given in units of ($W \cdot m^{-2}$). A flux is a measure of the amount of inflow or outflow such as through the side of a fixed volume, and thus is frequently used in Eulerian frameworks (Fig 2.5).

Because flow is associated with a direction, so is flux associated with a direction. We must account for fluxes $\Im_x$, $\Im_y$, and $\Im_z$ in the x, y, and z directions, respectively. A flux in the positive x-direction (eastward) is written with a positive value of $\Im_x$, while a flux towards the opposite direction (westward) is negative.

The total amount of heat or mass flowing through a plane of area A during time interval Δt is given by:

$$Amount = \Im \cdot A \cdot \Delta t \qquad (2.12)$$

For heat, $Amount \equiv \Delta Q_H$ by definition.

Fluxes are sometimes written in **kinematic** form, F, by dividing by the air density, ρ_{air}.

$$F = \frac{\Im}{\rho_{air}} \qquad (2.13a)$$

Kinematic mass flux equals the wind speed, M.

Heat fluxes $\Im_H$ can be put into kinematic form by dividing by both air density ρ_{air} and the specific heat for air C_p, which yields a quantity equal to temperature times wind speed (K·m·s^{-1}).

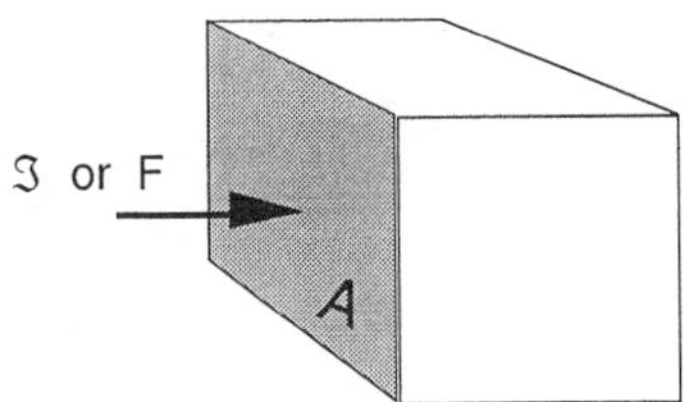

Figure 2.5
Flux F through an area A into one side of a volume.

> **Solved Example**
> The mass flux of air is 1 kg·m^{-2}·s^{-1} through a door opening that is 1 m wide by 2.5 m tall. What amount of mass of air passes through the door each minute, and what is the kinematic flux?
>
> **Solution**
> Given: A = (1 m) · (2.5 m) = 2.5 m^2
> $\Im$ = 1 kg·m^{-2}·s^{-1}
> Find: (a) $Amount$ = ? kg, and (b) F = ? m·s^{-1}
> Sketch: (see Fig 2.5)
>
> (a) Use eq. (2.12):
> $Amount$ = (1 kg·m^{-2}·s^{-1})·(2.5 m^2)·(1 min)·(60 s/min) = **150 kg**.
> (b) Assume: ρ = 1.225 kg·m^{-3} at sea-level
> Use eq. (2.13a):
> F = (1 kg·m^{-2}·s^{-1}) / (1.225 kg·m^{-3})
> = **0.82 m·s^{-1}**.
>
> **Check**: Units OK. Sketch OK. Physics OK.
> **Discussion**: The kinematic flux is equivalent to a very light wind speed of less than 1 m/s blowing through the doorway, yet it transports quite a large amount of mass each minute.

$$F_H = \frac{\Im_H}{\rho_{air} \cdot C_p} \quad \text{for heat only} \qquad (2.13b)$$

For dry air (subscript "d") at sea level:

$\rho_{air} \cdot C_{pd}$ = 1231 (W·m^{-2}) / (K·m·s^{-1})
= 12.31 mb·K^{-1}
= 1.231 kPa·K^{-1}.

The reason for sometimes expressing fluxes in kinematic form is that the result is given in terms of easily measured quantities. For example, while most people do not have "Watt" meters to measure the normal "dynamic" heat flux, they do have thermometers and anemometers. The resulting temperature times wind speed has units of a kinematic heat flux (K·m·s^{-1}). Similarly, for mass flux it is easier to measure wind speed than kilograms of air.

Heat fluxes can be caused by a variety of processes. Radiative fluxes are radiant energy (electromagnetic waves or photons) per unit area per unit time. This flux can travel through a vacuum. Advective flux is caused by wind blowing through an area, and carrying with it warmer or colder temperatures. For example a warm wind blowing toward the east causes a positive flux component F_x. A cold wind blowing toward the west also gives positive F_x. Turbulent fluxes are caused by eddy motions in the air, while conductive fluxes are caused by molecules bouncing into each other.

RADIATION PRINCIPLES

Propagation

Radiation can be modeled as electromagnetic waves, or as photons. Radiation propagates through a vacuum at a constant speed: c_o = 299,792,458 m·s^{-1}. For practical purposes, we can approximate this **speed of light** as $c_o \cong 3\text{x}10^8$ m·s^{-1}. Light travels slightly slower through air, at roughly c = 299,710,000 m·s^{-1} at standard sea-level pressure and temperature.

Using the wave model of radiation, the **wavelength** λ (m·cycle^{-1}) is related to the **frequency**, ν (Hz = cycles·s^{-1}) by:

$$\lambda \cdot \nu = c_o \qquad (2.14)$$

Wavelength units are sometimes abbreviated as m. Because the wavelengths of light are so short, they are often expressed in units of micrometers (μm).

Wavenumber is the number of waves per meter: σ (cycles·m^{-1}) = 1 / λ. Its units are sometimes abbreviated as m^{-1}. **Circular frequency** or **angular frequency** is ω (radians/s) = $2\pi \cdot \nu$. Its units are sometimes abbreviated as s^{-1}.

Solved Example
Red light has a wavelength of 0.7 μm. Find its frequency, circular frequency, and wavenumber in a vacuum.

Solution
Given: c_o = 299,792,458 m/s, λ = 0.7 μm
Find: ν = ? Hz, ω = ? s^{-1}, σ = ? m^{-1}.

Use eq. (2.14), solving for ν:
$\nu = c_o/\lambda$ = (3x10^8 m/s) / (0.7x10^{-6} m/cycle)
= **4.28x10^{14}** Hz.
ω = $2\pi\cdot\nu$ = 2·(3.14159)·(4.28x10^{14} Hz)
= **2.69x10^{15}** s^{-1}.
σ = $1/\lambda$ = 1 / (0.7x10^{-6} m/cycle)
= **1.43x10^6** m^{-1}.

Check: Units OK. Physics reasonable.
Discussion: Wavelength, wavenumber, frequency, and circular frequency are all equivalent ways to express the "color" of radiation.

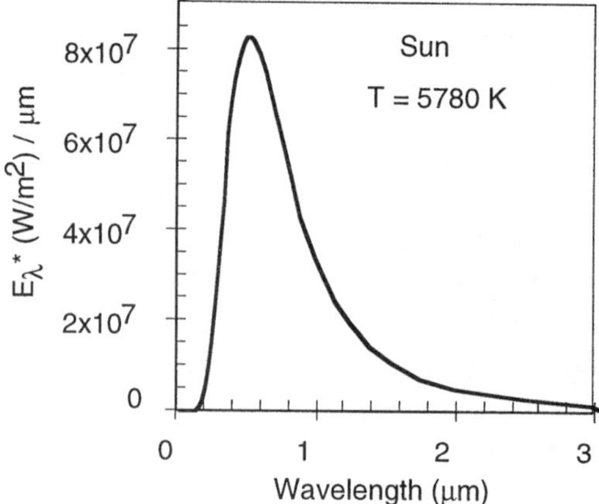

Figure 2.6
Planck blackbody irradiance, E_λ, from the sun.*

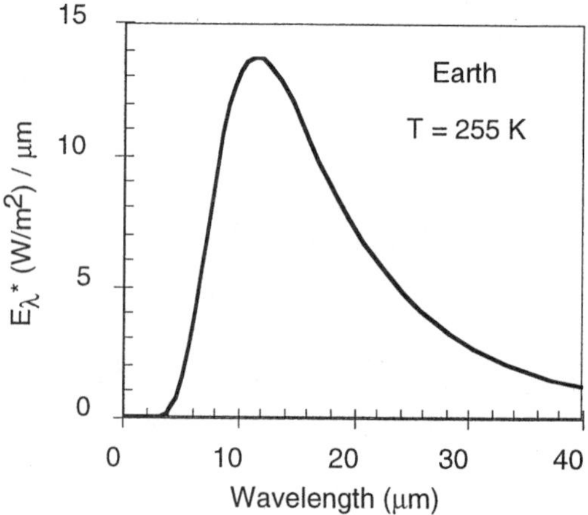

Figure 2.7
Planck blackbody irradiance, E_λ, from the earth.*

Emission

Objects warmer than absolute zero can emit radiation. An object that emits the maximum possible radiation for its temperature is called a **blackbody**. **Planck's law** gives the amount of blackbody monochromatic (single wavelength or color) radiative flux, called **irradiance**, E_λ*:

$$E_\lambda{}^* = \frac{c_1}{\lambda^5 \cdot \left[\exp(c_2 / \lambda \cdot T) - 1\right]} \qquad \bullet(2.15)$$

where T is absolute temperature, and the asterisk indicates blackbody. The two constants are

c_1 = 3.74 x 10^8 W·m^{-2} · µm^4
c_2 = 1.44 x 10^4 µm·K

To good approximation for typical temperatures on the earth or sun, Planck's law can be simplified to be:

$$E_\lambda{}^* = c_1 \cdot \lambda^{-5} \cdot \exp(-c_2 / \lambda \cdot T) \qquad (2.16)$$

Actual objects can emit less than the theoretical black body value: $E_\lambda = e_\lambda \cdot E_\lambda{}^*$, where $0 \le e_\lambda \le 1$ is **emissivity**, a measure of emission efficiency.

The Planck curve (eq. 2.15) for emission from the sun (T = 5780 K) is plotted in Fig 2.6. Peak emissions from the sun are in the visible range of wavelengths (0.4 µm for violet light through 0.7 µm for red light). Radiation from the sun is called **solar radiation** or **short-wave radiation.**

The Planck curve for emission from the whole earth-atmosphere system (T = 255 K) is plotted in Fig 2.7. Peak emissions from the average earth system are in the infrared range 8 to 18 µm. This radiation is called **terrestrial radiation, long-wave radiation**, or infrared (**IR**) radiation.

The wavelength of the peak emission is given by **Wien's law:**

$$\lambda_{max} = \frac{a}{T} \qquad (2.17)$$

where a = 2897 μm·K.

The total amount of emission (= area under Planck curve = total irradiance) is given by the

Stefan-Boltzmann law:

$$E^* = \sigma_{SB} \cdot T^4 \qquad \bullet(2.18)$$

where $\sigma_{SB} = 5.67\text{x}10^{-8}$ W·m^{-2}·K^{-4} is the **Stefan-Boltzmann constant,** and E* has units of W·m^{-2}.

Solved Example

Find the black-body monochromatic irradiance of green light of wavelength 0.53 µm from an object of temperature 3000 K.

Solution

Given: $\lambda = 0.53$ µm, $T = 3000$ K
Find: $E_\lambda^* = ?$ W·m^{-2}· µm^{-1} .

Use eq. (2.15):

$$E_\lambda^* = \frac{c_1}{\lambda^5 \cdot [\exp(c_2 / \lambda \cdot T) - 1]}$$

$$= \frac{(3.74 \times 10^8 \,\text{Wm}^{-2} \mu\text{m}^4) / (0.53 \mu\text{m})^5}{\left\{\exp\left[1.44 \times 10^4 \mu\text{mK} / (0.53 \mu\text{m} \cdot 3000\text{K})\right] - 1\right\}}$$

$= \underline{\mathbf{1.04 \times 10^6 \ W \cdot m}}^{-2} \cdot \mu m^{-1}$

Check: Units OK. Physics reasonable.
Discussion: Because 3000 K is cooler than the sun, about 50 times less green light is emitted. The answer is the watts emitted from each square meter of surface per µm wavelength increment.

Solved Example

What is the total irradiance (radiative flux) emitted from a blackbody earth at $T = 255$ K?

Solution

Given: $T = 255$ K
Find: E^* = ?W·m^{-2}

Sketch:

Use eq. (2.18):
E^*= (5.67x10^{-8} W·m^{-2}·K^{-4})·(255 K)^4 = **240 W·m^{-2}**.

Check: Units OK. Sketch OK. Physics OK.
Discussion: This doesn't seem like much radiation. One could create the same flux by placing a perfectly-efficient 240 W light bulb in front of a parabolic mirror that reflects the light and IR radiation into a beam that is 1.13 m in diameter.

For comparison, the surface area of the earth is about 5.1x10^{14} m^2, which when multiplied by the flux gives the total emission of 1.22x10^{17} W. Thus, the earth is quite a large-wattage IR light bulb.

ON DOING SCIENCE • Scientific Laws - The Myth

There are no scientific laws. Some theories or models have succeeded for every case tested so far, yet they may fail for other situations. Newton's "Laws of Motion" were accepted as laws for centuries, until they were found to fail in quantum mechanical and relativistic situations. It is better to use the word "relationship" instead of "law". In this textbook, the word "law" is used only for sake of tradition.

Einstein said "No amount of experimentation can ever prove me right; a single experiment can prove me wrong."

Because a single experiment can prove a relationship wrong, it behoves us as scientists to test theories and equations not only for reasonable values of variables, but also in the limit of extreme values, such as when the variables approach zero or infinity. These are often the most stringent tests of a relationship.

Example

Query: Eq. (2.16) was given as an approximation to Planck's law. Why is it not a perfect substitute?

Solution: If you compare the numerical answers from eqs. (2.15) and (2.16), you find that they agree very closely over the range of temperatures of the sun and the earth, and over a wide range of wavelengths. But...

a) What happens as temperature approaches absolute zero? For eq. (2.15), T is in the denominator of the argument of an exponential, which itself is in the denominator of eq. (2.15). If $T = 0$, then $1/T = \infty$. Exp(∞) = ∞ , and $1/\infty = 0$. Thus, $E_\lambda^* = 0$, as it should. Namely, no radiation is emitted at absolute zero (according to classical theory).

For eq. (2.16), if $T = 0$, then $1/T = \infty$, and exp($-\infty$) = 0. So it also agrees in the limit of absolute zero. Thus, both equations give the expected answer.

b) What happens as temperature approaches infinity? For eq. (2.15), if $T = \infty$, then $1/T = 0$, and exp(0) = 1. Then 1 – 1 = 0 in the denominator, and 1/0 = ∞ . Thus, $E_\lambda^* = \infty$, as it should. Namely, infinite radiation is emitted at infinite temperature.

For eq. (2.16), if $T = \infty$, then $1/T = 0$, and exp(–0) = 1. Thus, $E_\lambda^* = c_1 \cdot \lambda^{-5}$, which is not infinity. Hence, this approaches the wrong answer in the ∞ T limit.

Conclusion: Thus eq. (2.16) is not a perfect relationship, because it fails for this one case (b).

Science Graffito

"The great tragedy of science — the slaying of a beautiful hypothesis by an ugly fact." – Aldous Huxley

ON DOING SCIENCE • Seek Solutions

Most differential equations describing meteorological phenomena cannot be solved analytically. They cannot be integrated; they do not appear in a table of integrals; and they are not covered by the handful of mathematical tricks that you learned in math class.

But there is nothing magical about an analytical solution. **Any reasonable solution is better than no solution.** Be creative.

While thinking of creative solutions, also **think of ways to check your answer.** Is it the right order of magnitude, right sign, right units, does it approach a known answer in some limit, must it satisfy some other physical constraint or law or budget?

Example

Find the irradiance that can pass through an atmospheric "window" between λ_1 and λ_2.

Solution:

Method: Integrate Planck's law between the specified wavelengths. This is the area under the curve.

Check: The area under the whole spectral curve should yield the Stefan-Boltzmann (SB) law. Namely, the answer should be smaller than the SB answer, but should increase and converge to the SB answer as the lower and upper λ limits approach 0 and ∞.

Methods:

- Pay someone else to get the answer (Don't do this in school!), but be sure to check it yourself.
- Look up the answer in a Table of Integrals.
- Integrate it using the tricks you learned in math class.
- Integrate it using a symbolic equation solver on a computer, such as Mathematica or Maple.
- Find an approximate solution to the full equation. For example, integrate it numerically on a computer. (Trapezoid method, Gaussian integration, finite difference iteration, etc.)
- Find an exact solution for an approximation to the eq., such as a model or idealization of the physics. Most eqs. in this textbook have used this approach.
- Draw the Planck curve on graph paper. Count the squares under the curve between the wavelength bands, and compare to the value of each square, or to the area under the whole curve.
- Draw the curve, and measure area with a planimeter.
- Draw the Planck curve on cardboard or thick paper. Cut out the whole area under the curve. Weigh it. Then cut the portion between wavelengths, & weigh again.

BEYOND ALGEBRA • Incremental Changes

What happens to the total irradiance given by eq. (2.18) if temperature increases by 1°C? Such a question is important for climate change.

Solution using calculus:

Calculus allows a simple, elegant way to find the solution. By taking the derivative of both the left and right sides of eq. (2.18), one finds that:

$$dE^* = 4 \cdot \sigma_{SB} \cdot T^3 dT$$

assuming that σ_{SB} is constant. It can be written as

$$\Delta E^* = 4 \cdot \sigma_{SB} \cdot T^3 \cdot \Delta T$$

for small ΔT.

Thus, the same increase in temperature of ΔT = 1°C causes a much larger irradiance increase at high temperatures than at cold, because of the T^3 factor on the RHS.

Solution using algebra:

This particular problem could also have been solved using algebra, but with a more tedious and less elegant solution. First let

$$E_2 = \sigma \cdot T_2^4 \quad \text{and} \quad E_1 = \sigma \cdot T_1^4$$

Next, take the difference between these two eqs:

$$\Delta E \equiv E_2 - E_1 = \sigma \cdot \left[T_2^4 - T_1^4\right]$$

Recall from algebra that $(a^2 - b^2) = (a - b)\cdot(a + b)$

$$\Delta E = \sigma \cdot \left(T_2^{\,2} - T_1^{\,2}\right) \cdot \left(T_2^{\,2} + T_1^{\,2}\right)$$

$$\Delta E = \sigma \cdot (T_2 - T_1) \cdot (T_2 + T_1) \cdot \left(T_2^{\,2} + T_1^{\,2}\right)$$

But $(T_2 - T_1) / T_1 << 1$, then $(T_2 - T_1) = \Delta T$, but $T_2 + T_1 \cong 2T$, and $T_2^2 + T_1^2 = 2T^2$. This gives:

$$\Delta E = \sigma \cdot \Delta T \cdot 2T \cdot 2T^2$$

Or

$$\Delta E \approx \sigma \cdot \left[4T^3 \Delta T\right]$$

which is identical to the answer from calculus.

We were lucky this time, but it is not always possible to use algebra where calculus is needed.

Distribution

Radiation emitted from a spherical source decreases with the square of the distance from the center of the sphere:

$$E_2^* = E_1^* \cdot \left(\frac{R_1}{R_2}\right)^2 \qquad \bullet(2.19)$$

where R is the radius from the center of the sphere, and the subscripts denote two different distances from the center. This is called the **inverse square law**.

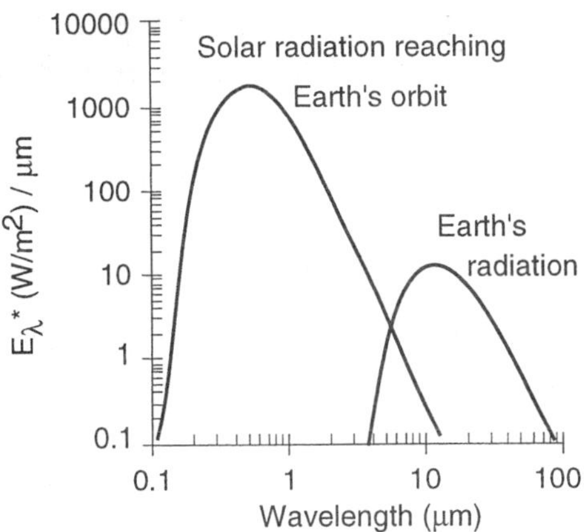

Figure 2.8
Black-body irradiance E reaching top of Earth's atmosphere from the sun and irradiance of terrestrial radiation leaving the top of the atmosphere, plotted on log-log graph.*

The reasoning behind eq. (2.19) is that as radiation from a small sphere spreads out radially, it passes through ever-larger conceptual spheres. If no energy is lost during propagation, then the total energy passing across the surface of each sphere must be conserved. Because the surface areas of the spheres increase with the square of the radius, this implies that the energy flux density must decrease at the same rate; i.e., inversely to the square of the radius.

From eq. (2.19) we expect that the radiative flux reaching the earth's orbit is greatly reduced from that at the surface of the sun. The solar emissions of Fig 2.6 must be reduced by a factor of 2.167×10^{-5}, based on the square of the ratio of solar radius to earth-orbital radius from eq. (2.19). This result is compared to the emission from earth in Fig 2.8.

The area under the solar-radiation curve in Fig 2.8 is the total (all wavelengths) **solar irradiance** reaching the earth's orbit. This quantity is commonly called the **solar constant**, S. The actual value of the solar irradiance measured at the top of the atmosphere by satellites is about $S = 1368 \pm 7$ W·m^{-2}, but it varies slightly. In kinematic units (based on sea-level density), the solar constant is roughly S = 1.125 K·m/s.

According to the inverse-square law, variations of distance between earth and sun cause changes of the solar constant:

$$S = S_o \cdot \left(\frac{\overline{R}}{R}\right)^2 \qquad (2.20)$$

where S_o = 1368 W·m^{-2} is the average solar constant measured at an average distance $\overline{R}$ = 149.6 Gm between the sun and earth. Remember that the solar constant is the flux across a surface that is **perpendicular** to the solar beam, measured at the top of the atmosphere.

Radiative flux (E = irradiance) such as the solar constant is the amount of energy crossing a unit area that is <u>perpendicular</u> to the path of the radiation. If this radiation strikes a surface that is not perpendicular to the radiation, then the radiation per unit surface area is reduced according to the **sine law**. The resulting flux, $\Im_{rad}$, at this surface is:

$$\Im_{rad} = E \cdot \sin(\Psi) \qquad (2.21a)$$

where Ψ is the **elevation angle** (the angle of the sun above the earth's surface). In kinematic form, this is

$$F_{rad} = \frac{E}{\rho \cdot C_p} \cdot \sin(\Psi) \qquad \bullet(2.21b)$$

Solved Example (§)

Using the results from an earlier solved example that calculated the true anomaly and sun-earth distance for several days during the year, find the solar constant for those days.

Solution

Given: R values from previously solved example
Find: S = ? W·m^{-2}

Sketch: (same as Fig 2.2)
Use eq. (2.20).

t (date)	R (Gm)	S (W/m^2)
3-Jan	**146.96**	**1418**
17-Jan	**147.03**	**1416**
31-Jan	**147.24**	**1412**
14-Feb	**147.57**	**1406**
28-Feb	**148.00**	**1398**
14-Mar	**148.53**	**1388**
28-Mar	**149.10**	**1377**
11-Apr	**149.70**	**1366**
25-Apr	**150.29**	**1355**

Check: Units OK. Physics OK.
Discussion: During Northern Hemisphere winter, the solar constant is up to 50 W·m^{-2} larger than average.

Solved Example

Estimate the value of the solar constant.

Solution

Given: $T_{sun} = 5780$ K

$R_{sun} = 6.96\text{x}10^5$ km = solar radius

$R_{earth} = 1.495\text{x}10^8$ km = earth orbit radius

Find: $S = ?$ W·m^{-2}

Sketch:

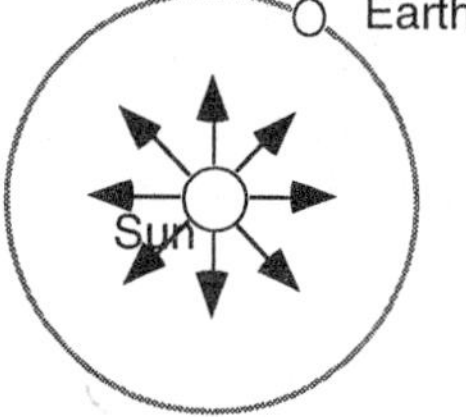

First, use eq. (2.18):

$E_1{}^* = (5.67\text{x}10^{-8}$ W·m^{-2}·K^{-4})·(5780 K)^4

$= 6.328\text{x}10^7$ W·m^{-2}.

Next, use eq. (2.19), with $R_1 = R_{sun}$ & $R_2 = R_{earth}$

$E_2{}^* = (6.328\text{x}10^7$ W·m^{-2})·

$(6.96\text{x}10^5 \text{ km}/1.495\text{x}10^8 \text{ km})^2$ = <u>1372 W·m</u>$^{-2}$.

Check: Units OK, Sketch OK. Physics OK.

Discussion: Answer is nearly equal to that measured by satellites.

Solved Example

During the equinox at noon, the solar elevation angle at a latitude of ϕ =60° is $\Psi = 90° - 60° = 30°$. If no solar radiation is absorbed by the atmosphere, then how much radiative flux is absorbed into a black asphalt parking lot?

Solution

Given: $\Psi = 30°$ = elevation angle

$E = S = 1368$ W·m^{-2}. solar constant

Find: $\Im_{rad} = ?$ W·m^{-2}

Sketch:

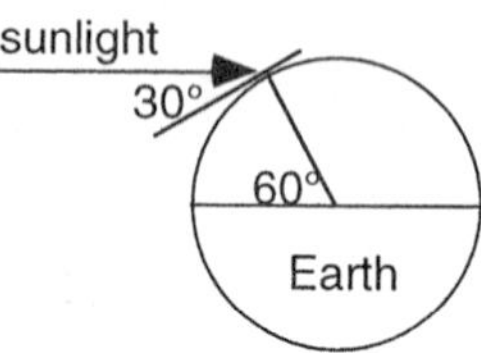

Use eq. (2.21a):

$\Im_{rad}$ = (1368 W·m^{-2})·sin(30°) = **<u>684 W·m</u>**$^{-2}$.

Check: Units OK. Sketch OK. Physics OK.

Discussion: Because the solar radiation is striking the parking lot at an angle, the radiative flux into the parking lot is half of the solar constant.

Average Daily Insolation

The incoming solar radiation (**insolation**) at the top of the atmosphere, averaged over 24 h, takes into account both the solar elevation angle, and the duration of daylight. For example, there is more total insolation at the poles in summer than at the equator, because the low sun angle near the poles is more than compensated by the long periods of daylight.

The average daily insolation $\overline{E}$ at any location is the solar constant that has been modified to account for the fact that the top of the atmosphere above any location is not usually perpendicular to the solar beam, and that the angle changes with time as the earth rotates.

$$\overline{E} = \frac{S_o}{\pi} \cdot \left(\frac{\overline{R}}{R}\right)^2 \cdot [h_o{}' \cdot \sin(\phi) \cdot \sin(\delta_s) + \cos(\phi) \cdot \cos(\delta_s) \cdot \sin(h_o)] \tag{2.22}$$

where S_o =1368 W/m^2 is the mean solar constant, $\overline{R}$ = 149.6 Gm is the mean sun-earth distance, R is the actual distance for any day of the year.

The hour angle h_o at sunrise and sunset is

$$\cos(h_o) = -\tan(\phi) \cdot \tan(\delta_s) \tag{2.23}$$

In eq. (2.22), h_o' is the hour angle in **radians.**

Solved Example

Find the average daily insolation over Vancouver during the summer solstice.

Solution:

Given: $d = d_r = 173$ at the solstice,

$\phi = 49.25°$N, $\lambda_e = 123.1°$W for Vancouver.

Find: $\overline{E}$ = ? W/m^2

Use eq. (2.8):

$\delta_s = \Phi_r = 23.45°$

Use eq. (2.5): $M = 167.55°$

Use eq. (2.7): $R = 151.892$ Gm

Use eq. (2.23):

$h_o = \arccos[-\tan(49.25°) \cdot \tan(23.45°)] = 120.23°$

$h_o{}' = h_o \cdot 2\pi/360° = 2.098$ radians

Use eq. (2.22):

$$\overline{E} = \frac{(1368 W \cdot m^{-2})}{\pi} \cdot \left(\frac{149.6 Gm}{151.892 Gm}\right)^2 \cdot$$

$$[2.098 \cdot \sin(49.25°) \cdot \sin(23.45°) + \cos(49.25°) \cdot \cos(23.45°) \cdot \sin(120.23°)]$$

$\overline{E}$ = (1327 W/m^2)·[2.098(0.3016)+0.5174]

$\overline{E}$ = **<u>486 W/m</u>**^{2}

Check: Units OK. Physics OK.

Discussion: At the equator on this same day, the average daily insolation is less than 400 W/m^2.

Absorption, Reflection & Transmission

The emissivity, e_λ, is the fraction of blackbody radiation that is actually emitted (see Table 2-2). The absorptivity, a_λ, is the fraction of radiation striking a surface that is absorbed. **Kirchhoff's law** states that the **absorptivity** and **emissivity** of a substance are equal at each wavelength, λ. Thus,

$$a_\lambda = e_\lambda \tag{2.24}$$

Some substances such as glass are semi-transparent. A fraction of the incoming (incident) radiation might also be reflected back:

$$r_\lambda = \frac{E_{\lambda\ reflected}}{E_{\lambda\ incident}} = \textbf{reflectivity} \tag{2.25}$$

A fraction is absorbed into the substance:

$$a_\lambda = \frac{E_{\lambda\ absorbed}}{E_{\lambda\ incident}} = \textbf{absorptivity} \tag{2.26}$$

And a fraction is transmitted through:

$$t_\lambda = \frac{E_{\lambda\ transmitted}}{E_{\lambda\ incident}} = \textbf{transmissivity} \tag{2.27}$$

The sum of these fractions must total 1, as 100% of the radiation at any wavelength must be accounted:

$$1 = a_\lambda + r_\lambda + t_\lambda \tag{2.28a}$$

or

$$E_{\lambda incoming} = E_{\lambda absorbed} + E_{\lambda reflected} + E_{\lambda transmitted} \tag{2.28b}$$

Table 2-2. Typical infrared emissivities.

Surface	e	Surface	e
soil, peat	0.97-0.98	snow, fresh	0.99
soils	0.9-0.98	snow, old	0.82
asphalt	0.95	ice	0.96
concrete	0.71-0.9	cloud, cirrus	0.3
gravel	0.92	cloud, alto	0.9
sandstone	0.98	cloud, low	1.0
desert	0.84-0.91	grass lawn	0.97
urban	0.85-0.95	forest, conif.	0.97
grass	0.9-0.95	forest, decid.	0.95
glass	0.87-0.94	sand, wet	0.98
paper	0.89-0.95	bricks, red	0.92
leaf 0.8μm	0.05-0.53	alfalfa	0.95
leaf 1 μm	0.05-0.6	shrubs	0.9
leaf 2.4μm	0.7-0.97	lumber, oak	0.9
leaf 10μm	0.97-0.98	silver	0.02
iron, galv.	0.13-0.28	human skin	0.95
aluminum	0.01-0.05	plaster, white	0.91

For opaque ($t_\lambda = 0$) substances such as the earth's surface, we find: $a_\lambda = 1 - r_\lambda$.

The reflectivity, absorptivity, and transmissivity usually vary with wavelength. For example, clean snow reflects about 90% of incoming solar radiation, but reflects almost 0% of IR radiation. Such behavior is crucial to temperature forecasts at the surface.

Instead of considering a single wavelength, it is also possible to examine the net effect over a range of wavelengths. The ratio of total reflected to total incoming solar radiation (i.e., averaged over all solar wavelengths) is called the **albedo**, A.

$$A = \frac{E_{reflected}}{E_{incoming}} \tag{•2.29}$$

Solved Example

What is the IR absorptivity of fresh snow?

Solution

Given: From Table 2-2: $e_{IRsnow} = 0.99$.
Find: $a_{IRsnow} = ?$
Use Kirchhoff's law (2.10): $a_{IRsnow} = e_{IRsnow} = 0.99$.

Check: Units OK. Physics reasonable.
Discussion: Snow absorbs virtually all of the downwelling IR radiation from the atmosphere.

Table 2-3. Typical albedos.

Surface	A	Surface	A
snow, fresh	75-95	road, asphalt	5-15
snow, old	35-70	road, dirt	18-35
ice, gray	60	concrete	15-37
water, deep	5-20	buildings	9
soil, dark wet	6-8	urban, mean	15
soil, light dry	16-18	field, fallow	5-12
soil, red	17	wheat	10-23
clay, wet	16	rice paddy	12
clay, dry	23	sugar cane	15
loam, wet	16	winter rye	18-23
loam, dry	23	corn	18
soil, sandy	20-25	tobacco	19
soil, peat	5-15	potatoes	19
lime	45	alfalfa	23-32
gypsum	55	cotton	20-22
lava	10	sorghum	20
granite	12-18	forest, conif.	5-15
stones	20-30	forest, decid.	10-25
tundra	15-20	grass, green	26
sand dune	20-45	meadow, grn	10-20
cloud, thick	70-95	savanna	15
cloud, thin	20-65	steppe	20

The average global albedo for solar radiation reflected from our planet is A = 30%. The actual global albedo at any instant varies with ice cover, snow cover, cloud cover, soil moisture, topography, and vegetation (Table 2-3).

The surface of the earth (land and sea) is a very strong absorber and emitter of radiation, making it relatively easy to use the relationships above to find the energy budget at the surface.

Within the air, however, the process is a bit more complicated. One approach is to treat the whole atmospheric thickness as a single object. Namely, we can compare the radiation at the top versus bottom of the atmosphere to examine the total emissivity, absorptivity, and reflectivity of the whole atmosphere. Over some wavelengths called **windows** there is little absorption, allowing the radiation to "shine" through. In other wavelength ranges there is partial or total absorption. Thus, the atmosphere acts as a filter.

Beer's Law

Sometimes we must examine radiative absorption across a short path length Δs within the atmosphere (Fig 2.9). Let n be the number density of absorbing particles in the air (#·m^{-3}), and b be the absorption cross section of each particle (m^2·#$^{-1}$), where this latter quantity gives the area of the shadow cast by each particle. **Beer's law** gives the relationship between incident radiative flux, $E_{incident}$, and outgoing transmitted radiative flux, $E_{transmitted}$, as

$$E_{transmitted} = E_{incident} \cdot e^{-n \cdot b \cdot \Delta s} \qquad (2.30)$$

Beer's law can also be written using an absorption coefficient, k:

$$E_{transmitted} = E_{incident} \cdot e^{-k \cdot \rho \cdot \Delta s} \qquad (2.31)$$

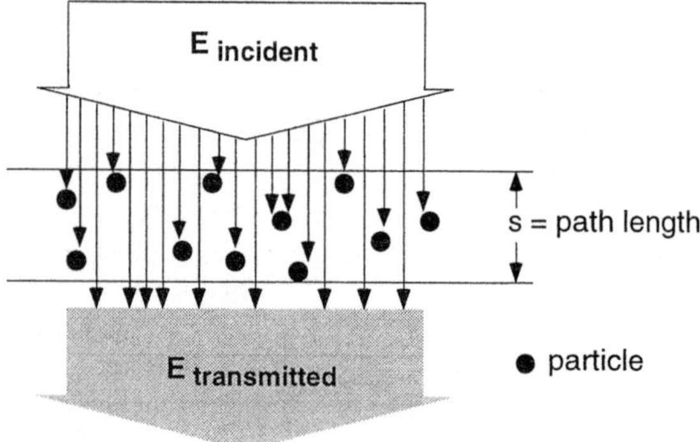

Figure 2.9
Reduction of radiation across a path due to absorption by particles, illustrating Beer's law.

Solved Example

Suppose that soot from a burning automobile tire has number density $n = 10^7$ m^{-3}, and an absorption cross section of $b = 10^{-9}$ m^2. Find the attenuation (i.e., reduction) of the solar constant across the 20 m diameter smoke plume.

Solution

Given: $n = 10^7$ m^{-3}, $b = 10^{-9}$ m^2
$E_{incident} = S = 1376$ W·m^{-2} solar const
Find: E_{tran} = ? W·m^{-2}

Sketch:

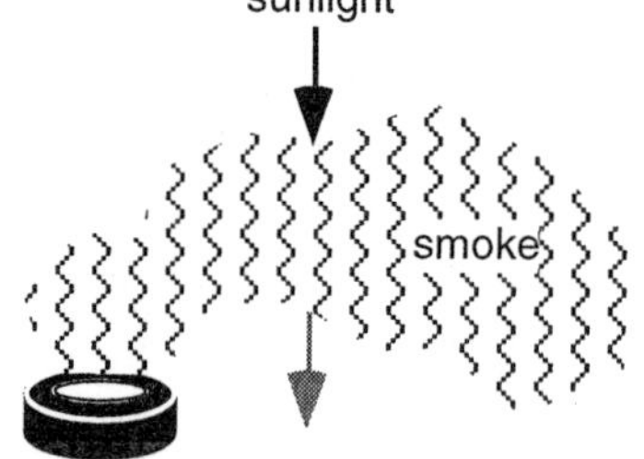

Use eq. (2.30)

$$E_{tran} = 1376\ \text{W·m}^{-2} \cdot \exp[-(10^7\ \text{m}^{-3})\cdot(10^{-9}\ \text{m}^2)\cdot(20\ \text{m})]$$
$$= 1376\ \text{W·m}^{-2} \cdot \exp[-0.2]$$
$$= \underline{\mathbf{1127\ W·m^{-2}}}$$

Check: Units OK. Sketch OK. Physics OK.
Discussion: Not much attenuation through this smoke plume: 1376 – 1127 = 249 W·m^{-2}.

where ρ is the density of air, and k has units of m$^2/g_{air}$. If the change in radiation is due only to absorption, the absorptivity across this layer is

$$a = \frac{E_{incident} - E_{transmitted}}{E_{incident}} \qquad (2.32)$$

SURFACE RADIATION BUDGET

Define $\mathfrak{I}^*$ as the **net radiative flux** (positive upward) perpendicular to the earth's surface. This net flux has contributions (Fig 2.10) from **downwelling solar** radiation $K\downarrow$, reflected **upwelling solar** $K\uparrow$, **downwelling longwave (IR)** radiation emitted from the atmosphere $I\downarrow$, and **upwelling longwave** emitted from the earth $I\uparrow$:

$$\mathfrak{I}^* = K\downarrow + K\uparrow + I\downarrow + I\uparrow \qquad \bullet(2.33)$$

where $K\downarrow$ and $I\downarrow$ are negative because they are downward.

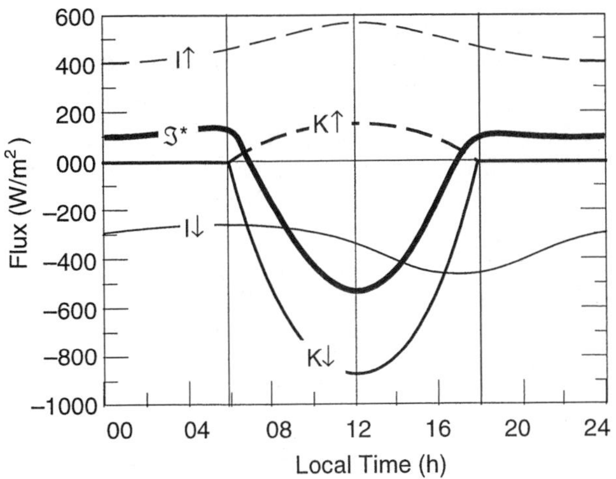

Figure 2.10
Typical diurnal variation of radiative fluxes at the surface. Fluxes are positive upward.

Solar

Recall that the solar irradiance (solar constant) is $S = 1368 \pm 7$ W·m^{-2} (equivalent to 1.125 K·m/s in kinematic form after dividing by ρC_p) at the top of the atmosphere. Some of this radiation is attenuated between the top of the atmosphere and the surface (Fig 2.11). Also, the sine law (eq. 2.21a) must be used to find the component of downwelling solar flux that is perpendicular to the surface $K\downarrow$. The result for daytime is

$$K\downarrow \ = \ -S \cdot T_r \cdot \sin\Psi \qquad (2.34)$$

where T_r is a **net sky transmissivity**. A negative sign is incorporated into eq. (2.34) because $K\downarrow$ is a downward flux. Eq. (2.9) can be used to find $\sin\Psi$. At night, the downwelling solar flux is zero.

Net transmissivity depends on path length through the atmosphere, atmospheric absorption characteristics, and cloudiness. One approximation for the net transmissivity is

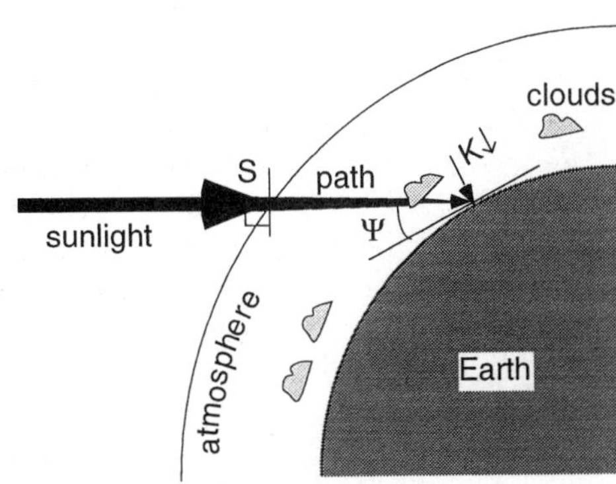

Figure 2.11
Fate of sunlight en route to the earth's surface.

$$T_r = (0.6 + 0.2\sin\Psi)(1 - 0.4\sigma_H)(1 - 0.7\sigma_M)(1 - 0.4\sigma_L) \qquad (2.35)$$

where cloud cover fractions for high, middle, and low clouds are σ_H, σ_M, and σ_L, respectively. These cloud fractions vary between 0 and 1, and the transmissivity also varies between 0 and 1.

Of the sunlight reaching the surface, a portion might be reflected:

$$K\uparrow \ = \ -A \cdot K\downarrow \qquad (2.36)$$

where the surface albedo is A.

Longwave (IR)

Upward emission of IR radiation from the earth's surface can be found from the Stefan-Boltzmann relationship:

$$I\uparrow = e_{IR} \cdot \sigma_{SB} \cdot T^4 \qquad (2.37)$$

where e_{IR} is the surface emissivity in the IR portion of the spectrum (=0.9 to 0.99 for most surfaces), and σ_{SB} is the Stefan-Boltzmann constant (= 5.67x10^{-8} W·m^{-2}·K^{-4}).

However, downward IR radiation from the atmosphere is much more difficult to calculate. As an alternative, sometimes a **net longwave flux** is defined by

$$I^* \ = \ I\downarrow \ + \ I\uparrow \qquad (2.38)$$

One approximation for this flux is

$$I^* \ = \ b \cdot (1 - 0.1\sigma_H - 0.3 \cdot \sigma_M - 0.6 \cdot \sigma_L) \qquad (2.39)$$

where parameter $b = 98.5$ W·m^{-2}, or $b = 0.08$ K·m·s^{-1} in kinematic units.

Net Radiation

Combining eqs. (2.33), 2.34), (2.35), (2.36) and (2.38) gives the **net radiation** ($\Im^*$, defined positive upward):

$$\Im^* \ = \ -(1-A) \cdot S \cdot T_r \cdot \sin(\Psi) \ + \ I^* \qquad \text{daytime} \quad \bullet(2.40a)$$

$$= I^* \qquad \text{nighttime} \quad \bullet(2.40b)$$

Solved Example

Find the net radiation at the surface in Vancouver, Canada at noon on 22 Jun. Low clouds are present with 30% coverage.

Solution

Assume: Grass lawns with albedo $A = 0.2$.
No other clouds.

Given: $\sigma_L = 0.3$

Find: $\mathcal{F}^* = ?\ \text{W·m}^{-2}$

Use $\Psi = 64.1°$ from an earlier solved example

Use eq. (2.35) to find the transmissivity

$$T_r = [0.6+0.2\cdot\sin\Psi]\cdot(1-0.4\cdot\sigma_L)$$
$$= [0.6 + 0.2\cdot\sin(64.1°)]\cdot(1-0.4\cdot0.3)$$
$$= [0.80]\cdot(0.88) = 0.686$$

Use eq. (2.39) to find net IR contribution

$$I^* = b\cdot(1 - 0.6\cdot\sigma_L)$$
$$= (98.5\ \text{W·m}^{-2})\cdot(1-0.6\cdot0.3) = 80.77\ \text{W·m}^{-2}$$

Use eq. (2.40a)

$$\mathcal{F}^* = -(1-A)\cdot S\cdot T_r\cdot\sin\Psi + I^*$$
$$= -(1-0.2)\cdot(1368\ \text{W·m}^{-2})\cdot0.686\cdot\sin(64.1°) + 80.77\ \text{W·m}^{-2}$$
$$=(-675.35 + 80.77)\ \text{W·m}^{-2} = \underline{-594.58}\ \text{W·m}^{-2}$$

Check: Units OK. Physics OK.

Discussion: The surface flux is only about 43% of that at the top of the atmosphere, for this case.

SUMMARY

The variations of temperature and humidity that we feel near the ground are driven by the diurnal cycle of solar heating during the day and infrared cooling at night. Both diurnal and seasonal heating cycles can be determined from the geometry of the earth's rotation and orbit around the sun. The same orbital mechanics describes weather satellite orbits.

Short-wave radiation is emitted from the sun and propagates to earth. It is distributed over the earth's surface where some is absorbed and some is reflected. This is the heat input to the earth-atmosphere system.

IR radiation is absorbed from the atmosphere, and IR radiation is also emitted. These radiative fluxes do not balance, leaving a net radiation term acting on the surface.

Threads

Just a reminder that the "threads" section at the end of each chapter is designed to highlight linkages to material in the other chapters. It is included to help you develop a coherent picture of atmospheric science, and to anticipate your question of "Why is this material important?".

Radiation affects motions on a wide range of scales, and it affects thermodynamics. Excess solar radiation near the equator, and excess IR cooling at the poles is the driving force for the global circulation (Chapter 11). It ultimately creates the synoptic weather patterns of high and low pressure systems and fronts (Chapter 12 & 13). Averaged over the whole globe, it controls the earth's climate (Chapter 18).

Radiative heating at different heights creates a vertical atmospheric structure quantified by a standard atmosphere (Chapter 1), and affects cloud growth and static stability (Chapter 7). On a smaller scale, IR radiative cooling drives downslope drainage winds at night (Chapter 10). Also, sunlight interacts with ice crystals and raindrops to create beautiful optical phenomena such as rainbows (Chapter 19).

Planck's law from Chapter 2 shows the peak wavelength of solar emissions to be in the visible portion of the spectrum. Sunlight drives the diurnal cycle of the surface heat budget described here in Chapter 3, which mostly affects air near the ground, in the atmospheric boundary layer (Chapter 4).

Orbital mechanics dictates weather satellite altitudes for observing such storms.

Solar elevation angle vs. latitude creates the equator-to-pole temperature difference that drives the global circulation (Chapter 11), and provides the baroclinicity that powers cyclones (Chapter 13). The budget of incoming and outgoing radiation might be sensitive to anthropogenic and natural factors that could modify the earth's climate (Chapter 18).

EXERCISES

Numerical Problems

(Students, don't forget to put a box around each answer.)

N1. Given distances R between the sun and planets compute the orbital periods of:
a. Mercury (R = 58 Gm)
b. Venus (R = 108 Gm)
c. Mars (R = 228 Gm)
d. Jupiter (R = 778 Gm)
e. Saturn (R = 1,427 Gm)
f. Uranus (R = 2,869 Gm)
g. Neptune (R = 4,498 Gm)
h. Pluto (R = 5,900 Gm)

N2. What is the value of the orbital time period for a polar orbiting satellite at altitude above the surface of: a. 850 km b. 450 km c. 250 km d. 1500 km e. 0 km (neglect air friction & mountains)?

N3. Calculate the precise distance between earth and sun today, using the equations in the book.

N4.(§) a. Use a spreadsheet (§) to compare the mean vs true anomalies for every day of the year.
b. For which day is the difference the greatest, and what is its value?

N5. What is the Julian day for
a. 15 March b. 4 July c. 25 December
d. 29 February e. 15 April f. 1 November
g. 2 January h. 5 September

N6. Find the solar declination angle for the days in the previous problem.

N7. How would Fig 2.4 be different if daylight (summer) time were used in place of standard time during the appropriate months?

N8(§) Plot local solar elevation angle vs. azimuth angle (such as in Fig 2.4) for a location on the Arctic Circle for a. 22 Dec b. 23 Mar c. 22 Jun
d, e, f. Repeat the plots for those same dates, but for a point at the North Pole.

N9. Compare the duration of evening civil twilight (difference between end of twilight and sunset times) during June 22 at: a. the equator b. 10°N
c. 30°N d. 43°N e. 55°N f. 65°N

N10. Plot the local solar elevation angle vs local time for 22 December, 23 March, and 22 June for three of the cities listed below.
a. Seattle, WA, USA
b. Corvallis, OR, USA
c. Boulder, CO, USA
d. Norman, OK, USA
e. Madison, WI, USA
f. Toronto, Canada
g. Montreal, Canada
h. Boston, MA, USA
i. New York City, NY, USA
j. University Park, PA, USA
k. Princeton, NJ, USA
l. Washington, DC, USA
m. Raleigh, NC, USA
n. Tallahassee, FL, USA
o. Reading, England
p. Toulouse, France
q. München, Germany
r. Bergen, Norway
s. Uppsala, Sweden
t. DeBilt, The Netherlands
u. Paris, France
v. Tokyo, Japan
w. Bejing, China
x. Warsaw, Poland
y. Madrid, Spain
z. Melbourne, Australia

N11. On 15 March for three of the cities listed in question N10, when is...? a. sunrise b. sunset
c. end of evening civil twilight

N12. Find the kinematic heat fluxes, given these regular fluxes ($W \cdot m^{-2}$): a. 1000 b. 500 c. 25 d. –100 .
Find regular fluxes, given these kinematic fluxes ($K \cdot m/s$): e. 0.1 f. 0.2 g. –0.05 h. 0.15

N13. Find the frequency, circular frequency, and wavenumber for light of color: a. red b. orange c. yellow d. green e. blue f. violet

N14.(§). Plot Planck curves for the following temperatures (K): a. 5000 b. 3000 c. 1000 d. 500
e. 300 f. 250 g. 100 h. 273

N15. What are the values of solar constant at the:
a. perihelion b. aphelion
c. summer solstice d. spring equinox
e. autumnal equinox f. winter solstice?

N16. During the equinox at noon, what is the solar elevation angle above the horizon at the following latitudes? a. 45°N b. 15°N c. 0°N d. 45°S e. 90°N f. 60°N g. 30°S h. 60°S

N17. Suppose polluted air reflects 30% and transmits 50% of the incoming solar radiation. How much is absorbed, and how much is emitted, assuming an irradiance equal to the solar constant?

N18. What product of number density times absorption cross section is needed in order for 50% of the incident radiation to be absorbed by airborne volcanic ash over a path length of 5 km?

N19. a. What is the value of solar downward direct radiative flux reaching the surface in your town at noon on 4 July, given 20% coverage of cumulus (low) clouds.

b. If the albedo is 0.5 in your town, what is the reflected solar flux at that same time?

N20. a. For a surface temperature of 20°C and emissivity of 1.0, find the emitted upwelling IR radiation.

b. Estimate the net (up & downwelling) IR radiation for clear skies.

c. Explain the difference between (a) and (b).

N21. Compare the net radiation reaching the surface at noon in Fairbanks, AK vs Miami, FL, USA, on 22 June, assuming clear skies and albedo 0.5 .

Understanding & Critical Evaluation

U1. At what time of year does the true anomaly equal: a. 90° b. 270°

U2(§) a. Calculate and plot the position (true anomaly and distance) of the earth around the sun for the first day of each month.

b. Verify Kepler's second law.

c. Compare the elliptical orbit to a circular orbit.

U3(§). Calculate and plot the noontime downwelling solar radiation every day of the year, assuming no clouds, and considering the change in solar irradiance due to changing distance between the earth and sun.

U4. What is the optimum angle for solar collectors at your location?

U5(§). Produce a graph with latitude on the ordinate, and day of the year on the abscissa. On this graph, draw isopleths of average daily insolation. Use increments of 5° latitude between both poles along the ordinate, and increments of 10 days along the abscissa.

U6. Design a device to measure the angular diameter of the sun when viewed from earth. (Hint, one approach is to allow the sun to shine through a pin hole on to a flat surface. Then measure the width of the projected image of the sun on this surface divided by the distance between the surface and the pin hole. What could cause errors in this device?)

U7. a. Compare the length of daylight in Fairbanks, AK, vs Miami, FL, USA.

b. Why do vegetables grow so large in Alaska?

c. Why are few fruits grown in Alaska?

U8. How many more minutes of direct sunlight occur at the spring equinox compared to the day before?

U9(§). Plot a diagram of sunrise times and of sunset times vs. day of the year, for your location.

U10. Explain the meaning of each term in eq. (2.22).

U11. Consider cloud-free skies at your town. If 50% coverage of low clouds moves over your town, how does net radiation change at noon? How does it change at midnight?

U12. On a clear day, record sunrise and sunset times, and the duration of twilight. a. Use that information to determine the day of the year. b. Based on your personal determination of the length of twilight, and based on your latitude and season, is your personal twilight most like civil, military, or astronomical twilight?

U13(§). Evaluate the quality of the approximation (2.5) against the exact Planck equation (2.4) by plotting both curves for a variety of typical sun & earth T.

U14. Find the irradiance that can pass through an atmospheric "window" between λ_1 and λ_2 . (See a previous "Beyond Algebra" box in this chapter for ways to do this without using calculus.)

U15. If the earth were to cool 5°C from its present radiative equilibrium temperature, by what percentage would the total emitted IR change?

U16. What solar temperature is needed for the peak intensity of radiation to occur at 0.2 micrometers? Remembering that humans can see light only between 0.4 and 0.7 microns, would the sun look brighter or dimmer at this new temperature?

U17. How much variation in earth orbital distance from the sun is needed to alter the solar constant by 10%?

U18. Solar radiation is a diffuse source of energy, meaning that it is spread over the whole earth rather than being concentrated in a small region. It has been proposed to get around the problem of the inverse square law of radiation by deploying very large mirrors closer to the sun to focus the light as columnated rays toward the earth. Assuming that all the structural and space-launch issues could be solved, would this be a viable method of increasing energy on earth?

U19. The "sine law" for radiation striking a surface at an angle is sometimes written as a "cosine law", but using the zenith angle instead of the elevation angle. Use trig to show that the two equations are physically identical.

U20. Using Table 2-2 of typical albedos, speculate how the average albedo will change if a pasture is developed into a residential neighborhood.

U21. **Visibility** has many different definitions, but one definition is that **visual range** is the distance where the intensity of transmitted light has decreased to 2% of the incident light. The product of particle number density n and absorption cross section b is sometimes known as the **extinction coefficient**. Use Beer's law to determine the relationship between visual range (km) and extinction coefficient (m^{-1}). (Note that extinction coefficient can be related to concentration of pollutants and relative humidity.)

U22. Suppose the optical capability on a geostationary weather satellite is good enough to resolve a feature on the earth's surface as small as 1 km. If an identical satellite were launched by the military to use as a spy satellite, but was placed in an orbit that is half the altitude above the earth's surface as current polar-orbiting weather satellites, what would be the observation resolution at the surface? Also, for how many minutes would such a satellite be within view of any object on the surface (assume objects are within view when the viewing angle is smaller than plus or minus 45° of vertical)?

Web-Enhanced Questions

(Students, don't forget to cite each URL web address that you use.)

W1. Access a full-disk visible satellite photo image of earth from the web. What visible clues can you use to determine the current solar declination angle? How does your answer compare with that expected for your latitude and time of year.

W2. Access "web cam" camera images from a city, town, ski area, mountain pass, or highway near you. Use visible shadows on sunny days, along with your knowledge of solar azimuth angles, to determine the direction that the camera is looking.

W3. Access from the web the exact time from military (US Navy) or civilian (National Institute of Standards and Technology) atomic clocks. Synchronize your clocks at home or school, utilizing the proper time zone for your location. What is the time difference between local solar noon (the time when the sun is directly overhead) and the official noon according to your time zone. Use this time difference to determine the number of degrees of longitude that you are away from the center of your time zone.

W4. Access orbital information about one planet (other than earth) that most interests your (or a planet assigned by the instructor). How elliptical is the orbit of the planet? Also, enjoy imagery of the planet if available.

W5. Access runway visual range reports from surface weather observations (METARs) from the web. Compare two different locations (or times) having different visibilities, and calculate the appropriate extinction coefficients. Also search the web to learn how runway visual range (RVR) is measured.

W6. Access both visible and infrared satellite photos from the web, and discuss why they look different. If you can access water vapor satellite photos, include them in your comparison.

W7. Search the web for information about the sun. Examine satellite-based observations of the sun made at different wavelengths. Discuss the structure of the sun. Do any of the web pages give the current value of the solar constant? If so, how has it varied recently?

W8. Access from the web daytime visible photos of the whole disk of the earth, taken from geostationary weather satellites. Discuss how variations in the apparent brightness at different locations (different latitudes; land v.s ocean, etc.) might be related to reflectivity.

W9. Some weather stations and research stations report hourly observations on the web. Some of these stations include radiative fluxes near the surface. Use this information to create surface net radiation graphs.

W10. Access photographs or artist drawings of those satellites currently in orbit. Compare the physical appearance and instrumentation differences between US geostationary satellites and polar orbiting satellites. Check that their orbital time agrees with their altitude. Determine the launch schedule for future weather satellites.

Synthesis Questions

(Students, don't forget to state and justify all assumptions.)

S1. What if the eccentricity of the earth's orbit around the sun changed to 0.2 ? How would the seasons be different than now?

S2. What if the tilt of the earth's axis relative to the ecliptic changed to 45° ? How would the seasons be different than now?

S3. What if the earth diameter decreased to half of its present value? How would sunrise and sunset time, and solar elevation angles change?

S4. Derive eq. (2.9) from basic principles of geometry and trigonometry. This is quite complicated. Show your work.

S5. What if the perihelion of the earth's orbit happened at the summer solstice, rather than near the winter solstice. How would noontime, clear-sky values of insolation change at the solstices compared to now?

S6. What if radiative heating was caused by the magnitude of the radiative flux, rather than by the radiative flux divergence. How would the weather or the atmospheric state be different, if at all?

S7. Suppose that Kirchhoff's law were to change such that $a_\lambda = 1 - e_\lambda$. What would be the implications?

S8. What if Wien's law were to be repealed, because it was found instead that the wavelength of peak emissions increases as temperature increases.
a. Write an equation that would describe this. You may name this equation after yourself.
b. What types of radiation from what sources would affect the radiation budget of earth?

S9. Consider Beer's law. If there are n particles per cubic meter of air, and if a path length in air is Δs, then multiplying the two gives the number of particles over each square meter of ground. If the absorption cross section b is the area of shadow cast by each particle, then multiplying this times the previous product would give the shadowed area divided by the total area of ground. This ratio is just the absorptivity a. Namely, by this reasoning, one would expect that $a = n \cdot b \cdot \Delta s$.

However, Beer's law is an exponential function. Why? What was wrong with the reasoning in the previous paragraph?

S10. Record the relevant weather information for one full day at your town, as specified by your instructor. After collecting that data, use it to calculate and plot the net radiation vs. time for a 24 h period. State and justify any assumptions. Compare your answer with the actual temperature variation that was observed. Discuss implications and limitations. Also, how could you modify the equations to improve the results over your town?

S11. What if the atmosphere were completely transparent to IR radiation. How would the surface net radiation budget be different?

S12. What if gravity on earth were twice as strong as now. What would be the altitude and orbital period of geostationary satellites?

S13. What if the earth were larger diameter, but had the same average density as the present earth. How large would the diameter have to be so that an orbiting geostationary weather satellite would have an orbit that is zero km above the surface? (Neglect atmospheric drag on the satellite.)

HEAT

CONTENTS

3 Energy cannot be created or destroyed, according to classical physics. However, it can change form. Heat is one form. Energy enters the atmosphere predominantly as short-wave radiation from the sun. Once here, the energy can change form many times while driving the weather. It finally leaves as long-wave terrestrial radiation. An energy conservation relationship known as the **First Law of Thermodynamics** includes all energy forms to balance the heat budget.

We can apply this budget within two frameworks: Lagrangian and Eulerian. Both methods must consider latent heat associated with phase changes of water.

Lagrangian means that we follow an air parcel as it moves about in the atmosphere. This is useful for determining the temperature of a rising air parcel, and whether clouds form.

Eulerian means that we examine a volume fixed in space, such as over a farm or town. This is useful for forecasting temperature at any location, such as at your house. Eulerian methods must consider the transport (flux) of energy to and from the fixed volume. Transport can be via radiation, advection, turbulence, and conduction.

Radiative inputs at the surface drive diurnally-varying turbulent heat and moisture fluxes. These sensible and latent heat fluxes cause the daily variations of temperature and humidity that we experience as part of daily life.

SENSIBLE AND LATENT HEATS

Sensible

Sensible heat ΔQ_H (in units of J) can be sensed by humans; namely, it is that portion of total heat associated with temperature change ΔT. This sensible heat per unit mass of air m_{air} is given by

$$\frac{\Delta Q_H}{m_{air}} = C_p \cdot \Delta T \qquad \bullet(3.1)$$

The specific heat at constant pressure, C_p, depends on the material being heated. For dry air, C_{pd} = 1004.67 J·kg^{-1}·K^{-1}. For moist air, the specific heat is

$$C_p = C_{pd} \cdot (1 + 0.84 \cdot r) \qquad (3.2)$$

where r is the mixing ratio of water vapor, in units of $g_{vapor}/g_{dry\ air}$. The specific heat of liquid water is much greater: $C_{liq} \cong 4200$ J·kg^{-1}·K^{-1}.

Solved Example

How much sensible heat is needed to warm 2 kg of dry air by 5°C?

Solution

Given: m_{air} = 2 kg , ΔT = 5°C ,
C_{pd} = 1004.67 J·kg^{-1}·K^{-1}.
Find: ΔQ_H = ? J

Rearrange eq. (3.1)
ΔQ_H = $m_{air} \cdot C_{pd} \cdot \Delta T$
= (2 kg)·(1004.67 J·kg^{-1}·K^{-1})·(5°C)
= <u>**10.046 kJ**</u>

Check: Units OK. Physics reasonable.
Discussion: This amount of air takes up a space of about 2.45 m^3, — the size of a small closet.

Latent

Latent heat is a hidden heat, until water **phase changes** occur. Evaporation of liquid water drops cools the air by removing sensible heat and storing it as latent heat. The following list summarizes those water phase changes that cool the air:

- **vaporization**: liquid to vapor
- **melting**: solid (ice) to liquid
- **sublimation**: solid to vapor

Phase changes of water in the opposite direction cause warming of the air by releasing latent heat:

- **condensation**: vapor to liquid
- **fusion**: liquid to solid (ice)
- **deposition**: vapor to ice

When air parcels or fixed volumes contain moist air, there is the possibility that water phase changes can alter the temperature, even without transport across the volume boundaries. The amount of heat ΔQ_E per unit mass of phase-changed water m_{water} is defined to be

$$\frac{\Delta Q_E}{m_{water}} \equiv L \qquad \bullet(3.3)$$

where L is known as the **latent heat**.

Values of latent heat are

L_v = ± 2.5x10^6 J·kg^{-1} $\equiv L_{condensation\ or\ vaporization}$

L_f = ± 3.34x10^5 J·kg^{-1} $\equiv L_{fusion\ or\ melting}$

L_d = ± 2.83x10^6 J·kg^{-1} $\equiv L_{deposition\ or\ sublimation}$

where the sign depends on the direction of phase change, as described above. See Appendix B for tables of geophysical constants.

Solved Example

How much dew must condense on the sides of a can of soda for it to warm the soda from 1°C to 16°C?

Hints: Neglect the heat capacity of the metal. The density of liquid water is 1025 kg·m^{-3}. Assume the density of soda equals that of pure water. Assume the volume of a can is 354 ml (milliliters), where 1 l = 10^{-3} m^3.

Solution

Given: ρ_{water} = 1025 kg·m^{-3}.
C_{liq} = 4200 J·kg^{-1}·K^{-1}
Vol in Can = 354 ml
L_{cond} = + 2.5x10^6 J·kg^{-1}
ΔT = 15 K
Find: *Vol of Condensate*

Sketch:

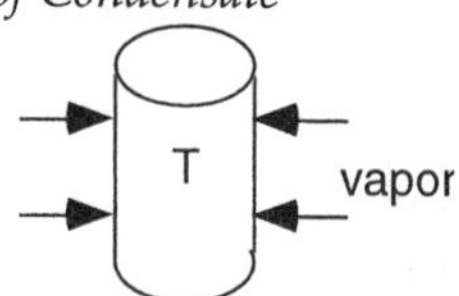

Equate the latent heat release by condensing water vapor (eq. 3.3) with the sensible heat gained by fluid in the can (eq. 3.1)

$\Delta Q_E = \Delta Q_H$
$\rho_{condensate} \cdot$ (*Vol of Condensate*)$\cdot L_{cond}$ =
$\rho_{soda} \cdot$ (*Vol of Can*)$\cdot C_{liq} \cdot \Delta T$

Assume the density of condensate and soda are equal, so they cancel. The equation can then be solved for *Volume of Condensate*.
Vol of Condensate =(*Vol of Can*)$\cdot C_{liq} \cdot \Delta T / L_{cond}$
= (354 ml)·(4200 J·kg^{-1}·K^{-1})·(15 K)/(2.5x10^6 J·kg^{-1})
= <u>**8.92 ml**</u>

Check: Units OK. Sketch OK. Physics OK.
Discussion: Latent heats are so large that an amount of water equivalent to only 2.5% of that in the can needs to condense on the outside to warm the can by 15°C. Thus, to keep your can cool, insulate the outside to prevent dew from condensing.

Solved Example

How much latent heat is released when 2 kg of water vapor condenses into liquid?

Solution

Given: $m_{vapor} = 2$ kg , $L_v = 2.5\text{x}10^6$ J·kg^{-1} .
Find: ΔQ_E = ? J

Rearrange eq. (3.3)

$$\Delta Q_E = (2.5 \times 10^6 \text{J} \cdot \text{kg}^{-1}) \cdot (2\text{kg}) = \underline{5{,}000} \text{ kJ}$$

Check: Units OK. Physics reasonable.
Discussion: This amount of liquid water would fill a typical bathtub to only a depth of 2 mm. Thus, a small amount of water can hold a large amount of latent heat. Water vapor is a very important source of energy to drive storms.

LAGRANGIAN HEAT BUDGET – PART 1: UNSATURATED

An **air parcel** is a hypothetical blob of air that we track as it moves about in the atmosphere. We can think of it as a balloon without any latex skin. As it moves we can compute its thermodynamic and dynamic state, based on work done on it, changes within it, fluxes across its surface, and mixing with its surrounding environment. This concept is often employed in **Lagrangian** studies.

In Part 1 of this Lagrangian description, we will study only those situations where there is NO condensation or evaporation of water; namely, **unsaturated** conditions. This does not mean that the air is dry — only that any water vapor within the air is not undergoing phase changes.

Historically, these processes with no phase changes are called "**dry**" processes, even though there may be water vapor in the air. In Chapter 5, after moisture variables are introduced, the analysis will continue in a section called " Lagrangian Heat Budget – Part 2: Saturated", where "**moist**" (saturated) processes are included.

First Law of Thermodynamics

The temperature of an air parcel of mass m_{air} changes by amount ΔT when heat (ΔQ_H) is added, and changes when work is done on or by the parcel. This relationship, called the **First Law of Thermodynamics**, can be written as:

$$\frac{\Delta Q_H}{m_{air}} = C_p \cdot \Delta T - \frac{\Delta P}{\rho} \qquad \bullet(3.4)$$

The first term on the right hand side is the sensible heat.

The last term represents the work per unit mass associated with parcel expansion and contraction due to pressure changes. To understand this, recall that pressure is force per unit area, and density is mass per unit volume. The ratio of these two variables is thus (force x distance)/mass, which is equivalent to work per unit mass. The pressure of the parcel usually equals that of its surrounding environment, which decreases exponentially with height.

The first law of thermodynamics can be reformulated using the hydrostatic equation to yield the temperature change of a rising or sinking air parcel

$$\Delta T = -\left(\frac{|g|}{C_p}\right) \cdot \Delta z + \frac{\Delta Q_H}{m_{air} \cdot C_p} \qquad \bullet(3.5)$$

The heat added term ($\Delta Q_H/m_{air}$) on the right can be caused by radiative heating, latent heating during condensation, dissipation of turbulence energy into heat, heat from chemical reactions, and convective or turbulent interactions between air inside and outside of the parcel. Advection and convection do not transport heat to/from the parcel, but they do move the parcel about (Fig 3.1).

The first law of thermodynamics is also called the **conservation of heat**. It says that the temperature of an air parcel will not change, unless heat is added or removed, or unless the parcel is raised or lowered. Thus, even without adding heat to a parcel, its temperature can change during vertical parcel motion. Vertical motion is very important in atmospheric circulations and cloud development.

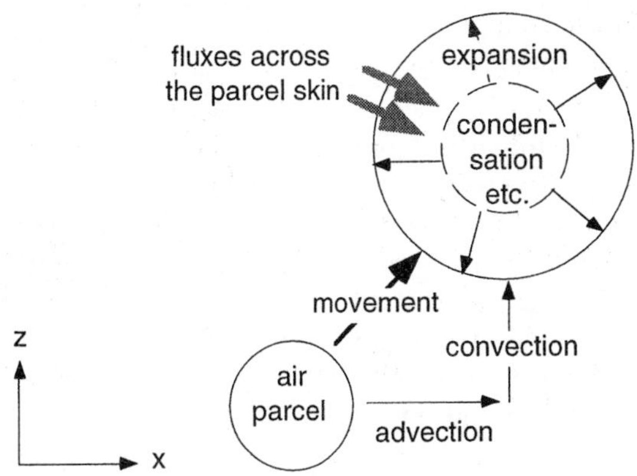

Figure 3.1
Processes affecting the temperature of an air parcel.

Solved Example

What is the temperature change of a 10 kg air parcel when heated at rate $H = 100$ W for 10 minutes? Water-vapor mixing ratio is $r = 0.01$ g_{vapor}/g_{air}, and the air parcel is stationary.

Solution

Given: $H = 100$ W, $m_{air} = 10$ kg, $\Delta z = 0$
$r = 0.01$ g_{vapor}/g_{air}, $\Delta t = 10$ min
Find: $\Delta T = ?$ K

First, calculate the specific heat from eq. (3.2):
$C_p = (1004.67\ \text{J·kg}^{-1}\text{·K}^{-1})\cdot[1 + 0.84 \cdot 0.01]$
$= 1013.11\ \text{J·kg}^{-1}\text{·K}^{-1}$.
Next, calculate the total heat added:
$\Delta Q_H = H\cdot\Delta t = (100\ \text{W})\cdot(10\ \text{min})$
$= (100\ \text{J/s})\cdot(600\ \text{s}) = 6\text{x}10^4\ \text{J}$.
Finally, use eq. (3.5):
$\Delta T = (6\text{x}10^4\ \text{J}) / [(10\ \text{kg}) \cdot (1013.11\ \text{J·kg}^{-1}\text{·K}^{-1})]$
$=$ <u>5.92 K</u>.

Check: Units OK. Physics reasonable.
Discussion: Air density is about 1.225 kg/m^3 at sea level, then we see that 10 kg is equivalent to 8.16 m^3, which is about the size of a large closet. According to the first law of thermodynamics, the air in this closet would be warmed by almost 6 °C if we turned on a 100 W electric light bulb, assuming no heat is lost to the walls of the closet.

Lapse Rate

The decrease of temperature with height is known as the **lapse rate,** Γ. Namely, $\Gamma = -\Delta T/\Delta z$. Two types of lapse rate are **environmental** lapse rates and **process** lapse rates.

When the existing temperature-state of the environment is measured with aircraft, weather balloons, or remote sensors, the resulting graph of temperature vs. height is called an **environmental temperature profile**, or an **environmental sounding**. The temperature decrease within this sounding gives the **environmental lapse rate** as a function of height. It is a static measure of the state of the environment.

Fig 1.9 is an example of an environmental sounding based on climatological average conditions. The actual environmental sounding on any day might differ. A wide range of soundings is possible.

A **process lapse rate** gives the temperature change associated with some action or process. Examples are rising air parcels and motions within clouds. The change of temperature with height for a process is usually defined by a physical relationship, such as the first law of thermodynamics. One common process is the adiabatic process, described next.

Adiabatic Lapse Rate

An **adiabatic** process is one where there is no heat transfer to or from the air parcel ($\Delta Q_H = 0$). An air parcel that rises adiabatically with no cloud formation (dry) experiences a temperature change given by eq. (3.5), which can be rearranged as:

$$\frac{\Delta T}{\Delta z} = -\left(\frac{|g|}{C_p}\right) = -9.8\ \text{K / km} \qquad \bullet(3.6)$$

Thus, the **"dry" adiabatic lapse rate** for a rising air parcel is $\Gamma_d = 9.8$ K / km, and it is a process lapse rate. As mentioned earlier, "dry" adiabatic processes also apply to humid air, but only if there is no condensation, clouds, or precipitation in it (i.e., only if the air is unsaturated).

The adiabatic lapse rate can also be found in terms of pressure instead of height. Substituting the ideal gas law into the first law of thermodynamics for an adiabatic process yields:

$$\frac{\Delta T}{T} = \left(\frac{\Delta P}{P}\right)^{\Re_d / C_p} \qquad (3.7)$$

or

$$\frac{T_2}{T_1} = \left(\frac{P_2}{P_1}\right)^{\Re_d / C_p} \qquad \bullet(3.8)$$

where $\Re_d/C_p = 0.28571$ (dimensionless) for dry air, and where temperatures are in degrees Kelvin.

A different adiabatic lapse rate ("moist" lapse rate) must be used if there is liquid or solid water in the air parcel. This is discussed later in Chapter 5.

Solved Example

Suppose an air parcel of initial temperature 15 °C rises adiabatically 2 km from the surface. What is the final temperature of the parcel?

Solution

Given: $T_{initial} = 15°\text{C}$, $\Delta z = 2$ km
Find: $T_{final} = ?$ °C

Using the definition of gradients (see Appendix A) along with eq. (3.6):

$$\frac{\Delta T}{\Delta z} = \frac{T_{final} - T_{initial}}{z_{final} - z_{initial}} = -9.8\ °\text{C/km}$$

Solving for T_{final} gives:
$T_{final} = T_{initial} - (9.8\ °\text{C/km}) \cdot \Delta z$
$= 15\ °\text{C} - (9.8\ °\text{C/km})\cdot(2\ \text{km}) =$ <u>-4.6 °C</u>.
Check: Units OK. Physics reasonable.
Discussion: Sufficient cooling to freeze water.

When calculating temperature differences, units of °C and K are interchangeable, because a change of 1°C equals a change of 1 K. However, when extracting a temperature from such a difference, always convert it to K before using it in a radiative or ideal gas law.

BEYOND ALGEBRA • Adiabatic Lapse Rate

Calculus can be used to get eq. (3.8) from (3.4). First, write the First Law of Thermo. in its more-accurate differential form, for an adiabatic process ($\Delta Q_H = 0$):

$$dP = \rho \cdot C_p \cdot dT_v$$

Then substitute for density using the ideal gas law $\rho = P / (\Re_d \cdot T_v)$, to give:

$$dP = \frac{P \cdot C_p \cdot dT_v}{\Re_d \cdot T_v}$$

Move pressures to the LHS, & temperatures to the RHS:

$$\frac{dP}{P} = \frac{C_p}{\Re_d} \cdot \frac{dT_v}{T_v}$$

Integrate between pressure P_1 where temperature is T_1, to pressure P_2 where temperature is T_2:

$$\int_{P_1}^{P_2} \frac{dP}{P} = \frac{C_p}{\Re_d} \cdot \int_{T_{v1}}^{T_{v2}} \frac{dT_v}{T_v}$$

where $C_p / \Re_d$ is relatively constant, and was moved out of the integral on the RHS. The result is:

$$\ln(P)\Big|_{P_1}^{P_2} = (C_p / \Re_d) \cdot \ln(T_v)\Big|_{T_{v1}}^{T_{v2}}$$

Plug in the limits, and use ln(a)–ln(b) = ln(a/b) :

$$\ln\left(\frac{P_2}{P_1}\right) = (C_p / \Re_d) \cdot \ln\left(\frac{T_{v2}}{T_{v1}}\right)$$

Move $C_p / \Re_d$ to the LHS:

$$(\Re_d / C_p) \cdot \ln\left(\frac{P_2}{P_1}\right) = \ln\left(\frac{T_{v2}}{T_{v1}}\right)$$

Recall that $a \cdot \ln(b) = \ln(b^a)$, thus:

$$\ln\left[\left(\frac{P_2}{P_1}\right)^{\Re_d / C_p}\right] = \ln\left(\frac{T_{v2}}{T_{v1}}\right)$$

Finally, take the anti-log of both sides ($e^{LHS} = e^{RHS}$):

$$\left(\frac{P_2}{P_1}\right)^{\Re_d / C_p} = \frac{T_{v2}}{T_{v1}} \qquad (3.8)$$

Potential Temperature

A **potential temperature** θ can be defined that removes the effects of "dry" adiabatic temperature changes experienced by air parcels during vertical motion. If height is used as the vertical coordinate, then:

$$\theta(z) = T(z) + \Gamma_d \cdot z \qquad •(3.9)$$

This equation can use either °C or K. For pressure coordinates:

$$\theta = T \cdot \left(\frac{P_o}{P}\right)^{\Re_d / C_p} \qquad •(3.10)$$

where P_o is a reference pressure, and where absolute temperatures (K) must be used. The word "potential" arises because θ corresponds to the temperature T that a parcel would potentially have if it were moved adiabatically to the ground or to a reference pressure.

It is common practice to use a reference level of P_o = 100 kPa in eq. (3.10), but this is not a strict definition. For example, boundary-layer meteorologists often use eq. (3.9) where z is height above local ground level, not above the 100 kPa level.

The potential temperature is a **conserved variable** (i.e., remains constant) for "dry" adiabatic processes (i.e., when $\Delta Q = 0$). However, it can change as a result of all other heat transfer processes such as mixing or radiative heating (i.e., when $\Delta Q \neq 0$), which are generically called **diabatic** processes.

To include the buoyant effects of water vapor and liquid water in air, we can define a **virtual potential temperature** for non-cloudy (3.11) and cloudy (3.12) air:

$$\theta_v = \theta \cdot (1 + 0.61 \cdot r) \qquad (3.11)$$

$$\theta_v = \theta \cdot (1 + 0.61 \cdot r_s - r_L) \qquad •(3.12)$$

where r is mixing ratio of water vapor, r_s is saturation mixing ratio (see Chapter 5) and r_L is the mixing ratio of liquid water (cloud and rain drops). θ_v is used to determine how high air parcels can rise.

Solved Example

What is the potential temperature of air at height z = 500 m and temperature T = 10°C?

Solution

Given: z = 500 m, T = 10°C

Find: θ = ? °C

Assume no liquid water, and use eq. (3.9)

θ = 10°C + (9.8 °C/km)· (0.5 km) = **14.9 °C**

Check: Units OK. Physics reasonable.

Discussion: This is the temperature that air would have when lowered dry adiabatically to the surface. In other words, the surface is used as the reference height here for the potential temperature. Potential temperatures are always greater than the actual temperature, for heights above the reference level.

Thermodynamic Diagrams –Part 1: Dry Adiabatic Processes

It is often necessary to compare environmental lapse rates with process or parcel lapse rates, in order to determine parcel buoyancy, cloudiness and storm growth. Sometimes it is easier to utilize a diagram that already indicates the proper thermodynamic relationships, rather than recalculating the thermodynamic equations for every situation. Such a diagram is called a **thermodynamic diagram,** or **thermo diagram** for short.

These diagrams usually include relationships between so many variables that they become quite cluttered with lines and numbers. We will develop our own thermodynamic diagrams little by little as the book progresses. In this first part, we will include only "dry"adiabatic processes.

Most thermodynamic diagrams have temperature along the horizontal axis, and pressure or height along the vertical axis. We will use pressure with a reversed logarithmic scale along the vertical, as an approximation to height. Fig 1.4 is an example of such a diagram.

To this background chart, we will add diagonal lines showing the temperature decrease with height associated with air parcels rising adiabatically from the surface. These lines are called "dry" **adiabats.** They are also sometimes called **isentropes.**

We can draw different lines for parcels starting at the bottom of the graph (bottom of the atmosphere) with different temperatures. Fig 3.2 shows the result.

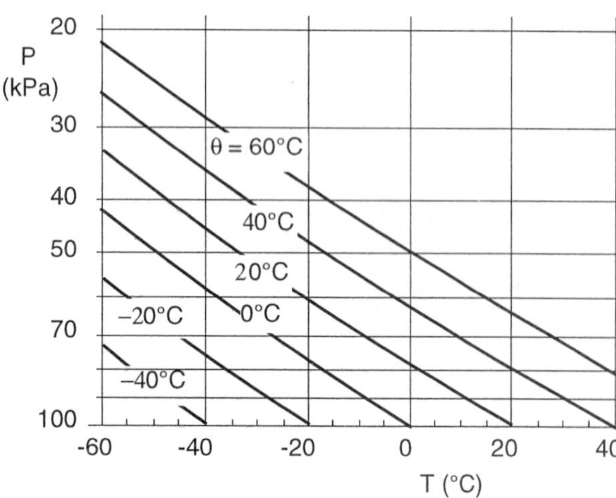

Figure 3.2
Thermodynamic diagram showing the temperature change of air parcels of several different initial surface temperatures, as they rise dry adiabatically. Recall from Chapt. 1 that a logarithmic pressure decrease corresponds to roughly a linear height increase. Thus, height increases upward in this diagram. The dark lines are called "dry" adiabats, and are labeled with potential temperature.

Potential temperature is constant along these diagonal lines, so the lines are usually labeled with θ. Sometimes the labeling is implicit; namely, by the temperature at the bottom of the graph where the adiabat intercepts P = 100 kPa. For example, the diagonal line third from the left in Fig 3.2 is the θ = 0°C adiabat.

FOCUS • Spreadsheet Thermodynamics

Thermodynamic diagrams can be produced with spreadsheet programs on personal computers. To calculate the dry adiabats, first type in the following row and column headers. Also shown in the figure below is the first set of temperature values representing the dry adiabats starting at four of the different temperatures. These temperatures were just typed in as numbers.

	A	B	C	D	E
1	Dry Adiabat Example				
2					
3	P (kPa)	T (°C)	T (°C)	T (°C)	T (°C)
4	100	60.0	40.0	20.0	0.0

Next, in cell A5, increment to the next pressure by typing equation: =A4-10 .

In cell B5, enter eq. (3.8) for the temperature along a dry adiabat: =((B$4+273)*($A5/A4)^0.28571)-273
The dollar sign implies an absolute reference; namely one that is not changed as the equation is auto-filled down or across.

Next select cell B5, and fill it across to E5:

	A	B	C	D	E
1	Dry Adiabat Example				
2					
3	P (kPa)	T (°C)	T (°C)	T (°C)	T (°C)
4	100	60.0	40.0	20.0	0.0
5	90	50.1	30.7	11.3	-8.1

Then, select cells A5 through E5, and fill down to row 12

	A	B	C	D	E
1	Dry Adiabat Example				
2					
3	P (kPa)	T (°C)	T (°C)	T (°C)	T (°C)
4	100	60.0	40.0	20.0	0.0
5	90	50.1	30.7	11.3	-8.1
6	80	39.4	20.7	1.9	-16.9
7	70	27.7	9.7	-8.4	-26.4
8	60	14.8	-2.5	-19.8	-37.1
9	50	0.2	-16.2	-32.6	-49.0
10	40	-16.7	-32.1	-47.5	-62.9
11	30	-36.9	-51.1	-65.3	-79.5
12	20	-62.7	-75.4	-88.0	-100.6

Finally, select all the cells holding numbers, and make a graph. Switch the axes such that pressure is on the vertical axis, reverse the order so pressure decreases upward, and finally make the vertical axis logarithmic. The result should look like Fig 3.2.

To use this diagram, plot the initial temperature and pressure of the air parcel as a point on the graph. If that air parcel rises or descends to a new pressure, draw a line parallel to the diagonal lines, starting at the initial point, and ending at the final pressure. This line shows how the parcel temperature changes, and what its final temperature is. This approach is easier than solving equations, but not as accurate.

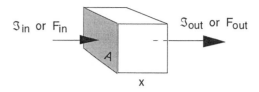

Figure 3.3
Change of flux F across distance Δx can cause changes within the volume. A is area of a face of the cube.

EULERIAN HEAT BUDGET

First Law of Thermodynamics – Revisited

If the heat flux into one side of a fixed volume is less than the flux out of the opposite side (Fig 3.3), then some amount of heat ΔQ is obviously removed from the volume. According to the first law of thermodynamics (3.5), a heat loss causes a temperature decrease.

Thus, it is not the flux itself, but the **gradient of flux** (i.e., the change of flux ΔF_x or $\Delta \mathfrak{F}_x$ across distance Δx) that causes temperature change within the volume. The gradient of flux is also known as a **flux divergence.** A positive value of $\Delta F_x/\Delta x$ gives a positive divergence (more flux leaving the volume than entering). Negative flux divergence is possible, which is sometimes called a **flux convergence.**

Considering the fluxes in all three dimensions, the first law of thermodynamics can be rewritten in Eulerian form as a **heat budget, heat balance**, or **heat conservation equation** for a stationary volume $Vol = A\cdot\Delta x$:

$$\frac{\Delta T}{\Delta t} = -\frac{1}{\rho\cdot C_p}\left[\frac{\Delta \mathfrak{F}_x}{\Delta x}+\frac{\Delta \mathfrak{F}_y}{\Delta y}+\frac{\Delta \mathfrak{F}_z}{\Delta z}\right]+\frac{\Delta So}{C_p\cdot\Delta t} \qquad (3.13a)$$

ΔSo is the internal source of heat per unit mass (e.g., $J\cdot kg^{-1}$), and would be negative for net cooling. Examples are **latent heating**, radioactive decay, and exothermal chemical reactions.

The kinematic-flux (F) form of the budget is:

$$\frac{\Delta T}{\Delta t} = -\left[\frac{\Delta F_x}{\Delta x}+\frac{\Delta F_y}{\Delta y}+\frac{\Delta F_z}{\Delta z}\right]+\frac{\Delta So}{C_p\cdot\Delta t} \qquad \bullet(3.13b)$$

In these equations, Δx, Δy, and Δz represent the length of sides of the volume, not movement of the volume. Finally, the heat budget can be rewritten using potential temperature:

$$\frac{\Delta \theta}{\Delta t} = -\left[\frac{\Delta F_x}{\Delta x}+\frac{\Delta F_y}{\Delta y}+\frac{\Delta F_z}{\Delta z}\right]+\frac{\Delta So}{C_p\cdot\Delta t} \qquad \bullet(3.13c)$$

As discussed in Appendix A, one must be careful when using gradients to insure that the numerator and denominator are in the same direction. Thus, the flux gradient in the x-direction is:

$$\frac{\Delta F_x}{\Delta x} = \frac{F_{x\ right} - F_{x\ left}}{x_{right} - x_{left}} \qquad (3.14)$$

Flux gradients are similar in the other directions. Gradients are like slopes, when plotted.

Equations (3.13) are important because they are used to make temperature forecasts. To solve these equations, we must determine the flux gradients. The flux gradients that appear in eq. (3.13) can be caused by **advection, conduction, turbulence,** and **radiation.**

$$\frac{\Delta F_x}{\Delta x} = \left.\frac{\Delta F_x}{\Delta x}\right|_{adv} + \left.\frac{\Delta F_x}{\Delta x}\right|_{cond} + \left.\frac{\Delta F_x}{\Delta x}\right|_{turb} + \left.\frac{\Delta F_x}{\Delta x}\right|_{rad} \qquad (3.15a)$$

$$\frac{\Delta F_y}{\Delta y} = \left.\frac{\Delta F_y}{\Delta y}\right|_{adv} + \left.\frac{\Delta F_y}{\Delta y}\right|_{cond} + \left.\frac{\Delta F_y}{\Delta y}\right|_{turb} + \left.\frac{\Delta F_y}{\Delta y}\right|_{rad} \qquad (3.15b)$$

$$\frac{\Delta F_z}{\Delta z} = \left.\frac{\Delta F_z}{\Delta z}\right|_{adv} + \left.\frac{\Delta F_z}{\Delta z}\right|_{cond} + \left.\frac{\Delta F_z}{\Delta z}\right|_{turb} + \left.\frac{\Delta F_z}{\Delta z}\right|_{rad} \qquad (3.15c)$$

We will describe each of these processes next. Finally, the source term will be estimated, and all the results will be put back into the Eulerian heat budget equation.

Solved Example

Given a cube of dry air at sea level, with side 20 m. An eastward moving heat flux of 3 $W\cdot m^{-2}$ flows in through the left, while a westward moving heat flux of 4 $W\cdot m^{-2}$ flows in through the right. There are no other fluxes or internal sources of heat. What is the kinematic heat flux through each side, and at what rate will temperature change?

(continued next column)

Solved Example *(continuation)*

Solution

Given: $\mathfrak{I}_{x\ right} = -4\ \mathrm{W \cdot m^{-2}}$ (negative because it is moving in the negative x direction)

$\mathfrak{I}_{x\ left} = 3\ \mathrm{W \cdot m^{-2}}$, $\Delta x = 20$ m

Find: $F_{x\ right} =$? K·m/s, $F_{x\ left}$ = ? K·m/s

$\Delta T/\Delta t$ = ? K/s

Sketch:

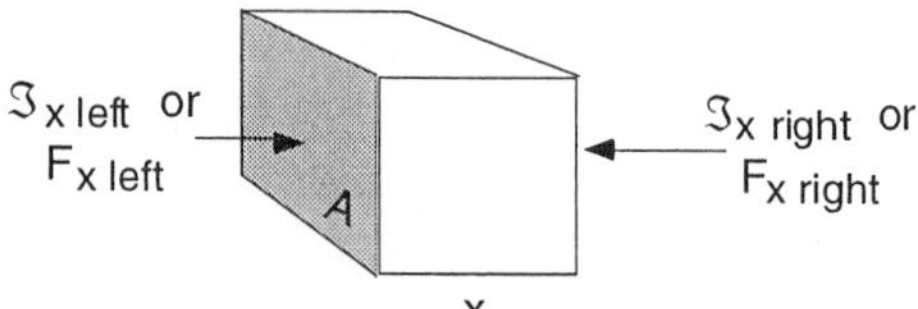

Use eq. (2.13b)

$F_{x\text{-}left} =$ (3 W·m^{-2}) / [1231 (W·m^{-2}) / (K·m·s^{-1})]
= **2.437x10^{-3} K·m·s^{-1}** .

$F_{x\text{-}right} =$ (–4 W·m^{-2}) / [1231 (W·m^{-2})/(K·m·s^{-1})]
= **– 3.249x10^{-3} K·m·s^{-1}** .

The flux gradient (eq. 3.14) is thus:

$$\frac{\Delta F_x}{\Delta x} = \frac{\left[(-3.249\times10^{-3})-(2.437\times10^{-3})\right]\ (\mathrm{K\cdot m\cdot s^{-1}})}{[20-0]\ (\mathrm{m})}$$

= – 2.843x10^{-4} K·s^{-1}.

Plugging this flux gradient into (3.13b) gives

$\Delta T/\Delta t$ = **+ 2.843x10^{-4} K·s^{-1}**.

Check: The sign is positive implying warming with time, because heat is flowing into both sides of the cube.

Discussion: The heating rate is equivalent to 1 K/h.

Advection

The word **advect** means "to be transported by the mean wind". **Temperature advection** relates to heat being blown to or from a region by the wind. The amount of **advective flux** of heat increases linearly with mean temperature, and with mean wind speed:

$$F_{x\ adv} = U \cdot T \tag{3.16a}$$

$$F_{y\ adv} = V \cdot T \tag{3.16b}$$

$$F_{z\ adv} = W \cdot T \tag{3.16c}$$

Vertical motion is also called **advection** if by the mean wind, but is called **convection** if by buoyancy.

The heat budget (3.13), however, uses the flux gradient, not the raw flux. As illustrated in Fig 3.4a, the temperature at a fixed point such as at the top of a tall tower might initially be 5 °C. However, if the air temperature is greater in the west than the east, and if the wind is blowing from the west, then after a while the warmer air will be over the tower (Fig 3.4b). Thus, the temperature increases with time at the fixed point for a situation with positive mean wind (blowing from west to east) and a negative horizontal temperature gradient (temperature decreasing from west to east).

If the mean wind speed varies little across the region of interest, then

$$\frac{\Delta F_{x\ adv}}{\Delta x} = \frac{U \cdot \left(T_{right} - T_{left}\right)}{x_{right} - x_{left}} = U \cdot \frac{\Delta T}{\Delta x} \tag{3.17a}$$

Similarly:

$$\frac{\Delta F_{y\ adv}}{\Delta y} = V \cdot \frac{\Delta T}{\Delta y} \tag{3.17b}$$

and

$$\frac{\Delta F_{z\ adv}}{\Delta z} = W \cdot \left[\frac{\Delta T}{\Delta z} + \Gamma_d\right] \tag{3.17c}$$

The dry adiabatic lapse rate Γ_d = 9.8 °C/km is added to the last term to compensate for the adiabatic temperature change that normally accompanies vertical motion as parcels move into regions of different pressure. In fact, this correction term must always be added to temperature whenever vertical motions are considered.

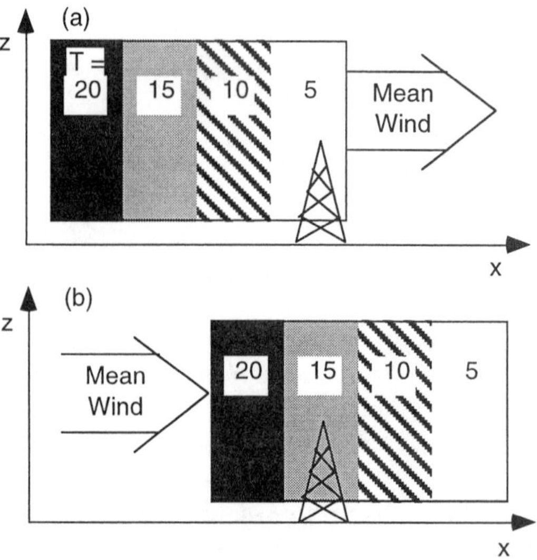

Figure 3.4
Advection illustration. (a) Initial state of the air with an east-west temperature gradient. (b) Later, after the air moves east.

Substituting (3.9) into (3.17) allows potential temperature to be used for advection:

$$\frac{\Delta F_{x\ adv}}{\Delta x} = U \cdot \frac{\Delta \theta}{\Delta x} \qquad \bullet(3.18a)$$

$$\frac{\Delta F_{y\ adv}}{\Delta y} = V \cdot \frac{\Delta \theta}{\Delta y} \qquad \bullet(3.18b)$$

$$\frac{\Delta F_{z\ adv}}{\Delta z} = W \cdot \frac{\Delta \theta}{\Delta z} \qquad \bullet(3.18c)$$

Solved Example

Suppose the average air temperature gets colder with height, such that $T = 15°C$ at $z = 200$ m and $T = 10°C$ at $z = 1000$ m, with a linear change in between. If the mean vertical velocity is blowing the cold air downward from aloft, then what is the advective cooling rate at $z = 600$ m, given that $W = -0.1$ m/s, and no other heat-transfer processes are present?

Solution

Given: $W = -0.1$ m/s, $z = 600$ m

$\Delta T/\Delta z = (10–15°C)/(1000–200 \text{ m})$

$= -0.00625$ °C/m

Find: $\Delta T/\Delta t = ?$ °C/s

This is a trick question. Although the air is colder aloft, we anticipate that it will warm as it descends due to the adiabatic compression discussed in the previous sections. When the air gets to the height of interest (600 m), it might be so hot as to cause warming rather than cooling. So we must be careful in solving this.

Plugging (3.17c) into (3.13b), and neglecting all other heat transfer processes, we find

$\Delta T/\Delta t = -W(\Delta T/\Delta z + \Gamma_d)$

$= -(-0.1 \text{ m/s})\cdot(-0.00625 + 0.0098°C/m)$

$= +$ <u>3.55×10^{-4} °C/s</u>

Check: Units OK. Physics reasonable.

Discussion: The positive sign indicates a warming rate, not cooling. This advective warming rate is the same at all heights between 200 and 1000 m, not just at 600 m. The warming rate is equivalent to 1.28 °C/hour, which could cause a significant temperature increase if it continues for many hrs.

Solved Example

Given a cube of air with winds as sketched in Fig 3.3. Let the mean wind speed be 10 m/s from west to east, but the temperature entering from the left is 20°C while that leaving on the right is 18°C. If the cube has side 20 m, then what is the advective flux gradient? (That is, what is the temperature tendency caused by advection.)

Solution

Given: $U = 10$ m/s, $\Delta T = 18 - 20°C = -2°C$

$\Delta x = 20$ m

Find: $\Delta F_{x\ adv} / \Delta x = ?$ °C/s

Use (3.17a): $\Delta F_{x\ adv} / \Delta x = U\cdot(\Delta T / \Delta x)$

$= (10 \text{ m/s})\cdot[-2°C / 20 \text{ m}] =$ <u>-1 °C/s</u>.

Check: Units OK. Sketch OK. Physics OK.

Discussion: If this flux gradient were used in the heat-budget equation (3.13), the warming rate would be +1 °C/s due to advection. This is called warm-air advection, because warmer air is entering the region than is leaving.

Conduction & Surface Fluxes

Heat can be transported by individual molecules bouncing into each other. This conduction process works in solids, liquids and gases, and works with or without any mean or turbulent wind. It is the primary way heat is transferred from the earth's surface to the atmosphere, and also the primary way that heat is transported down into the ground.

The amount of conductive heat flux in the vertical is

$$\mathfrak{I}_{z\ cond} = -k \cdot \frac{\Delta T}{\Delta z} \qquad (3.19)$$

where the molecular conductivity k is a property of the material through which heat flows. For air at standard conditions near sea level, $k = 2.53\times10^{-2}$ $W\cdot m^{-1}\cdot K^{-1}$.

Because temperature gradients $\Delta T/\Delta z$ in most interior parts of the troposphere are relatively small, molecular conduction is negligible. Thus, for simplicity we will use

$$\frac{\Delta F_{x\ cond}}{\Delta x} \approx \frac{\Delta F_{y\ cond}}{\Delta y} \approx \frac{\Delta F_{z\ cond}}{\Delta z} \approx 0 \qquad (3.20)$$

in the heat budget equation (3.13) everywhere except at the ground.

Solved Example

What vertical temperature difference is necessary across the bottom 1 mm of atmosphere to conduct 300 W·m^{-2} of heat flux?

Solution

Given: $\mathfrak{I}_{z\,cond}$ = 300 W·m^{-2}, Δz = 0.001 m
k = 2.53x10^{-2} W·m^{-1}·K^{-1}

Find: ΔT = ? °C

Sketch: (see Fig 3.5)

Rearrange eq. (3.19) to solve for ΔT:

$\Delta T = -\Delta z \cdot \mathfrak{I}_{z\,cond} / k$

ΔT = –(10^{-3} m)·(300 W·m^{-2}) / (2.53x10^{-2} W·m^{-1}·K^{-1})

ΔT = **– 11.9°K**

Check: Units OK. Physics OK.

Discussion: The air must be almost 12°C cooler than the surface to conduct the required heat flux. This temperature difference is often observed in the real atmosphere.

Within the bottom few millimeters of air near the surface, vertical temperature gradients can be quite large (Fig 3.5). For example, on a sunny summer day, the sun might heat the surface of an asphalt road to temperatures greater than 50°C, while the adjacent air temperatures are about 20 or 30°C. Such a large gradient is sufficient to drive a substantial heat flux.

When conduction and turbulence are considered together, we see that turbulence is zero at the ground where conduction can be important, while turbulence is important in the rest of the lower atmosphere where conduction is negligible. For this reason, turbulence and conduction are often combined into an **effective surface turbulent heat flux,** F_H.

In practice, it is virtually impossible to use eq. (3.19) directly, because the temperature gradient across the bottom few millimeters of air is rarely measured or predicted. However, on **windy** days the combination of conduction and turbulence causes the effective surface heat flux at the surface to be:

$$F_H = C_H \cdot M \cdot (\theta_{sfc} - \theta_{air}) \qquad \bullet(3.21a)$$

or

$$F_H \cong C_H \cdot M \cdot (T_{sfc} - T_{air}) \qquad \bullet(3.21b)$$

M is the mean wind speed at 10 m, T_{air} is the air temperature at a height of 10 m, T_{sfc} is the surface skin temperature, and θ_{air} and θ_{sfc} are the corresponding potential temperatures.

C_H is the **bulk heat transfer coefficient**. This coefficient is unfortunately not constant, but varies with surface conditions and turbulence intensity. It has a dimensionless value in the range of 2x10^{-3} over smooth surfaces to 2x10^{-2} over rough or forested surfaces. Eq. (3.21) is useful during strong advection.

The bottom 1 to 2 km of the troposphere is called the **atmospheric boundary layer** (**ABL**, see Chapt. 4). During **calm** days with strong solar heating and convection, warm rising **thermals** (buoyant air parcels) often occur within the ABL. This type of ABL is called a **mixed layer** (ML). The rising thermals cause a substantial effective surface heat flux. Instead of using eq. (3.21) on these calm days, the surface flux can be found from:

$$F_H = b_H \cdot w_B \cdot (\theta_{sfc} - \theta_{ML}) \qquad \bullet(3.22a)$$

or

$$F_H = a_H \cdot w_* \cdot (\theta_{sfc} - \theta_{ML}) \qquad \bullet(3.22b)$$

Solved Example

Find the effective surface heat flux over a forest when the wind speed is 5 m·s^{-1} at a height of 10 m, the surface temperature is 25°C, and the air temperature at 10 m is 20°C.

Solution

Given: M = 5 m/s, z = 10 m
$T_{sfc} - T_{air}$ = 25 – 20°C = 5 °C

Find: F_H = ? K·m·s^{-1}

Use eq. (3.21b):

F_H = (2x10^{-2})·(5 m/s)·(5°C) = **0.5 °C·m·s^{-1}**

Check: Units OK. Physics OK.

Discussion: Using conversion (2.13b), the dynamic heat flux is 615.5 W·m^{-2}, which is almost half of the solar constant of 1376 W·m^{-2}. This suggests that 50% of the incoming solar radiation heats the air, while the remainder causes evaporation, is conducted into the ground, and is reflected for this example.

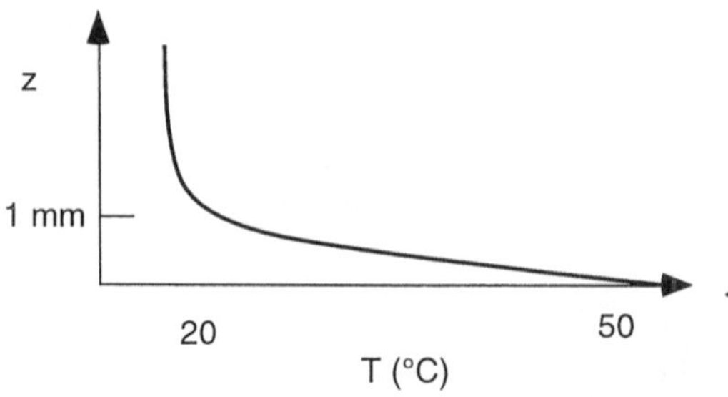

Figure 3.5
Strong temperature gradients within a few millimeters of the ground on a sunny day can drive strong conductive heat fluxes there, but higher in the atmosphere conduction is negligible.

where θ_{ML} is the potential temperature of the air at a height of about 500 m, in the middle of the mixed layer. The empirical **mixed-layer transport coefficient** is $a_H = 0.0063$, and the **convective transport coefficient** is $b_H = 5x10^{-4}$. Both are independent of sfc. roughness.

A **buoyancy velocity scale** w_B gives the effectiveness of thermals in producing vertical heat transport:

$$w_B = \left[\frac{|g| \cdot z_i}{\theta_{v\ ML}} \cdot (\theta_{v\ sfc} - \theta_{v\ ML}) \right]^{1/2} \qquad \bullet(3.23)$$

where z_i is the depth of the convective ABL (i.e., the mixed layer), and g is gravitational acceleration. Virtual potential temperatures (in absolute units of K) are used in this last expression, but not in eq. (3.22). Typical vertical velocities are about $0.02 \cdot w_B$.

The **Deardorff velocity** w_* is another convective velocity scale, and is defined as

$$w_* = \left[\frac{|g| \cdot z_i}{T_v} \cdot F_H \right]^{1/3} \qquad \bullet(3.24)$$

where T_v is an average absolute virtual temperature in the ABL. Typical values are 1 to 2 $m \cdot s^{-1}$.

Another way of finding F_H is with a surface heat budget or a Bowen ratio, as described later in this chapter. One could also write an expression similar to eqs. (3.22), except for surface moisture flux in terms of the difference in water vapor mixing ratio between the surface and the middle of the mixed layer.

Solved Example

On a calm sunny day, a 2 km thick mixed layer is dry with $\theta = 295$ K. If $T_{sfc} = 325$ K, find the effective surface heat flux.

Solution

Given: $\theta_{ML} = 295$ K, $\theta_{sfc} = 325$ K
$z_i = 2000$ m, $g = 9.8\ m \cdot s^{-2}$
Find: $F_{z\ eff.sfc.} = ?\ K \cdot m \cdot s^{-1}$

For dry air $\theta = \theta_v$ (see eq. 3.11).
Use eq. (3.23) first, then eq. (3.22a):

$$w_B = \left[\frac{|9.8 m / s^2| \cdot 2000 m}{295 K} \cdot (325 K - 295 K) \right]^{1/2}$$

$= (1993\ m^2/s^2)^{1/2} = 44.6\ m \cdot s^{-1}$
$F_H = (5x10^{-4}) \cdot (44.6\ m \cdot s^{-1}) \cdot (325\ K - 295\ K)$
$=$ **0.67** $K \cdot m \cdot s^{-1}$

Check: Units OK. Physics reasonable.
Discussion: Greater surface-atmos. temperature differences drive more vigorous thermals, and each thermal then transports more heat.

Turbulence

Turbulence is the quasi-random movement of air parcels by small (order of 2 mm to 2 km) swirls of motion called **eddies**. Although the air parcels leaving one region are always replaced by air parcels returning from somewhere else (conservation of air mass), the departing parcels might have a different temperature than the arriving parcels. By moving temperature around, the net result is a heat flux.

Because there are so many eddies of all sizes moving in such a complex fashion, meteorologists don't even try to describe the heat flux contribution from each individual eddy. Instead, meteorologists use statistics to describe the average heat flux caused by all the eddies within a region (see Chapter 4). The expressions below represent such statistical averages.

The net effect of turbulence is to mix together air parcels from different initial locations. Thus, turbulence tends to homogenize the air. Potential temperature, wind, and humidity gradually become mixed toward a more uniform state by the action of turbulence. The amount of mixing varies with time and location, as turbulence intensity changes.

During fair-weather, turbulence is strong during daytime over non-snowy ground, and can cause significant heat transport within the atmospheric boundary layer (ABL). Due to the turbulent homogenization, turbulent fluxes vary roughly linearly with height, and can be estimated from the fluxes at the top and bottom of the ABL.

In fair weather (**wx**) during daytime, the heat flux at the top of the ABL is often roughly 20% of the magnitude of the surface heat flux, but has opposite sign. The flux at the bottom of the ABL is associated with conduction, and was given in the previous subsection as an effective surface flux F_H. This gives a turbulent flux divergence of:

$$\frac{\Delta F_{z\ turb}}{\Delta z} \approx \frac{F_{z\ top} - F_{z\ bottom}}{z_i} \qquad (3.25a)$$

$0 < z < z_i$, fair wx

$$\frac{\Delta F_{z\ turb}}{\Delta z} \approx \frac{-1.2 \cdot F_H}{z_i} \qquad (3.25b)$$

where z_i is the depth of the ABL (typically of order 200 m to 2 km).

Solved Example

On a sunny fair-weather day with effective surface heat flux of 0.1 K·m·s^{-1}, what is the vertical flux divergence across an atmospheric boundary layer of depth 1 km?

Solution

Given: $F_H = 0.1$ K·m·s^{-1}, $z_i = 1000$ m

Find: $\Delta F_{z\ turb} / \Delta z = ?$ (K/s)

Use eq. (3.25b):

$$\frac{\Delta F_{z\ turb}}{\Delta z} \approx \frac{-1.2 \cdot F_H}{z_i}$$

$$\frac{\Delta F_{z\ turb}}{\Delta z} \approx \frac{-1.2 \cdot (0.1\ \mathrm{K \cdot m / s})}{1000\mathrm{m}}$$

$$\Delta F_{z\ turb} / \Delta z = \underline{\mathbf{-0.00012\ K \cdot s^{-1}}}$$

Check: Units OK. Physics OK.

Discussion: The negative sign on the gradient implies heating with time according to eq. (3.13), which makes sense during daytime.

The answer is equivalent to 0.43°C·h^{-1}, or about 5°C/12 h. This seems a bit small, compared to typical amounts of warming during the daylight hours. Thus, one would guess that daytime effective surface heat fluxes can be larger than 0.1 K·m·s^{-1}, and/or ABLs can be shallower than 1 km.

Above the ABL during fair weather, turbulence is often negligible:

$$\frac{\Delta F_{z\ turb}}{\Delta z} \approx 0 \qquad \text{for } z > z_i \text{, fair wx} \quad (3.26)$$

At night, turbulence is often much weaker. Thus, vertical flux divergence is often quite small during fair weather nights, and can be neglected everywhere except in the bottom 100 m or so.

During stormy weather, deep cumulus clouds and thunderstorms can cause vigorous turbulent mixing throughout the depth of the troposphere. Such convective clouds form in response to warm, buoyant air underlying cooler air (Fig 3.6). This type of pre-storm environment is said to be unstable, as will be discussed in more detail in the chapters on stability (Chapter 6) and thunderstorms (Chapter 15). The atmosphere tries to re-stabilize itself by letting the warm air rise and the cold air sink. This mixing process is called **moist convective adjustment**. Such mixing causes a vertical heat flux.

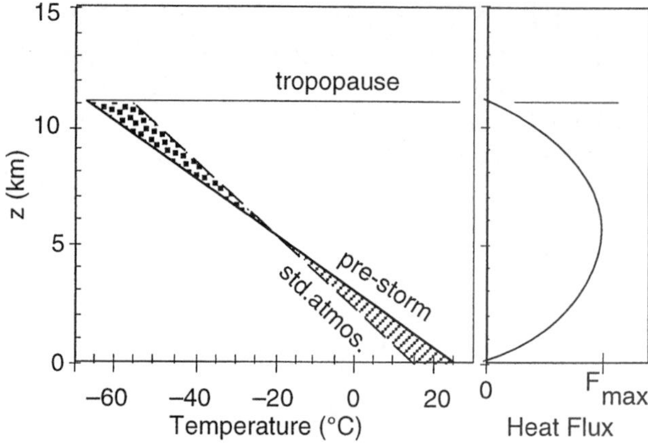

Figure 3.6

Pre-storm air at the bottom of the troposphere that is warmer (hatched region) than the standard atmosphere tends to rise and exchange places with cooler air aloft (stippled region) during a thunderstorm. The result is a vertical heat flux.

The mixing continues until some equilibrium profile of temperature is reached. As a first approximation, the **standard atmospheric lapse rate** ($\Gamma_{sa} = -\Delta T/\Delta z = +6.5$ K/km) is used here for the equilibrium profile. If the pre-storm lapse rate is $\Gamma_{ps} = -\Delta T/\Delta z$, then the vertical turbulent flux gradient is approximately:

$$\frac{\Delta F_{z\ turb}}{\Delta z} \approx \frac{z_T}{\Delta t} \cdot \left[\Gamma_{ps} - \Gamma_{sa}\right] \cdot \left(\frac{1}{2} - \frac{z}{z_T}\right) \qquad \text{stormy wx} \quad (3.27)$$

where Δt is the lifetime of the thunderstorms (order of 1 h), and z_T is depth of the troposphere (order of 11 km). The term in square brackets is positive for an unstable pre-storm environment.

As shown in Fig 3.6, the heat flux profile is shaped like a parabola. The turbulent vertical heat flux is zero at the top and bottom of the troposphere, and reaches a maximum of

$$F_{max} = z_T{}^2 \cdot \left[\Gamma_{ps} - \Gamma_{sa}\right] / (8 \cdot \Delta t) \qquad \text{stormy wx} \quad (3.28)$$

in the middle of the troposphere.

The convective adjustment effect of cooling at low levels and heating aloft must be added to the precipitation effect (see the source term) of latent heating at all heights, to find the net affect of the thunderstorm. An even more accurate accounting would include IR longwave radiation divergence from the top of the thunderstorm anvil cloud.

In those regions where turbulence exists, turbulent transport in any horizontal direction often cancels transport in the opposite direction, yielding negligible net flux gradients. Also, in non-turbulent regions there is obviously no turbulent flux gradient. Thus, as an idealized approximation, we can assume the following everywhere:

$$\frac{\Delta F_{x\ turb}}{\Delta x} \approx \frac{\Delta F_{y\ turb}}{\Delta y} \approx 0 \tag{3.29}$$

Solved Example

What is the vertical turb. flux gradient in stormy weather at 2 km altitude, if the pre-storm vertical temperature gradient was –10 °C/km?

Solution

Given: $\Gamma_{ps} = -\Delta T/\Delta z = +10$ K/km, $\Gamma_{sa} = 6.5$ K/km
Find: $\Delta F_{z\ turb}/\Delta z = ?$ (K/s)
Assume: $z_T = 11$ km, $\Delta t = 1$ h $= 3600$ s.

Use eq. (3.27):

$$\frac{\Delta F_{z\ turb}}{\Delta z} \approx \frac{z_T}{\Delta t}\cdot\left[\Gamma_{ps} - \Gamma_{sa}\right]\cdot\left(\frac{1}{2} - \frac{z}{z_T}\right)$$

$$\frac{\Delta F_{z\ turb}}{\Delta z} \approx \frac{11\text{ km}}{3600\text{s}}\cdot\left[(10-6.5)\frac{\text{K}}{\text{km}}\right]\cdot\left(\frac{1}{2} - \frac{2\text{km}}{11\text{km}}\right)$$

$\Delta F_{z\ turb} / \Delta z =$ **0.0034 K/s**

Check: Units OK. Physics OK.
Discussion: The positive flux gradient implies cooling over time, because of the negative sign on the flux gradient terms in eq. (3.13). This makes sense, because z = 2 km is in the bottom half of the troposphere, where the initially-warm air is becoming cooler with time due to mixing with the colder air from above.

Radiation

Air in the troposphere is relatively opaque to IR radiation, causing radiation emitted at one altitude to be reabsorbed into the air at a neighboring altitude, some of which is re-radiated back to the first level. The net result of this very complex process is a relatively small vertical IR flux gradient within the air, and a nearly zero horizontal flux gradient:

$$\frac{\Delta F_{x\ rad}}{\Delta x} \approx \frac{\Delta F_{y\ rad}}{\Delta y} \approx 0 \tag{3.30}$$

$$\frac{\Delta F_{z\ rad}}{\Delta z} \approx -0.1 \text{ to } -0.2 \text{ (K/h)} \tag{3.31}$$

In the absence of clouds, haze, or smoke, direct radiative warming of the air due to solar radiation is negligible. However, sunlight warms the air indirectly by heating the surface, which then transfers and distributes its heat to the air by conduction and convection. There can also be substantial solar warming of clouds during daytime.

Internal Source - Latent Heat

The amount of latent heat given to the air during condensation of $m_{condensing}$ grams of water vapor that are already contained in the Eulerian volume is $L_v \cdot m_{condensing}$. When distributed throughout the mass of air m_{air} within the volume, the net latent heating rate for the source term is:

$$\frac{\Delta So}{C_p \cdot \Delta t} = \frac{L_v}{C_p}\cdot\frac{m_{\text{condensing}}}{m_{\text{air}}\cdot\Delta t} \tag{3.32}$$

For evaporation of existing liquid water, the sign is negative (to cool the air).

Consider a column of air equal to the troposphere depth of roughly $z_{Trop} = 11$ km. The average condensational heating rate of the whole column is proportional to the amount of precipitation that falls out of the bottom. If *RR* denotes rainfall rate measured as the change of depth of water in a rain gauge with time, then the latent heating source term is:

$$\frac{\Delta So}{C_p \cdot \Delta t} = \frac{L_v}{C_p}\cdot\frac{\rho_{liq}}{\rho_{air}}\cdot\frac{RR}{z_{Trop}} \tag{3.33a}$$

where $\rho_{liq} = 1025$ kg·m^{-3} is the density of liquid water, ρ_{air} is air density averaged over the whole column, and $L_v / C_p = 2500$ K·kg$_{air}$·kg$_{liq}^{-1}$ is the ratio of latent heat of condensation to specific heat at constant pressure.

Utilizing the standard atmosphere with a tropopause depth of 11 km gives an **average tropospheric density** of $\rho_{air} = 0.689$ kg·m^{-3}. Thus, eq. (3.33a) can be rewritten as:

$$\frac{\Delta So}{C_p \cdot \Delta t} = a \cdot RR \tag{3.33b}$$

where $a = 0.338$ K/(mm of rain), and for *RR* in (mm of rain)/s. Divide by 3600 for *RR* in mm/h.

Solved Example

Find the tropospheric average latent heating for a rainfall rate of $5\ \text{mm·h}^{-1}$.

Solution

Given: $RR = 5\ \text{mm·h}^{-1}$.
Find: $\Delta So/(C_p\cdot\Delta t) = ?\ \text{K·m·s}^{-1}$

Use eq. (3.33b): $\Delta So/(C_p\cdot\Delta t) = 0.338\ \text{(K/mm)}\cdot(5\ \text{mm/h})\cdot(1\text{h}/3600\text{s}) = \underline{\mathbf{0.00047}}\ \text{K·m·s}^{-1}$

Check: Units OK. Physics OK.
Discussion: This is the same order of magnitude as other terms in the Eulerian heat budget.

Net Heat Budget

The net Eulerian heat budget is found by combining the first law of thermodynamics eq. (3.13b or c) with the definition of all the flux gradients (3.15). The long messy equation that results can be simplified for many atmospheric situations as follows.

Assume negligible vertical mean temperature advection. Assume negligible horizontal turbulent transport. Assume negligible conduction. Neglect direct solar heating of the troposphere, but approximate IR cooling as a small constant.

The resulting simplified Eulerian net heat budget equation is:

$$\left.\frac{\Delta T}{\Delta t}\right|_{x,y,z} = \underbrace{-\left[U\cdot\frac{\Delta T}{\Delta x}+V\cdot\frac{\Delta T}{\Delta y}\right]}_{\text{advection}} \underbrace{-0.1\ \frac{\text{K}}{\text{h}}}_{\text{radiation}} \underbrace{-\frac{\Delta F_{z\ turb}(\theta)}{\Delta z}}_{\text{turbulence}} + \underbrace{\frac{L_v}{C_p}\cdot\frac{m_{condensing}}{m_{air}\cdot\Delta t}}_{\text{latent heat}} \quad \bullet(3.34)$$

where the θ in the turbulence term is to remind us that this turbulence flux divergence is for heat. In later chapters we will see similar turbulence flux divergences, but for moisture or momentum. If there is no condensation or evaporation (i.e., for non-cloudy air), the source-sink term can be set to zero.

Solved Example

Find the temperature change during one hour at a fixed point, given $U = 5$ m/s, $V = 0$, $\Delta T/\Delta x = 1°\text{C}/10$ km, $m_{cond}/m_{air} = 2\ g_{water}/kg_{air}$, for a daytime 800 m thick ABL with effective surface kinematic flux of 0.2 K·m/s. Use $L_v/C_p = 2.5\ \text{K}/(g_{water}/kg_{air})$.

Solution

Given: (see above)
Find: $\Delta T = ?\ °\text{C}$

Use eq. (3.34). Do each term separately, times Δt:

$$\text{Adv}\cdot\Delta t = -\left[(5\text{m/s})\cdot\left(\frac{1°\text{C}}{10^4\text{m}}\right)\right]\cdot(3600\text{s}) = -1.8°\text{C}$$

$$\text{Source}\cdot\Delta t = \left(2.5\frac{\text{K}\cdot kg_{air}}{g_{water}}\right)\cdot\left(2\frac{g_{water}}{kg_{air}}\right) = +5.0°\text{C}$$

$$\text{Rad}\cdot\Delta t = \left(-0.1\ \frac{\text{K}}{\text{h}}\right)\cdot(1\text{h}) = -0.1°\text{C}$$

$$\text{Turb}\cdot\Delta t = -\frac{-1.2\cdot(0.2\ \text{K}\cdot\text{m/s})}{800\text{m}}\cdot(3600\text{s}) = +1.1°\text{C}$$

where (2.43b) was used for the turbulence term.

Thus: $\Delta T = (\text{Adv} + \text{Latent} + \text{Rad} + \text{Turb})\cdot\Delta t$
$= (-1.8 + 5.0 - 0.1 + 1.1)°\text{C} = \underline{\mathbf{4.2\ °C}}$

Check: Units OK. Physics OK.
Discussion: Many processes act simultaneously, some adding and some subtracting from the total. This is one of the complexities of weather forecasting.

During daytime, <u>direct</u> radiative heating of the air is small, and can be neglected. However, strong heating of the ground by solar radiation allows the warm ground to heat the air via the surface heat flux. Thus, radiative heating from the sun is getting to the air <u>indirectly</u> via conduction of sensible heat from the ground, and via surface evaporation of water that transports latent heat. Latent heating was quite important here.

SURFACE HEAT BUDGET

While the Lagrangian and Eulerian heat budgets give the temperature change and fluxes for a volume (moving or stationary, respectively), we can also examine the fluxes across the earth's surface. Because the surface has zero thickness, all fluxes must sum to zero. This holds for both heat and moisture.

Heat Budget

During daytime, the net radiative heat input to the surface via flux $\Im^*$ is balanced by three output fluxes from the surface. They are: turbulent sensible-heat transport into the air (**sensible heat flux** $\Im_H$, sometimes called the heat flux), turbulent latent-heat transport into the air (**latent heat flux** $\Im_E$ associated with evaporation from the surface), and **molecular conduction** flux into the ground $\Im_G$. At night, the directions (signs) of the fluxes are reversed. Regardless of the surface scenario (Fig 3.7), the fluxes into the surface must balance the fluxes away from the surface.

For an infinitesimally thin surface, this balance is

$$-\Im^* = \Im_H + \Im_E - \Im_G \qquad \bullet(3.35a)$$

or in kinematic form (after dividing by $\rho \cdot C_p$):

$$-F^* = F_H + F_E - F_G \qquad \bullet(3.35b)$$

where *all fluxes are defined positive upward*. All terms have units of $W \cdot m^{-2}$ in the first equation, and $K \cdot m \cdot s^{-1}$ in the second. Also, Fig 3.8 shows an example of the daily variation of these fluxes for a moist vegetated surface.

One can think of the net radiation as a forcing on the system, and the other fluxes in eq. (3.35) as the responses. Namely, all of the radiation absorbed at the surface must go somewhere.

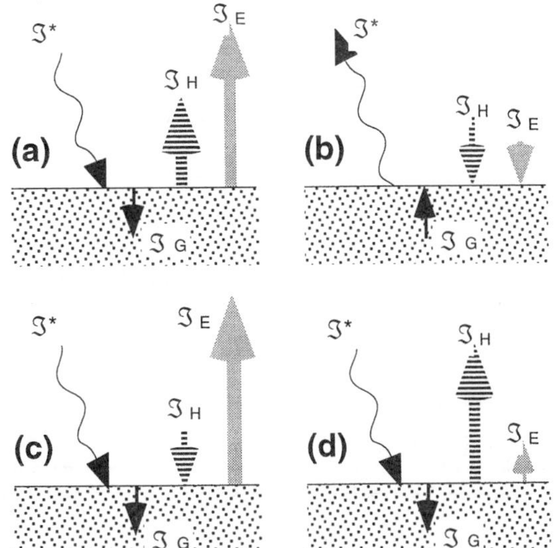

Figure 3.7
Fluxes affecting the surface heat budget. (a) Daytime over vegetated surface. (b) Nighttime over vegetated surface. (c) Oasis effect of warm, dry air advection over a cool, moist surface. (d) Daytime over a desert.

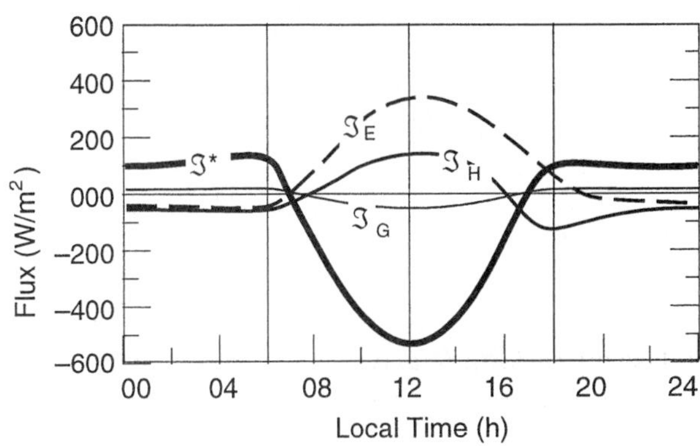

Figure 3.8
Typical diurnal variation of the surface heat budget. The $\Im^*$ curve comes from Fig 2.10 . Fluxes are positive upward.

To good approximation, the flux into the ground is proportional to the net radiative flux, which in dynamic and kinematic forms are:

$$\Im_G \approx X \cdot \Im^* \qquad (3.36a)$$

$$F_G \approx X \cdot F^* \qquad (3.36b)$$

where $X = 0.1$ during the day, and $X = 0.5$ at night.

Expressions for the effective sensible heat flux at the surface were presented in Chapter 2. Similar expressions can be used for the latent heat flux. A Bowen-ratio method for measuring the sensible and latent flux is described in the next subsection.

Note the difference between the heat budget across the surface (eq. 3.35) where the fluxes must balance, and the heat budget within a volume of air (eq. 3.13) where any imbalance of the fluxes causes a net heating or cooling within the volume.

Bowen Ratio

The ratio of the sensible to latent heat fluxes is called the **Bowen ratio**, B:

$$B = \frac{\Im_H}{\Im_E} = \frac{F_H}{F_E} \qquad (3.37)$$

The Bowen ratio can be 0.1 over the sea, 0.2 over irrigated crops, 0.5 over grassland, 5 over semi-arid regions, and 10 over deserts.

Many observations have confirmed that the sensible heat flux is proportional to the vertical gradient of potential temperature $\Delta\theta$ in the bottom 20 m of the atmosphere (see Appendix H). Similarly, the moisture flux is proportional to the vertical gradient of mixing ratio Δr (grams of water vapor in each kilogram of dry air, as will be described in detail in Chapt 5).

Solved Example

Given a net radiative surface flux of –600 W·m^{-2} over grassland, find the other terms of the surface heat budget.

Solution

Given: $\Im^* = -600$ W·m^{-2} (probably daytime)
$B = 0.5$ for grassland
Find: $\Im_G$, $\Im_H$, $\Im_E$ = ? W·m^{-2}

Use eq. (3.36) with $X = 0.1$ for day:
$\Im_G = 0.1 \cdot \Im^* = 0.1 \cdot (-600$ W·m$^{-2}) =$ **–60** W·m^{-2}

Combine eqs. (3.35a) & (3.37):
$\Im_H = B \cdot (\Im_G - \Im^*) / (1 + B)$
$\Im_E = (\Im_G - \Im^*) / (1 + B)$
Thus,
$\Im_H = 0.5 \cdot (-60 + 600$ W·m$^{-2}) / (1 + 0.5)$
= **180** W·m^{-2}
$\Im_E = (-60 + 600$ W·m$^{-2}) / (1 + 0.5)$
= **360** W·m^{-2}

Check: Units OK. Physics OK.
Discussion: Does the energy budget really balance:
$-\Im^* = \Im_H + \Im_E - \Im_G$
600 = 180 + 360 + 60 Yes.

The Bowen ratio is quite variable, even over a single type of surface such as grassland. For example, some grasses might be more moist or might be transpiring more than other grasses, depending on the maturity of the grass, insolation, air temperature, and soil moisture. For this reason, it is usually impossible to use the technique in this example to find the sensible and latent heat fluxes. A better alternative is to measure the Bowen ratio first.

By forming a ratio (the Bowen ratio) of these two fluxes, the constants of proportionality drop out, leaving:

$$B = \gamma \cdot \frac{\Delta\theta}{\Delta r} \tag{3.38}$$

where $\gamma = C_p/L_v = 0.4$ $(g_{water\ vapor}/kg_{air}) \cdot K^{-1}$ is called the **psychrometric constant.**

Thus, the Bowen ratio can easily be measured using **thermometers** and **hygrometers**. The potential temperature $\Delta\theta$ and mixing ratio Δr differences must be measured between the same pair of heights z to use this approach (see Fig 3.9). This gives, $\Delta\theta = T_2 - T_1 + (0.0098\ K/m) \cdot (z_2 - z_1)$, and $\Delta r = r_2 - r_1$.

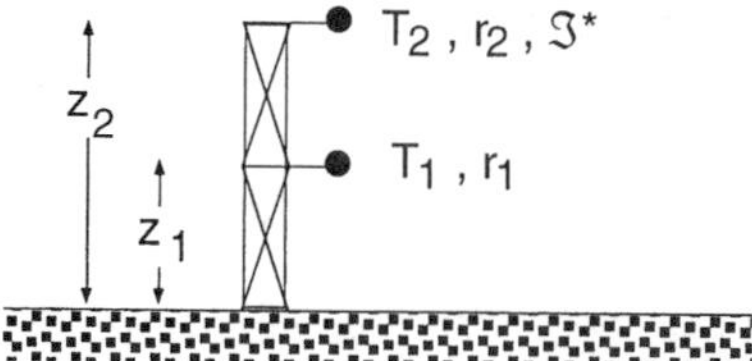

Figure 3.9
Instrumented tower for measuring temperature T and humidity r at two heights, z_1 and z_2. Net radiation is $\Im^*$.

When eq. (3.37) is combined with the surface energy balance eq. (3.35) and the approximation for ground flux eq. (3.36), the result for sensible heat flux is:

$$\Im_H = \frac{-0.9 \cdot \Im^*}{\dfrac{\Delta r}{\gamma \cdot \Delta\theta} + 1} \tag{3.39a}$$

or kinematically:

$$F_H = \frac{-0.9 \cdot F^*}{\dfrac{\Delta r}{\gamma \cdot \Delta\theta} + 1} \tag{3.39b}$$

Similarly, for latent heat flux:

$$\Im_E = \frac{-0.9 \cdot \Im^*}{\dfrac{\gamma \cdot \Delta\theta}{\Delta r} + 1} \tag{3.40a}$$

or

$$F_E = \frac{-0.9 \cdot F^*}{\dfrac{\gamma \cdot \Delta\theta}{\Delta r} + 1} \tag{3.40b}$$

Instead of calculating both equations, one can save time by calculating the first, and then using the surface energy balance to find the other:

$$\Im_E = -0.9 \cdot \Im^* - \Im_H \tag{3.41a}$$

or

$$F_E = -0.9 \cdot F^* - F_H \tag{3.41b}$$

Measurements of temperature and humidity at two heights (Fig 3.9) along with measurements of the net radiation near the surface are thus sufficient to determine the vertical effective fluxes of heat and moisture.

Surface heat flux can also be found from the potential temperature decrease near the surface, as was given in eqs. (3.21) or (3.22). A similar expression for surface moisture flux in terms of Δr is given in Chapter 5. Also in that chapter are equations for converting latent heat fluxes into water fluxes.

Solved Example

Given the following measurements from an instrumented tower, find the sensible and latent heat fluxes. Also, $\Im^* = -500\ \text{W·m}^{-2}$.

index	z (m)	T (°C)	r (g_vapor/kg_air)
2	10	15	8
1	2	18	10

Solution

Given: the table above.
Find: $\Im_H$ and $\Im_E$ = ? W·m^{-2}

First step is to find $\Delta\theta$:

$$\Delta\theta = T_2 - T_1 + (0.0098\ \text{K/m})\cdot(z_2 - z_1)$$
$$= 15\ \text{K} - 18\ \text{K} + (0.0098\ \text{K/m})\cdot(10\text{m} - 2\text{m})$$
$$= -3\ \text{K} + 0.0784\ \text{K} = -2.92\ \text{K}$$

Use eq. (3.39a)

$$\Im_H = \frac{-0.9\cdot(-500\ \text{W}\cdot\text{m}^{-2})}{\dfrac{(-2\ \text{g}_{vap}/\text{kg}_{air})}{[0.4\ (\text{g}_{vap}/\text{kg}_{air})\cdot\text{K}^{-1}]\cdot(-2.92\ \text{K})} + 1}$$

$\Im_H$ = **165.91** W·m^{-2}

Next, use eq. (3.41a):

$$\Im_E = -0.9\cdot\Im^* - \Im_H$$
$$= -0.9\cdot(-500\ \text{W·m}^{-2}) - 165.91\ \text{W·m}^{-2}$$
$$= \underline{284.09}\ \text{W·m}^{-2}$$

Check: Units OK. Physics OK.
Discussion: Sensible heat flux is roughly 50% of the latent heat flux (i.e., $B = 0.5$). Thus, we can guess that these observations were made over grassland.

APPARENT TEMPERATURES

Humans and livestock are warm blooded. Our metabolism generates heat, while our perspiration evaporates to keep us cool. Our bodies attempt to regulate our metabolism and perspiration to maintain a relatively constant internal temperature.

Whether we feel warm or cool depends not only on air temperature, but on wind speed and humidity. During winter, faster wind makes the air feel colder, because it removes heat from our bodies faster. Wind chill is a measure of this effect.

During summer, higher humidity makes the air feel hotter, because it reduces evaporation of our perspiration. The heat index measures this effect. Both of these indices are given as **apparent temperatures**; namely, how warm or cold it feels.

Wind Chill

Wind chill is the hypothetical air temperature in calm conditions ($M = 0$) that would cause the same heat flux from the skin as occurs for the true winds and true air temperature. The heat transfer equation (3.21) for flux across a surface can be slightly modified to give one version of the wind-chill equation:

$$T_{wind\ chill} = T_{skin} - \left(\frac{M + M_o}{M_o}\right)^{0.21}\cdot(T_{skin} - T_{air}) \qquad (3.42)$$

where $T_{skin} = 33°\text{C}$, and $M_o = 2$ m/s is the average speed that people walk. It is assumed that even in calm weather, there is an effective wind due to a person's movement.

Thus, the weather seems colder with colder air temperatures and with stronger winds. Table 3-1 shows wind chills computed using this equation on a computer spreadsheet, with the results plotted in Fig 3.10.

Table 3-1. Wind-chill temperatures (°C).

Wind Speed		Air Temperature (°C)					
km/h	m/s	10	0	-10	-20	-30	-40
0	0.0	10	0	-10	-20	-30	-40
10	2.8	5	-7	-19	-31	-43	-55
20	5.6	3	-11	-24	-37	-50	-64
30	8.3	1	-14	-28	-42	-56	-70
40	11.1	-1	-16	-31	-46	-61	-75
50	13.9	-3	-18	-33	-49	-64	-80
60	16.7	-4	-20	-36	-52	-68	-84

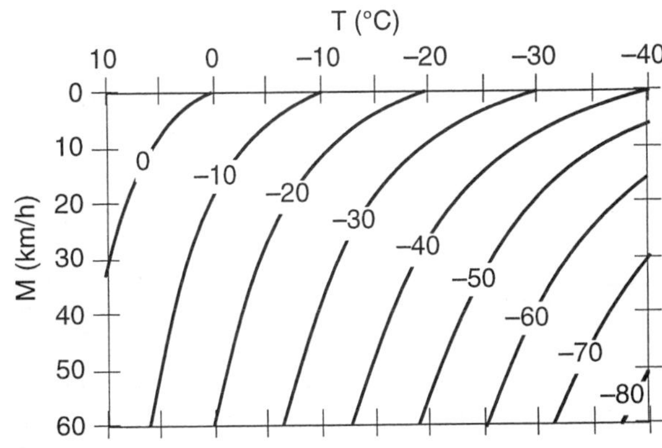

Figure 3.10
Curves are wind chill equivalent temperature (°C), as a function of air temperature (T) and wind speed (M).

> **Solved Example**
> What is the wind chill temperature in °F for an air temperature of –10°F and a wind of 20 mph.
>
> **Solution**
> Given: $T_{air} = -10°F$, $M = 20$ mph
> Find: $T_{wind\ chill} = ?$ °F.
>
> First, convert the units of the parameters:
> $T_{skin} = 33°C = 91.4°F$
> $M_o = 2\ m/s \approx 4$ mph
> Use eq. (3.42):
>
> $$T_{wind\ chill} = 91.4°F - \left(\frac{20+4}{4}\right)^{0.21} \cdot [91.4 - (-10)°F]$$
>
> $= \underline{\mathbf{-56°F}}$
>
> **Check**: Units OK. Physics OK.
> **Discussion**: This is close to the tabulated value.

Heat Index

More humid air feels warmer and more uncomfortable than the actual temperature. **Heat index** (or **apparent temperature** (AT) or **temperature-humidity index**) is one measure of heat discomfort and heat-stress danger.

The equation for heat index below is an inelegant regression to human-perceived apparent temperatures, but it works nicely. It gives heat index (HI) values in Fahrenheit, and is written in the form of a spreadsheet formula[1].

```
HI or AT = 16.923 +((1.85212*10^-1)*T)
          +(5.37941*RH)                              (3.43)
          -((1.00254*10^-1)*T*RH)
          +((9.41695*10^-3)*T^2)
          +((7.28898*10^-3)*RH^2)
          +((3.45372*10^-4)*T^2*RH)
          -((8.14971*10^-4)*T*RH^2)
          +((1.02102*10^-5)*T^2*RH^2)
          -((3.8646*10^-5)*T^3)
          +((2.91583*10^-5)*RH^3)
          +((1.42721*10^-6)*T^3*RH)
          +((1.97483*10^-7)*T*RH^3)
          -((2.18429*10^-8)*T^3*RH^2)
          +((8.43296*10^-10)*T^2*RH^3)
          -((4.81975*10^-11)*T^3*RH^3) .
```

where T is dry bulb temperature (°F), RH is relative humidity percentage (values of 0 to 100), and "^" means "raised to the power of". This equation works for $T > 70°F$.

[1] From web site http://www.zunis.org/16element_heat_index_equation.htm

Table 3-2 shows the solution to this equation. For low relative humidities, the air feels cooler than the actual air temperature because perspiration evaporates effectively, keeping humans cool. However, above 30% relative humidity, the apparent temperature is warmer than the actual air temperature.

Table 3-2. Heat Index Apparent Temperature °C

RH (%)	Actual Air Temperature (°C) 20	25	30	35	40	45	50
0	16	21	26	30	35	39	43
10	19	23	27	31	37	43	49
20	21	24	28	33	40	48	57
30	23	25	29	35	44	55	
40	24	25	30	38	49	63	
50	24	26	31	41	55		
60	25	26	33	45	62		
70	24	27	35	49			
80	24	27	37	55			
90	24	28	40	61			
100	24	29	43				

SUMMARY

Air parcels that move through the atmosphere while conserving heat are said to be adiabatic. When parcels rise adiabatically, they cool due to the change of pressure with height. The change of temperature with height is called the adiabatic lapse rate, and indicates a physical process. The ambient air through which the parcel moves might have a different lapse rate. Thermodynamic diagrams are convenient for determining how the temperature varies with height or pressure, and for comparing different lapse rates.

Once heat from the sun is in the earth-atmospheric system, it is redistributed by advection, radiation, turbulence, and conduction. Some of the sensible heat is converted between latent heat by water phase change. Eventually, heat is lost from the system as infrared radiation to space.

The net radiation is balanced by turbulent fluxes into the atmosphere and conduction into the ground. Turbulent fluxes consist of sensible heat flux, and latent heat flux associated with evaporation of water. The ratio of sensible to latent heat flux is the Bowen ratio. All of these fluxes vary with the diurnal cycle.

Warm blooded animals and humans feel heat loss rather than temperature. On cold, windy days, the heat loss is quantified as a wind chill temperature. During humid hot days, the apparent temperature is warmer than actual.

Threads

The conversion between latent and sensible heat provides energy that drives thunderstorms (Chapter 15) and hurricanes (Chapter 16). These thunderstorms stabilize the atmosphere by moist convective adjustment. The transport of moisture by the wind is thus equivalent to a transport of energy, and affects the global energy budget (Chapters 11 and 18).

Latent heat flux (i.e., evaporation driven by the incoming sunlight) provides the moisture (Chapter 5) for clouds (Chapter 6) and precipitation (Chapter 8), and the energy for thunderstorms (Chapter 15) and hurricanes (Chapter 16). Surface heat and moisture fluxes drive boundary layer evolution (Chapt. 4) and contribute to the Eulerian budget of moisture (Chapt. 6).

The Eulerian heat budget is one of the governing equations that is solved as part of a numerical weather forecast (Chapter 14). The Lagrangian heat budget associated with rising air parcels is used to determine cloud formation (Chapter 7), atmospheric turbulence (Chapter 4) and air pollutant dispersion (Chapter 17). Thermodynamic diagrams (Chapter 6) combine potential temperature, latent heat, and the Lagrangian heat budget. Moist convective adjustment is caused by thunderstorms (Chapter 15).

EXERCISES

Numerical Problems

N1. a. What is the lapse rate in the troposphere for the standard atmosphere?

b. What is the corresponding potential temperature gradient?

N2. Given an air parcel initially at P = 50 kPa and T = 0°C. If it were lowered in the atmosphere dry adiabatically to a height where P = 100 kPa, then what is its final temperature?

N3. Without doing any calculations, use Fig 3.2 to estimate the final temperature of an air parcel at P = 70 kPa that started with a temperature of T = 10°C at P = 100 kPa.

N4(§). Create a thermodynamic diagram on a spreadsheet like the one plotted in Fig 3.2, only print it with more lines.

N5. Using equations (and show your work) rather than a thermo diagram, what is the potential temperature of air at:

a.	z = 200 m	T = 20°C
b.	z = 1,000 m	T = 5°C
c.	z = 200 m	T = 5°C
d.	z = 1,000 m	T = 20°C
e.	z = 300 m	T = 10°C
f.	z = 6,000 m	T = –50°C
g.	z = 10,000 m	T = –100°C
h.	z = –100 m	T = 40°C

N6. What is the virtual potential temperature for the previous problem, assuming 0.005 g_{vapor}/g_{air} for temperatures over 15°C, and assuming 0.002 g_{vapor}/g_{air} for cooler temperatures? Assume no liquid water.

N7. Use the thermodynamic diagram (Fig 3.2) to answer the following.

	Given:	Find:
a.	T = 0°C, P = 70 kPa	θ (°C)
b.	T = 20°C, P = 100 kPa	θ (°C)
c.	T = –20°C, P = 40 kPa	θ (°C)
d.	θ = 40°C, P = 40 kPa	T (°C)
e.	θ = –20°C, P = 90 kPa	T (°C)
f.	θ = 40°C, P = 80 kPa	T (°C)
g.	θ = 40°C, T = 0 °C	P (kPa)
h.	θ = 60°C, T = –20 °C	P (kPa)

N8. a. A west wind of 8 m/s and a temperature gradient of $\Delta T/\Delta x$ = – 1°C/100 km causes what value of advective flux gradient? b. What if the temperature gradient were doubled? c. What if the wind speed were doubled?

N9. At a height of 10 m above the ground, typical vertical temperature gradients are –2°C per meter during the day.

a. What is the magnitude of conductive flux for this location?

b. What temperature gradient would be needed to create a conductive kinematic heat flux of 0.1 K·m/s, (which is of the order of turbulent fluxes)?

N10. Suppose the sea surface temperature is 30°C and the air temperature is 25°C at a height of 10 m, what is the effective surface heat flux for the following wind speeds (m·s^{-1})?

a. 2	b. 5	c. 10	d. 20
e. 50	f. 1	g. 15	h. hurricane force.

N11. Find the vertical turbulent flux gradient, for the following surface effective flux values and ABL depths, assuming fair-weather during daytime:

	F_H (K·m·s^{-1})	z_i (km)
a.	0.2	1.0
b.	0.1	2.0
c.	0.3	1.5
d.	0.05	0.2

N12. Just before thunderstorms form in a stormy troposphere, the air is 10°C warmer than standard near the ground, and is 5°C cooler than standard near the top of the troposphere, with a linear temperature variation in between. Find the vertical turbulent flux gradient at heights (km):

a. 0 b. 1 c. 2 d. 3
e. 4 f. 5.5 g. 10 h. 11

N13. Find the value of flux into the ground, given a net radiation of: a. –400 W·m^{-2} b. –150 W·m^{-2} c. –450 W·m^{-2}

N14. If the sensible heat flux is 300 W·m^{-2} and the latent flux is 100 W·m^{-2}, what is the Bowen ratio? What is the likely type of surface?

N15. Given the following measurements of temperature, and specific humidity, find the sensible and latent heat flux values. The net radiation is 100 W/m^2.

index	z(m)	T(°C)	r (g_{vap}/kg_{air})
2	20	18	9
1	10	14	8

N16. Assume you ride your bicycle at a speed given below, during a calm day of temperature given below. What wind chill temperature to your feel? of

a. 10 m/s, 20°C b. 5 m/s, 20°C c. 15 m/s, 20°C
d. 10 m/s, –5°C e. 5 m/s, –5°C f. 15 m/s, –5°C

N17(§). Use eq. (3.42) to create a wind-chill table or graph similar to Table 3-2 and Fig 3.10, but for wind speeds in miles per hour and temperatures in °F.

Understanding & Critical Evaluation

U1. Suppose you are given 1 kg of ice at 0°C. If the ice is placed in 1 kg of liquid water initially at 20°C, what will be the final temperature after all the ice melts?

U2. Derive eq. (3.5) from eq. (3.4), showing your steps. (Calculus is not needed to do this.) Show all your work and state any assumptions and limitations of the result.

U3. On a copy of Fig 3.2 plot the standard-atmosphere temperature profile. If an air parcel from this standard atmosphere at P = 100 kPa is lifted dry adiabatically to a height where P = 50 kPa, how does its final (process) temperature compare with the standard (environmental) temperature at the same height?

U4(§). Use the equations for the standard atmosphere from Chapter 1, and convert temperatures into potential temperatures every km within the troposphere and stratosphere. Plot the resulting θ vs z.

U5(§). Create a semi-log graph of log P vs. T similar to Fig 3.2, however do not plot the dry adiabats. Instead, solve the hypsometric equation to plot curves corresponding to the following heights: z = 1, 2, 3, 4, 5, 6, 7, 8, 9, and 10 km. In other words there will be one curve corresponding to z = 1 km, and another for z = 2 km, etc.

U6. Explain why a positive flux gradient is associated with cooling and not warming.

U7. Given a 1 km thick mixed layer over the ocean, with sea surface temperature 30°C and air temperature 25°C at z = 10 m, and calm winds. If the mixed layer potential temperature is 20°C and the water vapor mixing ratio is 0.015 g_{vapor}/g_{air}, then what is the convective effective surface heat flux?

U8. If no other heating process was acting except radiative divergence in the vertical, what would be the heating rate of air. This typically occurs on calm nights at altitudes roughly 500 m above ground.

U9. What are the limitations of the net heat budget equation (3.34)?

U10. Given a situation where a wind speed of U = 10 m/s is blowing air with horizontal (east-west) temperature gradient of 2°C/100 km. Also suppose that 2 g of water per 1 kg of air is condensing every 15 minutes. Neglect turbulence. Compare the magnitudes of each term in the net heat budget equation (3.34).

U11. Given a layer of air 1 km thick advecting over a sea surface of 20°C at wind speed $U = 20$ m/s. The initial air temperature is 10°C. The layer of air experiences bulk turbulent heat transfer into the bottom of it, but no heat transfer across the top, and no radiative heating. Assume the heat from the sea surface is mixed instantaneously in the vertical across the whole air layer. For a steady-state situation with only advection and the effective turbulent surface heat flux, derive an equation for the change of temperature with horizontal distance $\Delta T/\Delta x$.

U12. Use the flux estimates from eqs. (3.39 and 3.40) in the surface heat budget equation to show that they do indeed balance the budget. Also, what are the limitations of those equations?

U13. (§). An alternative equation for wind chill temperature is

$$T_{wc} = 91.4 + 0.0817 \cdot (T_{air} - 91.4) \cdot (3.71 \cdot M^{0.5} + 5.81 - 0.25 \cdot M)$$

for T_{air} in °F, and $M \geq 4$ mph. (Actually, this is the original equation that was used to create the wind-chill indices.) Create a table or graph of wind chills and compare to wind chills using the eq. in this chapter. Comment on which equation is more accurate, and on the limitations of both equations.

U14. Where in Fig 3.10 does a small increase in wind speed cause the greatest decrease in wind-chill temperature?

U15. Here is an alternative expression for heat index (HI) or apparent temperature having fewer parameters. Create tables or curves of heat index, and compare with the table in this chapter (after converting the results to Celsius) and with the table in Ahrens. This expression is designed for use in a spreadsheet, where the (Tf) and (RH) factors must be replace with the appropriate cell references for temperature in Fahrenheit and relative humidity (0 - 100) in percent.
HI (°F) = -42.379 + 2.04901523*(Tf) + 10.14333127*(RH) - 0.22475541*(Tf)*(RH) - (6.83783*10^(-3))*(Tf^(2)) - (5.481717*10^(-2))*(RH^(2)) + (1.22874*10^(-3))*(Tf^(2))*(RH) + (8.5282*10^(-4))*(Tf)*(RH^(2)) - (1.99*10^(-6))*(Tf^(2))*(RH^(2))
where "^" means "raised to the power of". This eq. is adapted from web site http://www.usatoday.com/weather/whumcalc.htm

U16. Sketch and discuss a figure such as Fig 3.7, except for nighttime over a desert.

U17. Speculate why the X factor for ground flux is different from day to night. Discuss limitations of that expression for ground flux.

Web-Enhanced Questions

W1(§). Access today's actual temperature sounding from the web for a rawinsonde station close to you (or for another sounding station specified by your instructor). Convert the resulting temperatures to potential temperatures, and plot the resulting θ vs z.

W2. Access upper-air soundings from the web for a rawinsonde launch site near you. What type of thermodynamic diagram is it plotted on? Can you identify which lines are the isobars; the isotherms, the dry adiabats? There are many different types of thermodynamic diagrams in use around the world, so it is important to learn how to identify the isopleths in each type.

W3. Access the current temperature and wind at your town or location. Then access the temperature from a town upwind of you. Calculate the portion of heating or cooling rate at your location that is based only on advection.

W4. During winter, access from the web reports of wind chill. Can you find on the web a weather map that plots the wind chill? In addition to checking official weather agencies, also check local and network TV web pages.

W5. During summer, access from the web reports of heat index, temperature-humidity index, or apparent temperature. Can you find on the web a weather map that plots any of these? In addition to checking official weather agencies, also check local and network TV web pages.

Synthesis Questions

S1. What if the sign of the latent heat of condensation or vaporization were opposite. Namely, when liquid water evaporates it heats the air, and when it condenses it cools the air. What would happen to the earth's oceans as water evaporates from the surface?

S2. Suppose the adiabatic lapse rate had the opposite sign, so that air parcels become warmer as they rise adiabatically. Assuming a standard atmosphere in the troposphere, what would happen to an air parcel from the middle of the troposphere that would be displace slightly up or down?

S3. What if the earth's crust and surface were made of a metal such a aluminum many hundreds of meters thick. Speculate about the importance of conduction on the surface heat budget, and other implications about the weather.

S4. Lagrangian frameworks follow air parcels as they move. Eulerian frameworks remain fixed relative to the earth's surface.

What if you were on a cruise ship moving at constant speed from east to west across the ocean. Devise a heat budget equation relative to this framework. You may name this equation after yourself. Check that it works for any wind direction, and in the limits of zero ship speed.

S5. For the real earth during cloud-free conditions, most visible light is absorbed by the earth's surface, while the troposphere is mostly transparent. Suppose that the troposphere were translucent to visible light, but opaque enough that the absorption of sunlight was spread uniformly over the depth of the troposphere, with no sunlight reaching the surface. How would atmospheric structure and weather be different, if at all?

S6. Suppose that the thermometers and hygrometers on the instrumented tower sketched in Fig 3.9 had random errors of 10% of their magnitudes. Would this error affect the surface flux estimate using the Bowen ratio method? If so, what is the sensitivity (what percentage error does the flux have, compared to the percentage error that the input measurements had)?

S7. Derive the wind-chill eq. (3.42) from the heat transfer eq. (3.21), stating all assumptions that are needed.

Science Graffito

A true weather forecast that appeared for Yukon Territory, Canada, during summer:
"Sunny today, sunny tonight, sunny tomorrow."

BOUNDARY LAYERS

CONTENTS

4 Sunrise, sunset, sunrise. The daily cycle of radiative heating causes a daily cycle of sensible and latent heat fluxes between the earth and the air. However, these fluxes cannot directly reach the whole atmosphere. They are confined by the troposphere to a shallow layer near the ground (Fig 4.1).

This layer is called the **atmospheric boundary layer** (ABL). It experiences a **diurnal** (daily) cycle of temperature, humidity, wind, and pollution variations. Turbulence is also routine, and is one of the causes of the unique nature of the ABL.

Because the boundary layer is where we live, where our crops are grown, and where we conduct our commerce, we have become familiar with its daily cycle. We perhaps forget that these variations are not experienced by the rest of the atmosphere. This chapter examines the formation and unique characteristics of the ABL.

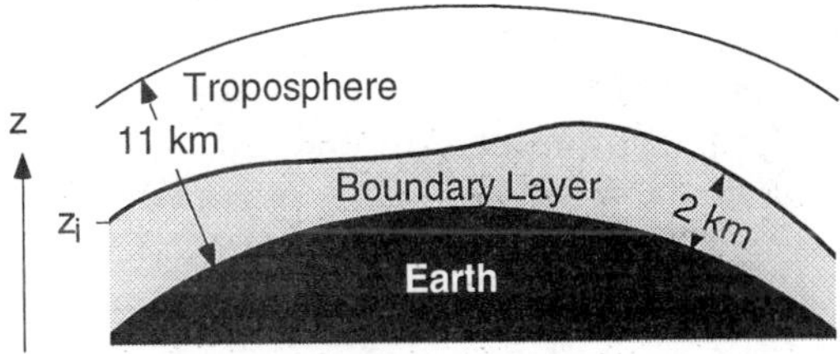

Figure 4.1
Location of the boundary layer, with top at z_i.

STATIC STABILITY — A PREVIEW

Explanation

Static stability controls formation of the ABL, and affects ABL wind and temperature profiles. Here is a preliminary explanation of static stability; Chapter 7 gives more details.

If a small blob of air (i.e., an **air parcel**) is warmer than its surroundings at the same height or pressure, the parcel is positively buoyant and rises. If cooler, it is negatively buoyant and sinks. A parcel with the same temperature as its surrounding environment experiences zero buoyant force.

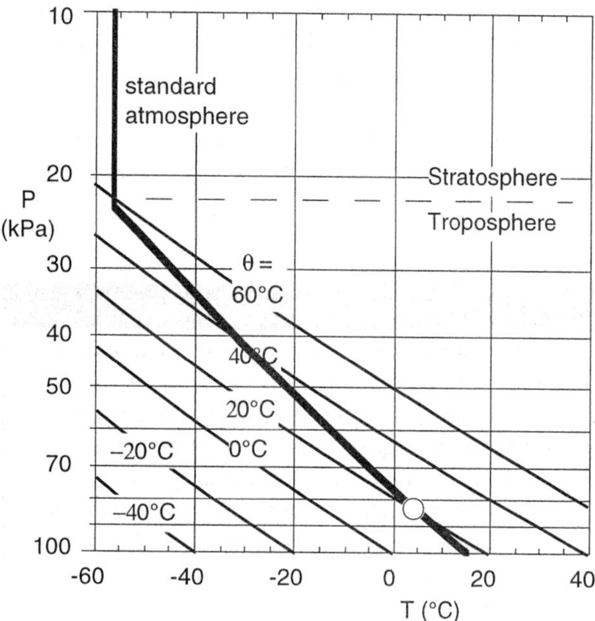

Figure 4.2
Standard atmosphere (heavy line) plotted on a thermodynamic diagram. The open circle represents a hypothetical air parcel. Thin lines are dry adiabats.

Figure 4.2 shows the standard atmosphere from Chapter 1, plotted on a thermodynamic diagram from Chapter 2. Let the standard atmosphere represent the environment or the background air. Consider an air parcel captured from one part of that environment (plotted as the circle). At its initial height, the parcel has the same temperature as the surrounding environment, and experiences no buoyant forces.

To determine static stability, you must ask what would happen to the air parcel if it were forceably displaced a small distance up or down. When moved from its initial capture altitude, the parcel and environment temperatures could differ, thereby causing buoyant forces.

If the buoyant forces on a displaced air parcel push it back to its starting altitude, then the environment is said to be **statically stable**. In the absence of any other forces, statically stable air is **laminar.** Namely, it is smooth and non-turbulent.

However, if the displaced parcel is pulled further away from its starting point by buoyancy, the portion of the atmosphere through which the air parcel continues accelerating is classified as **statically unstable**. Unstable regions are **turbulent** (gusty).

If the displaced air parcel has a temperature equal to that of its new surroundings, then the environment is **statically neutral**.

When an air parcel moves vertically, its temperature changes adiabatically, as described in previous chapters. Always consider such adiabatic temperature change before comparing parcel temperature to that of the surrounding environment. The environment is usually assumed to be **stationary**, which means it is relatively unchanging during the short time it takes for the parcel to rise or sink.

Looking at Fig 4.2, if an air parcel is captured at P = 83 kPa and T = 5°C, and is then is forceably lifted dry adiabatically, it cools following the θ = 20°C adiabat (one of the thin diagonal lines in that figure). If lifted to a height where the pressure is P = 60 kPa, its new temperature is about T = –20°C.

This air parcel, being colder than the environment (thick line in Fig 4.2) at that same height, descends back to its starting point. Similarly if displaced downward from its initial height, the parcel is warmer than its surroundings and would rise back to its starting point.

Air parcels captured from any initial height in the environment of Fig 4.2 always tend to return to their starting point. Therefore, the standard atmosphere is statically stable. This stability is critical for ABL formation.

Rules of Thumb for Stability in the ABL

Because of the daily cycle of radiative heating and cooling, there is a daily cycle of static stability in the ABL. ABL static stability can be anticipated as follows, without worrying about air parcels for now.

Unstable air adjacent to the ground is associated with light winds and a surface that is warmer than the air. This is common on sunny days in fair-weather. It can also occur when cold air blows over a warmer surface, day or night. In unstable conditions, thermals of warm air rise from the surface to heights of one to four kilometers, and turbulence within this layer is vigorous.

At the other extreme are **stable** layers of air, associated with light winds and a surface that is cooler than the air. This typically occurs at night in fair-weather with clear skies, or when warm air blows over a colder surface day or night. Turbulence is weak or sometimes nonexistent in stable layers adjacent to the ground. The stable layers of air are usually shallow (20 - 500 m) compared to the unstable daytime cases.

In between these two extremes are **neutral** conditions, where winds are moderate to strong and there is little heating or cooling from the surface. These occur during overcast conditions, often associated with bad weather.

BOUNDARY-LAYER FORMATION

Tropospheric Constraints

Because of buoyant effects, the vertical temperature structure of the troposphere limits the types of vertical motion that are possible. The standard atmosphere in the troposphere is not parallel to the dry adiabats (Fig 4.2), but crosses the adiabats toward warmer potential temperatures as altitude increases.

That same standard atmosphere is replotted as the thick hatched line in Fig 4.3, but now in terms of its potential temperature (θ) versus height (z). The standard atmosphere slopes toward warmer potential temperatures at greater altitudes. Such a slope indicates statically stable air; namely, air that opposes vertical motion.

The ABL is often turbulent. Because turbulence causes mixing, the bottom part of the standard atmosphere becomes homogenized. Namely, within the turbulent region, warmer potential-temperature air from the standard atmosphere in the top of the ABL is mixed with cooler potential-temperature air from bottom. The resulting mixture has a medium potential temperature that is uniform with height, as plotted by the black line in Fig 4.3. During situations when turbulence is particularly vigorous, the ABL is also called the **mixed layer (ML)**.

Above the mixed layer, the air is unmodified by turbulence, and retains the same temperature profile as the standard atmosphere in this idealized scenario. This air is known as the **free atmosphere (FA)**.

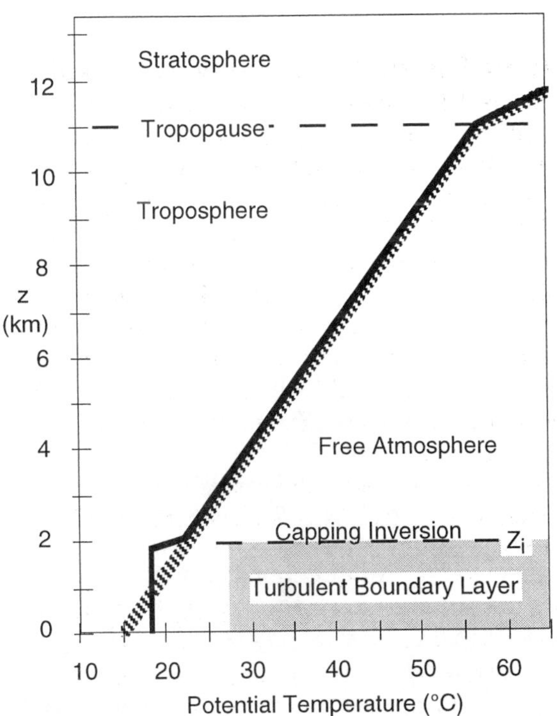

Figure 4.3
Restriction of ABL depth by tropospheric temperature structure in fair weather. The standard atmosphere is the thick hatched line. The modification of the standard atmosphere by turbulence near the ground is idealized with the thick black line.

FOCUS • Engineering Boundary Layers

In wind tunnel experiments, the layer of air that feels frictional drag against the bottom wall grows in depth indefinitely (Fig a). Boundary-layer thickness h grows proportional to the square root of downstream distance x, until hitting the top of the wind tunnel.

On an idealized rotating planet, the earth's rotation imposes a dynamical constraint on ABL depth (Fig b). This maximum depth is proportional to the ratio of wind drag (related to a friction velocity u_*) to earth's rotation (related to the Coriolis parameter f_c). Both those parameters will be defined later in the book.

For the real ABL on earth, the strong capping inversion at z_i makes the ABL unique compared to other fluid flows (Fig c). It constrains the ABL thickness and the eddies within it to a maximum size of order 200 m to 4 km. This stratification constraint supersedes the others. It means that the temperature structure is always very important for the ABL.

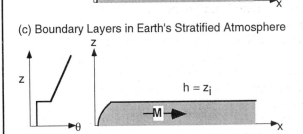

(a) Engineering Boundary Layers

z
$h \propto x^{1/2}$
B L
M
x

(b) Boundary Layers on a Rotating Planet

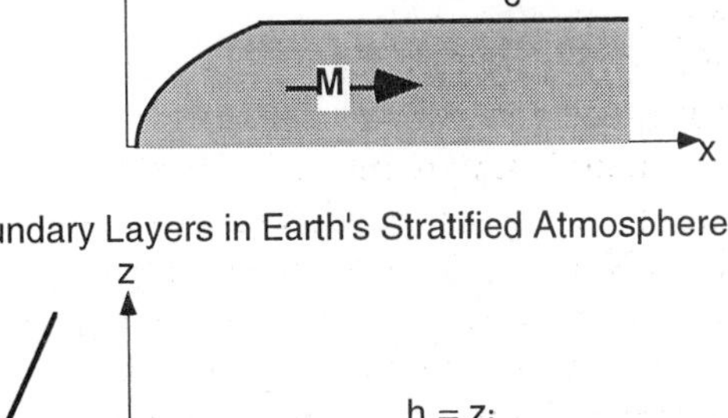

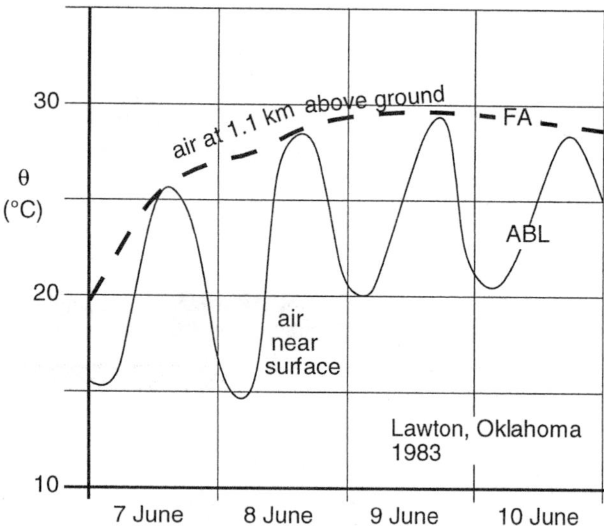

Figure 4.4
Observed variations of potential temperature in the ABL (solid line) and the free atmosphere (dashed line). The daily heating and cooling cycle that we are so familiar with near the ground does not exist above the ABL.

As a result of a turbulent mixed layer being adjacent to the unmixed free atmosphere, there is a sharp temperature increase at the mixed layer top. This transition zone is very stable, and is often a **temperature inversion**. Namely, it is a region where temperature increases with height. The altitude of this inversion is given the symbol z_i, and is a measure of the depth of the turbulent ABL.

The temperature inversion acts like a lid or cap to motions in the ABL. Picture an air parcel from the mixed layer in Fig 4.3. If turbulence were to try to push it out of the top of the mixed layer into the free atmosphere, it would be so much colder than the surrounding environment that a strong buoyant force would push it back down into the mixed layer. Hence, air parcels, turbulence, and air pollution are trapped within the mixed layer.

There is always a strong stable layer or temperature inversion capping the ABL. As we have seen, turbulent mixing in the bottom of the statically-stable troposphere creates this cap, and in turn this cap traps turbulence below it.

The capping inversion breaks the troposphere into two parts. Vigorous turbulence within the ABL causes the ABL to respond quickly to surface influences such as heating and frictional drag. However, the remainder of the troposphere does not experience this strong turbulent coupling with the surface, and hence does not experience frictional drag nor a daily heating cycle. Fig 4.4 illustrates this difference.

In summary, the bottom 200 m to 4 km of the troposphere is called the atmospheric **boundary layer**. ABL depth is variable with location and time. Turbulent transport causes the ABL to feel the direct effects of the earth's surface. The ABL exhibits strong diurnal (daily) variations of temperature, moisture, winds, pollutants, turbulence, and depth in response to daytime solar heating and nighttime IR cooling of the ground. The name "boundary layer" comes from the fact that the earth's surface is a boundary on the atmosphere, and the ABL is the part of the atmosphere that "feels" this boundary.

Synoptic Forcings

Weather patterns such as high (ℍ) and low (𝕃) pressure systems that are drawn on weather maps are known as **synoptic** weather. These large diameter (order of 1000 km) systems modulate the ABL. In the N. Hemisphere, ABL winds circulate clockwise and spiral out from high-pressure centers, but circulate counterclockwise and spiral in toward lows (Fig 4.6). Synoptic-scale winds will be discussed in detail in Chapter 9.

The outward spiral of winds around highs is called **divergence**, and removes ABL air horizontally from the center of highs. Conservation of air mass requires **subsidence** (downward moving air) over highs to replace the horizontally diverging air (Fig 4.5). Although this subsidence pushes free atmosphere air downward, it cannot penetrate into the ABL because of the strong capping inversion. Instead, the capping inversion is pushed downward closer to the ground as the ABL becomes somewhat thinner. This situation traps air pollutants in a shallow ABL, causing air stagnation and air pollution episodes.

Similarly, horizontally converging ABL air around lows is associated with upward motion (Fig 4.5). Often the synoptic forcings and storms associated with lows are so powerful that they easily lift the capping inversion or eliminate it altogether. This allows ABL air to be deeply mixed over the whole depth of the troposphere by thunderstorms and other clouds. Air pollution is usually reduced during this situation as it is diluted with cleaner air from aloft, and as it is washed out by rain.

Because winds in high-pressure regions are relatively light, ABL air lingers over the surface for sufficient time to take on characteristics of that surface. These include temperature, humidity, pollution, odor, and others. Such ABL air is called an **air mass**, and is discussed in Chapt. 12. When the ABLs from two different high-pressure centers are drawn toward each other by a low center, the zone separating those two air masses is called a **front**.

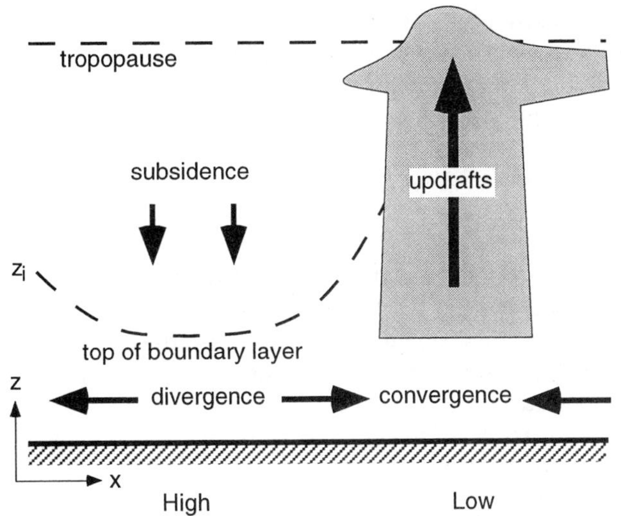

Figure 4.5
Influence of synoptic scale vertical circulations on the ABL.

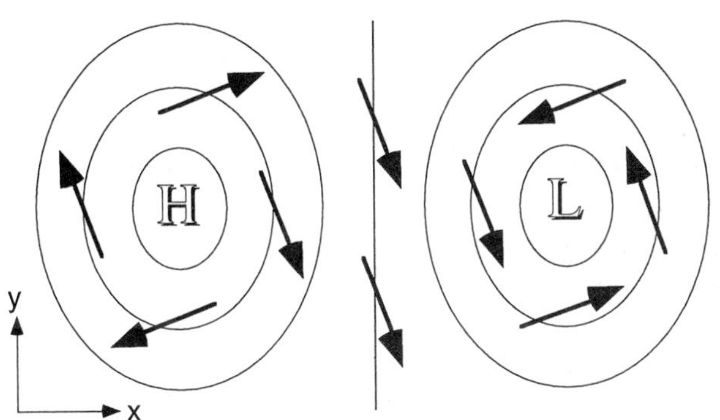

Figure 4.6
Synoptic scale horizontal winds (arrows) in the ABL. Thin lines are isobars around surface high (H) and low (L) pressure centers.

At a frontal zone, the colder, heavier air mass acts like a wedge under the warm air mass. As winds blow the cold and warm air masses toward each other, the wedge peels away the warm ABL from the ground, causing it to ride up over the colder air (Fig 4.7).

Although an ABL forms in the advancing air mass behind the front, the warm humid air that was pushed aloft is not called an ABL because it has lost contact with the surface. Instead, this rising warm air cools, allowing water vapor to condense and make the clouds that we often associate with fronts.

For synoptic-scale low-pressure systems, it is difficult to define a separate ABL, so boundary-layer meteorologists study the air below cloud base. The remainder of this chapter focuses on fair-weather ABLs associated with high-pressure systems.

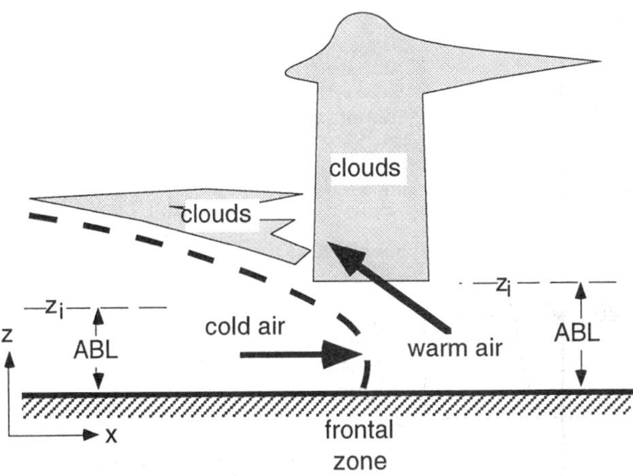

Figure 4.7
Idealized ABL modification near a frontal zone.

ABL STRUCTURE & EVOLUTION

The fair-weather ABL consists of the components sketched in Fig 4.8. During daytime there is a statically-unstable **mixed layer** (ML). At night, a statically **stable boundary layer** (SBL) forms under a statically neutral **residual layer** (RL). The residual layer contains the pollutants and moisture from the previous mixed layer, but is not very turbulent.

The bottom 20 to 200 m of the ABL is called the **surface layer** (not shown in Fig 4.8). Here frictional drag, heat conduction, and evaporation from the surface cause substantial changes of wind speed, temperature, and humidity with height. However, turbulent fluxes are relatively uniform with height; hence, the surface layer is known as the **constant flux layer**.

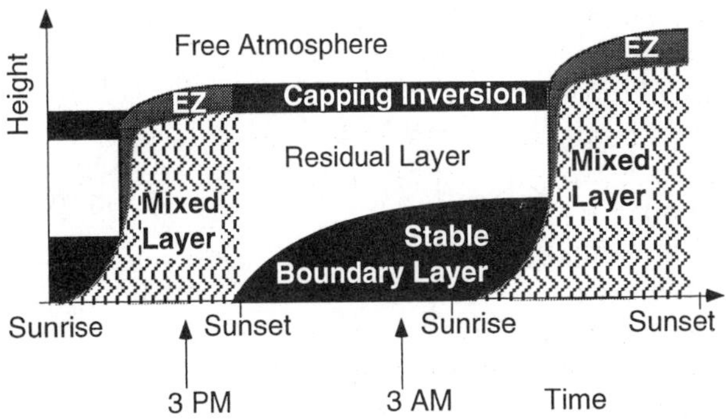

Figure 4.8
Components of the boundary layer during fair weather.

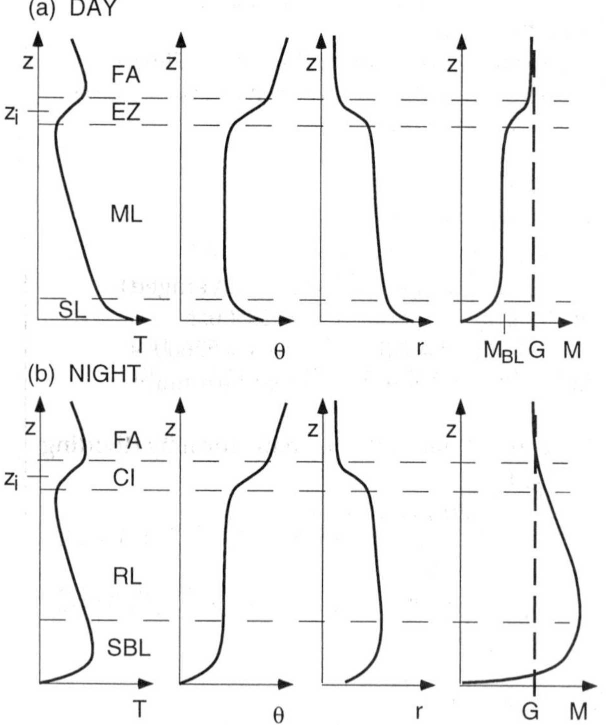

Figure 4.9
Typical vertical profiles of temperature (T), potential temperature (θ), mixing ratio (r, see Chapt. 5), and wind speed (M) in the ABL. Day is roughly 3 PM local time, and night is 3 AM. The dashed line labeled G is the geostrophic wind speed (a theoretical wind in the absence of surface drag, see Chapt. 9). M_{BL} is average wind speed in the ABL.

Separating the **free atmosphere** from the mixed layer is a strongly stable **entrainment zone** (EZ) of intermittent turbulence. Mixed-layer depth z_i is the distance between the ground and the middle of the EZ. At night, turbulence in the EZ ceases, leaving a nonturbulent separation layer called the **capping inversion** (CI), which is still strongly statically stable.

Typical vertical profiles of temperature, potential temperature, specific humidity, and wind speed are sketched in Fig 4.9. The "day" portion of Fig 4.9 corresponds to the 3 PM time indicated in Fig 4.8 below, while "night" is for 3 AM.

Next we look at ABL temperature, wind, and turbulence characteristics in more detail. Moisture and pollution are covered in Chapters 5 and 17.

TEMPERATURE

The capping inversion traps surface heat flux and evaporation in the ABL. As a result, heat accumulates within the ABL during day, or whenever the surface is warmer than the air. Cooling (actually, heat loss) accumulates during night, or whenever the surface is colder than the air. Thus, the temperature structure of the ABL depends on the accumulated heating or cooling.

Cumulative Heating or Cooling

During clear winter nights, the longer duration of nighttime and greater heat loss in the absence of clouds produce greater temperature decreases than for cloudy summer nights. Thus, the cumulative effect is more important than the instantaneous heat flux.

This cumulative heating or cooling Q_A equals the area under the curve of heat flux vs. time (Fig 4.10). We will examine cumulative nighttime cooling separately from cumulative daytime heating.

Nighttime

During clear nights, heat flux from the air to the cold surface is relatively constant with time (shaded portion of Fig 4.10). If we define t as the time since cooling began, then the accumulated cooling per unit surface area is:

$$Q_A = \Im_{H\ night} \cdot t \qquad (4.1a)$$

For night, Q_A is a negative number because $\Im_H$ is negative for cooling. Q_A has units of J/m^2.

Dividing eq. (4.1a) by $\rho_{air} \cdot C_p$ gives the kinematic form

$$Q_{Ak} = F_{H\ night} \cdot t \qquad (4.1b)$$

where Q_{Ak} has units of K·m, which is more convenient.

For a night with variable cloudiness that causes a variable surface heat flux (Chapt. 2), use the average value of $\Im_H$ or F_H.

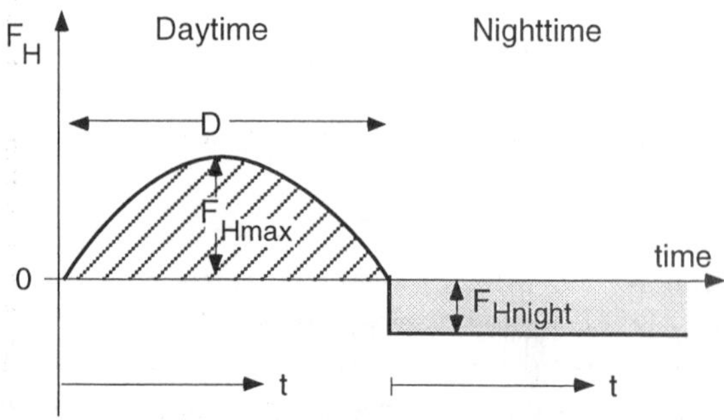

Figure 4.10
Idealization of the heat flux curve from Fig 3.8.

Daytime

On clear days, the nearly sinusoidal variation of solar elevation (Fig 3.4) causes nearly sinusoidal variation of surface net heat flux (Fig 3.8). Let t be the time since $\Im_H$ becomes positive in the morning, D be the total duration of positive heat flux, and $\Im_{Hmax}$ be the peak value of heat flux (Fig 4.10). These values can be found from data such as in Fig 3.8.

The accumulated daytime heating per unit surface area (units of J/m^2) is:

$$Q_A = \frac{\Im_{H\max} \cdot D}{\pi} \cdot \left[1 - \cos\left(\frac{\pi \cdot t}{D}\right)\right] \tag{4.2a}$$

In kinematic form (units of K·m), this equation is

$$Q_{Ak} = \frac{F_{H\max} \cdot D}{\pi} \cdot \left[1 - \cos\left(\frac{\pi \cdot t}{D}\right)\right] \tag{4.2b}$$

BEYOND ALGEBRA • Cumulative Heating

Derivation of Daytime Cumulative Heating

Assume that the kinematic heat flux is approximately sinusoidal with time:

$$F_H = F_{H\max} \cdot \sin(\pi \cdot t / D)$$

with t and D defined as in Fig 4.10 for daytime.
Integrating from time $t = 0$ to arbitrary time t :

$$Q_{Ak} \equiv \int_{t'=0}^{t} F_H \mathrm{d}t' = F_{H\max} \cdot \int_{t'=0}^{t} \sin(\pi \cdot t' / D)\mathrm{d}t'$$

where t' is a dummy variable of integration.

From a table of integrals, we find that:

$$\int \sin(a \cdot x) = -(1/a) \cdot \cos(a \cdot x)$$

Thus, the previous equation integrates to:

$$Q_{Ak} = \frac{-F_{H\max} \cdot D}{\pi} \cdot \cos\left(\frac{\pi \cdot t}{D}\right)\Bigg|_0^t$$

Plugging in the two limits gives:

$$Q_{Ak} = \frac{-F_{H\max} \cdot D}{\pi} \cdot \left[\cos\left(\frac{\pi \cdot t}{D}\right) - \cos(0)\right]$$

But the cos(0) = 1, giving the final answer (4.2b):

$$Q_{Ak} = \frac{F_{H\max} \cdot D}{\pi} \cdot \left[1 - \cos\left(\frac{\pi \cdot t}{D}\right)\right]$$

CAUTION. **Any answer is no better than the assumptions that go into it.** While F_H is indeed proportional to the $\sin(\psi)$ from eq. (3.19), $\sin(\psi)$ is NOT proportional to $\sin(t)$ [see eq. (3.7)].

Solved Example

Use Fig 3.8 to estimate the accumulated heating and cooling in kinematic form over the whole day and whole night.

Solution

By eye from Fig 3.8:

Day: $\Im_{Hmax} \cong 150$ W·m^{-2} , $D = 8$ h

Night: $\Im_{Hnight} \cong -50$ W·m^{-2} (averaged)

Given: Day: $t = D = 8$ h $= 28800.$ s

Night: $t = 24$h $- D = 16$ h $= 57600.$ s

Find: Q_{Ak} = ? K·m for day and for night

First convert fluxes to kinematic form by dividing by $\rho_{air} \cdot C_p$:

$F_{Hmax} = \Im_{Hmax} / \rho_{air} \cdot C_p$

$=(150$ W·m^{-2}) / [1231(W·m^{-2})/(K·m/s)]

= 0.122 K·m/s

F_{Hnight}=(–50W·m^{-2})/[1231(W·m^{-2})/(K·m/s)]

= –0.041 K·m/s

Day: use eq. (4.2b)

$$Q_{Ak} = \frac{(0.122\ \mathrm{K \cdot m/s}) \cdot (28800\ \mathrm{s})}{3.14159} \cdot [1 - \cos(\pi)]$$

= <u>**2237 K·m**</u>

Night: use eq. (4.1b)

Q_{Ak} = (–0.041 K·m/s)·(57600. s) = <u>**–2362 K·m**</u>

Check: Units OK. Physics OK.

Discussion: The heating and cooling are nearly equal for this example, implying that the daily average temperature is fairly steady.

Temperature-Profile Evolution

Idealized Evolution

A typical afternoon temperature profile is plotted in Fig 4.11a. During the daytime, the environmental lapse rate in the mixed layer is nearly adiabatic. The unstable surface layer (plotted but not labeled in Fig 4.11) is in the bottom part of the mixed layer. Warm blobs of air called **thermals** rise from this surface layer up through the mixed layer, until they hit the entrainment zone. Fig 4.12a shows a closer view of the surface layer.

These thermal circulations create strong turbulence, and cause pollutants, potential temperature, and moisture to be well mixed in the vertical (hence the name mixed layer). The whole mixed layer, surface layer, and bottom portion of the entrainment zone are statically unstable.

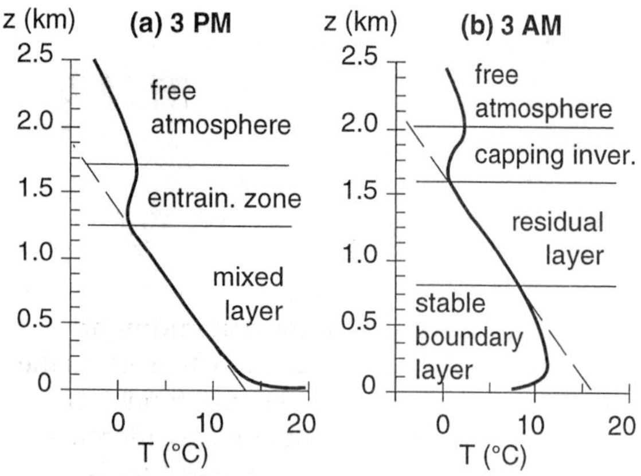

Figure 4.11
Examples of boundary-layer temperature profiles during day (left) and night (right). Adiabatic lapse rate is dashed. The heights shown here are illustrative only. In the real ABL the heights can be greater or smaller, depending on location, time, and season.

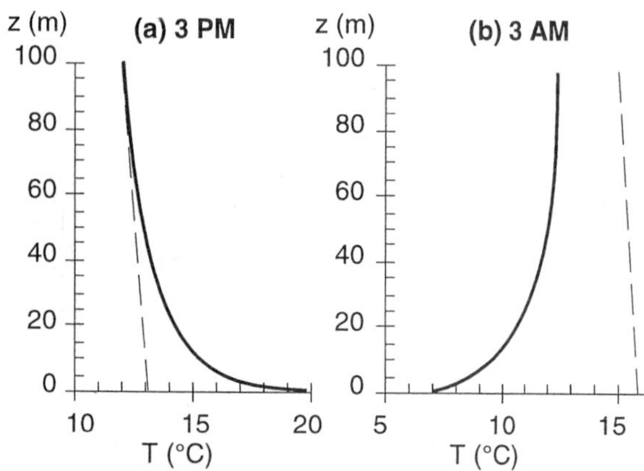

Figure 4.12
Examples of surface-layer temperature profiles during day (left) and night (right). Adiabatic lapse rate is dashed. Again, heights are illustrative only. Actual heights might differ.

In the entrainment zone, free-atmosphere air is incorporated or **entrained** into the mixed layer, causing the mixed-layer depth to increase during the day. Pollutants trapped in the mixed layer cannot escape through the EZ, although cleaner, drier free atmosphere air is entrained into the mixed layer. Thus, the EZ is a one-way valve.

At night, the bottom portion of the mixed layer becomes chilled by contact with the radiatively-cooled ground. The result is a stable ABL. The bottom portion of this stable ABL is the surface layer (again not labeled in Fig 4.11b, but sketched in Fig 4.12b).

Above the stable ABL is the residual layer. It has not felt the cooling from the ground, and hence retains the adiabatic lapse rate from the mixed layer of the previous day. Above that is the capping temperature inversion, which is the nonturbulent remnant of the entrainment zone.

Seasonal Differences

During summer at mid and high latitudes, days are longer than nights. More heating occurs during day than cooling at night. After a full 24 hours, the ending sounding is warmer than the starting sounding as illustrated in Fig 4.13 a & b. The convective mixed layer starts shallow in the morning, but rapidly grows through the residual layer. In the afternoon, it continues to rise slowly into the free atmosphere. If the air contains sufficient moisture, cumuliform clouds can exist. At night, cooling creates a shallow stable ABL near the ground, but leaves a thick residual layer in the middle of the ABL.

During winter at mid and high latitudes, more cooling occurs during the long nights than heating during the short days. Stable ABLs dominate, and

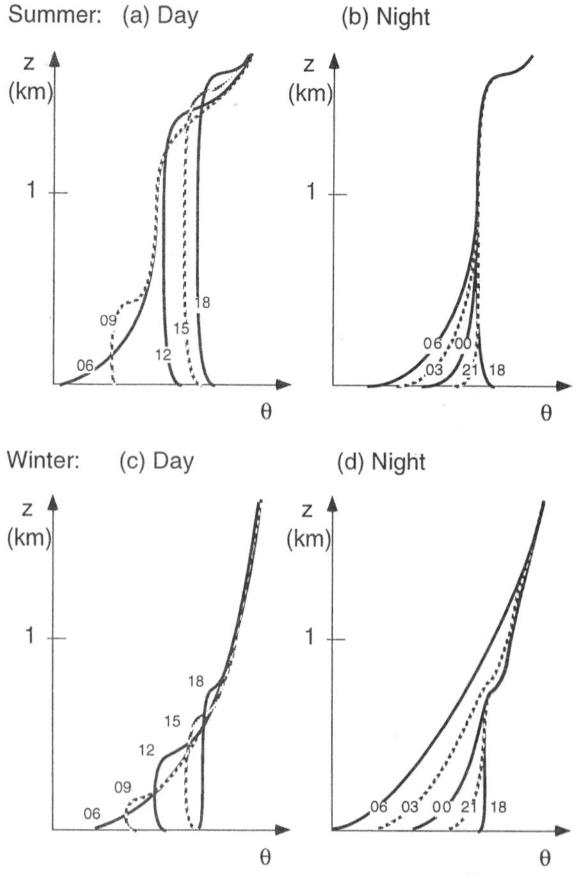

Figure 4.13
Evolution of potential temperature profiles. Curves are labeled with local time in hours. The nighttime curves begin at 18 local time, which corresponds to the ending sounding from the previous day.

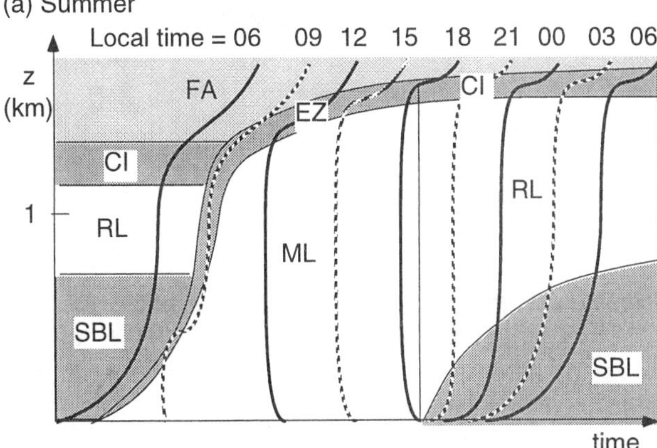

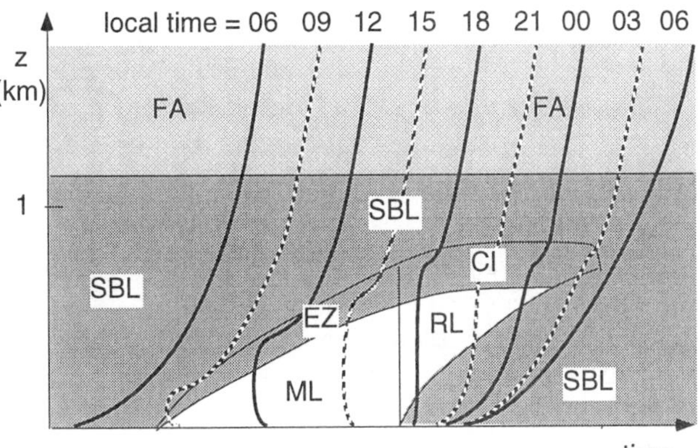

Figure 4.14
Potential temperature profiles identical to the previous figure, except offset equal increments to the right to show evolution with time. Boundary-layer structure is labeled: FA = free atmosphere, ML = mixed layer, RL = residual layer, SBL = stable ABL, CI = capping inversion, EZ = entrainment zone.

there is net temperature decrease over 24 hours (Fig 4.13 c & d). Any non-frontal clouds present are typically stratiform or fog. Any residual layer that forms early in the night is quickly overwhelmed by the growing stable ABL.

Fig 4.13 corresponds to fair weather conditions, such as in anticyclones. When fronts and cyclones pass, these cycles are disrupted as different air masses are advected into the region. Also, thunderstorms can cause mixing throughout a deep layer of the troposphere.

Figure 4.14 shows how the potential temperature soundings define the structure of the ABL. Weakly stable regions are lightly shaded. Strongly stable regions are shaded darker. Adiabatic regions are white. Although both the mixed layer and residual layer have nearly adiabatic temperature profiles, the mixed layer is nonlocally unstable, while the residual layer is neutral. This difference causes pollutants to disperse at different at rates in those two regions.

If the wind moves ABL air over surfaces of different temperatures, then ABL structures can evolve in space, rather than in time. For example, Fig 4.14 b could represent the structure of the ABL at midnight in mid-latitude winter for air blowing over snow-covered ground, except for an unfrozen lake in the center. If the lake is warmer than the air, it will create a mixed layer that grows as the column of air advects across the lake. This can occur even at midnight.

Don't be lulled into thinking the ABL evolves the same way at every location or at similar times. The most important factor is the temperature difference between the surface and the air. If the surface is warmer, a mixed layer will develop regardless of the time of day. Similarly, colder surfaces will create stable ABLs.

Stable-Boundary-Layer Temperature

Stable ABLs are quite complex. Turbulence can be intermittent, and coupling of air to the ground can be quite weak. In addition, any slope of the ground causes the cold air to drain downhill. Cold, downslope winds are called **katabatic winds**, and are discussed in Chapt. 10.

For a simplified case of a contiguously-turbulent stable ABL over a flat surface during light winds, the potential temperature profile is approximately exponential with height (Fig 4.15):

$$\Delta\theta(z) = \Delta\theta_s \cdot e^{-z/H_e} \tag{4.3}$$

where $\Delta\theta(z) = \theta(z) - \theta_{RL}$ is the potential temperature difference between the air at height z and the air in the residual layer. $\Delta\theta(z)$ is negative.

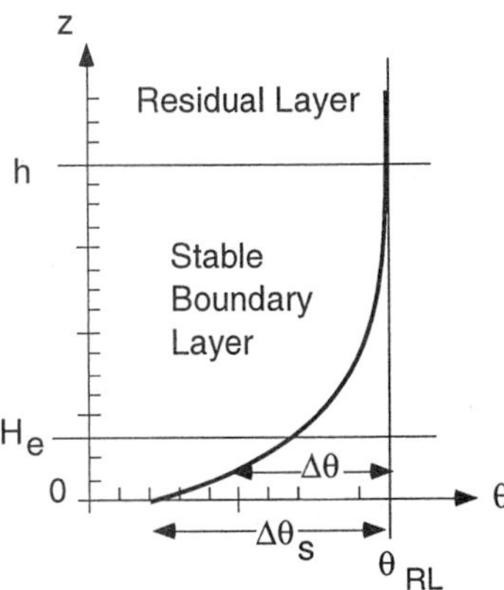

Figure 4.15
Idealized exponential-shaped potential temperature profile in the stable (nighttime) boundary layer.

The value of this difference near the ground is defined to be $\Delta\theta_s = \Delta\theta(z=0)$, and is sometimes called the **strength** of the stable ABL. H_e is an **e-folding height** for the exponential curve. The actual **depth** h of the stable ABL is roughly $h = 5 \cdot H_e$.

Depth and strength of the stable ABL grow as the cumulative cooling Q_{Ak} increases with time:

$$H_e \approx a \cdot M_{RL}{}^{3/4} \cdot t^{1/2} \qquad (4.4)$$

$$\Delta\theta_s = \frac{Q_{Ak}}{H_e} \qquad (4.5)$$

Solved Example (§)

Estimate the potential temperature profile at the end of a 12-hour night for two cases: windy (10 m/s) and less windy (5 m/s). Assume Q_{Ak} = –1000 K·m

Solution

Given: Q_{Ak} = –1000 K·m, t = 12 h
(a) M_{RL} = 10 m/s, (b) M_{RL} = 5 m/s
Find: θ vs. z

Assume: flat prairie.
Use eq. (4.4):
$H_e \approx (0.15\text{m}^{1/4} \cdot \text{s}^{1/4}) \cdot (10\text{m} \cdot \text{s}^{-1})^{3/4} \cdot (43200\text{s})^{1/2}$
(a) H_e = **175 m**
$H_e \approx (0.15\text{m}^{1/4} \cdot \text{s}^{1/4}) \cdot (5\text{m} \cdot \text{s}^{-1})^{3/4} \cdot (43200\text{s})^{1/2}$
(b) H_e = **104 m**

Use eq. (4.5):
(a) $\Delta\theta_s = (-1000\text{K} \cdot \text{m}) / (175\text{m})$ = **–5.71 °C**
(b) $\Delta\theta_s = (-1000\text{K} \cdot \text{m}) / (104\text{m})$ = **–9.62 °C**
Use eq. (4.3) in a spreadsheet to compute the potential temperature profiles:

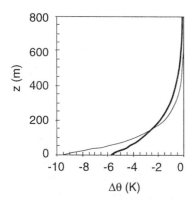

Check: Units OK, Physics OK. Graph OK.
Discussion: The windy case (thick line) is not as cold near the ground, but the cooling extends over a greater depth than for the less-windy case.

where $a \cong 0.15\ \text{m}^{1/4} \cdot \text{s}^{1/4}$ for flow over a flat prairie, and where M_{RL} is the wind speed in the residual layer. Because the cumulative cooling is proportional to time, both the depth H_e and strength $\Delta\theta_s$ of the stable ABL increase as the square root of time. Thus, fast growth of the SBL early in the evening decreases to a much slower growth by the end of the night.

Mixed-Layer (ML) Temperature

The shape of the potential temperature profile in the mixed layer is simple. To good approximation it is uniform with height. Of more interest is the evolution of mixed layer θ and z_i with time.

Use the potential temperature profile at the end of the night (early morning) as the starting sounding for forecasting daytime temperature profiles. In real atmospheres, the sounding might not be a smooth exponential as idealized in the previous subsection. The method below works for arbitrary shapes of the initial potential temperature profile.

A graphical solution is easiest. First, plot the early-morning sounding of θ vs. z. Next, determine the cumulative daytime heating Q_{Ak} that occurs between sunrise and some time of interest t_1, using eq. (4.2b). This heat warms the air in the ABL; thus the area under the sounding equals the accumulated heating (and also has units of K·m). Plot a vertical line of constant θ between the ground and the sounding, such that the area (cross hatched in Fig 4.16a) under the curve equals the cumulative heating.

This vertical line gives the potential temperature of the mixed layer $\theta_{ML}(t_1)$. The height where this vertical line intersects the early-morning sounding defines the mixed-layer depth $z_i(t_1)$. As cumulative heating increases with time during the day ($Area_2$ = total cross-hatched plus diagonal-ruled at time t_2),

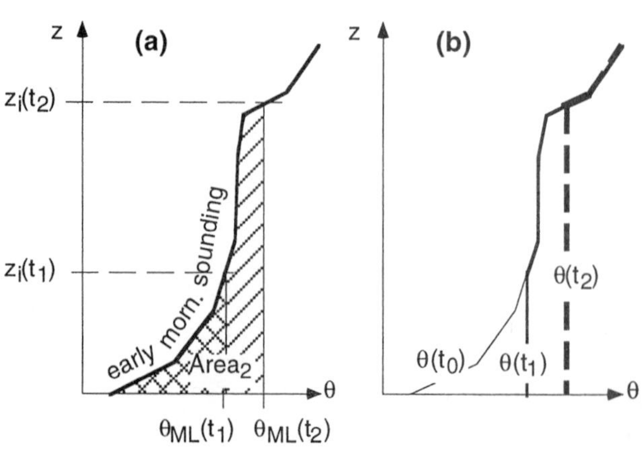

Figure 4.16
Evolution of the mixed layer with time, as cumulative heating increases the areas under the curves.

Solved Example

Given an early morning sounding with surface temperature 5°C and lapse rate $\Delta\theta/\Delta z$ = 3 K/km. Find the mixed-layer potential temperature and depth at 10 AM, when the cumulative heating is 500 K·m.

Solution

Given: θ_{sfc} = 5°C, $\Delta\theta/\Delta z$ = 3 K/km, Q_{Ak}=0.50 K·km

Find: θ_{ML} = ? °C, z_i = ? km

Sketch:

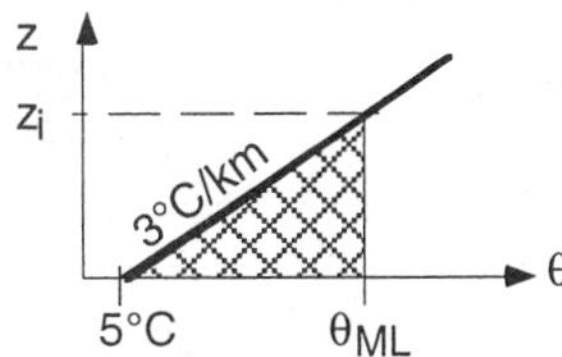

The area under this simple sounding is the area of a triangle: *Area* = 0.5·[*base*]·(*height*) ,

Area = $0.5 \cdot [(\Delta\theta/\Delta z) \cdot z_i] \cdot (z_i)$

0.50 K·km = $0.5 \cdot (3\ \text{K/km}) \cdot z_i^2$

Rearrange and solve for z_i:

z_i = **0.577 km**

Next, use the sounding to find the θ_{ML}:

$\theta_{ML} = \theta_{sfc} + (\Delta\theta/\Delta z) \cdot z_i$

θ_{ML} = (5°C) + (3°C/km)·(0.577 km) = **6.73 °C**

Check: Units OK. Physics OK. Sketch OK.

Discussion: Arbitrary soundings can be approximated by straight line segments. The area under such a sounding consists of the sum of areas within trapezoids under each line segment. Alternately, draw the early-morning sounding on graph paper, count the number of little grid boxes under the sounding, and multiply by the area of each box.

the mixed layer becomes warmer and deeper (Fig 4.16a). The resulting potential temperature profiles during the day are sketched in Fig 4.16b. This method of finding mixed layer growth is called the **encroachment method**, or **thermodynamic method**, and explains roughly 90% of typical mixed layer on sunny days with winds less than 10 m/s.

Entrainment

As was mentioned earlier, the turbulent mixed layer grows by entraining non-turbulent air from the free atmosphere. One can idealize the mixed layer as a **slab model** (Fig 4.17a), with constant potential temperature in the mixed layer, and a jump of potential temperature ($\Delta\theta$) at the EZ.

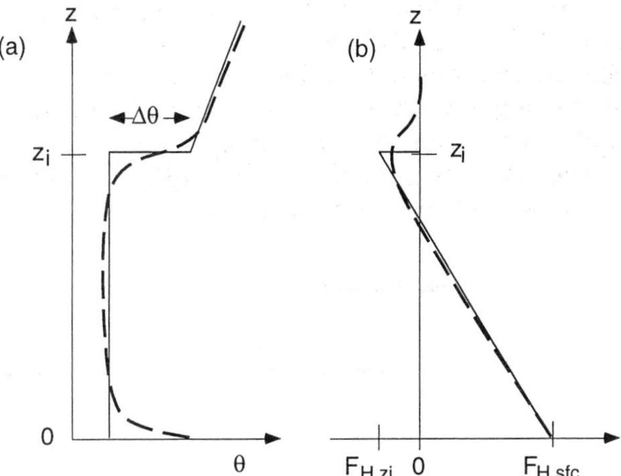

Figure 4.17
(a) Slab idealization of the mixed layer (solid line) approximates a more realistic potential temperature (θ) profile (dashed line). (b) Corresponding real and idealized heat flux (F_H) profiles.

Entrained air from the free atmosphere has warmer potential temperature than air in the mixed layer. Because this warm air is entrained downward, it corresponds to a negative heat flux F_{Hzi} at the top of the mixed layer. The heat-flux profile (Fig 4.17b) is often linear with height, with the most negative value marking the top of the mixed layer.

The entrainment rate of free atmosphere air into the mixed layer is called the **entrainment velocity**, w_e, and can never be negative. The entrainment velocity is the volume of entrained air per unit horizontal area per unit time. In other words it is a volume flux, which turns out to have the same units as velocity.

The rate of growth of the mixed layer is

$$\frac{\Delta z_i}{\Delta t} = w_e + w_s \qquad \bullet(4.6)$$

where w_s is the synoptic scale vertical velocity, and is negative for the subsidence that is typical during fair weather (recall Fig 4.5).

The kinematic heat flux at the top of the mixed layer is

$$F_{Hzi} = -w_e \cdot \Delta\theta \qquad \bullet(4.7)$$

where the sign and magnitude of the temperature jump is defined as $\Delta\theta = \theta$(just above z_i) – θ (just below z_i). Greater entrainment across stronger temperature inversions causes greater heat flux. Similar relationships describe entrainment fluxes of moisture, pollutants, and momentum as a function of jump of humidity, pollution concentration, or wind speed, respectively. As for temperature, the jump is defined as the value above z_i minus value below z_i. Entrainment velocity has the same value for all variables.

During free convection (when winds are weak and thermal convection is strong), the entrained kinematic heat flux is approximately 20% of the surface heat flux:

$$F_{H\ zi} \cong -A \cdot F_{H\ sfc} \tag{4.8}$$

where $A \cong 0.2$ is called the Ball ratio, and $F_{H\,sfc}$ is the surface kinematic heat flux. This special ratio works only for heat, and does not apply to other variables. During windier conditions, A can be greater than 0.2. Eq. (4.8) was used in Chapt. 2 to get the vertical heat-flux divergence (2.37), a term in the Eulerian heat budget.

Combining the two equations above gives an approximation for the entrainment velocity during free convection:

$$w_e \cong \frac{A \cdot F_{H\ sfc}}{\Delta\theta} \qquad \bullet(4.9)$$

Combining this with eq. (4.6) gives a mixed-layer growth equation called the **flux-ratio method**. From these equations, we see that stronger capping inversions cause slower growth rate of the mixed layer, while greater surface heat flux (e.g., sunny day over land) causes faster growth. The flux-ratio method and the thermodynamic methods usually give equivalent results for mixed layer growth during free convection.

Solved Example

Find the entrainment velocity for a potential temperature jump of 2°C at the EZ, and a surface kinematic heat flux of 0.2 K·m/s.

Solution

Given: $F_{H\,sfc}$ = 0.2 K·m/s, $\Delta\theta = 2\ °C = 2\ K$

Find: w_e = ? m/s

Use eq. (4.9):

$$w_e \cong \frac{A \cdot F_{H\ sfc}}{\Delta\theta} = \frac{0.2 \cdot (0.2\ K \cdot m/s)}{(2\ K)} = \underline{0.02\ m/s}$$

Check: Units OK. Physics OK.

Discussion: While 2 cm/s seems small, when applied over 12 h of daylight works out to z_i = 864 m, which is a reasonable mixed layer depth.

WIND

For any given weather condition, there is a theoretical equilibrium wind speed, called the **geostrophic wind** G, that can be calculated for frictionless conditions (see Chapt. 11). However, steady-state winds in the ABL are usually slower than geostrophic (i.e., **subgeostrophic**) because of frictional and turbulent drag of the air against the surface, as was illustrated in the top right of Fig 4.9.

Turbulence continuously mixes slower air from close to the ground with faster air from the rest of the ABL, causing the whole ABL to experience drag against the surface and to be subgeostrophic. This vertically averaged steady-state ABL wind M_{BL} is derived in Chapt. 11. Although the actual ABL winds are nearly equal to this theoretical M_{BL} speed over a large middle region of the ABL, the winds closer to the surface are even slower (Fig 4.9).

Wind Profile Evolution

Over land during fair weather, the winds often experience a diurnal cycle as illustrated in Fig 4.18. For example, a few hours after sunrise, say at 9 AM local time, there is often a shallow mixed layer, which is 300 m thick in this example. Within this shallow mixed layer the ABL winds are uniform with height, and winds next to the surface are nonzero.

As the day progresses, the mixed layer deepens, so by 3 PM a deep layer of subgeostrophic winds fills the ABL. Winds remain moderate near the ground as turbulence mixes down faster winds from higher in the ABL. After sunset, turbulence intensity usually

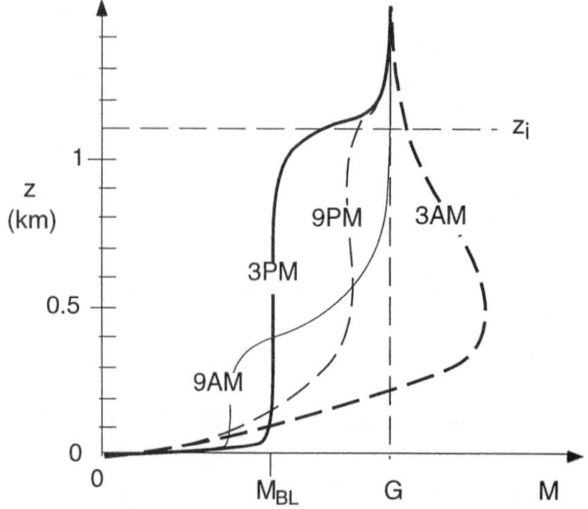

Figure 4.18
Typical ABL wind-profile evolution during fair weather over land. G is geostrophic wind speed, & M_{BL} is average ABL wind.

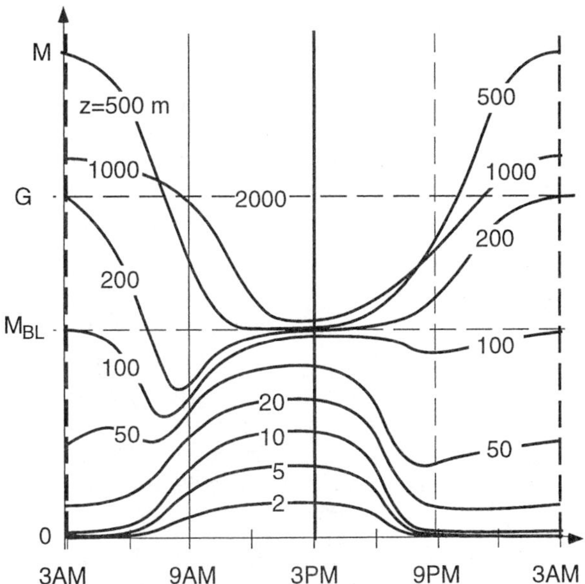

Figure 4.19
Typical ABL wind speed evolution at different heights. G is geostrophic wind speed, M_{BL} is average ABL wind, and the vertical time lines correspond to the profiles of Fig 4.18.

diminishes, allowing drag at the surface to reduce the winds at ground level. However, without turbulence, the air in the middle of the ABL no longer feels drag against the surface, and begins to accelerate. By 3 AM, the winds a few hundred meters above ground can be supergeostrophic, even though the winds at the surface might be calm. Then, after sunrise, turbulence begins vertical mixing again.

For measurements made at fixed heights on a very tall tower, the same wind speed evolution is shown in Fig 4.19. Below 20 m altitude, winds are generally calmer at night, and increase in speed during daytime. The converse is true above 100 m altitude, where winds are reduced during the day because of turbulent mixing with slower air from near the surface, but become faster at night when turbulence decays.

At the bottom of the ABL, near-surface wind speed profiles have been found **empirically** (i.e., based on experiment). In the bottom 5% of the statically neutral ABL is the **surface layer**, where wind speeds increase roughly logarithmically with height (Fig 4.20).

For the statically stable surface layer, this logarithmic profile changes to a more linear form (Fig 4.20). Winds close to the ground become slower than logarithmic, or near calm. Winds just above the surface layer are often not in steady state, and can temporarily increase to faster than geostrophic (**supergeostrophic**) in a process called an inertial oscillation (Fig 4.9 lower right, and Fig 4.20).

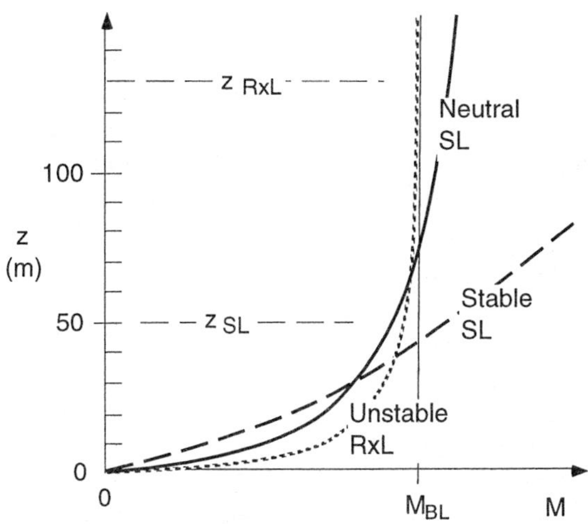

Figure 4.20
Typical wind speed profiles in the surface layer (SL, bottom 5% of the ABL) and radix layer (RxL, bottom 20%), for different static stabilities. z_{RxL} and z_{SL} give order-of-magnitude depths for the radix layer and surface layer.

The bottom 20% of the convective (unstable) ABL is called the **radix layer (RxL)**. Winds in the RxL have an exponential power-law relationship with height. The RxL has faster winds near the surface, but slower winds aloft than the neutral logarithmic profile. After a discussion of drag at the ground, these three wind cases at the bottom of the ABL will be described in more detail.

Drag, Stress, Friction Velocity, and Roughness Length

The frictional force between two objects such as the air and the ground is called **drag**. One way to quantify drag is by measuring the force required to push the object along another surface. For example, if you place your textbook on a flat desk, after you first start it moving you must continue to push it with a certain force (i.e., equal and opposite to the drag force) to keep it moving. If you stop pushing, the book stops moving.

Your book contacts the desk with a certain surface area. Generally, larger contact area requires greater force to overcome friction. The amount of friction force per unit surface contact area is called **stress**, τ, where for stress the force is parallel to the area. Contrast this with pressure, which is defined as a force per unit area that is perpendicular to the area. Units of stress are N/m^2, and could also be expressed as Pa.

Stress is felt by both objects that are sliding against each other. For example, if you stack two books on top of each other, then there is friction between both books, as well as between the bottom book and the table. In order to push the bottom book in one direction without moving the top book, you must apply a force to the top book in the opposite direction as the bottom book.

Think of air within the ABL as a stack of layers of air, much like a stack of books. Each layer feels stress from the layers above and below it. The bottom layer feels stress against the ground, as well as from the layer of air above. In turn, the surface tends to be pushed along by air drag. Over the ocean, this **wind stress** drives the **ocean currents**.

In the atmosphere, stress caused by turbulent motions is many orders of magnitude greater than stress caused by molecular viscosity. For that reason, we often speak of **turbulent stress** instead of frictional stress, and **turbulent drag** rather than frictional drag. This turbulent stress is also called a **Reynolds stress**, after Osborne Reynolds who related this stress to turbulent gust velocities in the late 1800s.

Because air is a fluid, it is often easier to study the stress per unit density ρ of air. This is called the **kinematic stress**. The kinematic stress against the earth's surface is given the symbol u_*^2, where u_* is called the **friction velocity**:

$$u_*^2 = |\tau / \rho| \qquad \bullet(4.10)$$

Typical values range from $u_* = 0$ during calm winds to $u_* = 1$ m/s during strong winds. Moderate-wind values are often near $u_* = 0.5$ m/s.

For fluid flow, turbulent stress is proportional to wind speed squared. Also stress is greater over rougher surfaces. A dimensionless **drag coefficient** C_D relates the kinematic stress to the wind speed M_{10} at z = 10 m.

$$u_*^2 = C_D \cdot M_{10}^2 \qquad \bullet(4.11)$$

The drag coefficient ranges from $C_D = 2\text{x}10^{-3}$ over smooth surfaces to $2\text{x}10^{-2}$ over rough or forested surfaces. It is similar to the bulk heat transfer coefficient of Chapter 2.

The surface roughness is usually quantified as an **aerodynamic roughness length** z_o. Table 4-1 shows typical values of the roughness length for various surfaces. Rougher surfaces such as sparse forests have greater values of roughness length than smoother surfaces such as a frozen lake. Roughness lengths in this table are not equal to the heights of the houses, trees, or other roughness elements.

Table 4-1. The Davenport-Wieringa roughness-length classification.

z_o (m)	Classi-fication	Landscape
0.0002	sea	sea, paved areas, snow-covered flat plain, tide flat, smooth desert
0.005	smooth	beaches, pack ice, morass, snow-covered fields
0.03	open	grass prairie or farm fields, tundra, airports, heather
0.1	roughly open	cultivated area with low crops & occasional obstacles (single bushes)
0.25	rough	high crops, crops of varied height, scattered obstacles such as trees or hedge-rows, vineyards
0.5	very rough	mixed farm fields and forest clumps, orchards, scattered buildings
1.0	closed	regular coverage with large size obstacles with open spaces roughly equal to obstacle heights, suburban houses, villages, mature forests
≥ 2	chaotic	centers of large towns and cities, irregular forests with scattered clearings

For statically neutral air, there is a relationship between drag coefficient and aerodynamic roughness length:

$$C_D = \frac{k^2}{\ln^2(z_R / z_o)} \qquad (4.12)$$

where $k = 0.4$ is the **von Kármán constant**, and z_R = 10 m is a reference height defined as the standard anemometer height for measuring "**surface winds**". The drag coefficient decreases as the air becomes more statically stable. For unstable air, roughness is less important, and alternative approaches are given in Chapt. 9.

Combining the previous two equations gives an expression for friction velocity in terms of surface wind speed and roughness length:

$$u_* = \frac{k \cdot M_{10}}{\ln[z_R / z_o]} \tag{4.13}$$

The physical interpretation is that faster winds over rougher surfaces causes greater kinematic stress.

Solved Example

Find the drag coefficient in statically neutral conditions to be used with standard surface winds of 5 m/s, over (a) villages, and (b) grass prairie. Also, find the friction velocity and surface stress.

Solution

Given: z_R = 10 m for "standard" winds
Find: C_D = ? (dimensionless), u_* = ? m/s, τ = ? N/m^2

Use Table 4-1:
(a) z_o = 1 m for villages. (b) z_o = 0.03 m for prairie
Use eq. (4.12) for drag coefficient:

(a) $C_D = \frac{0.4^2}{\ln^2(10\text{m}/1\text{m})} = \underline{0.030}$ (dimensionless)

(b) $C_D = \frac{0.4^2}{\ln^2(10\text{m}/0.03\text{m})} = \underline{0.0047}$ (dimensionless)

Use eq. (4.11) for friction velocity:

$u_*^2 = C_D \cdot M_{10}^2 = 0.03 \cdot (5\text{m/s})^2 = 0.75\ \text{m}^2\cdot\text{s}^{-2}$

(a) Thus u_* = **0.87 m/s**.
(b) Similarly, u_* = **0.34 m/s**.

Use eq. (4.10) for surface stress, and assume ρ = 1.2 kg/m^3:

$\tau = \rho \cdot u_*^2 = (1.2\text{kg/m}^3)\cdot(0.75\text{m}^2/\text{s}^2)$

(a) $\tau = 0.9\ \text{kg}\cdot\text{m}^{-1}\cdot\text{s}^{-2}$ = **0.9 Pa** (using Appen. A)
(b) $\tau = 0.14\ \text{kg}\cdot\text{m}^{-1}\cdot\text{s}^{-2}$ = **0.14 Pa**

Check: Units OK. Physics OK.
Discussion: The drag coefficient, friction velocity, and stress are smaller over smoother surfaces.

In this development we examined the stress for fixed wind speed and roughness. However, in nature, greater roughness and greater surface drag causes slower winds (see Chapt 9).

Log Profile in the Neutral Surface Layer

Wind speed M is zero at the ground (more precisely, at a height equal to the aerodynamic roughness length). Speed increases roughly logarithmically with height in the statically-neutral surface layer (bottom 50 to 100 m of the ABL), but the shape of this profile depends on the surface roughness:

$$M(z) = \frac{u_*}{k} \ln\left(\frac{z}{z_o}\right) \tag{•4.14a}$$

Alternately, if you know wind speed M_1 at height z_1, then you can calculate wind speed M_2 at any other height z_2 :

$$M_2 = M_1 \cdot \frac{\ln(z_2 / z_o)}{\ln(z_1 / z_o)} \tag{4.14b}$$

Many weather stations measure the wind speed at the standard height z_1 = 10 m.

An example of the **log wind profile** is plotted in Fig 4.21. A perfectly logarithmic wind profile (i.e., eq. 4.14) would be expected only for **neutral** static stability (e.g., overcast and windy) over a uniform surface. For other static stabilities, the wind profile varies slightly from logarithmic.

On a semi-log graph, the log wind profile would appear as a straight line. You can measure the roughness length by measuring the wind speeds at two or more heights, and then extrapolating the straight line in a semi-log graph to zero wind speed. The z-axis intercept gives the roughness length.

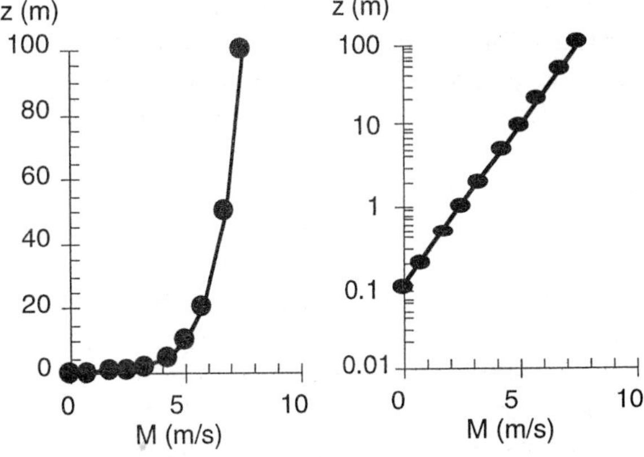

Figure 4.21
Wind profile in the statically-neutral surface layer, for a roughness length of 0.1 m. (a) linear plot. (b) semi-log plot.

Solved Example

On an overcast day, a wind speed of 5 m/s is measured with an anemometer located 10 m above ground within an orchard. What is the wind speed at the top of a 25 m smoke stack?

Solution

Given: $M_1 = 5$ m/s at $z_1 = 10$ m
Neutral stability (because overcast)
$z_o = 0.5$ m from Table 3-2 for an orchard
Find: $M_2 = ?$ m/s at $z_2 = 25$ m

Sketch:

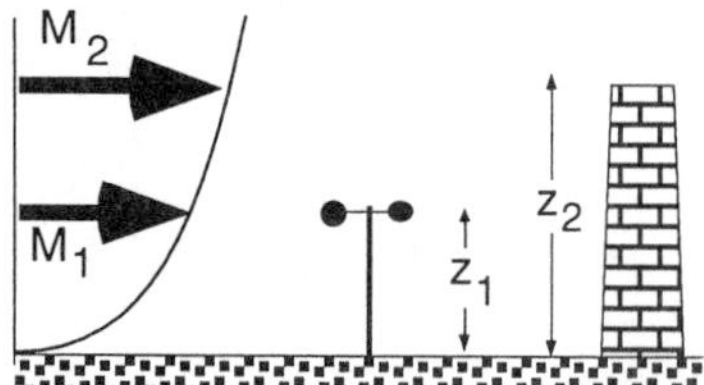

Use eq. (4.14b):

$$M_2 = 5(\mathrm{m/s}) \cdot \frac{\ln(25\mathrm{m}/0.5\mathrm{m})}{\ln(10\mathrm{m}/0.5\mathrm{m})} = \underline{6.53\ \mathrm{m/s}}$$

Check: Units OK. Physics OK. Sketch OK.
Discussion: Hopefully the anemometer is situated far enough from the smoke stack to measure the true undisturbed wind.

Log-Linear Profile in the Stable Surface Layer

During statically stable conditions, such as at nighttime over land, wind speed is slower near the ground, but faster aloft than that given by a logarithmic profile. This profile is empirically described by a formula with both a logarithmic and a linear term in z:

$$M(z) = \frac{u_*}{k}\left[\ln\left(\frac{z}{z_o}\right) + 6\frac{z}{L}\right] \qquad \bullet(4.15)$$

where M is wind speed at height z, $k = 0.4$ is the von Kármán constant, z_o is the aerodynamic roughness length, and u_* is friction velocity. As height increases, the linear term dominates over the logarithmic term, as sketched in Fig 4.20.

An **Obukhov length** L is defined as:

$$L \equiv \frac{-u_*^{\,3}}{k \cdot (g/T_v) \cdot F_{Hsfc}} \qquad \bullet(4.16)$$

where $g = 9.8$ m/s^2 is gravitational acceleration, T_v is the absolute virtual temperature, and F_{Hsfc} is the kinematic surface heat flux. L has units of m, and is positive during statically stable conditions (because F_{Hsfc} is negative then). The Obukhov length can be interpreted as the height in the stable surface layer below which shear production of turbulence exceeds buoyant consumption.

Solved Example(§)

For a friction velocity of 0.3 m/s, aerodynamic roughness length of 0.02 m, average virtual temperature of 300 K, and kinematic surface heat flux of –0.05 K·m/s at night, plot the wind-speed profile in the surface layer.

Solution

Given: $u_* = 0.3$ m/s, $z_o = 0.02$ m,
$T_v = 300$ K, $F_{Hsfc} = -0.05$ K·m/s
Find: $M(z) = ?$ m/s

Use eq. (4.16) to find $L = 41.3$ m
Use eq. (4.15) for M. (Also 4.14a for neutral M)

z (m)	M(m/s) n	M (m/s) stable
0.02	0.0	0.0
0.05	0.7	0.7
0.1	1.2	1.2
0.2	1.7	1.7
0.5	2.4	2.5
1	2.9	3.0
2	3.5	3.7
5	4.1	4.7
10	4.7	5.7
20	5.2	7.4
50	5.9	11.3
100	6.4	17.3

Check: Units OK. Physics OK. Plot OK.
Discussion: Open circles are for neutral, solid are for statically stable. The linear trend is clear.

Profile in the Convective Radix Layer

For statically unstable ABLs with vigorous convective thermals, such as occurs on sunny days over land, wind speed becomes uniform with height a short distance above the ground. Between that uniform wind-speed layer and the ground is the **radix layer (RxL)**. The wind speed profile in the radix layer is:

•(4.17a)

$$M(z) = M_{BL} \cdot \left(\zeta_*^D\right)^A \cdot \exp\left[A \cdot \left(1 - \zeta_*^D\right)\right] \quad \text{for } 0 \le \zeta^* \le 1.0$$

and (4.17b)

$$M(z) = M_{BL} \quad \text{for } 1.0 \le \zeta^*$$

where $\zeta^* = 1$ defines the top of the radix layer. In the bottom of the RxL, wind speed increases faster with height than given by the log wind profile for the neutral surface layer, but becomes tangent to the uniform winds in the mixed layer.

The dimensionless height in the eqs. above is

$$\zeta_* = \frac{1}{C} \cdot \frac{z}{z_i} \cdot \left(\frac{w_*}{u_*}\right)^B \quad (4.18)$$

where w_* is the **Deardorff velocity**, and the empirical coefficients are $A = 1/4$, $B = 3/4$, and $C = 1/2$. $D = 1/2$ over flat terrain, but increases to near 1.0 over hilly terrain.

The Deardorff velocity is defined as

$$w_* = \left[\frac{g}{T_v} \cdot z_i \cdot F_{Hsfc}\right]^{1/3} \quad •(4.19)$$

where $g = 9.8$ m/s^2 is gravitational acceleration, T_v is absolute virtual temperature, z_i is depth of the ABL (= depth of the mixed layer), and F_{Hsfc} is the kinematic sensible heat flux (units of K·m/s) at the surface. Typical values of w_* are on the order of 1 m/s. The Deardorff velocity and buoyancy velocity w_B (defined in Chapt. 2) are both convective velocity scales for the statically unstable ABL, and are related by: $w_* \cong 0.08\, w_B$.

To use eq. (4.17) you need to know the average wind speed in the middle of the mixed layer M_{BL}, as was sketched in Fig 4.9. Chapt 9 shows how to estimate this if it is not known from measurements.

For both the free-convection radix layer and the forced-convection surface layer, turbulence transports momentum, which controls the shape of the wind profile, which determines the shear (Fig 4.22). However, differences between the radix layer and surface layer are caused by differences in feedback.

Solved Example(§)

For a 1 km deep mixed layer with surface heat flux of 0.3 K·m/s and friction velocity of 0.2 m/s, plot the wind speed profile using a spreadsheet. Terrain is flat, and mid-ABL wind is 5 m/s.

Solution

Given: F_{Hsfc} =0.3 K·m/s, u_* = 0.2 m/s, z_i= 1000 m, M_{BL} = 5 m/s, D= 0.5 .

Find: $M(z)$ = ? m/s . First, find w_* = ? m/s.

Assume: g/T_v = 0.0333 m·s^{-2}·K^{-1} (typical)

Use eq. (4.19) first, to get Deardorff velocity:

$$w_* = \left[\left(0.0333\frac{\text{m}}{\text{s}^2\text{K}}\right) \cdot (1000\text{m}) \cdot \left(0.3\frac{\text{K}\cdot\text{m}}{\text{s}}\right)\right]^{1/3}$$

$$= (10\ \text{m}^3/\text{s}^3)^{1/3} = \underline{2.15\ \text{m/s}}$$

Use eq. (4.18) in the spreadsheet to get ζ_* at each height, and then use eq. (4.17) to find each M.

9	z (m)	ζ*	M (m/s)
10	0	0.000	0.00
11	0.1	0.001	2.74
12	0.2	0.002	2.98
13	0.5	0.006	3.32
14	1	0.012	3.59
15	2	0.024	3.87
16	5	0.059	4.24
17	10	0.119	4.51
18	15	0.178	4.66
19	20	0.237	4.75

etc.

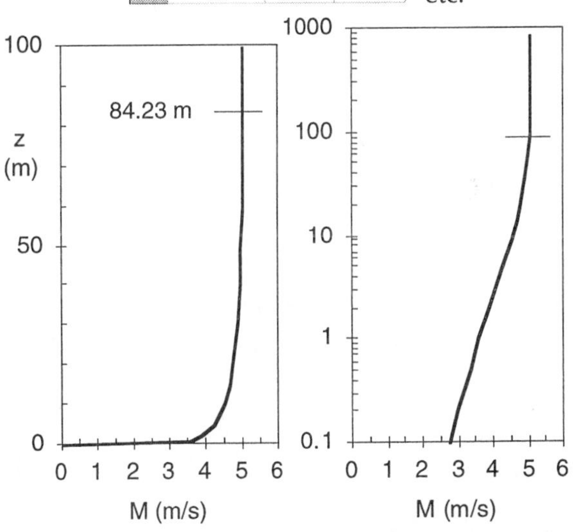

Check: Units OK. Physics OK. Sketch OK.

Discussion: This profile smoothly merges into the uniform wind speed in the mid-mixed layer, at height $\zeta_* = 1.0$, which is at z = 84.23 m.

In the neutral surface layer (Fig 4.22a) there is strong feedback because wind shear generates the turbulence, which in turn controls the wind shear. However, such feedback is broken for convective turbulence (Fig 4.22b), because turbulence is generated primarily by buoyant thermals, not by shear.

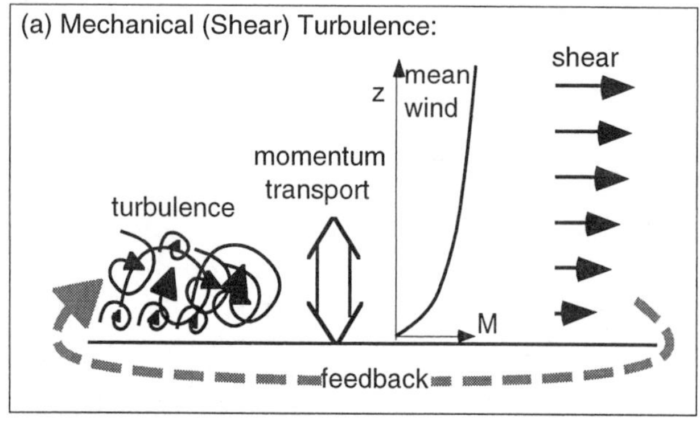

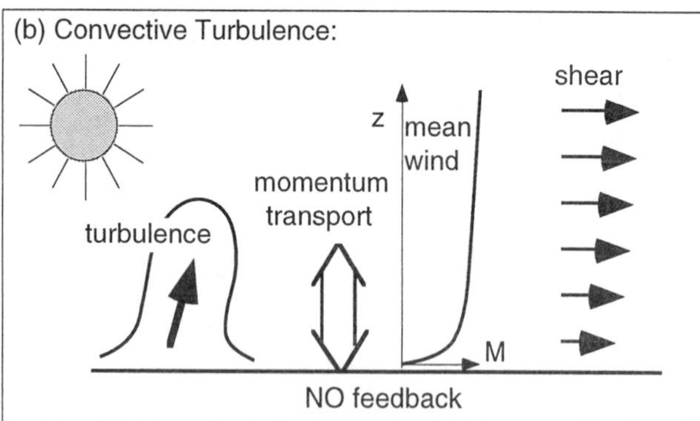

Figure 4.22
(a) Processes important for the log-wind profile in the surface layer dominated by mechanical turbulence. (b) Processes important for the radix-layer wind profile during convective turbulence.

TURBULENCE

Mean and Turbulent Parts

Wind can be quite variable. The total wind speed is the superposition of three types of flow:

mean wind – relatively constant, but varying slowly over the course of hours

waves – regular (linear) oscillations of wind, often with periods of ten minutes or longer

turbulence – irregular, quasi-random, nonlinear variations or gusts, with durations of seconds to minutes

These flows can occur individually, or in any combination. Waves are discussed in Chapters 7 and 10. Here, we focus on mean wind and turbulence.

Let $U(t)$ be the x-direction component of wind at some instant in time, t. Different values of $U(t)$ can occur at different times, if the wind is variable. By averaging the **instantaneous wind** measurements over a time period, P, we can define a **mean wind** $\overline{U}$, where the overbar denotes an average. This mean wind can be subtracted from the instantaneous wind to give the **turbulence** or gust part (Fig 4.23).

Similar definitions exist for the other wind components and temperature and humidity:

$$u'(t) = U(t) - \overline{U} \quad (4.20a)$$

$$v'(t) = V(t) - \overline{V} \quad (4.20b)$$

$$w'(t) = W(t) - \overline{W} \quad (4.20c)$$

$$T'(t) = T(t) - \overline{T} \quad (4.20d)$$

$$r'(t) = r(t) - \overline{r} \quad (4.20e)$$

Thus, the wind can be considered as a sum of mean and turbulent parts.

The averages in eq. (4.20) are defined over time or over horizontal distance. For example, the mean temperature is the sum of all individual temperature measurements, divided by the total number N of data points:

$$\overline{T} = \frac{1}{N}\sum_{k=1}^{N} T_k \quad \bullet(4.21)$$

where k is the index of the data point (corresponding to different times or locations). The averaging time in eq. (4.21) is typically about 0.5 h. If you average over space, typical averaging distance is 50 to 100 km.

Short term fluctuations (described by the primed quantities) are associated with small-scale swirls of motion called **eddies**. The superposition of many such eddies of various scales makes up the **turbulence** that is imbedded in the mean flow.

Molecular viscosity in the air causes friction between the eddies, tending to reduce the turbulence intensity. Thus, turbulence is NOT a conserved quantity, but is **dissipative**. Turbulence decays and

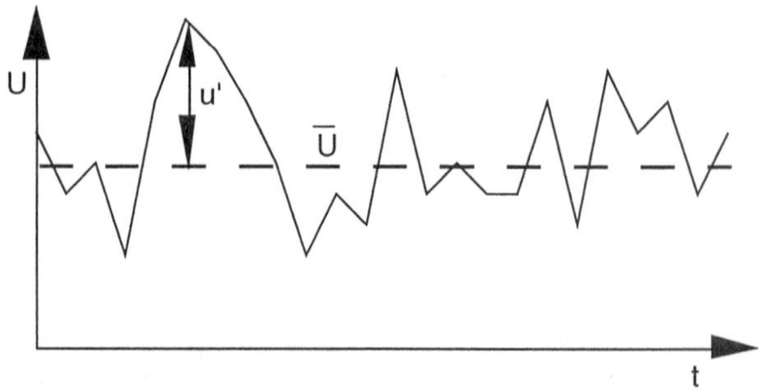

Figure 4.23
The instantaneous wind speed U shown by the zigzag thin line. The average wind speed $\overline{U}$ is shown by the thicker horizontal dashed line. A gust velocity u' is the instantaneous deviation of the instantaneous wind from the average.

Solved Example

Given the following measurements of total instantaneous temperature, T, find the average $\overline{T}$. Also, find the T' values.

t (min)	T (°C)	t (min)	T (°C)
1	12	6	13
2	14	7	10
3	10	8	11
4	15	9	9
5	16	10	10

Solution

As specified by eq. (4.21), adding the ten temperature values and dividing by ten gives the average $\overline{T}$ = **12.0°C**. Subtracting this average from each instantaneous temperature gives:

t (min)	T '(°C)	t (min)	T '(°C)
1	0	6	1
2	2	7	–2
3	–2	8	–1
4	3	9	–3
5	4	10	–2

Check: The average of these T' values should be zero, by definition. In fact, this is a good way to check for mistakes.

Discussion: If a positive T' corresponds to a positive w', then warm air is moving up. This contributes positively to the heat flux.

disappears unless there are active processes to generate it. Two such production processes are **convection**, associated with warm air rising and cool air sinking, and **wind shear**, the change of wind speed or direction with height.

Normally, weather forecasts are made for mean conditions, not turbulence. Nevertheless, the net effects of turbulence on mean flow must be included. Idealized average turbulence effects are given in Chapts. 2, 6, and 9 for the heat, moisture, and momentum budgets, respectively. Appendix H describes turbulence parameterization in detail.

Meteorologists use statistics to quantify the net effect of turbulence. Some statistics are described next. In this chapter we will continue to use the overbar to denote the mean conditions. However, we drop the overbar in subsequent chapters to simplify the notation.

Variance and Standard Deviation

The **variance** σ^2 of vertical velocity is an overall statistic of gustiness:

$$\sigma_w{}^2 = \frac{1}{N}\sum_{k=1}^{N}(W_k - \overline{W})^2 = \frac{1}{N}\sum_{k=1}^{N}(w_k{}')^2 = \overline{w'^2} \qquad \bullet(4.22)$$

Similar definitions can be made for $\sigma_u{}^2$, $\sigma_v{}^2$, $\sigma_\theta{}^2$, etc. Statistically, these are called "biased" variances. Velocity variances can exist in all three directions, even if there is a mean wind in only one direction.

The **standard deviation** σ is defined as the square-root of the variance, and can be interpreted as an average gust (for velocity), or an average turbulent perturbation (for temperatures and humidities, etc.). For example, standard deviations for vertical velocity, σ_w, and potential temperature, σ_θ, are:

$$\sigma_w = \sqrt{\sigma_w{}^2} = \overline{(w')^2}^{\,1/2} \qquad (4.23a)$$

$$\sigma_\theta = \sqrt{\sigma_\theta{}^2} = \overline{(\theta')^2}^{\,1/2} \qquad (4.23b)$$

Larger variance or standard deviation of velocity means more intense turbulence.

For statically **stable** air, standard deviations in an ABL of depth h have been empirically found to vary with height z as:

$$\sigma_u = 2 \cdot u_* \cdot [1-(z/h)]^{3/4} \qquad (4.24a)$$

$$\sigma_v = 2.2 \cdot u_* \cdot [1-(z/h)]^{3/4} \qquad (4.24b)$$

$$\sigma_w = 1.73 \cdot u_* \cdot [1-(z/h)]^{3/4} \qquad (4.24c)$$

where u_* is friction velocity. These equations work when the stability is weak enough that turbulence is not suppressed altogether.

For statically **neutral** air:

$$\sigma_u = 2.5 \cdot u_* \cdot \exp(-1.5 \cdot z/h) \qquad (4.25a)$$

$$\sigma_v = 1.6 \cdot u_* \cdot [1-0.5 \cdot (z/h)] \qquad (4.25b)$$

$$\sigma_w = 1.25 \cdot u_* \cdot [1-0.5 \cdot (z/h)] \qquad (4.25c)$$

For statically **unstable** air:

$$\sigma_u = 0.11 \cdot w_B \cdot (z/z_i)^{1/3} \cdot [1-0.7 \cdot (z/z_i)] \qquad (4.26a)$$

$$\sigma_v = 0.08 \cdot w_B \cdot \left[0.5 + 0.4 \cdot (1 - z/z_i)^2\right] \qquad (4.26b)$$

$$\sigma_w = 0.11 \cdot w_B \cdot (z/z_i)^{1/3} \cdot [1-0.8 \cdot (z/z_i)] \qquad (4.26c)$$

where z_i is the mixed-layer depth, w_B is buoyancy velocity (eq. 2.46). These relationships are important for air pollution dispersion, and will be used in Chapt. 17.

Isotropy

If turbulence has the same variance in all three directions, then turbulence is said to be **isotropic**. Namely:

$$\sigma_u{}^2 = \sigma_v{}^2 = \sigma_w{}^2 \qquad \bullet(4.27)$$

Don't confuse this word with "isentropic", which means adiabatic or constant entropy.

Turbulence is **anisotropic** (not isotropic) in many situations. During the daytime over bare land, rising thermals create stronger vertical motions than horizontal. Hence, a smoke puff becomes **dispersed** (i.e., spread out) in the vertical faster than in the horizontal. At night, vertical motions are very weak, while horizontal motions can be larger. This causes smoke puffs to fan out horizontally with only little vertical dispersion in statically stable air.

Solved Example (§)

(a) Given the following V-wind measurements. Find the mean wind speed, and standard deviation.
(b) If the vertical standard deviation is 1 m/s, is the flow isotropic?

t (h)	V (m/s)
0.1	2
0.2	–1
0.3	1
0.4	1
0.5	–3
0.6	–2
0.7	0
0.8	2
0.9	–1
1.0	1

Solution

Given: Velocities listed above, $\sigma_w = 1$ m/s.
Find: $\overline{V}$ = ? m/s, σ_v = ? m/s, isotropy = ?

(a) Use eq. (4.21), except for V instead of T:

$$\overline{V}(z) = \frac{1}{n}\sum_{i=1}^{n} V_i(z) = \frac{1}{10}(0) = \underline{0 \text{ m/s}}$$

Use eq. (4.22), but for V: $\sigma_v{}^2 = \frac{1}{n}\sum_{i=1}^{n}(V_i - \overline{V})^2$

$= (0.1)\cdot(4+1+1+1+9+4+0+4+1+1) = 2.6 \text{ m}^2/\text{s}^2$

$\sigma_v = \sqrt{2.6m^2 \cdot s^{-2}} = \underline{\mathbf{1.61\ m/s}}$

Check: Units OK. Physics OK.
Discussion: (b) **Anisotropic**, because $\sigma_v > \sigma_w$.

Turbulence Kinetic Energy

An overall measure of the intensity of turbulence is the turbulence kinetic energy per unit mass (*TKE*):

$$TKE = 0.5\cdot\left[\overline{(u')^2} + \overline{(v')^2} + \overline{(w')^2}\right] \qquad \bullet(4.28a)$$

$$TKE = 0.5\cdot\left[\sigma_u{}^2 + \sigma_v{}^2 + \sigma_w{}^2\right] \qquad \bullet(4.28b)$$

TKE is usually produced at the scale of the boundary-layer depth. The production is mechanically by wind shear and buoyantly by thermals.

Turbulent energy cascades through the **inertial subrange**, where the large size eddies drive medium ones, which in turn drive smaller eddies. Molecular viscosity continuously damps the tiniest (**microscale**) eddies, dissipating *TKE* into heat. *TKE* is not conserved.

Solved Example

Find u_*, the velocity standard deviations, and *TKE* in statically stable air at height 50 m in an ABL that is 200 m thick. Assume $C_D = 0.002$, and the winds at height 10 m are 5 m/s.

Solution

Given: $z = 50$ m, $h = 200$ m, $C_D = 0.002$,
$M = 5$ m/s. Statically stable.
Find: $u_*, \sigma_u, \sigma_v, \sigma_w$ = ? m/s. TKE = ? m²/s².

Use eq. (4.11):
$u_*{}^2 = 0.002\cdot(5\text{m/s})^2 = \text{m}^2/\text{s}^2$. $u_* = \underline{\mathbf{0.22\ m/s}}$

Use eqs. (4.24a-c):
$\sigma_u = 2\cdot(0.22\text{m/s})\cdot[1-(50\text{m}/200\text{m})]^{3/4} = \underline{\mathbf{0.35\ m/s}}$
$\sigma_v = 2.2\cdot(0.22\text{m/s})\cdot[1-(50\text{m}/200\text{m})]^{3/4} = \underline{\mathbf{0.39}}$
$\sigma_w = 1.73\cdot(0.22\text{m/s})\cdot[1-(50\text{m}/200\text{m})]^{3/4} = \underline{\mathbf{0.31}}$

Use eq. (4.28b):
$TKE = 0.5\cdot\left[0.35^2 + 0.39^2 + 0.31^2\right] = \underline{\mathbf{0.185}}\ \text{m}^2/\text{s}^2$.

Check: Units OK. Physics OK.
Discussion: In statically stable air, vertical turbulence is generally less than horizontal turbulence. Also, turbulence intensity increases with wind speed. If the atmosphere is too stable, then there will be no turbulence (see discussion of dynamic stability in Chapt. 7).

Science Graffito

"Big whirls have little whirls that feed on their velocity, And little whirls have lesser whirls and so on to viscosity – in the molecular sense." L.F. Richardson, 1922.

The tendency of *TKE* to increase or decrease is given by the following *TKE* budget equation:

$$\frac{\Delta TKE}{\Delta t} = A + S + B + Tr - \varepsilon \qquad \bullet(4.29)$$

where A is advection of *TKE* by the mean wind, S is shear generation, B is buoyant production or consumption, Tr is transport by turbulent motions and pressure, and ε is viscous dissipation rate.

Mean wind blows *TKE* from one location to another. The **advection** term is given by:

$$A = -U \cdot \frac{\Delta TKE}{\Delta x} - V \cdot \frac{\Delta TKE}{\Delta y} - W \cdot \frac{\Delta TKE}{\Delta z} \qquad (4.30)$$

Thus, turbulence can increase (or decrease) at any location if the wind is blowing in greater (or lesser) values of *TKE* from somewhere else.

Wind **shear** generates turbulence near the ground according to:

$$S = u_*^2 \cdot \frac{\Delta M}{\Delta z} \qquad (4.31a)$$

in the surface layer, where u_* is the friction velocity, and $\Delta M/\Delta z$ is the wind shear. To good approximation:

$$S \approx a \cdot M^3 \qquad (4.31b)$$

where $a \cong 2 \times 10^{-5}$ m^{-1} for wind speed M measured at a standard height of $z = 10$ m. Greater wind speeds near the ground cause greater wind shear, and generate more turbulence.

Buoyancy can either increase or decrease turbulence. When thermals are rising from a warm surface, they generate *TKE*. Conversely, when the ground is cold and the ABL is statically stable, buoyancy opposes vertical motion and consumes *TKE*.

The rate of **buoyant production or consumption** of TKE is:

$$B = \frac{g}{T_v} \cdot F_{H\ sfc} \qquad (4.32)$$

where g is gravity, T_v is the absolute virtual air temperature near the ground, and $F_{H\,sfc}$ is the effective surface heat flux (positive when the ground is warmer than the air). Over land, $F_{H\,sfc}$ and B are usually positive during the daytime, and negative at night.

Turbulence can advect or **transport** itself. For example, if turbulence is produced by shear near the ground (in the **surface layer**), then turbulence motions will tend to move the excess *TKE* from the surface layer to locations higher in the ABL. Pressure fluctuations can have a similar effect, because turbulent pressure forces can generate turbulence motions. This term is difficult to simplify, and will be grouped with the turbulent transport term, Tr, here.

Molecular viscosity dissipates turbulent motions into heat. The amount of heating is small, but the amount of damping of *TKE* is large. The **dissipation** is always a loss:

$$\varepsilon \approx \frac{(TKE)^{3/2}}{L_\varepsilon} \qquad (4.33)$$

where $L_\varepsilon \cong 50$ m is a dissipation length scale.

The ratio of buoyancy to shear terms of the TKE equation is called the **flux Richardson number,** R_f, which is approximately equal to the **gradient or bulk Richardson number**, to be discussed in Chapter 7. Generally, turbulence dies if $R_f > 1$.

$$R_f = \frac{-B}{S} \approx \frac{-(g/T_v) \cdot F_{H\ sfc}}{u_*^2 \cdot \dfrac{\Delta M}{\Delta z}} \approx \frac{-(g/T_v) \cdot F_{H\ sfc}}{a \cdot M^3} \qquad (4.34)$$

Solved Example

Assume steady state, and neglect advection and transport. What equilibrium *TKE* is expected in the surface layer with a mean wind of 5 m/s and surface heat flux of –0.02 K·m/s? The ambient temperature is 25°C, and the air is dry.

Solution

Given: $M = 5$ m/s, $F_{H\,sfc} = -0.02$ K·m/s,
$\Delta TKE/\Delta t = 0$ for steady state, $T_v = 298$ K.

Find: $TKE = ?$ m^2/s^2.

Rearrange eq. (4.29), using eqs. (4.30 - 4.32):

$$TKE = \left\{ L_\varepsilon \cdot \left[a \cdot M^3 + (g/T_v) \cdot F_{H\ sfc} \right] \right\}^{2/3} = \left\{ (50\text{m}) \cdot \left[\left(2 \times 10^{-5}\text{m}^{-1}\right)\left(5\frac{\text{m}}{\text{s}}\right)^3 + \frac{9.8\text{ms}^{-2}}{298\text{K}}(-0.02\text{Km/s}) \right] \right\}^{2/3}$$

$= \{0.125 - 0.033 \text{ m}^3/\text{s}^3\}^{2/3} = \underline{0.20} \text{ m}^2/\text{s}^2$

Check: Units OK. Physics OK.
Discussion: This turbulence intensity is weak, as is typical at night when heat fluxes are negative.

Free and Forced Convection

The nature of turbulence, and therefore the nature of pollutant dispersion, changes with the relative magnitudes of terms in the *TKE* budget. Two terms of interest are the shear S and buoyancy B terms.

When $|B| < |S/3|$, the atmosphere is said to be in a state of **forced convection** (Fig 4.24). These conditions are typical of windy overcast days, and are associated with near **neutral static stability**. Turbulence is nearly **isotropic**. Smoke plumes disperse at nearly equal rates in the vertical and lateral, which is called **coning**. The sign of B is not important here — only the magnitude.

When B is positive and $|B| > |3 \cdot S|$, the atmosphere is said to be in a state of **free convection**. Thermals are typical in this situation, and the ABL is **statically unstable** (in the nonlocal sense). These conditions often happen in the daytime over land, and during periods of cold-air advection over warmer surfaces. Turbulence is **anisotropic**, with more energy in the vertical, and smoke plumes loop up and down in a pattern called **looping**.

When B is negative and $|B| > |S|$, static stability is so strong that turbulence cannot exist. During these conditions, there is virtually no dispersion while the smoke blows downwind. **Buoyancy waves (gravity waves)** are possible, and appear as waves in the smoke plumes. For values of $|B| \cong |S|$, breaking Kelvin-Helmholtz waves can occur (see Chapter 7), which cause some dispersion.

For B negative but $|B| < |S|$, weak turbulence is possible. These conditions can occur at night. This is sometimes called **stably-stratified turbulence (SST)**. Vertical dispersion is much weaker than lateral, causing an **anisotropic** condition where smoke spreads horizontally more than vertically, in a process called **fanning**.

Fig 4.24 shows the relationship between different types of convection and the terms of the TKE equation. While the ratio of B/S determines the nature of convection, the sum $S + B$ determines the intensity of turbulence. A Pasquill-Gifford turbulence type (Fig 4.24) can also be defined from the relative magnitudes of S and B, and will be used in Chapt. 17 to help estimate pollution dispersion rates.

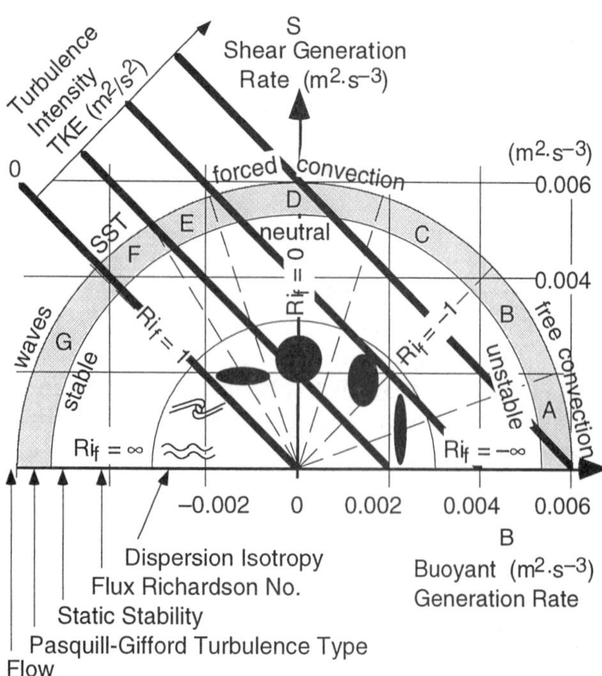

Figure 4.24
Rate of generation of TKE by buoyancy (abscissa) and shear (ordinate). Shape and rates of plume dispersion (dark spots or waves). Dashed lines separate sectors of different Pasquill-Gifford turbulence type. Isopleths of TKE intensity (dark diagonal lines). R_f is flux Richardson number. SST is stably-stratified turbulence.

Solved Example

For the previous solved example, determine the nature of convection (free, forced, etc.), the Pasquill-Gifford turbulence type, and the flux Richardson number. Assume no clouds.

Solution

Given: (see previous solved example)

Find: $S = ?\ \mathrm{m^2/s^3}$, $B = ?\ \mathrm{m^2/s^3}$, $R_f = ?$, PG =?

Use eq. (4.31b):

$$S \approx \left(2\times10^{-5}\mathrm{m}^{-1}\right)\left(5\frac{\mathrm{m}}{\mathrm{s}}\right)^3 = 0.0025\ \mathrm{m^2/s^3}$$

Use eq. (4.32):

$$B = \frac{9.8\mathrm{ms}^{-2}}{298\mathrm{K}}(-0.02\mathrm{Km/s}) = -0.00066\ \mathrm{m^2/s^3}$$

Use eq. (4.34): $R_f = -(-0.00066) / 0.0025 =$ <u>**0.264**</u>

Use Fig 4.24. Pasquill-Gifford Type = <u>**D**</u> (but on the borderline near E).

Convection is <u>**forced**</u>.

Check: Units OK. Physics OK.

Discussion: The type of turbulence is independent of the intensity. Intensity is proportional to S+B.

Turbulent Fluxes and Covariances

Rewrite eq. (4.20) for variance of w as

$$\mathrm{var}(w) = \frac{1}{N}\sum_{k=1}^{N}(W_k - \overline{W})\cdot(W_k - \overline{W}) \qquad (4.35)$$

By analogy, a **covariance** between vertical velocity w and potential temperature θ can be defined as:

$$\mathrm{covar}(w,\theta) = \frac{1}{N}\sum_{k=1}^{N}(W_k - \overline{W})\cdot(\theta_k - \overline{\theta})$$

$$= \frac{1}{N}\sum_{k=1}^{N}(w_k{}')\cdot(\theta_k{}') \qquad \bullet(4.36)$$

$$= \overline{w'\theta'}$$

where the overbar still denotes an average. Namely, one over N times the sum of N terms (see middle line of eq. 4.36) is the average of those items. Comparing eqs. (4.35) with (4.36), we see that variance is just the covariance between a variable and itself.

Covariance indicates the amount of common variation between two variables. It is positive where both variables increase or decrease together. Covariance is negative for opposite variation, such as when one variable increases while the other decreases. Covariance is zero if one variable is unrelated to the variation of the other.

The **correlation coefficient** is defined as the covariance normalized by the standard deviations of the two variables. Using vertical velocity and potential temperature for illustration:

$$r_{w,\theta} \equiv \frac{\overline{w'\theta'}}{\sigma_w \cdot \sigma_\theta} \qquad \bullet(4.37)$$

By normalized, we mean that $-1 \le r \le 1$. A correlation coefficient of positive one indicates a perfect correlation (both variables increase or decrease together proportionally), negative one indicates perfect opposite correlation, and zero indicates no correlation. Because it is normalized, the correlation coefficient gives no information on the absolute magnitudes of the variations.

In the ABL, many turbulent variables are correlated. For example, in the statically unstable ABL (Fig 4.25a), parcels of warm air rise and while other cool parcels sink in convective circulations. Warm air ($\theta' = +$) going up ($w' = +$) gives a positive product ($w'\theta'_{up} = +$). Cool air ($\theta' = -$) going down ($w' = -$) also gives a positive product ($w'\theta'_{down} = +$).

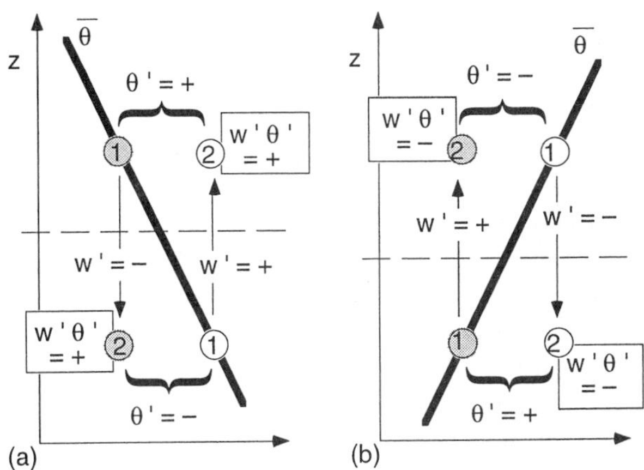

Figure 4.25
(a) Relationship between turbulent potential temperature θ and vertical velocity w for a statically unstable environment (e.g., daytime). (b) Same, but for a statically stable environment (e.g., nighttime). In both figures, the thick line represents the ambient environment, circles represent air parcels, with white being the warm air parcel and gray being cool. Numbers 1 and 2 indicate starting and ending positions of each air parcel during time interval Δt.

The average of those two products is also positive [$\overline{w'\theta'} = 0.5\cdot(w'\theta'_{up} + w'\theta'_{down}) = +$]. The result gives positive correlation coefficients $r_{w\theta}$ during convection, which is typical during daytime.

Similarly, for statically stable conditions (Fig 4.25b) where wind-shear-induced turbulence drives vertical motions against the restoring buoyant forces (see Chapt. 7), one finds cold air moving up, and warm air moving down. (Parcel warmness or coldness is measured relative to the ambient mean potential temperature $\overline{\theta}$ at the same ending height as the parcel.) This gives $\overline{w'\theta'} = -$, which is often the case during night.

More important than the statistics are the physical processes they represent. Covariances represent fluxes. Look at the air parcels crossing the horizontal dashed line in Fig 4.25a. Pretend that the dashed line is an edge view of a horizontal area that is 1 m^2.

During time interval Δt, the white (warm) air parcel moves warm air upward through that area. Heat flux is defined as heat moved per area per time. Thus this rising warm air parcel contributes to a positive heat flux. Similarly, the cold sinking air parcel contributes to a positive heat flux through that area (because negative w' times negative θ' is positive). Both parcels contribute to a positive heat flux.

This implies that covariance $\overline{w'\theta'}$ between vertical velocity and potential temperature is a turbulent kinematic heat flux, F_H [$= F_{z\ turb}(\theta)$ in the notation of Chapt. 2]:

$$\overline{w'\theta'} \equiv F_H \qquad \bullet(4.38a)$$

Similarly, the covariance between vertical velocity and water vapor mixing ratio, r (see Chapt. 5), is a kinematic moisture flux, $F_{z\ turb}(r)$:

$$\overline{w'r'} \equiv F_{z\ turb}(r) \qquad \bullet(4.38b)$$

Momentum flux is even more interesting. Recall from physics that momentum is mass times velocity. Units would be $kg \cdot m \cdot s^{-1}$. Therefore, momentum flux (momentum per area per time) would have units of $(kg \cdot m \cdot s^{-1}) \cdot m^{-2} \cdot s^{-1} = (kg \cdot m \cdot s^{-2}) \cdot m^{-2} = N \cdot m^{-2}$. Appendix A was used to find the equivalent units for a force of 1 Newton. But $N \cdot m^{-2}$ is a force per unit area, which is the definition of stress, τ. Thus, stress and momentum flux are physically the same in a fluid such as air.

A kinematic momentum flux is the momentum flux divided by air density ρ, which from the paragraph above is equal to τ / ρ. But this is just the definition of friction velocity squared u_*^2 (eq. 4.10).

As in Fig 4.25, if vertically moving air parcels (w') transport air with different horizontal velocities (u') across a horizontal area per unit time, then the covariance between vertical velocity and horizontal velocity is a kinematic momentum flux. From the paragraph above, the magnitude is also equal to the friction velocity squared. Thus:

$$\left| \overline{w'u'} \equiv F_{z\ turb}(\text{momentum}) \equiv \tau / \rho \right| \equiv u_*^2 \qquad \bullet(4.38c)$$

In Fig 4.25, air mass is conserved. Namely, each rising air parcel is compensated by a descending air parcel with the same air mass. Thus, as seen from the discussion above, turbulence can cause a net vertical transport of heat, moisture, and momentum, even though there is no net transport of air mass.

Turbulent fluxes given by eqs. (4.38) are called **eddy-correlation fluxes**. They can be measured with fast response velocity, humidity, and temperature sensors, sampling at about 10 Hz for 30 minutes. See Appendix H for parameterizations of eddy fluxes.

Solved Example(§)

Fast-response measurements of potential temperature θ, water vapor mixing ratio r, and u and w components of wind are given below as a function of time t. For θ and w, find their means, variances, and standard deviations. Also find the covariance, correlation coefficient, kinematic heat flux, and the heat flux (W/m^2).

Columns C and D are not used in this example, but will be used in some of the homeworks.

	A	B	C	D	E
1	Given:				
2	t (s)	θ (°C)	r (g/kg)	U (m/s)	W (m/s)
3	0	21	6.0	10	-5
4	0.1	28	9.5	6	4
5	0.2	29	10.0	7	3
6	0.3	25	8.0	3	4
7	0.4	22	6.5	5	0
8	0.5	28	9.5	15	-5
9	0.6	23	7.0	12	-1
10	0.7	26	8.5	16	-3
11	0.8	27	9.0	10	2
12	0.9	24	7.5	8	-4
13	1	21	6.0	14	-4
14	1.1	24	7.5	10	1
15	1.2	25	8.0	13	-2
16	1.3	27	9.0	5	3
17	1.4	29	10.0	7	5
18	1.5	22	6.5	11	2
19	1.6	30	10.5	2	6
20	1.7	23	7.0	15	-1
21	1.8	28	9.5	13	3
22	1.9	21	6.0	12	-3
23	2	22	6.5	16	-5
24	avg=	25			0

Solution:

Given: Data above in rows 2 through 23.

Find: $\overline{W}$ =? m/s , $\overline{\theta}$ = ? °C , $\sigma_w{}^2$ = ? m^2/s^2,
$\sigma_\theta{}^2$ = ? °C^2, σ_w =? m/s, σ_θ = ? °C,
$\overline{w'\theta'}$ = ? K·m/s, r = ? (dimensionless)
F_H = ? K·m/s, $\mathfrak{F}_H$ = ? W/m^2 .

First, use eq. (4.21) to find the mean values. These answers are shown in row 24 above.

$\overline{\theta}$ = **25 °C** , $\overline{W}$ = **0 m/s**

Next, use eq.s (4.20) to find the deviation from the mean, for each of the observations. The results are tabulated in columns G and H below. Then square each of those perturbation (primed) values, as tabulated in columns I and J below.

Use eq. (4.22), averaging the squared perturbations to give the variances (row 24, cols I and J): $\sigma_\theta{}^2$ = **8.48** °C^2, $\sigma_w{}^2$ = **12.38** m^2/s^2. The square root of those answers (eq. 4.23) gives the standard deviations in row 25, cols I and J:

σ_θ = **2.91** °C, σ_w = **3.52** m/s .

(continued next col.)

Solved Example(§) *(continued)*

Use eq. (4.36) and multiply each w' with θ' to give in col K the values of $\overline{w'\theta'}$. Average those to get the covariance: $\overline{w'\theta'} = F_H =$ **6.62** K·m/s, which is the kinematic heat flux by definition.

Use eq. (4.37) and divide the covar. by the std. deviations to give the correlation coef: $r =$ **0.65** .

Use eq. (2.2b) to give heat flux, with $\rho \cdot C_p$=1231 from Appendix B: $\Im_H = \rho \cdot C_p \cdot F_H =$ **8150** W/m^2 .

	G	H	I	J	K	L
1	θ'	w'	θ'^{2}	w'^{2}	w'θ'	
2	(°C)	(m/s)	(°C)^2	(m/s)^2	(Km/s)	
3	-4	-5	16	25	20	
4	3	4	9	16	12	
5	4	3	16	9	12	
6	0	4	0	16	0	
7	-3	0	9	0	0	
8	3	-5	9	25	-15	
9	-2	-1	4	1	2	
10	1	-3	1	9	-3	
11	2	2	4	4	4	
12	-1	-4	1	16	4	
13	-4	-4	16	16	16	
14	-1	1	1	1	-1	
15	0	-2	0	4	0	
16	2	3	4	9	6	
17	4	5	16	25	20	
18	-3	2	9	4	-6	
19	5	6	25	36	30	
20	-2	-1	4	1	2	
21	3	3	9	9	9	
22	-4	-3	16	9	12	
23	-3	-5	9	25	15	
24	0	var.=	8.48	12.38	6.62	=covar
25		st.dev=	2.91	3.52	0.65	= r

Check: Single primed values mean = 0. Units OK.
Discussion: $\Im_H$ magnitude is unrealistically big.

SUMMARY

We live in a part of the atmosphere known as the boundary layer (ABL), the bottom 200 m to 4 km of the troposphere. Tropospheric static stability and turbulence near the earth's surface combine to create this ABL, and cap it with a temperature inversion.

Above the boundary layer is the free atmosphere, which is not turbulently coupled with the ground (except during stormy weather such as near low pressure centers and fronts). Thus, the free atmosphere does not experience a strong diurnal cycle.

Within the ABL there are significant daily variations of temperature, winds, static stability, and turbulence. These variations are driven by the diurnal cycle of solar heating (exceeding IR cooling) during day, and IR cooling at night.

During daytime under fair weather (i.e., in anticyclonic or high-pressure regions), vigorous turbulence mixes potential temperature, humidity, wind speed, and pollutants to be nearly uniform with height. This turbulence creates a well-mixed layer that grows due to entrainment of free atmosphere air from above.

At night in fair weather, there is a shallow stable boundary layer near the ground, with a nearly neutral residual layer above. Turbulence is weak and sporadic. Winds often become calm near the surface, but can be very fast a few hundred meters above ground.

The bottom 5 to 10% of the ABL is called the surface layer. Surface drag causes the wind to be zero near the ground, and to increase with height. The shape of this wind profile depends on the roughness of the surface, and on convection.

Turbulence is a quasi-random phenomenon that can be described by statistics. Covariance of vertical velocity with another variable represents the vertical kinematic flux of that variable. Heat fluxes, moisture fluxes, and stress can be expressed as such an eddy correlation statistic.

Velocity variances represent turbulent kinetic energy (TKE) per unit mass, a measure of the intensity of turbulence. TKE is produced by wind shear and buoyancy, is advected from place to place by the mean and turbulent winds, and is dissipated into heat by molecular viscosity.

The relative magnitudes of the shear and buoyant production terms determine whether convection is free or forced. The sum of those terms is proportional to the intensity of turbulence. The ratio gives the flux Richardson number for determining whether turbulence can persist.

Threads

Formation and evolution of the ABL is intimately tied to the temperature structure of the troposphere, such as described with the standard atmosphere (Chapt. 1). The diurnal cycle in the ABL is created by the radiation budget at the surface (Chapts. 2 and 3).

Turbulence causes dispersion of air pollutants (Chapt. 17). Turbulence is a complex phenomena that can be parameterized (Appendix H). It mixes boundary layer air down to the ground where it takes on temperature and humidity characteristics of the underlying surface, thereby creating an air mass (Chapt. 12). The border between two such boundary layers is a front (Chapt. 12).

The ABL throttles the flow of heat (Chapt. 2) and moisture (Chapt. 6) from the surface into the troposphere, thereby controlling the formation of clouds and precipitation (Chapt. 8). The resulting sensible and latent heat can fuel cyclones (Chapt. 13), thunderstorms (Chapt. 15) and hurricanes (Chapt. 16). The creation of tornado rotation (Chapt. 15) is believed to be due to wind shear in the ABL. Thunder propagation depends on ABL temperature.

The north-south temperature gradient that drives the jet stream (Chapt. 11) is created by heat fluxes through the boundary layer. Mirages (Chapt. 19) form in ABL temperature gradients.

Additional boundary-layer results and equations are presented in the chapters covering budgets of heat (Chapt. 2), moisture (Chapt. 6), and momentum (Chapt. 9). Many of the aspects of atmospheric dispersion (Chapt. 17) are applications of boundary layers and turbulence. Thermals and katabatic winds of Chapt. 10 are also boundary layer phenomena.

EXERCISES

Numerical Problems

N1(§). Calculate and plot the increase of cumulative kinematic heat (cooling) during the night, for a case with kinematic heat flux (K·m/s) of:

a. –0.02 b. –0.05 c. – 0.01 d. –0.04
e. –0.03 f. – 0.06 g. – 0.07 h. –0.10

N2 (§). Calculate and plot the increase of cumulative kinematic heat during the day, for a case with daytime duration of 12 hours, and maximum kinematic heat flux (K·m/s) of:

a. 0.2 b. 0.5 c. 0.1 d. 0.4
e. 0.3 f. 0.6 g. 0.7 h. 0.9

N3(§). For a constant kinematic heat flux of –0.02 K·m/s during a 12-hour night, plot the depth and strength of the stable ABL vs. time. Assume a flat prairie, with a residual layer wind speed (m/s) of:

a. 2 b. 5 c. 8 d. 10
e. 15 f. 20 g. 25 h. 30

N4(§). For the previous problem, plot the vertical temperature profile at 1-h intervals.

N5. Find the entrainment velocity for a surface heat flux of 0.2 K·m/s, and a capping inversion strength of (°C): a. 0.1 b. 0.2 c. 0.5 d. 1.0
e. 2 f. 3 g. 5 h. 10

N6. For the previous problem, calculate the increase in mixed layer depth during a 6 h interval, assuming subsidence of – 0.02 m/s.

N7. Calculate the surface stress at sea level for a friction velocity (m/s) of:

a. 0.1 b. 0.2 c. 0.5 d. 1.0
e. 2 f. 0.3 g. 0.8 h. 0.9

N8. Find the roughness length and standard drag coefficient over: a. sea b. beach
c. tundra d. low crops e. hedgerows
f. orchards g. village h. city center

N9. Same as previous problem, but find the friction velocity given a 10 m wind speed of 10 m/s.

N10. Prepare a table similar to Table 4-1, except showing drag coefficient instead of roughness length.

N11. Find the friction velocity over a corn crop during wind speeds (m/s) of: a. 2 b. 5 c. 10 d. 20

N12(§). Given M_1 = 5 m/s at z = 10 m, plot wind speed vs height, for z_o (m) =

a. 0.001 b. 0.01 c. 0.1 d. 2

N13. Find the drag coefficients for problem N12.

N14(§). For a neutral surface layer, plot wind speed against height on linear and on semi-log graphs for friction velocity of 0.5 m/s and aerodynamic roughness length (m) of:

a. 0.001 b. 0.002 c. 0.005 d. 0.01
e. 0.02 f. 0.05 g. 0.1 h. 0.2

N15. An anemometer on a 10 m mast measures a wind speed of 8 m/s in a region of scattered hedgerows. Find the wind speed at height (m):

a. 0.5 b. 2 c. 5 d. 30

N16. The wind speed is 2 m/s at 1 m above ground, and is 5 m/s at 10 m above ground. Find the roughness length.

N17. Over a low crop with wind speed of 5 m/s at height 10 m, find the wind speed (m/s) at height (m), assuming overcast conditions:

a. 0.5 b. 1.0 c. 2 d. 5
e. 20 f. 30 g. 40 h. 50

N18. Same as previous problem, but during a clear night when friction velocity is 0.1 m/s and surface kinematic heat flux is –0.01 K·m/s. Assume g/T_v = 0.0333 m·s^{-2}·K^{-1}.

N19(§). Plot the vertical profile of wind speed in a stable boundary layer for roughness length of 0.2 m, friction velocity of 0.3 m/s, and surface kinematic heat flux (K·m/s) of:

a. –0.02 b. –0.05 c. – 0.01 d. –0.04
e. –0.03 f. – 0.06 g. – 0.07 h. –0.10

Assume g/T_v = 0.0333 m·s^{-2}·K^{-1}.

N20. For a 1 km thick mixed layer with g/T_v = 0.0333 m·s^{-2}·K^{-1}, find the Deardorff velocity (m/s) for surface kinematic heat fluxes (K·m/s) of:

a. 0.2 b. 0.5 c. 0.1 d. 0.4
e. 0.3 f. 0.6 g. 0.7 h. 0.9

N21§). For the previous problem, plot the wind speed profile, given u_* = 0.4 m/s, and M_{BL} = 5 m/s.

N22(§). For the following time series of temperatures (°C): 22, 25, 21, 30, 29, 14, 16, 24, 24, 20

a. Find the mean and turbulent parts
b. Find the variance
c. Find the standard deviation

N23(§). Using the data given in the last solved example in this Chapter:

a. Find the mean and variance for mixing ratio.
b. Find the mean and variance for U .
c. Find the covariance between r and W.
d. Find the covariance between U and W.
e. Find the covariance between r and U.

N24(§). Plot the standard deviations of U, V, and W with height given u_* = 0.5 m/s, w_B = 40 m/s, h = 600 m, and z_i = 2 km, for air that is statically:

a. stable b. neutral c. unstable

N25(§). Same as previous problem, but plot TKE with height.

N26(§). Plot wind standard deviation vs. height for all three wind components, for

a. stable air with h = 300 m, u_* = 0.1 m/s
b. neutral air with h = 1 km, u_* = 0.2 m/s
c. unstable air with z_i = 2 km, w_B = 0.3 m/s

N27(§). Plot the turbulence kinetic energy per unit mass vs. height for the previous problem.

N28. Given a wind speed of 20 m/s, surface kinematic heat flux of -0.1 K·m/s, TKE of 0.4 m^2/s^2, and g/T_v = 0.0333 m·s^{-2}·K^{-1}, find

a. shear production rate of TKE
b. buoyant production/consumption of TKE
c. dissipation rate of TKE
d. total tendency of TKE (neglecting advection and transport)
e. flux Richardson number
f. the static stability classification?
g. the Pasquill-Gifford turbulence type?
h. flow classification

Understanding & Critical Evaluation

U1. Can the ABL fill the whole troposphere? Could it extend far into the stratosphere? Explain.

U2. If the standard atmosphere was statically neutral in the troposphere, would there be a boundary layer? If so, what would be its characteristics.

U3. It is nighttime in mid winter with snow on the ground. This air blows over an unfrozen lake. Over the lake, what is the static stability of the air, and what type of ABL exists there.

U4. It is daytime in summer, and an eclipse blocks the sun. How will the ABL evolve during the eclipse?

U5. Given Fig 4.3, if turbulence were to become more vigorous and cause the ABL depth to increase, what would happen to the strength of the capping inversion? How would that result affect further growth of the ABL?

U6. The ocean often has a turbulent boundary layer within the top tens of meters of water. During a moderately windy, clear, 24 h period, when do you think stable and unstable (convective) ocean boundary layers would occur?

U7. Fig 4.4 shows the free atmosphere line touching the top peaks of the ABL curve. Is it possible during other weather conditions for the FA curve to cross through the middle of the ABL curve, or to touch the bottom valleys of the ABL curve? Discuss.

U8. It was stated that subsidence cannot penetrate the capping inversion, but pushes the inversion down in regions of high pressure. Why can subsidence not penetrate into the ABL?

U9. Fig 4.7 shows a cold front advancing. What if it were retreating (i.e., if it were a warm front). How would the ABL be affected differently, if at all?

U10. Copy Fig 4.9, and then trace the nighttime curves onto the corresponding daytime curves. Comment on regions where the curves are the same, and on other regions where they are different. Explain this behavior.

U11. Use the solar and IR radiation data to plot a curve of daytime sensible heat flux vs. time. Although the resulting curve looks similar to half of a sine wave (as in Fig 4.10), they are theoretically different. How would the actual curve change even further from a sine curve at higher or lower latitudes? How good is the assumption that the heat flux curve is a half sine wave?

U12. Similar to the previous question, but how good is the assumption that nighttime heat flux is constant with time?

U13. At nighttime it is possible to have well-mixed layers if wind speed is sufficiently vigorous. Assuming a well-mixed nocturnal boundary layer over a surface that is getting colder with time, describe the evolution (depth and strength) of this stable boundary layer. Also, for the same amount of cooling, how would its depth compare with the depth of the exponentially-shaped profile?

U14. Derive an equation for the strength of an exponentially-shaped nocturnal inversion vs. time, assuming constant heat flux during the night.

U15. For a linear early morning sounding (θ increases linearly with z), analytically compare the growth rates of the mixed layer calculated using the thermodynamic and the flux-ratio methods. For some idealized situations, is it possible to express one in terms of the other? Were any assumptions needed?

U16. Given an early morning sounding as plotted in Fig 4.16a. If the daytime heat flux were constant with time, sketch a curve of mixed-layer depth vs. time, assuming the thermodynamic method. Comment on the different stages of mixed layer growth rate.

U17. Assume that eq. (4.7) applies to moisture and momentum fluxes as well as to heat fluxes, but that (4.9) defines the entrainment velocity based only on heat. Combine those two equations to give the kinematic flux of moisture at the top of the mixed layer as a function of potential temperature and mixing ratio jumps across the mixed-layer top, and in terms of surface heat flux.

U18. If the ABL is a region that feels drag near the ground, why can the winds at night accelerate, as shown in Fig 4.18?

U19. Use eqs. (4.11) and (4.14a) to show how eq. (4.12) is derived from the log wind profile.

U20. Given the moderate value for u_* that was written after eq. (4.10), what value of stress (Pa) does that correspond to? How does this stress compare to sea-level pressure?

U21. Derive eq. (4.14b) from (4.14a).

U22. Given the wind speed profile of the solved example in the radix layer subsection, compare this profile to a log wind profile. Namely, find a best fit log-wind profile for the same data. Comment on the differences.

U23. For fixed w_B and z_i, plot eqs. (4.26) with height, and identify those regions with isotropic turbulence.

U24. Using eqs. (4.24 – 4.26), derive expressions for the *TKE* vs. z for statically stable, neutral, and unstable boundary layers.

U25. Derive an expression for the shear production term vs. z in the neutral surface layer, assuming a log wind profile.

U26. Knowing how turbulent heat flux varies with height (Fig 4.17) in a convective mixed layer, comment on the *TKE* buoyancy term vs. z for that situation.

U27. Given some initial amount of *TKE*, and assuming no production, advection, or transport, how would *TKE* change with time, if at all? Plot a curve of *TKE* vs. time.

U28. Use K-theory (from Appendix H) to relate the flux Richardson number to the gradient Richardson number (Chapt. 7).

U29. If shear production of *TKE* increases with time, how must the buoyant term change with time in order to maintain a constant Pasquill-Gifford stability category of E?

U30. Create a figure similar to Fig 4.25, but for the log wind profile. Comment on the variation of momentum flux $\overline{u'w'}$ with height in the neutral surface layer.

U31. In the explanation surrounding Fig 4.10, the accumulated heating and cooling were referenced to start times when the heat flux became positive or negative, respectively. We did not use sunrise and sunset as the start times because those times did not correspond to when the surface heat flux changes sign. Use all of the relevant previous equations in the chapter to determine the time difference at your town and for today, between:

a. sunrise and the time when surface heat flux first becomes positive.

b. sunset and the time when surface heat flux first becomes negative.

U32. What is the static stability now at your location?

U33. Draw a sketch similar to Fig 4.8, except indicating the static stabilities instead of the name of each domain.

U34. Use eq. (4.3) and the definition of potential temperature to replot Fig 4.15 in terms of actual temperature T vs. z. Discuss the difference in heights between the relative maximum of T and relative maximum of θ.

U35. Given an early morning sounding with surface temperature 5°C, find the mixed-layer potential temperature and depth at 11 AM, when the cumulative heating is 800 K·m. Assume the early-morning potential temperature profile is

a. $\Delta\theta/\Delta z = 2$ K/km = constant

b. $\Delta\theta(z) = (8°C)\cdot\exp(-z/150\text{m})$

U36. If the heating rate is proportional to the vertical heat-flux divergence, use Fig. 4.17b to determine the heating rate at each height in the mixed layer. How would the mixed-layer T profile change with time?

U37. A negative value of eddy-correlation momentum flux near the ground implies that the momentum of the wind is being lost to the ground. Is it possible to have a positive eddy-correlation momentum flux near the surface? If so, under what conditions, and what would it mean physically?

Web-Enhanced Questions

W1. Access the upper-air soundings every 6 or 12 h for a rawinsonde station near you (or other site specified by your instructor). For heights every 200 m (or every 2 kPA) from the surface, plot how the temperature varies with time over several days. The result should look like Fig 4.4, but with more lines. Which heights or pressure levels appear to be above the ABL?

W2. Access temperature profiles from the web for the rawinsonde station closest to you (or for some other sounding station specified by your instructor). Convert the resulting temperatures to potential temperatures, and plot the resulting θ vs z. Can you identify the top of the ABL? Consider the time of day when the sounding was made, to help you anticipate what type of ABL exists (e.g., mixed layer, stable boundary layer, neutral layer.)

W3. Same as the previous question, but first convert the temperatures to potential temperatures at those selected heights. This should look even more like fig 4.4.

W4. Access a weather map with surface wind observations. Find a situation or a location where there is a low pressure center. Draw a hypothetical circle around the center of the low, and find the average inflow velocity component across this circle. Using volume conservation, and assuming a 1 km thick ABL, what vertical velocity to you anticipate over the low?

W5. Same as previous question, but for a high-pressure center.

W6. Access a number of rawinsonde soundings for stations more-or-less along a straight line that crosses through a cold front. Identify the ABL both ahead of and behind the front.

W7. Find a rawinsonde station in the center of a clear high pressure region, and access the soundings every 6 h if possible. If they are not available over N. America, try Europe. Use all of the temperature, humidity, and wind information to determine the evolution of the ABL structure, and create a sketch similar to Fig 4.8, but for your real conditions. Consider seasonal affects, such as in Fig 4.13.

W8. Access from the web the current sunrise and sunset times, and estimate a curve such as in Fig 4.10.

W9. Access a rawinsonde sounding for nighttime, and find the best-fit parameters (height, strength) for an exponential potential temperature profile that best fits the sounding. If the exponential shape is a poor fit, explain why.

W10. Access an early morning sounding for a rawinsonde site near you. Calculate the anticipated accumulated heating, based on the eqs. in Chapts. 2 – 4, and predict how the mixed layer will warm and grow in depth as the day progresses.

W11. Compare two sequential rawinsonde soundings for a site, and estimate the entrainment velocity by comparing the heights of the ABL. What assumptions did you make in order to do this calculation?

W12. Access the wind information from a rawinsonde site near you. Compare the wind speed profiles between day and night, and comment on the change of wind speed in the ABL.

W13. Access the "surface" wind observations from a weather station near you. Record and plot the wind speed vs. time, using wind data as frequent as are available, getting a total of roughly 24 to 100 observations. Use these data to calculate the mean wind, variance, and standard deviation.

W14. Same as the previous exercise, but for temperature.

W15. Same as the previous exercise, except collect both wind and temperature data, and find the covariance and correlation coefficient.

Synthesis Questions

S1. At the end of one night, assume that the stable ABL profile of potential temperature has an exponential shape, with strength 10°C and e-folding depth of 300 m. Using this as the early-morning sounding, compute and plot how the potential temperature and depth of the mixed layer evolve during the morning. Assume that $D = 12$ hr, and $F_{Hmax} = 0.1$ K·m/s. (Hint: In the spirit of the "Seek Solutions" box in a previous chapter, feel free to use graphical methods to solve this without calculus. However, if you want to try it with calculus, be very careful to determine which are your dependent and independent variables.)

S2. Suppose that there was no turbulence at night, causing the radiatively-cooled surface to cool only the bottom 1 m of atmosphere by conduction. Given typical heat fluxes at night, what would be the resulting air temperature by morning? Also describe how the daytime mixed layer would evolve from this starting point?

S3. On a planet that does not have a solid core, but has a gas that increases in density as the core is approached, would there be a boundary layer? If so, how would it behave?

S4. What if you have been hired by an orchard owner to tell her how deep a layer of air needs to be mixed by electric fans to prevent frost formation. Create a suite of answers based on different scenarios of initial conditions, so that she can consult these results on any given day to determine the speed to set the fans (which we will assume can be used to control the depth of mixing). State and justify all assumptions.

S5. What if the earth's surface was a perfect conductor of heat, and had essentially infinite heat capacity (as if the earth were a solid sphere of aluminum, but had the same albedo as earth). Given the same solar and IR radiative forcings on the earth's surface, describe how ABL structure and evolution would differ from Fig 4.8, if at all?

S6. What if all the air in the troposphere were saturated and cloudy. Would the ABL be different, if so, how?

S7. Suppose that for some reason, the actual 2 km thick ABL in our troposphere was warm enough to have zero capping inversion on one particular day, but otherwise looked like Fig 4.3. Comment on the evolution of the ABL from this initial state. Also, how might this initial state have occurred?

S8. If the ABL were always as thick as the whole troposphere, how would that affect the magnitude of our diurnal cycle of temperature?

S9. It was stated in this chapter that entrainment happens one way across the capping inversion, from the nonturbulent air into the turbulent air. Suppose that the capping inversion was still there, but that both the ABL below the inversion and the layer of air above the inversion were turbulent. Comment on entrainment and the growth of the mixed layer.

S10. Suppose that TKE was not dissipated by viscosity. How would that change our weather and climate, if at all?

S11. Suppose that wind shear destroyed TKE rather than generated it. How would that change our weather and climate, if at all?

S12. Suppose there were never a temperature inversion capping the ABL. How would that change our weather and climate, if at all?

S13. Suppose that the winds felt no frictional drag at the surface of the earth. How would that change our weather and climate, if at all?

MOISTURE

CONTENTS

5 Water-vapor amount in the air is variable. Concentration of water vapor can be quantified by vapor pressure, mixing ratio, specific humidity, absolute humidity, relative humidity, dew point depression, saturation level, or wet-bulb temperature.

Warmer air can hold more water vapor at equilibrium than colder air. Air that holds this equilibrium amount is saturated. If air is cooled below the saturation temperature, some of the water vapor condenses into liquid, which releases latent heat and warms the air. Thus, temperature and moisture interact in a way that cannot be neglected.

SATURATION VAPOR PRESSURE

Vapor Pressure

Air is a mixture of gases. All of the gases contribute to the total pressure. The pressure associated with any one gas in a mixture is called the **partial pressure**. Water vapor is a gas, and its partial pressure in air is called the **vapor pressure**. Symbol e is used for vapor pressure. Units are pressure units, such as kPa.

Saturation

Air can hold any proportion of water vapor. However, for humidities greater than a threshold called the **saturation humidity**, water vapor tends to condense into liquid faster than it re-evaporates. This condensation process lowers the humidity toward the equilibrium (saturation) value. The process is so fast that humidities rarely exceed the equilibrium value. Thus, while air can hold any portion of water vapor, the threshold is rarely exceeded by more than 1% in the real atmosphere.

Air that contains this threshold amount of water vapor is **saturated**. Air that holds less than that amount is **unsaturated**. The equilibrium (saturation) value of vapor pressure over a flat surface of pure water is given the symbol e_s. For unsaturated air, $e < e_s$.

Air can be slightly **supersaturated** ($e > e_s$) when there are no surfaces upon which water vapor can condense (i.e., unusually clean air, with no cloud condensation nuclei, and no liquid or ice particles). Temporary supersaturation also occurs when the threshold value drops so quickly that condensation does not remove water vapor fast enough.

Even at saturation, there is a continual exchange of water molecules between the liquid water and the air. The evaporation rate depends mostly on the temperature of the liquid water. The condensation rate depends mostly on the humidity in the air. At equilibrium these two rates balance.

If the liquid water temperature increases, then evaporation will temporarily exceed condensation, and the number of water molecules in the air will increase until a new equilibrium is reached. Thus, the equilibrium (saturation) humidity increases with temperature. The net result is that warmer air can hold more water vapor at equilibrium than cooler air.

The **Clausius-Clapeyron** equation describes the relationship between temperature and saturation vapor pressure, which is approximately:

$$e_s = e_o \cdot \exp\left[\frac{L}{\Re_v} \cdot \left(\frac{1}{T_o} - \frac{1}{T}\right)\right] \qquad \bullet(5.1)$$

where $e_o = 0.611$ kPa and $T_o = 273$ °K are constant parameters, and $\Re_v = 461$ J·K^{-1}·kg^{-1} is the gas constant for water vapor. Absolute temperature in Kelvins must be used for T in eq. (5.1).

Because clouds can consist of liquid droplets and ice crystals suspended in air, we must consider saturations with respect to water and ice. Over a flat water surface, use the latent heat of vaporization $L = L_v = 2.5\text{x}10^6$ J·kg^{-1} in eq. (5.1), which gives $L/\Re_v = 5423$ K. Over a flat ice surface, use the latent heat of deposition $L = L_d = 2.83\text{x}10^6$ J·kg^{-1}, which gives $L/\Re_v = 6139$ K.

Solved Example
Find the sat. vapor pressure for $T = 21$°C?

Solution
Given: $T = 21$°C = 294 K
Find: $e_s = ?$ kPa
Use eq. (5.1) for liquid water.

$$e_s = (0.611\text{kPa}) \cdot \exp\left[(5423\text{K}) \cdot \left(\frac{1}{273\text{K}} - \frac{1}{294\text{K}}\right)\right]$$
$= (0.611$ kPa$)\cdot\exp(1.419)$ = **2.525 kPa**

Check: Units OK. Physics OK. Agrees Table 5-1.
Discussion: Average pressure at sea level is P = 101.3 kPa. Thus, water vapor accounts for roughly 2.5% of the total air pressure in this example.

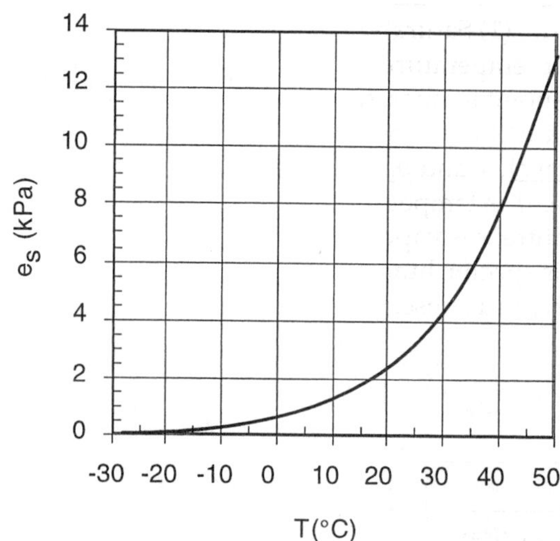

Figure 5.1
Saturation vapor pressure over a flat surface of pure water. A blowup of the portion of this curve colder than 0°C is given in the next figure.

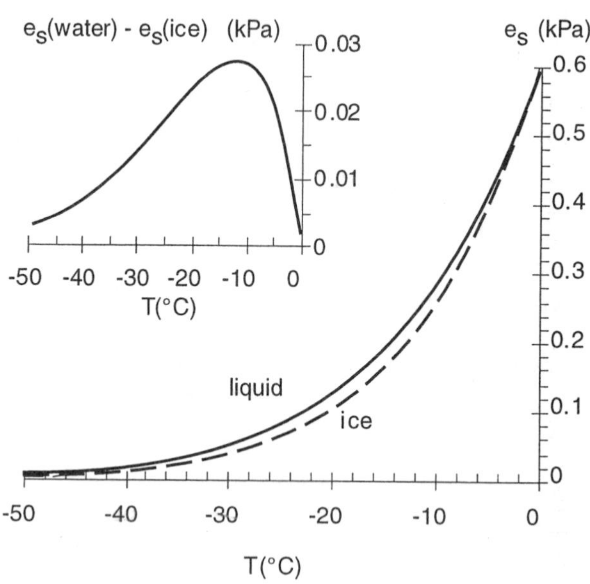

Figure 5.2
Saturation vapor pressure over flat surfaces of pure liquid water and ice, at temperatures below 0°C. Insert shows difference between saturation vapor pressures over water and ice.

Saturation vapor pressure is listed in Table 5-1 and plotted in Figs 5.1 and 5.2. These were produced by solving eq. (5.1) on a computer spreadsheet. The exponential-like increase of water vapor holding capacity with temperature has tremendous impact on clouds and storms. **Supercooled** (unfrozen) liquid water can exist between temperatures of 0 to –40°C, hence, the curve for saturation over water includes temperatures below 0°C.

Table 5-1. (1) Saturation values of humidity vs. actual air temperature. -or- (2) Actual humidities vs. dew-point temperature. Values are for over a flat surface of liquid water. r and q values are for sea level. e and ρ_v values are for any pressure. T is temperature, T_d is dew-point temperature, e is vapor pressure, r is mixing ratio, q is specific humidity, ρ_v is absolute humidity, and subscript s denotes a saturation value.

T	e_s	r_s	q_s	ρ_{vs}
		or		
T_d	e	r	q	ρ_v
(°C)	(kPa)	(g/kg)	(g/kg)	(kg/m³)
-20	0.127	0.78	0.78	0.00109
-18	0.150	0.92	0.92	0.00128
-16	0.177	1.09	1.09	0.00150
-14	0.209	1.28	1.28	0.00175
-12	0.245	1.51	1.51	0.00204
-10	0.287	1.77	1.76	0.00237
-8	0.335	2.07	2.06	0.00275
-6	0.391	2.41	2.40	0.00318
-4	0.455	2.80	2.80	0.00367
-2	0.528	3.26	3.25	0.00422
0	0.611	3.77	3.76	0.00485
2	0.706	4.37	4.35	0.00557
4	0.814	5.04	5.01	0.00637
6	0.937	5.80	5.77	0.00728
8	1.076	6.68	6.63	0.00830
10	1.233	7.66	7.60	0.00945
12	1.410	8.78	8.70	0.01073
14	1.610	10.05	9.95	0.01217
16	1.835	11.48	11.35	0.01377
18	2.088	13.09	12.92	0.01556
20	2.371	14.91	14.69	0.01755
22	2.688	16.95	16.67	0.01976
24	3.042	19.26	18.89	0.02222
26	3.437	21.85	21.38	0.02494
28	3.878	24.76	24.16	0.02794
30	4.367	28.02	27.26	0.03127
32	4.911	31.69	30.72	0.03493
34	5.514	35.81	34.57	0.03896
36	6.182	40.43	38.86	0.04340
38	6.921	45.61	43.62	0.04827
40	7.736	51.43	48.91	0.05362
42	8.636	57.97	54.79	0.05947
44	9.627	65.32	61.31	0.06588
46	10.717	73.59	68.54	0.07287
48	11.914	82.91	76.56	0.08051
50	13.228	93.42	85.44	0.08884

BEYOND ALGEBRA • Clausius-Clapeyron Eq.

Historical Underpinning

To build better steam engines during the industrial revolution, engineers were working to discover the thermodynamics of water vapor. Steam engineers B.-P.-E. Clapeyron in 1834 and R. Clausius in 1879 applied the principles of S. Carnot (early 1800s), studying isothermal compression of pure water vapor in a cylinder to find the saturation vapor pressure at the point where condensation occurred.

Variation of Vapor Pressure with Temperature

By repeating the experiment for various temperatures, they found that:

$$\frac{de_s}{dT} = \frac{L_v}{T}\left[\frac{1}{\rho_v} - \frac{1}{\rho_L}\right]^{-1}$$

where ρ_v is the density of water vapor, and ρ_L is the density of liquid water. Because $\rho_L >> \rho_v$ (see Appendix B and Table 1-2), thus $1/\rho_L << 1/\rho_v$, allowing us to neglect the ρ_L term:

$$\frac{de_s}{dT} \cong \frac{L_v}{T}\rho_v$$

The ideal gas law can be used to relate the water vapor pressure to the vapor density: $e_s = \rho_v \cdot \Re_v T$, where $\Re_v$ is the gas constant for water vapor. Solving for ρ_v and plugging the result into the eq. above gives:

$$\frac{de_s}{dT} \cong \frac{L_v \cdot e_s}{\Re_v \cdot T^2}$$

Use Separation of Variables: move all e_s terms to the left, all T terms to the right:

$$\frac{de_s}{e_s} \cong \frac{L_v}{\Re_v}\frac{dT}{T^2}$$

Then integrate, taking care with the limits:

$$\int_{e_o}^{e_s} \frac{de_s}{e_s} \cong \frac{L_v}{\Re_v}\int_{T_o}^{T} \frac{dT}{T^2}$$

where e_o is a known vapor pressure at reference temperature T_o. The integration result is:

$$\ln\left(\frac{e_s}{e_o}\right) \cong -\frac{L_v}{\Re_v}\left[\frac{1}{T} - \frac{1}{T_o}\right]$$

Taking the antilog (exp) of both sides gives eq. (5.1).

$$e_s = e_o \cdot \exp\left[\frac{L}{\Re_v} \cdot \left(\frac{1}{T_o} - \frac{1}{T}\right)\right]$$

See C. Bohren and B. Albrecht, 1998: *Atmospheric Thermodynamics.* Oxford. 402pp. for details, and for an interesting historical discussion.

Tetens' formula is an empirical expression for saturation vapor pressure with respect to liquid water that includes the variation of latent heat with temperature:

$$e_s = e_o \cdot \exp\left[\frac{b \cdot (T - T_1)}{T - T_2}\right] \tag{5.2}$$

where $e_o = 0.611$ kPa, $b = 17.2694$, $T_1 = 273.16$ K, and $T_2 = 35.86$ K. Absolute temperature (Kelvins) must be used in this formula.

Solved Example

Compare Tetens' formula with the Clausius-Clapeyron equation for a variety of temperatures

Solution

Use eqs. (5.1) and (5.2) on a spreadsheet.
The result below on a semi-log graph shows the Clausius-Clapeyron equation as a thin solid line, and Tetens' formula as the thicker dashed line.

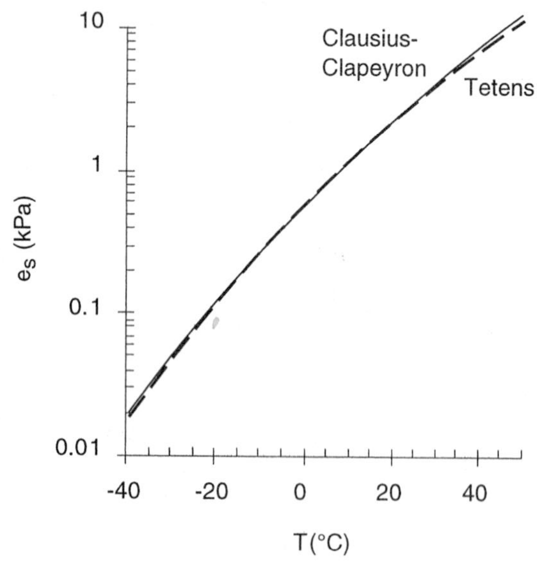

Check: Units OK. Physics OK. Agrees Table 5-1.
Discussion: Both formulae are extremely close over a wide range of temperatures.

FOCUS • Boiling

Definition:

Boiling is the state where ambient atmospheric pressure equals the equilibrium (saturation) vapor pressure:

$$P = e_s$$

Derivation of Boiling *T* vs. *z*:

Because ambient pressure decreases roughly exponentially with height, we can use (1.6b) to replace the left hand side (LHS) of the equation above by $P = P_o \cdot \exp(-z/H_p)$, where $P_o = 101.325$ kPa is sea level pressure (at $z = 0$), and $H_p \approx 7.29$ km is the scale height for pressure. The right hand side (RHS) can be replaced with the Clausius-Clapeyron eq. (5.1), which describes how this boiling pressure is related to T. The net result is:

$$P_o \cdot \exp\left[-\frac{z}{H_p}\right] = e_o \cdot \exp\left[\frac{L_v}{\Re_v} \cdot \left(\frac{1}{T_o} - \frac{1}{T}\right)\right]$$

where $L_v/\Re_v = 5423$ K. Dividing both sides by P_o, and using the trick that $e_o/P_o = \exp[\ln(e_o/P_o)]$, gives:

$$\exp\left[-\frac{z}{H_p}\right] = \exp\left[\ln\left(\frac{e_o}{P_o}\right) + \frac{L_v}{\Re_v}\left(\frac{1}{T_o}\right) - \frac{L_v}{\Re_v}\frac{1}{T}\right]$$

But the first two terms in square brackets on the RHS are constants, and can be grouped as a dimensionless constant a. Taking the ln of both sides gives:

$$\frac{z}{H_p} = \frac{L_v}{\Re_v}\frac{1}{T} - a$$

Rearranging to solve for temperature T (which equals $T_{boiling}$ because we had set $P = e_s$):

$$\boxed{T_{boiling} = \frac{L_v / \Re_v}{a + z/H_p}}$$

By definition of the Celsius temperature scale, $T = 100°C = 373.15$ K for boiling at sea level $z = 0$. Using these values of z and T in the eq. above allows us solve for a, yielding $a = 14.53$.

Discussion:

At $z = 1$ km, the eq. above gives $T_{boiling} = 96.6°C$, and at $z = 2$ km, $T_{boiling} = 93.2°C$. This is an average decrease of 3.4°C / km. Boiled foods take longer to cook at higher altitude because the cooking (boiling) temperature is lower.

HUMIDITY VARIABLES

Mixing Ratio

The ratio of mass of water vapor to mass of dry air is called the mixing ratio, r. It is given by

$$r = \frac{\varepsilon \cdot e}{P - e} \qquad \bullet(5.3)$$

where $\varepsilon = \Re_d / \Re_v = 0.622\ g_{vapor}/g_{dry\ air}$ is the ratio of gas constants for dry air to that for water vapor. Eq. (5.3) says that r is proportional to the ratio of partial pressure of water vapor (e) to partial pressure of the remaining gases in the air ($P - e$).

The **saturated mixing ratio,** r_s, is given by (5.3), except with e_s in place of e. Examples of mixing ratio values are given in Table 5-1, for air at sea level.

Although units of mixing ratio are g/g (i.e., grams of water vapor per gram of dry air), it is usually presented as g/kg (i.e., grams of water vapor per kilogram of dry air) by multiplying the g/g value by 1000. Also, note that g/g does *not* cancel to become 1, because the numerator and denominator represent mass of different substances.

Specific Humidity

The ratio of mass of water vapor to mass of total (*moist*) air is called the specific humidity, q, and to a good approximation is given by:

$$q = \frac{\varepsilon \cdot e}{P} \qquad (5.4)$$

It has units of g/g or g/kg. For **saturation specific humidity,** q_s, use e_s in place of e in eq. (5.4).

Both the saturation mixing ratio and specific humidity depend on ambient pressure, but the saturation vapor pressure does not. Thus, in Table 5-1 the vapor pressure numbers are absolute numbers that can be used anywhere, while the other humidity variables are given at sea-level pressure and density. The saturation values of mixing ratio and specific humidity can be calculated for any other ambient pressure using eqs. (5.3) and (5.4).

Absolute Humidity

The concentration, ρ_v, of water vapor in air is called the absolute humidity, and has units of grams of water vapor per cubic meter (g/m^3). Because the absolute humidity is essentially a partial density, it can be found from the partial pressure using the ideal gas law for water vapor:

$$\rho_v = \frac{e}{\Re_v \cdot T} = \frac{e}{P} \cdot \varepsilon \cdot \rho_d \qquad (5.5)$$

where ρ_d is the density of dry air. As discussed in Chapter 1, dry air density is roughly $\rho_d = 1.225$ kg/m^3 at sea level, and varies with altitude, pressure, and temperature according to the ideal gas law.

The **saturated** value of **absolute humidity,** ρ_{vs}, is found by using e_s in place of e in eq. (5.5).

Solved Example

Find the saturation values of mixing ratio, specific humidity, and absolute humidity for air of temperature 0°C and pressure 50 kPa.

Solution

Given: $T = 0°C$, $P = 50$ kPa

Find: $r_s = ?$ g/kg, $q_s = ?$ g/kg, $\rho_{vs} = ?$ g/m^3

$e_s = 0.611$ kPa was found directly from Table 5-1 because it does not depend on P.

Use eq. (5.3) for r_s:

r_s = [0.622·(0.611 kPa)] / [50 kPa - 0.611 kPa]

= 0.00770 g/g

= <u>7.70 g/kg</u>

Use eq. (5.4) for q_s:

q_s = 0.622·(0.611 kPa)/50 kPa = 0.00760 g/g

= <u>7.60 g/kg</u>

Use eq. (5.5) for ρ_{vs}:

ρ_{vs} = (0.611 kPa)/[(461 J·K^{-1}·kg^{-1})·(273 K)]

= <u>0.00485</u> kg·m^{-3} .

Check: Units OK. Physics OK.

Discussion: The first two values are roughly double those for sea-level pressure (see Table 5-1). The reason is that although the partial pressure of water vapor held in the air is the same, partial pressure of dry air is less. The last value equals the value in Table 5-1, because there is no P dependence.

Relative Humidity

The ratio of actual amount of water vapor in the air compared to the equilibrium (saturation) amount at that temperature is called the relative humidity, *RH*:

$$\frac{RH}{100\%} = \frac{e}{e_s} = \frac{q}{q_s} = \frac{\rho}{\rho_s} \approx \frac{r}{r_s} \qquad \bullet(5.6)$$

Relative humidity indicates the amount of net evaporation that is possible into the air, regardless of the temperature. At RH = 100% no net evaporation occurs because the air is already saturated. Variation of mixing ratio with RH is shown in Fig 5.3.

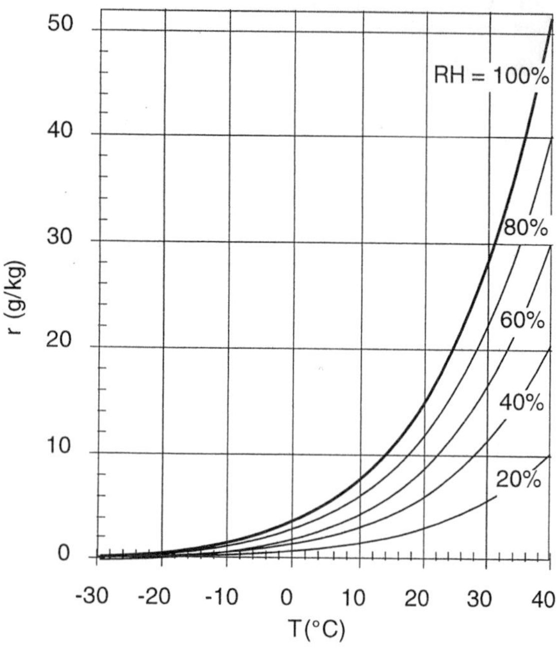

Figure 5.3
Mixing ratio at sea level vs. relative humidity and temperature.

> **Solved Example**
> Find the relative humidity for air of T = 20°C and e = 1 kPa.
>
> **Solution**
> Given: T = 20°C , e = 1 kPa
> Find: RH = ? %
>
> From Table 5-1 at T = 20°C : e_s = 2.371 kPa.
> Use eq. (5.6):
> RH = 100% · (1 kPa / 2.371 kPa) = <u>42%</u>.
>
> **Check**: Units OK. Physics OK.
> **Discussion**: The result does not depend on P.

Dew-Point Temperature

The temperature to which air must be cooled to become saturated at constant pressure is called the dew-point temperature, T_d. It is given by eq. (5.1) or found from Table 5-1 by using e in place of e_s and T_d in place of T. Making those substitutions and solving for T_d yields:

$$T_d = \left[\frac{1}{T_o} - \frac{\Re_v}{L} \cdot \ln\left(\frac{e}{e_o}\right)\right]^{-1} \qquad \bullet(5.7)$$

where e_o = 0.611 kPa, T_o = 273 K, and $\Re_v / L_v$ = 0.0001844 K^{-1}.

Saturation (equilibrium) with respect to a flat surface of liquid water occurs at a slightly colder temperature than saturation with respect to a flat ice surface. With respect to liquid water, use $L = L_v$ in the equation above, where T_d is called the **dew-point temperature**. With respect to ice, $L = L_d$, and T_d is called the **frost-point temperature** (see the "Saturation" subsection for values of L).

If $T_d = T$, the air is saturated. The **dew-point depression** or **temperature-dew-point spread** ($T - T_d$) is a relative measure of the dryness of the air. T_d is usually less than T (except during supersaturation, where it might be a fraction of a degree warmer). If air is cooled below the initial dew point temperature, then the dew point temperature drops to remain equal to the air temperature, and the excess water condenses or deposits as dew, frost, fog, or clouds (see Chapters 7 and 8).

Dew point temperature is easy to measure and provides the most accurate humidity value, from which other humidity variables can be calculated. **Dew-point hygrometers** measure humidity by reflecting a light beam off a chilled mirror. When the mirror temperature is cool enough for dew to form on it, the light beam scatters off of the dew drops instead of reflecting from the mirror. A photo detector records this change, and controls the electrical circuit that cools the mirror, in order to keep the mirror just at the dew point.

> **Solved Example**
> Given desert conditions with a temperature of 30°C and pressure of 100 kPa, find the dew point temperature for a relative humidity of 20%.
>
> **Solution**
> Given: T = 30°C , P = 100 kPa, RH = 20%
> Find: T_d = ? °C
>
> From Table 5-1, e_s = 4.367 kPa.
> Use eq. (5.6): e = 0.2·(4.367 kPa)= 0.8734 kPa
> We could use (5.7) to solve for T_d, but it is easier to look up the answers in Table 5-1. T_d = <u>5°C</u>.
>
> **Check**: Units OK. Physics OK.
> **Discussion**: The result does not depend on P. This dry air would need to be cooled 25°C before condensation would occur.

Saturation Level or Lifting Condensation Level (LCL)

When unsaturated air is lifted, it cools dry adiabatically. If lifted high enough, the temperature will drop to the dew-point temperature, and clouds will form. Dry air (air of low relative humidity) must be lifted higher than moist air. Saturated air needs no lifting at all.

The height at which saturation just occurs (with no supersaturation) is the **saturation level** or the **lifting condensation level** (LCL). **Cloud base** for convective clouds occurs there. Hence, LCL is a measure of humidity. LCL height (distance above the height where T and T_d are measured) for cumuliform clouds is very well approximated by:

$$z_{LCL} = a \cdot (T - T_d) \qquad \bullet(5.8)$$

where $a =$ 0.125 km/°C.

This expression does not work for stratiform (advective) clouds, because these clouds are not formed by air rising vertically from the underlying surface. Air in these clouds blows at a gentle slant angle from a surface hundreds to thousands of kilometers away.

Regardless of whether cumuliform clouds exist, the LCL or saturation level can be used as a measure of humidity. It also serves as a measure of total water content for saturated (cloudy) air. For this situation, the saturation level is the altitude to which one must lower an air parcel for all of the liquid or solid water to just evaporate. Eq. (5.8) does not apply to this situation.

Solved Example
Find the LCL if T = 20°C and T_d = 10°C.

Solution
Given: T = 20°C and T_d = 10°C.
Find: z_{LCL} = ? m

Use eq. (5.8)
z_{LCL} = (0.125km/°C)·(20 – 10°C) = **1.25 km**

Check: Units OK. Physics OK.
Discussion: Reasonable height for cumulus cloud base.

Wet-Bulb Temperature

When a thermometer bulb is covered by a cloth sleeve that is wet, it becomes cooler than the actual (**dry bulb**) air temperature T because of the latent heat associated with evaporation of water. Drier ambient air allows more evaporation, causing **wet-bulb** temperatures T_w to cool significantly below the air temperature. For saturated air there is no net evaporation and the wet bulb temperature equals that of the dry bulb. Humidity can be found from the difference (called **wet-bulb depression**) between the dry and wet bulb temperatures.

To work properly, the sleeve or wick must be wet with clean or distilled water. The wet bulb should be well ventilated by blowing air past it (**aspirated psychrometer**) or by moving the thermometer through the air (**sling psychrometer**). Usually psychrometers have both wet and dry bulb thermometers mounted next to each other.

Let T_w represent the wet-bulb temperature, and r_w represent the wet-bulb mixing ratio in the air adjacent to the wet-bulb after water has evaporated into it from the sleeve. Because the latent heat used for evaporation comes from the sensible heat associated with cooling, a simple heat balance gives:

$$C_p \cdot (T - T_w) = -L_v \cdot (r - r_w) \qquad (5.9)$$

where T and r are ambient air temperature and mixing ratio, respectively, C_p is the specific heat of air at constant pressure, and L_v is the latent heat of vaporization.

If the wet and dry bulb temperatures and ambient pressure P are measured or otherwise known, the corresponding mixing ratio can be found from

$$r_w = \frac{\varepsilon}{b \cdot P \cdot \exp\left(\frac{-c \cdot T_w}{T_w + \alpha}\right) - 1} \qquad (5.10)$$

$$r = r_w - \beta \cdot (T - T_w) \qquad (5.11)$$

where temperatures must have units of °C, and where $\varepsilon = 0.622$ g/g, $b = 1.631$ kPa^{-1}, $c = 17.67$, $\alpha =$ 243.5°C, and $\beta = 4.0224\text{x}10^{-4}$ (g/g)/°C. Knowing r, any other moisture variable is easily found.

Eqs. (5.10) and (5.11) are based on Tetens' formula. So if you use these equations to generate **psychrometric tables** of relative humidity as a function of wet-bulb temperature and wet-bulb depression, then you must also use Tetens' formula (eq. 5.2) for the saturation mixing ratio r_s in order to get the proper relative humidities $RH \cong r / r_s$.

Finding T_w from T_d is a bit trickier, but is possible using **Normand's Rule**:

- Step 1: Find z_{LCL} using eq. (5.8)
- Step 2: $T_{LCL} = T - \Gamma_d \cdot z_{LCL}$ (5.12)
- Step 3: $T_w = T_{LCL} + \Gamma_s \cdot z_{LCL}$ (5.13)

where Γ_s is the moist lapse rate described later in this chapter, Γ_d = 9.8 K/km is the dry rate, and T_{LCL} is the temperature at the LCL. Normand's Rule tells us that $T_d \le T_w \le T$.

The relationship between wet-bulb temperature and other moisture variables are given in tables, such as in Appendix D of Ahrens. Such **psychrometric tables** are the most common method for utilizing wet-bulb temperatures. **Wet-bulb depression** ($T - T_w$) is a measure of the relative dryness of the air.

Solved Example

At a pressure of 100 kPa, what is the mixing ratio if the dry and wet bulb temperatures of a psychrometer are 20°C and 14°C, respectively?

Solution

Given: $T = 20°C$, $T_w = 14°C$, $P = 100$ kPa
Find: $r = ?$ g/kg

First, solve eq. (5.10): $r_w = $

$$\frac{0.622\text{g/g}}{(1.631\text{kPa}^{-1})\cdot(100\text{kPa})\cdot\exp\left(\frac{-17.67\cdot 14°C}{14°C + 243.5°C}\right) - 1}$$

r_w = 0.01013 g/g.
Then solve eq. (5.11):
r=(0.01013g/g)–[4.0224x10^{-4} (g/g)/°C]·(20–14°C)
= 0.00772 g/g = <u>7.72 g/kg</u>.

Check: Units OK. Physics OK.
Discussion: The corresponding dew-point temperature is 10°C.

Solved Example

Find the wet bulb temperature for air at sea level with temperature 30°C and dew point 24°C. Assume Γ_s = 3.78 °C/km for this case.

Solution

Given: $T = 30°C$, $T_d = 24°C$, $P = 101.3$ kPa
Find: $T_w = ?°C$

Step 1, solve eq. (5.8):
z_{LCL} = (0.125 km/°C)·(30 – 24°C) = 0.75 km.
Step 2, solve eq. (5.12):
T_{LCL} = 30 – (9.8 K/km)·(0.75 km) = 22.65 °C.
Step 3, solve eq. (5.13):
T_w = 22.65 + (3.78 °C/km)·(0.75 km) = <u>25.5°C</u>.

Check: Units OK. Physics OK.
Discussion: The resulting wet bulb depression is $(T - T_w)$ = 4.5°C, which agrees with the value listed in Table D.1 of Ahrens.

More Relationships Between Moisture Variables

Units of g/g must be used for r and q in the relationships below:

$$q = \frac{r}{1+r} \tag{5.14}$$

$$e = \frac{r}{\varepsilon + r}\cdot P \tag{5.15}$$

$$q = \frac{\rho_v}{\rho_d + \rho_v} \tag{5.16}$$

$$r = \frac{\rho_v}{\rho_d} \tag{5.17}$$

For air that is not excessively humid:

$$r_s \approx q_s \tag{5.18}$$

$$q \approx r = \varepsilon\cdot e / P \tag{5.19}$$

FOCUS • Why So Many Moisture Variables?

Vapor pressure, mixing ratio, specific humidity, absolute humidity, relative humidity, dew point depression, saturation level, and wet bulb temperature are different ways to quantify moisture in the air. Why are there so many?

Some variables are useful because they can be measured. Others are useful in conservation equations for water substance, for describing physical characteristics of the air, or for describing how life is affected by humidity. Often we are given or can measure one moisture variable, but must convert it to a different variable to use it.

Directly and easily measurable variables include wet bulb temperature and dew point temperature. Wet bulb temperature is easy to measure by putting a wet cloth sleeve around the bulb of a thermometer, and ventilating the wet bulb. However, it is difficult to convert this into other humidity variables, which is why psychrometric tables have been developed (Ahrens Appendix D). Dew point is measured by chilling a mirror until dew forms on it. It is one of the most accurate measurements of humidity.

Absolute humidity can be measured by shining electromagnetic radiation (infrared, ultraviolet, or microwave) across a path of humid air. The attenuation of the signal is a measure of the number of water vapor molecules along the path.

(*continued next column*)

FOCUS • Why So Many Variables? (continued)

Relative humidity can be measured by: 1) expansion or contraction of organic fibers such as human hairs; 2) the change in electrical conductivity of a carbon powder emulsion on a glass slide; and 3) by how it affects the capacitance across a plastic dielectric. Relative humidity is extremely important because it affects the rate of evaporation from plants, animals, and the soil.

Mixing ratio or specific humidity are not usually measured directly, but are extremely useful because they are conserved within air parcels moving vertically or horizontally without mixing. Namely, heating, cooling, or changing the pressure of an air parcel will not change the mixing ratio or specific humidity in unsaturated air.

Lifting condensation level, or saturation level, can be used to estimate the altitude of cloud base for convective (cumulus) clouds.

Vapor pressure is neither easily measured nor directly useful. However, it is important theoretically because it describes from first principles how saturation humidity varies with temperature.

TOTAL WATER MIXING RATIO

Define **total-water mixing ratio** r_T as the grams of water of all phases (vapor, liquid, and ice) per gram of dry air

$$r_T = r + r_L + r_i \qquad \bullet(5.20)$$

where r is the **water-vapor mixing ratio,** r_L is the **liquid-water mixing ratio** (grams of liquid water per gram of dry air), and r_i is the **ice mixing ratio** (grams of ice per gram of dry air). Similar total water variables are defined in terms of specific humidity and absolute humidity.

For negligible precipitation, eq. (5.20) is approximately:

$$r_T = r \qquad \text{for unsaturated air} \quad \bullet(5.21a)$$

$$r_T = r_s + r_L + r_i \qquad \text{for saturated air} \quad \bullet(5.21b)$$

where the water-vapor mixing ratio is assumed to equal its saturation value r_s when liquid or ice is present. The word **saturated air** means cloudy air, and **unsaturated air** means non-cloudy air.

Total water is nearly conserved, in the sense that the amount of water lost or created via chemical reactions is small compared to the amounts advected with the wind or associated with precipitation. Thus, the change of total water in a volume can be calculated by the amount of water of all phases that enters or leaves the volume.

Recall that saturation water vapor mixing ratio decreases as temperature decreases (see Table 5-1). Thus, given some initial amount of total water r_T, eq. (5.21b) says that liquid or solid water must increase as temperature decreases, in order that the sum of all water phases remain constant in saturated air.

Clouds or fog form when the total water content of the air exceeds the saturation (equilibrium) value. Because total water is an important factor in cloud and precipitation formation, budget equations for total water are presented next.

Solved Example

Saturated air at sea level with temperature 10°C contains 3 g/kg liquid water. Find the total water mixing ratio.

Solution

Given: $T = 10°C$, $r_L = 3$ g/kg
Find: $r_T = ?$ g/kg.
Assume: No ice.

Use Table 5-1 because it applies for sea level. Otherwise, solve equations or use thermo diag.
At $T = 10°C$, $r_s = 7.66$ g/kg from the table.
Use eq. (5.21b):
$r_T = r_s + r_L + r_i = 7.66 + 3 + 0 = $ **10.66 g/kg**

Check: Units OK. Physics OK.
Discussion: According to Table 5-1, the air would need to warm to 15°C to evaporate all liquid water

LAGRANGIAN WATER BUDGET

Water Conservation

If we follow an air parcel as it moves, and assume no turbulent mixing with the environment, then the Lagrangian water budget is

$$\frac{\Delta r_T}{\Delta t} = S^{**} \qquad (5.22)$$

where S^{**} is a Lagrangian net source of water. This source could be evaporation from adjacent liquid water, or it could be a loss of water falling from the parcel as precipitation (causing negative S^{**}).

If there is no source or loss, then there is no change of total water content. Phase changes must compensate each other to maintain constant total water. Thus, for no net source:

$$(r + r_i + r_L)_{initial} = (r + r_i + r_L)_{final} \quad \bullet(5.23a)$$

For the special case of no ice, this reduces to:

$$(r + r_L)_{initial} = (r + r_L)_{final} \quad \bullet(5.23b)$$

On a thermodynamic diagram, conservation of total water corresponds to following an isohume of constant mixing ratio as a parcel rises or descends. Thus, it is total water mixing ratio that is conserved, not dew point temperature.

Solved Example

An air parcel at sea level has a temperature of 20°C and a mixing ratio of 10 g/kg. If the air is cooled to 10 °C, what is the liquid water content?

Solution

Given: $T = 20°C$, $r = 10$ g/kg
Find: $r_L = ?\ g_{liq}/g_{air}$ at $T = 10°C$

Use Table 5-1 because it applies for sea level. Otherwise, solve equations or use a thermo diagram. Assume no other sources or sinks.

Initially:
$r_s = 14.91$ g/kg from the table, at $T = 20°C$.
Air is initially unsaturated, because $r < r_s$.
Unsaturated air has no liquid water: $r_L = 0$.
Assume no ice at these warm temperatures.
$(r + r_L)_{initial}$ = (10 + 0 g/kg) = 10 g/kg

Finally:
After cooling the air to 10°C, Table 5-1 shows that the max amount of vapor that can be held in the air at equilibrium is $r = r_s = 7.66$ g/kg.
However, total water must be constant if it is conserved. Use eq. (5.23b):
10 g/kg = $(r + r_L)_{final}$ = 7.66 g/kg + r_{Lfinal}
Thus,
r_{Lfinal} = 10 g/kg – 7.66 g/kg = **2.34 g/kg**

Check: Units OK. Physics OK.
Discussion: Some liquid water has formed in the air parcel, so it is now a cloud or fog.

Thermo Diagrams – Part 2: Isohumes

Two classes of information are included on thermo diagrams: **states** and **processes.** State of the air is defined by **temperature, pressure,** and **moisture.** Two processes are unsaturated (**dry**) and saturated (**moist**) adiabatic vertical movement, such as experienced by air parcels rising through the atmosphere. Processes cause the state to change.

The chart background of vertical **isotherms** and horizontal **isobars** was introduced in Chapters 1 (Fig 1.4) and 3 (Fig 3.2). Here we will overlay **isohumes** of constant saturation-mixing-ratio state over the background.

Plugging eq. (5.15) into (5.7) yields:

$$T_d = \left[\frac{1}{T_o} - \frac{\Re_v}{L_v}\cdot \ln\left\{\frac{r\cdot P}{e_o\cdot(r+\varepsilon)}\right\}\right]^{-1} \quad (5.24)$$

where $e_o = 0.611$ kPa, $T_o = 273$ K, $\varepsilon = 0.622$ g/g, and $\Re_v/L = 0.0001844\ K^{-1}$. Be sure to use g/g and not g/kg for r. The answer is in Kelvin.

Thus, for any fixed value of r, eq. (5.24) gives curves of T_d vs. P that can be plotted on the thermo diagram. These same curves can be used for T vs. P along constant values of r_s. Solving eq. (5.24) on a spreadsheet for several values of r yields Fig 5.4.

Here are some examples of use of this diagram. Air of temperature 0°C and pressure 40 kPa has a saturation mixing ratio of about 10 g/kg. Air of dew point 0°C and pressure 40 kPa has an actual mixing ratio of 10 g/kg. Air of mixing ratio 20 g/kg and pressure 80 kPa has a dew point temperature of 20°C.

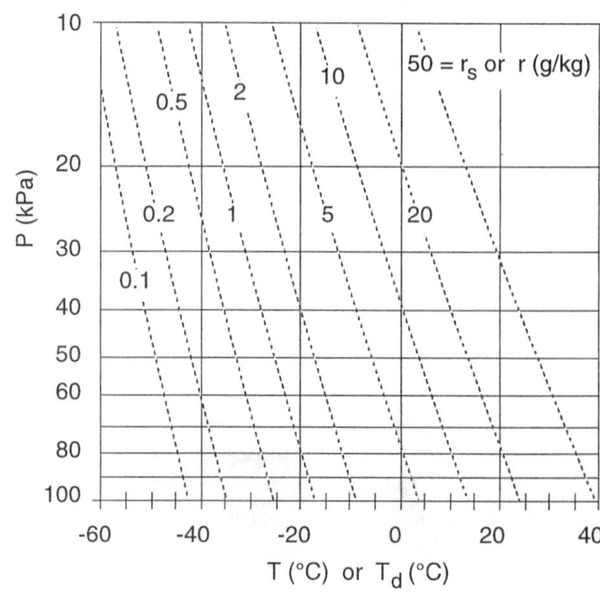

Figure 5.4
Thermodynamic diagram with isohumes of mixing ratio.

Solved Example
Calculate the dew point of air at pressure 80 kPa and mixing ratio 1 g/kg.

Solution
Given: P = 80 kPa, r = 1 g/kg
Find: T_d = ?°C

Use eq. (5.22):

$$T_d = \left[\frac{1}{273K} - 0.000184K^{-1} \cdot \ln\left\{\frac{(0.001g/g)\cdot(80kPa)}{0.611kPa\cdot(0.001+0.622g/g)}\right\}\right]^{-1}$$

= $[0.003663 - 0.0001844 \cdot \ln\{0.210\}]^{-1}$
= 253.12 K = **−20°C**

Check: Units OK. Physics OK.
Discussion: Agrees with Fig 5.4.

EULERIAN WATER BUDGET

Any change of total water content within a fixed Eulerian volume such as a cube must be explained by transport of water through the sides. Processes include advection by the mean wind, turbulent transport, and precipitation of liquid and solid water.

As was done for the heat budget, the moisture budget can be simplified for many atmospheric situations. In particular, the following terms are usually small except in thunderstorms, and will be neglected: vertical advection by the mean wind, horizontal turbulent transport, and conduction (except for effective moisture fluxes near the earth's surface). The net Eulerian total water (r_T) budget is:

$$\frac{\Delta r_T}{\Delta t} = \underbrace{-\left[U\cdot\frac{\Delta r_T}{\Delta x} + V\cdot\frac{\Delta r_T}{\Delta y}\right]}_{\text{advection}} + \underbrace{\left(\frac{\rho_L}{\rho_d}\right)\frac{\Delta Pr}{\Delta z}}_{\text{precipitation}} \underbrace{- \frac{\Delta F_{z\ turb}(r_T)}{\Delta z}}_{\text{turbulence}} \qquad \bullet(5.25)$$

where (U, V) are the horizontal wind components in the (x, y) Cartesian directions, z is height, Pr is precipitation rate, and $F_{z\ turb}(r_T)$ is the kinematic flux of total water. Most of these terms are similar to those in the heat budget eq. (3.34). The ratio of liquid-water density to dry-air density is ρ_L/ρ_d = 836.7 kg_{liq}/kg_{air} at STP. The liquid water density is ρ_L = 1025 kg_{liq}/m^3.

Each term is examined next. Surface moisture flux is also discussed because it contributes to the turbulence term for a volume of air on the surface.

Horizontal Advection

Horizontal advection of total water is similar to that for heat. Within a fixed volume, moisture increases with time if air blowing into the volume contains more water than air leaving. The advected water content includes water vapor (i.e., humidity), liquid, and solid cloud particles and precipitation.

Precipitation

Liquid water equivalent is the depth of liquid water that would occur if all solid precipitation were melted. For example, 10 cm depth of snow accumulated in a cylindrical container might be equivalent to only 1 cm of liquid water when melted.

Precipitation rate Pr has units of depth of liquid water equivalent per time (e.g., mm/hour, or m/s). It tells us how quickly the depth of water in a hypothetical rain gauge would change.

Solved Example
Consider a 9 km deep thunderstorm, and examine the bottom 1 km layer of the cloud. Rain from higher in the thunderstorm falls into the top of this layer at rate 0.9 cm/h. From the bottom of the layer, the precipitation rate is 1 cm/h. What is the rate of total water change in the layer?

Solution
Given: Pr_{top} = 0.9 cm/h = 2.5x10^{-5} m/s,
Pr_{bot} = 1.0 cm/h = 2.78x10^{-5} m/s,
Δz = 1 km = 1000 m
Find: $\Delta r_T/\Delta t$ = ? (g/kg)/s.
Assume: No advection or turbulence. Assume the average air density is 1 kg/m^3 in lower troposph.

Use eq. (5.25):
$\Delta r_T/\Delta t = (\rho_L/\rho_d)\cdot(Pr_{top}-Pr_{bot})/(z_{top}-z_{bot})$

$$= \frac{(1.025\times10^6 g_{water}/m^3)}{(1\ kg_{air}/m^3)} \cdot \frac{[(2.5-2.78)\times10^{-5} m/s]}{1000m}$$

= **−0.00287 (g/kg)/s**

Check: Units OK. Physics OK.
Discussion: Rate = −10.33 (g/kg)/h loss of water from this part of the cloud. If there is no advection of moisture to replace that lost as rain, then this part of the cloud will rain itself dry in several hours if loss continues at a constant rate.

Precipitation rate is a function of altitude, as if the rain gauge could be hypothetically mounted at any altitude in the atmosphere. Some precipitation might evaporate when falling through the air, or condensation might increase the precipitation rate on the way down. If more precipitation falls into the top of a volume than falls out of the bottom, then total water within the volume must increase. At the earth's surface, precipitation rate equals **rainfall rate** *RR* by definition.

Surface Moisture Flux

The latent heat flux described in Chapter 3 is, by definition, associated with a flux of water. Hence, the heat and moisture budgets are coupled. Methods to calculate the latent heat flux were described in that chapter under the topic of the surface heat budget. Knowing the surface latent heat flux, you can use the equations below to transform it into a water flux.

The vertical flux of water vapor $\Im_{water}$ ($kg_{water}\cdot m^{-2}\cdot s^{-1}$) is related to the latent heat flux $\Im_E$ (W/m^2) by:

$$\Im_{water} = \frac{\Im_E}{L_v} \qquad \bullet(5.26a)$$

$$\Im_{water} = \rho_{air} \cdot \frac{C_p}{L_v} \cdot F_E = \rho_{air} \cdot \gamma \cdot F_E \qquad (5.26b)$$

where the latent heat of vaporization is $L_v = 2.5\text{x}10^6$ J/kg, $\gamma = C_p/L_v = 0.4\ (g_{water}/kg_{air})\cdot K^{-1}$ is the **psychrometric constant**, and where the kinematic latent heat flux F_E has units of $K\cdot m\cdot s^{-1}$.

Eq. (5.26b) can be put into kinematic form by dividing by air density, ρ_{air}.

$$F_{water} = \frac{\Im_E}{\rho_{air} \cdot L_v} = \gamma \cdot F_E \qquad (5.27)$$

where F_{water} is like a mixing ratio times a vertical velocity, and has units of $(kg_{water}/kg_{air})\cdot(m\cdot s^{-1})$.

Sometimes, this moisture flux is expressed as an **evaporation rate** *Evap* in terms of depth of liquid water that is lost per unit time (e.g., mm/day):

$$Evap = \frac{\Im_{water}}{\rho_{liq}} = \frac{\Im_E}{\rho_{liq} \cdot L_v} = a \cdot \Im_E \qquad (5.28a)$$

$$Evap = \frac{\rho_{air}}{\rho_{liq}} \cdot F_{water} = \frac{\rho_{air}}{\rho_{liq}} \cdot \gamma \cdot F_E \qquad \bullet(5.28b)$$

where $a = 3.90\text{x}10^{-10}\ m^3\cdot W^{-1}\cdot s^{-1}$ or $a = 0.0337$ $(m^2/W)\cdot$ (mm/day), and $\rho_L = 1025\ kg_{liq}/m^3$.

Solved Example

Given a latent heat flux of 300 $W\cdot m^{-2}$, find the evaporation rate.

Solution

Given: $\Im_E$= 300 $W\cdot m^{-2}$

Find: *Evap* = ? mm/day

Use eq. (5.28a): $Evap = a \cdot \Im_E$

= [0.0337 $(m^2/W)\cdot$(mm/day)]$\cdot$(300 W/m^2)

= **10.11 mm/day**

Check: Units OK. Physics OK.

Discussion: Evaporation of about 1 cm of water per day could cause significant drying of soil and lowering of water levels in lakes unless replenished by precipitation.

If the latent heat flux is not known, you can estimate F_{water} from the mixing ratio difference between the surface (r_{sfc}) and the air (r_{air}). For windy, overcast conditions, use:

$$F_{water} = C_H \cdot M \cdot \left(r_{sfc} - r_{air}\right) \qquad (5.29)$$

where C_H is the bulk transfer coefficient (moisture and heat assumed to be identical, see eq. 3.21), and M is wind speed. Both r_{air} and M are measured at z = 10 m.

Solved Example

Find the kinematic water flux over a lake with saturation mixing ratio of r_{sfc} = 15 g/kg. The air is overcast with wind speed of 10 m/s and mixing ratio of 8 g/kg at z = 10 m

Solution

Given: r_{sfc} = 15 g/kg, M = 10 m/s,
r_{air} = 8 g/kg

Find: F_{water} = ? $(kg_{water}/kg_{air})\cdot(m\cdot s^{-1})$

Assume: smooth lake surface: $C_H = 2\text{x}10^{-3}$ (from eq. 3.21),

Use eq. (5.29): $F_{water} = C_H \cdot M \cdot \left(r_{sfc} - r_{air}\right)$

= ($2\text{x}10^{-3}$)$\cdot$(10 m/s)$\cdot$(15 – 8 g/kg)

= 0.14 $(g_{water}\cdot kg_{air})\cdot(m\cdot s^{-1})$

= **0.00014** $(kg_{water}/kg_{air})\cdot(m\cdot s^{-1})$

Check: Units OK. Physics OK.

Discussion: Eq. (5.28b) says this is equivalent to $1.64\text{x}10^{-7}$ m/s of evaporation, or **14.2 mm/day**, which is a strong evaporation rate.

During free convection of sunny days and light or calm winds:

$$F_{water} = b_H \cdot w_B \cdot \left(r_{sfc} - r_{ML}\right) \qquad (5.30)$$

where r_{ML} is the mixing ratio in the middle of the mixed layer, and $b_H = 5\text{x}10^{-4}$ is the **convective transport coefficient** for heat (assumed to be identical to that for moisture, see eq. 3.22). The buoyancy velocity scale was given earlier (eq. 3.23) for convective conditions.

These formula are difficult to use because of r_{sfc}. Over lakes and saturated ground, it is assumed equal to the saturation mixing ratio for the temperature of the surface skin. Over drier ground the value is less than saturated, but is not precisely known.

Turbulent Transport

The turbulent transport term tends to homogenize moisture by mixing moist and dry regions of air to create a larger region of average humidity. $\Delta F_{z\ turb}(r_T)$ is the change of turbulent flux of total water across distance Δz, and is analogous to the turbulence heat flux discussed in Chapt 3. $\Delta F / \Delta z$ is called the **flux divergence**.

In the fair-weather, daytime, convective boundary layer, turbulence causes a flux divergence in the vertical:

$$\frac{\Delta F_{z\ turb}(r_T)}{\Delta z} \equiv \frac{F_{z\ turb\ at\ zi} - F_{z\ turb\ at\ surface}}{z_i - z_{surface}} \qquad (5.31a)$$

or $\qquad$ for $0 < z < z_i$

$$\frac{\Delta F_{z\ turb}(r_T)}{\Delta z} = -\frac{\left[0.2\cdot(\Delta_{z_i} r_T / \Delta_{z_i}\theta)\cdot F_H\right] + F_{water}}{z_i} \qquad (5.31b)$$

where F_H is the kinematic effective surface heat flux (units of K·m/s), and F_{water} is the kinematic effective surface water vapor flux [units of $(\text{kg}_{water}/\text{kg}_{air})\cdot$ (m/s)]. The operator $\Delta_{zi} = (\ )_{zi+} - (\ )_{zi-}$ is the difference between the value of any variable ( ) just above the top of the boundary layer, minus the value just below the top (Fig 5.5), where the average top of the boundary layer is at height z_i.

Because surface moisture flux F_{water} is directly related to the surface latent heat flux F_E (eq. 5.27), one can rewrite (5.31b) as:

$$\frac{\Delta F_{z\ turb}(r_T)}{\Delta z} = -\frac{0.2\cdot(\Delta_{z_i} r_T / \Delta_{z_i}\theta)\cdot F_H + \gamma\cdot F_E}{z_i}$$

for $0 < z < z_i$ (5.31c)

for the convective boundary layer, where $\gamma = 0.4$ $(\text{g}_{water}/\text{kg}_{air})\cdot\text{K}^{-1}$ is the psychrometric constant.

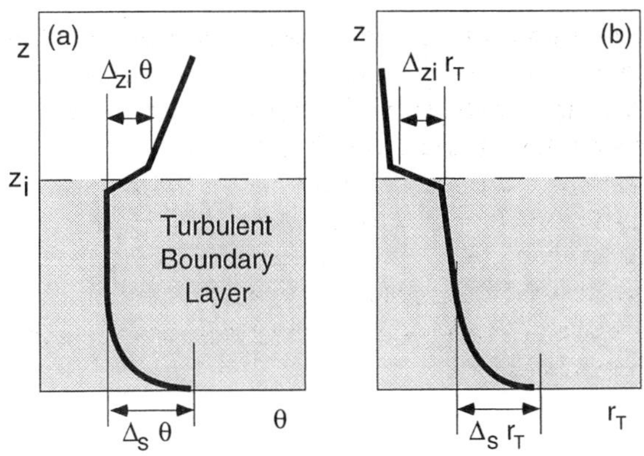

Figure 5.5
Method of estimating jumps in (a) potential temperature $\Delta\theta$ and in (b) total water Δr_T at the top and bottom of a turbulent boundary layer of depth z_i, given the sounding (thick lines). For moisture, the jump $\Delta_{zi}\ r_T$ can be positive or negative depending on whether the air aloft is moister or drier than the boundary layer. Only the case of drier air aloft is shown in (b).

Another version can be written using the effective surface heat flux parameterization of Chapt. 3:

$$\frac{\Delta F_{z\ turb}(r_T)}{\Delta z} = -\frac{b_H \cdot w_B}{z_i}\left\{\left[0.2\cdot\left(\Delta_{z_i} r_T\right)\cdot\left|\frac{\Delta_s\theta}{\Delta_{z_i}\theta}\right|\right] + \left|\Delta_s r_T\right|\right\}$$

for $0 < z < z_i$ (5.31d)

where b_H is the convective transport coefficient (see Chapt. 3), w_B is the buoyancy velocity scale, and operator $\Delta_s(\ ) = (\ )_{surface} - (\ )_{mid\ mixed\text{-}layer}$. The three equations above are equivalent — use the version that is most convenient for the values on hand.

Above the fair-weather boundary layer (i.e., for $z > z_i$), turbulence is often negligible:

$$\frac{\Delta F_{z\ turb}(r_T)}{\Delta z} \approx 0 \qquad (5.32)$$

Also, within the nighttime boundary layer, the equation above is valid at all heights.

For some situations, the moisture flux (F_{water}) at the earth's surface into or out of the bottom of a volume of air might be known. A cube of air sitting on a lake surface might gain total water by evaporation from the lake, resulting from an upward or positive flux of water at the surface. Alternately, a cube of humid air sitting over a snow surface might lose water by condensation onto the snow, corresponding to a negative or downward flux of water at the surface. If the flux at the top of the volume is near zero, then the vertical flux divergence is approximately $\Delta F_{z\ turb}(r_T)/\Delta z = -F_{bottom}(r_T)/d$, where d is depth of the volume.

Turbulent mixing of moisture by thunderstorms is vigorous and complex. Prior to the storm, a humid boundary layer underlies a dry layer of air in the middle troposphere. The thunderstorm lifts and consumes the humid air from the boundary layer, and replaces it with drier air from aloft. However, much of the moisture from the former boundary layer is squeezed out as precipitation, and is accounted for in the precipitation term of the Eulerian moisture budget. The net result is that the humid, prestorm boundary layer becomes drier after the storm.

Solved Example

Given measurements of a 1 km thick boundary layer as shown in Fig 5.5, with $|\Delta_s\theta| = 8°C$, $|\Delta_{zi}\theta| = 3°C$, $|\Delta_s r_T| = 5$ g/kg, and $\Delta_{zi} r_T = -4$ g/kg. The convective transport coefficient times buoyancy velocity is 0.025 m/s on this calm day. What is the vertical flux divergence of total water due to turbulence, and how does that contribute to moistening the boundary layer?

Solution

Given: $|\Delta_s\theta| = 9°C$, $|\Delta_{zi}\theta| = 3°C$,
$|\Delta_s r_T| = 5$ g/kg, $\Delta_{zi} r_T = -4$ g/kg,
$z_i = 1$ km $= 1000$ m, $b_H \cdot w_B = 0.025$ m/s.
Find: $\Delta r_T/\Delta t = -\Delta F_{z\ turb}(r_T)/\Delta z = ?$ (g/kg)/s
Assume: No advection or precipitation.

Use eq. (5.31d):

$$\frac{\Delta F_{z\ turb}(r_T)}{\Delta z} = -\frac{b_H \cdot w_B}{z_i}\left[0.2\cdot\left(\Delta_{z_i} r_T\right)\cdot\left|\frac{\Delta_s\theta}{\Delta_{z_i}\theta}\right| + \left|\Delta_s r_T\right|\right]$$

$$= -\frac{(0.025\text{m / s})}{(1000\text{m})}\left[0.2\cdot\left(-4\frac{\text{g}}{\text{kg}}\right)\cdot\left|\frac{9°\text{C}}{3°\text{C}}\right| + \left(5\frac{\text{g}}{\text{kg}}\right)\right]$$

$= \underline{-0.000065\ \text{(g/kg)/s}}$

Thus:

$\Delta r_T/\Delta t = -\Delta F_{z\ turb}(r_T)/\Delta z = \underline{+0.000065\ \text{(g/kg)/s}}$

Check: Units OK. Physics OK.

Discussion: This is equivalent to a rate of +0.234 (g/kg)/h increase of water in the boundary layer. The air is getting juicier with time.

Water evaporates from the surface to increase the humidity, but dry air entrained into the top of the boundary layer tends to decrease humidity. For this example, the moistening from the surface was greater than the drying from aloft, but it could be different in real life. Also, there can be other situations where moistening happens from both the top and bottom of the boundary layer, such as when a dry boundary layer is growing into a cloud layer aloft.

ON DOING SCIENCE • Look for Patterns

Suppose you know that $A = C$, and $C = E$. You can use this knowledge to yield new understanding, by various methods of reasoning.

Deductive reasoning combines known relationships in new ways to allow new applications. For example, using the knowns from the previous paragraph, you could deduce that $A = E$.

Inductive reasoning goes from the particular to the general. It does this by looking for patterns. By such reasoning, you could state that $E = G$. The pattern is that A equals a value that is two letters later in the alphabet, and C equals a value two letters later. By extension, you might hope that any letter of the alphabet equals a value that is two letters later in the alphabet.

Inductive reasoning is riskier than deductive. If the knowns you use as a starting point are unrepresentative of the whole situation, then you will get a biased (i.e., wrong) result. For example: given that I have a sister, and that my best friend has a sister, therefore by inductive reasoning everyone has a sister. To avoid such blunders, it helps to get a large sample of knowns before you extend the pattern into the unknown.

Inductive reasoning usually gives the greatest advances in science, while deductive reasoning gives the greatest utility to engineering. Both forms of reasoning are encouraged.

Regardless of the form of reasoning, you should test your conclusions against independent data. Isaac Newton used inductive reasoning to develop his concepts of motion and gravity, based on observations he made near his home in England. To test his theories with independent data, he built telescopes to watch the motion of planets. Because his theories were found to be **universal** (i.e., they worked elsewhere in the universe), Newton's theories were accepted as the Laws of Motion.

Similarly, engineers might use deductive reasoning to design a bridge, dam, or airplane. However, they first test their design with prototypes, scale models, or computer simulations to support their deductions before they construct the final products.

As an example, we can use reasoning with the Eulerian budgets for heat and moisture. From Chapter 3, the Eulerian heat budget was:

$$\left.\frac{\Delta T}{\Delta t}\right|_{x,y,z} = \underbrace{-\left[U\cdot\frac{\Delta T}{\Delta x} + V\cdot\frac{\Delta T}{\Delta y}\right]}_{\text{advection}} + \underbrace{\frac{L_v}{C_p}\cdot\frac{m_{condensing}}{m_{air}\cdot\Delta t}}_{\text{latent heat}}$$

$$\underbrace{-0.1\ \frac{\text{K}}{\text{h}}}_{\text{radiation}} \ \underbrace{-\frac{\Delta F_{z\ turb}}{\Delta z}}_{\text{turbulence}}$$

(*continued next column*)

ON DOING SCIENCE (continuation)

From the current chapter, the Eulerian moisture budget is:

$$\frac{\Delta r_T}{\Delta t} = \underbrace{-\left[U\cdot\frac{\Delta r_T}{\Delta x} + V\cdot\frac{\Delta r_T}{\Delta y}\right]}_{\text{advection}} + \underbrace{\left(\frac{\rho_L}{\rho_d}\right)\frac{\Delta Pr}{\Delta z}}_{\text{precipitation}} \underbrace{-\frac{\Delta F_{z\ turb}(r_T)}{\Delta z}}_{\text{turbulence}}$$

Both equations above have the same pattern; namely, a time-tendency term on the left, and advection and turbulence terms on the right. **Inductive** reasoning would allow us to state that conservation or budget equations in general should be of the form:

$$\frac{\Delta\Psi}{\Delta t} = \underbrace{-\left[U\cdot\frac{\Delta\Psi}{\Delta x} + V\cdot\frac{\Delta\Psi}{\Delta y}\right]}_{\text{advection}} \underbrace{-\frac{\Delta F_{z\ turb}(\Psi)}{\Delta z}}_{\text{turbulence}} + S_\Psi$$

where Ψ is any variable, and S_Ψ represents other terms. Mean vertical motion and horizontal turbulence flux divergence were neglected.

Given the equation above, we could use **deductive** reasoning to apply it to a new variable, such as the U component of velocity:

$$\frac{\Delta U}{\Delta t} = \underbrace{-\left[U\cdot\frac{\Delta U}{\Delta x} + V\cdot\frac{\Delta U}{\Delta y}\right]}_{\text{advection}} \underbrace{-\frac{\Delta F_{z\ turb}(U)}{\Delta z}}_{\text{turbulence}} + S_U$$

This would then need to be tested.

LAGRANGIAN HEAT BUDGET — PART 2: SATURATED

Saturated Adiabatic Lapse Rate

The unsaturated (dry) adiabatic lapse rate, Γ_d = 9.8 K/km, was discussed earlier. It is the temperature decrease ($\Delta T/\Delta z = -\Gamma_d$) an air parcel would experience as it moves vertically in the atmosphere into regions of lower pressure. No mixing or heat transfer across the skin of the parcel can occur for an adiabatic process. Although the word "dry" appears, this lapse rate applies to moist as well as dry air, but only if the air is unsaturated (i.e., not cloudy). Air cools adiabatically as it rises, and warms as it descends.

Saturated (cloudy) air also cools as it rises and warms as it descends, but at a lesser rate than a dry parcel. The reason is that, by definition, a saturated parcel is in equilibrium with the liquid drops. If the parcel then rises and cools, it is no longer in equilibrium, and some of the water molecules condense as a new equilibrium is approached. During condensation, latent heat is released into the air, partially compensating the adiabatic cooling.

The opposite is true during descent. Some of the cloud droplets evaporate and remove heat from the air, partially compensating the adiabatic warming.

For saturated air, the decrease of temperature with height ($-\Delta T/\Delta z = \Gamma_s$) in clouds and fog is given by the **saturated-adiabatic lapse rate,** Γ_s:

$$\Gamma_s = \frac{g}{C_p}\cdot\frac{\left(1+\dfrac{r_s\cdot L_v}{\Re_d\cdot T}\right)}{\left(1+\dfrac{L_v^2\cdot r_s\cdot\varepsilon}{C_p\cdot\Re_d\cdot T^2}\right)} \qquad (5.33a)$$

This is sometimes called the **moist lapse rate.**

If the variation of specific heat with humidity is neglected for simplicity, then (5.33a) simplifies to:

$$\Gamma_s = \Gamma_d\cdot\frac{[1+(a\cdot r_s/T)]}{[1+(b\cdot r_s/T^2)]} \qquad \bullet(5.33b)$$

where Γ_d = 9.8 K/km, a = 8711 K, and b = 1.35x10^7 K^2. In both equations the saturation mixing ratio r_s varies with temperature, and must be used in units of g/g. Also, temperature must be Kelvin.

Near the ground in humid conditions, the value of the moist adiabatic lapse rate is near 4 K/km, and often changes to 6 to 7 K/km in the middle of the troposphere. At high altitudes where the air is colder and holds less water vapor, the moist rate nearly equals the dry rate of 9.8 K/km.

The moist lapse rate can be written as:

$$\frac{\Delta T}{\Delta P} = \frac{\left[(\Re_d/C_p)\cdot T + (L_v/C_p)\cdot r_s\right]}{P\cdot\left(1+\dfrac{L_v^2\cdot r_s\cdot\varepsilon}{C_p\cdot\Re_d\cdot T^2}\right)} \qquad (5.34a)$$

The corresponding simplified form is:

$$\frac{\Delta T}{\Delta P} = \frac{[a\cdot T + c\cdot r_s)]}{P\cdot[1+(b\cdot r_s/T^2)]} \qquad \bullet(5.34b)$$

where a = 0.28571, b = 1.35x10^7 K^2, c = 2488.4 K.

Solved Example

Find the moist adiabatic lapse rate near sea level for $T = 26°C$, using (a) the actual specific heat, and (b) the specific heat for dry air.

Solution

Given: $T = 26°C = 299$ K , $P = 101.325$ kPa
Find: Γ_s = ? °C/km

In general, we would need to calculate the saturation mixing ratio using eqs. (5.3) and (5.1). We can use Table 5-1 because sea-level pressure was specified. Thus: $r_s = 21.85$ g/kg = 0.0219 g/g.

(a) Using eq. (3.2), the actual specific heat is:
$C_p = 1023\ \text{J·kg}^{-1}\text{·K}^{-1}$.
Thus: $g/C_p = 9.58$ K/km
$L_v/C_p = 2444$ K.
$L_v/\Re_d = 8711$ K.
Plugging these into eq. (5.33a) gives:

$$\Gamma_s = \frac{(9.58\frac{\text{K}}{\text{km}})\cdot\left[1+\frac{(0.0219\text{g/g})\cdot(8711\text{K})}{299\text{K}}\right]}{\left[1+\frac{(2444\text{K})\cdot(8711\text{K})\cdot(0.0219\text{g/g})\cdot 0.622}{(299\text{K})^2}\right]}$$

$= (9.58\ \text{K/km})\cdot[1.638]/[4.307] =$ **3.70 K/km**

(b) Eq. (5.33b) uses C_p for dry air:

$$\Gamma_s = \frac{(9.8\frac{\text{K}}{\text{km}})\cdot\left[1+\frac{(0.0219\text{g/g})\cdot(8711\text{K})}{299\text{K}}\right]}{\left[1+\frac{(1.35\times 10^7\text{K}^2)\cdot(0.0219\text{g/g})}{(299\text{K})^2}\right]}$$

$= (9.8\ \text{K/km})\cdot[1.638]\ /\ [4.307] =$ **3.73 K/km**.

Check: Units OK. Physics OK.
Discussion: As expected the rate of cooling of a rising cloudy air parcel is less than the 9.8 K/km of a dry parcel. Also, method (b) is accurate enough for many situations.

Thermo Diagrams – Part 3: Saturated Adiabats

Recall that two processes shown in thermo diagrams are unsaturated (**dry**) and saturated (**moist**) adiabatic vertical movement, such as experienced by air parcels rising through the atmosphere. Unsaturated processes were previously indicated by overlaying isentropes (**dry adiabats**) on the background (Chapt 3, Fig 3.2). Here we will add **saturated adiabats**, also known as **moist adiabats**.

Moist adiabats show how cloudy air changes temperature as it rises or sinks. Unfortunately, eq. (5.34) does not directly give the temperature of a parcel at any pressure height. Instead, one must solve the equation recursively, stepping upward from an initial temperature at $P = 100$ kPa. Each step must take a small-enough pressure-height increment that the moist lapse rate does not change significantly. For each step, one can solve:

$$T_2 = T_1 + \frac{\Delta T}{\Delta P}\cdot(P_2 - P_1) \tag{5.35}$$

where $(\Delta T/\Delta P)$ is from eq. (5.34).

We will use pressure height increments of $\Delta P = (P_2 - P_1) = -2$ kPa all the way from $P = 100$ kPa up to $P = 10$ kPa. Five starting temperatures at $P = 100$ kPa were used: $T_1 = 40, 20, 0, -20,$ and $-40°C$. Fig 5.6 below shows the results of such calculations on a spreadsheet. The calculations were done in units of Kelvin, and then converted to Celsius for plotting.

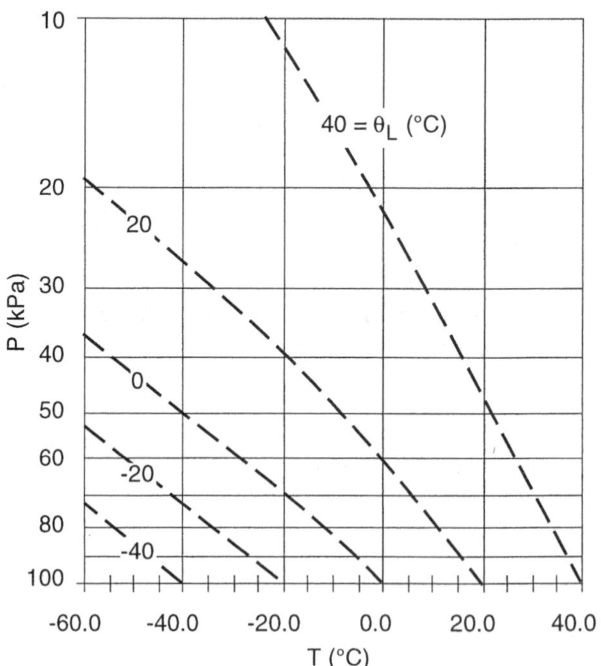

Figure 5.6
Thermodynamic diagram with moist adiabats, labeled by liquid-water potential temperature.

FOCUS • Spreadsheet Thermodynamics 2

Saturated (Moist) Adiabatic Lapse Rate

The iteration process of computing the θ_L = 20°C moist adiabat is demonstrated here.

First, set up a spreadsheet with the following row and column headers:

	A	B	C	D	E
1	**Moist Adiabat Example**				
2	ΔP(kPa)=	2			
3					
4	P (kPa)	T (°C)	es (kPa)	rs (g/g)	ΔT/ΔP
5	100	20.0			

where the starting pressure and temperature have been entered as numbers, and where the pressure increment of ΔP = 2 kPa has also been entered.

Because $\Delta T/\Delta P$ in eq. (5.34b) depends on r_S, which in turn depends on e_S, there are the two extra columns C and D in the spreadsheet above. Two typical mistakes are to forget to convert T into Kelvins, or to forget to use g/g for r.

In cell C5, you can enter eq. (5.1), which is:

```
=0.611*EXP(5423*((1/273)-(1/(B5+273))))
```

Then, in cell D5, enter eq. (5.3): `=0.622*C5/(A5-C5)`

Next in cell E5, enter eq. (5.34b):

```
=(0.28571*(B5+273)+(2488.4*D5)) /
(A5*(1+(13500000*D5/((B5+273)^2))))
```

In cell A6, increment the pressure: `=A5-$B$2`

Finally, in cell B6 enter eq. (5.35): `=B5-(E5*$B$2)`

where the previous 2 eqs referred to B2 as an absolute reference (B2 on some spreadsheets).

Finally, select cells C5:E5 and fill them down one row. Then, select all of row 6; namely cells A6:E6, and fill down about 40 or 45 rows. The result is:

	A	B	C	D	E
1	**Moist Adiabat Example**				
2	ΔP(kPa)=	2			
3					
4	P (kPa)	T (°C)	es (kPa)	rs (g/g)	ΔT/ΔP
5	100	20.0	2.371	0.0151	0.359
6	98	19.3	2.265	0.0147	0.369
7	96	18.5	2.162	0.0143	0.378
•••					
44	22	-52.2	0.006	0.0002	2.765
45	20	-57.7	0.003	0.0001	3.006

Discussion: Columns A and B can be plotted as the 20°C moist adiabat. For other moist adiabats, duplicate these cells, but with different starting T.

Liquid-Water and Equivalent Potential Temperatures

There are a variety of methods for identifying and labeling moist adiabats. In this book, we will label lines with values of **liquid-water potential temperature**, θ_L, which is the starting temperature of our calculations at P = 100 kPa. In other words, θ_L is the temperature that a *saturated* parcel would have when lowered or raised *moist* adiabatically to a pressure height of P = 100 kPa.

This is analogous to the labeling scheme for the dry adiabats, where potential temperature θ was used as the label. It is the temperature that an *unsaturated* parcel would have when lowered *dry* adiabatically to a height where P = 100 kPa.

Liquid-water potential temperature is found from:

$$\theta_L = \theta - \left(\frac{L_v \cdot \theta}{C_p \cdot T}\right) \cdot r_L \qquad \bullet(5.36)$$

or:

$$\theta_L \approx T + \frac{g}{C_p} \cdot z - \left(\frac{L_v \cdot \theta}{C_p \cdot T}\right) \cdot r_L \qquad (5.37)$$

where liquid water mixing ratio has units of $(g_{liq.water}/g_{air})$.

Solved Example

Compare on a thermo diagram dry and moist adiabats starting at T = 20°C at P = 100 kPa.

Solution

Given: T = 20°C at P = 100 kPa.

Plot: θ = 20°C and θ_L = 20°C.

Use curves from Fig 5.6 and Fig 3.2.

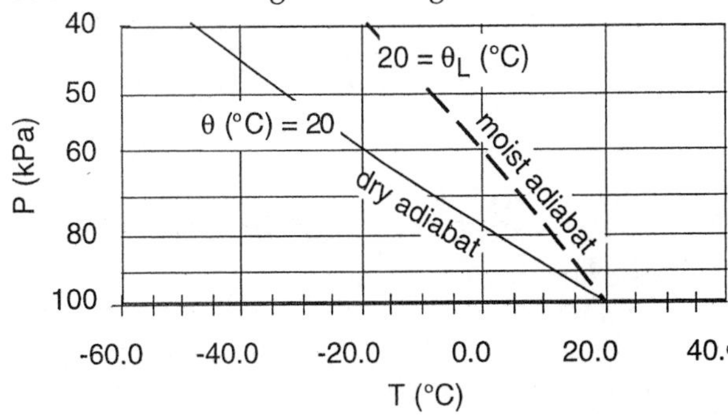

Check: Slopes look reasonable. Labels match.

Discussion: Temperature of the moist adiabat decreases much more slowly than that for the dry adiabat, because condensational latent heating partially compensates for the adiabatic cooling.

Historically, another common label for moist adiabats has been the **equivalent potential temperature** θ_e. This is the potential temperature that a saturated air parcel would have if raised moist adiabatically to the top of the atmosphere (where $P = 0$). Equivalent potential temperature is related to liquid-water potential temperature by:

$$\theta_e \approx \theta_L + \frac{L_v}{C_p} \cdot r_T \qquad (5.38)$$

where the total water mixing ratio (g/g) is $r_T = r + r_L$. Regardless how the line is labeled, it is still the same line given by eqs. (5.33) or (5.34).

Solved Example

Air at height 2 km and pressure 80 kPa has temperature 30°C. If the total water mixing ratio is 40 g/kg, find the liquid water and equivalent potential temperatures.

Solution

Given: P = 80 kPa, T = 30°C, z = 2 km, r_T = 40 g/kg

Find: θ_L = ?°C, θ_e = ?°C

First, determine if the air is saturated. From the thermo diagram of Fig 5.4 (or more accurate thermo diagrams in Chapt. 6, or from from eqs. 5.1 and 5.3), $r_s \approx 35$ g/kg. Given $r_T = 40$ g/kg, we see that $r_T > r_s$, therefore the air IS saturated, and the amount of liquid water is $r_L = r_T - r_s$ = (40 g/kg) – (35 g/kg) = 5 g/kg.

First, use eq. (3.9): $\theta = T + \Gamma_d \cdot z$

θ = 30°C + (9.8 K/km)·(2 km) = 49.6 °C

Then use eq. (3.2):

$C_p = C_{pd} \cdot (1 + 0.84 \cdot r) = (1004.67 \text{ J·kg}^{-1}\text{·K}^{-1}) \cdot [1 + 0.84 \cdot (0.036 \text{ g/g})] = 1035 \text{ J·kg}^{-1}\text{·K}^{-1}$

Next, use eq.(5.36): $\theta_L = \theta - (L_v / C_P) \cdot (\theta / T) \cdot r_L$

θ_L = 49.6°C–[(2.5x10^6 J·kg^{-1})/(1035 J·kg^{-1}·K^{-1})] ·[(49.6+273)/(30+273)]·(0.005 g/kg) = 49.6–12.9 °C

θ_L = **36.7°C**

Finally, use eq. (5.38):

θ_e = 36.7°C+2415·(0.040) = 133°C = **406K**

Check: Units OK. Physics OK.

Discussion: These same answers could have been found more easily using the thermo diagram.

Solved Example

Use Normand's rule on a thermodynamic diagram to find the mixing ratio and dew point temperature, given dry and wet-bulb observations of 40°C and 20°C, respectively, from a psychrometer near the surface (where P = 100 kPa).

Solution

Given: T = 40°C, T_w = 20°C, P = 100 kPa.

Find: T_d = ?°C, r = ? g/kg

Combining all the components of the thermo diagram (which will be covered in more detail in Chapter 6) yields the figure below. First, the given values of T and T_w at the given P are plotted as the two large points at the bottom of the Figure.

Starting from the T observation, follow a dry adiabat upward. Simultaneously, starting from the T_w observation, follow a moist adiabat upward. From the point where those two adiabats cross (also marked in this diagram), follow (or go parallel to) the isohumes back down to the starting pressure to find the T_d. Alternately, from the intersection point, follow the isohumes upward to find r (interpolating if necessary between the printed values).

The results are r = **8 g/kg**, and T_d = **10°C**.

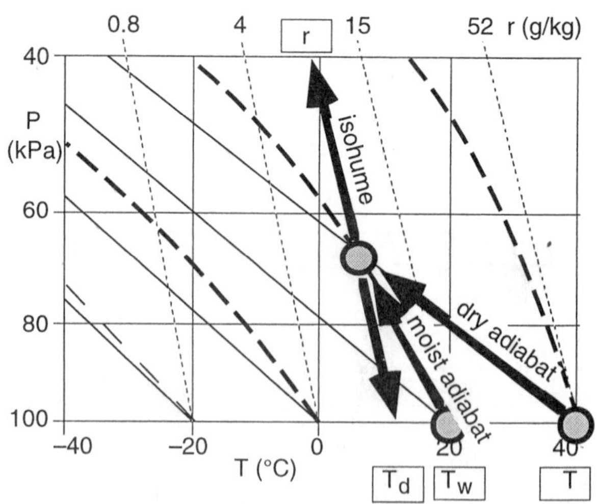

Check: Units OK. Physics OK.

Discussion: Using r_s = 52 g/kg from the starting T, the relative humidity is RH = 8/52 = 15%.

SUMMARY

The saturation (equilibrium) vapor pressure of warm air is greater than that of cool air. The Clausius-Clapeyron equation and Tetens' formula quantify this relationship.

Some relative moisture variables compare the actual amount of water vapor in the air to the saturation amount for that temperature: relative humidity, temperature dew-point spread, wet-bulb depression, and condensation level. Other variables measure the concentration of water in the air: vapor pressure, mixing ratio, specific humidity, and absolute humidity.

The amount of water in the air is described by a total water budget. Budgets can be formed following a cloud or air parcel (Lagrangian), and for a fixed volume (Eulerian). As saturated (cloudy) air rises, its temperature cools at the moist adiabatic lapse rate, which is less than the dry rate. These moist adiabats can be drawn on thermodynamic diagrams. Using such diagrams can often be quicker than calculating complex recursive equations for the moist lapse rate.

Isohumes can also be drawn on thermodynamic diagrams, to show the actual or maximum amount of moisture in the air. For many formulae and for thermodynamic diagrams, actual air temperature is related to equilibrium (saturation) amount of water vapor. Alternately, dew-point temperature is related to actual amount of water vapor in the air.

Threads

Isohumes and moist adiabats are used in Chapts. 6 and 7 to create more detailed thermodynamic diagrams to help understand cloud growth. Saturation mixing ratio and vapor pressure determine the growth of individual cloud particles leading to precipitation (Chapt. 8). Some of the energy for cyclones (Chapt. 13), thunderstorms (Chapt. 15) and hurricanes (Chapt. 16) comes from latent heat release, which again depends on the humidity and saturation values in the air. Global warming associated with climate change (Chapt. 18) affects the amount of condensation via the Clausius-Clapeyron equation.

Temperature and pressure of the standard atmosphere of Chapt. 1 were used to find boiling temperatures. The thermodynamic diagram of Chapt. 3 was extended here to include isohumes and moist adiabats.

Routine forecasts of humidity are made using numerical weather prediction (Chapt. 14). The Lagrangian water budget is used in thermodynamic diagrams to determine thunderstorm and hurricane development (Chapts. 15 and 16).

EXERCISES

Numerical Problems

N1. Which has the greatest r at P = 100 kPa?
- a. T = -10°C with RH = 100%
- b. T = +20°C with RH = 20%
- c. T = +5°C with RH = 50%
- d. T = -10°C with RH = 20%
- e. T = +20°C with RH = 50%
- f. T = +5°C with RH = 100%
- g. T = +50°C with RH = 100%
- h. T = +50°C with RH = 2%

N2. Find e_s (kPa), T_d (°C), r and r_s (g/kg), q and q_s (g/kg), ρ_v (g/m^3), RH (%), and T_w (°C) for the situations below:

	T (°C)	P (kPa)	e (kPa)
a.	30	101.3	1.5
b.	10	80.0	0.1
c.	0	90.0	0.5
d.	–10	50.0	0.1
e.	–20	101.3	0.1
f.	40	70.0	2.0
g.	10	85.0	0.2
h.	20	90.0	0.5

N3. Saturated outside air of T = 0°C is drawn into your house to replace hot air going up the chimney. If this induced air is heated in the house to 20°C, what is the relative humidity inside?

N4(§). Generate a table similar to Table 5-1, but for an altitude where P = 50 kPa.

N5. What is the LCL height for air with the following conditions at the surface. This is the cloud-base height for convective clouds.

	T (°C)	T_d (°C)
a.	30	10
b.	10	10
c.	0	–5
d.	30	0
e.	20	14
f.	40	35
g.	–20	–20
h.	–10	–40

N6. Psychrometer measurements give the (dry bulb) temperatures, T, and wet bulb temperatures, T_w, listed below. What is the mixing ratio, dew point and relative humidity?

	T (°C)	P (kPa)	T_w (°C)
a.	30	101.3	20
b.	10	80.0	10
c.	0	90.0	–5
d.	30	100.0	15
e.	20	90.0	14
f.	0	95.0	–5
g.	20	100.0	6
h.	40	100.0	20

N7. What is the total water mixing ratio at sea level, given

a. $r = 5$ g/kg
b. $r = 5$ g/kg, $r_L = 3$ g/kg
c. $T = 10$°C, $r_L = 2$ g/kg
d. $r = 10$ g/kg, $r_L = 5$ g/kg
e. $T = 5$°C, $r_L = 5$ g/kg

N8. What is the liquid water mixing ratio at sea level for

a. $T = 30$°C, $r_T = 25$ g/kg
b. $T = 20$°C, $r_T = 25$ g/kg
c. $T = 30$°C, $r_T = 30$ g/kg
d. $T = 20$°C, $r_T = 30$ g/kg
e. $T = 10$°C, $r_T = 20$ g/kg
f. $T = 0$°C, $r_T = 10$ g/kg

N9. Using Fig 5.4, find the dew point temperature for air at $P = 50$ kPa, given mixing ratios (g/kg) of

a. 30 b. 15 c. 8 d. 3
e. 20 f. 10 g. 5 h. 0

N10. Using Fig 5.4, what is the dew point temperature of air parcels ending at a pressure height of 50 kPa, if they start at 100 kPa with dew-point temperatures (°C) of

a. 25 b. 20 c. 15 d. 10
e. 5 f. 40 g. –40 h. –20

N11. Using Fig 5.4, find the relative humidity for air at

	P (kPa)	T (°C)	T_d (°C)
a.	70	20	0
b.	70	10	10
c.	70	30	–10
d.	90	30	30
e.	90	30	20
f.	90	30	10
g.	100	20	–20
h.	100	–20	–30

N12. a. An unsaturated air parcel from sea level has $r = 15$ g/kg and T = 30°C. If it rises to a height where it is saturated with $r_L = 2$ g/kg, what is its new value of r? What is its temperature?

b. Same as (a), but for $r_L = 5$ g/kg.

N13. a. Precipitation of rate 5 mm/h enters a volume of air at altitude 100 m, but only 4 mm/h reaches the ground. How does this precipitation gradient alter the total water mixing ratio within that 100 m-thick layer?

b. Same problem, but with top and bottom precipitation rates of 3 mm/h and 4 mm/h, respectively.

N14. What evaporation rate (mm/day) corresponds to a latent heat flux of: a. 100 W·m^{-2} b. 50 W·m^{-2}
c. 270 W·m^{-2} d. 15 W·m^{-2} e. 333 W·m^{-2}
f. 150 W·m^{-2} g. 75 W·m^{-2} h. 4 W·m^{-2}

N15. a. A wind of $U = 10$ m/s is advecting air of $r_T = 10$ g/kg at the west side of a region. The other side of the region is 100 km further east, and has a humidity of $r_T = 15$ g/kg. How does this advection alter the total water mixing ratio across that 100 km wide layer?

b. Same as (a), but with twice the wind speed.

c. Same as (a), but with $r_T = 20$ g/kg at the west side of the region.

N16. a. A radiosonde balloon launched during daytime fair-weather shows vertical profiles in a 1 km thick boundary layer as shown in Fig 5.5, with $|\Delta_s\theta| = 10$°C, $|\Delta_{zi}\theta| = 2$°C, $|\Delta_s r_T| = 4$ g/kg, and $\Delta_{zi} r_T = -6$ g/kg. The convective transport coefficient times buoyancy velocity is 0.025 m/s on this calm day.

What is the vertical flux divergence of total water due to turbulence, and how does that contribute to moistening the air?

b. Same as (a) but with $\Delta_{zi} r_T = +6$ g/kg.

N17. Using the equations, what is the value of the moist adiabatic lapse rate at $P = 100$ kPa for the following temperatures (°C)?

a. 30 b. 25 c. 15 d. 10 e. 5
Same, but at $P = 50$ kPa:
f. 30 g. 25 h. 15 i. 10 j. 5

N18. Using Fig 5.6, what is the final temperature of a saturated air parcel at $P = 70$ kPa if lifted moist adiabatically from $P = 100$ kPa with the following initial temperatures (°C):

a. –10 b. 30 c. 20 d. 10
e. 0 f. 40 g. –20 h. –30

N19. Same as previous exercise, but starting at P = 50 kPa and lowering moist adiabatically to P = 80 kPa.

N20. Identify the liquid water potential temperatures for the air parcels in the previous two exercises.

N21. At height 1 km, the temperature is 5°C, the water vapor mixing ratio is 6 g/kg, and liquid-water mixing ratio is 2 g/kg. Find the:
a. liquid water potential temperature
b. equivalent potential temperature

Understanding & Critical Evaluation

U1(§). a. Use a spreadsheet to calculate saturation vapor pressure over ice (not over water) for temperatures between –50 and 0°C. Plot the resulting ice curve together with the water curve on the same graph, as in Fig 5.2. Also, plot a curve of the difference between the ice and water curves, as shown in the inset in Fig 5.2.
b. Replot the results as semi-log and log-log.
c. Why are the saturation vapor pressures different over water vs. ice?

U2. The Clausius-Clapeyron equation is based on an assumption of air over a flat surface of pure liquid water, and assumes that the liquid water and the air are of the same temperature. However, sometimes in the atmosphere there is warm rain falling into a region of colder air. Explain what would happen in this situation. (Neglect droplet surface curvature.)

U3. Use the definitions of mixing ratio eq. (5.3), specific humidity eq. (5.4), and absolute humidity eq. (5.5) to verify relationships (5.14) - (5.19).

U4. a. Derive eq. (5.24) from the Clausius-Clapeyron equation.
b. Derive a similar equation for dew point temperature, but using Tetens' formula as the starting point.

U5. In Table 5-1, why can the same columns of numbers be used for actual temperature with saturation values of humidity (i.e., the top set of column labels), and for the dew point temperature with actual values of humidity (i.e., the bottom set of column labels)?

U6(§). It is often assumed that specific humidity is approximately equal to mixing ratio. How good is that assumption? Create a graph comparing the two values over a range of temperatures. Do this for a few different pressures.

U7. Vapor pressure e represents the partial pressure of water vapor in air. Namely, the total pressure in air can be thought of as consisting as the sum of partial pressures from each of the individual gases within the air.

If total air pressure remains constant, but if evaporation of liquid water into the air causes the humidity to increase (i.e., the vapor pressure increases), then what happens to the other gases in air? Where do they go, if total pressure remains constant?

U8. Create a table that lists the max and min values allowed for each humidity variable.

U9. Derive eq. (5.31d) from (5.31c).

U10. In arid regions (hot, dry), an inexpensive form of air conditioning (sometimes known as "**swamp coolers**") is possible where liquid water is evaporated into the dry air. The result is cooler air (due to latent heat effects) that is more humid.
a. For air of initial temperature T = 40°C, how effective are swamp coolers for different initial relative humidities?
b. At what initial humidity would a swamp cooler not cause the air to be more comfortable? (Hint, use the heat index info from Chapter 3.)

U11(§). Check whether eq. (5.8) is a reasonable approximation to the lifting condensation level (LCL). The exact definition is the height of a rising air parcel where the temperature and dew point become equal.
a. Do this check using the thermodynamic diagram, for a variety of T and T_d values at the same starting height. For T, follow a dry adiabat upward from the starting temperature and height. For T_d, follow an line of constant mixing ratio up from the starting T_d and height. Where the two lines cross is the LCL.
b. Do the check using the equations for adiabatic temperature change from an earlier chapter, and using the equation for dew point temperature as a function of height or pressure.

U12(§). On a thermodynamic diagram, **Normand's Rule** corresponds to the following graphical procedure. Starting with T and T_d, follow upward the dry adiabat and isohume, respectively, until they cross, and then follow a saturated adiabat back down to the starting height. The temperature at this point is T_w (see a previous solved example).

Verify that eqs. (5.12) and (5.13) describe the same process, and compare the numerical value of T_w using graphical vs. equation methods.

U13(§). Use a spreadsheet program to generate a table that gives dew point temperatures from measurements of wet and dry bulb temperatures. Compare your results to Table D.1 of Ahrens.

U14(§). Use a spreadsheet program to draw dry adiabats and mixing ratio lines on the same thermodynamic diagram, and plot as a semi-log graph. That is, combine Figs 3.2 and 5.4.

U15(§). Use a spreadsheet program to draw dry and moist adiabats on the same thermodynamic diagram, and plot as a semi-log graph.

U16(§). (Very time-consuming but important problem, resulting in a complete thermodynamic diagram.)

a. Use a spreadsheet program to draw dry and moist adiabats and mixing ratio lines on the same thermodynamic diagram. If you can draw graphs in color, use red lines for the dry adiabats, blue lines for the moist adiabats, and green lines for the mixing ratios. The other temperature and pressure grid lines can be black.

b. (Optional) Also, compute and plot on the same graph lines of constant altitude (m) vs. $T_{surface}$ and P, using the hypsometric equation. If you have color, draw these lines in brown or tan. Assume a dry adiabatic lapse rate from the surface.

U17. In eqs. (5.36) and (5.37) appear a factor θ / T. Why is it in those equations? Do they cancel?

U18. Let $r_T = r + r_L$ be the total water mixing ratio, which is the sum of the water vapor r plus any liquid water r_L (cloud droplets) present in the air. Consider a rising air parcel with initially no liquid water that is unsaturated, starting from near the surface. Thus the total water equals the initial water vapor mixing ratio.

a. As the air parcel rises toward its LCL, is θ_L conserved? Explain.

b. Above the LCL, a rising air parcel does not conserve θ, because condensation of water releases latent heat into the air. However, as the temperature of the rising air parcel continues to decrease with height, the saturation mixing ratio drops to values less than the initial mixing ratio. The difference is squeezed out of the air as liquid water. Is θ_L conserved for this rising air parcel that is above its LCL? Explain.

U19. What is the behavior of eq. (5.33b) in the limit of zero absolute temperature, and in the limit of very high temperature?

U20. Derive eq. (5.31b), assuming that $F_{water\ zi} = w_e \cdot \Delta_{zi} r_T$, and $F_{H\ zi} = 0.02 \cdot F_H$, and $F_{H\ zi} = w_e \cdot \Delta\theta$, where w_e is entrainment velocity.

U21. Without the simplifying assumptions, write the full Eulerian water budget equation. (Hint, look at the full heat budget equation from Chapter 3.)

U22. If air starting from near the surface (at $P = 100$ kPa) is initially just saturated, find the values of equivalent potential temperature θ_e for each of the moist adiabats drawn in Fig 5.6.

Explanation: $\theta_e > \theta_L$, because θ_e represents the potential temperature of the air after being heated by condensation of ALL the water vapor in the air, while θ_L represents the potential temperature of the air after being cooled by evaporation of all the liquid water present in the air. In other words, θ_L and θ_e are the potential temperatures at the bottom and top (at near infinite height) of any moist adiabat line in the thermo diagram.

U23. Rewrite eq. (5.25) to include vertical advection by mean wind W, and to include evaporation from the ocean. Assume that you are considering a fixed Eulerian volume (such as a hypothetical cube of air) that rests on the sea surface.

U24. Consider a Lagrangian, cloudy air parcel.

a. Suppose precipitation falls into the top of the air parcel, passes through the air parcel, and falls out of the bottom at the same rate that it entered. Discuss how eq. (5.23) would be affected.

b. What if no precipitation enters the top or sides of the air parcel, but there is some falling out of the bottom. Discuss how eq. (5.23) would be affected.

U25. In the Eulerian budget eq. for *total water* (5.25), there is a precipitation term. In the Eulerian budget eq. for *heat* (3.34), there is a latent heat term that depends on condensation. Condensation and precipitation are related, yet the terms in those two eqs. are different. Why must they be different?

U26. a. Rearrange the T_w equations to find the T_w value corresponding to $r = 0$, for arbitrary T and P.

b. Test these values from part (a) for a variety of temperatures at $P = 100$ kPa, and compare with the appendix in Ahrens.

Web-Enhanced Questions

W1. Access the surface weather data for a weather station close to your location (or at some other location specified by your instructor). What is the current humidity (value and units)?

W2. Find a weather map on the web that shows either colored or contoured values of humidity over your country for a recent observation time. Where are the humidities the greatest; the smallest?

W3. Do the humidity formulas presented in the USAToday web page agree with those in this textbook? Compare and discuss.
http://www.usatoday.com/weather/whumcalc.htm

W4. Search the web for weather calculator tools for calculating or converting humidities, temperatures, and/or pressures in or to various units.

W5. Using the web site listed below that lists manufacturers of meteorological instruments (or based on your own web searches if the site below doesn't work), create a catalog or table of the different types of instruments that can be used for measuring humidity. Don't give brand names here, but instead list the type of instrument (such as sling psychrometer). For each type of instrument, briefly explain the physical process that it utilizes to measure humidity, and state which humidity variable it measures.
http://www.ugems.psu.edu/~owens/WWW_Virtual_Library/instrument.html

W6. Search the web for tables or graphs or definitions of various indices that measure discomfort or heat-stress danger for humans and/or livestock as a function of air temperature and humidity. There are a variety of such indices: **temperature-humidity index (THI), discomfort index, apparent temperature, heat index** (also known as the R.G. **Steadman Heat Index), effective temperature, humiture, humisery, livestock weather safety index, summer simmer index, wet-bulb globe temperature, index of thermal stress.** (Hint, Two traditional sources of info are: Houghton, D.D., 1985: *Handbook of Applied Meteorology*, Wiley Interscience, p786. Weiss, M.H., 1983: Quantifying summer discomfort, *Bulletin American Meteor. Soc.*, **64**, 654-655.) (Hint 2: Try http://www.zunis.org/index.html).

W7. Search the web to find a current weather map analysis that shows isohumes (surface, or at 85 kPa, or at 70 kPa) over your state, province, or country (as assigned by the instructor). Locate your town or school on this map. Search for related current weather info that gives the wind speed and direction at that same altitude over your town.

a. Compute the moisture advection term of the Eulerian moisture budget, and discuss whether the result will cause the air at your location to become more or less humid with time.

b. What precipitation or evaporation rate would be needed to compensate for the advection from part (a), which would result in no change to total water mixing ratio (neglecting turbulence)?

W8. Access the current surface weather data near you every hour, and plot humidity vs. time. Alternately, a web site might already have a plot of humidity vs. time that you could use directly. Plots of weather variables vs. time are called **meteograms.**

W9. Search the web for web sites that plot upper-air soundings on thermodynamic diagrams. In particular, look for different types of thermo diagrams (not for different soundings on the same type of diagram). Print out a sample of each different type of thermo diagram.

In almost all thermo diagrams, the moist adiabats are curved lines that gradually become tangent to the dry adiabats at high altitudes. Use this behavior to identify the moist and dry adiabats in each type of diagram. Then, also identify the isotherms, isobars, and isohumes. You should be able to interpret any type of thermo diagram.

Science Graffito

"Once an explanation, no matter how faulty, has appeared in print three times it becomes an immutable truth, and a minor industry develops to feed it and keep it alive." — C. Bohren and B. Albrecht, 1998: *Atmospheric Thermodynamics.*

Synthesis Questions

S1. Suppose that saturation vapor pressure decreases as air temperature increases. Assume that clouds, fog, and precipitation can still form in air that becomes saturated. How would cloudiness and fogginess near the ground vary during a daily cycle? How would it vary seasonally?

S2. Suppose that dew point temperature did not depend on pressure or altitude. Would any of the lines in a thermo diagram be different? How would they look?

S3. Suppose you were to design a new humidity sensor that measured absorption of radiation across a path of humid air. Assume that the path length through which the radiation propagates, and the absorption coefficient for water vapor for the wavelength of interest, are both known. Which principle or law would you utilize to convert radiation absorption into a humidity, and which humidity variable would you get from this instrument?

S4. What if the amount of latent heat released in a rising, saturated air parcel was greater than the adiabatic cooling of the parcel as it expands into lower pressure.

a. What would be different in a thermo diagram?

b. How would this difference affect the weather and climate, if at all?

S5. What is the mass of an individual water molecule?

S6. Suppose the whole troposphere was saturated with water vapor (100% relative humidity), but there were no cloud droplets or ice crystals in the air.

a. Assuming a standard atmosphere temperature profile, how many kg of water vapor are there over each square meter of earth surface?

b. How many kg of water vapor would there be in the troposphere total, when summed around the whole surface of the earth?

c. If all this water were to rain out (leaving the atmosphere completely dry) and flow into the oceans, which cover roughly 70.8% of the earth's surface, how many meters would sea level rise?

d. Repeat questions a-c, but for the stratosphere only.

S7. a. Rewrite the Clausius-Clapeyron equation in the form $e_s = C \cdot \exp[-(L/\Re_v)\cdot(1/T)]$, where C is a constant. What is C equal to?

b. The gas constant for water vapor $\Re_v \equiv k_B/m_v$, where m_v is the mass of a water molecule, and $k_B = 1.3806\text{x}10^{-23}$ J·K^{-1}·molecule^{-1} is the **Boltzmann constant**. Rewrite the eq. from part (a) after substituting for the water vapor gas constant.

c. The result from (b) contains an exponent of an argument, where that argument is a ratio. The numerator represents the energy needed to separate one molecule from liquid water to allow it to become a vapor molecule in the neighboring air, while the denominator represents the average energy available (as a kinetic energy associated with temperature). Discuss the implications of this interpretation, and why it is relevant to determining equilibrium saturation vapor pressure.

STABILITY

CONTENTS

6 The day dawns, you awaken and notice how the atmosphere begins to change with time. At first, smoke from chimneys oscillates up and down in a smooth wavy pattern as it drifts downwind. Winds near the ground are light.

The sun rises and heats the ground, and you notice the first gentle gusts of wind. Later in the morning as thermals of warm air climb to greater heights, wind speeds near the ground suddenly increase. Cumulus clouds form in the updrafts, and you see birds soaring in circles. Smoke from tall stacks loops up and down, which you interpret as evidence of vigorous turbulence.

By afternoon, you observe that some of the clouds have risen to greater heights, forming cumulus congestus. Later, thunderstorms with intense up and down drafts form. An anvil is evident at cloud top, making a spectacular scene that you photograph. A few moments later, you see precipitation push downward from cloud base, and you feel a cold gust front as it advances over your neighborhood. After sunset, the storms weaken and dissipate.

Such an evolution from gentle turbulence and waves to violent updrafts, depends on the buoyancy of air. As the earth's surface is warmed or cooled diurnally, air near the ground becomes warmer or cooler than the air higher in the boundary layer and the troposphere. This changes the ability of air parcels to rise, for which **static stability** is one measure. We can utilize thermodynamic diagrams to help determine static stability.

Science Graffito

"True genius resides in the capacity for evaluation of uncertain, hazardous, and conflicting information." – Winston Churchill

THERMODYNAMIC DIAGRAMS – PART 4: APPLICATIONS

In previous chapters, we created various components of a thermodynamic diagram. These are reproduced here as Figs 6.1a-c, but with identical aspect ratios so that they can be overlaid. Fig 6.1d shows all the components combined into a complete, but simplified, thermodynamic diagram. It will be used throughout this chapter to study various atmospheric processes.

Figure 6.2 (next page)
Full T - log P thermodynamic diagram for the atmosphere. Thin vertical lines are isotherms T, labeled at bottom in °C. Thin horizontal lines are isobars P, labeled at left in kPa. Thin dashed lines nearly horizontal are height contours z, labeled at right in km. Thin diagonal dotted lines slightly-off-from-vertical are isohumes r, labeled at the top in g/kg. All these thin lines represent the state of the air. Thick lines represent adiabatic processes that can change the state. Heavy diagonal solid lines are dry adiabats, labeled by potential temperature θ at the bottom in °C. Heavy curved dashed lines are moist (saturated) adiabats, labeled by liquid-water potential temperature θ_L at the bottom in °C. [Potential temperature is defined using a reference pressure of 100 kPa, and height is set to zero there.]

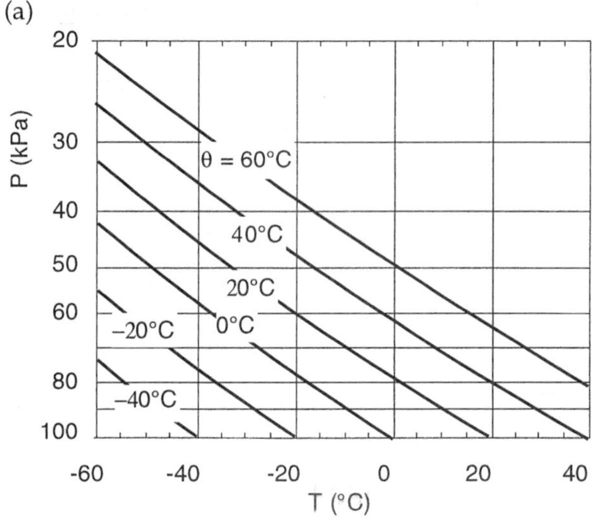

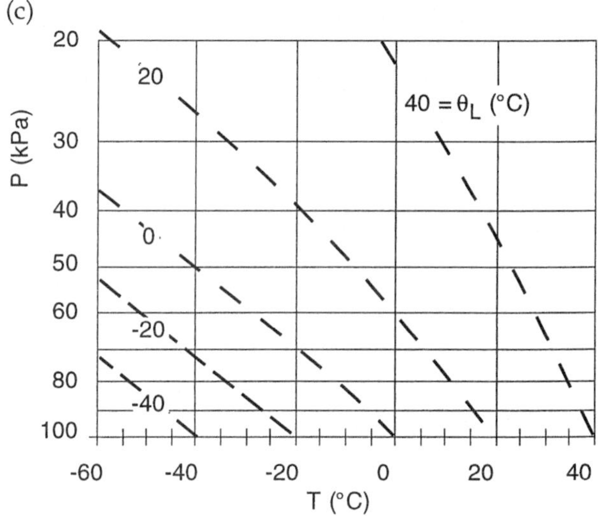

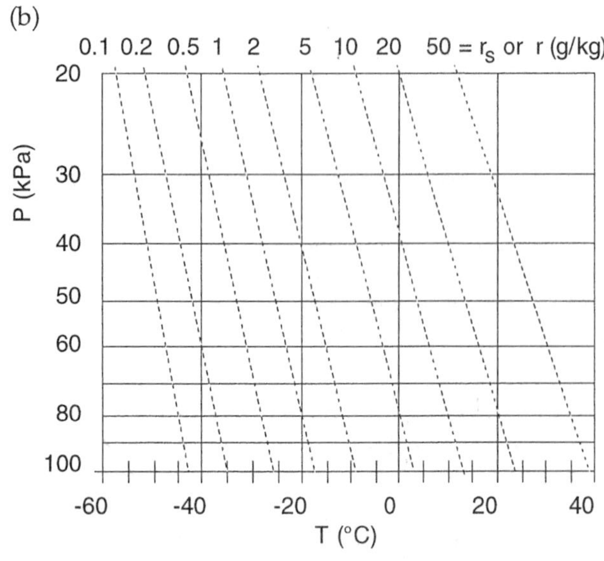

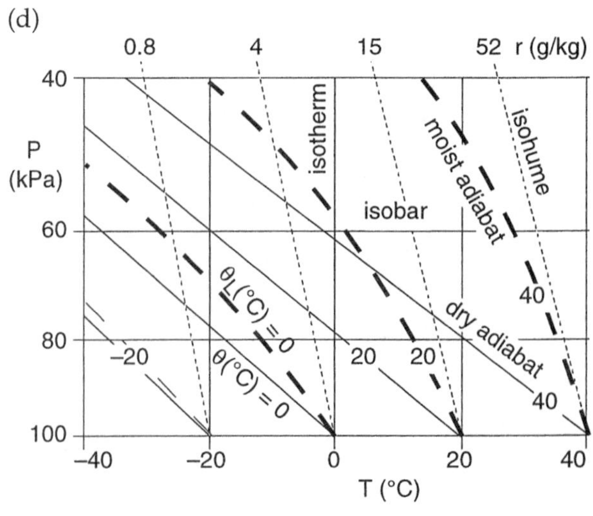

Figure 6.1
Construction of a thermodynamic diagram. The background on all the charts are isotherms (vertical thin solid lines), and isobars (horizontal thin solid lines). (a) dry adiabats (from Fig 3.2) are diagonal solid lines; (b) isohumes (from Fig 5.4) are nearly vertical dotted lines; (c) moist adiabats (from Fig 5.6) are curved dashed lines; and (d) complete diagram. P is pressure, T is temperature, r is mixing ratio, θ is potential temperature, and θ_L is liquid-water potential temperature.

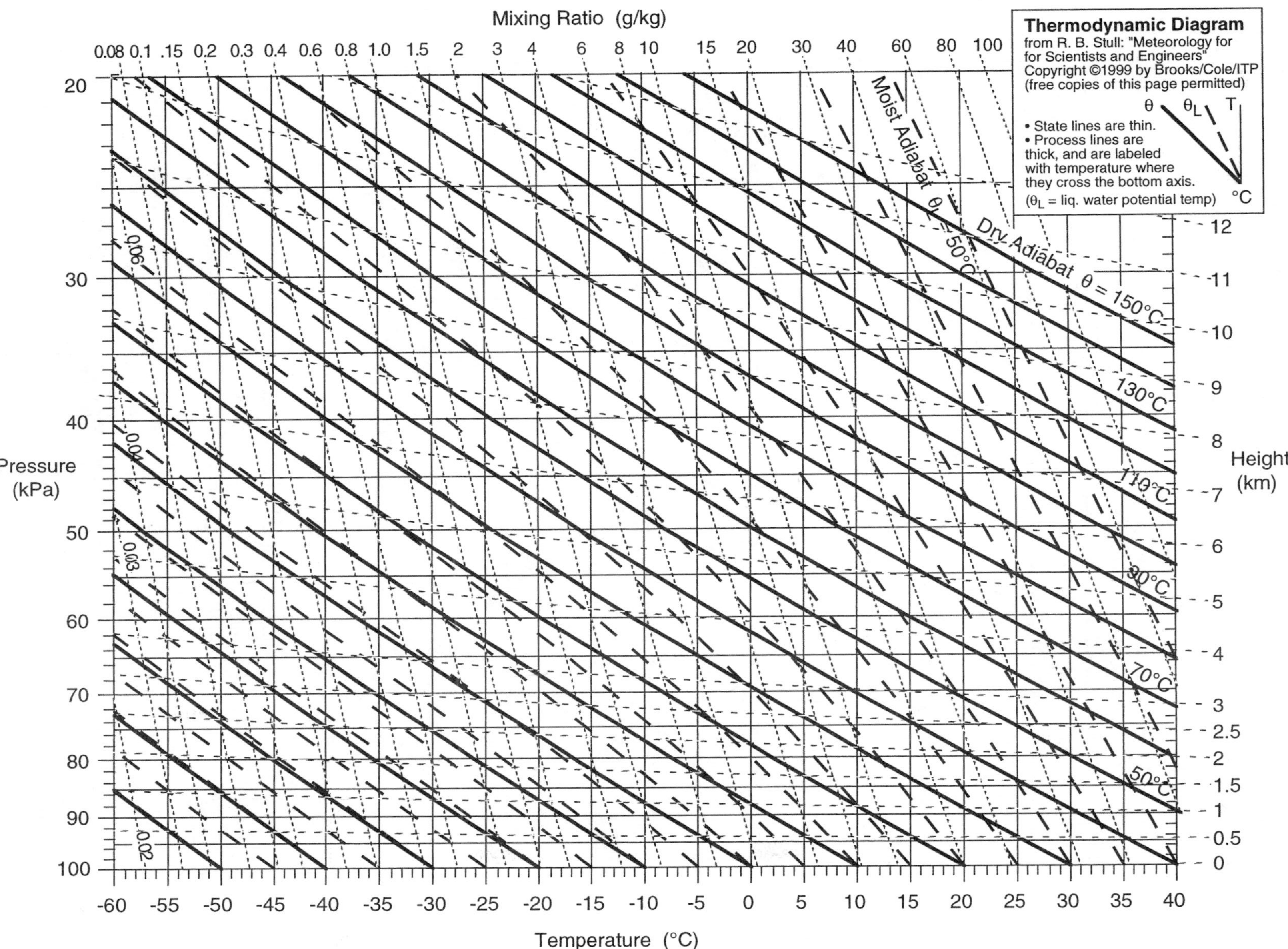
Thermodynamic Diagram
from R. B. Stull: "Meteorology for for Scientists and Engineers"
Copyright ©1999 by Brooks/Cole/ITP
(free copies of this page permitted)
θ
θL
T
°C
• State lines are thin.
• Process lines are thick, and are labeled with temperature where they cross the bottom axis.
(θL = liq. water potential temp)
Mixing Ratio (g/kg)
0.08 0.1 .15 0.2 0.3 0.4 0.6 0.8 1.0 1.5 2 3 4 6 8 10 15 20 30 40 60 80 100
Moist Adiabat θL = 50°C
Dry Adiabat θ = 150°C
130°C
110°C
90°C
70°C
50°C
0.06
0.04
0.03
0.02
Pressure (kPa)
20
30
40
50
60
70
80
90
100
Height (km)
12
11
10
9
8
7
6
5
4
3
2.5
2
1.5
1
0.5
0
Temperature (°C)
-60 -55 -50 -45 -40 -35 -30 -25 -20 -15 -10 -5 0 5 10 15 20 25 30 35 40

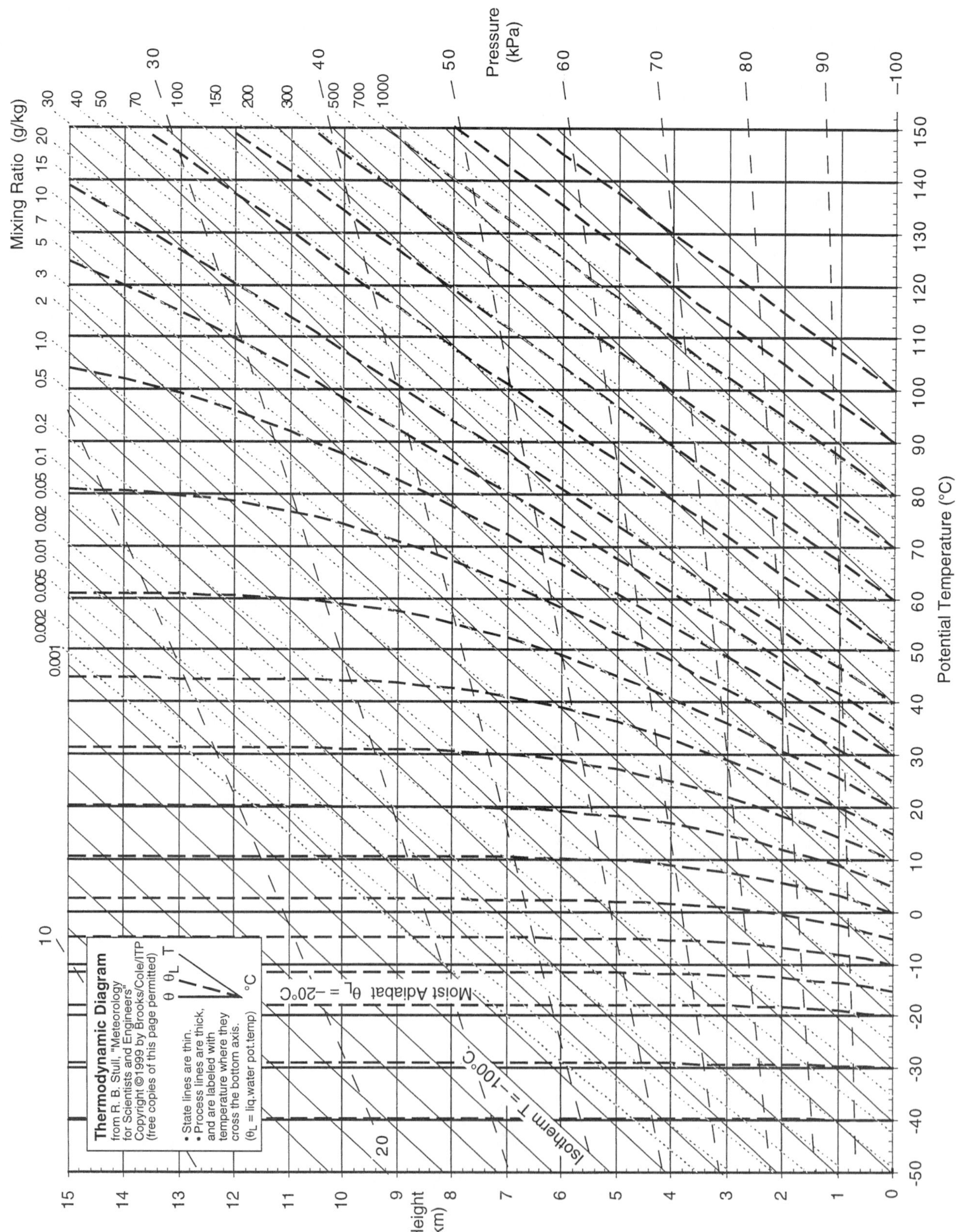
Thermodynamic Diagram
from R. B. Stull, "Meteorology for Scientists and Engineers"
Copyright ©1999 by Brooks/Cole/ITP
(free copies of this page permitted)
• State lines are thin.
• Process lines are thick, and are labeled with temperature where they cross the bottom axis.
(θ_L = liq.water pot.temp)
θ θ_L T
°C
Moist Adiabat θ_L = −20°C
Isotherm T = −100°C
Mixing Ratio (g/kg)
0.001 0.002 0.005 0.01 0.02 0.05 0.1 0.2 0.5 1.0 2 3 5 7 10 15 20 30 40 50 70 100 150 200 300 500 700 1000
Pressure (kPa)
10 20 30 40 50 60 70 80 90 −100
Height (km)
15 14 13 12 11 10 9 8 7 6 5 4 3 2 1 0
Potential Temperature (°C)
-50 -40 -30 -20 -10 0 10 20 30 40 50 60 70 80 90 100 110 120 130 140 150

Figure 6.3 (previous page)
Full θ vs. z thermodynamic diagram. Thick vertical lines are adiabats θ, labeled at bottom in °C. Thin horizontal lines are height z, labeled at left in km. Thin diagonal solid lines are isotherms T, labeled at bottom in °C. Thin nearly-horizontal dashed lines are isobars P, labeled at right in kPa. Thin diagonal dotted lines are isohumes of constant mixing ratio, r, labeled at top in g/kg. Thick dashed diagonal lines are moist (saturated) adiabats, labeled by liquid water potential temperature θ_L at bottom in °C. [Potential temperature is defined using a reference pressure of 100 kPa, and height is set to zero there.]

In Fig 6.1d are five types of lines: three giving the state of the air (**isobars, isotherms,** and **isohumes**) and two showing how the state of air changes during vertical movement of air (**dry and moist adiabats**). Additional state lines for **height contours** are left out to avoid clutter.

More detailed (and more useful) versions of thermodynamic diagrams are given in Figs 6.2, 6.3 6.12, and 6.13. These were produced on a computer spreadsheet. They look complicated at first, but with a bit of practice they can make meteorological analysis much easier than solving sets of coupled equations. We will call these figures **thermo diagrams**, for short.

State

For any one air parcel that is unsaturated, two points are needed to describe its state on a thermo diagram. One describes the temperature and the other describes the humidity (see Fig 6.4).

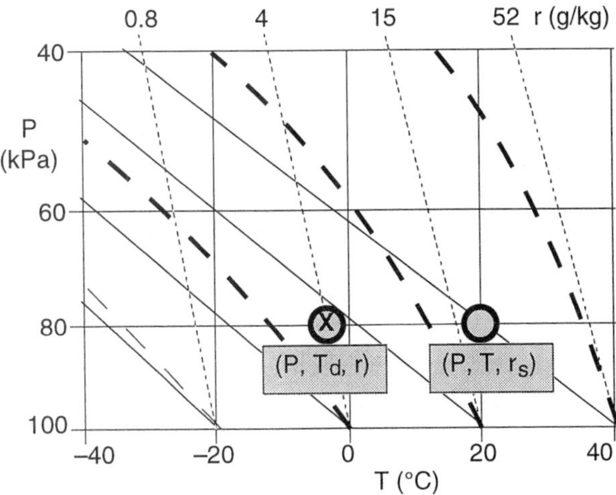

Figure 6.4
Two points represent one air parcel: right indicates temperature, and left (marked with the X) indicates moisture .

The first point is plotted at the actual air temperature (*T*) and pressure (*P*) for the parcel. The isohume associated with this point gives the saturation mixing ratio (r_s, the maximum equilibrium amount of moisture that could be held as vapor).

The second point is plotted at the dew-point temperature (T_d), at the same pressure (*P*) as the first point. The isohume associated with this point gives the actual mixing ratio (*r*) of water vapor currently in the air parcel.

Saturated (cloudy) air parcels need only one temperature point on the diagram, because the temperature equals the dew-point, and $r = r_s$. However, it is appropriate to keep another point on the diagram to represent total water, as described later.

Before using Figs 6.2 or 6.3, make many extra copies. That way, you can write on the copies and keep the original clean. Figs 6.2 and 6.3 are specially designed for black and white usage, so it should reproduce well. When you plot points on your copy, it helps to use colored pens or pencils, which stand out against the black and white background. Also, use a separate copy for each exercise to avoid confusion.

Solved Example
For the air parcel of Fig 6.4, find its temperature, dew point, pressure, mixing ratio, saturation mixing ratio, and relative humidity.

Solution
Given: Fig 6.4. Estimate points by eye.
Find: T = ? °C, T_d = ? °C, P = ? kPa
r = ? g/kg, r_s = ? g/kg, RH = ?%
For fog: $T_d - T$= ? °C, $r_s - r$ = ? g/kg

The point on the right is directly on the T = 20°C isotherm. The left point must be interpolated between the 0°C and -20°C isotherms. I estimate by eye roughly T_d = –2.5°C. The pressure, which must be the same for both points, is P = 80 kPa.

Interpolating the right point between isohumes gives roughly r_s = 19 g/kg, while the actual mixing ratio using the left point is r = 4 g/kg. Relative humidity is RH = (100%)·(4/19) = 21%.

Check: Units OK. Physics OK.
Discussion: I actually used Fig 6.2, because it is more accurate. The air is fairly dry.

Dry (Unsaturated) Processes

When an unsaturated air parcel rises, the temperature cools dry adiabatically due to the decreasing pressure. Any water molecules in the parcel stay in the parcel. Both potential temperature θ and mixing ratio r are conserved. On the thermodynamic diagram, this corresponds to moving the temperature point along the "dry" adiabat that passes through its initial position, and moving the moisture point (X) along the isohume of constant r (NOT along a T_d line). Always move both points together to the same pressure (see Fig 6.5).

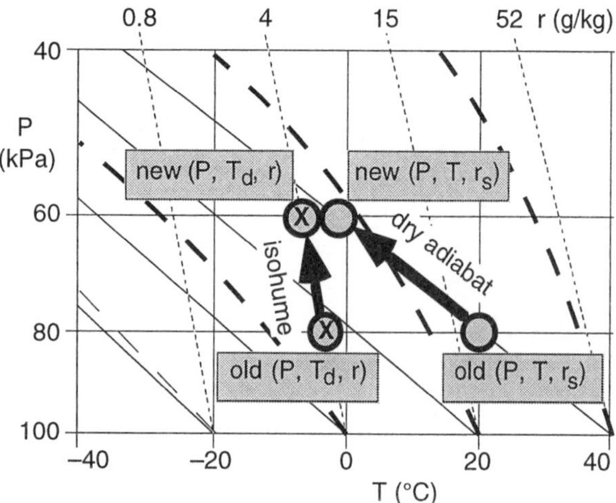

Figure 6.5
Potential temperature and mixing ratio are constant for rising unsaturated air. The new points still represent the one parcel. The new r equals the old r, although new r_s does not equal old r_s.

Solved Example
Given Fig 6.5, what is the new state of the air after being lifted dry adiabatically to P = 60 kPa?

Solution
Given: Fig 6.5 (or 6.2). Estimate points by eye.
Find: T & T_d = ? °C, r & r_s = ? g/kg, RH = ?%

The temperature, when moved along (or parallel to) the dry adiabats, has cooled to about T = <u>–3°C</u> at P = 60 kPa. The dew point temperature (X), when moved along (or parallel to) the isohumes, decreases slightly to T_d = <u>–6.5°C</u>.

Mixing ratio is still r = <u>4 g/kg</u>, but saturation mixing ratio has decreased to r_s = <u>5 g/kg</u>. The new relative humidity is: RH=(100%)·(4/5)= <u>80%</u>.

Check: Units OK. Physics OK.
Discussion: Although not yet saturated (i.e., not cloudy), the RH is much higher.

Likewise, for descending air parcels that are unsaturated, the temperature point follows the "dry" adiabat and the humidity point follows the isohume.

Moist (Saturated) Processes & Liquid Water

Adiabats and isohumes converge with increasing height. Thus, for a rising air parcel there will be some height where the temperature and moisture points coincide. At this altitude or pressure the air is just saturated, and corresponds to convective (cumulus) cloud base. It is called the **lifting condensation level (LCL)**, (Fig 6.6).

For air rising above its LCL, T and Td cool together moist adiabatically, maintaining $T_d = T$. Thus, on the thermo diagram we don't need two circles because there is only one temperature. However, we still need to keep track of total water, as indicated by the X in Fig 6.6. If all water molecules stay in the parcel (i.e., if cloud droplets move with the air and do not precipitate out), then total water mixing ratio r_T remains constant.

The difference between the actual mixing ratio in the cloudy air (using the isohume associated with the single point) and the original mixing ratio before the parcel was lifted gives the amount of liquid water that has condensed. Namely, $r_L = r_T - r_s$. In Fig 6.6, this is the difference between the X and the circle representing the new state of the parcel.

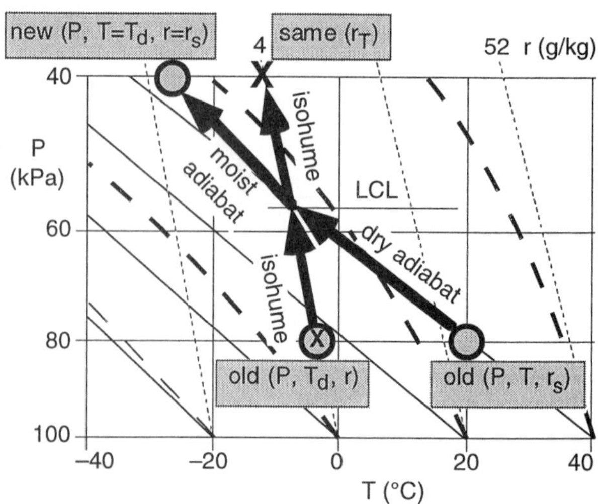

Figure 6.6
The height where the two points converge is the lifting condensation level (LCL). This is cloud base. As the saturated parcel continues to rise, the temperature point follows the moist adiabat. If no precipitation falls out of the parcel, then total water continues to follow the original isohume (indicated by X at same height as parcel).

For descending saturated air, follow the total water mixing ratio (from point X) down the isohume, and follow the temperature state down the moist adiabat until they intersect. This is cloud base for descending air. At the LCL, split the parcel point into two points representing temperature and unsaturated humidity, and follow them down the dry adiabat and isohume respectively. The result is exactly as sketched in Fig 6.6 except all arrows are in the opposite direction. The reason is that below the LCL, the descending air has warmed sufficiently to hold all the water as vapor.

Solved Example

Given the example in Fig 6.6:
(a) At what height is the LCL, and what is the state of the air parcel at that height?
(b) What is the state of the air parcel when it reaches a pressure height of 40 kPa?
(c) How much liquid water has been condensed out at that final height?

Solution

Given: Fig 6.6 (or 6.2). Estimate points by eye.
Find: (a) LCL = ?m, T=T_d= ?°C, r=r_s= ? g/kg
(b) T=T_d= ?°C, r=r_s= ? g/kg
(c) r_L = ? g/kg

Using the diagram:
(a) The LCL is at roughly P = **57 kPa**. The parcel is just barely saturated, with temperature and dew points of roughly $T = T_d$ = **–7°C**. The mixing ratio is still $r = r_s = r_T$ = **4 g/kg**.

(b) Moving parallel to the moist adiabat, from the LCL up to P = 40 kPa, the final state is $T = T_d$ = **–25.5°C**. The saturation mixing ratio is now only $r = r_s$ = **1.2 g/kg**. If no precipitation, then total water is unchanged at r_T = **4 g/kg**.

(c) $r_L = r_T - r_s$ = 4 – 1.2 = **2.8 g/kg**.

Check: Units OK. Physics OK.
Discussion: Liquid water amount not only indicates the amount of rainfall possible, but for conditions below freezing it indicates the rate of ice formation on aircraft wings, a hazard to flight. Most commercial jet aircraft heat the leading edge of the wing to prevent ice formation.

Precipitation

When liquid or solid water precipitate out, the X point representing total water moves to the left along an isobar, on the thermo diagram (Fig 6.7).

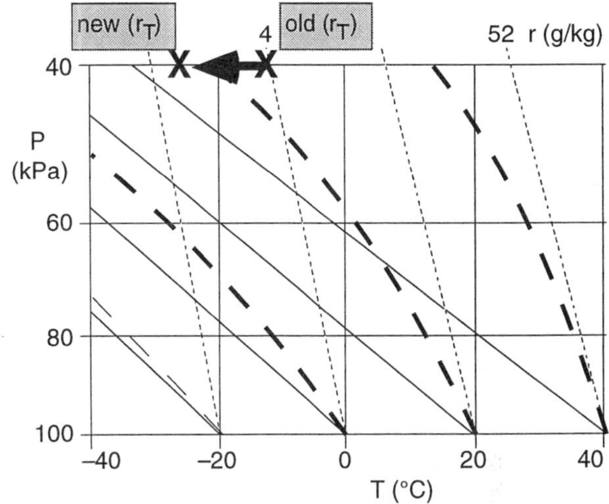

Figure 6.7
Reduction of total water due to precipitation. The new X point cannot be further left than the temperature state of the air parcel.

Solved Example

From the previous solved example, suppose that 1 g/kg of liquid water falls out, leaving the remaining liquid water as cloud droplets that stay with the parcel. If this parcel descends, what is its new LCL and thermo state back at P = 80 kPa?

Solution

Given: Fig 6.7 (or 6.2). r_{precip} = 1 g/kg
Find: LCL = ? m, r = ? g/kg, T & T_d = ? °C.

The amount of total water remaining in the cloud is $r_{T\,new} = r_{T\,old} - r_{precip}$ = 4 – 1 = 3 g/kg. Follow the r_T = 3 g/kg isohume down until it crosses the moist adiabat of the parcel. The intersection pressure height is the LCL: about **52 kPa**.

Starting from this intersection, create two points, where the temperature point follows the dry adiabat down to P = 80 kPa, and the dewpoint point follows the mixing ratio down. This gives: $r \cong$ **3 g/kg**, $T_d \cong$ **–6.5 g/kg**, $T \cong$ **23°C**.

Check: Units OK. Physics OK.
Discussion: The parcel is warmer and drier than when it started (in the first solved example), because the precipitation removed some water, and caused net latent heating due to condensation.

If all the liquid or solid water falls out, then the X on the thermo diagram would move all the way to the circle representing the parcel temperature. If the parcel were to then descend, it would do so dry adiabatically, with T following a dry adiabat, and total water and T_d following the isohume.

If only some of the rain or ice falls out, then the X would move only part way toward the parcel temperature. The new X must then be used as the starting point for total water if the parcel rises further or descends. Namely, the X-point stays on its new isohume during vertical parcel movement. If this parcel descends, it will reach its new LCL at a height greater than its original LCL.

The rain or snow that falls out of the parcel implies net heating. It is an irreversible process. Namely, the latent heating of condensation that occurred in the rising air is not balanced by evaporation in the descending air, because there is now less liquid or solid water to evaporate.

Radiative Heating/Cooling

Clouds are opaque to IR radiation, and translucent to visible light. This can warm or cool the cloudy air, but it does not alter the mixing ratio. Such warming or cooling is said to be **diabatic**, because there is heat transfer across the "skin" of the cloudy air parcel. (Recall that "adiabatic" means no heat transfer.)

There is net heat loss by IR radiation upward from cloud top toward space, both day and night. Also, cloud base exchanges IR radiation with the ground and with the underlying air. If the ground is warmer than cloud base, then cloud base will warm if the cloud is low enough.

Sunlight can penetrate into cloud top (and sides of cumuliform clouds) during daytime, with an e-folding extinction distance on the order of meters. Not all the sunlight is absorbed by the cloud droplets. Some scatters and rescatters off of the droplets, eventually leaving the cloud. This is why clouds look white during daytime.

At night, when IR cooling from cloud top dominates, cloud top becomes colder, and the cold parcels of cloudy air sink through the cloud and cause turbulence and mixing. The whole cloud becomes colder this way, even though only the top few centimeters experience the direct IR net radiative loss to space.

Fig 6.8 illustrates net IR cooling at cloud top. As temperature decreases, so must dew point, because the air is saturated and $T_d = T$. However, the colder air can hold less water vapor at equilibrium than before. Although total water remains constant, the liquid water mixing ratio increases.

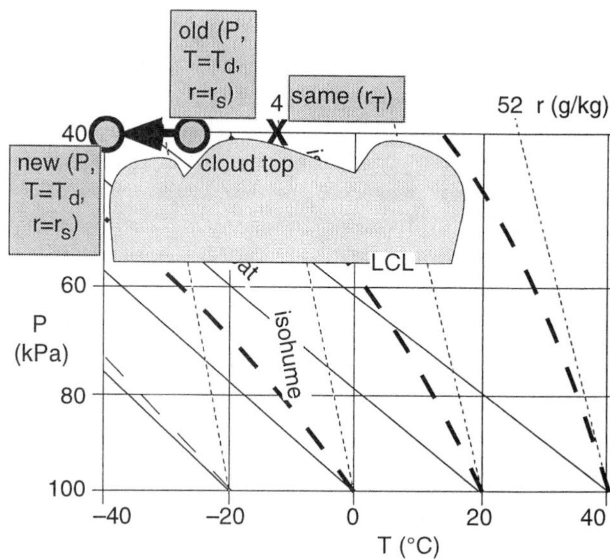

Figure 6.8
Radiative cooling of cloud top. The total water mixing ratio (X) does not change, but temperature and dew point decrease. As less of the total water can be held as vapor at equilibrium, the remainder condenses into additional liquid water.
Thus, $r_{L\,new} = r_T - r_{s\,new}$.

Solved Example
Given the example in Fig 6.8 with initial T = –25°C and r_T = 4 g/kg at cloud top (P = 40 kPa). What is the new state of the air in cloud top after IR radiation has cooled the air to –40°C? Also find the liquid water mixing ratio.

Solution
Given: Fig 6.8, T = –25°C, r_T = 4 g/kg
Find: T & T_d = ? °C, r & r_s = ? g/kg, r_L = ? g/kg

Initially:
Because the air is saturated, $T_d = T$ = **–25°C**.
Also, r_T = **4 g/kg** .
Using Fig 6.2, $r = r_s$ = **1.3 g/kg**.
Thus, $r_L = r_T - r_s$ = 4 – 1.3 = **2.7 g/kg**.

Finally:
The air is still saturated, $T_d = T$ = **–40°C**.
Total water is unchanged: r_T = **4 g/kg**.
Using Fig 6.2, $r = r_s$ = **0.3 g/kg**.
Thus, $r_L = r_T - r_s$ = 4 – 0.3 = **3.7 g/kg**.

Check: Units OK. Physics OK.
Discussion: Liquid water mixing ratio has increased as temperature decreased at cloud top.

Solved Example

Air initially at 100 kPa has temperature 40°C and dew point temperature of 20°C. It rises to a height where the pressure is 50 kPa. Precipitation reduces the total water by 5 g/kg, and the parcel radiatively cools by 11°C while at cloud top. Finally, the parcel descends back to 100 kPa. What is the final relative humidity?

Solution

Given: T = 40°C and T_d = 20°C at P = 100 kPa.
Find: RH = ?% at P = 100 kPa

Method: Use thermo diagram Fig 6.2. (Although Fig 6.1d is shown below for simplicity, Fig 6.2 was actually used for this exercise because it is much more accurate.)

The initial conditions are plotted as the two points initially at P = 100 kPa, in the Fig below. The initial temperature happens to be on the θ = 40°C dry adiabat, and the initial dew point happens to be on the r = 15 g/kg isohume.

Follow the dry adiabat (black arrow) and isohume (hatched arrow) up from their starting points, until those isopleths cross. This crossing point is the upward- bound LCL (i.e., cloud base) and is at P_{LCL} = 85 kPa at T_{LCL} = 15°C for this example.

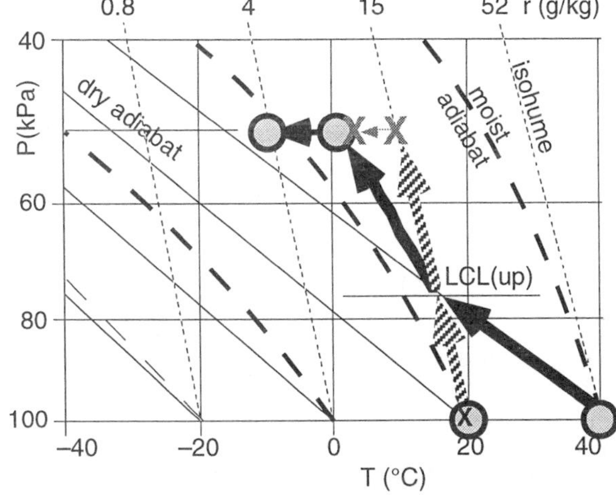

This LCL point happens to lie on the θ_L = 25°C moist adiabat . Above the LCL, follow this moist adiabat up to the pressure of 50 kPa. The parcel temperature is roughly T = 1°C. Also, continue tracking the total water (indicated by the X) as it follows the r_T = 15 g/kg isohume up.

(continued next column)

Solved Example *(continuation)*

At cloud top, the temperature drops by 11°C, from 1°C to –10°C (see diagram in previous column). This puts the temperature near the θ_L = 18°C moist adiabat (interpolating by eye between the θ_L = 15 and 20°C moist adiabats). Precipitation removes 5 g/kg, leaving the total water content at r_T = 15 – 5 = 10 g/kg (see gray X's).

Then the parcel descends. From the temperature point at cloud top (T = –10°C at P = 50 kPa), follow the new moist adiabat down (see Fig below). From the total water point (r_T = 10 g/kg at P = 50 kPa), follow the isohume down.

Where they cross is the LCL for the descending air. P_{LCL} = 80 kPa, which is different than the initial LCL because of the irreversible diabatic processes that acted on the parcel (i.e., radiative cooling, and loss of water). The temperature at this new LCL is roughly T = 10°C.

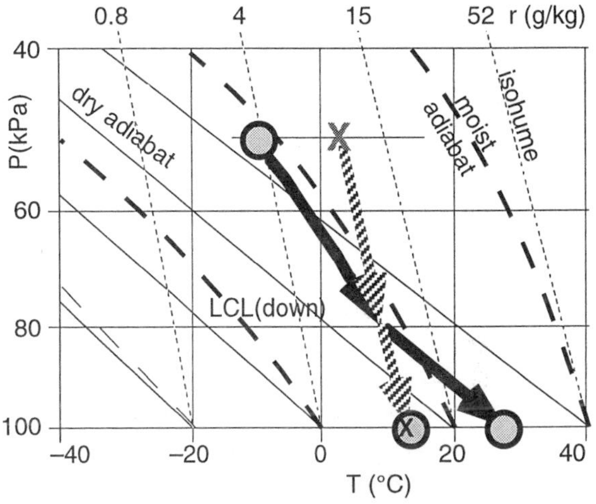

Continuing down below the LCL, temperature follows a dry adiabat down to P = 100 kPa. The final temperature is roughly T = 28°C. Mixing ratio follows the 10 g/kg isohume, which gives a final dew point of T_d = 14°C.

At the final temperature, the saturation mixing ratio is roughly r_s = 25 g/kg (interpolating by eye between the 20 and 30 g/kg isohumes that are in Fig 6.2). Thus, the final relative humidity is:

RH = r/r_s = (10 g/kg) / (25 g/kg) = <u>40%</u>.

Check: Units OK. Physics OK.
Discussion: It helps to use different colors for the temperature and the mixing ratio lines, to avoid confusion.

During day, the top few centimeters of cloud are still cooled by net IR radiative loss, but the top few meters are warmed by sunlight. The sum could be either positive or negative when averaged over the whole cloud. However, with cooling happening at the top of the cloud while warming occurs lower in the cloud, turbulence is set up by the cold sinking air and warm rising air.

In short lived cumulus clouds, the buoyancy of the thermal updraft rising from the ground often dominates, allowing radiative effects to be neglected. However, for stratiform clouds, and longer-lived thunderstorms and hurricanes, radiative effects can be important.

PARCELS vs. ENVIRONMENT

Air parcels do not move through a vacuum. They are moving through air that is already present at all heights and locations in the atmosphere. We call this omnipresent air the **environment.** Of course much of the air in the environment got there by moving from somewhere else; however, for simplicity we will assume in this chapter that environmental movement is small compared to the movement of parcels through it.

Environmental conditions are measured by making **soundings**. Traditionally, a sounding is collected from instruments hanging on a **weather balloon** that rises through the ambient environment and radios back its observations to a receiving station. **Remote sensors** are also used to observe the atmosphere from the ground or from satellite.

Regardless of the mechanics of acquiring a sounding, the purpose is to determine the state of the air at various heights at some instant in time. The end result is a **vertical profile** of temperature, pressure, humidity, and winds in the environment. Density is rarely measured directly, but can easily be computed using the ideal gas law. Constituents such as ozone and other pollutants can also be measured. At any height, the decrease of temperature with height is called the **environmental lapse rate.**

Plotting an environmental sounding is similar to that for individual parcels. Each measurement of environmental temperature and humidity at any one height (pressure) is plotted as two points on the thermo diagram for unsaturated air, and as a single point for cloudy air. Usually, a line is drawn connecting the temperature points, and a separate line is drawn for the humidity points for unsaturated air.

The "standard atmosphere" described in Chapter 1 is an example of an environmental sounding.

Solved Example

Plot the following sounding on a thermo diag.

P(kPa)	T_d(°C)	T (°C)
40	–40	–20
50	–30	–10
60	–5	–5
70	0	0
80	–5	10
90	9	10
99	9	19
100	11	25

Solution

Using black points and solid lines for T, and white points and dashed lines for T_d:

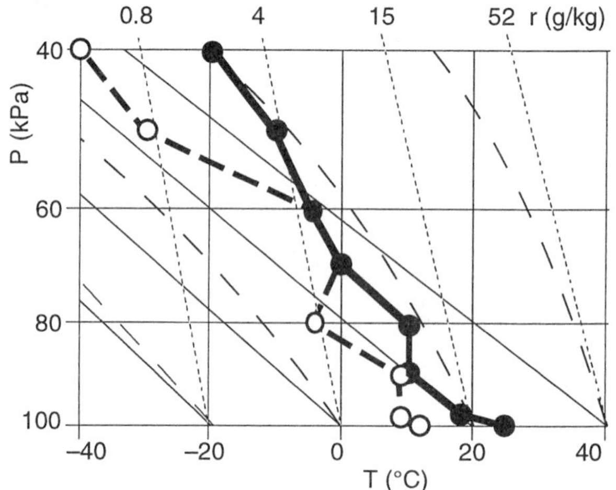

Check: Sketch OK. T_d always ≤ T.
Discussion: The layer of air between P = 70 and 60 kPa is probably cloudy, because the temperature and dew points are equal. Air aloft near 40 kPa is very dry.

BUOYANCY

As Archimedes discovered, an object that is emersed in a fluid displaces some of the fluid, resulting in buoyant forces on the object. The gravitational force minus the buoyant force gives the net force F acting on objects in a fluid:

$$\frac{F}{m} = \frac{\rho_o - \rho_f}{\rho_o} \cdot g \;\equiv\; g' \tag{6.1}$$

where m is the mass of the object, ρ_o and ρ_f are the densities of the object and fluid, respectively, and g = –9.8 m·s^{-2} is the gravitational acceleration, where the negative sign implies a downward acceleration. The right hand side of eq. (6.1) is called the **reduced gravity**, g', where the sign of g' can be different than the sign of g, depending on whether the net buoyant force is up or down.

The force F is essentially the weight of the object, and is defined to be positive in the same direction that g is positive. Negative F implies a downward force (sinking), and positive implies an upward force (floating). Zero force is said to be **neutrally buoyant**.

Air parcels can also be considered as objects within a fluid of the surrounding environmental air. Plugging the ideal gas law into eq. (6.1) and rearranging gives the reduced gravity for an air parcel:

$$\frac{F}{m} = \frac{T_{v\ e} - T_{v\ p}}{T_{v\ e}} \cdot g \equiv g' \qquad (6.2a)$$

where $T_{v\ e}$ and $T_{v\ p}$ are the absolute virtual temperatures (Kelvin) of the environment and parcel *at the same altitude or same pressure*. For relatively dry air, the approximation that $T_v \cong T$ is often used.

Warm air rises, as given in eq. (6.2). More specifically, a parcel that is warmer than its environment experiences an upward buoyant force.

On a thermodynamic diagram, buoyancy is determined by comparing the temperature of an air parcel against the temperature of the environment *at the same height or pressure*. An air parcel is assumed to move vertically in the direction of the buoyant force. Of course, as it moves its temperature changes according to the appropriate moist (cloudy) or dry (non-cloudy) process.

Thus, at each new height, the new parcel temperature must be compared to the environment temperature at its new altitude, and the new buoyant force can then be estimated to determine if the parcel continues to rise or not. Similar procedures must be used for downward moving parcels.

Virtual potential temperatures (see eqs. 2.21 and 2.22) can be used instead of virtual temperature for plotting soundings and determining buoyancy.

$$\frac{F}{m} = \frac{\theta_{v\ e} - \theta_{v\ p}}{T_{v\ e}} \cdot g \equiv g' \qquad (6.2b)$$

Don't forget that g is negative in eqs. (6.2). Thus, parcels that are warmer than their environment cause the ratio on the RHS to be negative, which when multiplied by the negative gravity gives a positive (upward) force (i.e., warm air wants to move upward).

Often gravitational acceleration is treated as a positive magnitude rather than as a vector. Thus:

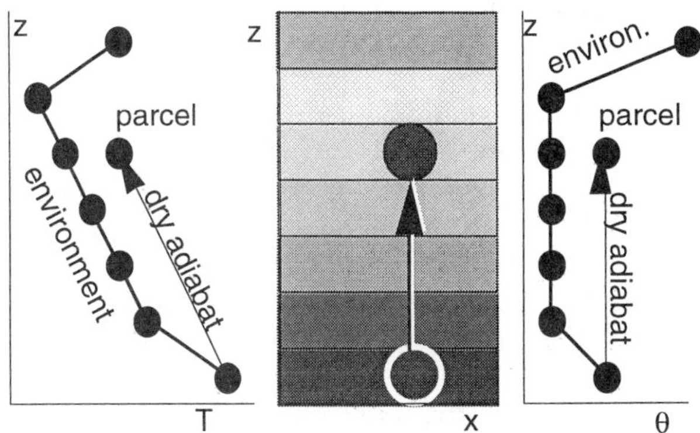

Figure 6.9
Comparison of temperature T(z) and potential temperature θ(z) methods of representing parcel movement through the environment. Darker shades in the center sketch represent warmer temperatures T. Horizontal distance is x.

$$\frac{F}{m} = \frac{\theta_{v\ p} - \theta_{v\ e}}{T_{v\ e}} \cdot |g| \equiv g' \qquad (6.2c)$$

An example is sketched in Fig 6.9. The left diagram represents the parcel and environment as it would be plotted on a thermo diagram (Fig 6.2). It cools as it rises. The right diagram represents the same situation, but in terms of potential temperature θ vs. height z (Fig 6.3). The center sketch shows the parcel rising through the environment, where the darker shading represents warmer temperatures T.

Solved Example

Find the buoyant force acting on an air parcel of temperature $T_p = 20°C$ in an environment of temperature $T_e = 15°C$.

Solution

Given: $T_p = 20+273 = 293$ K , $T_e = 15+273 = 288$ K
Find: $F/m = ?$ m·s^{-2}

Assume dry air, so $T_v = T$.
Use eq. (6.2):
F/m = (288 K – 293 K)·(–9.8 m·s^{-2}) / 288·K
= <u>0.17</u> m·s^{-2}

Check: Physics OK. Units don't look like a force per unit mass, but using conversions from Appendix A, the answer can also be written as 0.17 N/kg. Thus, acceleration is equivalent to force per unit mass in classical physics.
Discussion: The positive answer means an upward buoyant force. Hot air indeed rises.

Thermo diagram Fig 6.2 also gives θ, so instead of solving equations we can pick-off θ from the graph. The buoyancy of an air parcel at any height is given by the potential temperature difference between the parcel and the environmental sounding.

The advantage of using potential temperature is that the parcel maintains constant θ as it moves dry adiabatically. That is, it moves vertically on the graph. It is for this reason that the θ vs. z thermo diagram of Fig 6.3 was designed. Also, as a conserved variable it is convenient to use in numerical weather-forecast models (Chapt. 14).

Solved Example

Convert the following sounding to potential temperature, and plot on a linear graph. If a piece of the environment at P=100 kPa separates from the rest and rises as a parcel, what is the resulting potential temperature difference at P = 90 kPa?

P (kPa)	T (°C)
80	10
90	10
99	19
100	25

Solution

Given: Sounding above

Find: $\theta(z)$, and $\Delta\theta$ at P = 90 kPa

P (kPa)	θ (°C)
80	28.5
90	18.5
99	20
100	25

(using Fig 6.2 to find θ)

Plot θ vs z:

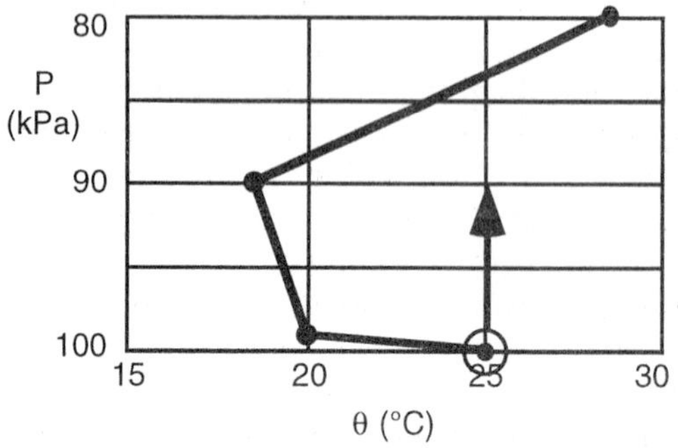

Air parcel retains θ = 25°C as it rises. Thus at P = 90 kPa: $\Delta\theta$ = 25 – 18.5°C = 6.5°C.

Check: Units OK. Physics OK.

Discussion: Hopefully you would find the same temperature difference using Fig 6.2, without any calculation of potential temperature.

STATIC STABILITY

Flow stability indicates whether the atmosphere will develop turbulence or growing waves. **Unstable** air becomes, or is, turbulent. **Stable** air becomes, or is, **laminar** (non-turbulent).

Static stability considers only buoyancy to estimate flow stability, and ignores shears in the mean wind. "Static" means not-moving. **Dynamic stability** considers both wind shears and buoyancy.

To determine static stability, a small portion of the environmental air can be conceptually displaced from its starting point and tracked as an air parcel, such as sketched in the middle part of Fig 6.9. When conceptually moving this air parcel, follow a dry adiabat while the air is unsaturated (i.e., below its LCL), otherwise follow a moist adiabat. At any height, the buoyant force is based on the temperature difference between the parcel and the environment.

Rule for determining Static Stability: •
First, find unstable domains using a nonlocal method described next. Then find neutral and stable layers using the local lapse rate, but only in the remaining regions that were not unstable.

Unstable

If the buoyant force on a displaced parcel is in the same direction as the displacement, then the air between the initial parcel height and the displaced height is unstable (see Fig 6.10). Because of this instability, the parcel will continue accelerating in the direction of its initial movement. This results in **convective** circulations, and perhaps convective clouds.

In some environments, it is possible for such an unstable parcel to move across large vertical distances before reaching a height where there is no net buoyancy force. In such a case, the whole vertical domain of buoyant movement of the air parcel is statically unstable. Thus, statically unstable air must be determined by **nonlocal** air parcel movement.

Rule for determining Unstable regions: •
Parcels from each relative θ maximum in the environmental sounding should be conceptually lifted adiabatically, and those from each relative minimum should be lowered adiabatically, until they hit the sounding again, or hit the ground (Fig 6.10). The domains spanned by these conceptual parcel movements define the statically unstable regions.

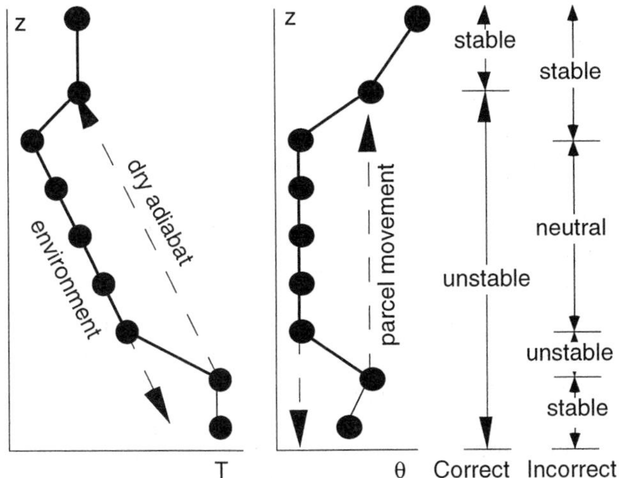

Figure 6.10
Example of correct static stability determination using parcel displacement. Incorrect determination utilizes lapse rates only. If the displaced parcel becomes saturated, follow a moist adiabat.

For this rule, remember to follow a dry adiabat below the LCL of the conceptual parcel, but follow the moist adiabat above the parcel's LCL. Each parcel from each relative max or min will have its own LCL.

By **relative maximum**, we mean a height with warmer θ than the neighboring environmental points just above and below it. Similarly, a **relative minimum** is a height with cooler θ than the neighboring points just above and below it. Either Fig 6.2 or Fig 6.3 can be used to determine static stability, because adiabats and θ are plotted on both.

Neutral

Those portions of the sounding where the environmental lapse rate approximately equals the adiabatic lapse rate, *and which are not otherwise nonlocally unstable*, are approximately statically neutral. A parcel displaced in this environment will feel no buoyant forces.

$$\frac{\Delta T}{\Delta z} \approx -\begin{cases} \Gamma_d & \text{if unsaturated} \\ \Gamma_s & \text{if saturated} \end{cases} \qquad \bullet(6.3a)$$

or

$$\frac{\Delta \theta}{\Delta z} \approx \begin{cases} 0 & \text{if unsaturated} \\ \Gamma_d - \Gamma_s & \text{if saturated} \end{cases} \qquad \bullet(6.3b)$$

Stable

Those portions of the sounding where the temperature-decrease with height is less than adiabatic, *and which are not otherwise nonlocally unstable*, are statically stable. A parcel displaced in this environment will experience buoyant forces opposite to the direction of displacement.

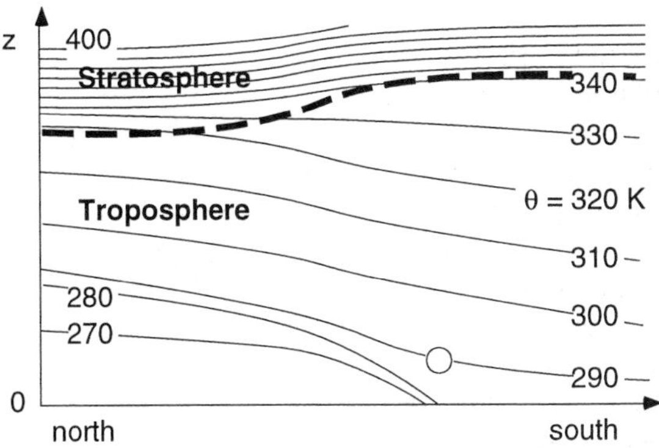

Figure 6.11
Simplified vertical slice through the atmosphere, showing isentropes (lines of equal potential temperature, θ). Heavy dashed line is the tropopause.

$$\frac{\Delta T}{\Delta z} > -\begin{cases} \Gamma_d & \text{if unsaturated} \\ \Gamma_s & \text{if saturated} \end{cases} \qquad \bullet(6.4a)$$

or

$$\frac{\Delta \theta}{\Delta z} > \begin{cases} 0 & \text{if unsaturated} \\ \Gamma_d - \Gamma_s & \text{if saturated} \end{cases} \qquad \bullet(6.4b)$$

Equation (6.4b) states that unsaturated air is statically stable if potential temperature increases with height (unless it was previously determined to be nonlocally unstable).

Fig 6.11 illustrates a hypothetical vertical slice through the atmosphere, showing the **isentropes** (lines of equal potential temperature). Let the circle in that figure represent a weather balloon. On the way up it measures increasing values of θ, resulting in a sounding such as plotted in Fig 4.3. Namely, the troposphere is statically stable. As the balloon rises into the stratosphere, θ increases even more rapidly with height, indicating greater static stability.

The spacing between isentropes is a measure of the static stability. Tight packing of isentropes indicates strong stability. Large spacing between isentropes indicates weak stability.

For example, if the balloon had been launched from the surface at the left part of the figure, it would have first risen through relatively cold air with large spacing between isentropes, indicating weak static stability (near neutral). However, as it rises across the 280 K and the 290 K isentropes, the tight packing there indicates stronger static stability, perhaps marking the top of a frontal surface or the inversion at the top of the mixed layer. Higher still in the troposphere, the medium spacing between isentropes indicates moderately stable air. Finally the balloon enters the stratosphere, where very tight packing of isentropes indicates very strong static stability.

Solved Example

Determine the static stability of the sounding in the solved example from the "Parcel vs. Environment" section.

Solution

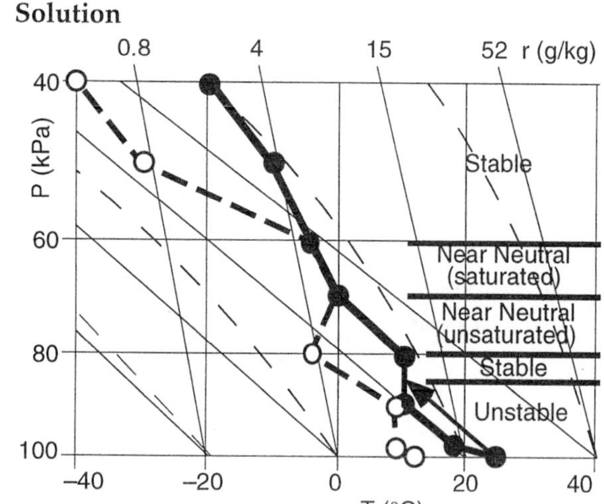

Check: Sketch OK.
Discussion: The unstable layer is turbulent. The near-neutral layers could become turbulent given the slightest wind shear.

Solved Example (§)

Given the following potential temperature measurements. Plot them on the T vs. log P diagram, and on the θ vs. z diagram.

Indicate the static stability of each layer, and find the location of z_i.

z (km)	θ (°C)
3.0	35
2.0	28
1.5	28
1.3	25
1.0	24
0.2	24
0	26

Solution

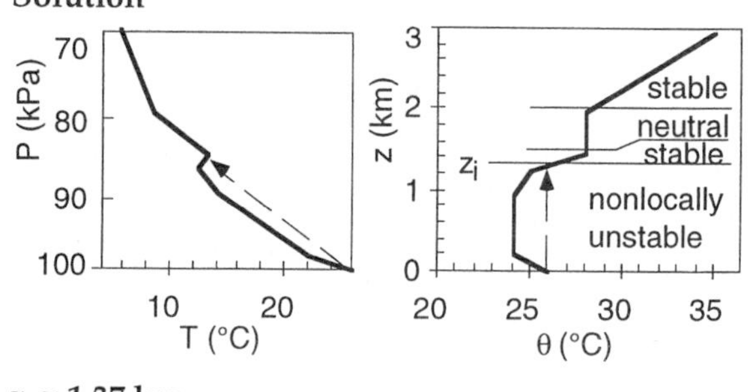

$z_i \cong$ **1.37 km**

Check: Physics OK. Sketch reasonable.
Discussion: θ vs. z is often easier to use than T vs. log(P). Note that z_i is not at the base of the subadiabatic region, but slightly above it due to nonlocal air-parcel rise.

THERMO DIAGRAMS FOR BOUNDARY LAYERS

Sometimes thermodynamic information is needed for the boundary layer, to study thermal rise, turbulence, low-cloud and fog formation, and smoke dispersion. An enlargement of the bottom portion of the two thermo diagrams is plotted as Figs 6.12 and 6.13.

Mixed-Layer Depth Determination

When the ground is warmer than the air, a convective **mixed layer** is created that is nonlocally unstable. The height z_i of the top of this mixed layer is defined as the top of the nonlocally unstable region. The easiest way to determine this height is to use a θ vs. z thermo diagram.

First, obtain a sounding of temperature or potential temperature vs. height. By plotting the sounding on this diagram and conceptually lifting an air parcel from the surface, z_i is found as the height where this rising parcel hits the sounding.

Other measures of stability, and applications of the mixed layer, are given in Chapts. 4 and 17.

Figure 6.12 **(next page)**
T vs. log(P) thermodynamic diagram for the boundary layer. Thin vertical lines are isotherms T, labeled at bottom in °C. Thin horizontal lines are isobars P, labeled at left in kPa. Thin dashed lines nearly horizontal are height contours z, labeled at right in km. Thin diagonal dotted lines slightly-off-from-vertical are isohumes of mixing ratio r, labeled at the top in g/kg. All these thin lines represent the state of the air. Thick lines represent adiabatic processes that can change the state. Heavy diagonal solid lines are dry adiabats, labeled by potential temperature θ at the bottom in °C. Heavy curved dashed lines are moist (saturated) adiabats, labeled by liquid-water potential temperature θ_L at the bottom in °C. [Potential temperature is defined using a reference pressure of 100 kPa, and height is set to zero there.]

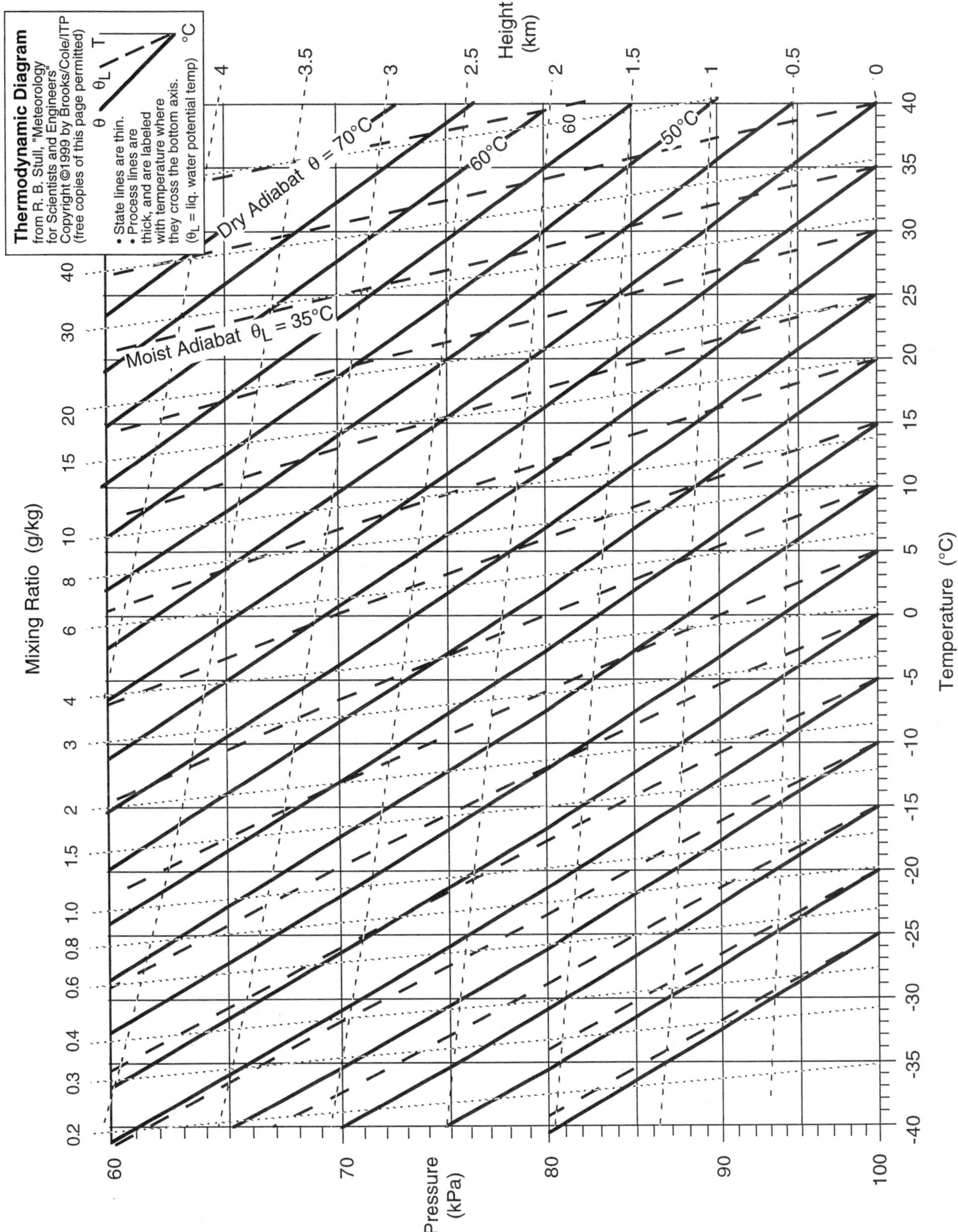
Thermodynamic Diagram
from R. B. Stull, "Meteorology for Scientists and Engineers"
Copyright ©1999 by Brooks/Cole/ITP
(free copies of this page permitted)
θ
θL
T
• State lines are thin.
• Process lines are thick, and are labeled with temperature where they cross the bottom axis.
(θL = liq. water potential temp)
°C
Dry Adiabat θ = 70°C
60°C
50°C
60
Moist Adiabat θL = 35°C
Height (km)
4
3.5
3
2.5
2
1.5
1
0.5
0
Mixing Ratio (g/kg)
0.2
0.3
0.4
0.6
0.8
1.0
1.5
2
3
4
6
8
10
15
20
30
40
Temperature (°C)
-40
-35
-30
-25
-20
-15
-10
-5
0
5
10
15
20
25
30
35
40
Pressure (kPa)
60
70
80
90
100

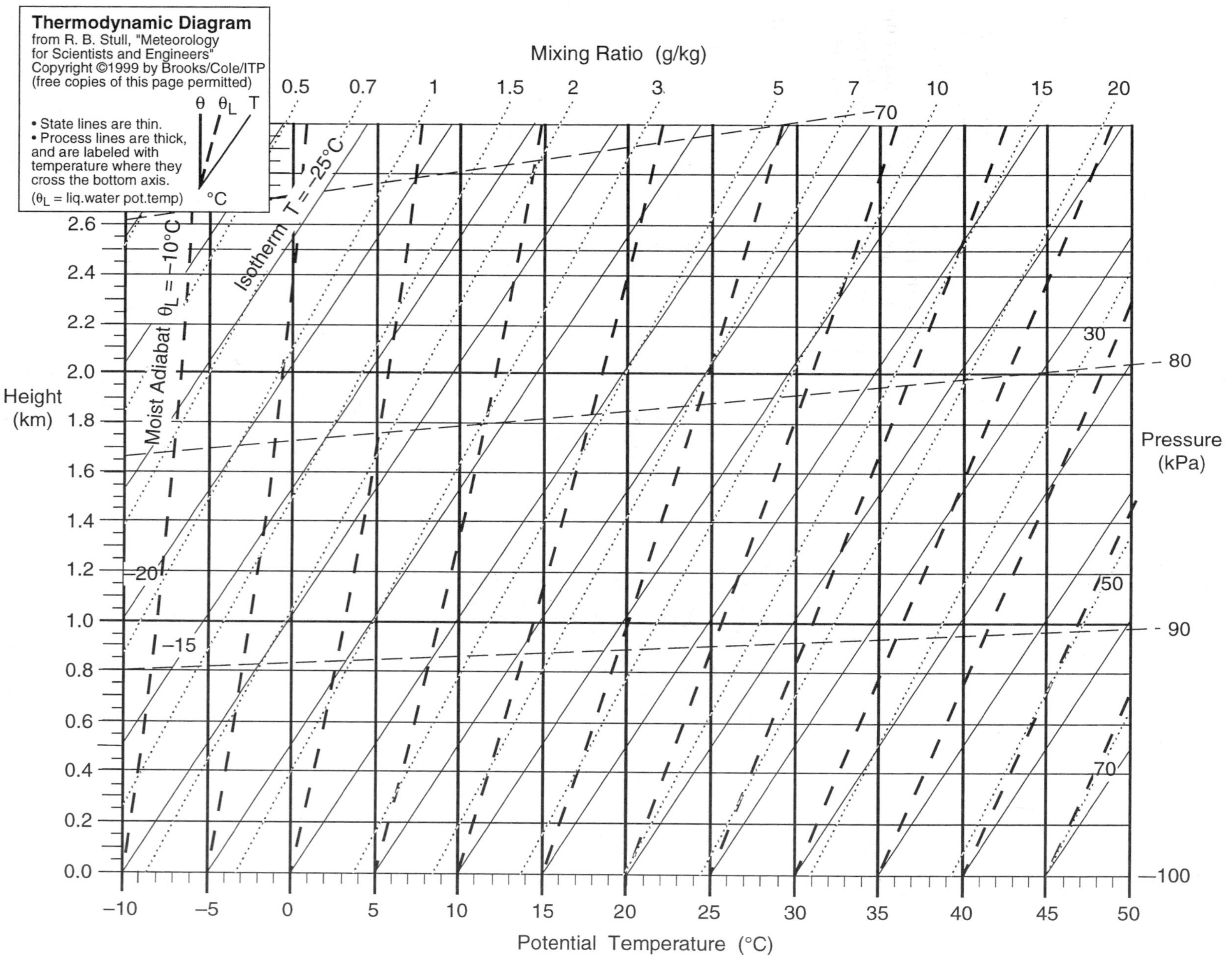
Thermodynamic Diagram
from R. B. Stull, "Meteorology for Scientists and Engineers"
Copyright ©1999 by Brooks/Cole/ITP
(free copies of this page permitted)
θ θL T
°C
• State lines are thin.
• Process lines are thick, and are labeled with temperature where they cross the bottom axis.
(θL = liq.water pot.temp)
Mixing Ratio (g/kg)
0.5 0.7 1 1.5 2 3 5 7 10 15 20
Height (km)
0.0 0.2 0.4 0.6 0.8 1.0 1.2 1.4 1.6 1.8 2.0 2.2 2.4 2.6
Pressure (kPa)
70 80 90 100
Potential Temperature (°C)
−10 −5 0 5 10 15 20 25 30 35 40 45 50
Isotherm T = −25°C
Moist Adiabat θL = −10°C
−20 −15 30 50 70

Figure 6.13 (previous page)
θ vs. z thermodynamic diagram for the boundary layer. Thick vertical lines are adiabats θ, labeled at bottom in °C. Thin horizontal lines are height z, labeled at left in km, with the top of the diagram at 3 km. Thin diagonal solid lines are isotherms T, labeled at bottom in °C. Thin nearly-horizontal dashed lines are isobars P, labeled at right in kPa. Thin diagonal dotted lines are isohumes of constant mixing ratio, r, labeled at top in g/kg. Thick dashed diagonal lines are moist (saturated) adiabats, labeled by liquid water potential temperature θ_L at bottom in °C. [Potential temperature is defined using a reference pressure of 100 kPa, and height is set to zero there.]

BRUNT-VÄISÄLÄ FREQUENCY

In a statically **stable** environment, an air parcel behaves somewhat like a mass hanging between two springs (Fig 6.14a). When displaced vertically from its rest position, the mass experiences a restoring force. As it accelerates back toward its equilibrium point, inertia causes it to overshoot. During this overshoot, the net spring force is in the opposite direction. The mass decelerates and changes direction. It overshoots again and again, the net result being an oscillation.

When an air parcel is lifted from its starting equilibrium height, it becomes colder than the environment due to adiabatic cooling (Fig 6.14b & c), and the buoyancy force acts downwards. The parcel then overshoots below its starting point, adiabatically warming during its descent to become hotter than the environment, thereby causing a buoyancy force that acts upward. Thus, buoyancy force acts to restore the parcel to its starting point, but inertia causes it to overshoot, the net result being an oscillation

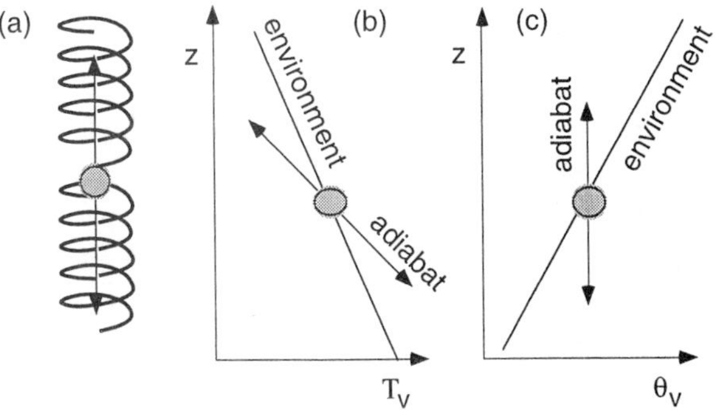

Figure 6.14
(a) A mass constrained between two springs will oscillate up and down if displaced from its equilibrium height. The same will happen for an air parcel displaced vertically in a statically stable environment, regardless of whether it is plotted as (b) virtual temperature T_v or (c) virtual potential temperature θ_v.

The frequency of oscillation is called the Brunt-Väisälä frequency, N_{BV}:

$$N_{BV} = \sqrt{\frac{|g|}{T_v} \cdot \left(\frac{\Delta T_v}{\Delta z} + \Gamma_d \right)} \qquad \bullet(6.5a)$$

or

$$N_{BV} = \sqrt{\frac{|g|}{T_v} \cdot \frac{\Delta \theta_v}{\Delta z}} \qquad \bullet(6.5b)$$

where virtual temperature must be in Kelvin in the denominator, and where the frequency has units of radians/s. For relatively dry air, it is often assumed that $T_v \cong T$, and $\theta_v \cong \theta$. Frequency increases as static stability increases, analogous to stiffening of the spring. The frequency is undefined for unstable air.

The period of oscillation (time to complete one full oscillation) is

$$P_{BV} = \frac{2\pi}{N_{BV}} \qquad (6.6)$$

As the static stability approaches neutral, the period of oscillation approaches infinity. In the real atmosphere, frictional drag on the parcel rapidly damps the oscillation.

Solved Example
Find the period of oscillation of a displaced air parcel in an isothermal environment of 300 K.

Solution
Given: $\Delta T/\Delta z = 0$, $T = 300$ K.
Find: $P_{BV} = ?$ s

Assume dry air. Combine eqs. (6.5a) and (6.6):

$$P_{BV} = \frac{2\pi}{\sqrt{\frac{|g|}{T} \cdot \left(\frac{\Delta T}{\Delta z} + \Gamma_d \right)}}$$

$$P_{BV} = \frac{2\pi}{\sqrt{\frac{(9.8\text{m} \cdot \text{s}^{-2})}{300\text{K}} \cdot (0 + 0.0098\text{K/m})}} = \underline{351 \text{ s}}$$

Check: Units OK. Physics OK.
Discussion: This oscillation period equals 5.9 minutes, which is greater than water-surface wave periods you might see from a ship. These atmospheric oscillations are so slow that you usually cannot see them by eye. However, a time-lapse camera captures these oscillations quite easily, using smoke as a tracer. Also, wave clouds are caused by these oscillations (see Chapt 10).

BEYOND ALGEBRA • Brunt-Väisälä Frequency

Setting up the Problem

An air parcel that moves up and down during an oscillation is governed by Newton's second law, which will be covered in detail in Chapt. 9. Newton's law is:

$$F = m \cdot a$$

where *F* is force, *m* is mass, and *a* is acceleration. Dividing both sides by *m*, and using the definition of acceleration as change of velocity *w* with time *t*:

$$F / m = \mathrm{d}w / \mathrm{d}t$$

Buoyant *F/m* is given by eq. (6.2b), and vertical velocity is defined as change of height *z* with time; namely, $w = \mathrm{d}z/\mathrm{d}t$. Thus:

$$\frac{\theta_{v\ e} - \theta_{v\ p}}{T_{v\ e}} \cdot g = \frac{\mathrm{d}^2 z}{\mathrm{d}t^2}$$

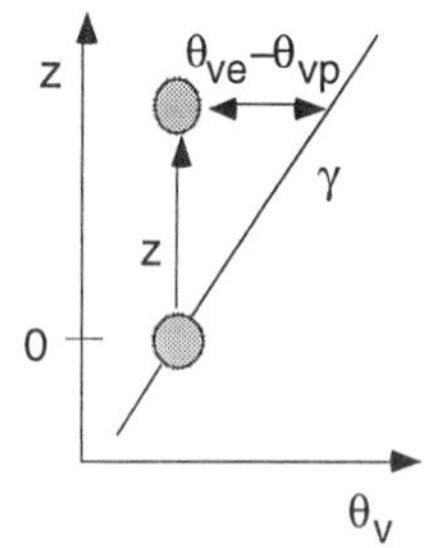

For simplicity, consider the environment shown at right, where environmental virtual potential temperature θ_{ve} increases linearly with height at rate $\gamma = \Delta\theta_v / \Delta z$. Define $z = 0$ at the initial (rest) height of the air parcel. From the geometry shown in the diagram, a parcel displaced upward by distance *z* will be cooler than the environ. by: $\theta_{v\ e} - \theta_{v\ p} = z \cdot \gamma$. Using this in the eq. above gives:

$$g \cdot \gamma \cdot z / T_{v\ e} = \mathrm{d}^2 z / \mathrm{d}t^2$$

Define $N_{BV}^2 \equiv |g| \cdot \gamma / T_{ve}$ (see eq. 6.5b). But $g = -|g|$. Thus:

$$\boxed{\mathrm{d}^2 z / \mathrm{d}t^2 = -N_{BV}^2 \cdot z} \qquad \text{(a)}$$

This is a simple second-order differential equation.

The Solution

From the bag of tricks learned in Diff. Eqs. class, assume a solution of the form: $z = A \cdot \sin(f \cdot t)$ (b) where *A* and *f* are unknown amplitude and oscillation frequency (in radians/s). Plugging (b) into (a) gives:

$$-f^2 \cdot A \cdot \sin(f \cdot t) = -N_{BV}^2 \cdot A \cdot \sin(f \cdot t)$$

Canceling the *A* and sine terms leaves: $f = N_{BV}$.

Thus: $$\boxed{z = A \cdot \sin(N_{BV} \cdot t)} \qquad \text{(c)}$$

Discussion

Thus, N_{BV} is the freq. of oscillation in radians/s. Use (6.6) to rewrite (c) as: $z = A \cdot \sin(2\pi \cdot t / P_{BV})$. As *t* increases from 0 to P_{BV}, the argument of the sine goes from 0 to 2π, causing the sine to make one full cycle.

Because of buoyant forces, air parcels tend to follow **isentropic surfaces** as they advect. In Fig 6.11, suppose that the circle represents an air parcel, not a weather balloon. Further suppose that the potential temperature of that parcel is 290 K. Because that air parcel is surrounded by an environment of the same potential temperature (as indicated by the 290 K isentrope), the parcel experiences no buoyant forces on it. Namely, it is at its equilibrium height.

If the parcel is blown horizontally to the north, it finds itself surrounded by cold air. Thus, the parcel rises until it reaches its equilibrium 290 K isentrope. Such rise happens concurrently with its horizontal motion, so that the parcel follows the isentrope. Similarly, if the parcel were blown south, it would follow the same isentrope and descend. The study of forces, motions, and thermodynamics along surfaces of constant potential temperature is called **isentropic analysis**.

DYNAMIC STABILITY

Richardson Number

Flow can become turbulent in statically stable air if the wind shear is strong enough. Such dynamic stability is indicated by the dimensionless **bulk Richardson number**, *Ri*:

$$Ri = \frac{|g| \cdot (\Delta T_v + \Gamma_d \cdot \Delta z) \cdot \Delta z}{T_v \cdot \left[(\Delta U)^2 + (\Delta V)^2\right]} \qquad \bullet(6.7a)$$

or

$$Ri = \frac{|g| \cdot \Delta\theta_v \cdot \Delta z}{T_v \cdot \left[(\Delta U)^2 + (\Delta V)^2\right]} \qquad \bullet(6.7b)$$

or

$$Ri = \frac{N_{BV}{}^2 \cdot (\Delta z)^2}{(\Delta U)^2 + (\Delta V)^2} \qquad (6.7c)$$

where ΔT_v, ΔU, and ΔV represent virtual temperature and wind differences across a height $\Delta z = z_2 - z_1$. The dry adiabatic lapse rate $\Gamma_d = 9.8$ ·K/km. Temperature in the denominator must be in Kelvin. For relatively dry air, it is often assumed that $T_v \cong T$, and $\theta_v \cong \theta$. Eq. (6.7c) can be used only for statically-stable situations, because N_{BV} is undefined otherwise.

To a good approximation when Δz is small, the flow is dynamically unstable and **turbulent** whenever

$$Ri < Ri_c \qquad \bullet(6.8)$$

The **critical Richardson number** is $Ri_c = 0.25$.

Solved Example

Given air at $z = 2$ km with $U = 13$ m/s, $V = 0$, and $T = 10°C$, while air at 1.8 km has $U = 5$ m/s, $V = 0$, and $T = 10°C$. Find the bulk Richardson number. Is the air turbulent?

Solution

Given: $\Delta T = 0$, $\Delta V = 0$, $\Delta U = 8$ m/s, $\Delta z = 0.2$ km
Find: $Ri = ?$ (dimensionless)

Assume relatively dry air.
Remembering to take all differences (Δ) in the same direction, use eq. (6.7a):

$$Ri = \frac{(9.8\text{m}\cdot\text{s}^{-2})\cdot[0+(9.8\text{K/km})\cdot(0.2\text{km})]\cdot(200\text{m})}{(283\text{K})\cdot\left[(8\text{m/s})^2+(0)^2\right]}$$

$= \underline{\mathbf{0.21}}$.

The flow is **dynamically unstable and turbulent** because $Ri < 0.25$.

Check: Units OK. Physics OK.
Discussion: This low value of Richardson number says that the wind shear is relatively strong, compared to the static stability. As turbulence causes mixing of temperature and winds, the denominator of Ri becomes smaller faster than the numerator. As a result, Ri increases above critical, and turbulence ceases. Thus, turbulence tends to eliminate itself (per LeChatelier's Principle), unless other processes continue to strengthen the wind shear to maintain turbulence.

FOCUS • LeChatelier's Principle

LeChatelier might have had his nose in chemistry beakers, but his notions of chemical equilibria are easily generalized to describe circulations in the atmosphere.

He started with a few reacting chemicals that were in equilibrium with each other. By adding an extra amount of one of the chemicals, he disturbed the equilibrium. However, reactions then occurred to partially eliminate the excess chemical, thereby bringing the reagents in the beaker to a new equilibrium.

An analogous chain of events occurs when solar heating induces instabilities in the atmosphere. The atmosphere generates circulations to eliminate the instability and reach a new equilibrium. Some of these thermal circulations result in clouds and turbulence.

For situations where the atmosphere is continuously being destabilized by solar heating of the ground, the atmosphere continuously reacts by creating thermal after thermal or cloud after cloud. Similarly, continuous dynamic destabilization creates continuous turbulence, as a reaction that reduces wind shear by mixing.

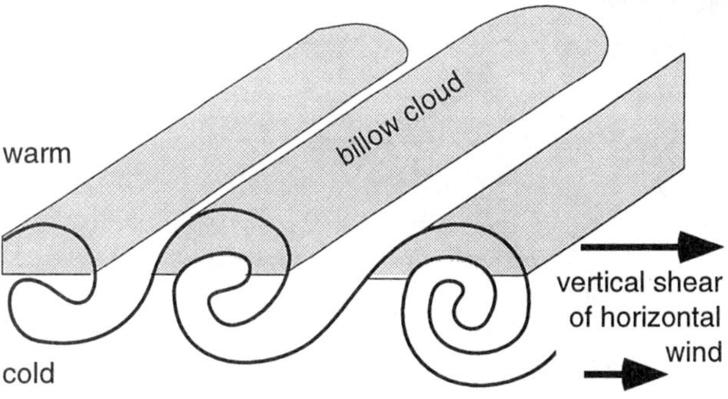

Figure 6.15
Kelvin-Helmholtz (K-H) waves with sufficient moisture to create rows of billow clouds perpendicular to the wind shear.

When condition (6.8) is met, the flow is called **dynamically unstable**. Otherwise, for larger Ri, the flow is **dynamically stable.** Statically unstable air results in negative Richardson numbers, which also implies dynamic instability.

Air that is just becoming dynamically unstable often forms **Kelvin-Helmholtz (K-H) waves** in the air, which look like ocean waves breaking on shore (Fig 6.15). When there is sufficient moisture in the air to make these waves visible, they appear as **billow clouds** (see Ahrens for photographs of these clouds).

Turbulence Determination

Both dynamic stability and static instability give incomplete measures of the existence of turbulence. Static stability does not include the effects of shear in turbulence generation. Dynamic stability does not include the nonlocal processes that can generate turbulence. Thus, to determine whether the flow is turbulent, you must compute both the static stability (using the nonlocal methods given in this chapter), and the dynamic stability.

Turbulence Determination Rule: •

1) To be turbulent, the flow must be either statically OR dynamically unstable.
2) To be laminar, the flow must be both statically AND dynamically stable.

SUMMARY

The **thermodynamic diagram** is a tool to simplify analyses of stability and thermodynamics. The state of the air, as well as processes that can change the state, are included on thermo diagrams. Processes include rising or sinking air, condensation, precipitation, and radiative cooling. With such a diagram, one can find the amount of liquid water in a cloud, the height of cloud base, and the cloud depth.

As winds move the air from place to place, an **air parcel** of one temperature can become immersed in an **environment** of different temperatures. Such a parcel experiences **buoyant forces**, which causes it to rise or sink. Parcels displaced from their starting location will accelerate further away in a statically **unstable** environment, but will tend to return to their starting height in a statically **stable** environment.

In a statically stable environment, air parcels can oscillate vertically at the Brunt-Väisälä frequency. When **wind shear**, the change of wind speed or direction with height, is considered, statically stable air can be dynamically unstable and turbulent.

Threads

The growth of thunderstorms (Chapt. 15) fueled by warm humid air, and the energy available to hurricanes (Chapt. 16) are analyzed using thermo diagrams. Thermo diagrams are used to study the vertical structure of the whole atmosphere (Chapt. 1), the boundary layer (Chapt. 4), dispersion of air pollutants (Chapt. 17), and frontogenesis (Chapt. 12).

The wavelength of mountain waves (Chapt. 10) and formation of lenticular clouds depend on the Brunt-Väisälä frequency. Anabatic and katabatic winds (Chapt. 10) are driven by buoyant forces. Boundary-layer turbulence (Chapt. 4) is triggered by both static and dynamic stability.

EXERCISES

Numerical Problems

Use the following environmental sounding to answer the exercises. Assume $V = 0$ everywhere.

P(kPa)	T(°C)	T_d(°C)	U (m/s)
60	–5	–30	40
70	–5	–20	40
80	2	0	10
87	7	4	5
98	16	6	5
100	20	9	0

N1. Plot the sounding on a thermodynamic diagram. Use solid dots (of red color if possible) for the temperature and mark an "X" (green color if available) for the dew points. Draw a line to connect the temperature dots, starting from the bottom up, and draw another lines for the X's.

a. Use a T vs. log(P) diagram

b. Use a θ vs. z diagram

N2. a. Find the potential temperature at each height, and plot it on a linear graph of θ vs P.

b. Calculate and plot θ_v vs P.

N3. Determine the static stability at all heights in the environment. Hint: neglect moisture & condensation.

N4. Determine if there is a mixed layer. If so, find the mixed layer depth z_i.

N5. Determine the Richardson number and the dynamic stability at all heights, and indicate which layers are turbulent.

N6. Using the thermo diagram, determine the pressure-height of the LCL for an air parcel rising from 100 kPa. How does this compare to the height using a formula for the LCL?

N7. For a parcel rising from 100 kPa, at what pressure-heights are cloud base and cloud top? What is the mixing ratio of water vapor in the cloud at cloud top? What is the value of liquid water mixing ratio in the cloud at cloud top? [Hint: see Chapt. 7.]

N8. a. Starting with the answer from the previous problem, assume that all of the liquid water falls out as precipitation. If the cloud-top parcel is then lowered to 100 kPa, what is its temperature and dew point?

b. Similar to (a) but only half of the liquid water falls out as precipitation.

N9. a. What is the relative humidity of the air parcel from the previous problem (part a) at 100 kPa?

b. What is the relative humidity of the air parcel from the previous problem (part b) at 100 kPa?

c. What is the value of relative humidity of the initial environment at 100 kPa?

N10. a. If a parcel taken from the $P = 60$ kPa environment is forced down to 70 kPa, find the buoyant force per mass, and the oscillation frequency.

b. Same as (a), but going from 70 to 98 kPa.

N11. What is the static stability of the different layers in the standard atmosphere?

N12. Using the standard atmosphere, what is the Brunt-Väisälä frequency in the:

a. Bottom layer of the stratosphere?

b. Top layer of the stratosphere?

c. Middle of the mesosphere?

Understanding & Critical Evaluation

U1. Why does an unsaturated, rising air parcel follow a line of constant mixing ratio in a thermo diagram, rather than a line of constant dew-point temperature?

U2. Under what conditions would a rising, unsaturated air parcel not follow a dry adiabat in a thermo diagram?

U3. In the full T vs. log$P$ thermodynamic diagram, why do the lines of constant height tilt up to the left, rather than up to the right? If warmer air has greater thickness than colder, then wouldn't the height lines be pushed up on the right side of the diagram? Explain.

U4. In the thermodynamic diagram, the labels at the top of the diagram for mixing ratio are not linear from left to right. Namely, as you look from left to right, the mixing ratio increment between neighboring lines gets larger and larger. Why?

U5. On the full T vs. log$P$ thermo diagram, notice that the dry adiabats are slightly curved. Why are they not straight? (Hint for viewing: hold the page so you can look almost along the page and along the line, rather than perpendicular to the page as is usually done for reading. This allows the curvature of lines to be more apparent. This is the same trick used by carpenters, sighting along a piece of lumber to detect warps and other imperfections.)

U6. a. Why do you need two points to represent one air parcel for unsaturated air?

b. Why do you need two points to represent one air parcel for saturated air?

U7. If you know the starting temperature and pressure of an unsaturated air parcel, and you know its LCL instead of its mixing ratio, is that sufficient information to represent its full thermodynamic state (regarding its temperature and humidity)?

U8. The method of moving air parcels up and down in a thermo diagram assumes that the background environment is steady. Is that a good assumption? When might that assumption be violated? Also, when an air parcel moves from its starting point, what happens to the hole it leaves behind?

U9. Suppose an air parcel is surrounded by an environment that has the same density. The net buoyant force on the parcel would be zero. Yet gravity is still acting on the parcel. Why doesn't gravity pull the parcel downward?

U10. Metal ships float in water. Is it possible for metal to float in air, even if it is not moving? Discuss.

U11. Eq. (6.1) gives buoyant force in terms of object (i.e., the air parcel) density minus surrounding fluid (i.e., environment) density. Yet eq. (6.2) gives the force in terms of fluid minus object. Why the sign change? Is it a mistake? Derive (6.2a) from (6.1).

U12. On the θ vs. z thermo diagram, why do the moist adiabats tilt to the right with increasing altitude, while the same moist adiabats tilt to the left on the T vs. log$P$ thermo diagram?

U13. Start with $T = 25°C$ and $T_d = 20°C$ for an air parcel at $P = 100$ kPa (assumed to be at $z = 0$). Plot this starting condition on copies of both versions of the boundary-layer thermodynamic diagram (T vs. log P, and θ vs. z).

a. Lift the parcel to the LCL on both diagrams, and compare the LCL heights. At this point, compare the values of T, r, and θ on both diagrams.

b. Continue lifting to a height of 2 km, and compare the values of T, r, r_T, and θ on both diagrams. Discuss.

U14. Fig 6.10 shows correct and incorrect methods of determining static stability. What is bad about using the incorrect method?

U15. The Brunt-Väisälä frequency is undefined for statically unstable air. Why? If θ decreases with increasing z, calculate the Brunt-Väisälä frequency.

U16. Why is the factor 2π in the numerator of eq. (6.6), rather than a factor of 1 ?

U17. a. Verify that eqs. (6.7) are indeed dimensionless.

b. For zero wind shear, the denominator of eqs. (6.7) are zero and Ri is infinite. Does that mean that there can never be any turbulence when the wind shear is zero?

U18. The thermo diagrams in this chapter indicate a surface pressure of 100 kPa for demonstration purposes. However, the average atmospheric pressure at sea-level is 101.325 kPa, and pressures on any given day can be greater than average. How can you plot a pressure greater than 100 kPa on the thermo diagram?

U19. Explain the last sentence in the caption for Fig 6.5.

U20(§). Create a new thermo diagram using a spreadsheet program, but utilize pressures in the range of 110 kPa to 10 kPa.

U21. Suppose the parcel in Fig 6.9 gradually mixes with its environment as it rises. Plot the path of such a parcel on a graph similar to that figure.

U22. What does it mean if the numerator of the Richardson number is negative?

U23. Compare the bulk Richardson number with the flux Richardson number from Chapter 4. (Hint: Consider Appendix H.)

Web-Enhanced Questions

W1. Search the web for a map of the locations of all the rawinsonde (i.e., upper-air) sites in your country. What are the three sites closest to your location, and what are the identifiers for those sites?

W2. Search the web for a site that allows you to find the name, abbreviation, latitude and longitude, altitude, and other characteristics of rawinsonde sites.

a. Type in the name of a major city, and determine if it has a rawinsonde launch site.

b. Compare the WMO (World Meteorological Organization) and ICAO (International Civil Aviation Organization) identification codes for several stations.

W3. Search the web for soundings plotted on other types of thermodynamic diagrams (e.g., **pseudoadiabatic diagrams, tephigrams, Stuve diagrams, skew-T log P diagrams**, etc.). Instead of focusing on the particular sounding, look at the background diagram, and identify each type of line in it (dry adiabats, moist adiabats, isobars, isotherms, isohumes, height contours, etc.). How many different types of thermo diagrams can you find on the web?

W4. Get from the web the current surface T and T_d for the weather station closest to you (or to a station specified by your instructor).

a. Plot those values on a thermo diagram, and determine the LCL.

b. If there are cumuliform clouds at the time of this data observation, also get cloud-base height from the web, and compare with the LCL. (Hint: search the web for METAR reports of cloud-base altitudes for the airport closest to you.)

W5. Get from the web the current sounding for the rawinsonde (upper-air) site closest to you (or to an upper-air site specified by your instructor). Plot this on a thermo diagram, using different colors for T and T_d. For air lifted from the surface, find the LCL. If this lifted air at the LCL is buoyant, continue lifting it until it becomes neutrally buoyant, and mark this height as cloud-top height.

W6. Get from the web a sequence of soundings (once every 12 h if available, otherwise once every 24 h) from the past few days for a rawinsonde (upper-air) site near you (or a site specified by your instructor).

a. Compare these soundings, to see how the weather evolves.

b. If you have appropriate software, put each sounding as a frame in a movie, and play the movie to see the soundings evolve.

c. Check the weather maps to see if frontal passage will happen during the next several days at your rawinsonde site. If so, download and save soundings from just before, during, and after frontal passage, and compare and discuss the results.

W7. Get from the web a sounding from just before a rain event in your area (or for a region specified by your instructor). From this sounding, conceptually lift an air parcel from the boundary layer to a reasonable height for cloud top. Calculate the liquid water mixing ratio for water that would have condensed during the rise of an air parcel from the surface. Compare this to the amount of precipitation that fell, and discuss why they are different.

W8. On some soundings plotted on the web, other information is also plotted (such as winds). Also, sometimes tables of information such as LCL are given. The web sites that serve such detailed soundings will often have an information page that you can select to learn the definitions of the other terms in such a table. Access and print those definitions, and use them to interpret the info table that is given with the sounding.

W9. Access the most recent sounding from the rawinsonde site closest to you (or a site specified by your instructor).

a. Determine the domains of different static stability in that sounding, and indicate them on or next to the sounding similar to that shown in Fig 6.10 (using the correct version only, please).

b. For the statically stable regions of the sounding, determine the Brunt-Väisälä frequency for each of those regions. (For simplicity, you can neglect the moisture contribution to virtual temperature for this exercise, and just use the air temperature in the calculation. However, normally you should not neglect the moisture contribution to virtual temperature when you calculate stabilities and Brunt-Väisälä frequencies.)

c. For each layer in the sounding, calculate the Richardson number, using the winds and temperatures at the top and bottom of each layer. The resulting Richardson numbers would apply across the whole layer between the top and bottom heights from which temperature and winds were obtained. But each different layer probably has a different Richardson number.

d. Using both (b) and (c), determine the ranges of heights that are likely to be turbulent.

W10. Access the most recent early morning sounding from the rawinsonde site closest to you (or a site specified by your instructor). How warm must the surface temperature be later in the day in order to allow cumuliform cloud tops to reach an altitude where the pressure is 50 kPa? Do you think that such warming is possible on this day?

W11. Access the most recent sounding from the rawinsonde site closest to you (or a site specified by your instructor), and determine the depth of the boundary layer.

Synthesis Questions

S1. The denominator of eq. (6.7c) looks something like a kinetic energy per unit mass (ignoring the missing factor of 1/2). Could the numerator be interpreted as a potential energy per unit mass? Explain.

S2. Why did Archimedes run naked through the streets of Syracuse when he discovered the principle of displacement and buoyancy? [Note: Archimedes also devised city defenses using lenses to blind the enemy and cranes to capsize their ships, but the Romans conquered Syracuse anyway, and killed him.]

S3. Sketch lines in a thermo diagram for a world where condensation of water causes cooling of the air instead of warming. How would the weather and climate be different, if at all? Why?

S4. What if buoyancy depended on the temperature of an air parcel instead of the temperature difference between the parcel and its environment. How would weather and climate be different, if at all? Why?

S5. The Brunt-Väisälä frequency implies oscillations, and oscillations imply waves in the air. Could there really be waves in the air? If so, where do they "come ashore" and break?

S6. Why isn't vertical velocity included in the definition of the Richardson number?

S7. What if global warming caused the atmosphere to warm, but the earth's surface stayed about the same temperature. How would static stability and vertical cloud growth be affected, and how would the resulting weather and climate be different, if at all?

S8. In a nuclear winter scenario (after nuclear bombs have destroyed the earth, and had sprayed bits of earth into the atmosphere), the stratosphere would be filled with fine dust and dark soot particles. Given the short and long wave radiation characteristics that you have studied in earlier chapters, how might the static stability of the troposphere change, and what effect would it have on weather and climate? (Hint: Assume that you are a cockroach, because all other life forms would have been destroyed, so you are the only species left to do this important calculation – as if you would care.)

CLOUD FORMATION

CONTENTS

7 Clouds have immense beauty, variety, and complexity. They show weather patterns on a global scale, as viewed by satellites. Yet they are made of tiny droplets that fall gently through the air. Clouds can have richly complex fractal shapes, and a wide distribution of sizes.

Clouds form when air becomes saturated. Saturation can occur by adding water, by cooling, or by mixing; hence, Lagrangian water and heat budgets are useful. The buoyancy of the cloudy air and the static stability of the environment determine the vertical extent of the cloud.

While fogs are clouds that touch the ground, their location in the boundary layer means that turbulent transport of heat and moisture from the underlying surface is important. Knowing such transport, the formation, growth, and dissipation of fog can be estimated in certain situations.

CLOUD DEVELOPMENT

Active – Cumuliform

Cumuliform clouds form in statically unstable air. These are dynamically **active**, in the sense that buoyant forces enhance their vertical growth. Once triggered, they continue to grow and evolve somewhat independently.

These look like fluffy puffs of cotton, cauliflower, rising castle turrets, or in extreme instability as anvil-topped thunderstorms that look like big mushrooms. These clouds have diameters roughly equal to their heights above ground (Fig 7.1). **Cloud morphology** is the study and classification of cloud shapes.

Clouds along and behind cold fronts are typically cumuliform. These include: **stratocumulus**, **cumulus humilis**, **cumulus mediocris**, **cumulus congestus**, and **cumulonimbus** (thunderstorms). On a thermo diagram, **cloud top** is where the buoyant cloud parcel crosses the environmental sounding, and loses its positive buoyancy (Fig 7.1).

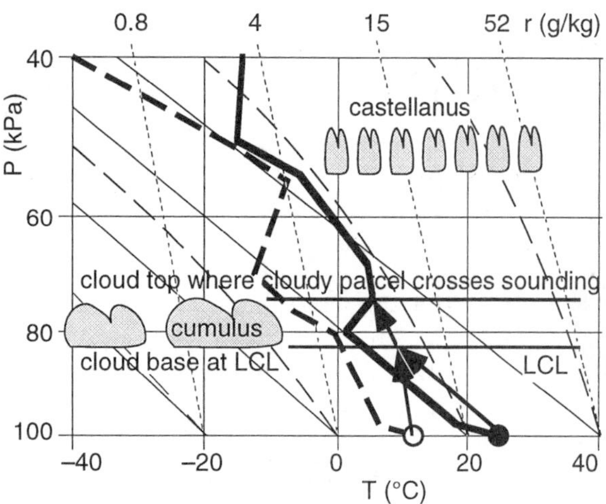

Figure 7.1
Characteristic sounding for cumulus humilis and altocumulus castellanus clouds.

Almost all active cumuliform clouds form within convective updrafts rising from the ground. The reason is that unstable air tends to turbulently stabilize itself, unless there is continued destabilization being forced externally. The ground is the most common location for continued destabilization, because of the solar heating of the surface.

Solved Example

Use the sounding in the solved example of the "Parcel vs. Environment" section of Chapter 6. Where is cloud base and top for a parcel rising from the surface?

Solution

Given: previous sounding
Find: P_{base} = ? kPa, P_{top} = ? kPa

First, plot the sounding on a copy of Fig 6.2. Then conceptually lift a parcel dry adiabatically from the surface until it crosses the isohume from the surface. That LCL is at P_{base} = 80 kPa.

However, the parcel never gets there. It crosses the environment below the LCL, at roughly P = 83 kPa. Neglecting any inertial overshoot of the parcel, it would have zero buoyancy and stop rising 3 kPa below the LCL. Thus, **there is no cloud**.

Check: Units OK. Physics OK.
Discussion: Trick question.

However, a type of cloud known as **castellanus** can form in layers of unstable air aloft. These layers are forced by **differential advection**; namely, by wind blowing in air of different temperatures from different directions at different heights. For castellanus clouds, differential advection creates a layer of warm air under colder air, which is statically unstable. Castellanus clouds are distinctive because their diameters are small compared to their height above ground, making them look like castle turrets (Fig 7.1).

Passive – Stratiform

Stratiform clouds form in statically stable air. These are dynamically **passive**, in the sense that buoyant forces suppress vertical growth. They exist only while some external process causes lifting to overcome the buoyancy.

These are layered clouds that look like sheets or blankets covering wide areas (Fig 7.2). Clouds along warm fronts are typically stratiform, including: **cirrus, cirrostratus, cirrocumulus, altostratus, altocumulus, stratus**, and **nimbostratus**.

All of the stratiform clouds listed above are not convectively coupled with the ground underneath them. Such clouds are formed by advection of humid air up the gently-inclined warm-frontal surface from moisture sources hundreds to thousands of kilometers distant.

This process is illustrated in Fig 6.11. Let the circle in that figure represent a humid air parcel with potential temperature 290 K. If a southerly wind were tending to blow the parcel toward the north, the parcel would follow the 290 K isentrope like a train on tracks, and would ride up over the colder surface air in the northern portion of the domain.

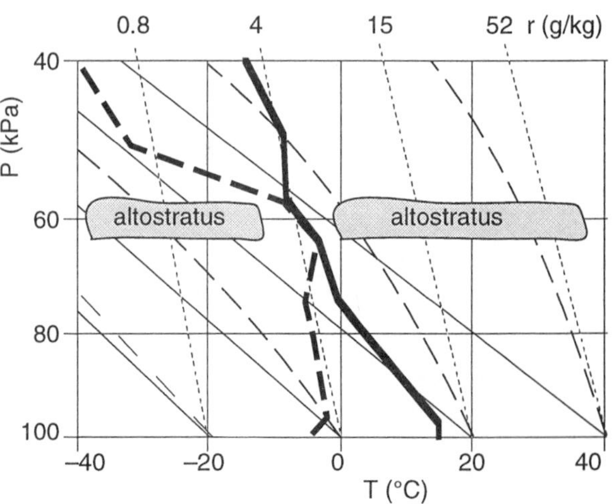

Figure 7.2
Characteristic sounding for stratiform clouds.

The gentle rise of air along the isentropic surface creates sufficient cooling to cause the condensation.

Cirrostratus and altostratus are particularly smooth looking, which implies little or no turbulence within them. The cirrocumulus and altocumulus clouds are lumpy, but the lumps are small. We infer from these small sizes that the turbulent eddies causing these lumps are locally generated within the thin cloud layer, and are not associated with updrafts from the ground. In spite of their lumpiness, cirrocumulus and altocumulus are formed primarily by advection, and are passive layer clouds.

Stratus is a thick, smooth, cloud layer at low altitudes, but this type of cloud is not turbulently coupled with the underlying surface. Nimbostratus clouds are thick enough to allow drizzle or rain to form and fall out.

Other dynamically passive clouds are **cap clouds** on the tops of mountains, and **lenticular** clouds downwind of mountains. These are caused by updrafts forced when the horizontal wind hits sloping terrain and mountains (Chapt. 10). **Arc** or **shelf** clouds near thunderstorms form in air forced upward by undercutting cold air flowing out from the thunderstorms. **Pileus** clouds form in mid-tropospheric humid layers of statically stable air that are forced upward by tops of cumulus congestus clouds rising rapidly from below.

Stratocumulus

Stratocumulus clouds are somewhat different from either active and passive clouds. Unlike passive clouds, they can be formed by updrafts by air rising from near the surface. Unlike active clouds, these updrafts are often not caused by buoyancy, but can be caused by turbulence generated by wind shear. The larger eddy circulations in the turbulence can lift air parcels from the ground up to the LCL, allowing the clouds to form. Unlike stratus clouds, stratocumulus clouds are turbulently coupled with the underlying surface.

In real atmospheres, often there is both buoyancy and shear that contribute to the updrafts into the base of stratocumulus clouds. Also, IR radiation emitted upward from cloud top can cool the cloud top, creating cool air parcels that sink as upside-down thermals. The resulting turbulent circulation can contribute to the lumpiness of the stratocumulus deck (cloud layers are sometimes called **cloud decks**).

CLOUD SIZES

Cumuliform clouds typically have diameters roughly equal to their depths, as was mentioned in the previous section. For example, a fair weather cumulus cloud typically averages about 1 km in size, while a thunderstorm might be 10 km.

Not all clouds are created equal. At any given time the sky contains a spectrum of cloud sizes that has a **lognormal distribution** (Fig 7.3, eq. 7.1)

$$f(X) = \frac{\Delta X}{\sqrt{2\pi} \cdot X \cdot S_X} \cdot \exp\left[-0.5 \cdot \left(\frac{\ln(X / L_X)}{S_X} \right)^2 \right] \qquad (7.1)$$

where X is the cloud diameter or depth, ΔX is a small range of cloud sizes, $f(X)$ is the fraction of clouds of sizes between $X-0.5\Delta X$ and $X+0.5\Delta X$, L_x is a location parameter, and S_x is a dimensionless spread parameter. These parameters vary widely in time and location.

According to this distribution, there are many clouds of nearly the same size, but there also are a few clouds of much larger size. This causes a skewed distribution with a long tail to the right (see Fig 7.3).

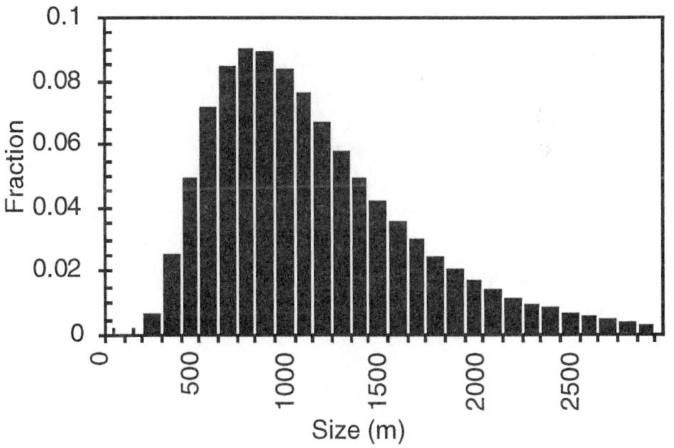

Figure 7.3
Lognormal distribution of cloud sizes.

> **Solved Example (§)**
> Use a spreadsheet to find and plot the fraction of clouds ranging from X = 50 to 2950 m width, given ΔX = 100 m, S_X =0.5, L_X = 1000 m.
>
> **Solution**
> Given: ΔX = 100 m, S_X =0.5, L_X = 1000 m.
> Find: $f(X)$= ?
>
> *(continued next column)*

Solved Example (§) *(continuation)*

Solve eq. (7.1) on a spreadsheet:

X (m)	f(X)
50	2.558x10-8
150	0.0004
250	0.0068
350	0.0252
450	0.0495
550	0.0710
etc.	etc.

Sum of all f = 0.986

Check: Units OK. Physics almost OK, but the sum of all f should equal 1.0, representing 100% of the clouds. The reason for the error is that we should have considered clouds even larger than 2950 m, because of the tail on the right of the distribution. Also smaller ΔX would help.

Discussion: Although the dominant cloud width is about 800 m for this example, the long tail on the right of the distribution shows that there are also a small number of large-diameter clouds.

FRACTAL CLOUD SHAPES

Fractals are patterns made of the superposition of similar shapes having a range of sizes. An example is a dendrite snow flake. It has arms protruding from the center. Each of those arms has smaller arms attached, and each of those has even smaller arms. Other aspects of meteorology exhibit fractal geometry, including lightning, turbulence, and clouds.

Fractal Dimension

Euclidian geometry includes only integer dimensions; for example 1-D, 2-D or 3-D. Fractal geometry allows a continuum of dimensions; for example $D = 1.35$.

Fractal dimension is a measure of space-filling ability. Common examples are drawings made by children, and newspaper used for packing cardboard boxes.

A straight line (Fig 7.4a) has fractal dimension $D = 1$; namely, it is one-dimensional both in the Euclidian and fractal geometry. When toddlers draw with crayons, they fill areas by drawing tremendously wiggly lines (Fig 7.4b). Such a line might have fractal dimension $D = 1.7$, and gives the impression of almost filling an area. Older children succeed in filling the area, resulting in fractal dimension $D = 2$.

A different example is a sheet of newspaper. While it is flat and smooth, it fills only two-dimensional area, hence $D = 2$. However, it takes up more space if you crinkle it, resulting in a fractal dimension of, say, $D = 2.2$. By fully wadding it into a tight ball of $D = 2.7$, it begins to behave more like a three-dimensional object, which is handy for filling empty space in cardboard boxes.

A **zero set** is a lower-dimensional slice through a higher-dimensional shape. Zero sets have fractal dimension one less than that of the original shape. Sometimes it is easier to measure the fractal dimension of a zero set, from which we can calculate the dimension of the original object.

For example, start with the wad of paper having fractal dimension $D = 2.7$. It would be difficult to measure the dimension for this wad of paper. Instead, carefully slice that wad into two halves, dip the sliced edge into ink, and create a print of that inked edge on a flat piece of paper. The wiggly line that was printed (Fig 7.4b) has fractal dimension of $D = 2.7 - 1 = 1.7$, and is a zero set of the original shape. It is easier to measure.

To continue the process, draw a straight line across the middle of the print. The wiggly printed line crosses the straight line at a number of points (Fig 7.4c); those points have a fractal dimension of $D = 1.7 - 1 = 0.7$, and represent the zero set of the print. Thus, while any one point has a Euclidian dimension of zero, the set of points appears to partially fill a 1-D line.

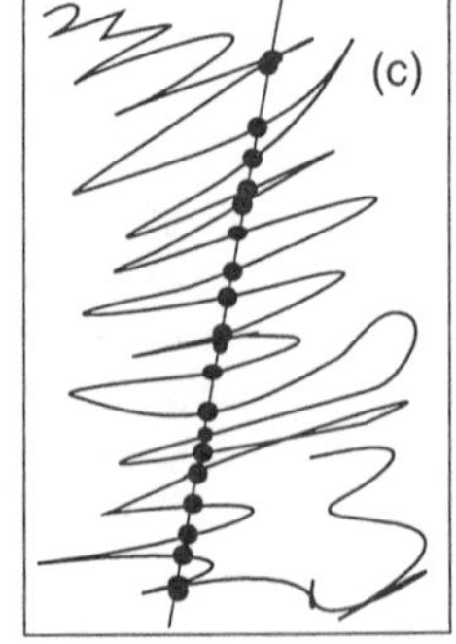

Figure 7.4
(a) A straight line has fractal dimension $D = 1$. (b) A wiggly line has fractal dimension $1 < D < 2$. (c) The zero-set of the wiggly line is a set of points with dimension $0 < D < 1$.

Measuring Fractal Dimension

Consider an irregular-shaped area, such as the shadow of a cloud. The perimeter of the shadow is a wiggly line, so we should be able to measure its fractal dimension. Based on observations of real clouds, $D = 1.35$. If we assume that this cloud shadow is a zero set of the surface of the cloud, then the cloud surface dimension is $D = 2.35$.

A **box-counting** method can be used to measure fractal dimension. Put the cloud-shadow picture within a square domain. Then tile the domain with smaller square boxes, with M tiles per side of domain (Fig 7.5). Count the number N of tiles through which the perimeter passes. Repeat the process with smaller tiles.

Get many samples of N vs. M, and plot these as points on a log-log graph. The fractal dimension D is the slope of the best fit straight line through the points. Define subscripts 1 and 2 as the two end points of the best fit line. Thus:

$$D = \frac{\log(N_1 / N_2)}{\log(M_1 / M_2)} \qquad (7.2)$$

This technique works best when the tiles are small. If the line curves, try to find the slope near large M.

Figure 7.5
Cloud shadow overlaid with small tiles. The number of tiles M per side of domain is: (a) 8, (b) 10, (c) 12, (d) 14.

Solved Example
Use Fig 7.5 to measure the fractal dimension of the cloud shadow.

Solution
For each figure, the number of boxes is:

M	N
8	44
10	53
12	75
14	88

(You might get a slightly different count.)
Plot the result, and fit a straight line.

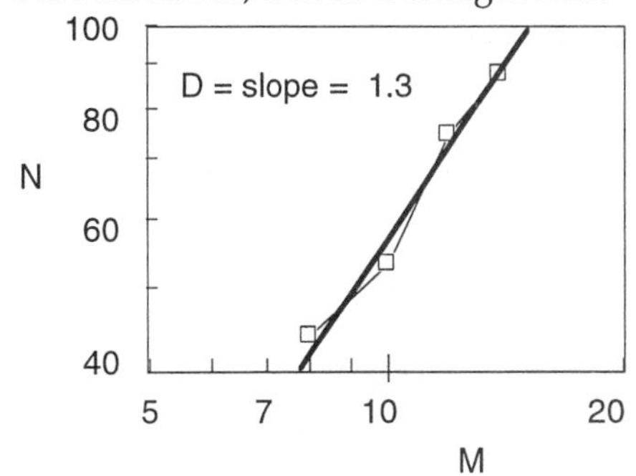

Use the end points of the best-fit line in eq. (7.2):

$$D = \frac{\log(100/40)}{\log(15/7.5)} = \underline{1.32}$$

Check: Units OK. Physics OK.
Discussion: About right for a cloud. The original domain size relative to cloud size makes no difference. Although you will get different M and N values, the slope D will be almost the same.

PROCESSES CAUSING SATURATION

Cooling and Moisturizing

Unsaturated air parcels can reach saturation by three processes: cooling, adding moisture, or mixing. The first two processes are shown in Fig 7.6. where saturation is reached by either **cooling** until the temperature equals the dew point temperature, or **adding moisture** until the dew point temperature is raised to the actual ambient temperature.

The temperature change necessary to saturate an air parcel is:

$$\Delta T = T_d - T \qquad (7.3)$$

Whether this condition is met can be determined by finding the actual temperature change based on the

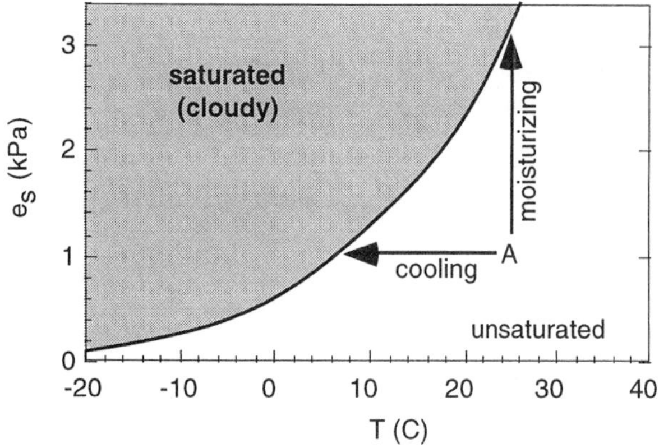

Figure 7.6
Unsaturated air parcel A can become saturated by the addition of moisture, or by cooling. The curved line is the saturation vapor pressure from Chapter 5.

first law of thermodynamics (eq. 3.5 for a moving air parcel, or eq. 3.34 at a fixed location).

The moisture addition necessary to reach saturation is

$$\Delta r = r_s - r \tag{7.4}$$

Whether this condition is met can be determined by using the moisture budget to find the actual humidity change.

In the real atmosphere, sometimes both cooling and moisturizing happen simultaneously. Schematically, this would correspond to an arrow from parcel *A* diagonally to the saturation line of Fig 7.6.

Solved Example

Air at sea level has a temperature of 20°C and a mixing ratio of 5 g/kg. How much cooling or moisturizing is necessary to reach saturation?

Solution

Given: $T = 20°C$, $r = 5$ g/kg
Find: $\Delta T = ?$ °C, $\Delta r = ?$ g/kg.

Use Table 5-1 because it applies for sea level. Otherwise, we can solve equations or use a thermo diagram.
At $T = 20°C$, $r_s = 14.91$ g/kg from the table.
At $r = 5$ g/kg, $T_d = 4$ °C from the table.

Use eq. (7.3): $\Delta T = 4 - 20 =$ **<u>–16°C</u>** needed.
Use eq. (7.4): $\Delta r = 14.91 - 5 =$ **<u>9.9 g/kg</u>** needed.

Check: Units OK. Physics OK.
Discussion: This air parcel is fairly dry. Much cooling or moisturizing is needed.

Mixing

Mixing of two unsaturated parcels can result in a saturated mixture, as shown in Fig 7.7. Jet contrails and your breath on a cold winter day are examples of clouds that form by the mixing process.

Mixing essentially occurs along a *straight* line in this graph connecting the thermodynamic states of the two original air parcels. However, the saturation line (given by the Clausius-Clapeyron equation) is curved, so a mixture can be saturated even if the original parcels are not.

Let m_B and m_C be the original masses of air in parcels *B* and *C*, respectively. The mass of the mixture (parcel *X*) is obviously :

$$m_X = m_B + m_C \tag{7.5}$$

The temperature and vapor pressure of the mixture are the weighted averages of the corresponding values in the original parcels:

$$T_X = \frac{m_B \cdot T_B + m_C \cdot T_C}{m_X} \tag{7.6}$$

$$e_X = \frac{m_B \cdot e_B + m_C \cdot e_C}{m_X} \tag{7.7}$$

Specific humidity or mixing ratio can be used in place of vapor pressure in eq. (7.7).

Instead of using the actual masses of the air parcels in eqs. (7.5) - (7.7), one can use the relative portions that mix. For example, if the mixture consists of 3 parts *B* and 2 parts *C*, then we can use $m_B = 3$ and $m_C = 2$ in the equations above.

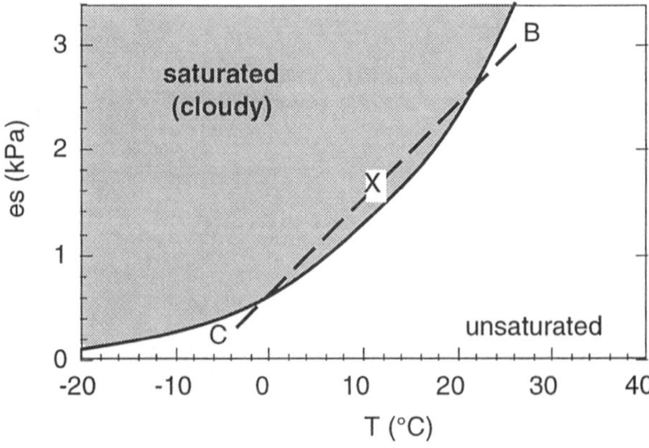

Figure 7.7
Mixing of two unsaturated air parcels B and C, which occurs along a straight line (dashed), can cause a saturated mixture X. The curved line is the saturation vapor pressure from Chapt. 5.

Solved Example

Suppose that parcel *B* has T = 30°C and e = 3.4 kPa, while parcel *C* has T = -4°C and e = 0.2 kPa. If each parcel contains 1 kg of air, then will the mixture be saturated at P = 100 kPa?

Solution

Given: *B* has T = 30°C and e = 3.4 kPa
C has T = -4°C and e = 0.2 kPa.
Find: T = ? °C and e = ? kPa at *X*.

Use eq. (7.5): $m_X = 1 + 1 =$ **2 kg**
Use eq. (7.6): T_X =
[(1kg)·(30°C) + (1kg)·(–4°C)]/(2kg) = **13°C.**
Use eq. (7.7): e_X =
[(1kg)·(3.4 kPa) + (1kg)·(0.2 kPa)]/(2kg) = **1.8 kPa**

Check: Units OK. Physics OK.
Discussion: At T = 13°C, Table 5-1 gives e_s = 1.5 kPa. Thus, the mixture **is saturated** because its vapor pressure exceeds the saturation vapor pressure. This mixture would be cloudy/foggy.

CLOUDS AND UPSLOPE FOG

Clouds usually form by adiabatic cooling of rising air. Air can be rising due to its own buoyancy, or can be forced up over hills or frontal boundaries. Once formed, infrared radiation from cloud top can cause additional cooling to help maintain the cloud.

Upslope fog is formed by adiabatic cooling in rising air that is forced up sloping terrain by the wind. Basically, this fog is a cloud that touches the ground (or more precisely, the ground rises to touch the cloud). As already discussed in Chapter 5, air parcels must rise or be lifted to their lifting condensation level (LCL, eq. 5.8) to form a cloud or upslope fog.

OTHER FOGS

Processes

Fog can form by cooling, moisturizing, and/or mixing. **Radiation** and **advection fogs** are formed by cooling of the air via conduction from the cold ground. **Precipitation** or **frontal fog** is formed by adding moisture, via the evaporation from warm rain drops falling through cooler air.

Steam fog occurs when cold air moves over warm humid surfaces such as unfrozen lakes during early winter. The lake warms the air near it by conduction, and adds water by evaporation. However, this thin layer of moist warm air near the surface is unsaturated. As turbulence causes it to mix with the colder air higher above the surface, the mixture becomes saturated, which we see as steam fog.

Idealized models for formation, evolution, and dissipation of some fog types are given next.

Solved Example

A layer of air adjacent to the surface (where P = 100 kPa) is initially at temperature 20°C and relative humidity 68%. (a) To what temperature must this layer be cooled to form radiation or advection fog? (b) To what altitude must this layer be lifted to form upslope fog? (c) How much water must be evaporated into each kilogram of dry air from falling rain drops to form frontal fog?

Solution

Given: P = 100 kPa, T = 20°C, RH = 68%
Find: a) T_d= ?°C, b) z_{LCL}= ?m, c) r_s=?g/kg

Using Table 5-1: e_s = 2.371 kPa.
Using eq. (5.6): $e = (RH/100\%)\cdot e_s$
$= (68\%/100\%)\cdot(2.371 \text{ kPa}) = 1.612$ kPa

(a) Knowing e and using Table 5-1: T_d = **14°C**.

(b) Using eq. (5.8):
$z_{LCL} = a\cdot[T - T_d] = (0.125 \text{ m/K})\cdot$
[(20+273)K – (14+273)K] = **0.75 km**

(c) Using eq. (5.3), the initial state is:
$r = \varepsilon\cdot e/(P-e) = 0.622\cdot(1.612 \text{ kPa}) / (100 \text{ kPa} - 1.612 \text{ kPa}) = 0.0102 \text{ g/g} = 10.2 \text{ g/kg}$
The final mixing ratio at saturation is:
$r_s = \varepsilon\cdot e_s/(P-e_s) = 0.622\cdot(2.371 \text{ kPa}) / (100 \text{ kPa} - 2.371 \text{ kPa}) = 0.0151 \text{ g/g} = 15.1$ g/kg.
The amount of additional water needed is
$\Delta r = r_s - r = 15.1-10.2 =$ **4.9** g_{water}/kg_{air}

Check: Units OK. Physics OK.
Discussion: For many real fogs, cooling of the air and addition of water via evaporation from the surface happen simultaneously. Thus, a fog might form in this example at temperatures warmer than 14°C.

Solved Example

A 100 m thick layer of air adjacent to a warm lake surface (where P = 100 kPa) is initially at temperature 20°C and relative humidity 68%. How much evaporation from the lake is necessary to form steam fog throughout the whole layer?

Solution

Given: T=20°C, RH=68%, P=100 kPa, z=100 m
Find: mm of lake water evaporated into air

This is an extension of the previous solved example . The initial absolute humidity using eq. (5.5) and saturation absolute humidity are:

$\rho_v = \varepsilon \cdot e \cdot \rho_d / P$
$= (0.622)\cdot(1.612 \text{ kPa})\cdot(1.275 \text{ kg·m}^{-3})/(100 \text{ kPa})$
$= 0.01278 \text{ kg·m}^{-3}$

$\rho_{vs} = \varepsilon \cdot e_s \cdot r_d / P$
$= (0.622)\cdot(2.371 \text{ kPa})\cdot(1.275 \text{ kg·m}^{-3})/(100 \text{ kPa})$
$= 0.01880 \text{ kg·m}^{-3}$

The difference must be added to the air to reach saturation:

$\Delta\rho = \rho_{vs} - \rho_v$
$= (0.01880–0.01278) \text{ kg·m}^{-3} = 0.00602 \text{ kg·m}^{-3}$

Over A =1 m^2 of surface area, volume is $V_{air} = A \cdot z = 100 \text{ m}^3$.

The mass of water needed in this volume is $m = \Delta\rho \cdot V = 0.602$ kg

But knowing liquid water density = 1000 kg·m^{-3}: $V_{liq} = m / \rho_{liquid} = 0.000602 \text{ m}^3$

The depth of liquid water under the 1 m^2 area is $d = V_{liq} / A = 0.000602 \text{ m} =$ **0.602 mm**

Check: Units OK. Physics OK.
Discussion: A little evaporation goes a long way.

Advection Fog

For formation and growth of advection fog, suppose a fogless mixed layer of thickness z_i advects with speed M over a cold surface such as snow covered ground or a cold lake. If the surface potential temperature is θ_{sfc}, then the air potential temperature θ evolution is

$$\theta = \theta_{sfc} + (\theta_o - \theta_{sfc}) \cdot \exp\left(-\frac{C_H}{z_i} \cdot x\right) \quad (7.8)$$

where θ_o is the initial air potential temperature, C_H is the heat transfer coefficient (see eq. 3.21), and x is travel distance over the cold surface. This assumes that there is sufficient turbulence caused by the wind speed to keep the boundary layer well mixed.

BEYOND ALGEBRA • Advection Fog

Derivation of eq. (7.8):

Start with the Eulerian heat balance, neglecting all contributions except for turbulent flux divergence:

$$\frac{\partial\theta}{\partial t} = -\frac{\partial F_{z\ turb}(\theta)}{\partial z}$$

But for a mixed layer of fog, F is linear with z, thus:

$$\frac{\partial\theta}{\partial t} = -\frac{F_{z\ turb\ zi}(\theta) - F_{z\ turb\ sfc}(\theta)}{z_i - 0}$$

If entrainment at the top of the fog layer is small, then $F_{z\ turb\ zi}(\theta) = 0$, leaving:

$$\frac{\partial\theta}{\partial t} = \frac{F_{z\ turb\ sfc}(\theta)}{z_i} \equiv \frac{F_H}{z_i}$$

Use eq. (3.21): $F_H = C_H \cdot M \cdot (\theta_{sfc} - \theta)$. Thus:

$$\frac{\partial\theta}{\partial t} = \frac{C_H \cdot M \cdot (\theta_{sfc} - \theta)}{z_i}$$

If the wind speed is approximately constant with height, then let the Eulerian volume move with speed M:

$$\frac{\partial\theta}{\partial t} = \frac{\partial\theta}{\partial x}\frac{\partial x}{\partial t} \equiv \frac{\partial\theta}{\partial x} \cdot M$$

Plugging this into the LHS of the previous eq. gives:

$$\frac{\partial\theta}{\partial x} = \frac{C_H \cdot (\theta_{sfc} - \theta)}{z_i}$$

To help integrate this, define a substitute variable $s \equiv \theta - \theta_{sfc}$, for which $\partial s = \partial\theta$, thus:

$$\frac{\partial s}{\partial x} = -\frac{C_H \cdot s}{z_i}$$

By separation of variables:

$$\frac{ds}{s} = -\frac{C_H}{z_i} dx$$

Which can be integrated (using the prime to denote a dummy variable of integration):

$$\int_{s'=s_o}^{s} \frac{ds'}{s'} = -\frac{C_H}{z_i} \int_{x'=0}^{x} dx'$$

Yielding:

$$\ln(s) - \ln(s_o) = -\frac{C_H}{z_i} \cdot (x - 0)$$

or

$$\ln\left(\frac{s}{s_o}\right) = -\frac{C_H}{z_i} \cdot x$$

Taking the antilog of each side (i.e., exp):

$$\frac{s}{s_o} = \exp\left(-\frac{C_H}{z_i} \cdot x\right)$$

Upon rearranging, and substituting for s:

$$\theta - \theta_{sfc} = (\theta - \theta_{sfc})_o \cdot \exp\left(-\frac{C_H}{z_i} \cdot x\right)$$

But θ_{sfc} is assumed constant, thus:

$$\boxed{\theta = \theta_{sfc} + (\theta_o - \theta_{sfc}) \cdot \exp\left(-\frac{C_H}{z_i} \cdot x\right)} \quad (7.8)$$

Advection fog forms when the temperature drops to the dew-point temperature T_d. At the surface (more precisely, at z = 10 m), $\theta \cong T$. Thus, setting $\theta \cong T = T_d$ and solving the equation above for x gives the distance over the lake at which fog first forms:

$$x = \frac{z_i}{C_H} \cdot \ln\left(\frac{T_o - T_{sfc}}{T_d - T_{sfc}}\right) \tag{7.9}$$

Surprisingly, neither the temperature evolution nor the distance to fog formation depends on wind speed.

Advection fog, once formed, experiences radiative cooling from fog top. Such cooling makes the fog more dense and longer lasting as it can evolve into radiation fog, described in the next subsection.

Solved Example (§)

Air of initial temperature 5°C and depth 200 m flows over a frozen lake of surface temperature –3°C. If the initial dew point of the air is –1°C, how far from shore will advection fog first form?

Solution

Given: T_o = 5°C, T_d = –1°C,
T_{sfc} = –3 °C, z_i =200 m

Find: x = ? km. Assume smooth ice: C_H = 0.002 .

Use eq. (7.9): x = (200 m/0.002)·
ln[(5–(–3))/(–1–(–3))] = **138.6 km**

Sketch, where eq. (7.8) was solved for T vs. x:

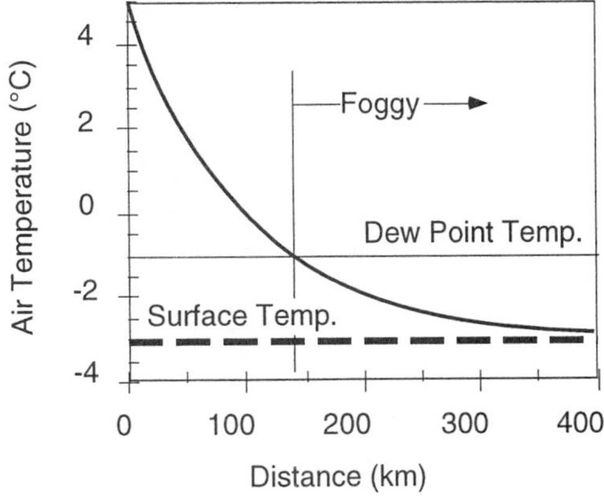

Check: Units OK. Physics OK.

Discussion: If the lake is smaller than 138.6 km in diameter, then no fog forms. Also, if the dew-point temperature needed for fog is colder than the surface temperature, then no fog forms.

Dissipation of advection fog is usually controlled by the synoptic and mesoscale weather patterns. If the surface becomes warmer (e.g., all the snow melts), or if the wind changes direction, then the conditions that originally created the advection fog might disappear. At that point, dissipation depends on the same factors that dissipate radiation fog. Alternately, frontal passage or change of wind direction might blow out the advection fog, and replace the formerly-foggy air with cold dry air that might not be further cooled by the underlying surface.

Radiation Fog

For formation and growth of radiation fog, assume a stable boundary layer forms and grows, as given by eqs. (4.3 - 4.5). If the surface temperature T_s drops below the dew point temperature T_d then fog can form (Fig 7.8). The depth of fog is estimated by the height where the nocturnal temperature profile crosses the initial dew-point temperature.

The time t_o between when nocturnal cooling starts and the onset of fog is

$$t_o = \frac{a^2 \cdot M^{3/2} \cdot (T_{RL} - T_d)^2}{(-F_H)^2} \tag{7.10}$$

where $a = 0.15\ \text{m}^{1/4} \cdot \text{s}^{1/4}$, T_{RL} is the residual layer temperature (extrapolated adiabatically to the surface), M is wind speed in the residual layer, and F_H is the average surface kinematic heat flux. Faster winds and drier air delay the onset of fog. For most cases, fog never happens because night ends first.

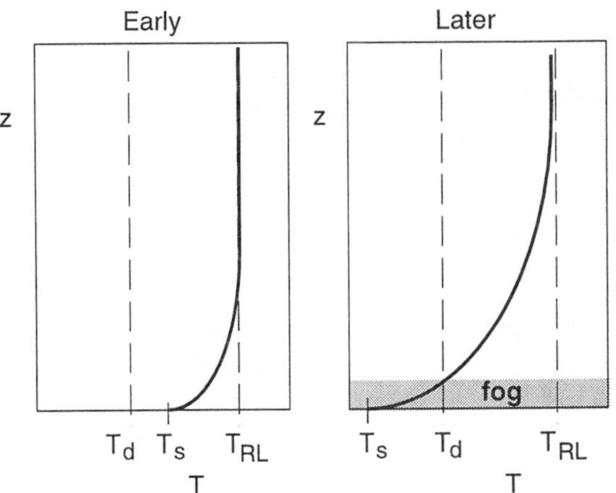

Figure 7.8
Stable boundary layer evolution at night leading to radiation fog onset, where T_s is near-surface air temperature, T_d is original dew point temperature, and T_{RL} is the original temperature.

Once fog forms, evolution of its depth is approximately

$$z = a \cdot M^{3/4} \cdot t^{1/2} \cdot \ln\left[\left(\frac{t}{t_o}\right)^{1/2}\right] \quad (7.11)$$

where t_o is the onset time from the previous equation. This equation is valid for $t > t_o$.

Liquid water content increases as a saturated air parcel cools. Also, visibility decreases as liquid water increases. Thus, the densest (lowest visibility) part of the fog will generally be in the coldest air, which is initially at the ground.

Solved Example (§)

Given a residual layer temperature of 20°C, dew point of 10°C, and wind speed 1 m/s. If the surface heat flux is constant during the night at –0.02 K·m/s, then what is the onset time and height evolution of radiation fog?

Solution

Given: T_{RL}= 20°C, T_d= 10°C, M = 1 m/s
F_H= –0.02 K·m/s
Find: t_o = ? h, and z vs. t.

Use eq. (7.10):

$$t_o = \frac{(0.15 m^{1/4} \cdot s^{1/4})^2 \cdot (1 m/s)^{3/2} \cdot (20 - 10°C)^2}{(0.02°C \cdot m/s)^2}$$

= **1.563 h**

The height evolution for this wind speed, as well as for other wind speeds, is calculated using eq. (7.11):

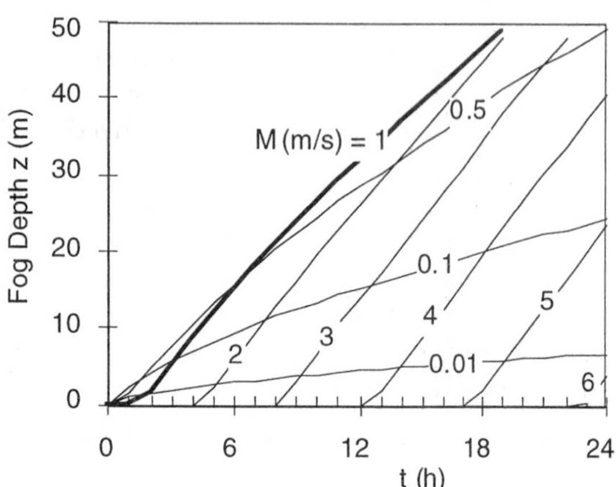

Check: Units OK. Physics OK.
Discussion: From the graph, windier nights cause later onset of fog, but stimulates rapid growth of the fog depth.

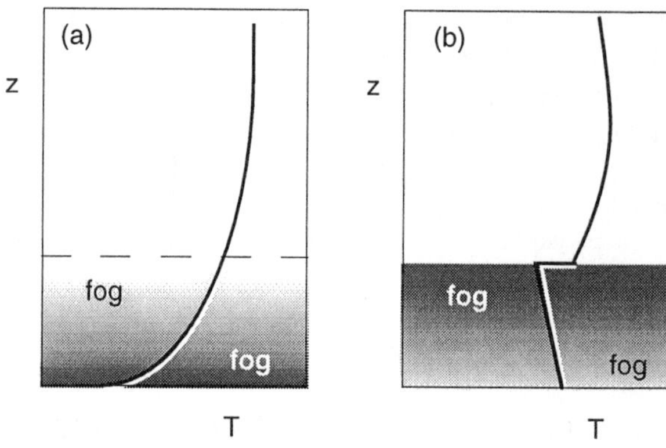

Figure 7.9
(a) Stratified fog that is more dense and colder at ground. (b) Well-mixed fog that is more dense and colder at the top due to IR radiative cooling.

Initially, fog density decreases smoothly with height because temperature increases smoothly with height (Fig 7.9a). As the fog layer becomes optically thicker and more dense , it reaches a point where the surface is so obscured that it can no longer cool by direct IR radiation to space.

Instead, the height of maximum radiative cooling moves upward into the fog away from the surface.

Solved Example

Initially, total water is constant with height at 10 g/kg, and liquid water potential temperature increases with height at rate 2°C/100m in a stratified fog. Later, a well-mixed fog forms with depth 100 m. Plot the total water and θ_L profiles before and after mixing.

Solution

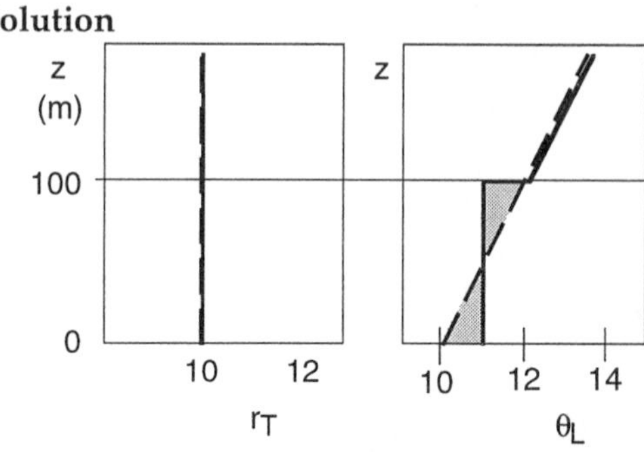

Check: Units OK. Physics OK.
Discussion: Dashed lines show initial profiles, solid are final profiles. For θ_L, the two shaded areas must be equal for heat conservation. This results in a temperature jump of 1°C across the top of the fog layer for this example. Recall from Chapter 5 that actual temperature decreases with height along lines of constant θ_L .

Cooling of air within the nocturnal fog causes air to sink as cold thermals. Convective circulations then turbulently mix the fog.

Very quickly, the fog changes into a well-mixed fog with liquid-water-potential temperature θ_L and total-water content r_T that are uniform with height. During this rapid transition, total heat and total water averaged over the whole fog layer are conserved. As the night continues, θ_L decreases and r_L increases with time due to continued radiative cooling.

In this fog the actual temperature decreases with height at the moist adiabatic lapse rate (Fig 7.9b), and liquid water content increases with height. Continued IR cooling at fog top can strengthen and maintain this fog.

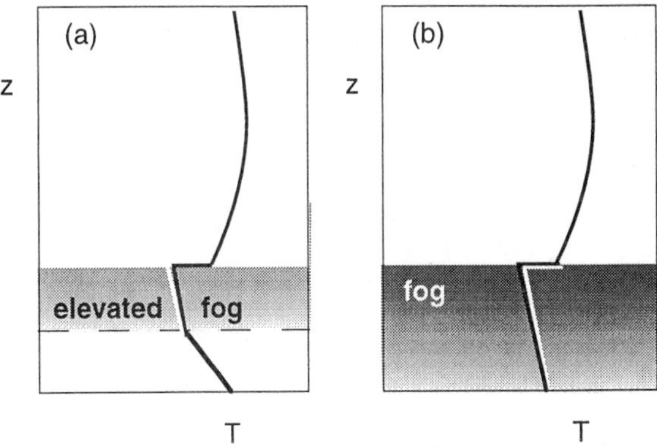

Figure 7.10
(a) Heating during the day can modify the fog of Fig 7.9b, causing the base to lift and the remaining elevated fog to be less dense (b) If solar heating is insufficient to totally dissipate the fog, then nocturnal radiative cooling can re-strengthen it.

Dissipation of Well-Mixed (Radiation and Advection) Fogs

During daytime, solar heating and IR cooling are both active. Fogs can become less dense, can thin, can lift, and can totally dissipate due to warming by the sun.

Stratified fogs (Fig 7.9a) are optically thin enough that sunlight can reach the surface and warm it. The fog albedo can be in the range of A = 0.3 to 0.5 for thin fogs. This allows rapid warming of the fog layer, evaporation of the liquid drops, and dissipation of the fog.

For optically-thick well-mixed fogs (Fig 7.9b), albedoes can be A =0.6 to 0.9. What little sunlight is not reflected off the fog is absorbed in the fog itself. However, IR radiative cooling continues, and can compensate the solar heating. One way to estimate whether fog will totally dissipate is to calculate the sum of accumulated cooling and heating (Chapt. 4) since time t_o when the fog first formed:

$$Q_{Ak} = F_{H.night} \cdot (t - t_o) + (1-A) \cdot \frac{F_{H.max} \cdot D}{\pi} \cdot \left[1 - \cos\left(\frac{\pi \cdot (t - t_{SR})}{D}\right)\right] \tag{7.12}$$

The first term on the right should be included only when $t > t_o$, which includes not only nighttime when the fog originally formed, but the following daytime also. This approach assumes that the rate $F_{H.night}$ of IR cooling during the night is a good approximation to the continued IR cooling during the day.

The second term should be included only when $t > t_{SR}$, where t_{SR} is sunrise time. (See Chapter 4 for definitions of other variables). When Q_{Ak} becomes positive, fog dissipates (neglecting other factors such as advection, and assuming no fog initially).

While IR cooling happens from the very top of the fog, solar heating occurs over a greater depth in the top of the fog layer. Such heating underneath cooling statically destabilizes the fog, creating convection currents of cold thermals that sink from fog top. These upside-down cold thermals continue to mix the fog even though there might be net heating when averaged over the whole fog. Thus, any radiatively heated or cooled air is redistributed and mixed vertically throughout the fog by convection.

Such heating can cause the bottom part of the fog to evaporate, which appears to an observer as lifting of the fog (Fig 7.10a). Sometimes the warming during the day is insufficient to evaporate all the fog. When night again occurs and radiative cooling is not balanced by solar heating, the bottom of the elevated fog lowers back down to the ground (Fig 7.10b).

In closing this section on fogs, please be aware that the equations above for formation, growth, and dissipation of fogs were based on very idealized situations, such as flat ground and horizontally uniform heating and cooling. In the real atmosphere, even gentle slopes can cause katabatic drainage of cold air into the valleys and depressions, which are then the favored locations for fog formation. The equations above were meant only to illustrate some of the physical processes, and should not to be used for operational fog forecasting.

Science Graffito

"Science is the belief in the ignorance of experts." – Richard Feynman.

Solved Example (§)

When will fog dissipate if it has an albedo of A = (a) 0.4; (b) 0.6; (c) 0.8 ? Assume daylight duration of D = 12 h. Assume fog forms at t_o = 3 h after sunset. Given: $F_{H\,night}$ = –0.02 K·m/s, and $F_{H\,max\,day}$ = 0.2 K·m/s. Also plot the cumulative heating.

Solution

Given: (see above) Let t = time after sunset.
Use eq. (7.12), & solve on a spreadsheet:
(a) t = 16.33 h = 4.33 h after sunrise = **10:20 AM**
(b) t = 17.9 h = 5.9 h after sunrise = **11:54 AM**
(c) **fog never dissipates**

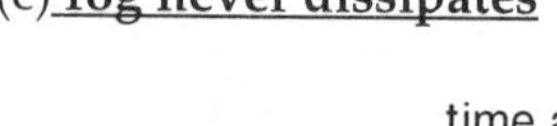

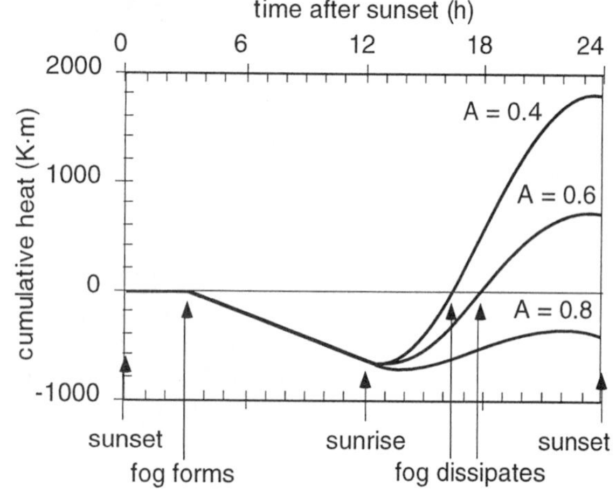

Check: Units OK. Physics OK.
Discussion: Albedo makes a big difference for fog dissipation. There is some evidence that albedo depends on the type and number concentration of fog nuclei.

SUMMARY

Condensation of water vapor occurs by cooling, adding moisture, or mixing. Cooling and moisturizing are governed by the Eulerian heat and water budgets, respectively. Turbulent mixing of two nearly-saturated air parcels can yield a saturated mixture. We see the resulting droplet-filled air as clouds or fog.

If clouds are buoyant (i.e., the cloud and subcloud air is statically unstable), thermals of warm air actively rise to form cumuliform clouds. These have a lognormal distribution of sizes, and have fractal shapes. Active clouds are coupled to the underlying surface, allowing thermo diagrams to be used to estimate cloud base and top altitudes.

If the clouds are not buoyant (i.e., the cloud layer is statically stable), the clouds remain on the ground as fog, or they are forced into existence as stratiform clouds by advection along an isotropic surface.

While almost all clouds are created in rising air and the associated adiabatic cooling, fogs can form other ways. IR radiative cooling creates radiation fogs, while cooling associated with advection of humid air over a cold surface causes advection fogs. Precipitation and steam fog form by adding moisture to the air that touches a liquid water surface (e.g., raindrops, or a lake), followed by mixing with the surrounding cooler air to reach saturation.

Although fog forms in a layer of cold air resting on the ground, that cold-air can be either statically stable or unstable depending on whether continued cooling is being imposed at the bottom or top of the layer, respectively. Heat and moisture budget equations can be used to forecast fog onset, development, and dissipation for some idealized situations.

Threads

Clouds of many different sizes and fractal shapes shade the ground, thereby altering the earth's radiation budget and climate (Chapts. 2 and 18). In turn, heat (Chapt. 3) and moisture (Chapt. 5) budgets are used to determine if saturation is reached.

Cloud depth depends on static stability (Chapt. 6), for which thermo diagrams are useful. Radiation (Chapt. 2) causes radiation fog. Droplet growth (Chapt. 8) determines fog visibility, and precipitation rates from clouds. Condensation of water vapor can occur in rising plumes from smoke stacks and cooling towers (Chapt. 17).

Adiabatic cooling process also causes condensation in tornado funnels (Chapt. 15). Advection fog is one example of air-mass transformation (Chapt. 12). All fogs are within the boundary layer (Chapt. 4).

EXERCISES

Numerical Problems

N1. Use a thermo diagram to plot the following environmental sounding:

P (kPa)	T (°C)
20	–15
25	–25
35	–25
40	–15
45	–20
55	–15
70	0
80	9
85	6
95	15

Determine cloud activity (active, passive, fog, none), cloud-base height, and cloud top height, for the (T, T_d) °C combinations of near-surface (P = 100 kPa) air parcels given below:

a. (20, 14) b. (20, 7) c. (20, 17) d. (20, 19)
e. (15, 15) f. (30, 0) g. (30, 10) h. (30, 24)

N2. Given saturated air of temperature and liquid water mixing ratio as listed below. What is its virtual temperature, and how would it compare to the virtual temperature with no liquid water?

	T (°C)	r_L (g/kg)
a.	20	5
b.	10	7
c.	0	2
d.	30	0

N3.(§). Use a spreadsheet to find and plot the fraction of clouds vs. size. Use ΔX = 100 m, with X in the range 0 to 5000 m.

a. For fixed S_x = 0.5, plot curves for L_X =
(1) 250 (2) 500 (3) 1000 (4) 2000 m

b. For fixed L_X = 1000 m, plot curves for S_X =
(1) 0.1 (2) 0.2 (3) 0.5 (4) 0.8

N4. In Fig 7.5, divide each tile into 4 equal quadrants, and count N vs. M for these new smaller tiles to add data points to the solved-example figure for measuring fractal dimension. Do these finer-resolution tiles converge to a different answer? Do the box-counting for Fig 7.5 part:

a. (a) b. (b) c. (c) d. (d)

N5. On a winter day, suppose your breath has T = 30°C with T_d = 28°C, and it mixes with the ambient air of temperature T = –10°C and T_d = –11°C. Will you see your breath? Assume you are at sea level, and that your breath and the environment mix in proportions of:

a. 2 parts breath to 1 part environment.
b. 5 parts breath to 1 part environment.
c. 1 parts breath to 1 part environment.
d. 1 parts breath to 2 part environment.
e. 1 parts breath to 5 part environment.

N6(§). a. Suppose jet exhaust has T = 200°C and T_d = 150°C. The jet is flying at an altitude where the ambient conditions are P = 30 kPa, T = -40°C, and T_d = -42°C. Will a visible contrail form? (Hint, assume that all possible mixing proportions occur.)

b. Same as (a), but with jet exhaust of T = 400°C and T_d = 350°C.

N7. a. Air at sea level is initially of temperature 10°C and relative humidity of 50%.

1) How high must it be lifted to form upslope fog?
2) How much water must be added to cause precipitation fog?
3) How much cooling is necessary to cause radiation fog?

b. Same as (a), but for air with 80% initial relative humidity.

N8(§). a. In spring, humid tropical air of initial temperature 20°C and dew point 15°C flows over colder land of surface temperature 2 °C. At what downwind distance will advection fog form? Also plot the air temperature vs. distance. Assume z_i = 200 m, and C_H = 0.005 .

b. Same as (a), but for a surface temperature of –10°C.

N9(§). Given a residual layer of 15°C, dew point 10 °C, and wind speed 3 m/s, and F_H = –0.02 K·m·s^{-1} .

a. When will radiation fog form? Plot fog depth vs. time.

b. Same as (a), but with dew point 13°C and wind speed 2 m/s.

N10(§). When will fog dissipate? Assume: albedo is 0.5, fog forms 6 h after sunset, daylight duration is 12 h, $F_{H\ night}$ = –0.02 K·m/s, and $F_{H\ max\ day}$ = 0.15 K·m/s. Also, plot the cumulative heat vs. time.

Understanding & Critical Evaluation

U1(§). On a spreadsheet, type the cloud-size parameters (S_X, L_X, Δ_X) from the solved example into separate cells. Create a graph of the lognormal cloud distribution, referring to these parameter cells.

b. Next, change the values of each of these parameters to see how the shape of the curve changes. Can you explain why L_x is called the location parameter, and S_X is called the spread parameter? Is this consistent with an analytical interpretation of the factors in eq. (7.5)? Explain.

U2. The box-counting method can also be used for the number of points on a straight line, such as sketched in Fig 7.4c. In this case, a "box" is really a fixed-length line segment, such as increments on a ruler or meter stick.

a. Using the straight line and dots plotted in Fig 7.4c, use a centimeter rule to count the number of dots in each successive centimeter along the line. Repeat for half cm increments. Repeat with ever smaller increments. Then plot the results as in the fractal-dimension solved example, and calculate the average fractal dimension. Use the zero-set characteristics to find the fractal dimension of the original wiggly line.

b. In Fig 7.4c, draw a different, nearly vertical, straight line, mark the dots where this line crosses the underlying wiggly line, and then repeat the dot-counting procedure as described in part (a). Repeat this for a number of different straight lines, and then average the resulting fractal dimensions to get a more statistically-robust estimate.

U3. a. Crumple a sheet of paper into a ball, and carefully slice it in half. This is easier said than done. (A head of cabbage or lettuce could be used instead.) Place ink on the cut end using an ink pad, or by dipping the paper wad or cabbage into a pan with a thin layer of red juice or diluted food coloring in the bottom. Then make a print of the result onto a flat piece of paper, to create a pattern such as shown in Fig 7.4. Use the box counting method to find the fractal dimension of the wiggly line that was printed. Using the zero-set characteristic, estimate the fractal dimension of the crumpled wad of paper.

b. Repeat the experiment, using crumpled paper wads that are more tightly packed, or more loosely packed. Compare the fractal dimensions of the wads. (P.S. Don't throw the crumpled wads at the instructor. They tend to get annoyed easily.)

U4. Derive eq. (7.10), based on the exponential temperature profile for a stable boundary layer (Chapt. 4). State and justify all assumptions. [This requires calculus.]

U5. Derive eq. (7.11) from (7.10), and justify all assumptions.

U6. Derive eq. (7.12), and justify all assumptions.

U7. Use the data in the solved example for fog dissipation, but find the critical albedo at which fog will just barely dissipate.

Web Enhanced Questions

W1. Search the web for thermo diagrams showing soundings for cases where different cloud types were present (e.g., active, passive, fog). Estimate cloud base and cloud top altitudes from the sounding, and compare with observations. Cloud base observations are available from the METAR surface reports made at airports. Cloud top observations are reported by aircraft pilots in the form of PIREPS.

W2. Search the web for cloud-classification images. Copy the best example of each cloud type to your own page. Be sure to also record and cite the original web site for each cloud photo.

W3. a. Search the web for fractal images that you can print. Use box counting methods on these images to find the fractal dimension, and compare with the fractal dimension that was specified by the creator of that fractal image.

b. Make a list of the URL web addresses of the ten fractal images that you like the best (try to find ones from different research groups or other people).

W4. Search the web for a discussion about which satellite channels can be used to discriminate between clouds and fog. Summarize your findings.

W5. a. Search the web for very high resolution visible satellite images of cumulus clouds or cloud clusters over an ocean. Display or print these images, and then trace the edges of the clouds. Find the fractal dimension of the cloud edge, similar to the procedure that was done with Fig 7.5.

W6. Search the web or consult engineering documents that indicate the temperature and water vapor content of exhaust from aircraft jet engines. What ambient temperature and humidity of the atmosphere near the aircraft would be needed such that the mixture of the exhaust with the air would just become saturated and show a jet contrail in the sky. (You might need to utilize a spreadsheet to experiment with different mixing proportions to find the one that would cause saturation first.)

W7. Search the web for methods that operational forecasters use to predict onset and dissipation times of fog. Summarize your findings. (Hint: Try web sites of regional offices of the national weather service.)

W8. Search the web for satellite images that show fog. (Hints: Try near San Francisco, or try major river valleys in morning, or try over snowy ground in spring, or try over unfrozen lakes in Fall). For the fog that you found, determine and justify the type of fog (radiation, advection, steam, etc.) in that image.

Synthesis Questions

S1. Utilize the information in (a) and (b) below to explain why cloud sizes might have a lognormal distribution.

a. The **central-limit theorem** of statistics states that if you repeat an experiment of adding *N* random numbers (using different random numbers each time), then there will be more values of the sum in the middle of the range than at the extremes. That is, there is a greater probability of getting a middle value than of getting a small or large value. This probability distribution has the shape of a **Gaussian curve** (i.e., a bell curve or a "normal" distribution; see Chapt 17 for an example).

For anyone who has rolled **dice**, this is well known. Namely, if you roll one die (consider it to be a **random number generator** with a uniform distribution) you will have an equal chance of getting any of the numbers on the die (1, 2, 3, 4, 5, or 6).

However, if you roll two dice and sum the numbers from each die, you have a much greater chance of getting a sum of 7 than of any other sum (which is exactly half way between the smallest possible sum of 2, and the largest possible sum of 12). You have slightly less chance of rolling a sum of 6 or 8. Even less chance of rolling a sum of 5 or 9, etc.

The reason is that 7 has the most ways (6 ways) of being created from the two dice (1+6, 2+5, 3+4, 4+3, 5+2, 6+1). The sums of 6 and 8 have only 5 ways of being generated. Namely, 1+5, 2+4, 3+3, 4+2, and 5+1 all sum to 6, while 2+6, 3+5, 4+4, 5+3, 6+2 all sum to 8. The other sums are even less likely.

b. The logarithm of a product of numbers equals the sum of the logarithms of those numbers.

S2. Build an instrument to measure relative sizes of cumulus clouds as follows. On a small piece of clear plastic, draw a fine-mesh square grid like graph paper. Or take existing fine-mesh graph paper and make a transparency of it using a copy machine. Cut the result to a size to fit on the end of a short tube, such as a toilet-paper tube.

Hold the open end of the tube to your eye, and look through the tube toward cumulus clouds. Do this over relatively flat, level ground, perhaps from the roof of a building or from a window just at tree-top level. Pick clouds of medium range, such that the whole cloud is visible through the tube.

For each cloud, record the relative diameter (i.e., the number of grid lines spanned horizontally by the cloud), and the relative height of each cloud base above the horizon (also in terms of number of grid lines). Then, for each cloud, divide the diameter by the cloud-base height to give a normalized diameter. This corrects for perspective, assuming that cumulus cloud bases are all at the same height.

Do this for a relatively large number of clouds that you can see during a relatively short time interval (such as half an hour), and then count the clouds in each bin of normalized cloud diameter.

a. Plot the result, and compare it with the lognormal distribution of Fig 7.3.

b. Find the L_X and S_X parameters of the lognormal distribution that give the best fit to your data. (Hint: Use trial and error on a spreadsheet that has both your measured size distribution and the theoretical distribution on the same graph. Otherwise, if you have had a statistics course, you can use a method such a Maximum Likelihood to find the best-fit parameters.)

S3. Suppose that you extend Euclidian space to 4 dimensions, to include time as well as the 3 space dimensions. Speculate and describe the physical nature of something that has fractal dimension of 3.4 .

S4. What if the saturation curve in Fig 7.6 was concave down instead of concave up, but that saturation vapor pressure still increased with increasing temperature. Describe how the cooling, moisturizing, and mixing processes to reach saturation would be different.

S5. For advection fog, eq. (7.8) is based on a well-mixed fog layer. However, it is more likely that advection fog would initially have a temperature profile similar to that for a stable boundary layer (Chapt. 4). Derive a substitute equation in place of eq. (7.8) that includes this better representation of the the temperature profile. Assume for simplicity that the wind speed is constant with height, and state any other assumptions you must make to get an answer. Remember that any reasonable answer is better than no answer, so be creative.

ON DOING SCIENCE • Cargo Cult Science

"In the South Seas there is a cargo cult of people. During the war they saw airplanes land with lots of good materials, and they want the same thing to happen now. So they've arranged to make things like runways, to put fires along the sides of the runways, to make a wooden hut for a man to sit in, with two wooden pieces on his head like headphones and bars of bamboo sticking out like antennas — he's the controller — and they wait for the airplanes to land. They're doing everything right. The form is perfect. It looks exactly the way it looked before. But it doesn't work. No airplanes land. So I call these things cargo cult science, because they follow the apparent precepts and forms of scientific investigation, but they're missing something essential, because the planes don't land."

"Now it behooves me, of course, to tell you what they're missing. ... It's a kind of scientific integrity, a principle of scientific thought that corresponds to a kind of utter honesty — a kind of leaning over backwards. For example, if you're doing an experiment, you should report everything that you think might make it invalid — not only what you think is right about it: other causes that could possibly explain your results; and things you thought of that you've eliminated by some other experiment, and how they worked — to make sure the other fellow can tell they have been eliminated."

"Details that could throw doubt on your interpretation must be given, if you know them. You must do the best you can — if you know anything at all wrong, or possibly wrong — to explain it. If you make a theory, for example, and advertise it, or put it out, then you must also put down all the facts that disagree with it, as well as those that agree with it. There is also a more subtle problem. When you have put a lot of ideas together to make an elaborate theory, you want to make sure when explaining what it fits, that those things it fits are not just the things that gave you the idea for the theory; but that the finished theory makes something else come out right, in addition."

"In summary, the idea is to try to give all the information to help others to judge the value of your contribution; not just the information that leads to judgement in one particular direction or another."

– Richard P. Feynman, 1985: *"Surely You're Joking, Mr. Feynman!". Adventures of a Curious Character.* Bantam. 322pp

PRECIPITATION

CONTENTS

8 **Precipitation** is a generic term for all liquid water or solid ice particles that are large enough to fall to the ground. **Hydrometeors** include liquid and ice particles of all sizes, from the tiniest cloud droplets and ice crystals to the largest precipitation including hail. Hydrometeors that are large and heavy enough to fall out of the cloud, but which evaporate before reaching the ground, create a weather element called **virga**.

Precipitation particles are much larger than cloud particles, as illustrated in Fig 8.1. One "typical" raindrop holds as much water as a million "typical" cloud droplets. How do such large precipitation particles form?

Three processes control the growth of precipitation: nucleation, diffusion, and collision. **Nucleation** involves the formation of new liquid or solid hydrometeors from vapor. These hydrometeors form on tiny contaminant particles called cloud or ice condensation nuclei, which are carried inside the air parcel.

Diffusion is the somewhat random migration of water vapor molecules through the air toward existing hydrometeors. Conduction of heat away from the droplet is also necessary to compensate the latent heating of condensation or deposition.

Collision between two hydrometeors allows them to combine into larger particles. These processes affect water and ice differently.

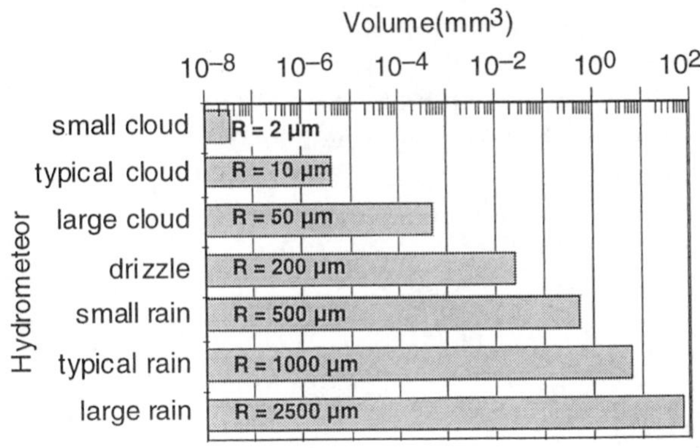

Figure 8.1
Drop volumes and radii, R.

FOCUS • Meteors and Meteorology

Ancient Greeks defined "**meteors**" as anything in the sky. They were particularly concerned about missiles the gods might toss down, such as bits of rock, ice, or lightning bolts.

Only much later did scientists discriminate between missiles from space (bits of rock called **meteoroids**) and missiles from the atmosphere (bits of just about anything else from the sky). But by then "**meteorology**" was firmly entrenched as the name for **atmospheric science**.

According to the *Glossary of Meteorology*, meteorologists study the following meteors:

- **hydrometeors** – wet: clouds, rain, snow, fog, dew, frost, etc.
- **lithometeors** – dry: dust, sand, smoke, haze
- **igneous meteors** -lightning, corona
- **electrometeors** – lightning (again), thunder
- **luminous meteors** – rainbows, halos, sun dogs

Except for "hydrometeors", these terms are seldom used any more.

NUCLEATION OF LIQUID DROPLETS

Nucleation in clean air is called **homogeneous**, while that involving impurities in air is called **heterogeneous**. We will show that homogeneous nucleation is virtually impossible in the real atmosphere and can be neglected. Even with heterogeneous nucleation, there is a barrier to droplet formation.

Cloud Condensation Nuclei (CCN)

Clouds form by heterogeneous nucleation when water vapor condenses on tiny dust and aerosol particles in the air called **cloud condensation nuclei**. Small nuclei are much more abundant than larger ones (Table 6.1 of Ahrens). Over continental regions, the **number density** (n = count of particles per volume of air) as a function of particle radius R is approximately:

$$n(R) = c \cdot R^{-4} \qquad (8.1)$$

for particles larger than 0.2 µm. Constant c depends on the total concentration of particles. This distribution is called the **Junge distribution** (Fig 8.2).

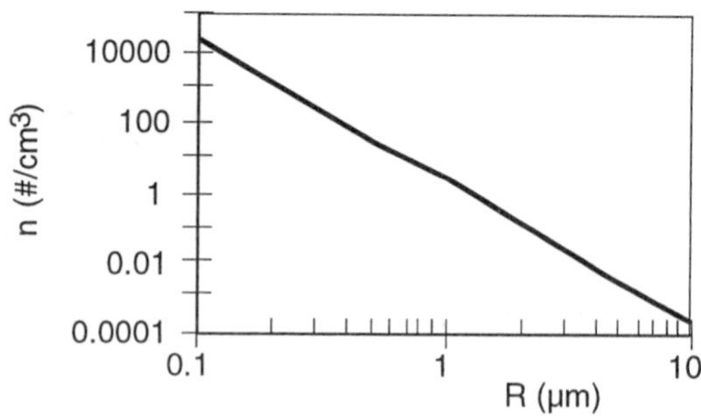

Figure 8.2
CCN particle count n vs. radius R, for c = 2,000,000 $\mu m^4 \cdot m^{-3}$.

Solved Example

If $c = 5\times10^6\ \mu m^4 \cdot m^{-3}$, then how many CCN are expected of radii (a) 0.5 µm and (b) 1 µm?

Solution

Given: R = 0.5 µm & 1 µm, $c = 5\times10^6\ \mu m^4 \cdot m^{-3}$
Find: n = ? #/m^3

Use eq. (8.1):
(a) $n = (5\times10^6\ \mu m^4 \cdot m^{-3})\cdot(0.5\ \mu m)^{-4}$ = **8×10^7** / m^3.
(b) $n = (5\times10^6\ \mu m^4 \cdot m^{-3})\cdot(1\ \mu m)^{-4}$ = **5×10^6** / m^3.

Check: Units OK. Physics OK.
Discussion: A doubled particle radius has more than a ten-fold decrease in number density. If each CCN is the nucleus of a cloud droplet, then there are tens of millions of cloud droplets in each cubic meter of cloudy air. In real clouds, the concentration can be tens to thousands times greater.

Curvature and Solute Effects

Both droplet curvature and chemical composition affect the evaporation rate. Evaporation rate from the curved surface of a droplet is greater than that from a flat water surface. This is a barrier to droplet formation, because the tiniest, insipient droplets have the tightest curvature, which tends to destroy the droplet by net evaporation rather than allowing the droplet to grow by net condensation.

Solutions evaporate water molecules at a slower rate than does pure water. Solutions occur when condensation occurs on impurities such as certain cloud condensation nuclei (CCN) that dissolve in the nascent water droplet. This can partially compensate the curvature effect.

These two opposing effects can be combined into one equation (**Köhler equation**) for the ratio of actual saturation vapor pressure $e_s{}^*$ in equilibrium over a solution with a curved surface, to vapor pressure over flat pure water e_s:

$$\frac{e_s^*}{e_s} \approx \frac{\exp\left(\frac{c_1}{T \cdot R}\right)}{1+\frac{c_2 \cdot i \cdot m_s}{M_s \cdot R^3}} \qquad \bullet(8.2)$$

where T is absolute temperature, R is drop radius, i is number of ions per molecule in solution (called the **van't Hoff** factor), m_s is mass of solute in the droplet, and M_s is molecular weight of solute. The two parameters are: c_1 = 0.3335 K·µm, and $c_2 = 4.3\text{x}10^{12}$ µm^3·g^{-1} .

Relative humidity can be defined as RH = 100%·e_s^*/e_s, which can be greater than 100%. **Supersaturation** can be defined as a fraction:

$$S = (e_s^*/e_s) - 1 \qquad \bullet(8.3a)$$

or as a percentage:

$$S = RH\% - 100\% = 100\% \cdot [(e_s^*/e_s) - 1] \qquad \bullet(8.3b)$$

Thus, the left hand side of eq. (8.2) is easily rewritten as supersaturation.

The numerator of eq. (8.2) describes the **curvature effect,** and together with the left hand side is called the **Kelvin equation**. The denominator describes the **solute effect** of impurities in the water. Table 8-1 gives parameters for common solutes in the atmosphere. These are used with eq. (8.2) in a spreadsheet to produce the **Köhler curves** in Fig 8.3. From these we can study nucleation.

First, the curve for pure water increases exponentially as drop radius becomes smaller. The evaporation from tiny droplets of pure water is so large that any such droplets will vaporize unless the relative humidity is significantly over 100%.

Second, solutions of some chemicals can form small droplets even at humidities of less than 100%. Such **hygroscopic** pollutants in the air will grow into droplets by taking water vapor out of the air.

Third, if humidities become even slightly greater than the peaks of the Köhler curves, or if droplet radius becomes greater than the radius at the peak, then droplets can grow unimpeded. CCN reaching this state are said to be **activated**. Growth of activated nuclei continues until enough vapor is consumed to reduce the supersaturation back toward 100%. Pure droplets cannot form or co-exist in an environment with neighboring solution droplets because of the low supersaturation remaining in the air. That is why homogeneous nucleation can be neglected for practical purposes.

Fourth, although not shown in these curves, the equations allow droplets to form at lesser supersaturations if the mass of dissolved solute is greater. Hence, larger CCN can grow into droplets earlier and faster than smaller CCN.

Table 8-1. Properties of some solutes. M_s is molecular weight, i is approximate ion count.

Solute	Chemistry	M_s	i
salt	NaCl	58.44	2
ammonium sulfate	$(NH_4)_2SO_4$	132.13	3
hydrogen peroxide	H_2O_2	34.01	2
sulfuric acid	H_2SO_4	98.07	3
nitric acid	HNO_3	63.01	2

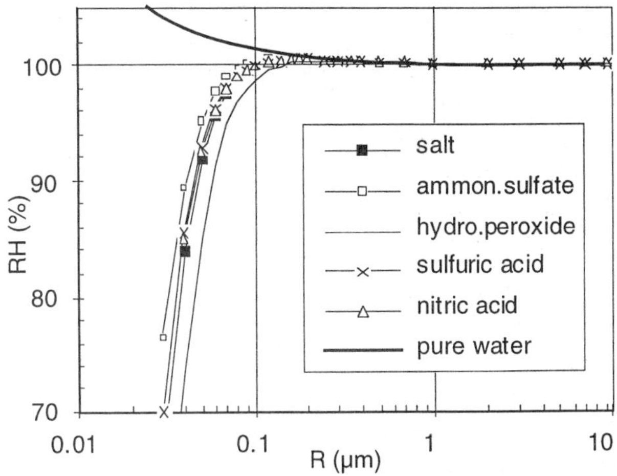

Figure 8.3a
Equilibrium relative humidities over droplets of different radius with various solutes. T = 0°C. Solute mass = 10^{-16} g.

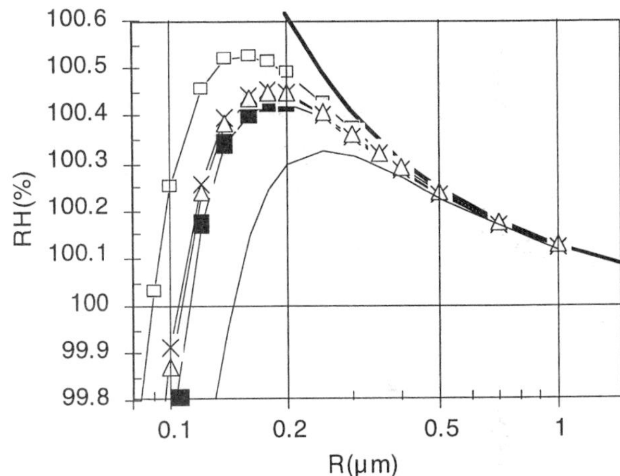

Figure 8.3b
Blow-up of Fig 8.3a.

Solved Example

Find the equilibrium relative humidity over a droplet of radius 0.2 µm, temperature 20°C, containing 10^{-16} g of ammonium sulfate.

Solution

Given: $R = 0.2$ µm, $T = 293$ K, $m_s = 10^{-16}$ g

Find: $RH = 100\% \cdot (e_s^*/e_s) = ?$ %

From Table 8-1 for ammonium sulfate:

$M_s = 132.13$, $i \cong 3$

Use eq. (8.2):

$$\frac{e_s^*}{e_s} \approx \frac{\exp\left(\frac{0.3335\text{K}\cdot\mu\text{m}}{(293\text{K})\cdot(0.2\mu\text{m})}\right)}{1+\frac{(4.3\times10^{12}\mu\text{m}^3\cdot\text{g}^{-1})\cdot 3\cdot(10^{-16}\text{g})}{(132.13)\cdot(0.2\mu\text{m})^3}}$$

$= 1.00571 / (1+0.00122) = 1.00448$

$RH = 100\% \cdot (e_s^*/e_s) =$ **<u>100.448%</u>**

Check: Units OK. Physics OK.

Discussion: Fig 8.3b gives a value of about 100.49% for a temperature of 0°C. Thus, warmer temperatures require less supersaturation of water vapor in the air to prevent the droplet from vaporizing.

Critical Radius

The location of the peak of the Köhler curves marks the barrier between the larger, activated droplets that can continue to grow, from the smaller haze droplets that reach an equilibrium at small size. The drop radius R^* at this peak is called the **critical radius**, and the corresponding **critical supersaturation** fraction is $S^* = e_s^*/e_s - 1$. They are given by:

$$R^* = \sqrt{\frac{c_3 \cdot i \cdot m_s \cdot T}{M_s}} \qquad •(8.4)$$

and

$$S^* = \sqrt{\frac{c_4 \cdot M_s}{i \cdot m_s \cdot T^3}} \qquad •(8.5)$$

where $c_3 = 3.8681\text{x}10^{13}$ µm^2·K^{-1}·g^{-1}·, and $c_4 = 1.278\text{x}10^{-15}$ K^3·g.

Critical conditions are plotted in Fig 8.4 for various masses of different chemicals. Obviously S^* is inversely related to R^*, so as solute mass increases, smaller supersaturations are necessary to reach the critical point, and at that point the droplets will be larger. Also, notice that different chemicals will grow to different sizes, which is one factor causing a range of drop sizes in the cloud.

Solved Example

Find the critical radius and supersaturation value for 10^{-15} g of ammonium sulfate at 0°C.

Solution

Given: $m_s = 10^{-15}$ g, $T = 273$ K

Find: $R^* = ?$ µm, $S^* = ?$ %.

Use eq. (8.4) & Table 8-1. $R^* =$

$$\sqrt{\frac{(3.8681\times10^{13}K^{-1}\cdot g^{-1}\cdot \mu m^2)\cdot 3\cdot(10^{-15}g)\cdot(273K)}{132.13}}$$

$R^* =$ **<u>0.49 µm</u>**

Use eq. (8.5) & Table 8-1. $S^* =$

$$S^* = \sqrt{\frac{(1.278\times10^{-15}\text{K}^3\cdot\text{g})\cdot 132.13}{3\cdot(10^{-15}\text{g})\cdot(273\text{K})^3}} = 0.00166$$

$S^* =$ **<u>0.166%</u>**

Check: Units OK. Physics OK.

Discussion: Agrees with the point plotted in Fig 8.4 for a.s. Thus, the larger mass nucleus needs less supersaturation to become activated.

For a parcel of rising air with increasing supersaturation, the larger nuclei will become activated first, followed by the smaller nuclei if the parcel keeps cooling and if the excess vapor is not removed by the larger nuclei first.

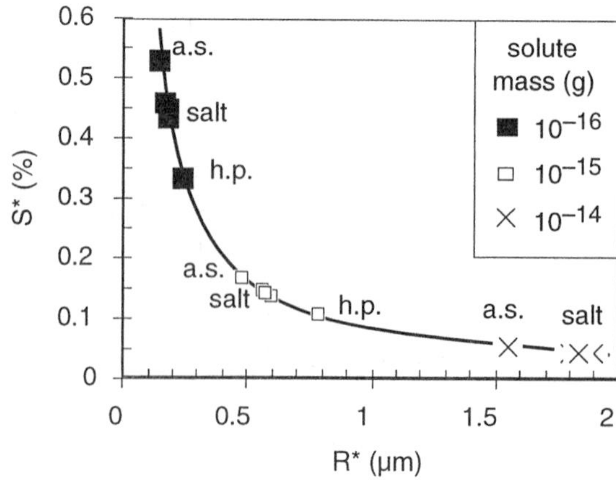

Figure 8.4

Critical values of supersaturation vs. drop radius, for various solute masses and chemicals. a.s. = ammonium sulfide, salt = sodium chloride, h.p. = hydrogen peroxide. Nitric acid and sulfuric acid are near the salt data points.

BEYOND ALGEBRA • Critical Radius

Derivation of the critical radius, eq. (8.4).

The critical radius is at e_s^*/e_s = maximum. But at the maximum, the slope is zero: $d(e_s^*/e_s) / dR = 0$. By finding this derivative of eq. (8.2) with respect to R, and setting it to zero, we can solve for R at the maximum. This is R^* by definition.

The RHS of eq. (8.2) is of the form a/b. A rule of calculus is:

$$d(a/b)/dR = [b\cdot(da/dR) - a\cdot(db/dR)] / b^2.$$

Also, a is of the form $a = \exp(f)$, for which another rule is:

$$da/dR = a\cdot(df/dR).$$

Combining these 2 rules and setting the whole thing to zero gives:

$$0 = \frac{a}{b}\left[\frac{df}{dR} - \frac{1}{b}\frac{db}{dR}\right]$$

But (a/b) is just the original RHS of eq. (8.2), which we know is close to 1.0 at the max, not close to 0. Thus, the eq. above equals 0 only if:

$$\frac{df}{dR} = \frac{1}{b}\frac{db}{dR}$$

Plugging in for f and b and differentiating yields:

$$\frac{-c_1}{T\cdot R^2} = \frac{1}{\left[1 + \frac{c_2\cdot i\cdot m_s}{M_s\cdot R^3}\right]}\frac{(-3)\cdot c_2\cdot i\cdot m_s}{M_s\cdot R^4}$$

Multiply both sides by $(-R^4\cdot T/c_1)$:

$$R^2 = (3\cdot T/c_1)\cdot\frac{\left[\frac{c_2\cdot i\cdot m_s}{M_s}\right]}{\left[1 + \frac{c_2\cdot i\cdot m_s}{M_s\cdot R^3}\right]}$$

Mult. the numerator and denominator of the RHS by: $M_s / c_2\cdot i\cdot m_s$, which gives:

$$R^2 = (3\cdot T/c_1)\cdot\frac{1}{\left[\frac{M_s}{c_2\cdot i\cdot m_s} + \frac{1}{R^3}\right]}$$

By plugging in typical values, we can show that the $1/R^3$ term is small enough to be negligible compared to the other term in square brackets. This leaves:

$$R^2 \cong \left(\frac{3\cdot c_2}{c_1}\right)\cdot\left(\frac{T\cdot i\cdot m_s}{M_s}\right)$$

Define $c_3 = 3\cdot c_2/c_1$, set $R = R^*$, and take the square root of both sides to get the final answer:

$$R^* \cong \sqrt{\frac{c_3\cdot i\cdot m_s\cdot T}{M_s}} \qquad (8.4)$$

Haze

For conditions left of the peak on any Köhler curve (i.e., $R < R^*$), CCN rapidly grow into small droplets that stop growing at an equilibrium size determined by the humidity, temperature, and solute. These small droplets are called **haze** droplets. Thus, tiny droplets can exist even at relative humidities below 100%.

An **aerosol** is any tiny solid or liquid particle suspended in the air, such as CCN and the smallest cloud droplets. Haze droplets are also aerosols. When a tiny, dry CCN grows to its equilibrium size by the condensation of water molecules, this process is called **aerosol swelling**. Aerosol swelling is responsible for reducing **visibility** in polluted air as humidities increase above 70 to 80%.

Even haze particles contain many water molecules. Liquid water contains about ρ_m = 3.3×10^{28} molecules/m^3. Thus, the smallest haze particles of radius 0.02 µm contain roughly n = $\rho_m\cdot[(4/3)\cdot\pi\cdot R^3]$ = 1.1 million molecules.

The word **smog** is a contraction of "smoke" and "fog", which is a reasonable lay description of haze. Many urban smogs are a stew of ingredients including ozone, volatile hydrocarbons such as evaporated gasoline, and various oxides of nitrogen. These react in the atmosphere, particularly in the presence of sunlight, to create sulfates, nitrates, and hydrogen peroxide CCNs. Aerosol swelling and reduced visibilities are quite likely in such urban smogs, particularly when the air is humid.

Solved Example

For 10^{-16} g of ammonium sulfate at T = 0°C, how does haze droplet radius change as RH increases from 70 to 80%? Also, how many molecules are in each aerosol?

Solution

Assume: Same conditions as in Fig 8.3a.
Given: RH = 70%, 80%.
Find: R = ? µm, n = ? molecules

Solve eq. (8.2), or use Fig 8.3a. I will use the Fig.
$R \cong$ <u>**0.027 µm**</u> at 70%; $R \cong$ <u>**0.032 µm**</u> at 80%.

The number of molecules is $n = \rho_m\cdot[(4/3)\cdot\pi\cdot R^3]$
n = <u>**2.72x10^6**</u> molecules; n = <u>**4.53x10^6**</u> molecules

Check: Units OK. Physics OK.
Discussion: Haze particles indeed become larger as relative humidity increases, thereby reducing visibility. Scattering of light by this size of particles is called **Mie scattering** (see Chapt. 19).

Activated Nuclei

For conditions right of the peak on any Köhler curve (i.e., $R > R^*$), CCN are activated and can continue growing. There is no equilibrium which would stop their growth, assuming sufficient water vapor is present. These droplets can become larger than haze droplets, and are called **cloud droplets.**

Because atmospheric impurities consist of a variety of chemicals with a range of masses, we anticipate from the Köhler curves that different CCN will become activated at different amounts of supersaturation. The number density n_{CCN} (# of CCN per m^3) activated as a function of supersaturation fraction S is roughly:

$$n_{CCN} = c \cdot (100 \cdot S)^k \tag{8.6}$$

but varies widely. In maritime air, where evaporated sea spray generates salt CCN, the parameters are $c \cong 1\text{x}10^8\ m^{-3}$ and $k \cong 0.7$. In continental air $c \cong 6\text{x}10^8\ m^{-3}$ and $k \cong 0.5$. The number of activated nuclei is in the range of 10^8 to $10^9\ m^{-3}$, which is usually just a small fraction of the total number of particles in the air.

The distance x between cloud droplets is on the order of 1 mm, and is given by

$$x = n_{CCN}^{-1/3} \tag{8.7}$$

Solved Example

How many nuclei would be activated in continental air of supersaturation percentage 0.5%? Also, how much air surrounds each droplet, and what is the distance between drops?

Solution

Given: $S = 0.005$

Find: n_{CCN} = ? particles/m^3,
V = ? mm^3, x = ? mm.

(a) Use eq. (8.6):

$n_{CCN} = (6 \times 10^8 m^{-3}) \cdot (0.5)^{0.5}$ = $\underline{\mathbf{4.24x10^8}}$ /m^3

(b) Also:

$V = 1/n_{CCN}$ = $\underline{\mathbf{2.36}}$ mm^3

(c) Using eq. (8.7): $x = V^{1/3}$ = **1.33 mm**.

Check: Units OK. Physics OK.

Discussion: Assuming all these nuclei become cloud droplets, that's over 40 million droplets within each cubic meter of cloud. But there is a relatively large distance between each drop.

NUCLEATION OF ICE CRYSTALS

Ice cannot exist at temperatures above 0°C. Below that temperature, ice crystals can exist in equilibrium with air that is supersaturated with respect to ice. The saturation curve for ice was plotted in Fig 5.2, and is close to, but slightly below, the curve for supercooled liquid water.

There is a thermodynamic barrier to ice formation, analogous to the barrier for droplet growth. This barrier can be overcome with either very cold temperatures much below 0°C, or by the presence of ice nuclei.

Processes

Homogeneous freezing nucleation is the name for the spontaneous freezing that occurs within liquid water droplets near –40°C. No impurity is needed for this. Instead, ice embryos form by chance when clusters of water molecules happen to come together with the correct orientations. At –40°C, the cluster needs only about 250 molecules. At slightly warmer temperatures the critical embryo size is much larger, and thus less likely to occur. Because larger droplets contain greater numbers of molecules, they are more likely to form an ice embryo and freeze.

Deposition nucleation occurs when water vapor deposits directly on an ice nucleus. While the solute effect was important for liquid droplet nucleation, it does not apply to ice nucleation because salts are excluded from the ice-crystal lattice as water freezes. Thus, the size and crystal structure of the ice nucleus is more important. Deposition is unlikely on particles of 0.1 µm or less. Colder temperatures and greater supersaturation increases deposition nucleation.

Immersion freezing occurs for those liquid drops already containing an undissolved ice nucleus. Each nucleus has a critical temperature below 0°C at which it triggers freezing. Thus, as external processes cause such a contaminated drop to cool, it can eventually reach the critical temperature and freeze.

Larger drops contain more ice nuclei, and have a greater chance of containing a nucleus that triggers freezing at warmer temperatures (although still below 0°C). To freeze half of the drops of radius R, the temperature must drop to T, given statistically by

$$T \approx T_1 + T_2 \cdot \ln(R / R_o) \tag{8.8}$$

where R_o = 5 µm, T_1 = 235 K, and T_2 = 3 K.

Condensation freezing is a cross between deposition nucleation and immersion freezing. In this scenario, which occurs below 0°C, nuclei are more attractive as condensation nuclei than as deposition nuclei. Thus, supercooled liquid water starts to condense around the nucleus. However, this liquid water immediately freezes due to the immersion-nucleation properties of the nucleus.

Contact freezing occurs when an uncontaminated supercooled liquid drop happens to hit an external ice nucleus. If the droplet is cooler than the critical temperature of the ice nucleus, then it will freeze almost instantly. This also happens when supercooled (**freezing**) **rain** hits and instantly freezes on trees and power lines. Ice crystals in the air are also good contact nuclei for supercooled water.

Solved Example

How cold must a cloud become so that half of the 100 µm radius droplets would likely freeze due to immersed nuclei?

Solution

Given: R = 100 µm
Find: T = ? K

Use eq. (8.8)
$T = 235\text{K} + (3\text{K})\cdot\ln(100\mu\text{m} / 5\mu\text{m}) = 244\text{K} = \underline{-29°\text{C}}$

Check: Units OK. Physics OK.
Discussion: Smaller droplets can remain unfrozen at much colder temperatures than larger drops, which are thus available to participate in the Wegener-Bergeron-Findeisen (WBF) precipitation growth process described later.

Nuclei

Only substances with similar molecular structure as ice can serve as ice nuclei. Such substances are said to be **epitaxial** with ice.

Natural ice nuclei include fine particles of clay such as kaolinite stirred up by the wind. Certain bacteria and amino acids such as *l*-leucine and *l*-tryptophan from plants also can nucleate ice. Combustion products from forest fires contain many ice nuclei. Also, ice crystals from one cloud can fall or blow into a different cloud, triggering continued ice formation.

Other substances have been manufactured specifically to seed clouds. Silver iodide has been a popular chemical for cloud seeding. Table 8-2 lists some ice nuclei and their critical temperatures. Because contact nucleation occurs at the warmest temperatures for most substances, it is the most likely process causing ice nucleation.

Table 8-2. Warmest ice nucleation threshold temperatures (°C). Processes are: **1** contact freezing, **2** condensation freezing, **3** deposition, and **4** immersion freezing.

Substance	Process			
	1	2	3	4
silver iodide	–3	-4	-8	–13
cupric sulfide	–6	x	–13	–16
lead iodide	–6	–7	–15	x
cadmium iodide	–12	x	–21	x
metaldehyde	–3	–2	–10	x
1,5-dihydroxynaphlene	–6	–6	–12	x
phloroglucinol	x	–5	–9	x
kaolinite	–5	–10	–19	-32

DROPLET GROWTH BY DIFFUSION

Max Droplet Radius via Diffusion

Before examining the details of diffusion, it is worthwhile to find the bounds on diffusive growth. Droplets and ice crystals stop growing by diffusion (condensation and deposition) when they have consumed all the available supersaturation. By partitioning the available liquid water (r_L, from previous chapter thermo diagrams) equally between all activated CCN, we can estimate a radius for the hydrometeor:

$$R = \left[\frac{3}{4\pi}\cdot\frac{\rho_{air}}{\rho_{water}}\cdot\frac{r_L}{n_{CCN}}\right]^{1/3} \qquad \bullet(8.9)$$

where r_L should be in kg_{water}/kg_{air}. Typical values are R = 2 to 50 µm, which is small compared to the 1000 µm separation between droplets, and is too small to be precipitation.

This is an important consideration. Namely, even if we ignore the slowness of the diffusion process (described next), the droplets stop growing by diffusion before they become precipitation. Thus, there must be another physical mechanism that causes precipitation particles to grow; namely, the Wegener-Bergeron-Findeisen (WBF) cold-cloud process to be discussed later.

Solved Example

Within a cloud, suppose $\rho_{air} = 1$ kg/m^3 and r_L = 4 g/kg. Find the final drop radius for n_{CCN} = (a) 10^8/m^3, and (b) 10^9/m^3.

Solution

Given: $r_L = 0.004$ kg$_{water}$/kg$_{air}$, $\rho_{air} = 1$ kg/m^3

Find: R = ? µm

Use eq. (8.9). Part (a):

$$R = \left[\frac{3}{4\pi} \cdot \frac{(1 kg_{air} / m^3)}{(10^3 kg_{water} / m^3)} \cdot \frac{(0.004 kg_{water} / kg_{air})}{10^8} \right]^{1/3}$$

$= 2.12\text{x}10^{-5}$ m = **21.2 µm**

(b) Similarly, R = **9.8 µm**.

Check: Units OK. Physics OK.

Discussion: Both of these numbers are well within the range of "typical" cloud droplets. Thus, the final drop size is NOT large enough to become precipitation.

Growth Rate

Diffusion is the process whereby individual water-vapor molecules meander through the air via Brownian motion. The direction of diffusion is always down the humidity gradient toward drier air.

The diffusive moisture flux $\Im$ in kg$_{water}$·m^{-2}·s^{-1} is

$$\Im = -D \cdot \frac{\Delta \rho_v}{\Delta x} \qquad \bullet(8.10a)$$

where x is distance, D is diffusivity, and ρ_v is absolute humidity (water vapor density) in kg$_{water}$/m^3. In kinematic form, divide by dry air density to give:

$$F = -D \cdot \frac{\Delta r}{\Delta x} \qquad \bullet(8.10b)$$

where r is water-vapor mixing ratio, and kinematic flux F has units of mixing ratio times velocity [(kg$_{water}$/kg$_{air}$)·(m/s)]. Larger gradients cause larger fluxes, which cause droplets to grow faster.

In a supersaturated environment, growing droplets remove water vapor by condensation from the adjacent air (Fig 8.5). This lowers the humidity near the drop, creating a humidity gradient down which water vapor can diffuse.

Larger drops create a more gentle gradient than smaller drops. The humidity profile is given by:

$$S \approx S_\infty + \frac{R}{|x|} \cdot (S_R - S_\infty) \qquad (8.11)$$

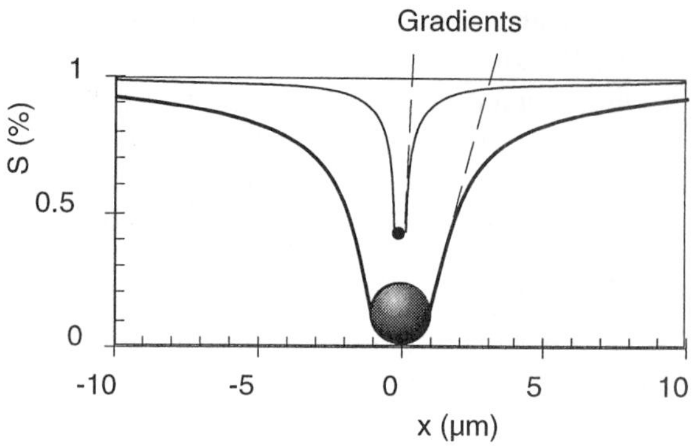

Figure 8.5
Humidity gradients in supersaturated environment near growing droplets. S is supersaturation, x is distance. Nucleus for both droplets is salt of mass 10^{-16} g. Background supersaturation is 1%, the small droplet has radius 0.2 µm, and the large drop 1.0 µm. Equilibrium supersaturation adjacent to the droplets is taken from Fig 8.3b. Nearest neighbor droplets are roughly 1000 µm distant.

where S is supersaturation fraction at distance x from the center of the drop, S_∞ is background supersaturation at a large distance from the droplet, S_R is equilibrium supersaturation adjacent to the drop, and R is droplet radius. Eq. (8.11) was solved on a spreadsheet to create Fig 8.5.

The rate of diffusion is called the **diffusivity** D, and is approximately

$$D = c \cdot \frac{P_o}{P} \cdot \left(\frac{T}{T_o}\right)^{1.94} \qquad (8.12)$$

where $c = 2.11\text{x}10^{-5}$ m^2·s^{-1} is an empirical constant, P_o = 101.3 kPa, and T_o = 273.15 K. This molecular diffusivity for moisture is similar to the thermal conductivity for heat, discussed in Chapter 3.

Solved Example

Find the water vapor diffusivity at P=100 kPa and T = –10°C.

Solution

Given: P=100 kPa and T = 263 K.

Find: D = ? m^2·s^{-1}.

Use eq. (8.12):

$$D = (2.11 \times 10^{-5} \text{m}^2\text{s}^{-1}) \cdot \left(\frac{101.3\text{kPa}}{100\text{kPa}}\right) \cdot \left(\frac{263\text{K}}{273.15\text{K}}\right)^{1.94}$$

= **1.99x10^{-5}** m^2·s^{-1}.

Check: Units OK. Physics OK.

Discussion: Such a small diffusivity means that a large gradient is needed to drive the vapor flux.

During droplet growth, not only must water vapor diffuse through the air toward the droplet, but heat must conduct away from the drop. This is the latent heat that was released during condensation. Without conduction of heat away from the drop, it would become warm enough to prevent further condensation, and would stop growing.

Droplet radius R increases with the square-root of time t, as governed by the combined effects of water diffusivity and heat conductivity:

$$R \approx c \cdot (D \cdot S_\infty \cdot t)^{1/2} \tag{8.13}$$

where S_∞ is the background supersaturation fraction far from the drop, and $c = (2 \cdot r_\infty \cdot \rho_{air} / \rho_{liq.water})^{1/2}$, for background mixing ratio r_∞. Small droplets grow by diffusion faster than larger droplets, because of the greater humidity gradients near the smaller drops (Fig 8.5 and eq. 8.11). Thus, the small droplets will tend to catch up to the larger droplets.

The result is a drop size distribution that tends to become **monodisperse** with time, where most of the drops have approximately the same radius. Also, eq. (8.13) suggests that time periods of many days would be necessary to grow rain-size drops by diffusion alone. But real raindrops form in much less time (minutes), and are known to have a wide range of sizes. Hence, diffusion cannot be the only physical process contributing to rain formation.

Solved Example(§)

Find and plot drop radius vs. time for diffusive growth under the same conditions as above. Assume 1% supersaturation.

Solution

Given: P=100 kPa, T = 263 K, $D = 2\text{x}10^{-5}\ \text{m}^2 \cdot \text{s}^{-1}$.
Find: R(µm) vs. t

First get ρ_{air} from ideal gas law:

$$\rho = \frac{P}{\Re_d \cdot T} = \frac{100\text{kPa}}{(0.287\text{kPa} \cdot \text{K}^{-1} \cdot \text{m}^3 \cdot \text{kg}^{-1}) \cdot (263\text{K})}$$

$= 1.325\ \text{kg} \cdot \text{m}^{-3}$.

$r_\infty = 2$ g/kg from the thermo diagram. I guess a slight supersaturation. For eq. (8.13) use:

$$c = \left[2 \cdot \left(0.002 \frac{\text{kg}_{water}}{\text{kg}_{air}}\right) \cdot \frac{1.325\text{kg/m}^3}{1000\text{kg/m}^3}\right]^{1/2}$$

$= 0.0023$ (dimensionless)

Finally solve eq. (8.13) on a spreadsheet:

$$R \approx 0.0023 \cdot \left[\left(2 \times 10^{-5}\text{m}^2/\text{s}\right) \cdot 0.01 \cdot t\right]^{1/2}$$

(continued next col)

Solved Example(§) *(continuation)*

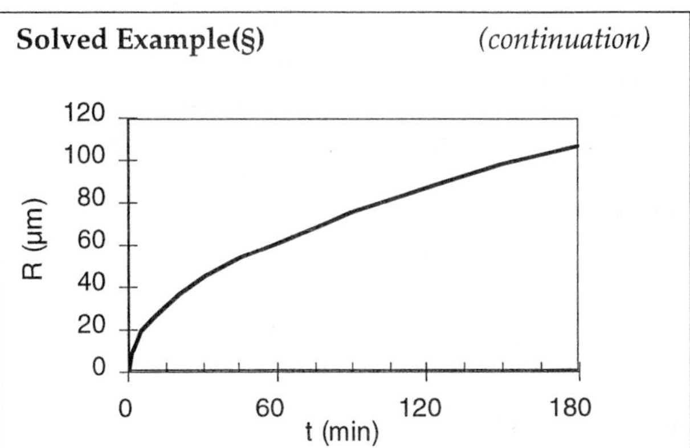

Check: Units OK. Physics OK.
Discussion: The droplet radius increases with the square root of time — fast initially and slower later. Even after 3 hours, it has a size on the borderline between cloud and rain drops. Thus, it is virtually impossible to grow full-size rain drops solely by diffusion.

ICE GROWTH BY DIFFUSION

Ice Crystal Habits

Because of the hexagonal lattice structure of ice, ice crystals grow in a variety of hexagonal shapes, depending on temperature and supersaturation. These shapes are called **habits** (Fig 8.6).

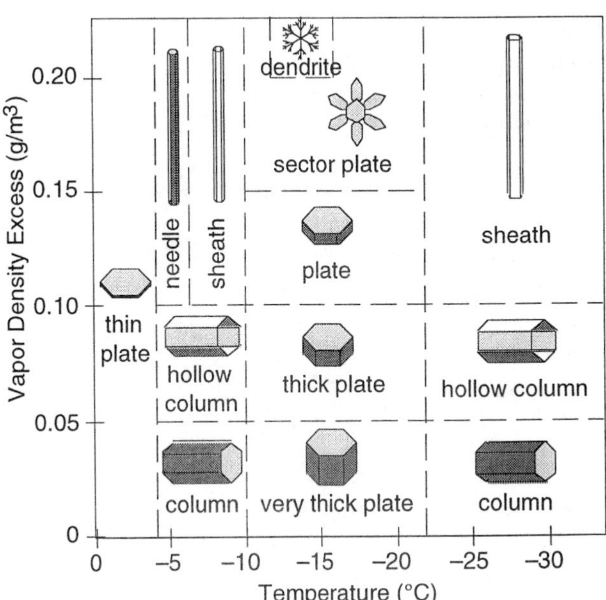

Figure 8.6
Ice crystal habits (idealized).

Growth Rates

Because of the diversity of shapes, it is better to measure crystal size by its mass m rather than by some not-so-representative radius. Rate of growth by diffusion depends on crystal habit.

Columns and very thick plates have **aspect ratio** (height to width ratio) of roughly 1. If the aspect ratio remains constant during growth, then the growth equation is:

$$m \propto (D \cdot S \cdot t)^{3/2} \tag{8.14}$$

where D is diffusivity, S is supersaturation fraction, and t is time. If the crystal were spherical with radius R, then its mass would be $m = \rho_{liq.water} \cdot (4 \cdot \pi/3) \cdot R^3$. Taking the cube root of both sides gives an equation similar to eq. (8.13). Thus growth rate of a 3-D crystal is very similar to growth of a liquid droplet.

For 2-D growth, such as dendrites or plates of constant thickness, the growth equation changes to

$$m \propto (D \cdot S \cdot t)^2 \tag{8.15}$$

For 1-D growth of needles and sheaths of constant diameter, the growth equation is

$$m \propto \exp\left[(D \cdot S \cdot t)^{1/2}\right] \tag{8.16}$$

These three growth rates are sketched in Fig 8.7.

Evidently 2-D crystals increase mass faster than 3-D ones, and 1-D crystals increase mass faster still. Those crystals that gain the most mass are the ones that will precipitate first.

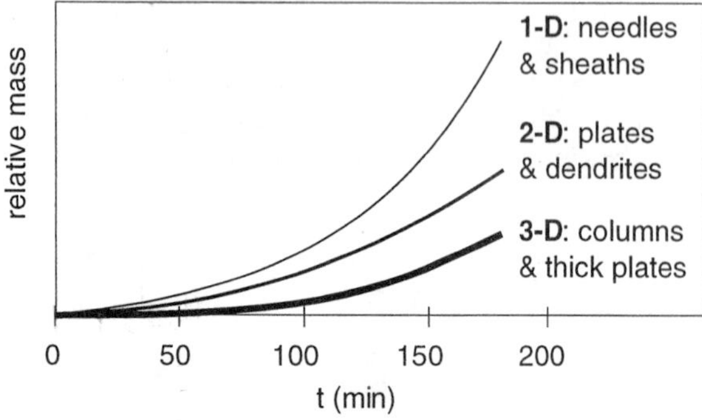

Figure 8.7
Relative growth rates of crystals of different habits.

The Wegener-Bergeron-Findeisen (WBF) Process

Recall from Chapter 5 that ice has a lower saturation vapor pressure than liquid water at the same temperature. Fig 8.8 shows an enlargement of the saturation vapor pressure curves for liquid water and ice.

Suppose that initially (time 1, on the time line in Fig 8.8) there are only supercooled liquid water droplets in a cloudy air parcel. These droplets exist in a supersaturated environment and therefore grow. As the air parcel rises and cools within the cloud, some ice nuclei might become activated (time 2), causing ice crystals to form and grow.

Both the ice crystals and liquid droplets continue to grow, because both are in a supersaturated environment (time 3). However, the ice crystal grows a bit faster because it is further from its ice saturation line (i.e., more supersaturated) than the liquid droplet is from liquid saturation line.

As both hydrometeors grow, water vapor is removed from the air, reducing the supersaturation. Eventually, near point 4 on the time line, so much vapor has been consumed that the relative humidity has dropped below 100% with respect to water. Hence, the liquid droplet begins to evaporate into the unsaturated air. However, at point 4 the ice crystal continues to grow because the air is still supersaturated with respect to ice.

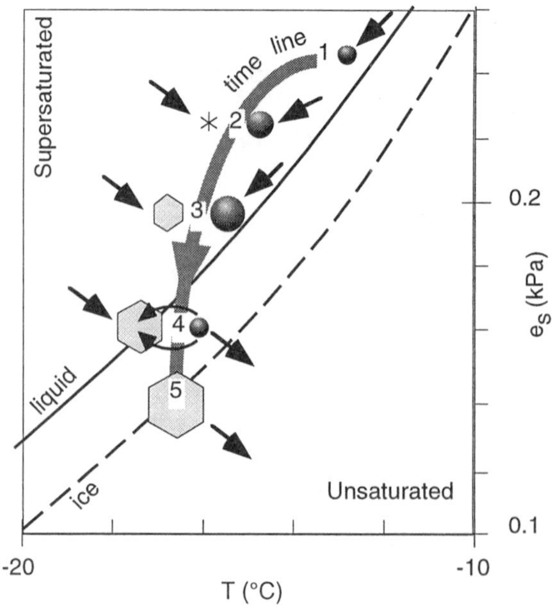

Figure 8.8
Enlargement of a portion of Fig 5.2, illustrating the WBF ice-growth process in a rising, cooling air parcel. Shown are saturation vapor pressure vs. temperature over liquid water and ice. Spheres represent cloud droplets, and hexagons represent ice crystals. Small arrows indicated movement of water vapor.

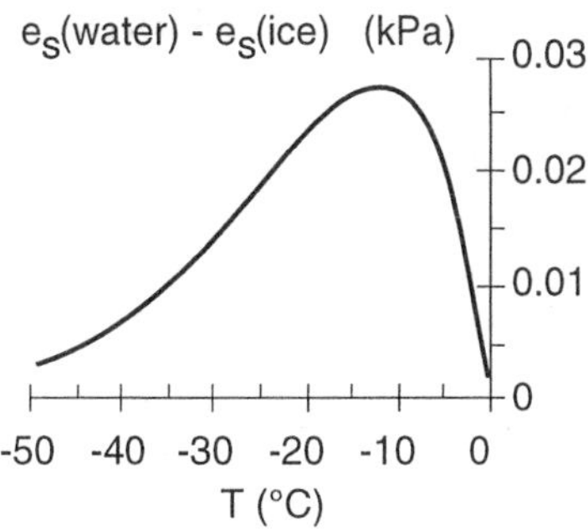

Figure 8.9
Saturation vapor pressure difference over water vs. over ice.

The net result is that the ice crystals grow at the expense of the evaporating liquid droplets, until the droplets disappear (point 5).

The difference between ice and liquid saturation vapor pressures is greatest in the range –8°C to –16°C, as shown in Fig 8.9 (from the insert in Fig 5.2). This is the temperature range where we expect the maximum effect from the WBF growth process, also known as the **cold cloud** process because temperatures below freezing are needed.

If a large number of ice nuclei exist in the air, then a large number of ice crystals will form that are each too small to precipitate. For a very small number of ice nuclei, those few ice crystals will rapidly grow large and precipitate out, leaving behind many small liquid cloud droplets in the cloud. Both of these scenarios lead to relatively little precipitation.

Only when there is between 1 to 10 ice nuclei per liter (compared to about a million liquid droplets in the same volume) will the ice nuclei be able to scavenge most of the condensed water before precipitating out. This scenario causes the maximum precipitation for the WBF processes. But a restriction on this precipitation formation process is that it happens only in **cold clouds** (clouds colder than 0°C).

COLLISION AND COLLECTION

Those hydrometeors that have the most mass fall the fastest. As a result, different hydrometeors move at different speeds, allowing some to collide. Not all collisions result in the merging of two hydrometeors. Those particles that do merge form a particle that is even heavier, falls faster, and collides with even more particles. Hence, this chain reaction can cause hydrometeors to rapidly grow large enough to precipitate.

Terminal Velocity Of Droplets

Everything including cloud and rain drops are pulled by gravity. The equilibrium velocity resulting when gravity balances frictional drag is called the **terminal velocity**.

Cloud Droplets

For particles of radius $R < 40$ µm, which includes most cloud droplets and aerosols, **Stokes Drag Law** gives the terminal velocity w as

$$w \approx k_1 \cdot R^2 \qquad (8.17)$$

where $k_1 = -1.19\text{x}10^8 \text{ m}^{-1}\cdot\text{s}^{-1}$. The negative sign on k_1 indicates the droplets are falling.

When drops fall at their terminal velocity, the gravitational pull on the drops is transmitted by frictional drag to the air. In other words, the weight of the air includes the weight of the drops within it. Hence, droplet-laden air is heavier than cloud-free air, and behaves as if it were colder (see virtual temperature, eq. 1.14). Falling rain also tends to drag air with it.

Solved Example
What updraft wind is needed to keep a typical cloud droplet (R = 10 µm) from falling?

Solution
Given: R = 10 µm
Find: w = ? m/s

Use eq. (8.17):

$$w \approx (-1.19\times10^8 \text{m}^{-1}\cdot\text{s}^{-1})\cdot(10\times10^{-6}\text{m})^2$$
$$= -0.012 \text{ m/s} = \underline{\mathbf{-1.2 \text{ cm/s}}}$$

Check: Units OK. Physics OK.
Discussion: The required updraft velocity is positive 1.2 cm/s, which is a very gentle movement of air. Recalling that most clouds form by adiabatic cooling within updrafts of air, these updrafts also keep cloud droplets afloat.

Rain Drops

Rain drops are sufficiently large and fall fast enough that Stokes drag law is not appropriate. If raindrops were perfect spheres, then

$$w \approx k_2 \cdot \left(\frac{\rho_o}{\rho_{air}} \cdot R \right)^{1/2} \tag{8.18}$$

where $k_2 = -220 \text{ m}^{1/2} \cdot \text{s}^{-1}$ (negative means a downward velocity), $\rho_o = 1.225 \text{ kg} \cdot \text{m}^{-3}$ is air density at sea level, and ρ_{air} is air density at the drop altitude.

However, the larger raindrops become flattened as they fall due to the drag. They do not have a teardrop shape. This flattening increases air drag even further, and reduces their terminal velocity from that of a sphere. Fig 8.10 shows raindrop terminal velocities. For the smallest drops, the curve has a slope of 2, corresponding to Stokes law. At larger sizes of about R = 500 to 1000 µm, the slope is 0.5, which corresponds to eq. (8.18). At the largest sizes (just under **drop-breakup**-size of about R = 2500 µm), the terminal velocity curve has near zero slope.

Let R be the **equivalent radius** of a sphere having the same volume as the deformed drop. An empirical curve for terminal velocity over range $20 \le R \le 2500$ µm is:

$$w = -c \cdot \left[w_o - \exp\left(\frac{R_o - R}{R_1} \right) \right] \tag*{•(8.19)}$$

where w_o = 12 m/s, R_o = 2500 µm, and R_1 = 1000 µm. This curve gives a maximum terminal velocity of 11 m/s for the largest drops. Density correction factor $c = (\rho_{70kPa}/\rho_{air})^{1/2} \cong (70 \text{ kPa}/P)^{1/2}$, where P is ambient pressure. Rain falls faster where the air is thinner.

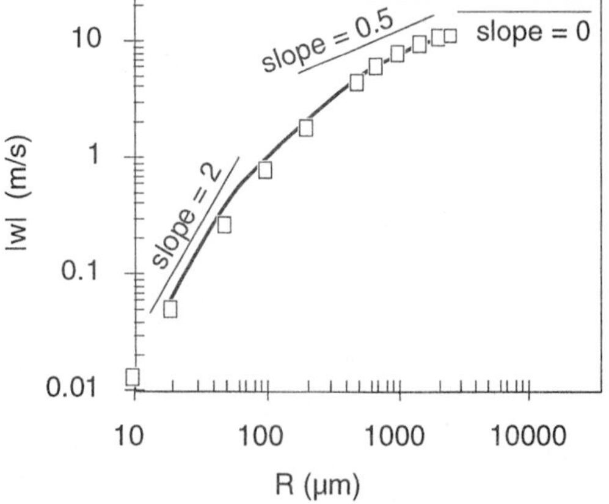

Figure 8.10
Raindrop terminal velocities w at P = 70 kPa and T = 0°C. R is the equivalent radius of a spherical drop. Curve is eq. (8.19).

Solved Example

Find the terminal velocity of a droplet of equivalent radius 1500 µm, at P = 70 kPa.

Solution

Given: R = 1500 µm, c = 1 at P = 70 kPa.
Find: w = ? m/s.

Use eq. (8.19):

$$w = -1 \cdot \left[(12 \text{m/s}) - \exp\left(\frac{2500\mu\text{m} - 1500\mu\text{m}}{1000\mu\text{m}} \right) \right]$$

= **−9.3 m/s**

Check: Units OK. Physics OK. Agrees with Fig 8.10. Negative sign means falling downward.
Discussion: This terminal velocity is not very fast – roughly equivalent to 21 mph. Updrafts in thunderstorms can easily exceed this velocity, and keep even these large drops afloat.

Processes

The merging of two liquid droplets is called **coalescence**. This is the only process for making precipitation-size hydrometeors that can happen in **warm clouds** (clouds warmer than 0°C), and is thus called the **warm cloud process**.

When ice particles collide and stick to other ice particles, the process is called **aggregation**. The growth of ice particles by collection and instant freezing of supercooled liquid droplets is called **accretion** or **riming**. Hydrometeors that become so heavily rimed as to completely cover and mask the original habit are called **graupeln** (a single such particle is a **graupel**). If the collected water does not freeze instantly upon contacting the ice particle, but instead flows around it before freezing, then **hail** can form. Graupel is much less dense than hail, because of the air trapped between the frozen droplets on the graupel.

Droplet Size Distributions

Small raindrops outnumber large ones. Classically, the rain-drop spectrum has been fit by an exponential function, known as the **Marshall-Palmer distribution**:

$$N = \frac{N_o}{\Lambda} \exp(-\Lambda \cdot R) \tag*{•(8.20)}$$

where N is the number of drops of radius greater than R within each cubic meter of air, and $N_o \cong 1.6\text{x}10^7 \text{ m}^{-4}$. Parameter $\Lambda(\text{m}^{-1}) = 8200 \cdot (RR)^{-0.21}$, where RR is rainfall rate in mm/h.

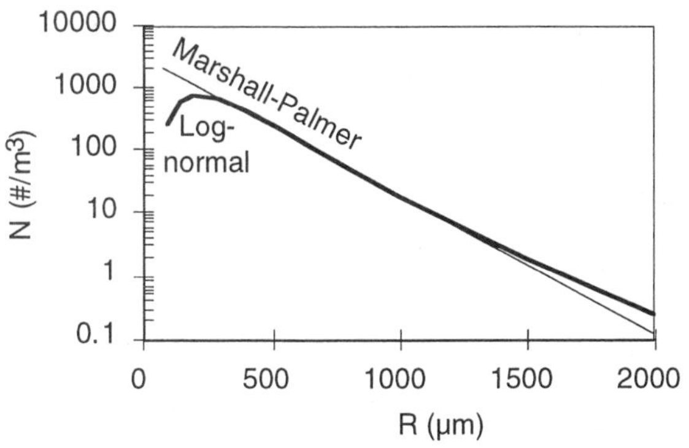

Figure 8.11
Rain drop size distributions.

Although eq. (8.20) might apply to raindrops within the cloud, the smallest drops might not reach the ground, due to either evaporation or suspension in updrafts. A lognormal distribution describes this decrease of the small droplets:

$$N = \frac{N_1}{(2 \cdot \pi^{0.5} \cdot \sigma_1 \cdot R)} \exp\left[-\frac{\ln^2(R/R_1)}{2 \cdot \sigma_1^2} \right] \qquad (8.21)$$

where R_1 is average drop radius, σ_1 is the standard deviation of drop radii, and N_1 is an abundance parameter. Fig 8.11 compares the Marshall -Palmer and lognormal distributions.

Solved Example
Compare Marshall-Palmer and lognormal raindrop distributions, using the parameters given below.

Solution
Given: RR = 10 mm/h, N_o = 1.6x10^7 m^{-4}
R_1 = 300 µm, σ_1 = 0.55, N_1 = 4x10^5.
Find and plot N vs. R.

Use eq. (8.20) for Marshall-Palmer, with
$\Lambda = 8200 \cdot RR^{-0.21} = 5056 \text{ m}^{-1}$.
Use eq. (8.21) for lognormal. The results were already plotted in Fig 8.11.

Check: Units OK. Physics OK. Fig reasonable.
Discussion: While there are on the order of 1000 drizzle drops in each cubic meter of air, there are only tens of drops of typical raindrop size.
Beware that these curves vary significantly from storm to storm. Many thunderstorms have much greater numbers of large 1000 to 2500 µm radius drops.

PRECIPITABLE WATER

If all the vapor, liquid, and solid water within a column of air in the atmosphere were to precipitate out, the resulting depth of water on the earth's surface is called the **precipitable water**, d_W. It is given by

$$d_W = \frac{r_T}{g \cdot \rho_{liq}} \cdot (P_B - P_T) \qquad \bullet(8.22)$$

where r_T is average total-water mixing ratio, P_B and P_T are ambient air pressures at the bottom and top of the column segment, g = 9.8 m·s^{-2} is gravitational acceleration, and ρ_{liq} = 1000 kg·m^{-3} is density of liquid water.

If the mixing ratio varies with height, then the column can be broken into smaller columns of near uniform mixing ratio, and the resulting depths can be summed to give the total depth. In the real atmosphere, winds can advect moisture into a column to replace that lost from precipitation. Consequently, the total rainfall amount observed at the surface after storm passage is often much greater than the precipitable water at any instant. Some microwave and infrared sensors on satellites measure precipitable water averaged over almost the whole depth of the troposphere.

Solved Example
If an average of 5 g/kg of water existed in the troposphere, between 100 kPa and 30 kPa, what is the precipitable water depth?

Solution
Given: r_T=0.005 kg/kg, P_B=100 kPa, P_T=30 kPa
Find: d_W = ? m

Use eq. (8.22): and recall 1 kPa = 1000 kg$_{air}$·m·s^{-2}

$$d_W = \frac{(0.005\text{kg}_{liq}/\text{kg}_{air})}{(9.8\text{m}\cdot\text{s}^{-2})\cdot(1000\text{kg}_{liq}\cdot\text{m}^{-3})} \cdot (100 - 30\text{kPa})$$

= 0.036 m = <u>3.6 cm</u> ≅ 1.4 inches.

Check: Units OK. Physics OK.
Discussion: Why so much precipitation? Because we used an unrealistically large mixing ratio over the whole depth of the troposphere. Although 5 g/kg is reasonable near the surface, higher in the troposphere the air is much colder and can hold much less water vapor.

RAINFALL RATE ESTIMATED BY RADAR

Heavier rain reflects more microwave energy back to a radar than lighter rain. However, more distant rain also gives a weaker return signal. A range-corrected and equipment-calibrated measure of reflectivity from rain is given by

$$\log(Z) = \log(\textit{received power}) + 2\cdot\log(\textit{range}) + \textit{const} \tag{8.23}$$

where Z is the **radar reflectivity factor**. Because Z has such a wide range of values, the reflectivity is usually expressed as **decibels** *dB* of Z:

$$dBZ = 10\cdot\log(Z) \tag{8.24}$$

Larger and more numerous drops reflect more radar energy:

$$Z = \frac{\sum D^6}{V} \tag{8.25}$$

where *D* is drop diameter, *V* is volume of air holding the drops, and the sum is over all raindrops within the volume. But the number and diameter of drops also determines the rainfall rate. When the above three equations are combined and empirically tuned to the observations, the result is a formula for converting radar echo intensity in *dBZ* to rainfall rate *RR*:

$$RR = c\cdot 10^{(0.0625\cdot dBZ)} \tag{•(8.26)}$$

where $c = 0.036$ mm/h.

Six discrete "**levels**" of radar echo intensity have been specified, corresponding to descriptive **rainfall categories**. These are easier for the general public to understand than *dBZ*. The definitions of these levels with respect to *dBZ* are given in Fig 8.12. Color-enhanced weather radar displays on TV show these six levels as different colors

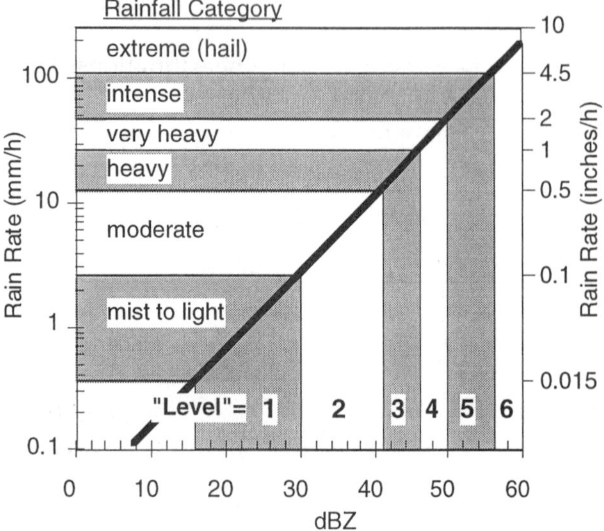

Figure 8.12
Rainfall intensity chart.

Solved Example
What is the rainfall rate for an echo of 43 *dBZ*?

Solution
Given: *dBZ* = 43
Find: *RR* = ? mm/h.

Use eq. (8.26)
$RR = (0.036\text{mm/h})\cdot 10^{(0.0625\cdot 43)}$ = **17.5 mm/h**.

Check: Units & Physics OK. Agrees with Fig 8.11.
Discussion: This would be displayed as a level 3 echo on a radar display, corresponding to heavy rain. Pilots utilize intensity levels for flight planning. They try to avoid the intensity levels 3 and higher due to the chance of strong turbulence in thunderstorms.

Science Graffito

"Bemused locals [in Stroud, Gloucestershire, UK] were not only stunned to witness frogs raining from the sky [in October 1987], but these frogs were also pink. Scientists identified them as a rare albino strain indigenous to southern Spain and North Africa."

"Minnows and sticklebacks fell over Glamorganshire in 1959, and sand eels rained down on Sunderland in August 1918. Crabs, starfish, flounder, and periwinkles have also fallen on the peaceful English countryside."

"Falls of fish are common in India,...[and have included] shrimps, squid, and herring." "At Acapulco, Mexico, a heavy October rain shower during the 1968 Olympics was accompanied by falling maggots."

"Mr. Alfred Wilson of Bristol [was] caught in a shower of hazel nuts." "In Gorky, USSR, in 1940, a treasure chest was whipped into the air by a whirlwind, and the ... residents were treated to an inundation of silver coins."

"Other reports of curious offerings from above include water lizards, snails, snakes, worms, cannon balls, turtles, ducks, and an alligator which fell on Charleston, Mississippi, in 1893. More gruesomely, it rained skulls, skeletons, and coffins when a tornado ripped up the vaults of a cemetery in New Orleans in the nineteenth century."

{ Roberts, S.K., 1999: Raining cats and frogs, *Weather*, **54**, p126. }

SUMMARY

Cloud droplets that form on cloud condensation nuclei overcome a formation barrier caused by the surface tension of the curved surface. These drops grow slowly by diffusion, and usually remain too small to rain.

In clouds colder than 0°C, ice nuclei trigger ice crystals to grow. Ice crystals exist in the air along with supercooled liquid drops. Because of the difference between liquid and ice saturation humidities, the ice can grow at the expense of evaporating liquid droplets. If the ratio of water to ice hydrometeors is about a million to one, then most of the water will be transferred to ice crystals, which are then heavy enough to fall as precipitation. Most rain at midlatitudes results from melted snow that formed from this "cold cloud" process.

As a spectrum of hydrometeors of different sizes forms, they fall at different terminal velocities and hit each other. Some collisions result in merging of the hydrometeors to form rain, snow aggregates, rimed ice crystals, graupeln, and possibly hail. Precipitation can be detected by weather radar, and rainfall rates can be estimated from echo intensity.

Threads

Larger precipitation particles form in deeper clouds, where cloud depth (Chapt. 7) depends on atmospheric stability (Chapt. 6). Thunderstorms (Chapt. 15) produce the largest rain droplets and ice crystals, including hail.

Air pollution (Chapt. 17) near urban areas often becomes cloud and ice nuclei, causing precipitation over and downwind of large cities to be altered (Chapt. 18) from the background climatological average. Precipitation and clouds are important feedback factors in global climate (Chapt. 18).

Total rainfall out of severe thunderstorms (Chapt. 15) and hurricanes (Chapt. 16) is often much greater than calculated precipitable water, primarily due to the strong inflow of humid air that fuels these storms.

Ice crystals of various shapes cause beautiful halos and other optical phenomena (Chapt. 19). Rain drops cause rainbows, and cloud droplets cause corona and glory (Chapt. 19). Droplet size distribution is very uniform in lenticular clouds (Chapt. 10), leading to iridescence optical phenomena (Chapt. 19). Haze increases scattering of light, and reduces visibility (Chapt. 19).

The supersaturation (Chapt. 5) needed to cause droplet and ice-crystal formation occurs in air that rises and cools adiabatically (Chapt. 3), due to the decrease of atmospheric pressure with height (Chapt. 1). However, the condensation that forms droplets also releases latent heat (Chapt. 3), which causes rising air parcels to cool at the moist rather than the dry adiabatic rate (Chapt. 5).

EXERCISES

Numerical Problems

N1. Using Fig 8.1, how many of the following are needed to fill a large rain drop.
a. small cloud droplet
b. typical cloud droplet
c. large cloud droplet
d. drizzle droplet
e. small rain droplet
f. typical rain droplet

N2. Given $c = 4\text{x}10^6\ \mu m^4 \cdot m^{-3}$, find the number-density of CCN of radius (µm):

a. 0.2 b. 0.5 c. 1.0 d. 2.0
e. 10. f. 2.5 g. 1.5 h. 5.0

N3(§) Produce Köhler curves such as in Fig 8.3b, but only for salt of the following masses (g) at 0°C:

a. 10^{-16} b. 10^{-15} c. 10^{-14} d. 10^{-13}
e. 10^{-18} f. 10^{-17} g. 10^{-12} h. 10^{-11}

N4(§). Produce Köhler curves for a solute mass of 10^{-16} g of salt for the following temperatures:

a. 20°C b. 0°C c. –20°C d. –39°C
e. –5°C f. –10°C g. –15°C h. 10°C

N5. Find the critical radii and supersaturations at a temperature of –10°C, for
a. 10^{-16} g of hydrogen peroxide
b. 10^{-14} g of sulfuric acid
c. 10^{-15} g of nitric acid
d. 10^{-13} g of ammonium sulfate
e. 10^{-13} g of hydrogen peroxide
f. 10^{-12} g of sulfuric acid
g. 10^{-12} g of nitric acid
h. 10^{-15} g of ammonium sulfate

N6. For the nuclei of the previous exercise, find the equilibrium haze droplet radius for the following relative humidities:

a. 80% b. 90% c. 100% d. 70%
e. 60% f. 95% g. 98% h. 100.1%

N7. How many CCN will be activated in maritime air at supersaturations of

a. 0.2% b. 0.5% c. 1%? d. 1.5%
e. 0.1% f. 0.3% g. 0.4% h. 0.7%

N8. Find the average separation distances between cloud droplets for the previous problem.

N9. What temperature is needed to immersion-freeze half the droplets of radius

a. 10 μm b. 100 μm c. 500 μm d. 1000 μm
e. 5 μm f. 50 μm g. 200 μm h. 700 μm

N10. What chemical will form ice crystals at the warmest temperature by the following processes?

a. contact freezing b. condensation freezing
c. deposition freezing d. immersion freezing

N11. For air at pressure altitude P = 70 kPa containing 5 g/kg of liquid water, find the maximum drop size via diffusive growth for the following CCN concentrations (# /m^3):

a. 10^7 b. 10^{10} c. 10^8 d. 10^9
e. 10^6 f. 10^5 g. 10^{11} h. 10^{12}

N12. Find the diffusivity for water vapor, given

a. P = 80 kPa, T = 10°C
b. P = 70 kPa, T = 0°C
c. P = 50 kPa, T = -20°C
d. P = 70 kPa, T = 10°C
e. P = 50 kPa, T = 0°C
f. P = 80 kPa, T = -20°C
g. P = 80 kPa, T = –5°C
h. P = 70 kPa, T = –5°C

N13. For a supersaturation gradient of 1% per 2 μm, find the kinematic moisture flux due to diffusion. Use D of:

a. 10^{-5} $m^2 \cdot s^{-1}$ b. 10^{-4} $m^2 \cdot s^{-1}$
c. 10^{-3} $m^2 \cdot s^{-1}$ d. 10^{-6} $m^2 \cdot s^{-1}$
e. 10^{-7} $m^2 \cdot s^{-1}$ f. 10^{-8} $m^2 \cdot s^{-1}$

N14(§). For conditions of the previous problem, plot droplet radius vs. time for diffusive growth.

N15(§). Compute and plot supersaturation vs. distance away from drops of the following radii, given a background supersaturation of 0.5%

a. 0.15 μm containing 10^{-16} g of ammon.sulfate
b. 0.3 μm containing 10^{-16} g of hydrogen perox.
c. 1 μm containing 10^{-16} g of salt
d. 0.15 μm containing 10^{-16} g of sulfuric acid
e. 0.3 μm containing 10^{-16} g of nitric acid
f. 0.5 μm containing 10^{-15} g of salt
g. 0.15 μm containing 10^{-16} g of hydrogen perox.
h. 0.3 μm containing 10^{-16} g of ammon.sulfate

N16. What crystal habit could be expected for

a. $\rho_{v\ excess}$ = 0.03 $g \cdot m^{-3}$, T = –25°C
b. $\rho_{v\ excess}$ = 0.22 $g \cdot m^{-3}$, T = –15°C
c. $\rho_{v\ excess}$ = 0.15 $g \cdot m^{-3}$, T = –3°C
d. $\rho_{v\ excess}$ = 0.12 $g \cdot m^{-3}$, T = –15°C
e. $\rho_{v\ excess}$ = 0.08 $g \cdot m^{-3}$, T = +5°C
f. $\rho_{v\ excess}$ = 0.20 $g \cdot m^{-3}$, T = –30°C
g. $\rho_{v\ excess}$ = 0.02 $g \cdot m^{-3}$, T = –8°C
h. $\rho_{v\ excess}$ = 0.12 $g \cdot m^{-3}$, T = –5°C

N17. Suppose the following ice crystals were to increase mass at the same rate. Find the rate of increase with time of the requested dimension.

a. effective radius of column growing in 3-D
b. diameter of plate growing in 2-D
c. length of needle growing in 1-D

N18. Find the terminal velocity for drops of radius:

a. 5 μm b. 20 μm c. 30 μm d. 40 μm
e. 2 μm f. 10 μm g. 35 μm h. 25 μm

N19. For a Marshall-Palmer distribution, how many droplets are expected of size greater than:

a. 200 μm b. 500 μm c. 1000 μm d. 2500 μm
assuming a rainfall rate of 10 mm/h.

e. 1500 μm f. 500 μm g. 1000 μm h. 2000 μm
assuming a rainfall rate of 20 mm/h.

N20. Find the terminal velocities for the drops in the previous exercise.

N21. a. Given 10 g/kg total water mixing ratio between 100 kPa and 80 kPa, and 3 g/kg up to 30 kPa, find precipitable water.

b. Given 20 g/kg total water mixing ratio between 100 kPa and 90 kPa, and 5 g/kg up to 20 kPa, find precipitable water.

N22. Find the rainfall rate, intensity "level", and rainfall category for these radar echo dBZ values:

a. 10 b. 35 c. 20 d. 45
e. 58 f. 48 g. 52 h. 25

N23. If weather radar indicates the following reflectivity values during the periods listed, what is the total rainfall at the end of the 40 minutes?

Time (min)	dBZ
0 - 20	13
20-23	25
23-27	39
27-32	42
32-35	21
35-39	6

Understanding & Critical Evaluation

U1. Once formed, does the volume of a falling raindrop remain constant? Discuss.

U2. Compare CCN size spectra with raindrop size spectra.

U3. Discuss the differences in nucleation between liquid water droplets and ice crystals, assuming the air temperature is below freezing.

U4. Using the full Köhler equation, discuss how supersaturation varies with:

a. temperature.
b. molecular weight of the nucleus chemical.
c. mass of solute in the insipient droplet.

U5. Discuss the differences in abundance of cloud vs. ice nuclei.

U6. If the atmosphere were to contain absolutely no CCN, discuss how clouds and rain would form, if at all.

U7. Can some chemicals serve as both water and ice nuclei? For these chemicals, describe how cloud particles would form and grow in a rising air parcel.

U8. If saturation is the maximum amount of water vapor that can be held by air at equilibrium, how is supersaturation possible?

U9. What is so special about the critical radius, that droplets larger than this radius continue to grow, while smaller droplets remain at a constant radius?

U10. In air parcels that are rising toward their LCL, aerosol swelling increases and visibility decreases as the parcels get closer to their LCL. If haze particles are at their equilibrium radius by definition, why could they be growing? Explain.

U11. Considering Fig 8.3, which type of CCN chemical would allow easier formation of cloud droplets: (a) a CCN with larger critical radius but lower peak supersaturation in the Köhler curve; or (b) a CCN with smaller critical radius but higher supersaturation? Explain.

U12. If droplets grow by diffusion to a final radius given by eq. (8.9), why do we even care about the diffusion rate?

U13. The droplet growth-rate equation (8.13) considers only the situation of constant background supersaturation. However, as the droplet grows, water vapor would be lost from the air causing supersaturation to decrease with time. Modify eq. (8.13) to include such an effect, assuming that the temperature of the air containing the droplets remains constant. Does the droplet still grow with the square root of time?

U14. Considering the different mass growth rates of different ice-crystal shapes, which shape would grow fastest in length of its longest axis?

U15. The Marshall-Palmer and log-normal raindrop distributions are very similar over a wide range. What are the advantages and disadvantages of each?

U16. If ice particles grew as spheres, which growth rate equation would best describe it? Why?

U17. Large ice particles can accumulate smaller supercooled liquid water drops via the aggregation process known as accretion or riming. Can large supercooled liquid water drops accumulate smaller ice crystals? Discuss.

U18. Can ice crystals still accrete smaller liquid water droplets at temperatures greater than freezing? Discuss.

U19. Compare the terminal velocity of rain vs. cloud droplets.

U20. Derive equation (8.22) for precipitable water. (Hint: Recall from Chapt. 1 that pressure difference between two heights is a measure of the amount of mass in that portion of the atmospheric column.) State and justify all assumptions.

U21. In eq. (8.21), why is the log(*range*) multiplied by 2 ?

U22. How does radar reflectivity factor Z change when *dBZ* doubles?

U23. For a weather radar in good working order, under what conditions might the radar not be able to see cells of heavy rain?

U24. Derive eq. (8.9), stating and justifying all assumptions. (Hint: consider the volume of a spherical drop.)

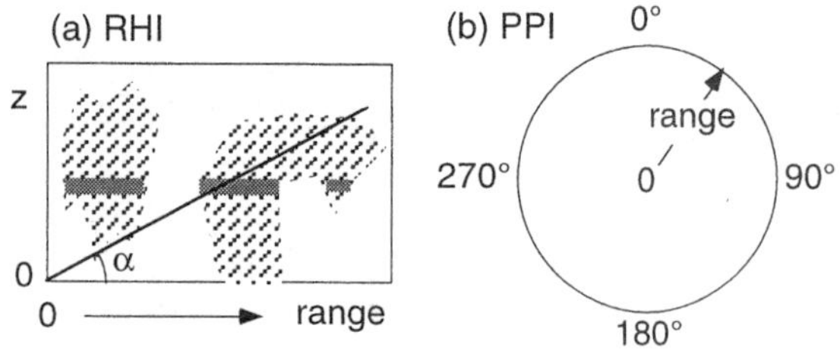

Figure 8.13
(a) Range height indicator (RHI) radar display, where hatching indicates precipitation, and the darker gray horizontal regions are the "bright bands". (b) Plan position indicator (PPI) display, made by scanning in azimuth a full 360°, but at fixed elevation angle α.

U25. Sometimes the weather conditions are such that ice crystals fall through a warm layer of air on the way to the ground. As the ice crystals start to melt, there is an altitude where each crystal has a solid core, but is coated with a thin layer of liquid water. Such a precipitation particle causes enhanced reflection of radar energy, and appears brighter on a radar display.

For a **range-height indicator** (**RHI**) radar display showing a vertical cross section through the precipitation area (Fig 8.13a), the altitude containing these water-coated ice crystals appears as a **bright band**. How would the bright-band effect look in a **plan-position indicator** (**PPI**) display (Fig 8.13b), which is when the radar scans 360° in azimuth while at a fixed elevation angle?

U26. If both supercooled liquid droplets and ice crystals were present in sinking cloudy air that is warming adiabatically, describe the evolutions of both types of hydrometeors relative to each other.

U27. How high above the LCL must air be lifted to cause sufficient supersaturation to activate CCN?

U28. a. Use the "typical" number of particles given in Ahrens' Table 6.1 to determine the value and units of parameter c in the CCN number density eq.

b Plot the resulting curve on a linear graph, and also on a log-log graph.

c. Why does eq. (8.1) appear as a straight line in Fig 8.2? What is the slope of the straight line in Fig 8.2? [Hint: for **log-log graphs**, the **slope** is the number of decades along the vertical axis (ordinate) spanned by the line, divided by the number of decades along the horizontal axis (abscissa) spanned by the line. **Decade** means a factor of ten; namely, the interval between major tic marks on a logarithmic axis.]

d. For number densities less than $1/\text{cm}^3$, what does it mean to have less than one particle (but greater than zero particles)? What would be a better way to quantify such a number density?

e. As discussed in Chapt. 19, air molecules range in diameter between roughly 0.0001 and 0.001 μm. If the line in Fig 8.2 were extended to that small of size, would the number density indicated by the Junge distribution agree with the actual number density of air molecules? If they are different, discuss why they should or should not be so.

U29. a. Find a relationship between the number density of CCN particles n, and the corresponding mass concentration c ($\mu\text{g}\cdot\text{m}^{-3}$), using the curve in Fig 8.2 and assuming that the molecular weight is m_w.

b. Consult a reference book such as the *Handbook of Chemistry and Physics* [Lide, D.R. (Ed.), 2000: CRC Press] to determine the molecular weight of sulfuric acid (H_2SO_4) and nitric acid (HNO_3), which are two contributors to acid rain. Assuming tiny droplets of these acids are the particles of interest for Fig 8.2, find the corresponding values or equations for mass concentration of these air pollutants.

Web-Enhanced Questions

W1. Use the web to access weather radar imagery for your local area (or other area specified by your instructor).

a. What different types of weather radar information are available?

b. How recent are each of these types of radar images?

c. If radar movie loops are available, comment on how precipitation cell movement is different or similar to movement of the whole storm system or front.

W2. Some radar imagery available on the web includes a precipitation estimate from the past 24 h. For a current precipitation event, compare these rainfall estimates to those available from conventional rain gauges in the same area.

W3. During certain conditions, weather radar can see insects, birds, and non-precipitating clouds. Can you find on the web radar sites displaying any of these conditions? What is the URL web address, and what does it show?

W4. For your country, search the web for a map or list of all weather radar locations. How much of your country is covered by such radar?

W5. Search the web for any journal articles, conference papers, or other technical reports that have pictures of droplet or ice-crystal growth or fall processes (such as photos taken in vertical wind tunnels).

W6. Can you find any satellite photos on the web showing haze? If so, which satellites and which channels on those satellites show haze the best? Do you think that satellites could be used to monitor air pollution in urban areas?

W7. Search the web for photos of typical weather radar antenna and/or enclosures. Also, can you find a table listing the scan rate, elevation angles, and transmitted power for normal operation of these weather radars?

W8. Search the web for photos of ice crystals and snow flakes with different habits. What habits in those photos were not given in the idealized Fig. 8.6?

W9. Find on the web a clear microphotograph of a dendrite snow flake. Print it out, and determine the fractal dimension of the snow flake. (Hint, see Chapt. 7 for a discussion of fractals.)

W10. Search the web for weather maps showing precipitable water.

a. Describe what info was used to create such a map (i.e., which weather forecast model, satellite, or radar, etc.).

b. What is the current and forecast amount of precipitable water over your location?

c. Compare model forecasts of precipitable water for your area, with radar estimates of 24 h precipitation that fell. Discuss differences in the results.

W11. Different radar displays and weather web information providers use different color schemes to indicate rainfall intensity or reflectivity dBZ. Print out or make note of the color scheme used in web images in your area, at your favorite web site.

W12. Search various government air pollution web sites (such as the U.S. Environmental Protection Agency: http://www.epa.gov/) for sizes of aerosol pollutants. Based on typical concentrations (or on concentrations specified in the air quality standards) of these pollutants, determine the number density, and compare with Fig 8.2.

Synthesis Questions

S1. What if the saturation vapor pressure over water and ice were equal.

a. Discuss the formation of cloud droplets and precipitation.

b. Contrast with those processes in the real atmosphere.

c. Discuss how the weather and climate might change.

S2. Suppose that the saturation vapor pressure over ice were greater, not less, than that over water. How would the WBF process change, if at all? How would the weather and climate change, if at all?

S3. What if you were hired to seed warm ($T > 0°C$) clouds, in order to create or enhance precipitation. Which would work better: (a) seeding with 10^5 salt particles/cm^3, each with identical of radius 0.1 µm; or (b) seeding with 1 salt particles/cm^3, each with identical radius of 0.5 µm; or (c) seeding with a range of salt particles sizes? Discuss, and justify.

S4. What if you were hired to seed cold ($T < 0°C$) clouds, in order to create or enhance precipitation. Would seeding with water or ice nuclei lead to the most precipitation forming most rapidly? Explain and justify.

S5. Is it possible to seed clouds in such a way as to reduce or prevent precipitation? Discuss the physics behind such weather modification.

S6. What if all particles in the atmosphere were **hydrophobic** (i.e., repelled water). How would the weather and climate be different, if at all?

S7. What if the concentration of nuclei that could become activated were only one-millionth of what currently exists in the atmosphere. How would the weather and climate change, if at all?

S8. Suppose a rising air parcel has a fixed amount of total water. If there is a high number-density of CCN in the parcel, then there will be a large number of small drops. For fewer CCN, there will be a smaller number of larger drops. Which combination gives the greatest radar reflectivity? Create a curve or equation of Z vs. n_{CCN} to describe this relationship.

S9. Eq. (8.17) indicates that smaller droplets and aerosol particles fall slower. Does Stoke's law apply to particles as small as air molecules? What other factors do air molecules experience that would affect their motion, in addition to gravity?

S10. What if Stoke's law indicated that smaller particles fall faster than larger particles. Discuss the nature of clouds for this situation, and how earth's weather and climate might be different.

S11. What if rain droplet size distributions were such that there were more large drops than small drops. Discuss how this could possibly happen, and describe the resulting weather and climate.

S12. Suppose that large rain drops did not break up as they fell. That is, suppose they experienced no drag, and there was no upper limit to rain drop size. How might plant and animal life on earth have evolved differently? Why?

S13. What if cloud and rain drops of all sizes fell at exactly the same terminal velocity. Discuss how the weather and climate might be different.

S14. Weather modification is as much a social issue as a scientific/technical issue. Consider a situation of cloud seeding to enhance precipitation over arid farm land in county *X*. If you wanted to make the most amount of money, would you prefer to be the:

a. meteorologist organizing the operation,
b. farmer employing the meteorologist,
c. company insuring the farmer's crop,
d. company insuring the meteorologist, or
e. lawyer in county *Y* downwind of county *X*, suing the meteorologist, farmer, and insurance companies?

Justify your preference.

S15. Suppose that you discovered how to control the weather via a new form of cloud seeding. Should you ...

a. keep your results secret and never publish or utilize them, thereby remaining impoverished and unknown?

b. publish your results in a scientific journal, thereby achieving great distinction?

c. patent your technique and license it to various companies, thereby achieving great fortune?

d. form your own company to create tailored weather, and market weather to the highest bidders, thereby becoming a respected business leader?

e. modify the weather in a way that you feel is best for the people on this planet, thereby achieving great power?

f. allow a government agency to hold hearings to decide who gets what weather, thereby achieving great fairness and inefficiency?

g. give your discovery to the military in your favorite country, thereby expressing great patriotism? (Note: they will probably take it anyway, regardless of whether you give it willingly.)

Discuss and justify your position. (Hint: See the "On Doing Science" box at the end of this chapter before you answer this question.)

ON DOING SCIENCE • Consequences

The scenario of exercise S15 is not as far-fetched as it might appear. Before World War II, American physicists received relatively little research funding. During the war, the U.S. Army offered a tremendous amount of grant money and facilities to physicists and engineers willing to help develop the atomic bomb as part of the Manhattan Project.

While the work they did was scientifically stimulating and patriotic, many of these physicists had second thoughts after the bomb was used to kill thousands of people at the end of the war. These concerned scientists formed the "Federation of Atomic Scientists", which was later renamed the "Federation of American Scientists" (FAS).

The FAS worked to discourage the use of nuclear weapons, and later addressed other environmental and climate-change issues. While their activities were certainly worthy, one has to wonder why they did not consider the consequences before building the bomb.

As scientists and engineers, it is wise for us to think about the moral and ethical consequences before starting each research project.

DYNAMICS

CONTENTS

9 Forces cause objects such as air parcels to accelerate or decelerate, thereby creating or altering winds. The relationship between forces and winds is called **atmospheric dynamics**. Newtonian physics describes atmospheric dynamics quite well.

Pressure, drag, and advection are some of the atmospheric forces that act in the horizontal. Other forces, called apparent forces, are caused by the Earth's rotation, and turning of the wind around a curve. The most important horizontal force is pressure-gradient force, without which there would be no wind.

Different forces dominate in different parts of the atmosphere, and during different weather conditions. Fig. 9.1 illustrates winds near the ground, and the isobaric pattern that drives those winds, for one of those special weather conditions.

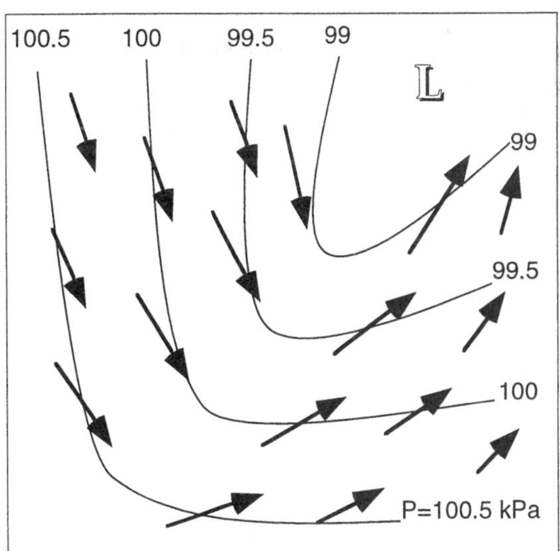

Figure 9.1
Sketch of sea-level pressure (thin lines are isobars) and the resulting near-surface winds (arrows). "L" marks a low pressure center.

NEWTON'S SECOND LAW OF MOTION

Lagrangian Momentum Budget

Forces, winds, and acceleration are vectors possessing both magnitude and direction. Newton's second law of motion states that a vector force $\vec{F}$ acting on an object such as an air parcel of mass m causes it to accelerate $\vec{a}$ in the same direction as the applied force:

$$\vec{F} = m \cdot \vec{a} \qquad \bullet(9.1)$$

Acceleration is defined as the rate of change of velocity with time t

$$\vec{a} = \frac{\Delta \vec{V}}{\Delta t} \qquad (9.2)$$

where $\vec{V}$ is the vector wind velocity.

Combining eqs. (9.1) and (9.2) give

$$\vec{F} = m \cdot \frac{\Delta \vec{V}}{\Delta t} \qquad (9.3a)$$

If mass is constant, then the equation can be written as:

$$\vec{F} = \frac{\Delta (m \cdot \vec{V})}{\Delta t} \qquad (9.3b)$$

Solved Example

A 1500 kg car accelerates from 0 to 60 mph in 9 seconds, heading south. (a) Find its average acceleration. (b) What force acted on it?

Solution

Given: $\vec{V}_{initial} = 0$, $\vec{V}_{final} = 60$ mph $= 27$ m/s
$t_{initial} = 0$, $t_{final} = 9$ s. Direction is south.
$m = 1500$ kg.

Find: $\vec{a} = ?$ m·s^{-2} , $\vec{F} = ?$ N

(a) Use eq. (9.2)

$$\vec{a} = \frac{\Delta \vec{V}}{\Delta t} = \frac{(27-0)\text{m/s}}{(9-0)\text{s}} = \underline{3\ \text{m·s}^{-2}\ \text{to the south}}$$

(b) Use eq. (9.1):

$\vec{F} = (1500 kg) \cdot (3 m \cdot s^{-2}) =$ __4500 N to the south__

(From Appendix A, recall that 1 N = 1 kg·m·s^{-2})

Check: Units OK. Physics OK.

Discussion: If the car falls off a cliff, gravity would accelerate it at $g = -9.8$ m·s^{-2}, where the negative sign denotes the downward direction.

But mass times velocity equals momentum. Thus, eq. (9.3b) describes the change in momentum with time following the air parcel; namely, it is the **Lagrangian momentum budget**.

ON DOING SCIENCE • Be Creative

Isaac Newton grew up on his mother's farm, Woolsthorpe, and built model windmills, clocks, and sundials for fun. His grades in school were OK, and like most other young students he got into occasional fights with his classmates, and carved his name in his desk at school. However, his schoolmaster saw a spark of talent in Isaac, and suggested to his mother that Isaac should go to college, because he would likely never make a good farmer.

So at age 18, Isaac Newton went to Cambridge University in England in 1661, and after working at odd jobs to pay his way, finally graduated with a B.A. in 1665. Later that year the plague hit, killing 10% of the London population within three months. For fear that the plague would spread, the dons at Cambridge decided to close the university until 1667. Isaac and the other students were sent home.

He continued his scientific studies in seclusion at his mother's farmhouse during the 18 months that school was closed. During this period, at age 23 - 24, he made many of his major discoveries. This included the laws of motion, the study of optics, the invention of the reflection telescope, the explanation for the orbits of planets, the understanding of gravity, and the invention of calculus to help describe his physical laws.

Often the most creative science, music, literature, and art are done by young men and women who have not been biased and (mis)directed by studying the works of others too much. Such knowledge of past work will often subconsciously steer one's research in the directions that others have already taken, which unfortunately discourages novel ideas.

Be creative, and learn from your mistakes. So what if you "re-invent the wheel", and "discover" something that was already discovered a century ago. So what if you make the same mistakes made by some famous scientist 50 years ago. The freedom to make personal discoveries and mistakes and the knowledge you gain by doing so allows you to be much more creative than if you had just read about the end result in a journal.

So we have a paradox. On one hand, I wrote this book to help you learn the advances in meteorology that were made by others. On the other hand, I discourage you from studying the works of others. As a scientist or engineer, you must make your own decision about the best balance of these two philosophies that will guide your future work.

FOCUS • In Newton's Own Words

Newton's laws of motion, in his own words, are given below. (Actually, his original words were Latin, but here is the translation.)

"**Law I**. Every body perseveres in its state of resting or moving uniformly straight on, except inasmuch as it is compelled by impressed forces to change that state.

Law II. Change in motion is proportional to the motive force impressed and takes place following the straight line along which that forced is impressed.

Law III. To any action there is always a contrary, equal reaction; in other words, the actions of two bodies each upon the other are always equal and opposite in direction.

Corollary 1. A body under the joint action of forces traverses the diagonal of a parallelogram in the same time as it describes the sides under their separate actions."

Rearranging eq. (9.3a) gives a forecast equation for the wind velocity:

$$\frac{\Delta \vec{V}}{\Delta t} = \frac{\vec{F}_{net}}{m} \qquad (9.3c)$$

The subscript "net" implies that there might be many forces acting on the air parcel, and we need to consider the vector sum of all forces in eq. (9.3c).

Eulerian Momentum Budget

For wind forecasts over a fixed location such as a town or lake, use an Eulerian reference frame. Define a local Cartesian coordinate system with x increasing toward the local East, y toward the local North, and z up.

Horizontal winds (parallel to the earth's surface) can be found by rewriting the vector equation (9.3c) as two separate scalar equations: one for the west-to-east wind component (U) and one for the south-to-north component (V):

$$\frac{\Delta U}{\Delta t} = \frac{F_{x\ net}}{m} \qquad \bullet(9.4a)$$

$$\frac{\Delta V}{\Delta t} = \frac{F_{y\ net}}{m} \qquad \bullet(9.4b)$$

where subscripts x and y indicate the component of the net vector force toward the east and toward the north, respectively. Relationships between the "speed and direction" method of representing a vector wind, vs. the "U and V component" method, were given in Chapter 1.

Recall that $\Delta U/\Delta t = [U(t+\Delta t) - U(t)]/\Delta t$.
Thus, we can rewrite eqs. (9.4) as forecast equations:

$$U(t+\Delta t) = U(t) + \frac{F_{x\ net}}{m} \cdot \Delta t \qquad \bullet(9.5a)$$

$$V(t+\Delta t) = V(t) + \frac{F_{y\ net}}{m} \cdot \Delta t \qquad \bullet(9.5b)$$

These equations are often called the **equations of motion**. Together with the continuity equation (later in this chapter), the equation of state (ideal gas law from Chapter 1), and the budget equations for heat and moisture (Chapters 3 and 5), they describe the dynamic and thermodynamic state of the air.

To forecast winds at some time in the future [$U(t+\Delta t)$ and $V(t+\Delta t)$], we need to know the winds now [$U(t)$ and $V(t)$], and the forces acting on the air. Generically, this is called an **initial value problem**, because we must know the initial winds in order to forecast the future winds. Even computerized weather forecast models (see Chapt. 14) need to start with an **analysis** of the current weather observations.

Solved Example

If $F_{x\ net}/m = 1\text{x}10^{-4}$ m·s^{-2} acts on air initially at rest, then what will be the final wind speed after 10 minutes?

Solution

Given: $U(0)=0$, $F_{x\ net}/m=1\text{x}10^{-4}$ m·s^{-2}, Δt=600s
Find: $U(\Delta t)$ = ? m/s.
Assume: no V-component forces or winds. V=0.

Use eq. (9.5a):
$U(t+\Delta t) = U(t) + \Delta t \cdot (F_{x\ net}/m)$
= 0 + (600s)·(1x10-4 m·s-2) = <u>**0.06 m/s**</u>.

Check: Units OK. Physics OK.
Discussion: Not very fast, but over many hours it becomes large. Because U is positive, the wind direction is toward the east.

Science Graffito

"If I have been able to see further than others, it was because I stood on the shoulders of giants." - Sir Isaac Newton.

FORCES

To solve the equations of motion for horizontal winds in an Eulerian framework, we need the horizontal forces acting on the air. The net "force per unit mass" consists of contributions from **advection** (AD), the **pressure-gradient force** (PG), the **Coriolis force** (CF), and **turbulent drag** (TD). In addition, imbalances between forces sometimes balance **centrifugal force** (CN):

$$F_{x\ net} = F_{x\ AD} + F_{x\ PG} + F_{x\ CF} + F_{x\ TD} \qquad (9.6a)$$

$$F_{y\ net} = F_{y\ AD} + F_{y\ PG} + F_{y\ CF} + F_{y\ TD} \qquad (9.6b)$$

Units of force per mass are N/kg, which is identical to units of $m \cdot s^{-2}$ (see Appendix A). We will use these latter units.

Advection

Wind can not only blow air of different temperature or humidity into a region, but it can also blow air of different momentum. But momentum per unit mass is just the wind itself. Thus, the wind can blow different winds into a region. Namely, winds can change due to advection, in an Eulerian framework.

While advection is not usually considered a force in the traditional Lagrangian sense, it has the same effect as a force. Namely, it can accelerate the wind. Advection terms always appear when Lagrangian budget equations are rewritten for fixed Eulerian frameworks, as we have seen earlier for the Eulerian heat and moisture budgets.

The components of advective force per unit mass are:

$$\frac{F_{x\ AD}}{m} = -U \cdot \frac{\Delta U}{\Delta x} - V \cdot \frac{\Delta U}{\Delta y} \qquad \bullet(9.7a)$$

$$\frac{F_{y\ AD}}{m} = -U \cdot \frac{\Delta V}{\Delta x} - V \cdot \frac{\Delta V}{\Delta y} \qquad \bullet(9.7b)$$

where $\Delta U/\Delta x$ is the horizontal gradient of U-wind in the x-direction, and the other gradients are defined similarly. We see that advection needs gradients. Without a change of wind with distance, there can be no momentum advection.

Solved Example

Vancouver, Canada, is roughly 250 km north of Seattle, Washington. The winds (U, V) are (8, 3) m/s in Vancouver, and (5, 5) m/s in Seattle. Find the advective force per unit mass.

Solution

Given: (U, V) = (8, 3) m/s in Vancouver,
(U, V) = (5, 5) m/s in Seattle
Δy = 250 km, Δx = not relevant (unknown)
Find: $F_{x\ AD}/m$ =? $m \cdot s^{-2}$, $F_{y\ AD}/m$ =? $m \cdot s^{-2}$

Use the definition of a gradient:
$\Delta U/\Delta y = (8 - 5\ m/s)/250{,}000\ m = 1.2\times10^{-5}\ s^{-1}$
$\Delta U/\Delta x = 0$ (unknown in this problem)
$\Delta V/\Delta y = (3 - 5\ m/s)/250{,}000\ m = -0.8\times10^{-5}\ s^{-1}$
$\Delta V/\Delta x = 0$ (unknown)
Average $U = (8 + 5\ m/s)/2 = 6.5\ m/s$
Average $V = (3 + 5\ m/s)/2 = 4\ m/s$

Use eq. (9.7a):

$$\frac{F_{x\ AD}}{m} = -(6.5 m/s)\cdot 0 - (4 m/s)\cdot(1.2\times10^{-5} s^{-1})$$
$$= \underline{\mathbf{-4.8x10^{-5}}}\ m \cdot s^{-2}$$

Use eq. (9.7b):

$$\frac{F_{y\ AD}}{m} = -(6.5 m/s)\cdot 0 - (4 m/s)\cdot(-0.8\times10^{-5} s^{-1})$$
$$= \underline{\mathbf{3.2x10^{-5}}}\ m \cdot s^{-2}$$

Check: Units OK. Physics OK.
Discussion: The U winds are slower in Seattle than Vancouver, but are being blown toward Vancouver by the southerly flow. Thus, advection is decreasing the U-wind, hence, the negative sign. The V-wind is faster in Seattle, and these faster winds are being blown toward Vancouver, causing a positive acceleration there.

Pressure-Gradient Force

Pressure gradient force always acts perpendicular to the isobars (or height contours) on a weather map, from high to low pressure (or heights). This force exists regardless of the wind speed, and does not depend on the wind speed. It starts the horizontal winds, and can accelerate, decelerate, or change the direction of existing winds. On a weather map, more-closely-space isobars indicate a greater force (Fig 9.2).

The components of this force are given by

$$\frac{F_{x\ PG}}{m} = -\frac{1}{\rho} \cdot \frac{\Delta P}{\Delta x} \qquad \bullet(9.8a)$$

$$\frac{F_{y\ PG}}{m} = -\frac{1}{\rho} \cdot \frac{\Delta P}{\Delta y} \qquad \bullet(9.8b)$$

where ρ is air density, and ΔP is the change of pressure across distance Δx or Δy. The negative sign makes the force act from high toward low pressure. (See p 188 for eqs. 9.8c & d.)

Pressure gradient force is the ONLY force that can drive the horizontal winds in the atmosphere. The other forces, such as Coriolis, drag, centrifugal, and even advection, disappear for zero wind speed. Hence, they can change the direction and speed of an existing wind, but they cannot create a wind out of calm conditions.

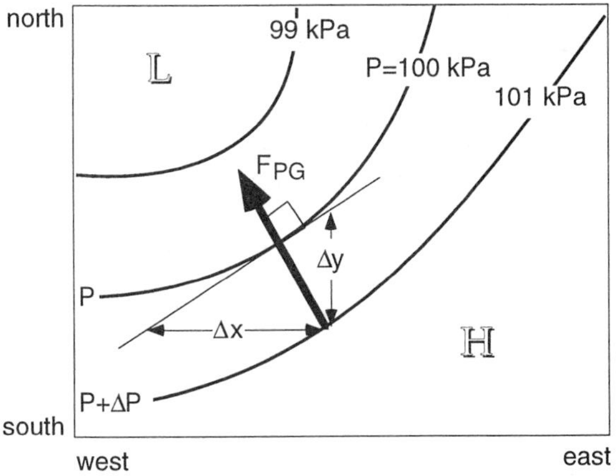

Figure 9.2
Pressure gradient force (heavy line) is perpendicular to isobars (medium lines) from high to low pressure. H and L indicate regions of high and low pressure, respectively.

Solved Example

If Milwaukee is 100 km east of Madison, Wisconsin, and the surface pressure at Milwaukee is 100.1 kPa and at Madison is 100 kPa, then what is the pressure gradient force per mass? Assume $\rho = 1.2\ \text{kg·m}^{-3}$.

Solution

Define: Cartesian coord. with $x = 0$ at Madison.
Given: $P = 100.1$ kPa at $x = 100$ km,
$P = 100.0$ kPa at $x = 0$ km. $\rho = 1.2\ \text{kg·m}^{-3}$.
Find: $F_{x\ PG}/m = ?\ \text{m·s}^{-2}$

Use eq. (9.8a):

$$\frac{F_{x\ PG}}{m} = -\frac{1}{(1.2\text{kg}\cdot\text{m}^{-3})}\cdot\frac{(100{,}100-100{,}000)\text{Pa}}{(100{,}000-0)\text{m}}$$

$= \underline{\mathbf{-8.33x10^{-4}}}\ \text{m·s}^{-2}$.

where $1\ \text{Pa} = 1\ \text{kg·m}^{-1}\text{·s}^{-2}$ was used (Appen. A).

Check: Units OK. Physics OK.
Discussion: The negative answer implies that the force is in the negative x-direction; that is, from Milwaukee toward Madison.

Centrifugal Force

Newton's laws of motion state that an object tends to move in a straight line unless acted upon by a force. Such a force, called **centripetal force**, causes the object to change direction and bend its trajectory. Centripetal force is caused by an imbalance of other forces.

Centrifugal force is an apparent force that is opposite to centripetal force, and pulls outward from the center of the turn. The components of centrifugal force are

$$\frac{F_{x\ CN}}{m} = +s\cdot\frac{V\cdot|V|}{R} \qquad \bullet(9.9a)$$

$$\frac{F_{y\ CN}}{m} = -s\cdot\frac{U\cdot|U|}{R} \qquad \bullet(9.9b)$$

where s is a sign coefficient given in Table 9-1, and R is the radius of curvature. The sign depends on whether air is circulating around a high or low pressure center, and whether it is in the northern or southern hemisphere. The total magnitude is:

$$\left|\frac{F_{CN}}{m}\right| = \frac{M^2}{R} \qquad (9.9c)$$

Table 9-1. Sign s for centrifugal-force equations.

Hemisphere	For flow around a Low	For flow around a High
Northern	1	–1
Southern	–1	1

Solved Example

On the back side of a low pressure center in the northern hemisphere, winds are from the north at 10 m/s at distance 250 km from the low center. Find the centrifugal force.

Solution

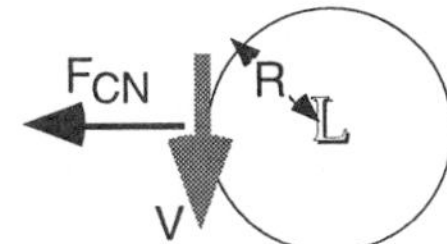

Given: $R = 2.5\text{x}10^5$ m,
$V = -10$ m/s
Find: $F_{x\ CN}/m = ?\ \text{m·s}^{-2}$.

Use eq. (9.9a):

$$\frac{F_{x\ CN}}{m} = +1\cdot\frac{-10\text{m/s}\cdot 10\text{m/s}}{2.5\times10^5} = \underline{\mathbf{-4x10^{-4}}}\ \text{m·s}^{-2}.$$

Check: Units OK. Physics OK. Sketch OK.
Discussion: The negative sign indicates a force toward the west, which is indeed outward from the center of the circle.

Coriolis Force

A **Coriolis parameter** is defined as

$$f_c = 2 \cdot \Omega \cdot \sin(\phi) \qquad \bullet(9.10)$$

where $2 \cdot \Omega = 1.458 \times 10^{-4}\ s^{-1}$, and ϕ is latitude. This parameter is constant at any fixed location. At mid-latitudes, the magnitude is of order of $f_c = 1 \times 10^{-4}\ s^{-1}$.

Coriolis force is an apparent force caused by the rotation of the earth. It acts perpendicular to the wind direction, to the right in the N. Hemisphere, and to the left in the Southern (Fig 9.3). In the N. Hemisphere, it is:

$$\frac{F_{x\ CF}}{m} = f_c \cdot V \qquad \bullet(9.11a)$$

$$\frac{F_{y\ CF}}{m} = -f_c \cdot U \qquad \bullet(9.11b)$$

Thus, there is no Coriolis force when there is no wind. Coriolis force cannot cause the wind to blow; it can only change its direction.

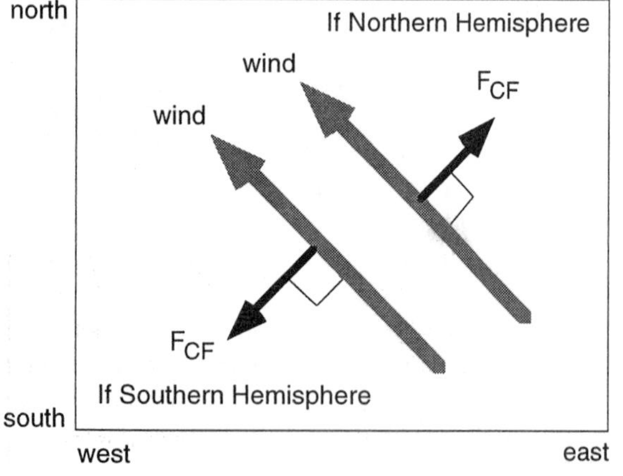

Figure 9.3
Coriolis force (dark lines).

Solved Example
Find the Coriolis force at Norman, Oklahoma, given a wind of $U = 10$ m/s.
Solution
Given: $U = 10$ m/s, $\phi =$ 35.2°N at Norman.
Find: $F_{x\ CF}/m = ?$ m·s^{-2}.
First, find the Coriolis parameter using eq. (9.10):
$f_c = (1.458 \times 10^{-4}\ s^{-1}) \cdot \sin(35.2°) = 8.4 \times 10^{-5}\ s^{-1}$.
Coriolis force in the y-direction (eq. 9.11b) is:

$$\frac{F_{y\ CF}}{m} = -(8.4 \times 10^{-5} s^{-1})\left(10\frac{m}{s}\right) = \underline{-8.4 \times 10^{-4}}\ m \cdot s^{-2}.$$

Check: Units OK. Physics OK.
Discussion: The – sign means force is north to south.

FOCUS • Coriolis Force

In 1835, Gaspar Gustave Coriolis used kinetic energy conservation to explain the apparent force that now bears his name. The following clarification was provided by Anders Persson in 1998.

Background

Coriolis force can be interpreted as the difference between two other forces: centrifugal force and gravitational force.

As discussed previously, **centrifugal force** is $F_{CN}/m = M_{tan}^2/R$, where M_{tan} is the tangential velocity of an object moving along a curved path with radius of curvature R (see Fig a). This force increases if the object moves faster, or if the radius becomes smaller. The symbol X marks the center of rotation of the object, and the small black circle indicates the object.

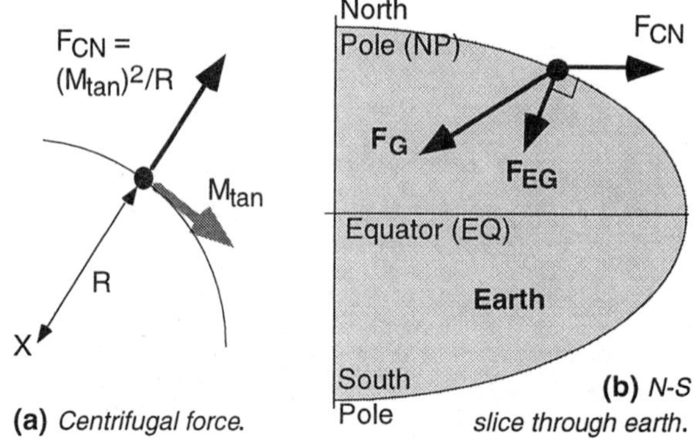

(a) *Centrifugal force.* **(b)** *N-S slice through earth.*

Because the earth is plastic (i.e., deformable), centrifugal force due to the earth's rotation and **gravitational force** F_G have shaped the surface into an ellipsoid, not a sphere. This is exaggerated in Fig b. The vector sum of F_G and F_{CN} is the effective gravity, F_{EG}. This effective gravity acts perpendicular to the local surface, and defines the direction we call **down**. Thus, a stationary object feels no net force (the downward force is balanced by the earth holding it up).

Now that "up" and "down" are identified, we can split centrifugal and actual gravitational forces into local horizontal (subscript $_H$) and vertical (subscript $_V$) components (Fig c). Because F_{CN} is always parallel to the equator, trigonometry gives $F_{CNH} \cong F_{CN} \cdot \sin(\phi)$, where ϕ is latitude.

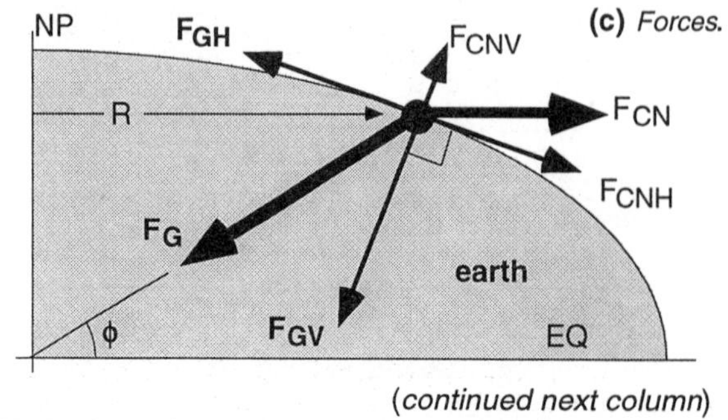

(c) *Forces.*

(*continued next column*)

FOCUS • Coriolis Force (continuation)

Objects at Rest

The earth rotates counterclockwise when viewed from above the north pole (NP). During time interval Δt, any single meridian (a longitude line, such as labeled with distance R in Fig d) will rotate by angle $\Omega \Delta t$, where Ω is the angular velocity of earth ($\cong$ 360°/24h).

Suppose an object (the dark circle) is at rest on this meridian. Then during the same time interval Δt, it will move as shown by the gray arrow, at speed $M_{tan} = \Omega \cdot R$. Because this movement follows a parallel (latitude line), and parallels encircle the earth's axis, the stationary object is turning around a circle. This creates centrifugal force. The horizontal component F_{CNH} balances F_{GH}, giving zero net apparent force on the object (Fig d).

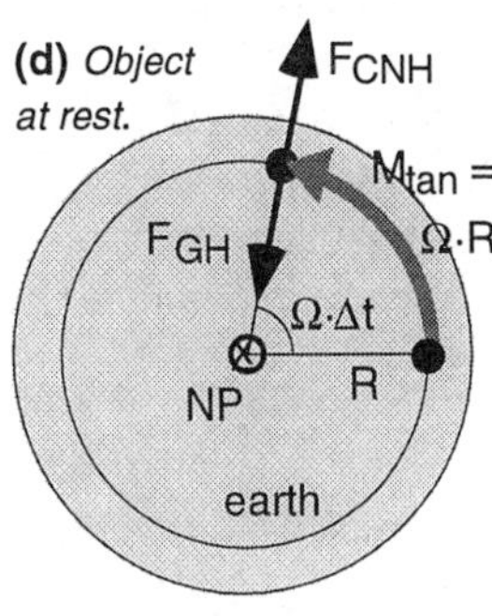

Objects Moving East or West

Next, we can ask what happens if the object moves eastward with velocity M relative to the earth's surface (shown by the thin white arrow in Fig e). The earth is rotating as before, as indicated by the thin meridian lines in the figure. Thus, the total movement of the object is faster than before, as shown by the gray arrow. This implies greater total centrifugal force, which results in a greater horizontal component F_{CNH}. However, the gravity component F_{GH} is unchanged.

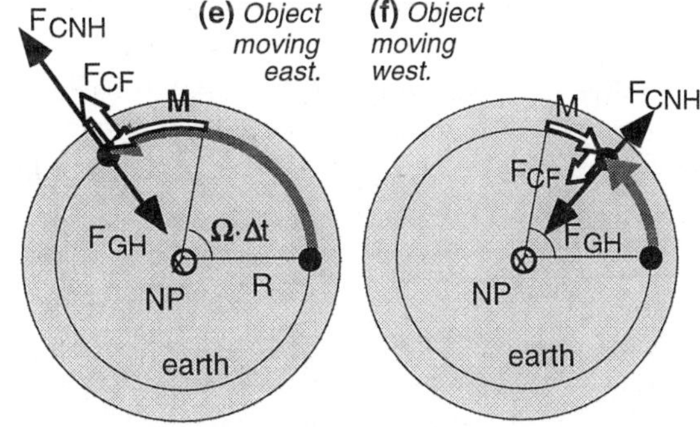

Thus, those two forces (F_{CNH} & F_{GH}) do NOT balance. The difference between them is a net force to the right of the relative motion M. This force difference is called **Coriolis force**, F_{CF}, and is indicated by the thick white arrow in Fig e.

Similarly, for an object moving westward (thin white arrow in Fig f), the net tangential velocity (gray arrow) is slower, giving an imbalance between F_{CNH} & F_{GH} that acts to the right of M. This is identified as Coriolis force.

(continued next column)

FOCUS • Coriolis Force (continuation)

Objects Moving North or South

For a northward moving object, the rotation of the earth (dashed thick gray line) and the relative motion of the object (M, thin white arrow) combine to cause a path shown with the solid thick gray line (Fig g). This has a smaller radius of curvature (R) about a center of rotation (X) that is NOT on the north pole (o). The smaller radius causes a greater horizontal component of centrifugal force (F_{CNH}), which points outward from X.

We can conceptually divide this horizontal centrifugal force into a north-south component ($F_{CNH\text{-}ns}$) and an east-west component ($F_{CNH\text{-}ew}$). The southward component $F_{CNH\text{-}ns}$ balances the horizontal gravitational component F_{GH}, which hasn't changed. However, the east-west component is acting to the right of the relative object motion M, and is identified as Coriolis force $F_{CF} = F_{CNH\text{-}ew}$.

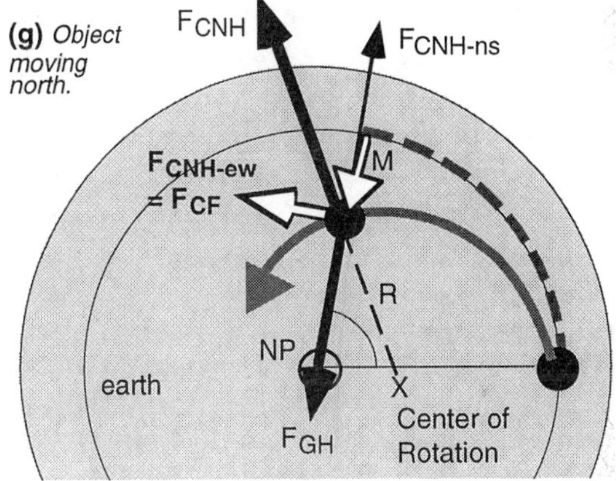

Similarly, a southward moving object has a larger radius of curvature, giving a Coriolis force to the right. In fact, an object moving in any arbitrary direction has Coriolis force acting to the right in the N. Hemisphere.

Magnitude of Coriolis Force

For an object at rest (Figs c & d), $F_{GH} = F_{CNH} \equiv F_{CNHR}$, where subscript $_R$ denotes "rest". At rest, $M_{tan\ rest} = \Omega \cdot R$. For an eastward moving object (Fig e), Coriolis force is defined as:

$$F_{CF} \equiv F_{CNH} - F_{GH} \qquad \text{(definition)}$$
$$= F_{CNH} - F_{CNHR} \qquad \text{(line 1 of this subsection)}$$
$$= \sin(\phi)\cdot[F_{CN} - F_{CNR}] \qquad \text{(from Fig c)}$$
$$F_{CF}/\,m = \sin(\phi)\cdot[(M_{tan})^2/R - (M_{tan\ rest})^2/R]$$
$$= \sin(\phi)\cdot[(\Omega\cdot R + M)^2/R - (\Omega\cdot R)^2/R]$$
$$= \sin(\phi)\cdot[(2\cdot\Omega\cdot M) + (M^2/R)]$$

But the last term is so small, it can be neglected compared to the first term. Thus:

$$F_{CF}/m \cong 2\cdot\Omega\cdot\sin(\phi)\cdot M$$
$$\equiv f_c\cdot M \qquad \text{(from eq. 9.10)}$$

The same answer is found for motion in any direction.

Turbulent Drag Force

At the earth's surface the air experiences drag against the ground. This turbulent drag force increases with wind speed, and is always in a direction opposite to the wind direction. Namely, drag slows the wind (Fig 9.4).

Only the **boundary layer** (see Chapter 4) experiences this drag. It is not felt by the air in the remainder of the troposphere (except for wave drag, see Ch. 10). The drag force acting on a boundary layer of depth z_i is:

$$\frac{F_{x\ TD}}{m} = -w_T \cdot \frac{U}{z_i} \qquad \bullet(9.12a)$$

$$\frac{F_{y\ TD}}{m} = -w_T \cdot \frac{V}{z_i} \qquad \bullet(9.12b)$$

where w_T is a turbulent **transport velocity**.

During windy conditions of near **neutral** static stability, turbulence is generated primarily by the **wind shear** (change of wind speed or direction with height). This turbulence transports frictional information upward from the ground to the air at rate:

$$w_T = C_D \cdot M \qquad (9.13)$$

where wind speed M is always positive, and C_D is a dimensionless **drag coefficient** in the range of $2\text{x}10^{-3}$ over smooth surfaces to $2\text{x}10^{-2}$ over rough or forested surfaces. It is similar to the bulk heat transfer coefficient of Chapter 3.

During statically **unstable** conditions of light winds and strong surface heating (e.g., daytime), buoyant thermals transport the frictional information, upward at rate:

$$w_T = b_D \cdot w_B \qquad (9.14)$$

where w_B is the **buoyancy velocity scale** (always positive (Chapt. 3), and $b_D = 1.83\text{x}10^{-3}$ is dimensionless.

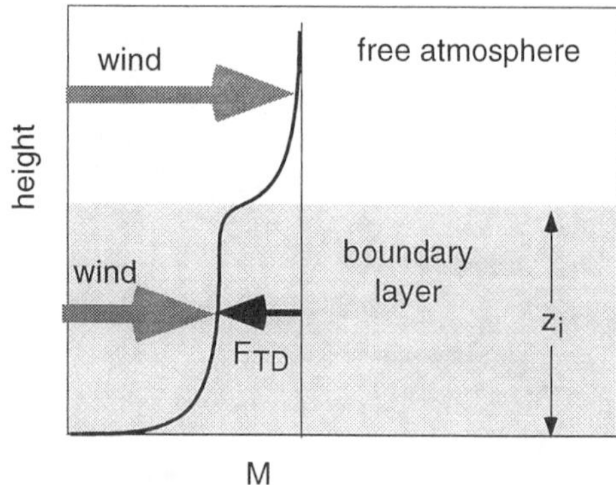

Figure 9.4
Turbulent drag force opposes the wind in the boundary layer.

Solved Example

Find the drag force per unit mass on a wind of U = 10 m/s, V = 0 for (a) statically neutral over smooth ground; and (b) statically unstable with w_B = 45 m/s. The boundary layer is 1 km thick.

Solution

Given: $U = M = 10$ m/s, $z_i = 1000$ m,
$C_D = 2\text{x}10^{-3}$, $w_B = 45$ m/s.
Find: $F_{x\ TD}/m$ = ? m·s^{-2}.

(a) Combine eqs. (9.13) and (9.12a):

$$\frac{F_{x\ TD}}{m} = -C_D \cdot M \cdot \frac{U}{z_i} = -(0.002) \cdot \frac{(10\text{m/s})^2}{1000\text{m}}$$

= **$-2\text{x}10^{-4}$** m·s^{-2}.

(b) Combine eqs. (9.14) and (9.12a)

$$\frac{F_{x\ TD}}{m} = -b_D \cdot w_B \cdot \frac{U}{z_i}$$

$$= -(0.00183) \cdot (45m/s) \cdot \frac{(10\text{m/s})}{1000\text{m}} =$$

= **$-8.24\text{x}10^{-4}$** m·s^{-2}.

Check: Units OK. Physics OK.
Discussion: The negative sign means that the drag force is toward the west, which is opposite the wind direction. Both mechanical and buoyant turbulence are equally effective at transporting frictional information to the air.

FULL EQUATIONS OF MOTION

Combining eqs. (9.4) and (9.6 - 9.12) gives the full equations of motion:

$$\frac{\Delta U}{\Delta t} = -U\frac{\Delta U}{\Delta x} - V\frac{\Delta U}{\Delta y} - \frac{1}{\rho}\cdot\frac{\Delta P}{\Delta x} + f_c \cdot V - w_T \cdot \frac{U}{z_i} \qquad \bullet(9.15a)$$

$$\frac{\Delta V}{\Delta t} = -U\frac{\Delta V}{\Delta x} - V\frac{\Delta V}{\Delta y} - \frac{1}{\rho}\cdot\frac{\Delta P}{\Delta y} - f_c \cdot U - w_T \cdot \frac{V}{z_i} \qquad \bullet(9.15b)$$

tendency — *advection* — *pressure gradient* — *Coriolis* — *turbulent drag*

Centrifugal force is not included because it is an imbalance of the forces already included above.

As was shown in the solved examples, each of the terms can be of similar magnitude: $1\text{x}10^{-4}$ to $10\text{x}10^{-4}$ m·s^{-2}. For some situations, some of the terms are small enough to be neglected compared to the others. For example, above the boundary layer the turbulent drag term is near zero. Near the equator Coriolis force is near zero. At the center of high or low pressure regions, pressure gradient is near zero.

There are other physical processes that have been neglected in the simplified equations just presented. **Molecular friction** is significant in the bottom few millimeters of the ground. **Mountain-wave drag** can be large in mountainous regions (see Chapter 10). Cumulus clouds can cause turbulent **convective mixing** above the boundary layer (Chapt. 4). Also, mean vertical motions (e.g., large-scale subsidence) have been neglected so far. They will be examined later.

HEIGHT CONTOURS ON ISOBARIC SURFACES

The equations of motion (9.15) were given in geometric Cartesian coordinates (x, y, z), where z is geometric distance above ground level, or above sea level in some cases, or is geopotential height in other cases. However, an alternative coordinate system can use pressure P as the vertical coordinate, because pressure decreases monotonically with increasing height. **Pressure coordinates** consist of (x, y, P).

A **monotonic** variable is one that changes only in one direction. For example, it increases or is constant, but never decreases. Or it decreases or is constant, but never increases. In the case of pressure in the atmosphere, it never increases with increasing height.

A surface connecting points of equal pressure is an **isobaric** surface. In low-pressure regions, this surface is closer to the ground than in high-pressure regions. Thus, this surface curves up and down through the atmosphere (Fig 9.5).

Low pressures on a constant height map correspond to low heights of a constant pressure surface. Similarly, high pressures correspond to high heights. Although isobaric surfaces are not flat, we draw them as flat weather maps on the computer screen or paper.

It is impossible for two different isobaric surfaces to cross each other. Also, these surface never fold back on themselves, because pressure decreases monotonically with height. However, they can intersect the ground, such as frequently happens in mountainous regions.

Constant pressure charts are used for three reasons. First, the old radiosondes measured pressure instead of altitude, so it was easier to plot their measurements of temperature, humidity and wind on an isobaric surface. Second, pressure is a measure of mass in the air. Third, some weather forecast models use pressure coordinate systems in the vertical, because the equations of motion that can be derived for this system have some advantages.

The hydrostatic equation (1.17) can be used to convert the pressure gradient terms from height to pressure coordinates. On isobaric surfaces, the pressure gradient terms become:

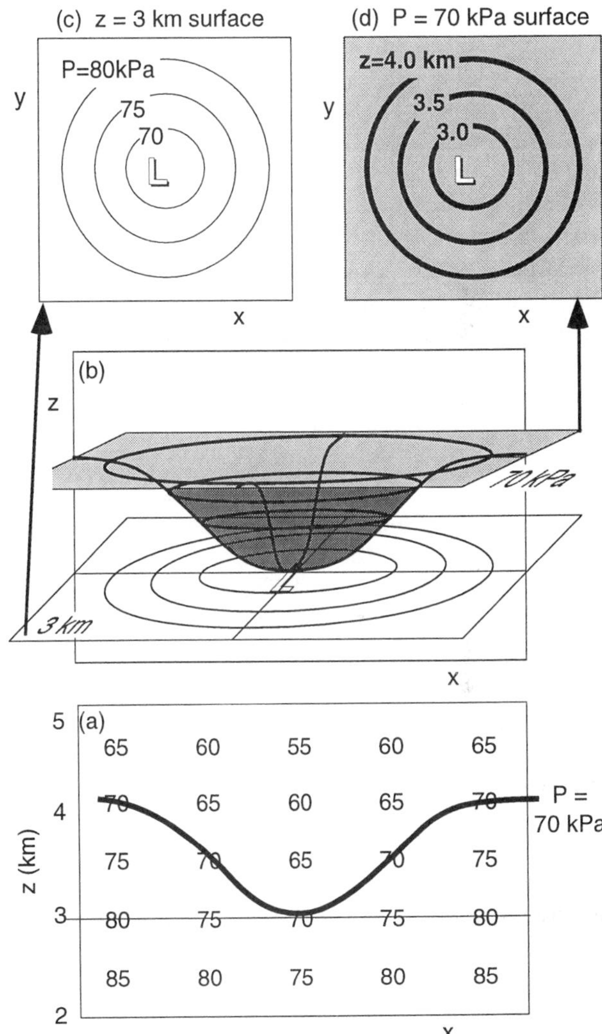

Figure 9.5
(a) Vertical slice through atmosphere, showing pressure values (kPa). Thick curved line is the 70 kPa isobar. Thin straight line is the 3 km height contour. The location of lowest pressure on the height contour corresponds to the location of lowest height of the isobar. (b) 3-D sketch of the same 70 kPa isobaric surface (shaded), and 3 km height surface (white). (c) Pressures intersected by the 3 km constant height surface. (d) Heights crossed by the 70 kPa surface. The low pressure center (L) in (c) matches the low height center in (d).

$$\frac{F_{x\ PG}}{m} = -g \cdot \frac{\Delta z}{\Delta x} \tag{9.8c}$$

$$\frac{F_{y\ PG}}{m} = -g \cdot \frac{\Delta z}{\Delta y} \tag{9.8d}$$

where $\Delta z / \Delta x$ and $\Delta z / \Delta y$ are the slopes of the isobaric surfaces (i.e., change of height with distance), and g = 9.8 m·s^{-2} is gravitational acceleration. If you could place a hypothetical ball on the isobaric surface plotted in Fig 9.5b, the direction that it would roll downhill is the direction of the pressure-gradient force.

Isobaric charts will be used extensively in the remainder of this book when describing upper air features. The reason is that most previous upper-air observations were made with radiosonde balloons that measured pressure but not height. However, more and more routine upper air soundings around the world are being made with **GPS** (**Global Positioning System**, satellite triangulation method) sondes that can measure geometric height as well as pressure. So in the future we might see the large government weather data centers starting to produce upper air weather maps on constant height surfaces.

In the section that follows, the equations of motion are simplified for some special cases, to yield theoretical winds. Where appropriate, the forces and winds will also be given in isobaric coordinates.

Solved Example

If the height of the 50 kPa pressure surface decreases by 10 m northward across a distance of 500 km, what is the pressure gradient force?

Solution

Given: Δz = –10 m, Δy = 500 km, g = 9.8 m·s^{-2}.
Find: F_{PG}/m = ? m·s^{-2}

Use eqs. (9.8 c & d):
$F_{x\ PG}/m$ = 0 m·s^{-2}, because $\Delta z/\Delta x = 0$.

$$\frac{F_{y\ PG}}{m} = -g \cdot \frac{\Delta z}{\Delta y} = -\left(9.8 \frac{\text{m}}{\text{s}^2}\right) \cdot \left(\frac{-10\text{m}}{500{,}000\text{m}}\right)$$

= **0.000196** m·s^{-2}

Check: Units OK. Physics OK. Sign OK.
Discussion: For our example here, height decreases toward the north, thus a hypothetical ball would roll downhill toward the north. A northward force is in the positive y direction, which explains the positive sign of the answer.

WINDS

When winds accelerate, forces such as Coriolis and drag change too, because they depend on the wind speed. This, in turn, changes the acceleration via eqs. (9.15), so there is a **feedback process**. This feedback continues until the forces balance each other. At that point, there is no net force, and no further acceleration.

This condition is called **steady state**:

$$\frac{\Delta U}{\Delta t} = 0, \qquad \frac{\Delta V}{\Delta t} = 0 \tag{•(9.16)}$$

In steady-state conditions, wind speeds do not change with time, but are not necessarily zero. Only the acceleration is zero.

Some special steady-state winds are examined next. These are theoretical winds that apply in idealized situations. The real winds for these special situations are often close to the theoretical winds.

Geostrophic Wind

The geostrophic wind (U_g, V_g) is a theoretical wind that results from a steady-state balance between pressure-gradient force and Coriolis force. After setting the other forces to zero, eqs. (9.15) become:

$$0 = -\frac{1}{\rho} \cdot \frac{\Delta P}{\Delta x} + f_c \cdot V \tag{9.17a}$$

$$0 = \underbrace{-\frac{1}{\rho} \cdot \frac{\Delta P}{\Delta y}}_{\text{pressure gradient}} \underbrace{- f_c \cdot U}_{\text{Coriolis}} \tag{9.17b}$$

Solving these equations for U and V, and then defining $U \equiv U_g$ and $V \equiv V_g$, gives:

$$U_g = -\frac{1}{\rho \cdot f_c} \cdot \frac{\Delta P}{\Delta y} \tag{•(9.18a)}$$

$$V_g = +\frac{1}{\rho \cdot f_c} \cdot \frac{\Delta P}{\Delta x} \tag{•(9.18b)}$$

In regions of straight isobars above the top of the boundary layer and away from the equator, the actual winds are approximately geostrophic. These winds blow parallel to the isobars or height contours, with low pressure to the left in the Northern Hemisphere (Fig 9.6). The wind is faster in regions

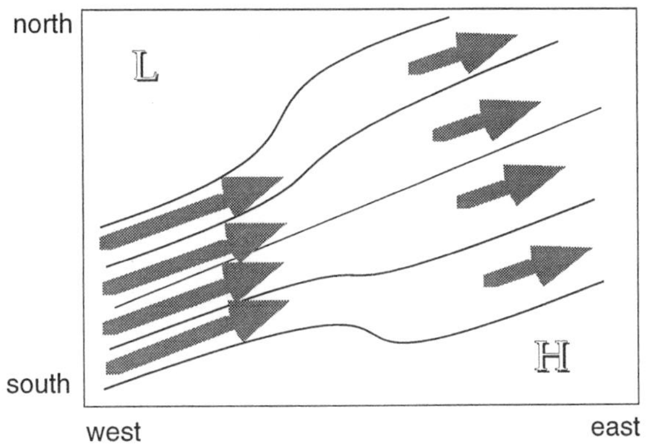

Figure 9.6
Geostrophic winds (grey arrows) are faster where isobars (thin dark lines) are closer together. (For Northern Hemisphere.)

> **Solved Example**
> If pressure increases by 1 kPa eastward across a distance of 500 km, then what is the geostrophic wind speed, given $\rho = 1$ kg/m^3 and $f_c = 10^{-4}$ s^{-1} ?
>
> **Solution**
> Given: $\Delta P = 1$ kPa, $\Delta x = 500$ km, $\rho = 1$ kg/m^3, $f_c = 10^{-4}$ s^{-1}.
> Find: $G = ?$ m/s
>
> $U_g = 0$, thus $G = V_g$. Use eq. (9.18b):
> $$V_g = +\frac{1}{(1\text{kg/m}^3)\cdot(10^{-4}\text{s}^{-1})}\cdot\frac{(1\text{kPa})}{(500\text{km})} = \underline{\textbf{20 m/s}}$$
>
> **Check**: Units OK. Physics OK.
> **Discussion**: The "kilo" in the numerator & denominator cancel. Given that $\Delta P/\Delta x = 0.002$ kPa/km, the answer agrees with Fig 9.7. The wind is toward the north.

where the isobars are closer together (i.e., where the isobars are tightly packed), and at higher latitudes (Fig 9.7). The total geostrophic wind speed G is:

$$G = \sqrt{U_g^{\,2} + V_g^{\,2}} \qquad (9.19)$$

The geostrophic wind as a function of horizontal distances between height contours on a constant pressure chart is:

$$U_g = -\frac{g}{f_c}\cdot\frac{\Delta z}{\Delta y} \qquad \bullet(9.20a)$$

$$V_g = +\frac{g}{f_c}\cdot\frac{\Delta z}{\Delta x} \qquad \bullet(9.20b)$$

where $g = 9.8$ m·s^{-2} is gravitational acceleration, and f_c is the Coriolis parameter. Figs 9.8 and 9.9 show examples of winds and heights on a 30 kPa isobaric surface.

> **Solved Example**
> If height increases by 100 m eastward across a distance of 500 km, then what is the geostrophic wind speed, given $f_c = 10^{-4}$ s^{-1} ?
>
> **Solution**
> Given: $\Delta z = 100$ m, $\Delta x = 500$ km, $f_c = 10^{-4}$ s^{-1}.
> Find: $G = ?$ m/s
>
> $U_g = 0$, thus $G = V_g$. Use eq. (9.20b):
> $$V_g = +\frac{g}{f_c}\cdot\frac{\Delta z}{\Delta x} = \left(\frac{9.8\text{m}\cdot\text{s}^{-2}}{0.0001\text{s}^{-1}}\right)\cdot\left(\frac{100\text{m}}{500{,}000\text{m}}\right)$$
> $= \underline{\textbf{19.6 m/s}}$
>
> **Check**: Units OK. Physics OK.
> **Discussion**: This is nearly the same answer as before. Thus, a height change of 100 m on an isobaric surface corresponds to a pressure change of roughly 1 kPa on a constant pressure surface. A hypothetical ball rolling downhill would start moving toward the west, but would be deflected toward the north by Coriolis force in the N. Hem.

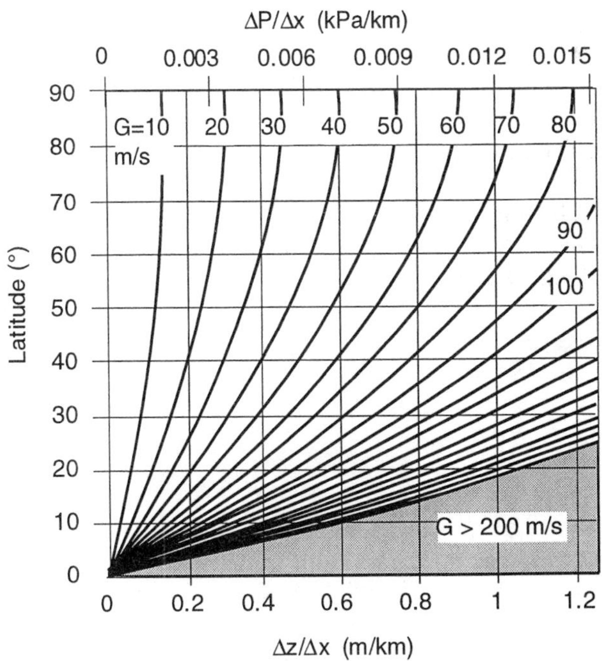

Figure 9.7
Geostrophic wind speed G vs. latitude and height gradient (Δz/Δx) on a constant pressure surface. Top scale is pressure gradient at sea level.

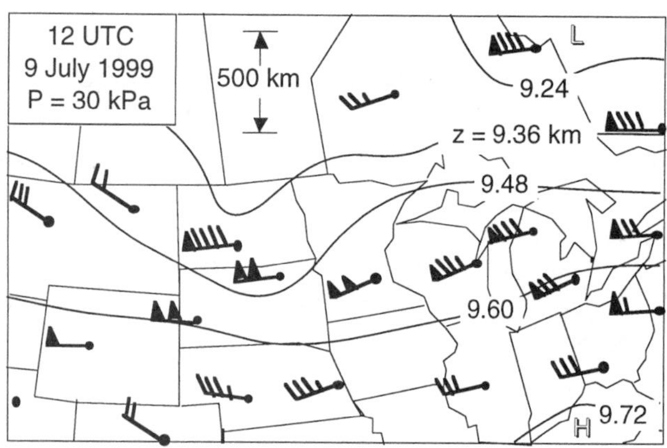

Figure 9.8
Weather map for a 30 kPa constant pressure surface over central N. America. Height of this surface above mean sea level (MSL) is contoured as the solid lines. Hence, this is called a 30 kPa height chart. Rawinsonde observations of wind are also plotted, where a pennant is worth 25 m/s, a full barb is 5 m/s, and a half barb is 2.5 m/s. Wind speed is the sum of all pennants and barbs.

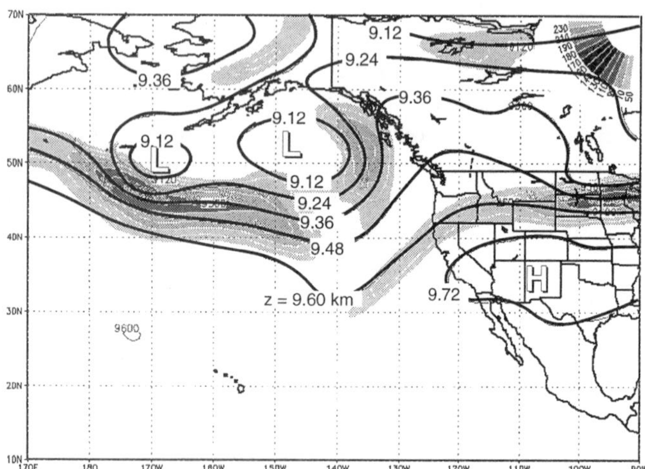

Figure 9.9
Upper air 30 kPa height chart valid at 12 UTC on 9 July 99. Shading indicates isotachs in the jet stream, where light gray denotes roughly 25 m/s or greater winds, and darker shading is roughly 50 m/s or greater. Thin lines are height contours. Faster winds occur where contours are packed. [Adapted from a US Navy Fleet Numerical Meteorological and Oceanographic Center (FNMOC) chart.]

Gradient Wind

Around a high or low pressure center, the steady-state wind follows the curved isobars, with low pressure to the left in the Northern Hemisphere. Around lows, the wind is slower than geostrophic, called **subgeostrophic**, regardless of the hemisphere. Around highs, the steady-state wind is faster than geostrophic, or **supergeostrophic**. The curved steady-state wind is called the **gradient wind**.

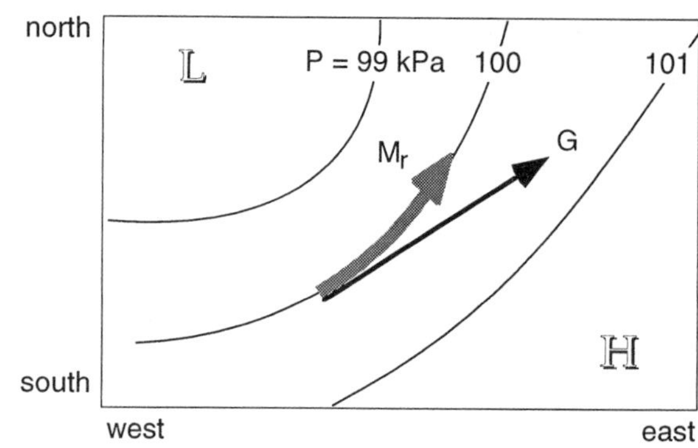

Figure 9.10
Geostrophic wind G and gradient wind M_r around a low pressure center in the Northern Hemisphere.

The gradient wind occurs because of an imbalance between pressure gradient and Coriolis forces of proper magnitude to balance the centrifugal force:

$$0 = -\frac{1}{\rho}\cdot\frac{\Delta P}{\Delta x} + f_c \cdot V + s\cdot\frac{V\cdot|V|}{R} \tag{9.21a}$$

$$0 = \underbrace{-\frac{1}{\rho}\cdot\frac{\Delta P}{\Delta y}}_{\text{pressure gradient}} \underbrace{- f_c \cdot U}_{\text{Coriolis}} \underbrace{- s\cdot\frac{U\cdot|U|}{R}}_{\text{centrifugal}} \tag{9.21b}$$

Define $U_r \equiv U$ and $V_r \equiv V$ as the components of gradient wind, with a total gradient wind speed of $M_r = [U_r^2 + V_r^2]^{1/2}$. One solution to eqs. (9.21) is

$$M_r = G \pm \frac{M_r^{\,2}}{f_c \cdot R} \tag{9.22}$$

where M_r is the gradient wind speed, and where the negative sign is used for flow around low pressure centers, and the positive sign for highs. This solution demonstrates that wind is "slow around lows" (Fig 9.10), meaning slower than geostrophic G. However, the solution is implicit because the desired wind M_r is on both sides of the equal sign.

Solving the quadratic equation (9.22) for the cyclonic flow around a low yields:

$$M_r = 0.5\cdot f_c \cdot R \cdot\left[-1 + \sqrt{1 + \frac{4\cdot G}{f_c\cdot R}}\right] \tag{•9.23a}$$

and anticyclonic flow around a high:

$$M_r = 0.5\cdot f_c \cdot R \cdot\left[1 - \sqrt{1 - \frac{4\cdot G}{f_c\cdot R}}\right] \tag{•9.23b}$$

A "curvature" **Rossby number** can be defined that uses the radius of curvature as the relevant length scale:

$$Ro_c = \frac{G}{f_c \cdot R} \qquad (9.24)$$

Neither R nor Ro_c are the Rossby deformation radius (see eq. 9.43, and Chapters 11 and 12). Small values of **Rossby number** indicate flow that is nearly in geostrophic balance.

The cyclonic and anticyclonic gradient winds are:

$$M_r = \frac{G}{2 \cdot Ro_c} \cdot \left[-1 + \sqrt{1 + 4 \cdot Ro_c}\right] \qquad (9.25a)$$

$$M_r = \frac{G}{2 \cdot Ro_c} \cdot \left[1 - \sqrt{1 - 4 \cdot Ro_c}\right] \qquad (9.25b)$$

For high-pressure centers, steady-state physical (non-imaginary) solutions exist only for $Ro_c \le 1/4$. Thus, around anticyclones, isobars cannot be both closely spaced and sharply curved. In other words, the pressure cannot decrease rapidly away from high centers. There is no analogous restriction on cyclones, because any value of Ro_c is possible. Thus, pressure gradients and winds must be gentle in highs, but can be vigorous near low centers (Figs 9.11 & 9.12).

By combining the definition of the Rossby number with that for geostrophic wind, and setting $Ro_c = 1/4$, we find that the maximum pressure P or height z variation near anticyclones is

$$z = z_c - \frac{f_c^2 \cdot R^2}{8 \cdot g} \qquad \bullet(9.26a)$$

$$P = P_c - \frac{\rho \cdot f_c^2 \cdot R^2}{8} \qquad \bullet(9.26b)$$

where z_c and P_c are the reference height or pressure at the center of the anticyclone, and R is distance from the center of the anticyclone (Fig 9.11).

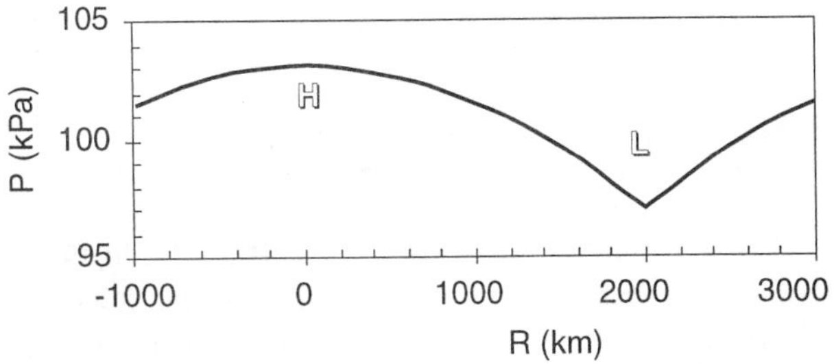

Figure 9.11
Variation of surface pressure across an anticyclone (H) and cyclone (L), showing that the curve can have a cusp at the low, but not at the high. Arbitrary center pressures of 103 kPa and 97 kPa were chosen for the anticyclone and cyclone, respectively.

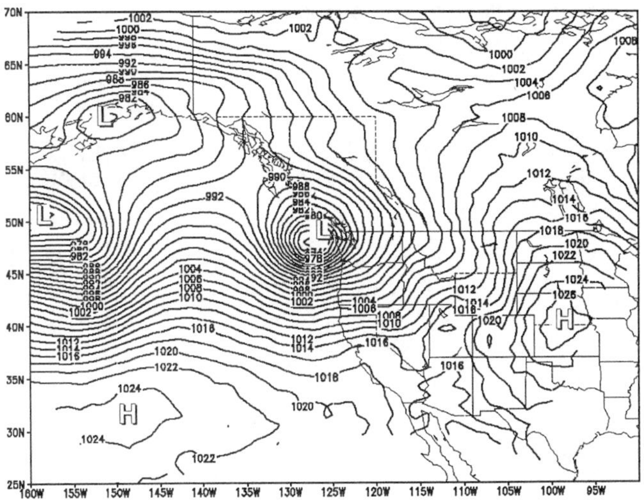

Figure 9.12
Sea-level pressure (contoured every 2 mb= 0.2 kPa), for 00 UTC on 24 Nov 1998. Notice the tight packing of isobars around lows, but looser spacing near high pressure centers.

Solved Example

If the geostrophic wind around a low is 10 m/s, then what is the gradient wind speed, given $f_c = 10^{-4}\ s^{-1}$ and a radius of curvature of 500 km? Also, what is the curvature Rossby number?

Solution

Given: $G = 10$ m/s, $R = 500$ km, $f_c = 10^{-4}\ s^{-1}$
Find: $M_r = ?$ m/s, $Ro_c = ?$ (dimensionless)

Use eq. (9.23a)

$$M_r = 0.5 \cdot (10^{-4} s^{-1}) \cdot (500000 m) \cdot \left[-1 + \sqrt{1 + \frac{4 \cdot (10 m/s)}{(10^{-4} s^{-1}) \cdot (500000 m)}}\right]$$

$= \underline{8.54\ m/s}$

Use eq. (9.24):

$$Ro_c = \frac{(10 m/s)}{(10^{-4} s^{-1}) \cdot (5 \times 10^5 m)} = \underline{0.2}$$

Check: Units OK. Physics OK.
Discussion: The small Rossby number indicates that the flow is in geostrophic balance. The gradient wind is indeed slower than geostrophic around this low.

Boundary-Layer Wind

Turbulent drag in the boundary layer slows the wind below the geostrophic value, and turns the wind to point at a small angle across the isobars toward low pressure (Fig 9.13). For flow along straight isobars, the steady-state equations of motion become:

$$0 = -\frac{1}{\rho}\cdot\frac{\Delta P}{\Delta x} + f_c \cdot V - w_T \cdot \frac{U}{z_i} \qquad (9.27a)$$

$$0 = \underbrace{-\frac{1}{\rho}\cdot\frac{\Delta P}{\Delta y}}_{\text{pressure gradient}} \underbrace{- f_c \cdot U}_{\text{Coriolis}} \underbrace{- w_T \cdot \frac{V}{z_i}}_{\text{turbulent drag}} \qquad (9.27b)$$

Define $U_{BL} = U$ and $V_{BL} = V$ as components of the boundary layer wind. An implicit solution of eqs. (9.27) is:

$$U_{BL} = U_g - \frac{w_T \cdot V_{BL}}{f_c \cdot z_i} \qquad (9.28a)$$

$$V_{BL} = V_g + \frac{w_T \cdot U_{BL}}{f_c \cdot z_i} \qquad (9.28b)$$

with $M_{BL}^2 = U_{BL}^2 + V_{BL}^2$. Eqs. (9.28) can be solved by iteration (guess V_{BL} and plug into the first equation, solve for U_{BL} and plug into second equation, solve for V_{BL} and repeat the process).

If the boundary layer is statically **neutral** with strong wind shear, then an approximate explicit solution is:

•(9.29a)

$$U_{BL} \approx (1 - 0.35 \cdot a \cdot U_g) \cdot U_g - (1 - 0.5 \cdot a \cdot V_g) \cdot a \cdot V_g \cdot G$$

•(9.29b)

$$V_{BL} \approx (1 - 0.5 \cdot a \cdot U_g) \cdot a \cdot G \cdot U_g + (1 - 0.35 \cdot a \cdot V_g) \cdot V_g$$

for $a \cdot G < 1$, where $a = C_D / (f_c \cdot z_i)$.

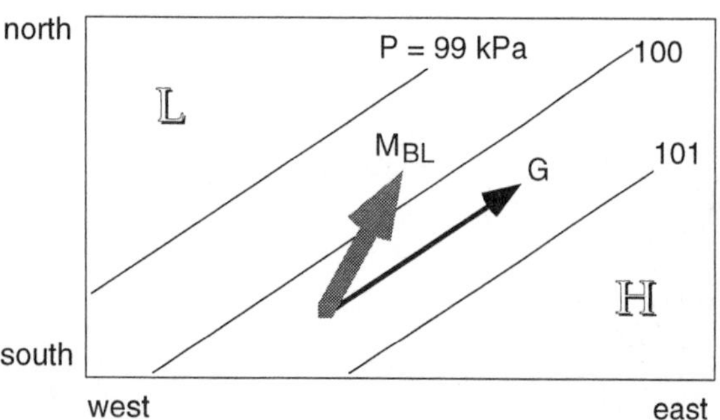

Figure 9.13
Boundary layer wind M_{BL} and geostrophic wind speed G.

Solved Example

Find the boundary layer winds given U_g = 10 m/s, V_g = 0, z_i = 1 km, C_D = 0.002, and $f_c = 10^{-4}$ s^{-1}. Also, what angle do the winds cross the isobars? This is a statically **neutral** boundary layer.

Solution

Given: (see above)
Find: U_{BL} = ? m/s, V_{BL} = ? m/s, M_{BL} = ? m/s, α = ? °

Use eqs. (9.29): with $G = (U_g^2 + V_g^2)^{1/2} = 10$ m/s

$$a = \frac{0.002}{(10^{-4}\text{s}^{-1})\cdot(1000\text{m})} = 0.02 \text{ s/m}$$

Check: $a \cdot G$ = (0.02s/m)·(10 m/s) = 0.2 < 1. Good.

$U_{BL} \approx [1 - 0.35 \cdot (0.02\text{s/m}) \cdot (10\text{m/s})] \cdot (10\text{m/s})$
$\cong$ **9.3 m/s**

$V_{BL} \approx [1 - 0.5 \cdot (0.02\text{s/m}) \cdot (10\text{m/s})] \cdot (0.02\text{s/m}) \cdot (10\text{m/s}) \cdot (10\text{m/s})$ = **1.8 m/s**

$M_{BL} = \sqrt{U_{BL}^2 + V_{BL}^2} = \sqrt{9.3^2 + 1.8^2}$ = **9.47 m/s**

The geostrophic wind is parallel to the isobars. The angle between BL wind and geostrophic is

$\alpha = \tan^{-1}(V_{BL}/U_{BL}) = \tan^{-1}(1.8/9)$ = **11°**

Check: Units OK. Physics OK.
Discussion: The boundary layer wind speed is indeed slower than geostrophic (9.47 vs. 10 m/s), but only slightly slower because the drag coefficient for this example was very small. Also, it crosses the isobars slightly toward low pressure. (The geostrophic wind toward the east means low pressure is to the north.)

For a statically **unstable** boundary layer with light winds, an exact explicit solution is:

$$U_{BL} = c_2 \cdot [U_g - c_1 \cdot V_g] \qquad \bullet(9.30a)$$

$$V_{BL} = c_2 \cdot [V_g + c_1 \cdot U_g] \qquad \bullet(9.30b)$$

where $c_1 = \dfrac{b_D \cdot w_B}{f_c \cdot z_i}$, and $c_2 = \dfrac{1}{[1 + c_1^2]}$.

The net result for both neutral and unstable static stabilities is that boundary-layer winds cross the isobars at a small angle toward low pressure. This cross-isobaric flow occurs for both straight and curved isobars.

Solved Example

Find the boundary layer winds given U_g = 10 m/s, V_g = 0, z_i = 1 km, w_B = 45 m/s, b_D = 1.83x10^{-3} , and $f_c = 10^{-4}$ s^{-1}. Also, at what angle does the wind cross the isobars? This is a statically **unstable** boundary layer.

Solution

Given: (see above)

Find: U_{BL} = ? m/s, V_{BL} = ? m/s, M_{BL} = ? m/s

Use eqs. (9.26):

$$c_1 = \frac{(1.83\times10^{-3})\cdot(45\text{m/s})}{(10^{-4}\text{s}^{-1})\cdot(1000\text{m})} = 0.824 \text{ (dimensionless)}$$

$c_1 = 1/[1+0.824^2] = 0.60$ (dimensionless)

$U_{BL} \approx 0.60\cdot[(10\text{m/s})-0] =$ **6.0 m/s**

$V_{BL} \approx 0.60\cdot[0+(0.824)\cdot(10\text{m/s})] =$ **4.94 m/s**

$M_{BL} = \sqrt{U_{BL}{}^2 + V_{BL}{}^2} \cong$ **7.77 m/s**

$\alpha = \tan^{-1}(4.94/6.0) =$ **39.5°**

Check: Units OK. Physics OK.

Discussion: As before, the boundary layer winds are subgeostrophic, and cross the isobars toward low pressure. Turbulent drag has similar effects, regardless of whether the turbulence is generated mechanically by wind shear, or by buoyant rising thermals.

Boundary-Layer Gradient (BLG) Wind

In regions of curved isobars at the bottom of cyclones and anticyclones, drag force exists in the boundary layer (BL) in addition to pressure gradient and Coriolis force. The imbalance of these forces creates a centripetal force that causes the air to spiral in towards low-pressure centers, and spiral out from high-pressure centers. Figs 9.1 and 4.6 show sketches of the BL gradient winds in the N. Hemisphere, and the associated isobars.

Assume steady state, neglect advection, and require an imbalance of forces equal to the centrifugal force, to reduce the equations of motion to:

$$0 = -\frac{1}{\rho}\cdot\frac{\Delta P}{\Delta x} + f_c\cdot V - w_T\cdot\frac{U}{z_i} + s\cdot\frac{V\cdot|V|}{R} \qquad (9.31a)$$

$$0 = \underbrace{-\frac{1}{\rho}\cdot\frac{\Delta P}{\Delta y}}_{\textit{pressure gradient}} \underbrace{- f_c\cdot U}_{\textit{Coriolis}} \underbrace{- w_T\cdot\frac{V}{z_i}}_{\textit{turbulent drag}} \underbrace{- s\cdot\frac{U\cdot|U|}{R}}_{\textit{centrifugal}} \qquad (9.31b)$$

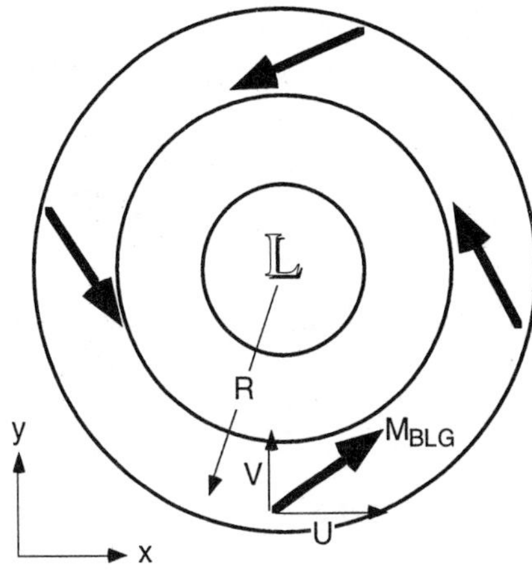

Figure 9.14
Tangential (U) and radial (V) components of the BLG wind in the N. Hemisphere.

Without actually solving these equations, we can anticipate the following from our previous understanding of geostrophic, gradient, and boundary-layer winds. Boundary-layer gradient (BLG) wind speed is slower than the gradient wind speed due to the drag. The BLG winds flow counterclockwise around lows in the N. Hemisphere, and clockwise around highs. Instead of blowing parallel to the curved isobars like the gradient wind, BLG winds cross the isobars at a small angle (tens of degrees) toward low pressure.

The turbulent drag term is different for highs and lows. Winds are strong around lows and skies are often overcast, hence the transport velocity is best represented by the statically neutral parameterization:

$$w_T = C_D\cdot\sqrt{U^2+V^2} \qquad (9.12 \text{ again})$$

which adds even more nonlinearity to eqs. (9.31). In highs, winds are light and skies are clear, suggesting that transport velocity should be given by the statically unstable parameterization during daytime by:

$$w_T = b_D\cdot w_B \qquad (9.13 \text{ again})$$

where w_B is not a function of wind speed. At night in fair weather, one cannot even assume steady state, so eqs. (9.31) are invalid.

While the equations of motion for geostrophic and gradient winds were simple enough to allow an analytical solution, and we could devise an approximate analytical solution for the BL wind, we are not so lucky with the BLG wind. The set of coupled equations (9.31) are nonlinear and quite nasty to solve.

Nonetheless, we can numerically iterate towards the answer by including the tendency term on the LHS of each equation. Also, rewrite eqs. (9.31) in cylindrical coordinates, letting U be the tangential component, and V be the radial component (Fig 9.14). Use geostrophic wind G as a surrogate for the radial pressure gradient.

For BLG winds around a low in the N. Hemisphere (i.e., $s = +1$), eqs. (9.31) can be rewritten as the following set of coupled equations, which are valid day or night:

$$M = \sqrt{U^2 + V^2} \qquad \text{(1.1 again)}$$

$$U(t+\Delta t) = U(t) + \Delta t \cdot \left[f_c \cdot V - \frac{C_D \cdot M \cdot U}{z_i} + s \frac{V \cdot |V|}{R} \right] \qquad \text{(9.32a)}$$

$$V(t+\Delta t) = V(t) + \Delta t \cdot \left[f_c \cdot (G - U) - \frac{C_D \cdot M \cdot V}{z_i} - s \frac{U \cdot |U|}{R} \right] \qquad \text{(9.32b)}$$

Similar equations can be derived for convective boundary layers in high-pressure regions.

To use this approach:
(1) Make a first guess that $(U, V) = (0, 0)$.
(2) Plug in these values everywhere that U, $U(t)$, V, or $V(t)$ appears in eqs. (1.1) and (9.32).
(3) Solve eqs. (9.32) for new values of (U, V).
(4) Repeat steps (2) and (3), but using the new winds.
(5) Continue until the (U, V) wind components converge to steady values, which equal (U_{BLG}, V_{BLG}) by definition.

This iterative solution is tedious when done on a hand calculator, so use a spreadsheet or computer program instead. An example of a spreadsheet solution is shown in the solved example. In this example, notice that the iterations don't proceed directly to the final solution, but initially spiral away from the solution. This spiral is called a **damped inertial oscillation**.

Equations (9.32) are valid for even unsteady (time varying) solutions, such as occurs at night. At night when drag is weak, the winds many never reach steady state, and may continue as undamped or weakly-damped **inertial oscillations**. Such an oscillation can temporarily cause winds to be greater than geostrophic in regions of straight isobars, or greater than gradient in regions of curved isobars. This is one explanation for the supergeostrophic **nocturnal jet** that was mentioned in Chapter 4.

Solved Example(§)

Find the BLG winds around a low pressure center in the N. Hemisphere, given $G = 20$ m/s at radius $R = 500$ km from the low center, $z_i = 1$ km, $C_D = 0.01$, and $f_c = 10^{-4}$ s^{-1}. Also, find the speed.

Solution

Given: (see above)
Find: U_{BLG} = ? m/s, V_{BLG} = ? m/s, M_{BLG} = ? m/s

Use eqs. (1.1) and (9.32) in a spreadsheet. Set the time step to $\Delta t = 1$ h. Choose $U = 0$ and $V = 0$ as the first guess. Make 20 iterations.

	A	B	C	D	E	F
1	**Boundary Layer Gradient Wind**					
2						
3	G (m/s) =	20	zi (km) =	1		
4	fc (1/s) =	0.0001	CD =	0.01		
5	R (km) =	500	delta t(s)=	3600		
6						
7						
8	Iteration	U (m/s)	V (m/s)	M (m/s)	delta U	delta V
9	0	0.00	0.00	0.00	0.000	7.200
10	1	0.00	7.20	7.20	2.965	5.334
11	2	2.97	12.53	12.88	4.268	0.258
12	3	7.23	12.79	14.70	1.956	-2.548
13	4	9.19	10.24	13.76	-0.110	-1.791
14	5	9.08	8.45	12.41	-0.498	-0.437
15	6	8.58	8.02	11.74	-0.280	0.192
16	7	8.30	8.21	11.67	-0.050	0.266
17	8	8.25	8.47	11.83	0.053	0.131
18	9	8.31	8.60	11.96	0.055	0.009
19	10	8.36	8.61	12.00	0.022	-0.035
20	11	8.38	8.58	11.99	-0.001	-0.027
21	12	8.38	8.55	11.97	-0.008	-0.009
22	13	8.37	8.54	11.96	-0.005	0.002
23	14	8.37	8.54	11.96	-0.001	0.004
24	15	8.37	8.55	11.96	0.001	0.002
25	16	8.37	8.55	11.96	0.001	0.000
26	17	8.37	8.55	11.97	0.000	-0.001
27	18	8.37	8.55	11.97	0.000	0.000
28	19	8.37	8.55	11.96	0.000	0.000
29	20	8.37	8.55	11.96	0.000	0.000

U_{BLG} = **8.37 m/s**
V_{BLG} = **8.55 m/s**
M_{BLG} = **11.96 m/s**

Check: Units OK. Physics OK. Check to see if the computed winds agree with: (1) the geostrophic wind when R approaches ∞, and $C_D = 0$; (2) the gradient winds when $C_D = 0$; (3) and the boundary layer winds when R approaches ∞.

I performed these checks for the scenarios of the previous solved examples, and found: (1) for geostrophic: $U_{BLG} = G$, and $V_{BLG} = 0$; (2) for gradient: $U_{BLG} = 8.54$ m/s, $V_{BLG} = 0$; and (3) for BL: $U_{BLG} = 9.63$ m/s, $V_{BLG} = 1.89$ m/s. This BL solution is the exact solution; it is better than the approximate analytical solution presented in the BL wind section.

(continued next column)

Solved Example *(continuation)*
Discussion: The solution converges toward the BLG wind, as if an air parcel started from rest and began accelerating. The Fig below illustrates this convergence, using smaller time steps (of Δt = 1000 s, rather than Δt = 1 h). Starting at zero wind speed, each dot shows the wind forecast for the next time step in the iteration.

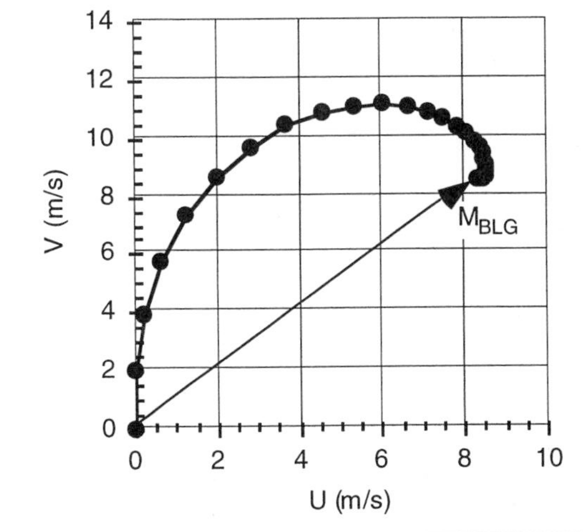

Cyclostrophic Wind

In intense vortices such as tornadoes, waterspouts, and near the eye wall of hurricanes, strong winds rotate around a very tight circle. For these phenomena, centrifugal force and pressure-gradient force dominate. While Coriolis force and surface drag can still be present, they are small compared to the other terms and can be neglected. For steady state winds, the equations of motion reduce to:

$$0 = -\frac{1}{\rho}\cdot\frac{\Delta P}{\Delta x} + s\cdot\frac{V\cdot|V|}{R} \qquad (9.33a)$$

$$0 = \underbrace{-\frac{1}{\rho}\cdot\frac{\Delta P}{\Delta y}}_{\text{pressure gradient}} \underbrace{- s\cdot\frac{U\cdot|U|}{R}}_{\text{centrifugal}} \qquad (9.33b)$$

Because of the cylindrical nature of these flows as they rotate around intense low-pressure centers, it is easier to write and solve the equations in cylindrical form:

$$M_{cs} = M_{\tan} = \sqrt{\frac{R}{\rho}\cdot\frac{\Delta P}{\Delta R}} \qquad (9.34)$$

where R is radial distance outward from the center of rotation, $\Delta P/\Delta R$ is the local radial pressure gradient, and the resulting tangential speed around the vortex is M_{tan}. This is defined equal to the cyclostrophic speed M_{cs}.

Cyclostrophic winds never occur around high pressure centers because the strong pressure gradients needed to drive such winds are not possible there, as discussed in the gradient wind section. Around lows, cyclostrophic winds turn counterclockwise in the northern hemisphere, and clockwise in the southern. Magnitudes are typically of order 50 m/s in water spouts and hurricanes, and 100 m/s in tornadoes. A solved example is given in the hurricane chapter (Chapt. 16).

FULL EQUATIONS OF MOTION - REVISITED

Use geostrophic wind as a surrogate measure of the pressure gradient. We can then group this term with the Coriolis term, which in eqs. (9.15) gives:

$$\frac{\Delta U}{\Delta t} = -U\frac{\Delta U}{\Delta x} - V\frac{\Delta U}{\Delta y} + f_c\cdot\left(V - V_g\right) - w_T\cdot\frac{U}{z_i} \qquad \bullet(9.35a)$$

$$\underbrace{\frac{\Delta V}{\Delta t}}_{\text{tendency}} = \underbrace{-U\frac{\Delta V}{\Delta x} - V\frac{\Delta V}{\Delta y}}_{\text{advection}} \underbrace{- f_c\cdot\left(U - U_g\right)}_{\text{geostrophic departure}} \underbrace{- w_T\cdot\frac{V}{z_i}}_{\text{turbulent drag}} \qquad \bullet(9.35b)$$

where a centrifugal term could also be added for winds associated with curved isobars.

Framed in this way, we see that Coriolis force alters the wind direction only if the wind is not equal to its geostrophic value (i.e., if not balanced by other forces such as pressure-gradient force). Hence, the third term on the right is sometimes called the **geostrophic departure** term. The wind difference is also called the **ageostrophic wind** (U_{ag}, V_{ag}):

$$U_{ag} \equiv U - U_g \qquad \bullet(9.36a)$$

$$V_{ag} \equiv V - V_g \qquad \bullet(9.36b)$$

MASS CONSERVATION

Barring any nuclear reactions, air molecules are not converted into energy, and air mass is conserved. In an Eulerian framework, mass flowing into a fixed volume minus the mass flowing out gives the change of mass within the volume. The equation describing this mass balance is called the **continuity equation.**

Continuity Equation

Recall that mass within a unit volume is defined as density, ρ. The **mass budget** is:

$$\frac{\Delta \rho}{\Delta t} = -\rho \left[\frac{\Delta U}{\Delta x} + \frac{\Delta V}{\Delta y} + \frac{\Delta W}{\Delta z} \right] \tag{9.37}$$

which is also called the continuity equation. This can be rearranged (using calculus) to be:

$$\frac{\Delta \ln(\rho)}{\Delta t} = -\left[\frac{\Delta U}{\Delta x} + \frac{\Delta V}{\Delta y} + \frac{\Delta W}{\Delta z} \right] \tag{9.38}$$

Solved Example

Just before a tornado strikes your garage, the air density inside is 1 kg/m^3. Because you accidently left your garage door open, winds of 100 m/s enter the garage from the west, but nothing leaves from the east. Also, your garage is temporarily intact, so the other walls, floor, and ceiling prevent winds in those other directions. The east end of your garage is 8 m from the open west end. What is the density change in your garage during the first 1 s?

Solution

Given: U_{door} = 100 m/s, U_{end} = 0 m/s, Δx = 8 m, ρ = 1 kg/m^3.

Find: $\Delta\rho/\Delta t$ = ? kg·m^3·s^{-1} .

Use eq. (9.37), with $V = W = 0$:

$$\frac{\Delta \rho}{\Delta t} = -\rho \frac{U_{door} - U_{end}}{x_{door} - x_{end}} = -\left(1 \frac{\text{kg}}{\text{m}^3}\right) \frac{(100 - 0)\text{m/s}}{(0 - 8)\text{m}}$$

$\Delta\rho/\Delta t$ = +<u>12.5</u> kg·m^3·s^{-1} .

Check: Units OK. Physics OK.

Discussion: Thanks to Bernoulli's principle (Chapt. 10), you won't have to worry about this high density for very long. The pressure inside your garage will increase so fast that it will blow out your walls and roof as if a bomb exploded.

Be careful when calculating the wind gradients; calculate them as wind at location 2 minus wind at location 1, divided by distance at location 2 minus distance at location 1. Don't accidently subtract 1 from 2 in the numerator, and then do 2 from 1 in the denominator, because it will give the wrong sign.

Incompressible Continuity Equation

Density at any fixed altitude changes only little with temperature and humidity for most non-violent weather conditions. Therefore, we can neglect mass changes within the volume, compared to the inflow and outflow. The air is said to be **incompressible** when the density does not change. This approximation fails in strong thunderstorm updrafts and tornadoes.

For incompressible flow, the left hand side of eq. (9.37) is zero. This requires inflow to balance outflow:

$$\frac{\Delta U}{\Delta x} + \frac{\Delta V}{\Delta y} + \frac{\Delta W}{\Delta z} = 0 \tag{•9.39}$$

This continuity equation is illustrated in Fig 9.15 for an example of inflow and outflow across the faces of a cube. The length of the grey arrows represents the strength of the wind components. Also, for this example $\Delta x = \Delta y = \Delta z$.

In the x-direction, there is less wind (U) entering the cube than leaving ($U + \Delta U$). This is called **divergence**. For this case, ΔU is positive.

In the z-direction (vertical), there is more air entering (W) than leaving ($W + \Delta W$). This is called **convergence**, and ΔW is negative.

In the y-direction, the air entering (V) and leaving ($V + \Delta V$) are equal, so there is no convergence or divergence. For this example, ΔV is zero.

For this example, the continuity equation is

$$\text{(positive)} + (0) + \text{(negative)} = 0$$

so we anticipate that mass is conserved. In the real atmosphere, the directions having convergence or divergence might differ from this example, but the sum must always equal zero. Namely convergence in one or two directions must be balanced by divergence in the other direction(s).

For situations such as circular isobars around a low pressure center, cylindrical coordinates are easier to use (Fig 9.16). The continuity equation is then

$$\frac{2 \cdot V_{in}}{R} = \frac{\Delta W}{\Delta z} \tag{•9.40}$$

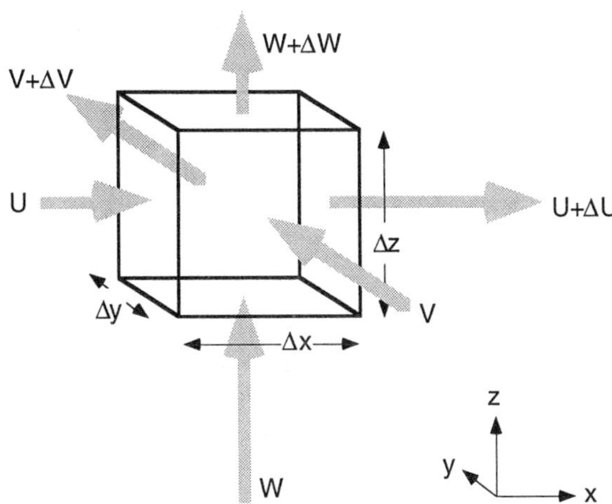

Figure 9.15
Inflow and outflow from a fixed volume.

where R is the radius of the cylinder, and Δz is the cylinder depth. We assume that the inflow velocities V_{in} through the sides of the cylinder are everywhere equal. When V_{in} is positive (indicating horizontal inflow), then ΔW is also positive (indicating vertical outflow).

Around low pressure regions near the surface, turbulent drag causes horizontal inflow (convergence) within the boundary layer. There is no vertical air motion ($W = 0$) at the bottom of the boundary layer because of the ground. Thus, the horizontal inflow must be balanced by vertical outflow from the boundary-layer top.

This mechanism for creating mean upward motion is called **Ekman pumping** or **boundary-layer pumping**. The upward motion carries water vapor, which then condenses in the adiabatically cooled air, causing clouds and precipitation. Thus, low-pressure regions generally have foul weather.

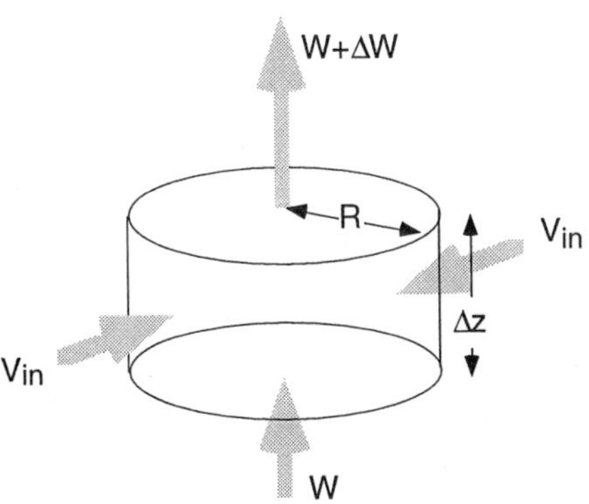

Figure 9.16
Mass conservation in cylindrical coordinates.

Around high pressure centers, turbulent drag causes horizontal outflow (divergence). This is balanced by downward motion called **subsidence** at the top of the boundary layer. Subsidence warms air adiabatically, thereby evaporating most clouds and causing fair weather.

Solved Example
At a radius of 400 km from a low center, boundary layer winds have a 2 m/s component that crosses the circular isobars inward. If the boundary layer is 1 km thick, estimate the average vertical velocity out of the top of the boundary layer.

Solution
Given: $V_{in} = 2$ m/s, $R = 500$ km
Find: $W + \Delta W = ?$ m/s

Use eq. (9.40):

$$\Delta W = 2 \cdot V_{in} \cdot \frac{\Delta z}{R} = 2 \cdot (2\text{m/s}) \cdot \frac{1\text{km}}{400\text{km}} = \underline{\mathbf{0.01\ m/s}}$$

Check: Units OK. Physics OK.
Discussion: For non-thunderstorm conditions, this magnitude of about 1 cm/s is typical for vertical velocities in the atmosphere. Such small velocities allow use of the hydrostatic assumption (see Chapter 1).

Boundary-Layer Pumping

Turbulent drag against the ground causes wind to be a bit slower than the gradient wind, and to cross the isobars at a small angle toward low pressure, as was discussed in the section on the BLG wind. In that section, we defined V_{BLG} as the radial component of wind, which by definition equals the inflow velocity V_{in} for eq. 9.40 (see Fig 9.17).

Unfortunately, we were unable to find an analytical solution for V_{BLG}. However, there is an analytical solution available for V_{BL} from an earlier section, which is slightly greater than V_{BLG}. For the subsequent analysis, we will use $V_{in} \cong V_{BL}$, knowing that the result will be an upper limit on what are likely slower velocities in the real atmosphere.

Assuming that the winds are relatively strong around lows, and that cloudy conditions preclude strong surface heating or cooling, we can assume that the boundary layer is statically **neutral**. For this situation, the inflow velocity across the isobars is given by eq. (9.29b).

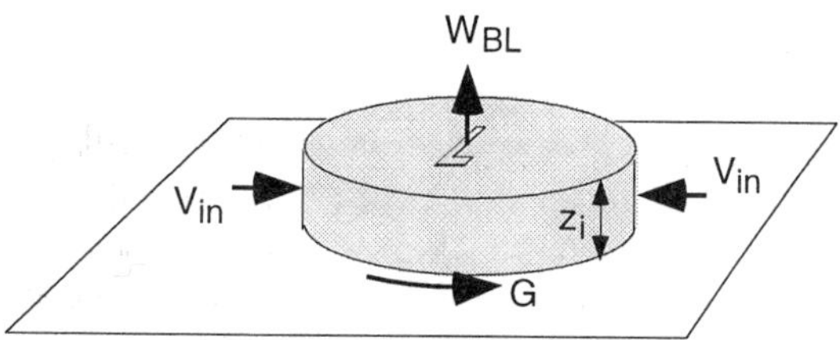

Figure 9.17
Convergence and updrafts in a cyclone.

Combining this inflow velocity with the cylindrical form of the continuity equation (9.40) allows us to calculate the vertical velocity W_{BL} at the top of the boundary layer (Fig 9.17):

$$W_{BL} = \frac{2 \cdot b \cdot C_D}{f_c} \cdot \frac{G^2}{R} \qquad \bullet(9.41)$$

where R is the radius of curvature of the isobars around the low center, G is the geostrophic wind speed, f_c is the Coriolis parameter, and C_D ($\cong 0.005$ over land) is the drag coefficient for a neutral boundary layer.

The factor $b = \{ 1 - 0.5 \cdot [C_D \cdot G / (f_c \cdot z_i)]\}$ is from eq. (9.29b),where boundary-layer depth is z_i. For simplicity, the statically neutral **boundary-layer depth within a cyclone** can be approximated as

$$z_i \approx \frac{G}{N_{BV}} \qquad (9.42)$$

where N_{BV} is the Brunt-Väisälä frequency (see Chapter 6) above the boundary layer. This allows b to be rewritten as: $b = \{ 1 - 0.5 \cdot [C_D \cdot N_{BV} / f_c]\}$. Note that the approximation for boundary-layer cross-isobaric flow is valid only when $[C_D \cdot N_{BV} / f_c] < 1$.

The updraft velocity in eq. (9.41) depends on both the size and rotation speed of the cyclone. Vorticity is one measure of the combined effects of rotation speed and size. **Geostrophic relative vorticity** can be defined as

$$\zeta_g = \frac{2 \cdot G}{R} \qquad (9.43)$$

which is a measure of the rotation of the air. More details of vorticity will be covered in Chapter 11.

Using the geostrophic vorticity in eq. (9.41) yields an alternative expression for the vertical velocity pumped out of the top of a cyclone:

$$W_{BL} = C_D \cdot \frac{G}{f_c} \cdot \zeta_g \cdot \left[1 - 0.5 \frac{C_D \cdot N_{BV}}{f_c} \right] \qquad (9.44)$$

The terms outside of the brackets suggest that greater vertical velocities (and hence nastier storms) occur for stronger geostrophic winds (i.e., tighter packing of isobars) over rougher terrain, and where there is greater curvature of the cyclonic flow. Also, lower latitudes give smaller Coriolis parameters, which allows greater vertical velocity.

The term in square brackets shows that the fastest vertical velocity is expected in statically neutral flow (where $N_{BV} = 0$). Greater static stabilities cause weaker vertical velocities.

By utilizing the approximation for mixed-layer depth (eq. 9.42), an "internal" **Rossby radius of deformation** can be approximated as

$$\lambda_R \approx \frac{G}{f_c} \cdot \frac{z_T}{z_i} \qquad (9.45)$$

where z_i is the boundary-layer depth and z_T is the depth of the troposphere. The internal Rossby radius of deformation is discussed in more detail in Chapter 11, and an external Rossby radius is given in Chapter 12.

Thus, an alternative form for the vertical velocity equation can be written using this Rossby radius of deformation:

$$W_{BL} = C_D \cdot \frac{z_i}{z_T} \cdot \lambda_R \cdot \zeta_g \cdot \left[1 - 0.5 \cdot C_D \cdot \frac{\lambda_R}{z_T} \right] \qquad (9.46)$$

Solved Example

A geostrophic wind of 10 m/s blows cyclonically around a low-center with radius of curvature of 1000 km. The latitude is such that $f_c = 0.0001\ s^{-1}$, and the drag coefficient is 0.005. The tropospheric lapse rate is standard above the BL.

What is the vertical velocity out of the top of the boundary layer? Also, estimate the boundary layer depth, geostrophic relative vorticity, and internal Rossby radius.

Solution

Given: $G = 10$ m/s, $R = 10^6$ m, $f_c = 0.0001\ s^{-1}$, $C_D = 0.005$, $N_{BV} = 0.0113\ s^{-1}$ (from a previous solved example using the standard atmos.)

Find: z_i = ? m, λ_R = ? km, W = ? m/s

Assume: tropospheric depth $z_T = 11$ km

Use eq. (9.42):

$z_i \cong G/N_{BV} = (10\ m/s)/(0.0113\ s^{-1}) =$ **885 m**

(continued next column)

Solved Example *(continuation)*

Use eq. (9.43):

$$\zeta_g = \frac{2\cdot(10\text{m/s})}{10^6\text{m}} = \underline{\mathbf{2x10^{-5}}\ \mathrm{s^{-1}}}$$

Use eq. (9.45):

$$\lambda_R \approx \frac{(10\text{m/s})}{(0.0001\text{s}^{-1})}\cdot\frac{11\text{km}}{0.885\text{km}} = \underline{\mathbf{1243\ km}}$$

First check that $\cdot[C_D \cdot N_{BV}/f_c] < 1$.

$[0.005\cdot(0.0113\text{s}^{-1})/(0.0001\text{s}^{-1})] = 0.565 < 1$. OK

Use eq. (9.46): $W_{BL} =$

$$0.005\cdot\frac{(0.885\text{km})}{(11\text{km})}\cdot(1.243\times10^6\text{m})\cdot(2\times10^{-5}\text{s}^{-1})$$

$$\cdot\left[1-0.5\cdot(0.005)\cdot\frac{1243\text{km}}{11\text{km}}\right]$$

$= (0.01\text{ m/s})\cdot[0.718] = \underline{\mathbf{0.0072\ m/s}}$

Check: Units OK. Physics OK.

Discussion: This vertical velocity of 7.2 mm/s is typical of synoptic circulations. Although weak, it is sufficient to lift air to cause condensation, releasing latent heat which allows stronger buoyant updrafts within the clouds.

SUMMARY

Pressure-gradient force can start winds moving, and can change wind direction and speed. Once the air is moving, other forces such as turbulent drag or Coriolis force also act on the air. Coriolis force is an apparent force that accounts for our moving frame of reference on the rotating earth. Turbulent drag is important only near the ground, in the boundary layer. The relationship between forces and acceleration of the winds is given by Newton's second law of motion.

When all forces balance, the winds are steady. In regions of straight isobars above the boundary layer, pressure-gradient and Coriolis force balance to cause winds that are geostrophic. Around highs or lows, there are slight imbalances associated with centrifugal force, which causes steady-state gradient winds. In the boundary layer, winds are slower than either geostrophic or gradient because of turbulent drag. In tornadoes and hurricanes, centrifugal and pressure gradient forces balance to create the intense cyclostrophic wind.

Outside of thunderstorms, vertical winds are well described by incompressible mass continuity. Thus horizontal gradients of horizontal wind cause weak vertical motion.

Threads

Forces cause winds, and winds flow over mountains to form mountain waves and lenticular clouds (Chapt. 10). Winds embody the general circulation (Chapt. 11), which moves air masses around to create fronts (Chapt. 12), cyclones (Chapt. 13) and our global climate (Chapt. 18). The equations of motion are solved every day on large computers to make the daily weather forecasts (Chapt. 14). The interplay between buoyancy (Chapt. 6) and dynamics creates phenomena such as thunderstorms, tornadoes (Chapt. 15) , and hurricanes (Chapt. 16). Winds advect air pollutants from place to place (Chapt. 17).

Not only do winds advect temperature (Chapt. 3), but horizontal temperature variations cause horizontal pressure variations via the hypsometric equation (Chapt. 1), thereby driving the jet-stream winds. Winds riding over colder air masses carry water vapor (Chapt. 5), some of which can condense to make clouds (Chapt. 7) and precipitation (Chapt. 8). The complexity of the atmosphere is becoming apparent.

EXERCISES

Numerical Problems

N1. Find the acceleration of a 75 kg person when pushed by a force of

a. 1 N b. 10 N c. 100 N d. 1000 N
e. 2 N f. 30 N g. 500 N h. 2000 N

N2. If the initial velocity is zero, find the final velocity after 100 s for a force per unit mass of

a. 10 N/kg b. 10 m·s^{-2}
c. 30 N/kg d. 20 m·s^{-2}
e. 50 N/kg f. 5 m·s^{-2}
g. 20 N/kg h. 15 m·s^{-2}

N3. What is the pressure-gradient force per unit mass between Seattle, WA, and Corvallis, OR (about 340 km south of Seattle), if sea-level pressure is 98 kPa in Seattle and 100 kPa in Corvallis?

N4. For a neutral boundary layer, find the turbulent drag force per unit mass over a forest for

a. $U = 5$ m/s and $V = 25$ m/s
b. $U = -10$ m/s and $V = 5$ m/s
c. $U = 5$ m/s and $V = -15$ m/s
d. $U = -5$ m/s and $V = -5$ m/s
e. $U = -40$ m/s and $V = 5$ m/s
f. $U = 5$ m/s and $V = 35$ m/s
g. $U = 25$ m/s and $V = -2$ m/s
h. $U = 0$ m/s and $V = 10$ m/s

N5. Calculate the Coriolis parameter for
a. Munich, Germany
b. Oslo, Norway
c. Madrid, Spain
d. Chicago, USA
e. Buenos Aires, Argentina
f. Melbourne, Australia
g. Vancouver, Canada
h. Bejing, China
i. Moscow, Russia
j. Tokyo, Japan
k. (your town)

N6. For Chicago, find the Coriolis force per unit mass in the N. Hemisphere for:
a. $U = 10$ m/s b. $V = 5$ m/s c. $U= 3$ m/s
d. $U = -10$ m/s e. $V = -5$ m/s f. $U = 8$ m/s
g. $U = -3$ m/s h. $V = -8$ m/s

N7. Draw a northwest wind of 5 m/s in the S. Hemisphere on a graph, and show the directions of forces acting on it.

N8. For a statically unstable boundary layer, find the turbulent drag force per unit mass, given a buoyant velocity scale of 60 m/s and
a. $U = 5$ m/s and $V = 1$ m/s
b. $U = -1$ m/s and $V = 3$ m/s
c. $U = 2$ m/s and $V = -4$ m/s
d. $U = -2$ m/s and $V = -1$ m/s
e. $U = -4$ m/s and $V = 0$ m/s
f. $U = 5$ m/s and $V = 3$ m/s
g. $U = 5$ m/s and $V = -2$ m/s
h. $U = 0$ m/s and $V = 2$ m/s

N9. Given $U = -5$ m/s and $V = 10$ m/s, find the components of centrifugal force around a 800 km diameter:
a. low in the southern hemisphere
b. high in the northern hemisphere
c. high in the southern hemisphere
d. low in the northern hemisphere

N10. What is the geostrophic wind at a height where $\rho = 0.7$ kg/m^3, $f_c = 10^{-4}$ s^{-1}, and the pressure gradient (kPa/100 km) is:
a. 0.5 b. 1.0 c. 2.0 d. 2.5
e. 0.2 f. 1.5 g. 0.7 h. 2.2

N11. At a latitude of 60°N, find the geostrophic wind given a height gradient (m/km on a constant pressure surface) of:
a. 0.5 b. 1.0 c. 2.0 d. 2.5
e. 0.2 f. 1.5 g. 0.7 h. 2.2

N12. If the geostrophic wind around a high is 10 m/s, then what is the gradient wind speed, given $f_c = 10^{-4}$ s^{-1} and a radius of curvature of:
a. 800 km b. 500 km c. 600 km
d. 1000 km e. 2000 km f. 1500 km

N13. Find the boundary layer winds given $U_g = 5$ m/s, $V_g = 5$ m/s, $z_i = 500$ m, $C_D = 0.002$, and $f_c = 10^{-4}$ s^{-1}. Also, what angle do the winds cross the isobars? This is a statically neutral boundary layer.

N14. Same as previous problem, but for an unstable boundary layer with w_B (m/s) of:
a. 75 b. 100 c. 50 d. 200
e. 150 f. 250 g. 125 h. 250

N15(§). Recompute M_{BLG} as in the solved example in the Boundary Layer Gradient Wind section, but with the following changes:
a. $G = 10$ m/s b. $z_i = 2$ km c. $C_D = 0.002$
d. $R = 1000$ km e. $f_c = 2\text{x}10^{-4}$ s^{-1}

N16. Given a pressure gradient of 1 kPa/km, compute the cyclostrophic wind at the following radii (km): a. 0.1 b. 0.2 c. 0.3 d. 0.5
e. 1 f. 2 g. 5 g. 10

N17. If the geostrophic wind around a low is 50 m/s, and $f_c = 10^{-4}$ s^{-1}, then what is the cyclostrophic wind speed, given a radius of curvature of:
a. 2 km b. 5 km c. 3 km
d. 100 m e. 20 m f. 10 m

N18. At a radius of 200 km from a low center, boundary layer winds have a 2 m/s component that crosses the circular isobars inward.
a. If the boundary layer is 2 km thick, estimate the average vertical velocity through the top of the boundary layer (BL).
b. Same as (a) but for a 1 km thick BL.

N19. Estimate boundary layer depth within a cyclone given an isothermal environment above the boundary layer, and a geostrophic wind near the surface of (m/s) a. 5 b. 10 c. 15 d. 2
e. 20 f. 25 g. 30 h. 8

N20(§). For a 1 km thick boundary layer over a surface of drag coefficient 0.003, plot the vertical velocity due to boundary-layer pumping as a function of geostrophic wind speed. Plot separate curves for the following radii (km) of curvature.
Assume a latitude of 45°:
a. 500 b. 1000 c. 2000 d. 4000
Assume a latitude of 60°:
e. 500 f. 1000 g. 2000 h. 4000

N21. Estimate the value of internal Rossby radius of deformation at latitude 60°N for a tropospheric depth of 11 km and geostrophic wind speed of 15 m/s. Assume a boundary layer depth of

a. 500 m b. 1 km c. 2 km d. 800 m
e. 200 m f. 1.5 km g. 2.5 km h. 3 km

Understanding & Critical Evaluation

U1. Compare eq. (9.1) with (1.16), and discuss.

U2. Can eqs. (9.5) be used to make a forecast if the initial conditions (i.e., the current winds) are not known? Discuss.

U3. If all of the net forces (eq. 9.6) are zero, does that mean that the wind speeds (eq. 9.4) are zero? Explain.

U4. Eqs. (9.7) suggest that winds can advect winds. How is that possible?

U5. In eq.s (9.7), why does advection depend on the gradient of winds (e.g., $\Delta U/\Delta x$) across the Eulerian domain, rather than on just the value of wind that is being blown into the domain?

U6. In Fig 9.1, and on weather maps, what is the relationship between packing of isobars (i.e., the number of isobars that cross through a square cm of weather map) and the pressure gradient?

U7. In the N. Hemisphere, the pressure gradient points from high to low pressure. Which way does it point in the S. Hemisphere?

U8. Eqs. (9.8) give the horizontal components of the pressure-gradient force. Combine those equations vectorally to show that the vector direction of the pressure-gradient force is indeed perpendicular to the isobars and pointing toward lower pressure, for any arbitrary isobar direction such as shown in Fig 9.1.

U9. An air parcel at rest (relative to the earth) near the equator experiences greater tangential velocity due to the earth's rotation than do air parcels at higher latitudes. Yet the Coriolis parameter is zero at the equator. Why?

U10. Eqs. (9.11) give the horizontal components of the Coriolis force. Combine those equations vectorally to show that the vector direction of the Coriolis force is indeed 90° to the right of the wind direction, for any arbitrary wind direction such as the two shown in Fig 9.2.

U11. Rewrite eqs. (9.0 - 9.11) for the S. Hemisphere.

U12. Eqs. (9.12) give the horizontal components of the turbulent drag force.

a. Combine those equations vectorally to show that the vector direction of the drag force is indeed opposite to the wind direction, for any arbitrary wind direction.

b. Show that the magnitude of the drag force is proportional to the square of the wind speed, M.

U13. Compare the values of the turbulent transport velocity during windy (statically neutral) and convective (statically unstable) conditions. Discuss.

U14. Eqs. (9.9) give the horizontal components of centrifugal force. Combine those equations vectorally to show that the vector direction of centrifugal force is indeed outward from the center of rotation, and is proportional to the square of the tangential velocity.

U15. Plug eqs. (9.18) into (9.19) to find the vector speed and direction of the geostrophic wind as a function of the vector pressure gradient.

U16. Re-derive the geostrophic wind eqs. (9.18) for the S. Hemisphere.

U17. Derive eqs. (9.20) from eqs. (9.18).

U18. The geostrophic wind approaches infinity as the equator is approached (see Fig 9.6), yet the winds in the real atmosphere are not infinite there. Why?

U19. Verify that eq. (9.22) is indeed a solution to the gradient wind eqs. (9.21).

U20. Verify that eqs. (9.23) are solutions to eq. (9.22).

U21. Imagine an idealized weather map that had a single high pressure center next to a single low pressure center. Further, suppose that if you were to draw a line through those two centers, that the pressure variation along that line would be the same as Fig 9.11. Given that information, and assuming circular cyclones and anticyclones, draw isobars on the weather map at ±0.5 kPa increments, starting at 100 kPa. Comment on the packing of isobars near the centers of the cyclone and anticyclone.

U22. a. Is there any limit on the strength of the pressure gradient that can occur just outside of the center of cyclones? (Hint: consider Fig 9.11)

b. What controls this limit?

c. What max winds are possible around cyclones?

U23. What is implicit about the implicit solution (eq. 9.28) for the boundary layer wind?

U24. How accurate are the approximate boundary layer wind solutions (eq. 9.29)? Under what conditions are the approximate solutions least accurate? (Hint: compare with an iterative solution to the implicit equations 9.28).

U25. Why is an exact explicit solution possible for the steady-state winds in the unstable boundary layer, but not for the neutral boundary layer?

U26. Verify that eqs. (9.30) are indeed exact solutions to (9.28) or (9.27).

U27. Verify that the cyclostrophic winds are indeed a solution to the governing equations (9.33).

U28. a. Derive eq. (9.41).
 b. For horizontal winds, we know that an increased drag coefficient will reduce wind speed. Why in eq. (9.41) does an increased drag coefficient cause increased vertical wind speed?
 c. When considering that the factor b in eq. (9.41) contains a negative function of the drag coefficient, does your answer to part (b) change? (Hint: Consider eq. 9.44).

U29. The paragraph after eq. (9.44) gives a physical interpretation of the equation. Show how that interpretation was reached, by examining each term in the equation and discussing its impact on W.

U30. Given eq. (9.45), determine the physical meaning of the internal Rossby radius of deformation by how it affects the various terms in (9.46).

U31. Regarding balances of forces, if only Coriolis and turbulent drag forces were acting, speculate on the nature of the wind.

U32. Derive an equation for the steady-state wind that occurs when pressure-gradient, Coriolis, turbulent drag, and centrifugal forces are all active.

U33. Modify eqs. (9.41) through (9.46) if necessary, to describe the boundary-layer pumping around highs (anticyclones) rather than around lows (cyclones). Discuss the significance of your equations.

U34. Re-derive the equations for cyclostrophic wind, but in terms of height gradient on a constant pressure surface, instead of pressure gradient along a constant height surface. [Hint: Compare eqs. (9.20) to (9.18).]

U35. Calculate the geostrophic and gradient winds, as appropriate, at a number of locations for the height contours plotted in Fig. 9.7. Compare them to the observed winds and comment.
 b. Same, but for Fig 9.12, and don't do the comparison with observations.

U36(§). a. Recreate on a spreadsheet the solved example for the boundary-layer gradient winds.
 b. Do and print the checks of that equation for the special cases where it reduces to the geostrophic wind, gradient wind, and boundary layer wind.
 c. Can the analytical solutions for the gradient wind and the (neutral) boundary layer wind be combined to yield an approximate analytical solution to the BLG winds. If so, what are the limitations, and the magnitude of the errors. (Hint: Try substituting M_r in place of G in the equations for boundary layer wind.)

Web-Enhanced Questions

W1. Search the web for historical discussions of Isaac Newton, and summarize what you find. What did you find most unusual or interesting?

W2. Search the web for "Coriolis Force", to find sites that show animations of the movement of objects in a rotating coordinate system. Tabulate a list of the best sites.

W3. a. Find a current sea-level weather map on the web that covers your area, and which includes isobars. Measure the distance (km) between isobars, and compute the pressure gradient force.
 b. Same as (a), but do this for several days in a row. Then plot a graph of how the pressure gradient changes with time over your location.

W4. Search the web for 50 kPa (500 mb) weather maps that cover part of the S. Hemisphere. Compute the geostrophic and gradient wind speeds and directions in regions of straight and curved isobars, respectively, and compare with upper-air observations. Discuss the differences between these winds and corresponding winds in the N. Hemisphere.

W5. Find a current 50 kPa (= 500 mb) weather map (or other map as specified by your instructor) that has height contours already drawn on it. Look for a region in the map where the isobars are nearly straight.

a. Compute the geostrophic wind speed components, and the total geostrophic wind speed vector (speed and direction).

b. If upper air (rawinsonde) observations are available near your area, compare the measured winds with the geostrophic value computed from part (a).

c. If there are multiple regions of nearly straight isobars at different latitudes in your weather map, see how the observed winds in these regions vary with latitude, and compare with the expected latitudinal variation of geostrophic wind (as from Fig 9.6).

W6. a. Same as W5, except look for a region on the map where the isobars are curving in a cyclonic (counterclockwise in the N. Hemisphere) direction when following along with the wind direction. Compute the gradient wind for this case.

b. Same as (a) but for anticyclone (clockwise) turning winds.

c. Compare the measured winds from rawinsondes to the theoretical winds from (a) and (b). Confirm that the gradient winds (both theoretical and observed) are slower than geostrophic around lows, and faster around highs.

W7. a. Find a sea-level weather map from the web that has isobars drawn on it. Look for a day or a region where there are neighboring regions of strong high and low pressure. Print this map, and draw a straight line through the center of the high and the low. Extend this line well past the centers of the high and low. Using the isobars that cross your drawn line, find how pressure varies with distance between the high and low, and plot the results similar to Fig 9.8.

b. If you set the location of the center of the high as the origin of your coordinate system, check whether the shape of the pressure curve agrees with eq. (9.26b). Confirm that the pressure variation across a low pressure center can have a cusp, while that across a high cannot.

W8. Search the web for a weather map at 50 kPa (500 mb) or any other altitude above the boundary layer, that includes the equator. Find a region of relatively straight isobars near the equator, and compute the geostrophic wind speed based on the plotted pressure gradient. Compare this theoretical wind with observed upper-air winds for the same altitude. Why don't the observations agree with the theory?

W9. Find a current sea-level weather map from the web, that shows both wind direction (such as the wind symbol on surface station observations), and isobars.

a. For a region of the map where the isobars are relatively straight, and hopefully over non-mountainous and non-coastal terrain, confirm that the observed boundary layer winds indeed cross the isobars at some small angle from high to low pressure. What is the average angle?

b. For a region where the isobars are curved around a high or low, confirm that the winds spiral in towards the center of the low, and out from the center of a high.

c. Around either the high or low, estimate the average value of the component of wind that represents inflow or outflow from the low or high. Use that value of V_{in} (or V_{out}) in eq. (9.40) to compute the vertical velocity at the top of the boundary layer.

d. Based on the inward or outward component of velocity from (c), estimate the drag force acting on the air. If this calculation is for flow around a low where the winds are relatively fast, find the value of the drag coefficient C_D for statically neutral conditions.

W10. a. If it is hurricane season, search the web for a weather map that is just above the top of the boundary layer (85 kPa or 850 mb might be good enough), but which is well below the altitude of the 60 kPa pressure. Look for a map (either observed or forecast) that has the height contours around the hurricane on this pressure surface (85 kPa). Find the location just outside of the eye where the height contours are packed closest together, and calculate the pressure gradient there. Then use that pressure gradient to compute the cyclostrophic wind, and compare your theoretical value with the observed upper-air values at that height.

b. Do the same as (a), but using a sea-level weather map showing the isobars. Also comment on the effect of boundary layer drag.

Synthesis Questions

S1. Suppose Newton's second law of motion was not a function of mass. How would the motion of bullets, cannon balls, and air parcels be different, if at all?

S2. What if Newton's second law of motion stated that velocity was proportional to force/mass. How would weather and climate be different, if at all?

S3. What if wind could not advect itself. How would the weather and climate be different, if at all?

S4. Suppose that pressure gradient force was along isobars, rather than perpendicular to them. Describe how winds would be different, how weather maps would be different, and how this might affect the weather and climate, if at all.

S5. There is some debate in the literature that our understanding of Coriolis force might be incorrect. We think that Coriolis force is an apparent force. Can an apparent force change the momentum and kinetic energy of air parcels? If so, would this violate Newton's laws when viewed from a non-rotating frame? Discuss.

S6. Fig. 9.3 shows that there is wind shear across the top of the boundary layer, where subgeostrophic winds in the boundary layer change to geostrophic winds above the boundary layer. The shear can exist because of the strongly statically stable layer of air that caps the boundary layer (recall Chapt. 4). If such a stable capping inversion did not exist, how might the wind profile be different over the depth of the troposphere?

S7. Suppose that turbulent drag force acted 90° to the right of the wind direction in the boundary layer. Discuss how the boundary layer winds would work around highs and lows, and in regions of straight isobars. How would the weather and climate be different, if at all?

S8. Suppose that the boundary layer drag force did not increase with velocity (in the case of an unstable boundary layer) or with velocity squared (in the case of a neutral boundary layer), but was constant regardless of wind speed. How would the boundary layer winds, weather, and climate change, if at all?

S9. What if there were no Coriolis force. How would winds, weather, and climate be different, if at all?

S10. On our present world where we think there is Coriolis force, are there situations where it is possible for wind to blow directly high to low pressure, rather than more-or-less parallel to the isobars? Describe such scenarios.

S11. What if geostrophic winds could turn around high or low pressure systems without feeling centripetal or centrifugal force. How would the weather and climate be different, if at all?

S12. Suppose that cusps in pressure were allowed at high-pressure centers as well as at low pressure centers (see Fig 9.8) so that strong pressure gradients could exist near the center of both types of pressure centers. Describe how the winds, weather, and climate might be different, if at all?

S13. Suppose that the earth's surface were frictionless. How would the weather and climate be different, if at all?

S14. Suppose that the tropopause acted like a rigid lid on the troposphere, and that the air at the top of the troposphere felt frictional drag against this rigid lid. How would the winds, weather, and climate be different, if at all?

S15. What if the earth were a flat spinning disk instead of a spinning sphere. How would the weather and climate be different, if at all?

S16. Suppose the earth's rotation were twice as fast as now. How would the weather and climate change, if at all?

S17. Suppose that the axis of the earth's rotation were along a radial line drawn from the sun (i.e., is in the plane of the ecliptic), rather than being more or less perpendicular to the plane of the ecliptic. How would the weather and climate be different, if at all?

S18. Suppose the earth did not rotate. How would the winds, weather, and climate be different, if at all?

S19. Why is incompressibility such a good approximation for the real atmosphere? How does the atmosphere react to density changes, that might help ensure little density change? (Hint: consider the first solved example in the continuity equation section.)

S20. Extend the discussion of the Coriolis Focus Box by deriving the magnitude of the Coriolis force for an object moving

a. westward b. northward

LOCAL WINDS

CONTENTS

10 Each locale has a unique landscape that creates or modifies the wind. These local winds affect where we choose to live, how we build our buildings, what we can grow, and how we are able to travel.

During synoptic high pressure (i.e., fair weather), some winds are generated locally by temperature differences. These gentle circulations include thermals, anabatic winds, and katabatic winds.

During synoptically windy conditions, mountains can modify the winds. Examples are gap winds, boras, hydraulic jumps, foehns, and mountain waves.

SCALES OF MOTION

In the atmosphere are superimposed motions of many scales, from small turbulent eddies through thunderstorms and cyclones to large planetary-scale circulations such as the jet stream. Scales of horizontal motion are classified in Table 10-1.

Small atmospheric phenomena of horizontal dimension less than about 2 km are frequently **isotropic**; namely, their vertical and horizontal dimensions are roughly equal. Horizontally-larger phenomena are somewhat pancake-like, because the vertical dimension is generally limited by the depth of the troposphere.

Table 10-1. Scales of horizontal motion.

Size	Scale	Name
	macro α	planetary scale
20,000 km	macro β	synoptic scale
2,000 km	meso α	mesoscale
200 km	meso β	
20 km	meso γ	
2 km	micro α	boundary-layer turbulence
200 m	micro β	surface-layer turbulence
20 m	micro γ	inertial subrange turb.
2 m	micro δ	
200 mm		fine-scale turbulence
20 mm		
2 mm	viscous	dissipation subrange

Time scales τ of most phenomena are approximately proportional to horizontal scales λ:

$$\tau \approx a \cdot \lambda \qquad (10.1)$$

where $a \cong 1$ s/m. For example, microscale turbulence about 1 m in diameter might last about a second. Boundary layer thermals of diameter 1 km have circulation lifetimes of about 15 min. Thunderstorms of size 10 km might last a couple hours, while cyclones of size 1000 km might last a week.

Many microscale motions occur in the atmospheric boundary layer, and were discussed in Chapt. 4. Some approximations for turbulence are given in Appendix H. Other local winds are discussed here.

Solved Example

Suppose a tornado makes a damage path on the ground roughly 100 m wide. Classify its length scale, and estimate its lifetime.

Solution

Given: λ = 100 m horizontal scale
Find: Scale name, and τ

Using Table 10-1. Scale = <u>**micro β**</u>.
Using eq. (10.1) τ = <u>**1.7 min**</u>

Check: Units OK. Physics reasonable.
Discussion: While most tornado touchdowns are usually short lived, they are spawned from mesocyclones that could be 10 km in diameter with lifetimes of a few hours or more. Some tornadoes have been observed to stay on the ground for 30 min or more.

WIND-SPEED PROBABILITY

Wind speeds are rarely constant. During a year, there might be a few days of strong mean winds, some days of moderate winds, many days of light winds, and few calms (Fig 10.1). This distribution of mean wind speeds can be modeled with the **Weibull distribution**:

$$Pr = \frac{\alpha \cdot \Delta M \cdot M^{\alpha-1}}{M_o{}^{\alpha}} \cdot \exp\left[-\left(\frac{M}{M_o}\right)^{\alpha}\right] \qquad (10.2)$$

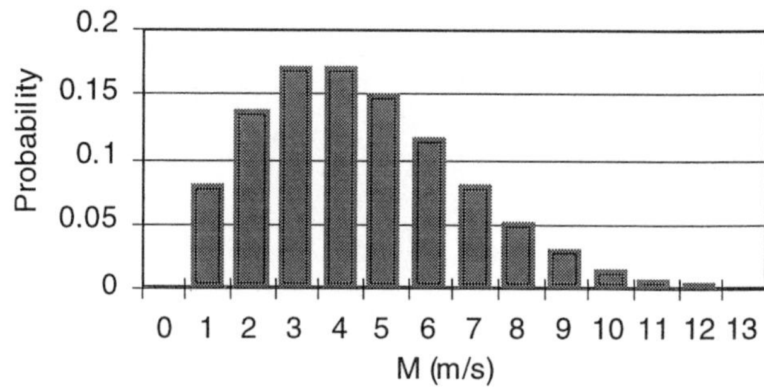

Figure 10.1
Weibull distribution of wind-speed probability, for α = 2 and M_o = 5 m/s.

where Pr is the probability of having a wind speed $M \pm \Delta M$. Such wind speed variations are caused by synoptic, mesoscale, and boundary-layer processes.

Parameter M_o is proportional to the mean annual wind speed, and is called a "location" parameter. Parameter α is a "spread" parameter – smaller α causes a wider spread of winds about the mean. Values of the parameters and the resulting distribution shape vary from place to place.

The bin size or resolution is ΔM. For example, the probability plotted in Fig 10.1 for M = 3 m/s is really the probability of finding a wind between 2.5 and 3.5 m/s. The width of each bar in the bar graph is ΔM = 1 m/s.

The sum of probabilities for all wind speeds should equal 1, meaning there is a 100% chance that

Solved Example

Given M_o = 5 m/s and α = 2, what is the probability that the wind speed will be between 5.5 and 6.5 m/s?

Solution

Given: M_o= 5 m/s, α= 2, ΔM= 1 m/s, M= 6 m/s
Find: Pr = ?

Use eq. (10.2):

$$Pr = \frac{2 \cdot (1\text{m/s}) \cdot (6\text{m/s})^{2-1}}{(5\text{m/s})^2} \cdot \exp\left[-\left(\frac{6\text{m/s}}{5\text{m/s}}\right)^2\right]$$

= 0.114 = <u>**11.4%**</u>

Check: Units OK. Physics OK.
Discussion: This agrees with Fig 10.1 at M = 6 m/s, which had the same parameters as this example. To get a sum of probabilities that is very close to 100%, use a very small bin size and be sure not to cut off the tail of the distribution at high wind speeds.

the wind is between zero and infinity. This is a good way to check the distribution for errors. Eq. (10.2) is only approximate, so the sum of probabilities almost equals 1.

Wind-speed distributions are used to estimate the ability to generate electrical power using wind turbines. They are also used in the design of buildings and bridges, to give the probability of extreme wind events.

WIND TURBINES AND POWER GENERATION

Although kinetic energy of the wind is proportional to air mass times wind-speed squared, the rate at which this energy is blown through a wind turbine is equal to the wind speed M. Thus, the power of the wind is proportional to wind speed cubed.

The theoretical power available from the wind is:

$$Power = (\pi / 2) \cdot \rho \cdot E \cdot R^2 \cdot M^3 \qquad (10.3)$$

where R is the turbine-blade radius and ρ is air density. Turbine efficiencies are about E = 30% to 40%.

To prevent destruction of the turbine in winds greater than about 10 m/s, the blades are designed to gradually feather (reduce their angle of attack) as wind speed increases. In this regime, the turbine spins at constant speed regardless of the wind speed; power generation is constant; and a smaller percentage of the theoretically available power is extracted. For winds in excess of about 30 m/s, the blades are feathered completely and power production ceases. Synoptic, mesoscale, and boundary-layer scales affect wind power generation.

As shown in Chapt. 4, wind speeds increase nearly logarithmically with height near the ground. On the disk spanned by the rotating blades, stronger winds push against the top than the bottom. The angle of attack of the turbine blades can be automatically varied with rotation angle to compensate.

Solved Example

A wind turbine at sea level uses a 30 m radius blade to convert a 10 m/s wind into electrical power at 40% efficiency. What is the theoretical power output?

Solution

Given: $\rho = 1.225$ kg·m^{-3}, R = 30 m, M = 10 m/s
$E = 0.4$

Find: $Power$ = ? kW

Use eq. (10.3): $Power$ =

$$(\pi / 2) \cdot (1.225\text{kg} \cdot \text{m}^{-3}) \cdot (0.4) \cdot (30\text{m})^2 \cdot (10\text{m / s})^3$$

$= 6.93\text{x}10^5$ kg·m^2·s^{-3} = **693 kW**

Check: Units OK. Physics OK.

Discussion: Because of a physical limitation of practical turbine blade sizes, the way to generate more power is to build more wind turbines. See Appendix A for the definition of Watts.

VERTICAL EQUATION OF MOTION

Analogous to the horizontal equations of motion described in Chapt. 9, Newton's second law can be used to develop an equation of motion for vertical velocity W:

•(10.4a)

$$\underbrace{\frac{\Delta W}{\Delta t}}_{tendency} = \underbrace{-U\frac{\Delta W}{\Delta x} - V\frac{\Delta W}{\Delta y}}_{advection} \underbrace{-\frac{1}{\rho}\frac{\Delta P}{\Delta z}}_{pressure\ gradient} \underbrace{+g}_{gravity} \underbrace{-\frac{F_{z\ TD}}{m}}_{turbulent\ drag}$$

where the left hand side is the acceleration, and the right hand side lists the vertical forces per mass. (U, V, W) are the three Cartesian velocity components in the (x, y, z) directions, P is pressure, ρ is air density, t is time, and $F_{z\ TD}/m$ is the turbulent drag force per unit mass m. Gravitational acceleration g = –9.8 m·s^{-2} appears in this equation, but not Coriolis force (which is negligible).

The pressure gradient and gravity terms of eq. (10.4a) can be simplified. First rewrite them as:

$$\frac{1}{\rho}\left[-\frac{\Delta P}{\Delta z} - \rho|g|\right]$$

Density fluctuations are important in the vertical, because they cause buoyancy (see Chapt. 6). Split density into a mean and fluctuating part ($\rho = \bar{\rho} + \rho'$) similar to what was done in Chapt. 4. The terms above become:

$$\frac{1}{(\bar{\rho} + \rho')}\left[-\frac{\Delta P}{\Delta z} - \bar{\rho}|g| - \rho'|g|\right]$$

Using the **hydrostatic assumption** (Chapt. 1), the first two terms in the square brackets cancel out.

Next, a **Boussinesq approximation** is made that $\rho'/(\bar{\rho}+\rho') \cong \rho'/\bar{\rho}$, which implies that density fluctuations are important in the gravity term, but negligible for all other terms. This is reasonable because $\rho' << \bar{\rho}$. In Chapt. 6, it was shown that density fluctuations can be described by virtual temperature T_v fluctuations (with a sign change because low density corresponds to high temperature):

$$-\frac{\rho'}{\bar{\rho}}\cdot|g| = \frac{\theta_v'}{\bar{T}_v}\cdot|g| = \frac{\theta_{v\,p}-\theta_{v\,e}}{\bar{T}_{v\,e}}\cdot|g|$$

where subscript p denotes parcel and e environment. T_{ve} in the denominator must be in Kelvin.

Plugging this back into eq. (10.4a) gives:

$$\underbrace{\frac{\Delta W}{\Delta t}}_{tendency} = \underbrace{-U\frac{\Delta W}{\Delta x} - V\frac{\Delta W}{\Delta y}}_{advection} + \underbrace{\frac{\theta_{v\,p}-\theta_{v\,e}}{\bar{T}_{v\,e}}\cdot|g|}_{buoyancy} \underbrace{-\frac{F_{z\,TD}}{m}}_{turbulent\ drag} \qquad \bullet(10.4b)$$

Turbulent drag is the resistance of vertically moving air against other surrounding (environmental) air. Air with no vertical movement relative to its environment has no drag.

Solved Example

An updraft of 8 m/s exists 2 km west of your location, and there is a west wind of 5 m/s. At your location there is zero vertical velocity, but the air is 3°C warmer than the surrounding environment of 25°C. What is the initial vertical acceleration of the air over your location?

Solution

Given: $\Delta\theta =$ 3°C, T_e = 273+25 = 298 K, U= 5m/s
$\Delta W/\Delta z$= (8m/s – 0) / (–2,000m – 0)
Assume: $V = 0$. Drag = 0 initially, given W = 0.
Dry air, thus $T_v = T$.
Find: $\Delta W/\Delta t$ = ? m·s^{-2}

Use eq. (10.4b): $\frac{\Delta W}{\Delta t} = -U\frac{\Delta W}{\Delta x} + \frac{\theta_p-\theta_e}{\bar{T}_e}\cdot|g|$

= –(5 m/s)·(–8m·s^{-1}/2,000m)+(3/298)·(9.8m·s^{-2})
= 0.020 + 0.099 = **0.119** m·s^{-2}

Check: Units OK. Physics OK.
Discussion: Extrapolated over a minute, this initial acceleration gives W = 7.1 m/s. However, this vertical velocity would not be achieved because as soon as the velocity is nonzero, the drag term also becomes nonzero and tends to slow the vertical acceleration.

THERMALLY DRIVEN CIRCULATIONS

Thermals

Thermal updrafts are warm rising air parcels in the boundary layer. Thermal diameters are nearly equal to their depth, z_i (Fig 10.2).

Use eq. (10.4b), and neglect advection. A rising thermal feels drag from the surrounding environment (not against the ground). This drag is proportional to the square of the thermal updraft velocity relative to its environment. The equation of vertical motion reduces to:

$$\underbrace{\frac{\Delta W}{\Delta t}}_{tendency} = \underbrace{\frac{\theta_{v\,p}-\theta_{v\,e}}{\bar{T}_{v\,e}}\cdot|g|}_{buoyancy} \underbrace{-C_w\frac{W^2}{z_i}}_{turbulent\ drag} \qquad (10.5)$$

where z_i is the mixed layer (boundary layer) depth, and vertical drag coefficient $C_w \cong 5$.

At steady state, the acceleration is near zero. Eq. (10.5) can be solved for the updraft speed of buoyant thermals (i.e., of warm air parcels):

$$W = \sqrt{\frac{|g|\cdot z_i}{C_w}\frac{(\theta_{v\,p}-\theta_{v\,e})}{\bar{T}_{v\,e}}} \qquad (10.6)$$

Thus, warmer thermals in deeper boundary layers have greater updraft speeds.

This equation also applies to deeper convection, such as thunderstorms that rise to the top of the troposphere (synoptic and mesoscales). For that case, z_i is the depth of the troposphere, and the temperature difference is that between the middle of the cloud and the surrounding environment at the same height.

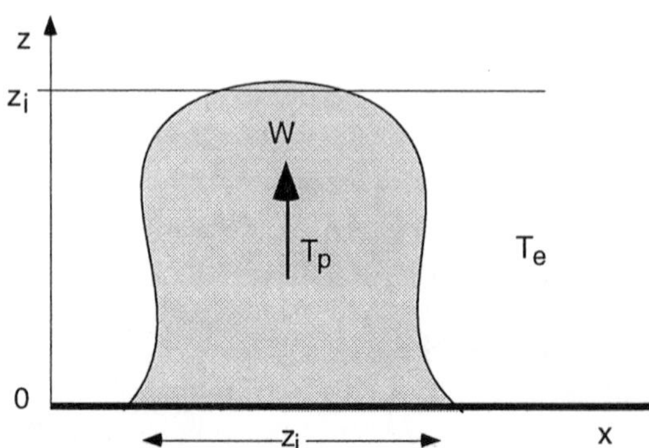

Figure 10.2
Thermal characteristics.

Solved Example

Find the steady-state updraft speed in the middle of (a) a thermal in a boundary layer that is 1 km thick; and (b) a thunderstorm in a 11 km thick troposphere. The virtual temperature excess is 2°C for the thermal and 5°C for the thunderstorm, and $|g|/T_{ve} = 0.0333$ m·s^{-2}·K^{-1}.

Solution

Given: $|g|/T_{ve} = 0.0333$ m·s^{-2}·K^{-1},
(a) $z_i = 1000$ m, $T_{vp}–T_{ve} = 2$°C
(b) $z_i = 11{,}000$ m, $T_{vp}–T_{ve} = 5$°C,

Find: $W = ?$ m/s

Use eq. (10.6): (a) For the thermal:

$$W = \sqrt{\frac{(0.0333 m \cdot s^{-2} K^{-1})(1000 m)(2K)}{5}} = \underline{3.65 \text{ m/s}}$$

(b) For the thunderstorm:

$$W = \sqrt{\frac{(0.0333 m \cdot s^{-2} K^{-1})(11000 m)(5K)}{5}} = \underline{19.1 \text{ m/s}}$$

Check: Units OK. Physics OK.
Discussion: Actually, neither thermal nor thunderstorm updrafts maintain a constant speed. However, this gives us an order of magnitude estimate. In convection, these updrafts must have downdrafts between them, but usually of larger diameter and slower speeds. Air mass continuity requires that mass flow up must balance mass flow of air down, across any arbitrary horizontal plane. This would cause quite a bumpy ride in an airplane, which is why most aircraft pilots try to avoid areas of convection.

Anabatic Wind

During daytime when sunlight heats mountain slopes, the mountain surface heats the neighboring air, which then rises. However, instead of rising vertically like thermals, the flow hugs the slope as it rises. This warm rising air is called an **anabatic wind** (Fig. 10.3). Anabatic wind is the updraft portion of a **cross-valley circulation.**

When the warm air reaches ridge top, it breaks away from the mountain and rises vertically, often joined by the updraft from the neighboring slope. If there is sufficient moisture in the air, cumulus clouds called **anabatic clouds** form just above ridge top in this updraft.

The dotted line in Fig. 10.3 is at a constant height above sea level. Following the line from left to right in Fig 10.3a, potential temperatures of about 19°C are constant until reaching close to the mountain slope,

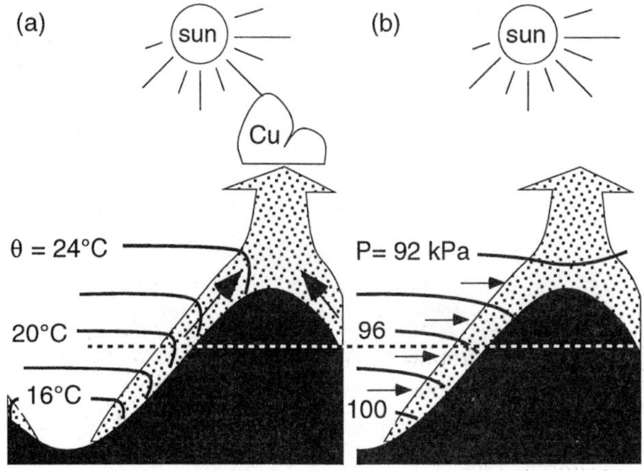

Figure 10.3
Anabatic winds (stippled). (a) Isentropes. (b) Isobars.

where the potential temperature rises past 20°C to nearly 22°C in this illustration. This is the warm anabatic air.

As the warm air tries to rise, it leaves slightly lower pressure adjacent to the slope, as sketched in Fig 10.3b. This causes a horizontal pressure gradient, as can be seen following the dashed line from left to right in Fig 10.3b. The pressure gradient force has a component towards the mountain (shown by the horizontal arrows in Fig 10.3b) that holds the warm rising air against the mountain.

Katabatic Wind

Cold air is heavier than warm, and thus flows downhill as a katabatic wind (Fig 10.4). An equation of motion for this tilted coordinate system is:

$$\underbrace{\frac{\Delta U}{\Delta t}}_{tendency} \underbrace{+ U\frac{\Delta U}{\Delta x} + V\frac{\Delta U}{\Delta y}}_{advection} = \underbrace{|g| \cdot \frac{\Delta \theta_v}{T_{ve}} \sin(\alpha)}_{buoyancy} \underbrace{+ f_c \cdot V}_{Coriolis} \underbrace{- C_D \cdot \frac{U^2}{h}}_{turbulent\ drag} \quad (10.7)$$

where the x-axis points down the fall line of the slope, α is the slope angle, h is the depth of cold air, and U is the downslope wind, not the horizontal component of wind. A cross-slope equation for V is similar, but without the buoyancy term.

The virtual potential temperature in a katabatic flow is generally coldest at the ground, and smoothly increases with height (Fig 10.4). Thus, the difference $\Delta\theta_v = (\theta_{ve} - \theta_{vkat})$ is the average environmental temperature minus the average temperature of the colder layer of air that is draining downhill.

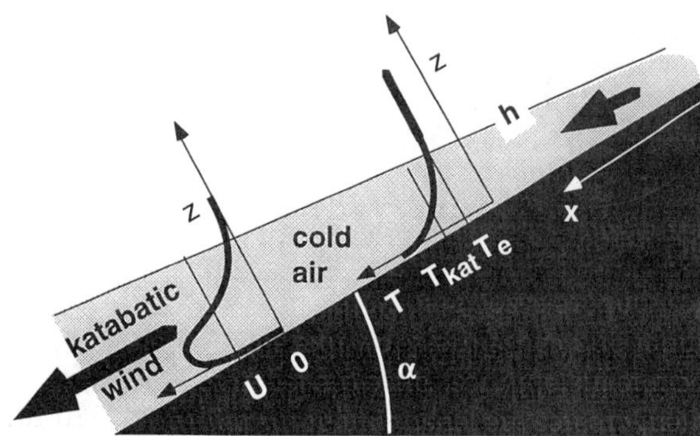

Figure 10.4
Katabatic winds (shaded light gray).

In Antarctica, downslope distances (hundreds of kilometers), speeds (order of 3 - 20 m/s), and durations (many days) are sufficiently large that Coriolis force turns the equilibrium wind direction 30 to 50° left of the fall line. For most smaller valleys and slopes, Coriolis force and the V-wind can be neglected and eq. (10.7) can be solved for some steady-state situations.

If there were no friction against the surface, then the fastest downslope winds would be where the air is the coldest; namely, closest to the ground. However, winds closest to the ground are reduced due to frictional drag, leaving a nose of fast winds just above ground level (Fig 10.4).

Initially the wind (averaged over the depth of the katabatic flow) is influenced mostly by buoyancy and advection. It accelerates with distance downslope:

$$U = \left[|g| \cdot \frac{\Delta\theta_v}{T_{ve}} \cdot x \cdot \sin(\alpha) \right]^{1/2} \qquad (10.8)$$

It eventually approaches an equilibrium where drag balances buoyancy

$$U_{eq} = \left[|g| \cdot \frac{\Delta\theta_v}{T_{ve}} \cdot \frac{h}{C_D} \cdot \sin(\alpha) \right]^{1/2} \qquad (10.9)$$

where C_D is the total drag against both the ground and against the slower air aloft.

Katabatic winds are boundary-layer scale.

Solved Example(§)

Air adjacent to a 10° slope averages 10°C cooler over its 20 m depth than the surrounding air of virtual temperature 10°C. Find and plot the wind speed vs. downslope distance, and the equilibrium speed. C_D = 0.005 .

Solution

Given: $\Delta\theta_v$= 10°C, T_{ve}=283 K, α= 10°, C_D= 0.005
Find: U (m/s) vs. x (km)

Use eq. (10.9) to find final equilibrium value:

$$U_{eq} = \left[\frac{(9.8\text{m}\cdot\text{s}^{-2})\cdot(10°\text{C})}{283\text{K}} \cdot \frac{(20\text{m})}{0.005} \cdot \sin(10°) \right]^{1/2}$$

= **<u>15.5 m/s</u>**

Use eq. (10.8) for the initial variation.

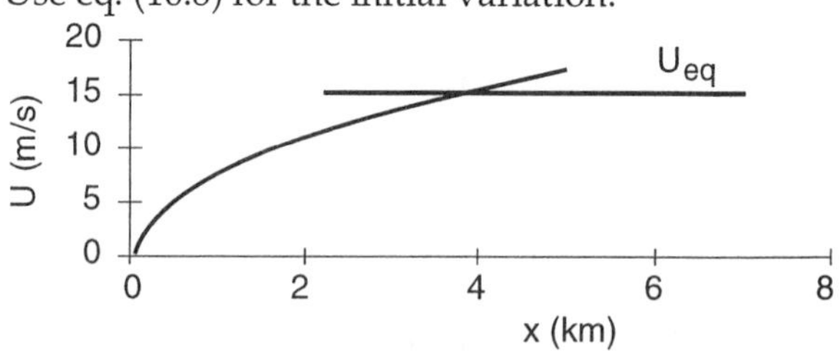

Check: Units OK. Physics OK.
Discussion: Although these two curves cross, the complete solution to eq. (10.7) smoothly transitions from the initial curve to the final equilibrium value.

STREAMLINES, STREAKLINES, AND TRAJECTORIES

Streamlines are everywhere parallel to the flow at some instant (i.e., a snapshot). This is an Eulerian point of view. Fig 10.5 shows an example of streamlines on a weather map. Streamlines never cross each other except where the speed is zero, and the wind never crosses streamlines. Streamlines can start and end anywhere, and can change with time. They are often not straight lines.

Streaklines are lines deposited in the flow during a time interval by continuous emission of a tracer from a fixed point. An example would be an aerial photograph of the smoke plume that was emitted from a smokestack.

Trajectories, also called **path lines**, trace the route traveled by an air parcel during a time interval. This is the Lagrangian point of view. For stationary (not changing with time) flow, streamlines, streaklines, and trajectories are identical.

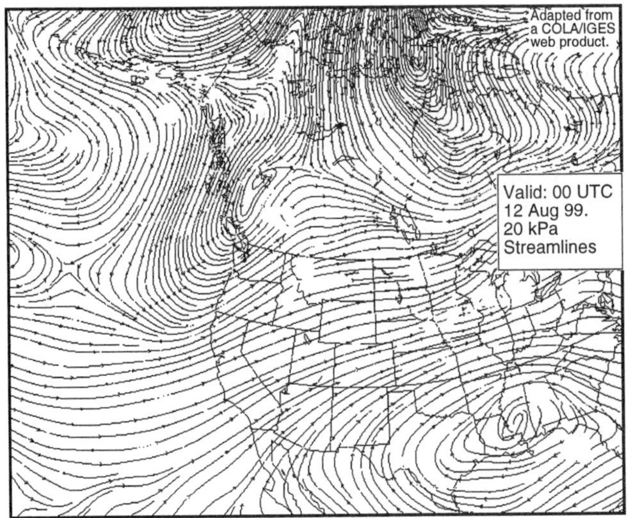

Figure 10.5
Streamlines near the tropopause, over N. America.

For nonstationary flow, there can be significant differences between them. For example, suppose that initially the flow is steady and from the north. Later, the wind suddenly shifts to come from the west. Fig 10.6 shows the situation shortly after this wind shift.

Streamlines (thin solid lines) are everywhere from the west in this example. The streakline caused by emission from a smokestack is shown as a thick hatched line. The dashed line shows the path followed by one of the air parcels in the smoke plume.

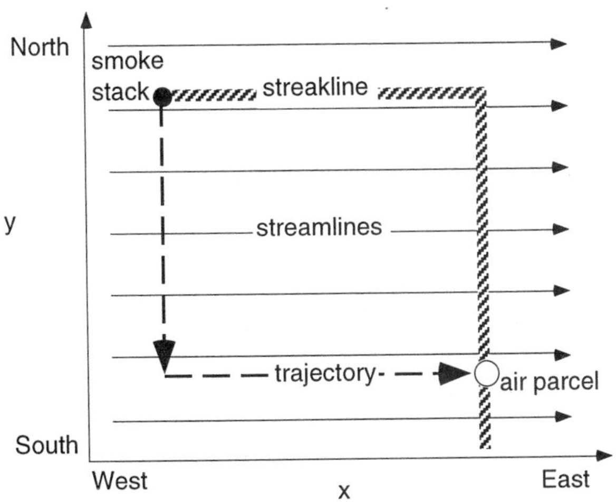

Figure 10.6
Streamlines, streaklines, and trajectories in nonstationary flow.

BERNOULLI'S EQUATION

Start with the vertical equation of motion (10.4a), and consider a steady-state situation ($\Delta W / \Delta t = 0$) with no drag:

$$0 = \underbrace{-U\frac{\Delta W}{\Delta x} - V\frac{\Delta W}{\Delta y}}_{advection} \underbrace{- \frac{1}{\rho}\frac{\Delta P}{\Delta z}}_{pressure\ gradient} \underbrace{+ g}_{gravity} \tag{10.10}$$

This equation, together with the corresponding horizontal equations of motion for steady state with no drag, can be solved for flow along a streamline. After much manipulation, the result is **Bernoulli's equation**:

$$\frac{1}{2}v^2 + \frac{P}{\rho} + |g| \cdot z = C_B \qquad \bullet(10.11)$$

where v is the total velocity along the streamline, and C_B is an arbitrary constant called **Bernoulli's constant** or **Bernoulli's function**. C_B is constant along any one streamline, but can vary from streamline to streamline.

Bernoulli's equation focuses on energy conservation along a streamline. The first term on the left is the kinetic energy per unit mass, the middle term is the work done on the air that has been stored as pressure, and the last term on the left is the potential energy per mass. Along any one streamline, energy can be converted from one form to another, providing the sum of these energies is constant.

In hydraulics, the gravity term is given by the change in depth of the water, especially when considering a streamline along the water surface. In meteorology, a similar situation occurs when cold air rises into a warmer environment; namely, it is a dense fluid rising against gravity. However, the gravity force felt by the rising cold air is reduced because of its buoyancy within the surrounding air. To compensate for this, the gravity term in Bernoulli's equation can be replaced with a **reduced gravity** $g' \equiv |g| \cdot \Delta\theta_v / T_v$, yielding:

$$\frac{1}{2}v^2 + \frac{P}{\rho} + |g|\frac{\Delta\theta_v}{T_v} \cdot z = C_B \qquad \bullet(10.12)$$

where $\Delta\theta_v$ is the virtual potential temperature difference between the warm air aloft and the cold air below. Thus, the gravity term is nonzero only when the streamline of interest is surrounded by air of different virtual potential temperature.

To use eq. (10.12), if any one or two terms increase in the equation, the other term(s) must decrease so that the sum remains constant. In other words, the sum of changes of all the terms must equal zero:

$$\Delta\left[\frac{1}{2}v^2\right] + \Delta\left[\frac{P}{\rho}\right] + \Delta\left[|g|\cdot\frac{\Delta\theta_v}{T_v}\cdot z\right] = 0 \qquad (10.13)$$

Dynamic vs. Static Pressure

When the wind approaches an obstacle, much flows around, as shown in Fig 10.7a. However, one of the streamlines that hits the obstacle decelerates from the upstream velocity of v_s to an ending velocity of zero. This ending point is called the **stagnation point**.

The pressure (called the **dynamic pressure**, P_d) at this stagnation point is higher than the free stream pressure (called the **static pressure**, P_s) well away from the obstacle. The dynamic pressure can be calculated from Bernoulli's equation.

For flow at constant altitude, the only two terms that change in Bernoulli's equation are the kinetic energy and pressure. Defining $\Delta P = P_d - P_s$ in eq. (10.13) gives the dynamic pressure:

$$P_d = P_s + \frac{\rho}{2}{v_s}^2 \qquad \bullet(10.14)$$

Dynamic effects make it difficult to measure static pressure in the wind. When the wind hits the pressure sensor, it decelerates and causes the pressure to increase. For this reason, static pressure instruments are designed to minimize flow deceleration and dynamic errors.

Temperature instruments have similar problems. When the flow decelerates and the pressure increases, the temperature will also increase because of the ideal gas law. This effect is called **dynamic warming**, and also must be avoided when designing temperature sensors that are in the wind.

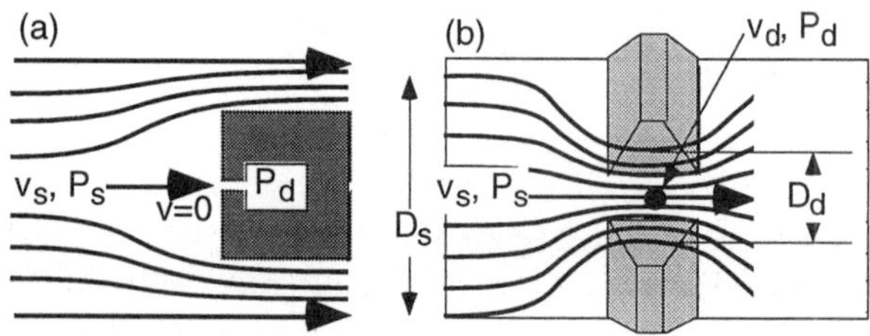

Figure 10.7
Streamlines (thick solid lines). (a) Stagnation. (b) Gap flow.

Dynamic pressure can be used to measure wind speed. An instrument that does this is the **pitot tube**. Aircraft instruments measure dynamic pressure with the pitot tube facing forward into the flow, and static pressure with another port facing sideways to the flow to minimize dynamic effects. The instrument then computes airspeed from the **pitot – static** pressure difference.

During tornadoes and hurricanes, if strong winds encounter an open garage door or house window, the wind trying to flow into the building causes pressure inside the building to increase dynamically. As was discussed in a previous solved example, the resulting pressure difference across the roof and walls of the building can cause them to blow out so rapidly that the building appears to explode.

Solved Example

For tornadic winds of 100 m/s that blow into a garage and stagnate, what force pushes against the inside of a wall of dimension 3 x 5 m?

Solution

Given: v_s = 100 m/s, Wall area A = 3 x 5 = 15 m^2
Find: F = ? N

Assume: ρ = 1 kg/m^3
Use eq. (10.14):

$$\Delta P = \frac{\rho}{2}{v_s}^2 = \frac{1\ \text{kg}\cdot\text{m}^{-3}}{2}\left(100\frac{\text{m}}{\text{s}}\right)^2 = 5\ \text{kPa}$$

$F = P \cdot A$ = (5000 N/m^2)·(15 m^2) = <u>**75 kN**</u>

Check: Units OK. Physics OK.
Discussion: This force is equivalent to the weight of 1000 people. Hide in the basement. Quickly.

Venturi Effect and Gap Winds

When winds are forced through a narrow opening such as a gap in a mountain range (Fig 10.7b), the winds accelerate to conserve the amount of air mass flowing. If there is a strong temperature inversion preventing the depth of the flow from changing through the gap, then air volume conservation gives:

$$v_d = \frac{D_s}{D_d}\cdot v_s \qquad \bullet(10.15)$$

where D is the width of the flow, subscript s represents the upstream flow, and subscript d represents flow in the narrowest part of the mountain pass.

Thus, if $D_d < D_s$, then $v_d > v_s$ to preserve the equality. Such strong winds are called **gap winds**, or **gap flow**.

Bernoulli's equation says that if the velocity increases in the region of flow constriction, then the pressure decreases. This is called the **Venturi effect**.

For gap winds of constant depth, eqs. (10.14 and 10.15) can be combined to give:

$$P_d - P_s = \frac{\rho}{2} \cdot v_s^2 \cdot \left[1 - \left(\frac{D_s}{D_d} \right)^2 \right] \qquad (10.16)$$

Solved Example

If a 20 km wide band of winds of 5 m/s must contract to pass through a 2 km wide gap, what is the wind speed and pressure decrease in the gap.

Solution

Given: D_s = 20 km, D_d = 2 km, v_s = 5 m/s.
Find: v_d = ? m/s, ΔP = ? kPa
Assume: $\rho = 1.2$ kg/m^3

Use eq. (10.15):

$v_d = (D_s / D_d) \cdot v_s = (20 / 2) \cdot (5 \text{m / s})$ = **50 m/s**

Use eq. (10.16): $\Delta P = (\rho / 2)(v_s^2 - v_d^2)$

$= (0.6 \text{ kg} \cdot \text{m}^{-3})(25 - 2500 \text{m}^2 / \text{s}^2)$

$= -1485 \text{ kg} \cdot \text{m}^{-1} \cdot \text{s}^{-2}$ = **–1.5 kPa**

Check: Units OK. Physics OK.
Discussion: The actual winds would be slower, because turbulence would cause significant drag.

Downslope Winds – Bora and Foehn

Consider a wintertime situation of a statically stable atmosphere (cold air under warm), and with a synoptic weather pattern that forces strong winds toward a mountain range. Flow over ridge-top depends on the depth of cold air, and on the strength of the static stability.

If the mountain height H is greater than the thickness z_i of cold air upstream, then the cold air is dammed behind the mountain and doesn't flow over (Fig. 10.8a). The strong warm winds aloft can flow over the ridge top, and can warm further upon descending adiabatically on the lee side. The result is a warm wind called the **Foehn**. (A second mechanism for generating Foehn winds by precipitation is given at the end of this chapter.)

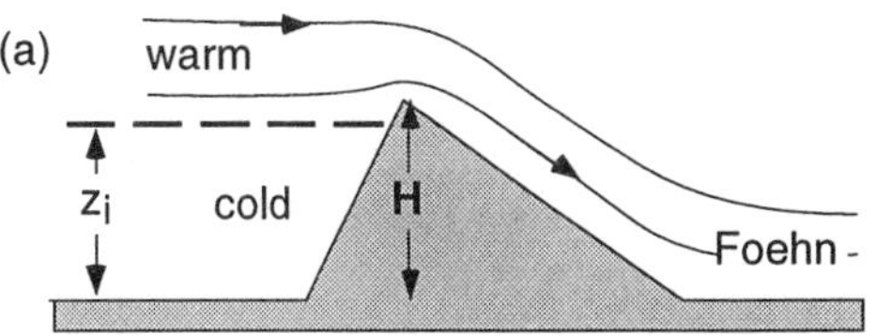

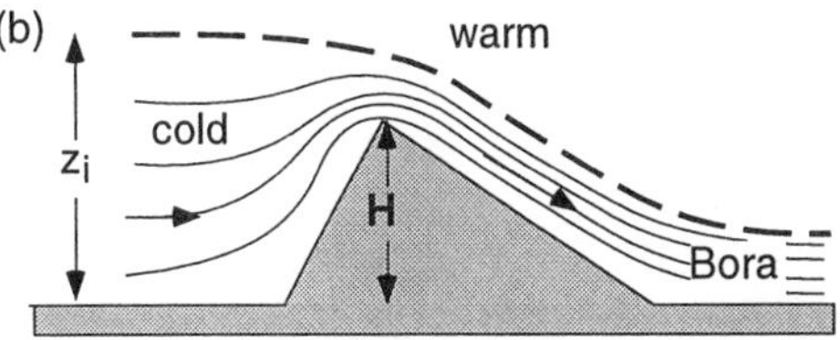

Figure 10.8
(a) Warm Foehn winds, or (b) cold Bora winds, during synoptic weather patterns where strong winds are forced toward the ridge from upstream. Thin lines are streamlines, and the thick dashed line is a temperature inversion.

If the fast-moving cold air upstream is deeper than the ridge top (Fig 10.8b), then very fast (hurricane force) cold winds can descend down the lee side. This phenomena is called a **Bora**. The winds first accelerate in the constriction between mountain and the overlying inversion, and pressure drops according to the Venturi effect. This lower pressure upsets hydrostatic balance, and draws the cold air layer downward, causing the fast winds to hug the slope.

The overlying warmer air is also drawn down by this same pressure drop. Because work must be done to lower this warm air against buoyancy, Bernoulli's equation tells us that the Bora winds decelerate slightly on the way down. Once the winds reach the lee valley floor, they are still destructive and much faster than the winds upstream of the mountain, but are slightly slower than the winds at ridge top.

The difference between katabatic and Bora winds is significant. Katabatic winds are driven by the local thermal structure, and form during periods of weak synoptic forcing such as in high pressure areas of fair weather and light winds. Boras are driven by the inertia of strong upstream winds that form in regions of low pressure and strong horizontal pressure gradient. Although both phenomena are cold downslope winds, they are driven by different dynamics.

Science Graffito

"With enough parameters I can fit an elephant." - An unknown scientist, commenting on an inelegant method of developing an equation to fit observations.

Solved Example

For the Bora situation of Fig 10.7b, the inversion of strength 6°C is 1200 m above the upstream valley floor. Ridge top is 1000 m above the valley floor. If upstream winds are 10 m/s, find the Bora wind speed in the lee valley.

Solution

Given: $H = 1$ km, $z_i = 1.2$ km, $\Delta\theta_v = 6°C$,
$v_s = 10$ m/s. Assume $g/T_v = 0.0333$ m·s^{-2}·K^{-1}

Find: $v_{Bora} = ?$ m/s

Volume conservation similar to eq. (10.15) gives ridge top winds v_d:

$$v_d = \frac{z_i}{z_i - H} \cdot v_s$$

Assume Bora thickness = constant = $z_i - H$.
Follow the streamline indicated by the thick dashed line in Fig 10.8b.
Assume ending pressure equals starting pressure on this streamline.

Use Bernoulli's eq. (10.12):

$$\left[\frac{1}{2}v^2 + |g|\frac{\Delta\theta_v}{T_v}\cdot z\right]_{ridgetop} = \left[\frac{1}{2}v^2 + |g|\frac{\Delta\theta_v}{T_v}\cdot z\right]_{Bora}$$

Combine the above eqs., set $z = H$, and solve for v_{Bora}:

$$v_{Bora} = \left[\left(\frac{z_i}{z_i - H}\right)^2 \cdot v_s^2 - 2\cdot\frac{g}{T_v}\cdot\Delta\theta_v \cdot H\right]^{1/2}$$

Finally, we can plug in the numbers:

$v_{Bora} =$

$$\sqrt{\left(\frac{1.2\text{km}}{0.2\text{km}}\right)^2\left(10\frac{\text{m}}{\text{s}}\right)^2 - 2\cdot\left(0.033\frac{\text{m}}{\text{s}^2\text{K}}\right)\cdot(6\text{K})\cdot(1000\text{m})}$$

$$= \sqrt{(3600 - 400)\text{m}^2\text{s}^{-1}} = \underline{56.6 \text{ m/s}}$$

Check: Units OK. Physics OK.

Discussion: Winds at ridge top were 60 m/s. Although the Bora winds at the valley floor are weaker than ridge top, they are still strong and destructive. Most of the wind speed up was due to volume conservation, with only a minor decrease given by Bernoulli's equation.

Hydraulic Jump

Consider the Bora situation, redrawn in Fig 10.9, with strong winds being forced synoptically toward a mountain ridge. Upstream of the mountain, where pressure decrease with height is hydrostatic, we know that $P_2 < P_1 < P_{sfc}$, where P_2 is the pressure at the dashed streamline that separates the warm and cold air. However, in the Bora solved example just presented, we assumed that the pressure is constant along the dashed streamline.

Thus, just downwind of the mountain, in the region of the strong Bora, pressure P_2 is much closer to the ground than it was upstream of the mountain (Fig 10.9). Further downstream of the mountain, we can assume that the pressure is again hydrostatic. Thus, at the constant height level shown by the dotted line in Fig 10.9, there is a <u>horizontal</u> pressure gradient between P_2 at the Bora and P_1 further downstream.

According to Bernoulli's equation, this pressure increase in the lee (downwind) valley causes the wind speed to decrease and the temperature inversion to rise back toward its starting altitude. This height increase is often sudden and turbulent, and is called an **hydraulic jump**.

Another interpretation of the hydraulic jump is that it is a wave or density current (**bore**) propagating upstream on the density interface between the cold and warm air. As it nears the Bora, its propagation speed is counteracted by increasing downslope winds, until the two balance and the wave or bore becomes stationary.

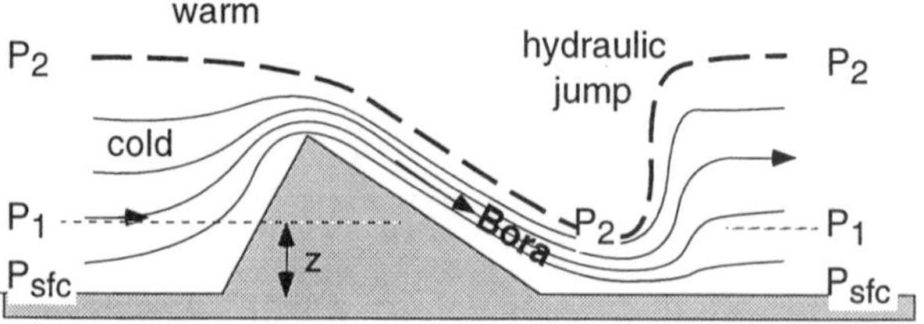

Figure 10.9
Hydraulic jump.

ON DOING SCIENCE • Simple is Best

Fourteenth century philosopher William of Occam suggested that the simplest scientific explanation is the best. This tenant is known as **Occam's Razor**, because with it you can cut away the bad theories and complex equations from the good.

But why should the simplest or most elegant be the best? There is no law of nature that says it must be so. It is just one of the philosophies of science, as is the scientific method of Descartes. Ultimately, like any philosophy or religion, it is a matter of faith.

MOUNTAIN WAVES

The Bora and hydraulic jump were associated with a sharp temperature interface separating a layer of cold air below from warm air aloft. Now, we will examine a different situation, where there is a gradual potential temperature increase with height. Again, we will assume that winds are being forced synoptically to approach a mountain ridge.

Natural Wavelength

When statically stable air flows with speed M over a hill or ridge, it is set into oscillation at the Brunt-Väisälä frequency, N_{BV}. The natural wavelength λ is

$$\lambda = \frac{2\pi \cdot M}{N_{BV}} \qquad \bullet(10.17)$$

Longer wavelengths occur in stronger winds, or weaker static stabilities.

These waves are known as **mountain waves**, gravity waves, buoyancy waves, or **lee waves**. They can cause damaging winds, and interesting clouds.

Friction and turbulence damp the oscillations with time (Fig 10.10). The resulting path of air is a damped wave:

$$z = z_1 \cdot \exp\left(\frac{-x}{b \cdot \lambda}\right) \cdot \cos\left(\frac{2\pi \cdot x}{\lambda}\right) \qquad (10.18)$$

where z is the height of the air above its starting equilibrium height, z_1 is the initial amplitude of the wave (based on height of the mountain), x is distance downwind of the mountain crest, and b is a damping factor. Wave amplitude reduces to $1/e$ at a downwind distance of b wavelengths (that is, b is the e-folding distance).

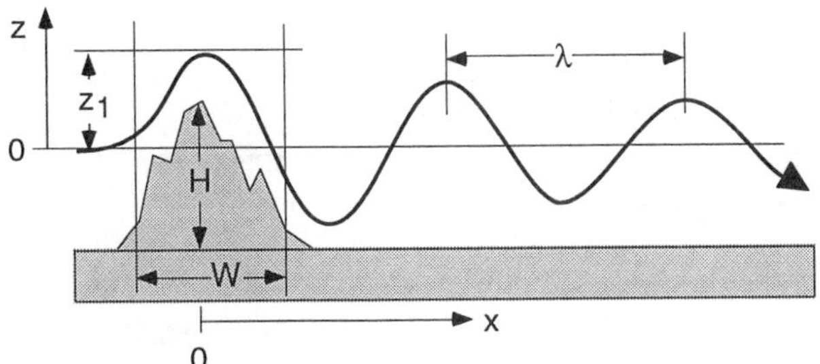

Figure 10.10
Mountain wave characteristics.

Solved Example(§)
Find and plot the path of air over a mountain, given: $z_1 = 500$ m, $M = 30$ m/s, $b = 3$, $\Delta T/\Delta z = -0.005$ K/m, $T = 10°C$, and $T_d = 8°C$.

Solution
Given: (see above). Thus $T = 283$ K
Find: $N_{BV} = ?\ s^{-1}$, $\lambda = ?$ m, and plot z vs. x

Use eq. (6.5a):

$$N_{BV} = \left[\frac{9.8\text{m}\cdot\text{s}^{-2}}{283\text{K}} \cdot (-0.005 + 0.0098)\right]^{1/2} = \underline{\mathbf{0.0129s^{-1}}}$$

Use eq. (10.17)

$$\lambda = \frac{2\pi \cdot (30\text{m/s})}{0.0129\text{s}^{-1}} = \underline{\mathbf{14.62\ km}}$$

Use eq. (10.18)

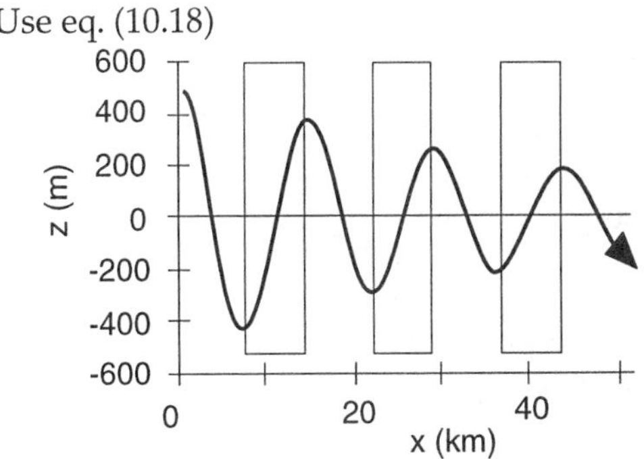

Check: Units OK. Physics OK. Sketch OK.
Discussion: Glider pilots can soar in the updraft portions of the wave, highlighted with white boxes.

Lenticular Clouds

In the updraft portions of mountain waves, the rising air cools adiabatically. If sufficient moisture is present, clouds can form, called **lenticular clouds**. The first cloud, which forms over the mountain crest, is usually called a **cap cloud**.

The droplet sizes in these clouds are often quite uniform, because of the common residence times of air in the clouds. This creates interesting optical phenomena such as **corona** and **iridescence** when the sun or moon shines through them.

Knowing the temperature and dew point of air at the starting altitude before blowing over the mountain, a lifting condensation level (LCL) can be calculated using eq. (5.8). Clouds will form in the crests of those waves for which $z > z_{LCL}$.

Solved Example (§)

Replot the results from the previous solved example, indicating which waves have lenticular clouds.

Solution

Given: (see previous solved example)
Find: z_{LCL} = ? m.

Use eq. (5.8):
$z_{LCL} = (125m/°C) \cdot (10°C - 8°C) = 250$ m
From the sketch below, **1 cap cloud and 2 lenticular clouds** form.

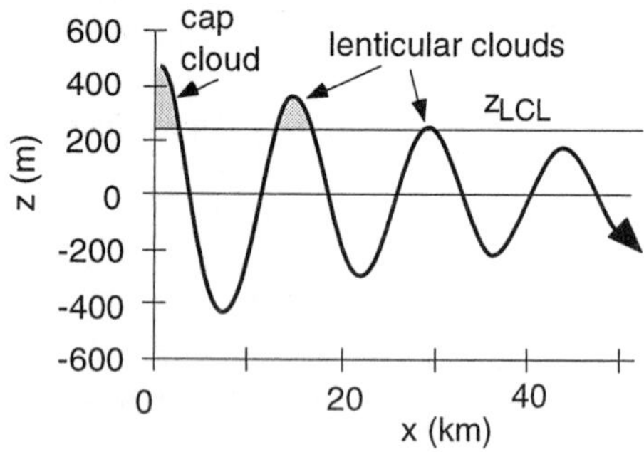

Check: Units OK. Physics OK. Sketch OK.
Discussion: Most clouds blow with the wind. Not **standing lenticular clouds**. They are stationary while the wind blows through them.

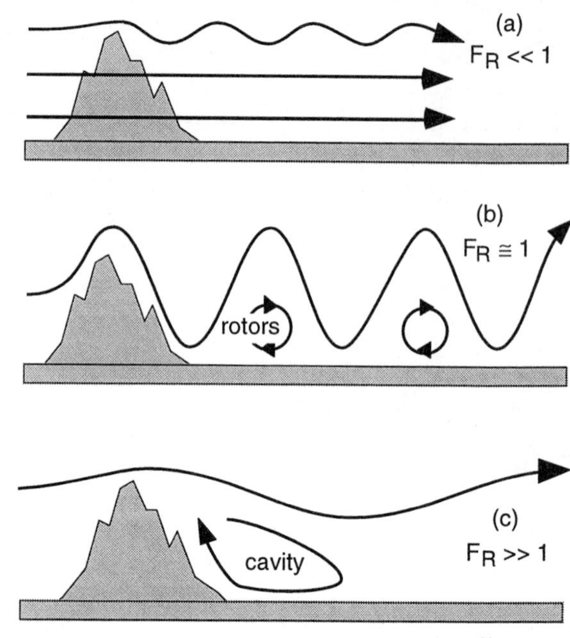

Figure 10.11
Mountain wave behavior vs. Froude number, Fr.

Froude Number

For individual hills not part of a continuous ridge, some air can flow around the hill rather than over the top. When less air flows over the top, shallower waves form.

The Froude Number Fr is a measure of the ability of waves to form over hills. It is given by

$$Fr = \frac{\lambda}{2 \cdot W} \qquad \bullet(10.19)$$

where W is the hill width, and λ is the natural wavelength. Fr is dimensionless.

For strong static stabilities or weak winds, $Fr << 1$. The natural wavelength of air is much shorter than the width of the mountain, resulting in only a little air flowing over the top of the hill, with small waves (Fig 10.11a). If H is the height of the hill (Fig 10.10), then $z_1 < H$ for this case. Most of the air is **blocked** in front of the hill, or flows around the sides.

For moderate stabilities where the natural wavelength is nearly equal to twice the hill width, $Fr \cong 1$. The air **resonates** with the terrain, causing very intense waves (Fig 10.11b). These waves have the greatest chance of forming lenticular clouds, and pose the threat of violent turbulence to aircraft. Wave amplitude roughly equals hill height: $z_1 \cong H$. Sometimes **rotor** circulations and **rotor clouds** will form near the ground under the wave crests (Fig 10.11b).

Solved Example

For a natural wavelength of 10 km and a hill width of 15 km, describe the type of waves.

Solution

Given: λ = 10 km, W = 15 km
Find: Fr = ?

Use eq. (10.19):

$$Fr = \frac{(10\text{km})}{2 \cdot (15\text{km})} = 0.333$$

Waves as in Fig 10.11a form off the top of the hill. Some air also flows around sides of hill.

Check: Units OK. Physics OK.
Discussion: Most natural hills are quite irregular, and can be thought of as small hills on top of larger hills. There are a wide variety of hill widths that apply, although only one natural wavelength.

When static stability is weak and winds strong, the natural wavelength is much greater than the hill width, $Fr >> 1$. Again the wave amplitude is weak, $z_1 < H$. For this situation, a turbulent **wake** will often form downwind of the mountain, sometimes with a **cavity** of reverse flow near the ground (Fig 10.11c). The cavity and rotor circulations are driven by the wind like a bike chain turning a gear.

Mountain-Wave Drag

For $Fr < 1$, wave crests tilt upwind with altitude (Fig 10.12). The angle α of tilt relative to vertical is

$$\cos(\alpha) = Fr \qquad (10.20)$$

For this situation, slightly lower pressure develops on the lee side of the hill, and higher pressure on the windward side. This pressure gradient opposes the mean wind, and is called wave drag. Such wave drag adds to the skin drag. The wave drag (WD) force per unit mass near the ground is:

$$\frac{F_{x\ WD}}{m} = -\frac{H^2 \cdot N_{BV}{}^2}{8 \cdot h_w} \cdot Fr \cdot \left[1 - Fr^2\right]^{1/2} \qquad (10.21)$$

where h_w is the depth of air containing waves. Not surprisingly, higher hills cause greater wave drag.

Solved Example

For $Fr = 0.8$, find the angle of the wave crests and the wave drag over a hill of height 800 m. The Brunt-Väisälä frequency is 0.01 s^{-1}, and the waves fill the 11 km thick troposphere.

Solution

Given: $Fr = 0.8$, $H = 800$ m, $N_{BV} = 0.01$ s^{-1}.
Find: $\alpha = ?°$, $F_{x\ WD}/m = ?$ m·s^{-2}

Use eq. (10.20):

$$\alpha = \cos^{-1}(0.8) = \underline{36.9°}$$

Use eq. (10.21):

$$\frac{F_{x\ WD}}{m} = -\frac{(800\text{m} \cdot 0.01\text{s}^{-1})^2}{8 \cdot (11{,}000\text{m})} \cdot 0.8 \cdot \left[1 - 0.8^2\right]^{1/2}$$

$$= \underline{3.5\text{x}10^{-4}} \text{ m·s}^{-1}.$$

Check: Units OK. Physics OK.
Discussion: This is of the same order of magnitude as the other forces in the equations of motion.

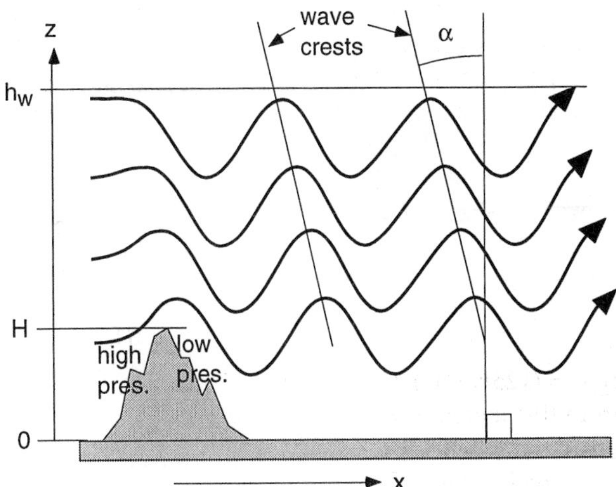

Figure 10.12
Vertical wave propagation, tilting crests, and wave drag.

FOEHNS (again) AND CHINOOKS

Air that is not statically stable will not form waves when it flows over mountains. However, thermodynamic changes in the air will occur to make it warmer and drier.

Consider an air parcel before it flows over the mountain, indicated as point (1) in Fig 10.13. Suppose that the temperature is 20°C and dew point is 10°C initially, as indicated by the black circles in Fig 10.14. This corresponds to about 50% relative humidity.

As the air begins to rise along the windward slopes, it will cool dry-adiabatically while conserving mixing ratio until the LCL is reached (2). Further lifting is moist adiabatic (3) within the orographic cloud. Suppose that most of the condensed water falls out as precipitation.

Over the summit (4), suppose that the air has risen to a height of 60 kPa. It now has a temperature of about –8°C. As it begins to descend down the lee side, any residual cloud droplets will quickly evaporate in the adiabatically warming air. The trailing edge of the cloud is called a **Foehn wall**, because it looks like a wall when viewed from downwind valleys.

Continued descent will be dry adiabatic (5) because there are no liquid water drops to cause evaporative cooling. By the time the air reaches its starting altitude on the lee side (6), its temperature has warmed to about 35°C, with a dew point of about –2°C. This is roughly 10% relative humidity.

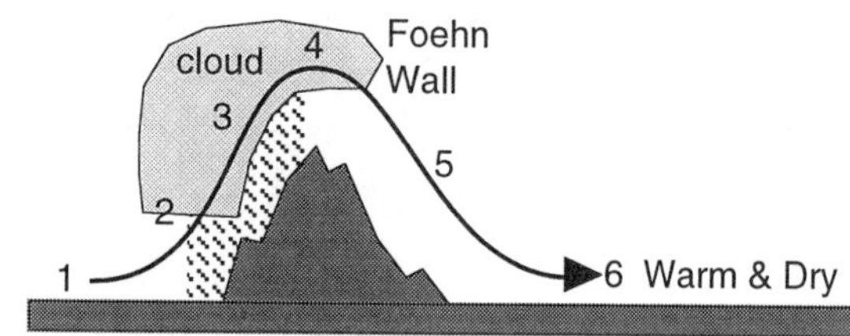

Figure 10.13
Foehn weather.

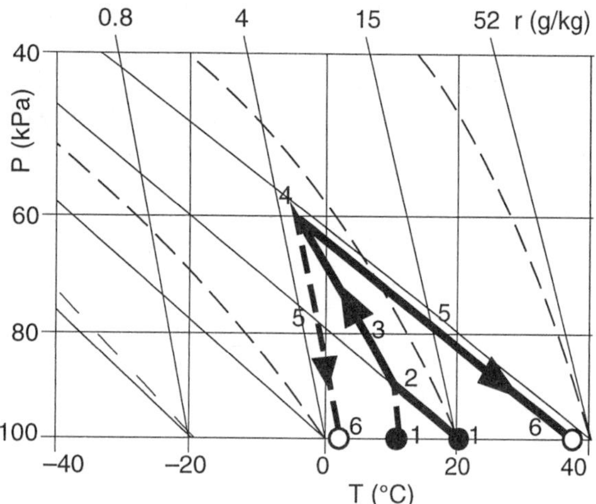

Figure 10.14
Foehn thermodynamics.

Solved Example

Air from the Pacific Ocean (T =5°C, T_d = 3°C, z = 0) flows over the Coast Mountains ($z \cong 3000$ m), and descends toward the interior plateau of British Columbia, Canada ($z \cong 1000$ m). Find the final T and T_d.

Solution

Given: $T = 5$ °C, $T_d = 3$ °C, $z = 0$ m initially.
Find: T = ? °C, T_d = ? ° at $z = 1000$ m finally.

Use the thermo diagram from Fig 7.8:

The initial mixing ratio is $r \cong 5$ g/kg. Clouds form as the air rises over the mountains, with base at z_{LCL} = (0.125 km/°C)·(5–3°C) = 0.25 km. From there to z = 3000 m, the air follows a moist adiabat, reaching $T = T_d \cong -17$°C.

Assuming all condensate falls out as precipitation, the air then descends dry-adiabatically to Calgary. The final state is:
$T \cong$ <u>**4°C**</u> at z = 1000 m, and $T_d \cong$ <u>**–13°C**</u>.

Check: Units OK. Physics OK.
Discussion: The air is much drier ($RH \cong 25$%), but nearly the same temperature as initially. This is typical of a foehn wind.

Hence, the net result of this process is: clouds and precipitation form on the windward slopes of the mountain, a Foehn wall forms just downwind of the mountain crest, and there is warming and drying in the lee valley. Such warm dry winds are called **Foehns** in the Alps, and **Chinooks** in the Rocky Mountains. Other types of Foehns not requiring precipitation were described earlier in this Chapter.

SUMMARY

Forces cause winds on all scales. At synoptic scales, changing weather patterns cause fast winds on some days and slower winds on others. The probability of any wind speed can be described by a Weibull distribution. Those regions with greater probability of strong winds are ideal for siting wind turbines to generate electrical power.

At mesoscales, statically stable air forms lee waves as it flows over mountains. These mountain waves cause turbulence, and are a form of drag against air movement. Winds accelerate according to Bernoulli's equation when channeled through mountain gaps.

Of the three types of downslope winds discussed, only the katabatic wind is driven by local cooling, and occurs in the absence of synoptic-scale winds. The Bora is also a cold wind, but is driven dynamically by the synoptic scale flow. Foehn winds are also driven dynamically, but are warm. Hydraulic jumps occur downstream of bora winds as the flow re-adjusts to hydrostatic equilibrium.

At boundary-layer scales, thermals often rise from the heated surface through the boundary layer. Near a valley wall, these warm updrafts hug the slope, forming anabatic winds. Wind and temperature instruments are constructed to minimize dynamic pressure and heating errors.

Threads

All scales of motion are covered in this book, from turbulent (Appendix H) and boundary layer (Chapt. 4) scales, through mesoscales such as thunderstorms (Chapt. 15) and fronts (Chapt. 12), synoptic scales including hurricanes (Chapt. 16) and cyclones (Chapt. 13), to planetary scales of the general circulations (Chapt. 11). All of these motions are described by Newton's second law of motion (Chapt. 9).

Mountain waves, Boras, and Foehns require a statically stable environment (Chapt. 6) such as is in the standard atmosphere (Chapt. 1). Warm foehn winds can occur by latent heating (Chapts. 5 and 6), or by adiabatic warming (Chapt. 3) of downslope air.

Wind turbines are often sited near hilly regions where wind speed increases. Turbines are designed to withstand the wind shear and gusts typical of the atmospheric boundary layer (Chapt. 4).

EXERCISES

Numerical Problems

N1. What is the scale classification and time duration for the following phenomena, given their horizontal size?

a. dust devils 5 m
b. cumulus clouds 1 km
c. gust fronts 20 km
d. sea breezes 100 km
e. convective complexes 200 km
f. hurricanes 500 km
g. jet streams 800 km
h. low pressure systems 1500 km

N2.(§) Plot the probability of wind speeds using a Weibull distribution with a resolution of 0.5 m/s, and M_o = 8 m/s, for α =

a. 1.5 b. 2 c. 3 d. 10
e. 1.0 f. 1.8 g. 5 h. 8

N3. A wind turbine of blade radius 20 m runs at 30% efficiency. At sea level, find the theoretical power for winds of: a. 2 m/s b. 8 m/s c. 15 m/s d. 40 m/s
e. 1 m/s f. 4 m/s g. 10 m/s h. 20 m/s

N4. Find the equilibrium updraft speed of a thermal in a 2 km boundary layer at 15°C. The thermal temperature is (°C): a. 17 b. 20 c. 23 d. 16
e. 18 f. 22 g. 25 h. 30

N5(§). Plot katabatic wind speed vs downslope distance if the environment is 20°C and the cold katabatic air is 15°C. The slope angle (°) is:

a. 1 b. 2 c. 5 d. 10
e. 12 f. 18 g. 20 h. 45

N6. Find the equilibrium downslope speed for the previous problem, if the katabatic air is 5 m thick and the drag coefficient is 0.002.

N7. Assuming standard sea-level density and streamlines that are horizontal, find the pressure change given the following velocity (m/s) change:

a. 2 b. 3 c. 5 d. 10
e. 15 f. 20 g. –5 h. –10

N8. Assume that pressure is constant along the streamline, and that the air in the streamline is 5°C warmer than the surrounding environment, which is of uniform potential temperature. For the previous problem, find the change in height of the streamline.

N9. Winds of 10 m/s are flowing in a valley of 10 km width. Further downstream, the valley narrows to the width (km) given below. Find the wind speed and pressure change in the constriction, assuming constant depth flow.

a. 8 b. 10 c. 6 d. 5
e. 3 f. 2 g. 1 h. 0.5

N10. For the bora solved example, redo the calculation assuming that the initial inversion height (km) is: a. 1.5 b. 1.7 c. 2.0 d. 2.5
e. 1.1 f. 2.2 g. 3.0 h. 4.0

N11. Find the natural wavelength given

a. $M = 2$ m/s, $\Delta T/\Delta z = 5$ °C/km
b. $M = 20$ m/s, $\Delta T/\Delta z = -8$ °C/km
c. $M = 5$ m/s, $\Delta T/\Delta z = -2$ °C/km
d. $M = 20$ m/s, $\Delta T/\Delta z = 5$ °C/km
e. $M = 5$ m/s, $\Delta T/\Delta z = -8$ °C/km
f. $M = 2$ m/s, $\Delta T/\Delta z = -2$ °C/km
g. $M = 5$ m/s, $\Delta T/\Delta z = 5$ °C/km
h. $M = 2$ m/s, $\Delta T/\Delta z = -8$ °C/km

N12. For a mountain of width 25 km, find the Froude number for the previous problem.

N13. For the previous problem, find the angle of the wave crests, and the wave-drag force per unit mass. Assume $H = 1000$ m and $h_w = 11$ km.

N14(§). Plot the wavy path of air as it flows past a mountain, given an initial vertical displacement of 300 m, an wavelength of 12.5 km, and a damping factor of

a. 4 b. 1 c. 2 d. 10
e. 5 f. 0 g. 8 h. 15

N15(§) Given a temperature dew-point spread of 1.5°C at the initial height of air in the previous problem, identify which wave crests contain lenticular clouds.

N16. Use a thermodynamic diagram. Air of initial temperature 10°C and dew point 0°C start at a height of 95 kPa. This air rises to height 70 kPa as it flows over a mountain, during which all liquid and solid water precipitate out. Air descends on the lee side of the mountain to an altitude of 100 kPa. What is the temperature, dew point, and relative humidity of the air at its final altitude? How much precipitation occurred on the mountain?

Understanding & Critical Evaluation

U1. Anabatic and lenticular clouds were described in this chapter. Compare these clouds and their formation mechanisms. Is it possible for both clouds to occur simultaneously over the same mountain?

U2. Find the horizontal length scale identifier for:

a. thunderstorms	b. hurricanes	d. fronts
f. chinook	g. tornadoes	h. thermals
i. gap winds	j. hydraulic jump	k. Bora

U3. The scaling of Table 10-1 was applied to the horizontal. Comment on why it was not applied to the vertical.

U4. Identify some meteorological phenomena that differ considerably from the space-time scale relationship of eq. (10.1).

U5. To double the amount of electrical power produced by a wind turbine, wind speed must increase by what percentage, or turbine radius increase by what percentage?

U6(§). Create a computer spreadsheet with location and spread parameters in separate cells. Create and plot a Weibull frequency distribution for winds by referencing those parameters. Then try changing the parameters to see if you can get the Weibull distribution to look like other well-known distributions, such as Gaussian (symmetric, bell shaped), exponential, or others.

U7. Why was an asymmetric distribution such as the Weibull distribution chosen to represent winds?

U8. For a wind turbine radius of 8 m and hub height of 10 m, derive an equation for the average power produced as a function of u_* and z_o, assuming a neutral surface-layer log wind profile.

U9. Term by term, compare the full horizontal (eq. 9.15) and vertical equations (10.4) of motion, and discuss similarities and differences.

U10. Both the "full" horizontal (eq. 9.15) and vertical equations (10.4) of motion are missing one advection term. Can you anticipate what it is, generalizing from the advection terms that are already listed? Also, why do you think it was not given for those equations? For what situations might this missing term be important?

U11. What is required for the Boussinesq approximation to be valid?

U12. If thermals with average updraft velocity of W = 5 m/s occupy 40% of the horizontal area in the boundary layer, find the average downdraft velocity.

U13. What factors might affect rise rate of the thermal, in addition to the ones already given in this chapter?

U14. What factors control the shape of the katabatic wind profile, as plotted in Fig. 10.4?

U15. The solved example for katabatic wind shows the curves from eqs. (10.8) and (10.9) as crossing. Given the factors that appear in those equations, is a situation possible where the curves never cross? Describe.

U16. If during the course of a day, the wind speed is constant but the wind direction gradually changes direction by a full 360°, draw a graph of the resulting streamline, streakline, and path line at the end of the period. Assume continuous emissions from a point source during the whole period.

U17. Identify the terms of Bernoulli's equation that form the hydrostatic approximation. According to Bernoulli's equation, what must happen or not happen in order for hydrostatic balance to be valid?

U18. Describe how the terms in Bernoulli's equation vary along a mountain-wave streamline as sketched in Fig 10.10.

U19. If a cold air parcel is given an upward push in a warmer environment of uniform potential temperature, describe how the terms in Bernoulli's equation vary with parcel height.

U20. In Fig. 10.7a, would it be reasonable to place a static pressure port on the top center of the darkly shaded block? Comment on potential problems with a static port at that location.

U21. Comment on the differences and similarities of the two mechanisms shown in this Chapter for creating Foehn winds.

U22. Explain in terms of Bernoulli's equation the horizontal pressure gradient force acting on anabatic winds.

U23. Why are lenticular clouds called "standing lenticular"?

U24. It is known from measurements of the ionosphere that the vertical amplitude of mountain waves increases with altitude. Explain this using Bernoulli's equation.

U25. Is there any max limit to the angle α of mountain wave crests (see Fig 10.12)? Comment.

U26. If the air in Fig. 10.13 went over a mountain but there was no precipitation, would there be a Foehn wind?

U27. Relate the amount of warming of a Foehn wind to the average upstream wind speed and the precipitation rate in mm/h .

Web-Enhanced Questions

W1. Download a weather map from the web, and comment on the different scales of motion that it contains.

W2. Find on the web climatological maps giving locations of persistent, moderate winds. These are favored locations for wind turbine farms. Also search the web for locations of existing turbine farms.

W3. Search the web for a weather map showing vertical velocities over your country or region. Sometimes, these vertical velocities are given as ω rather than w, where ω is the change of pressure with time experienced by a vertically moving parcel, and is defined in Chapt. 13. What is the range of vertical velocities on this particular day, in m/s?

W4. Search the web for the highest resolution (hopefully 0.5 km or better resolution) visible satellite imagery for your area. Which parts of the country have rising thermals, based on the presence of cumulus clouds at the top of the thermals?

W5. Search the web for the highest resolution (hopefully 0.5 km or better resolution) visible satellite imagery for your area. Also search for an upper air sounding (i.e., thermo diagram) for your area. Does the depth of the mixed layer from the thermo diagram agree with the diameter of thermals (clouds) visible in the satellite image? Comment.

W6. Access IR high resolution satellite images over cloud-free regions of the Rocky Mountains (or Cascades, Sierra-Nevada, Appalachians, or other significant mountains) for late night or early morning during synoptic conditions of high pressure and light winds. Identify those regions of cold air in valleys, as might have resulted from katabatic winds. Sometimes such regions can be identified by the fog that forms in them.

W7. Access high-resolution visible satellite images from the web during clear skies, that show the smoke plume from a major source (such as Gary, Indiana, or Sudbury, Ontario, or a volcano, or a forest fire). Assume that this image shows a streakline. Also access the current winds from a weather map corresponding roughly to the altitude of the smoke plume, from which you can infer the streamlines. Compare the streamlines and streaklines, and speculate on how the flow has changed over time, if at all. Also, draw on your printed satellite photo the path lines for various air parcels within the smoke plume.

W8. From the web, access weather maps that show streamlines. These are frequently given for weather maps of the jet stream near the tropopause (at 20 to 30 kPa). Also access from the web weather maps that plot the actual upper air winds from rawinsonde observations, valid at the same time and altitude as the streamline map. Compare the instantaneous winds with the streamlines.

W9. From the web, access a sequence of weather maps of streamlines for the same area. Locate a point on the map where the streamline direction has changed significantly during the sequence of maps. Assume that smoke is emitted continuously from that point. On the last map of the sequence, plot the streakline that you would expect to see. (Hint, from the first streamline map, draw a path line for an air parcel that travels until the time of the next streamline map. Then, using the new map, continue finding the path of that first parcel, as well as emit a new second parcel that you track. Continue the process until the tracks of all the parcels end at the time of the last streamline map. The locus of those parcels is a rough indication of the streakline.)

W10. Access from the web information for aircraft pilots on how the pitot tube works, and/or its calibration characteristics for a particular model of aircraft.

W11. Access from the web figures that show the amount of destruction for different intensities of tornado winds. Prepare a table giving the dynamic pressures and forces on the side of a typical house for each of those different wind categories.

W12. Access from the web news stories of damage to buildings or other structures caused by Boras, mountain waves, or downslope windstorms.

W13. Access visible high-resolution satellite photos of mountain wave clouds downwind of a major mountain range. Measure the wavelength from these images, and compare with the wind speed accessed from upper air soundings in the wave region. Use those data to estimate the Brunt-Väisälä frequency.

W14. Access from the web photographs taken from ground level of lenticular clouds. Also, search for iridescent clouds on the web, to find if any of these are lenticular clouds.

W15. Access from the web pilot reports of turbulence, chop, or mountain waves in regions downwind of mountains. Do this over several days, and show how these reports vary with wind speed and static stability.

Synthesis Questions

S1. Suppose that a commercial wind farm has wind turbines deployed along a ridge. These turbines have radius R, and hub height h. You were hired to increase the hub height in order to install longer turbine blades, to produce more electricity.

Knowing what you know about the log wind profile in the boundary layer under windy, neutral conditions, and knowing that the drag on the turbines is proportional to $R^2 \cdot M^2$, find and plot a relationship between power production and drag. Based on this, what is an optimum design for the turbines?

S2. Suppose that in year 2100 everyone is required by law to have their own wind turbine. Since wind turbines take power from the wind, the wind becomes slower. What effect would this have on the weather and climate, if any?

S3. Suppose that all air parcels felt an electrostatic attraction force toward the ionosphere. Revise the vertical equation of motion to include this effect.

S4. What if fair-weather thermals routinely rose as high as the tropopause without forming clouds. Comment on changes to the weather and climate, if any.

S5. Modify the katabatic wind equation to work for anabatic winds. Justify your assumptions.

S6. Suppose that katabatic winds were frictionless. Namely, no turbulence, no friction against the ground, and no friction against other layers of air. Speculate on the shape of the vertical wind profile of the katabatic winds, and justify your arguments.

S7. Is it possible for a moving air parcel to not be traveling along a streamline? Comment.

S8. Suppose that Bernoulli's equation says that pressure <u>de</u>creases as velocity <u>de</u>creases along a streamline of constant height. How would the weather and climate be different, if at all? Start by commenting how Boras would be different, if at all.

S9. If mountain wave drag causes the winds to be slower, does that same drag force cause the earth to spin faster? Comment.

S10. Suppose that mountain wave drag worked oppositely, and caused winds to accelerate aloft. How would the weather and climate be different, if at all?

GLOBAL CIRCULATION

CONTENTS

11 Solar radiation absorbed in the tropics exceeds infrared loss, causing heat to accumulate. The opposite is true at the poles, where there is net radiative cooling. Such **radiative differential heating** between poles and equator creates an imbalance in the atmosphere/ocean system (Fig 11.1a).

As expected by LeChatelier's Principle, this unstable system reacts to undo the imbalance (Fig 11.1b). Warm tropical air rises due to buoyancy and flows toward the poles, and the cold polar air sinks and flows toward the equator.

Because the radiative forcings are unremitting, a ceaseless movement of wind and ocean currents results in what we call the **general circulation** or **global circulation**. The circulation cannot keep up with the continued destabilization, causing the tropics to remain slightly warmer than the poles.

The atmosphere is said to be **baroclinic** where there is a horizontal temperature gradient. This gradient creates the jet stream at midlatitudes. Baroclinic instabilities cause the jet stream to meander and cause the eddy-like storm systems and waves that are associated with midlatitude weather.

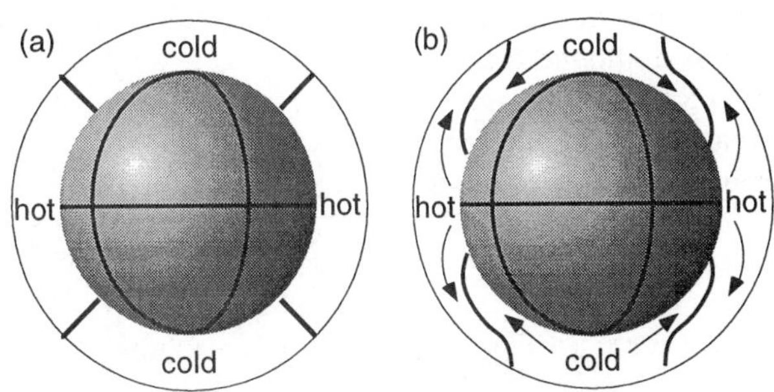

Figure 11.1
Earth/atmosphere system, showing that (a) differential heating by radiation causes (b) atmospheric circulations.

NOMENCLATURE

Latitude lines are **parallels**, and east-west winds are called **zonal** flow (Fig 11.2). Longitude lines are **meridians**, and north-south winds are called **meridional** flow.

Midlatitudes are the region between about 30° and 60° latitude. The **subtropical** zone is at roughly 30° latitude, and the **subpolar** zone is at 60° latitude, both of which partially overlap midlatitudes. **Tropics** span the equator, and **polar** regions are near the earth's poles. **Extratropical** refers to everything poleward from roughly 30°N and 30°S. Similar terminology is used in the southern hemisphere.

For example, extratropical cyclones are low-pressure centers, that are typically called lows and labeled with L, found in midlatitudes or high latitudes (> 60° N or S, roughly). Tropical cyclones include hurricanes and typhoons, and other lows in tropical regions.

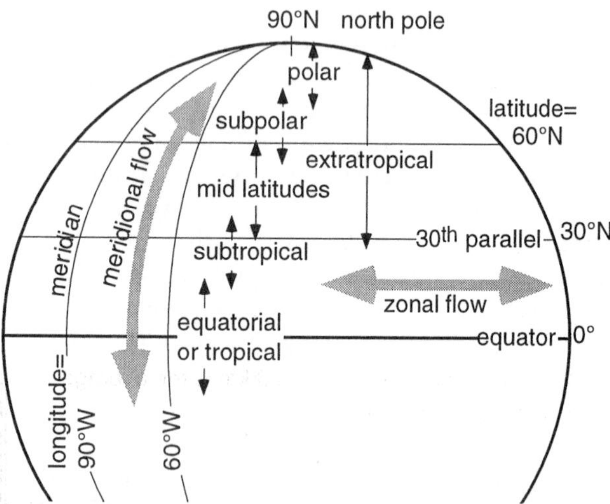

Figure 11.2
Global nomenclature.

DIFFERENTIAL HEATING

The balance between incoming and outgoing radiation drives the global circulation. Outgoing infrared radiation depends on the earth's temperature, as discussed in Chapter 2. Because the general circulation cannot completely eliminate the temperature disparity across the globe, there is a residual north-south temperature gradient that we will examine first.

Meridional Temperature Gradient

Sea-level temperatures are warmer near the equator than the poles, when averaged over a whole year and averaged along lines of constant latitude (Fig 11.3a). Although the Northern Hemisphere is slightly cooler than the Southern, an idealization of the meridional temperature variation at sea level is

$$T_{sea\ level} \approx a + b \cdot \left[\frac{3}{2} \cdot \left(\frac{2}{3} + \sin^2 \phi \right) \cdot \cos^3 \phi \right] \qquad (11.1)$$

where $a \cong -12°C$, $b \cong 40°C$, and ϕ is latitude. Parameter b represents a temperature difference between equator and pole, so it could also have been written as $b \cong 40$ K.

In the tropics there is little temperature variation – it is hot everywhere. However, in midlatitudes, there is a significant north-south temperature gradient that supports a variety of storm systems. Although eq. (11.1) might seem unnecessarily complex, it is designed to give sufficient uniformity in the tropics to support midlatitude jet streams.

The gradient associated with eq. (11.1) is

$$\frac{\Delta T}{\Delta y} \approx -b \cdot c \cdot \left[\sin^3 \phi \cdot \cos^2 \phi \right] \qquad (11.2)$$

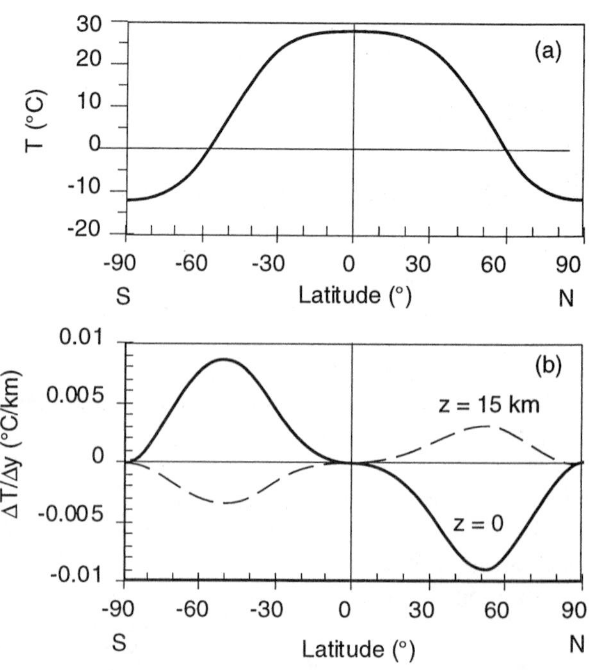

Figure 11.3
Annually and zonally averaged (a) temperature at sea level (idealized), and (b) north-south temperature gradient at sea level and at 15 km altitude.

where $c = 1.18\text{x}10^{-3}$ km^{-1}, and $b \cong 40°C$ at sea level, as before. The sea-level meridional temperature gradient is plotted in Fig 11.3b.

The magnitude of the north-south temperature gradient decreases with altitude, and changes sign in the stratosphere. Namely, in the stratosphere it is cold in the tropics and warmer in the polar regions. For this reason, parameter b in eqs. (11.1) and (11.2) can be approximated by

$$b \approx b_1 \cdot \left(1 - \frac{z}{z_T}\right) \qquad (11.3)$$

where $b_1 = 40°C$, z is height above sea level, and $z_T \cong 11$ km is the average depth of the troposphere. With this change to eq. (11.2), the horizontal temperature gradient at $z = 15$ km is also plotted in Fig 11.3b.

Solved Example

Find the meridional temperature and temperature gradient at 45°N latitude, at sea level.

Solution

Given: $\phi = 45°$

Find: $T = ?$ °C, $\Delta T/\Delta y = ?$ °C/km

Use eq. (11.1):

$$T_{sl} \approx -12°C + (40°C) \cdot \left[\frac{3}{2} \cdot \left(\frac{2}{3} + \sin^2 45°\right) \cdot \cos^3 45°\right]$$

$= \underline{12.75\ °C}$

Use eq. (11.2):

$$\frac{\Delta T}{\Delta y} \approx -(40°C) \cdot (1.18 \times 10^{-3}) \cdot \left[\sin^3 45° \cdot \cos^2 45°\right]$$

$= \underline{-0.0083\ °C/km}$

Check: Units OK. Physics OK. Agrees with Figs.
Discussion: Temperature decreases toward the north in the northern hemisphere, which gives the negative sign for the gradient. As a quick check, from Fig 11.3a the temperature decreases by about 9°C between 40°N to 50°N latitude. But each 1° of latitude equals 111 km of distance, y. This temperature gradient of –9°C/(1110 km) agrees with the answer above.

BEYOND ALGEBRA • Temperature Gradient

Problem: Derive eq. (11.2) from eq. (11.1).

Solution: Given: $T \approx a + b \cdot \left[\frac{3}{2} \cdot \left(\frac{2}{3} + \sin^2 \phi\right) \cdot \cos^3 \phi\right]$

Find: $\frac{\partial T}{\partial y} = \frac{\partial T}{\partial \phi} \cdot \frac{\partial \phi}{\partial y}$ (a)

The first factor on the RHS is: $\frac{\partial T}{\partial \phi} = b \cdot \left(\frac{3}{2}\right) \cdot$

$$\left[(2\sin\phi \cdot \cos\phi)\cos^3\phi - 3\left(\frac{2}{3} + \sin^2\phi\right)\cos^2\phi \cdot \sin\phi\right]$$

Taking $\sin\phi \cdot \cos^2\phi$ out of the square brackets:

$$\frac{\partial T}{\partial \phi} = b\left(\frac{3}{2}\right)\sin\phi \cdot \cos^2\phi \cdot \left[2\cos^2\phi - 2 - 3\sin^2\phi\right]$$

But $\cos^2\phi = 1 - \sin^2\phi$, thus $2\cos^2\phi = 2 - 2\sin^2\phi$:

$$\frac{\partial T}{\partial \phi} = b\left(\frac{3}{2}\right)\sin\phi \cdot \cos^2\phi \cdot \left[-5\sin^2\phi\right] \qquad (b)$$

Inserting eq. (b) into (a) gives:

$$\frac{\partial T}{\partial y} = -b \cdot \left(\frac{15}{2} \cdot \frac{\partial \phi}{\partial y}\right) \cdot \sin^3\phi \cdot \cos^2\phi \qquad (c)$$

Define: $\left(\frac{15}{2} \cdot \frac{\partial \phi}{\partial y}\right) \equiv c$ Thus, the final answer is:

$$\boxed{\frac{\partial T}{\partial y} = -b \cdot c \cdot \sin^3\phi \cdot \cos^2\phi} \qquad (11.2)$$

All that remains is to find c. Note that $\partial\phi / \partial y$ is the change of latitude per distance traveled north. The total change in latitude to circumnavigate the earth from pole to pole is 2π radians, and the circumference of the earth is $2\pi R$ where $R = 6356.766$ km is the average radius of the earth. Thus:

$$\frac{\partial \phi}{\partial y} = \frac{2\pi}{2\pi \cdot R} = \frac{1}{R}$$

The constant c is then:

$$c = \left(\frac{15}{2} \cdot \frac{1}{R}\right) = 1.18 \times 10^{-3}\,\text{km}^{-1} \qquad (e)$$

Check: The neighboring solved example verified that eq. (11.2) is a reasonable answer, given the temperature defined by eq. (11.1).

Caution: Eq. (11.1) is only a crude approximation to nature, designed to be convenient for analytical calculations of temperature gradients, thermal winds, IR emissions, etc. While it serves an education purpose here, more accurate models should be used elsewhere.

Radiative Forcings

Idealize the zonally- and annually-averaged outgoing radiative flux E_{out} by a truncated curve:

$$E_{out} \approx \varepsilon \cdot \sigma \cdot T_{sea\ level}^4 \qquad (11.4a)$$

with

$$E_{out} \le 260 \quad \mathrm{W/m^2} \qquad (11.4b)$$

where $\sigma = 5.67\text{x}10^{-8}$ W·m^{-2}·K^{-4} is the Stefan-Boltzmann constant, and $\varepsilon \cong 0.7$ is an effective emissivity that crudely compensates for emission from altitudes above the surface. The reason for truncating the top of the curve is because extensive cloudiness in the tropics limits the emissions there.

Incoming radiative flux E_{in}, averaged zonally and annually, is idealized as

$$|E_{in}| \approx |E_1| \cdot \cos(\phi) \qquad (11.5a)$$

with

$$|E_{in}| \le 310 \quad \mathrm{W/m^2} \qquad (11.5b)$$

where $|E_1| \cong 315$ W/m^2.

Solved Example(§)

Plot zonally-averaged incoming and outgoing radiation vs. latitude, for each 5° of latitude.

Solution

Find: E_{in} & E_{out} (W/m^2) vs. ϕ (°)

Use eqs. (11.1), (11.4), and (11.5) in a spreadsheet, a portion of which is shown below:

ϕ (°)	ϕ (rad)	E_{in} (W/m^2)	T_{sfc} (°C)	E_{out} (W/m^2)
90	1.571	0.0	-12.0	184.2
85	1.484	27.5	-11.9	184.4
80	1.396	54.7	-11.5	185.6
75	1.309	81.5	-10.3	188.9
etc.				

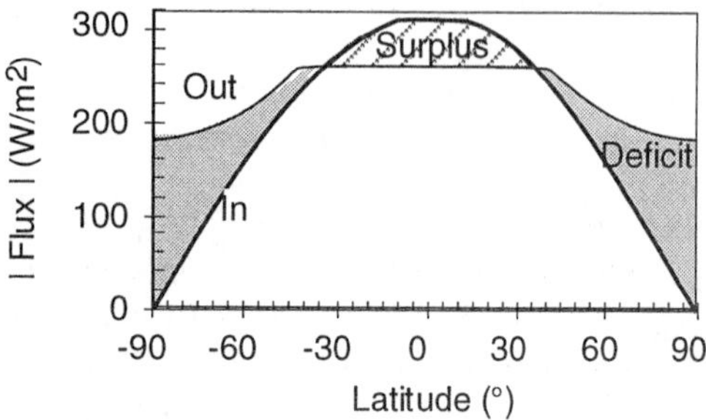

Check: Units OK. Physics OK.
Discussion: In tropical regions where incoming radiation exceeds outgoing, there is a surplus of energy. Similarly, near the poles there is a deficit. This is the "differential heating" between the equator and poles.

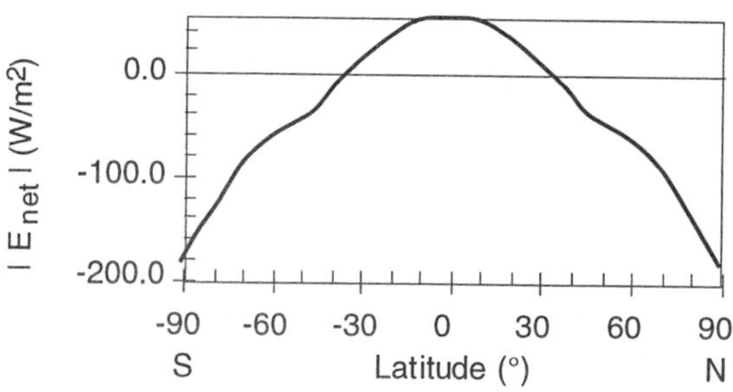

Figure 11.4
Magnitude of net radiative flux into the troposphere.

The net radiative heat flux per vertical column of atmosphere is simply

$$E_{net} = |E_{in}| - E_{out} \qquad (11.6)$$

which is plotted in Fig 11.4.

Eq. (11.6) and Fig 11.4 can be deceiving, because latitude belts have shorter circumference near the poles than near the equator. Namely, there are fewer square meters near the poles that experience the net deficit than there are near the equator that experience the net surplus.

To compensate for the shrinking latitude belts, multiply the radiative flux by the circumference of the belt $2\pi \cdot R_{earth} \cdot \cos(\phi)$, to give a more appropriate measure of heating vs. latitude:

$$E_\phi = 2\pi \cdot R_{earth} \cdot \cos(\phi) \cdot E \qquad (11.7)$$

where E is the magnitude of incoming or outgoing radiative flux, and R_{earth} = 6357 km is the average earth radius. These E_ϕ values are plotted in Fig 11.5.

The difference D_ϕ between incoming and outgoing values of E_ϕ is plotted in Fig 11.6.

$$D_\phi = E_{\phi\ in} - E_{\phi\ out} \qquad (11.8)$$

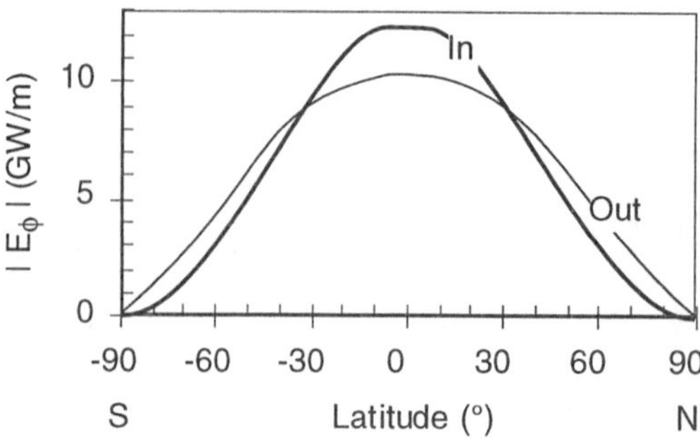

Figure 11.5
Magnitude of radiation accumulated around latitude belts.

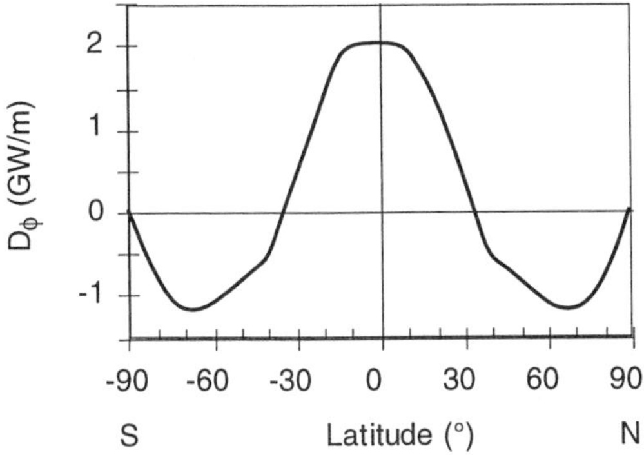

Figure 11.6
Difference between incoming and outgoing radiation, accumulated around latitude belts. This is the "differential heating."

To interpret this curve, picture a sidewalk built around the world along a latitude line. If this sidewalk is 1 m wide, then Fig 11.6 gives the number of gigawatts of net radiative power absorbed by the sidewalk. This curve shows the radiative **differential heating**. The areas under the positive and negative portions of the curve in Fig 11.6 balance, leaving the earth in overall equilibrium.

ON DOING SCIENCE • Toy Models

Some problems in meteorology are so complex, and involve so many interacting variables and processes, that they are intimidating if not impossible to solve. However, it is sometimes possible to gain insight into fundamental aspects of the problem by approximating the true physics by idealized, simplified physics. Such an approximation is sometimes called a **toy model**.

The meridional variation of temperature given by eq. (11.1) is an example of a toy model. It was designed to capture only the dominant temperature variations in a way that could be used to analytically calculate temperature gradients, energy balances and geostrophic winds. It is a model of the atmosphere — a toy model, not a complete model.

Toy models are used extensively to study climate change. For example, the greenhouse effect is examined using toy models in the last chapter of this book. Toy models capture only the dominant effects, and neglect the subtleties. They should never be used to infer the details of a process, particularly in situations where two large but opposite processes nearly cancel each other.

For other examples of toy models applied to the environment, see John Harte's 1988 book *Consider a Spherical Cow*, University Science Books. 283 pp.

Heat Transport Needed

The radiative imbalance in Fig 11.6 must be compensated by atmospheric and oceanic circulations. The net meridional transport is zero at the poles by definition. Starting at the north pole, and summing over all the "sidewalks" down to any latitude Φ of interest gives the total atmospheric and oceanic northward transport Tr needed to compensate the radiation:

$$Tr = \sum_{\phi=90°}^{\Phi} (-D_\phi) \cdot \Delta y \qquad (11.9)$$

where Δy is the width of the sidewalk.

Winds transport some of this heat. These winds are generated by the horizontal temperature gradient, as described next.

Solved Example(§)

Plot the net atmospheric and oceanic northward transport Tr vs. latitude Φ.

Solution

Find: Tr (W) vs. Φ (°).

Start with the spreadsheet results from the previous solved example. Then include eqs. (11.7 to 11.9). A sidewalk width of 5° latitude was used in the previous spreadsheet, which corresponds to Δy = 555 km.

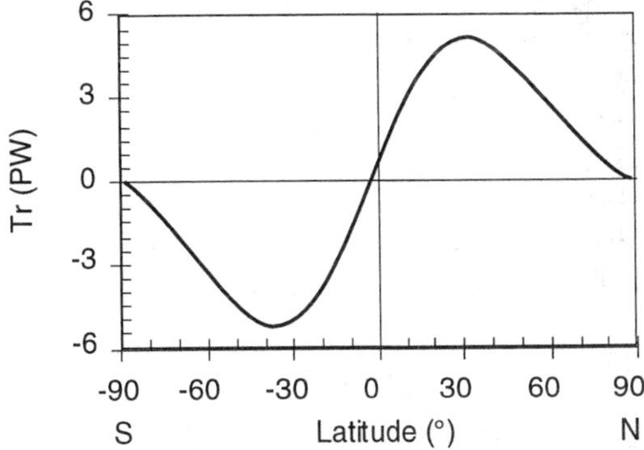

Check: Units OK. Physics OK. Curve good.
Discussion: (1 PW = 1 petawatt = 10^{15} W).

Interpretation: At 30°N, the northward transport by fluid circulations is roughly 5 PW, while at 60°N it is about 2.5 PW. Hence, more energy is flowing into the midlatitudes than out, causing a net heating that compensates radiative cooling.

THERMAL WIND RELATIONSHIP

As sketched in Fig 11.7, there is a greater thickness between pressure surfaces in warm air than in cold (see eq. 1.18, the hypsometric equation). If a horizontal temperature gradient is present, it changes the tilt of pressure surfaces with increasing altitude. But geostrophic wind is proportional to the tilt of the pressure surfaces (eq. 9.20). Thus, geostrophic wind changes with altitude when temperature changes in the horizontal.

The relationship between horizontal temperature gradient and vertical gradient of geostrophic wind (U_g, V_g) is called the **thermal wind relationship**:

$$\frac{\Delta U_g}{\Delta z} \approx \frac{-g}{T_v \cdot f_c} \cdot \frac{\Delta T_v}{\Delta y} \qquad \bullet(11.10a)$$

$$\frac{\Delta V_g}{\Delta z} \approx \frac{g}{T_v \cdot f_c} \cdot \frac{\Delta T_v}{\Delta x} \qquad \bullet(11.10b)$$

where $g = 9.8 \text{ m·s}^{-2}$ is gravitational acceleration, T_v is the virtual temperature (in Kelvins, and nearly equal to the actual temperature if the air is fairly dry), and f_c is the Coriolis parameter. Note that the north-south temperature gradient alters the east-west geostrophic winds with height, and vice versa.

As it turns out, most of the atmosphere is nearly in geostrophic equilibrium. Thus, the change of actual wind speed with height is nearly equal to the change of the geostrophic wind.

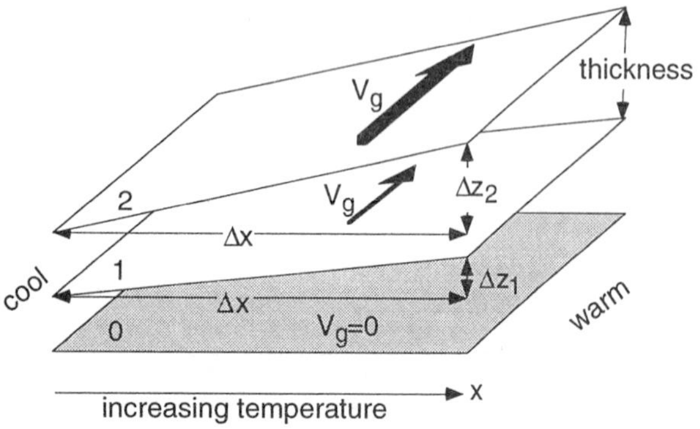

Figure 11.7
The three planes are surfaces of constant pressure (i.e., isobaric surfaces). Surface #2 has lower pressure than surface #1, etc. A horizontal temperature gradient tilts the pressure surfaces and causes the geostrophic wind to increase with height.

Solved Example

Suppose temperature increases from 8°C to 12°C toward the east, across a distance of 100 km (such as plotted in Fig 11.7). Find the vertical gradient of geostrophic wind, given $f_c = 10^{-4} \text{ s}^{-1}$.

Solution

Assume: Dry air. Thus, $T_v = T$

Given: $\Delta T = 12 - 8°C = 4°C$, $\Delta x = 100$ km,
$T = 0.5·(8+12°C) = 10°C = 283$ K, $f_c = 10^{-4} \text{ s}^{-1}$.

Find: $\Delta U_g/\Delta z$ & $\Delta V_g/\Delta z$ = ? (m/s)/km

Use eq. (11.10a):

$$\frac{\Delta U_g}{\Delta z} \approx \frac{-(9.8\text{m/s}^2)}{(283K)\cdot(10^{-4}\text{s}^{-1})} \cdot (0°\text{C/km}) = \underline{0}$$

Thus, U_g = uniform with height.

Use eq. (11.10b):

$$\frac{\Delta V_g}{\Delta z} \approx \frac{(9.8\text{m/s}^2)}{(283K)\cdot(10^{-4}\text{s}^{-1})} \cdot \frac{(4°\text{C})}{(100\text{km})}$$

$= \underline{13.9 \text{ (m/s)/km}}$

Check: Units OK. Physics OK. Agrees with Fig.
Discussion: Each kilometer gain in altitude gives a 13.9 increase in northward geostrophic wind speed. For example, if the wind at the surface is –3.9 m/s (i.e., light from the north), then the wind at 1 km altitude is 10 m/s (strong from the south).

Thickness

The **thickness** between two different pressure surfaces is a measure of the average virtual temperature within that layer. For example, consider the two different pressure surfaces colored white in Fig 11.7. The air is warmer in the east than in the west. Thus, the thickness of that layer is greater in the east than in the west.

Maps of thickness are used in weather forecasting. One example is the "100 to 50 kPa thickness" chart, such as shown in Fig 11.8. This is a map with contours showing the thickness between the 100 kPa and 50 kPa isobaric surfaces. For such a map, regions of low thickness correspond to regions of cold temperature, and vise versa. It is a good indication of average temperature in the bottom half of the troposphere, and is useful for identifying air masses and fronts (discussed in the next chapter).

Define thickness TH as

$$TH = z_{P1} - z_{P2} \qquad (11.11)$$

where z_{P2} and z_{P1} are the heights of the P_2 and P_1 isobaric surfaces.

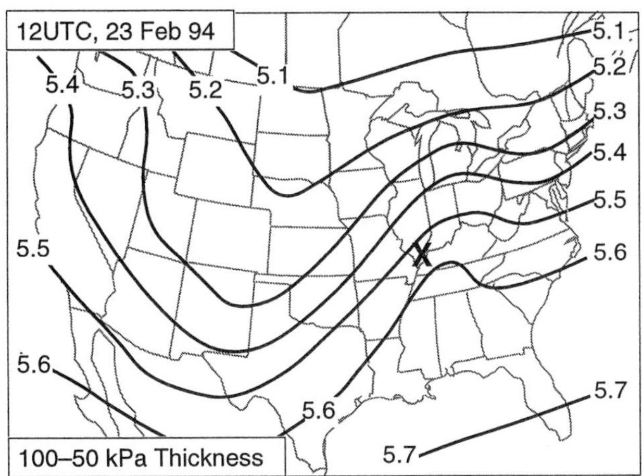

Figure 11.8
100–50 kPa thickness (km), valid at 12 UTC on 23 Feb 94. X marks sfc. low.

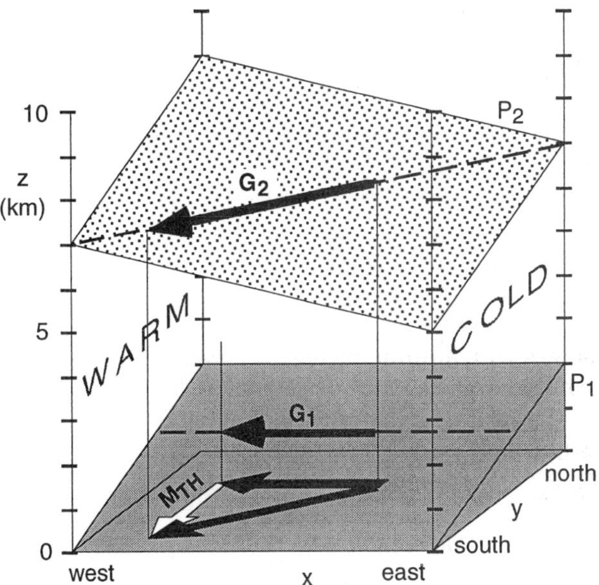

Figure 11.9
Given an isobaric surface P_1 (gray), with a height contour shown as the dashed line. The geostrophic wind G_1 on the surface is parallel to the height contour, with low heights to the left (N. Hem.). Isobaric surface P_2 (for $P_2 < P_1$) is also plotted (stippled). Cold air to the east is associated with a thickness of TH = 5 km between the two pressure surfaces. To the west, warm air has thickness 7 km. These thicknesses are added to the bottom pressure surface, to give the corners of the upper surface. On that upper surface are shown a height contour (dashed line) and the geostrophic wind vector G_2. We see that $G_2 > G_1$ because the top surface has greater slope than the bottom. The thermal wind is parallel to the thickness contours with cold to the left (white arrow). This thermal wind is the vector difference between the geostrophic winds at the two pressure surfaces, as shown in the projection on the ground.

Thermal Wind

The thermal wind relationship can be applied over the same layer of air bounded by isobaric surfaces as was used to define thickness. This gives:

$$U_{TH} \equiv U_{g2} - U_{g1} = -\frac{g}{f_c}\frac{\Delta TH}{\Delta y} \qquad \bullet(11.12a)$$

$$V_{TH} \equiv V_{g2} - V_{g1} = +\frac{g}{f_c}\frac{\Delta TH}{\Delta x} \qquad \bullet(11.12b)$$

where subscripts $g2$ and $g1$ denote the geostrophic winds on the P_2 and P_1 pressure surfaces, g is gravitational acceleration, and f_c is the Coriolis parameter.

The variables U_{TH} and V_{TH} are known as the **thermal wind** components. Taken together (U_{TH}, V_{TH}) they represent the vector difference between the geostrophic winds at the top and bottom pressure surfaces. Thermal wind magnitude M_{TH} is:

$$M_{TH} = \sqrt{U_{TH}^{\,2} + V_{TH}^{\,2}} \qquad (11.13)$$

In Fig 11.7, this vector difference happened to be in the same direction as the geostrophic wind. But in general, this is not usually the case, as illustrated in Fig 11.9.

Eqs. (11.12) imply that the thermal wind is parallel to the thickness contours, with cold temperatures (low thickness) to the left in the N. Hemisphere (Fig 11.10). Closer packing of the thickness lines gives stronger thermal winds, because the horizontal temperature gradient is larger there.

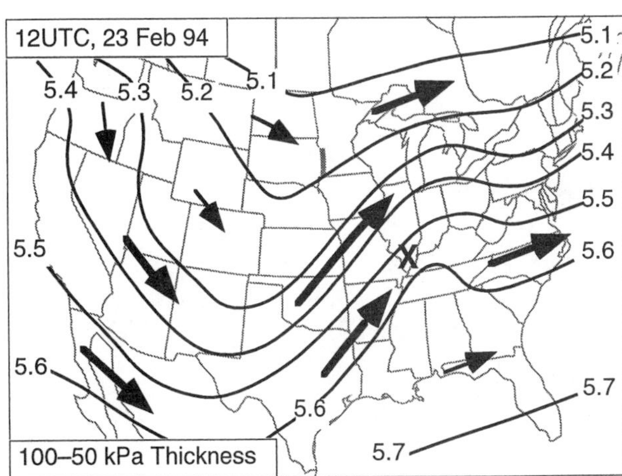

Figure 11.10
100–50 kPa thickness (km), with thermal wind vectors added. Larger vectors qualitatively denote stronger thermal winds.

Thermal winds on a thickness map behave analogously to geostrophic winds on a constant pressure or height map, making their behavior a bit easier to remember. However, while it is possible for actual winds to equal the geostrophic wind, there is no real wind that equals the thermal wind. The thermal wind is just the vector difference between geostrophic winds at two different heights or pressures.

Solved Example

Suppose the thickness of the 100 - 70 kPa layer is 2.9 km at one location, and 3.0 km at a site 500 km to the east. Find the components of the thermal wind vector, given $f_c = 10^{-4}\ s^{-1}$.

Solution

Assume: No north-south thickness gradient.
Given: $TH_1 = 2.9$ km, $TH_2 = 3.0$ km,
$\Delta x = 500$ km, $f_c = 10^{-4}\ s^{-1}$.
Find: $U_{TH} = ?$ m/s, $V_{TH} = ?$ m/s

Use eq. (11.12a): $U_{TH} = \underline{\mathbf{0\ m/s}}$.
Use eq. (11.12b):

$$V_{TH} = \frac{g}{f_c}\frac{\Delta TH}{\Delta x} = \frac{(9.8\text{ms}^{-2})\cdot(3.0-2.9)\text{km}}{(10^{-4}\text{s}^{-1})\cdot(500\text{km})}$$
$$= \underline{\mathbf{19.6\ m/s}}$$

Check: Units OK. Physics OK.
Discussion: There is no east-west thermal wind component because the thickness does not change in the north-south direction. The positive sign of V_{TH} means a wind from south to north, which agrees with the rule that the thermal wind is parallel to the thickness contours with cold air to the left (west, in this example).

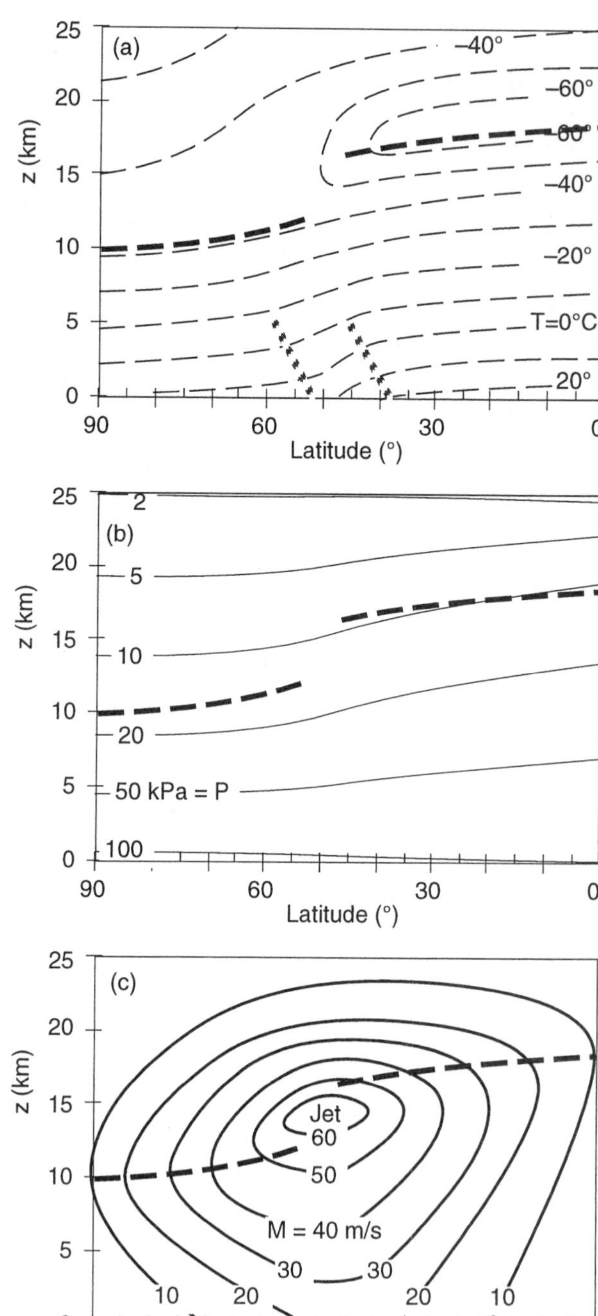

Figure 11.11
Vertical slices through the N. Hemisphere atmosphere: (a) isotherms, (b) isobars, (c) isotachs. Heavy dashed line marks the tropopause, and a frontal zone is between the dotted lines in (a). Wind direction in (c) is from the reader into the page. These figures can be overlain.

JET STREAM

Baroclinicity (i.e., the north-south temperature gradient) drives the west-to-east winds near the top of the troposphere, via the thermal wind relationship. Angular momentum issues will also be explored.

Baroclinicity

Fig 11.11a shows a typical isotherm pattern. Air near the ground is warmer near the equator, colder at the poles, and there is a frontal zone at mid-latitudes where temperature decreases rapidly toward the north. This north-south temperature gradient exists throughout the troposphere.

However, the tropopause is lower near the poles than near the equator. Thus, temperature begins increasing with height at a lower altitude near the poles than near the equator. This causes a temperature reversal in the stratosphere, where the air is colder over the equator and warmer over the poles. The spatial distribution of temperature is called the **temperature field**.

Apply the thermal wind equation to the temperature field to give the **pressure field** (the spatial distribution of pressures, Fig 11.11b). In the troposphere, greater thickness between pressure surfaces in the warmer equatorial air than the colder polar air causes the isobars to become more tilted at mid-latitudes as the tropopause is approached. Above the tropopause, tilt decreases because the north-south temperature gradient is reversed.

Regions with the greatest tilt have the greatest south-to-north pressure gradient, which drives the fastest geostrophic wind (Fig. 11.11c). This maximum of westerly winds, known as the **jet stream**, occurs at the tropopause in mid-latitudes. The center of the jet stream, known as the **core**, occurs in a region that can be idealized as a break or fold in the tropopause, as will be discussed in Chapter 12. Moderately strong winds also occur near the frontal zone.

The temperature model presented in eq. (11.1) and Fig 11.3 is a starting point for quantifying the nature of the jet stream. Equations (11.2), (11.3) and (11.10) can be combined to give the wind speed as a function of latitude ϕ and altitude z, where it is assumed for simplicity that the winds near the ground are zero:

$$U_{jet} \approx \frac{g \cdot c \cdot b_1}{2\Omega \cdot T_v} \cdot z \cdot \left(1 - \frac{z}{2 \cdot z_T}\right) \cdot \cos^2(\phi) \cdot \sin^2(\phi) \qquad (11.14)$$

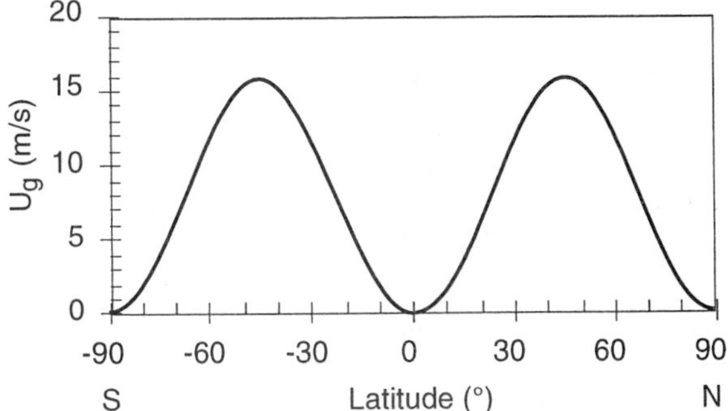

Figure 11.12
Geostrophic winds at 11 km altitude (idealized).

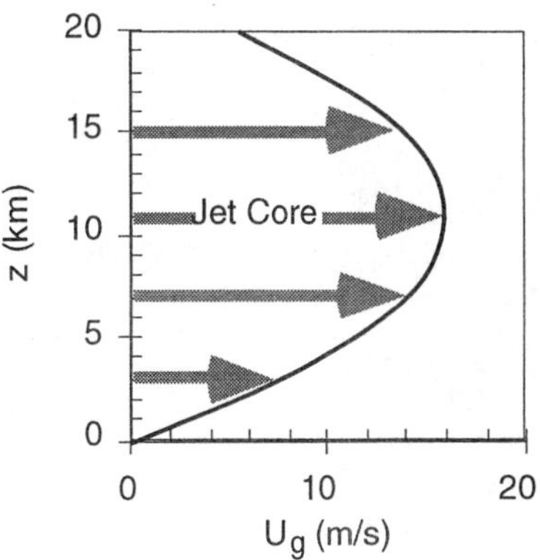

Figure 11.13
Vertical profile of (idealized) geostrophic wind at 45° latitude.

where U_{jet} is used in place of U_g to indicate a specific application of geostrophic wind to the polar jet stream, and where $c = 1.18\text{x}10^{-3}$ km^{-1} and $b_1 \cong 40$ K as before. The definition of the Coriolis parameter eq. (9.10) $f_c = 2 \cdot \Omega \cdot \sin\phi$ was used, where $2 \cdot \Omega =$ $1.458\text{x}10^{-4}$ s^{-1}. The average troposphere depth is $z_T \cong 11$ km, gravitation acceleration is $g = 9.8$ m/s^2, and T_v is the average virtual temperature (in Kelvins).

From Fig 11.3b we see there are two extrema of north-south temperature gradient, one in the northern hemisphere and one in the southern. Those are the latitudes where we can anticipate to find the strongest jet velocities (Fig 11.12). Although the temperature gradient in the southern hemisphere has a sign opposite to that in the north, the sign of the Coriolis parameter also changes. Thus, the jet stream velocity is positive (west to east) in both hemispheres.

A spreadsheet solution of eq. (11.11) is shown in Figs 11.12 to 11.14. At the tropopause (z = 11 km), Fig 11.12 shows the jet-stream velocity, with maxima at 45° north and south latitudes. For 45° latitude, Fig 11.13 shows the vertical profile of geostrophic wind speed, with the fastest speeds (i.e., the core of the jet) at the tropopause. Fig 11.14 shows a vertical cross-section (altitude vs. latitude).

The gross features shown in this figure are indeed observed in the atmosphere. However, actual spacing between jets in the northern and southern hemispheres is less than 90°; it is roughly 70° or 75° latitude difference. Also, the centers of both jets shift slightly northward during northern-hemisphere summer, and southward in winter. Jet speeds are not zero at the surface, and the jets tend to spread slightly toward the poles. Nevertheless, the empirical model presented here via eq. (11.1) serves as an instructive first introduction.

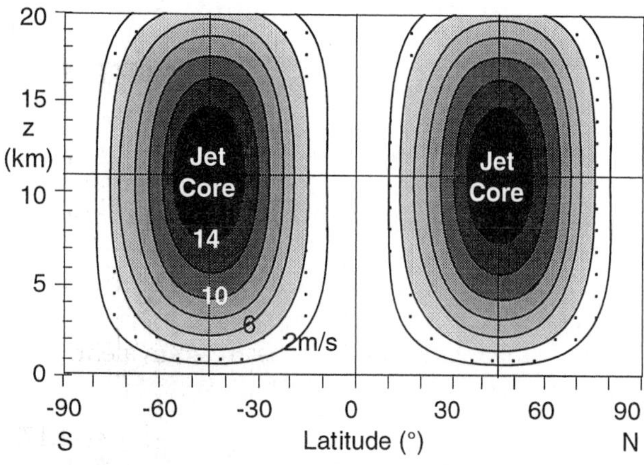

Figure 11.14
Idealized vertical cross-section of the atmosphere, showing isotachs.

Solved Example

Find the geostrophic wind speed at the tropopause at 45° latitude, assuming an idealized temperature structure as plotted in Fig 11.3. Assume an average temperature of 0°C, and neglect moisture.

Solution

Given: $z = z_T = 11$ km, $\phi = 45°$, $T_v = 273$ K.
Find: $U_g = ?$ m/s

Use eq. (11.14):

$$U_{jet} \approx \frac{(9.8\text{m}\cdot\text{s}^{-2})\cdot(1.18\times10^{-3}\text{km}^{-1})\cdot(40\text{K})}{(1.458\times10^{-4}\text{s}^{-1})\cdot(273\text{K})}\cdot$$

$$(11\text{km})\cdot\left(1-\frac{1}{2}\right)\cdot\cos^2(45°)\cdot\sin^2(45°)$$

= <u>16 m/s</u>

Check: Units OK. Physics OK.
Discussion: Actual average wind speeds of about 40 m/s are observed in the jet stream of the winter hemisphere over a three-month average. Sometimes jet velocities as great as 100 m/s are observed on individual days. In the winter hemisphere, the temperature gradient is often greater than that given by eq. (11.3), which partially explains our low wind speed.

Angular Momentum

Another factor affecting the jet stream is conservation of angular momentum. This approach has serious deficiencies, which we will discuss.

Consider air initially at rest at some initial latitude such as the equator; namely, it is moving eastward with the earth's surface as the earth rotates about its axis. If there were no forces acting on the air, then it would preserve its eastward **angular momentum** as it moves northward from a source latitude ϕ_s to a destination latitude ϕ_d (Fig 11.15), then:

$$m\cdot U_s\cdot R_s = m\cdot U_d\cdot R_d \qquad \bullet(11.15)$$

where m is the mass of air, U_s is the tangential velocity of the earth at the source latitude, and U_d is the tangential velocity of the air from the source after it gets to the destination. The radius from the earth's axis to the surface at latitude ϕ is $R_\phi = R_e\cdot\cos(\phi)$, where R_ϕ represents source or destination radius (R_s or R_d), and $R_e = 6357$ km is the earth radius.

The tangential velocity of the earth at latitude ϕ is $U_\phi = \Omega\cdot R_\phi = \Omega\cdot R_e\cdot\cos(\phi)$, where the rate of rotation is $\Omega = 0.729\text{x}10^{-4}\ \text{s}^{-1}$. When an air parcel moves from a source to a destination, its velocity U' relative to the earth is

$$U' \approx \Omega\cdot R_e\cdot\left[\frac{\cos^2\phi_s}{\cos\phi_d}-\cos\phi_d\right] \qquad (11.16)$$

For example, Fig 11.16 shows the U' values of air reaching a destination of 45° latitude from sources at other latitudes.

Jet-stream winds of 100 - 300 m/s predicted by this method are not routinely observed on earth. In the real atmosphere, angular momentum is <u>not</u> conserved because pressure gradient, Coriolis, and turbulent drag forces act on the air.

For example, as a poleward moving parcel accelerates toward the east, increasing Coriolis force would turn it back toward the equator. Such turning causes wind to converge and alter the pressure

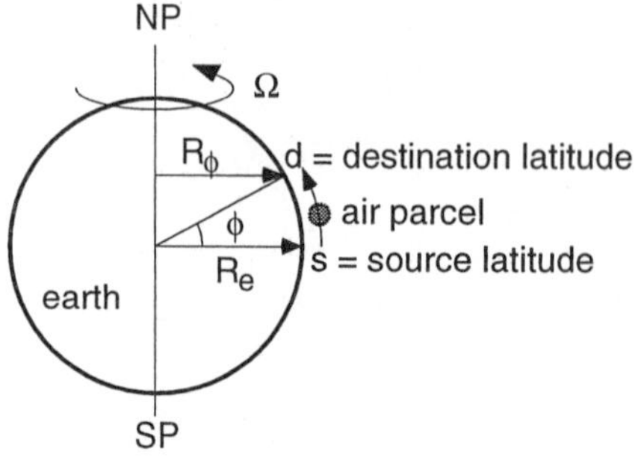

Figure 11.15
Northward air parcel movement.

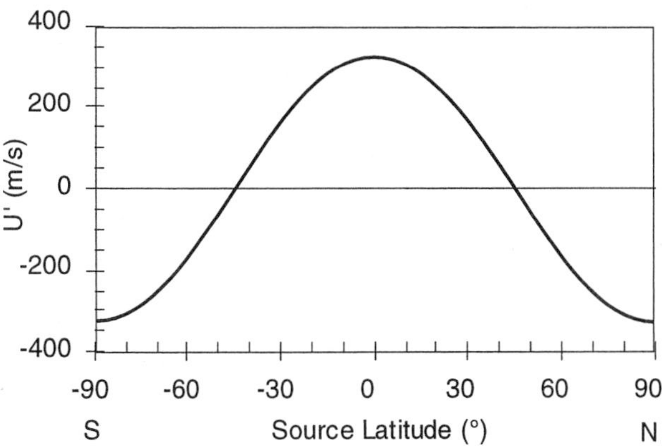

Figure 11.16
Relative wind speed at destination 45° latitude of air that came from various source latitudes.

gradient. The action of pressure-gradient, Coriolis, and turbulent drag forces prevents angular momentum from being conserved, and is one reason why the equator-to-pole circulation pattern of Fig 11.1 does not happen in the real atmosphere. The actual vertical circulation pattern extends only to about 30° north and south and is called the **Hadley cell**, as will be explained at the end of this chapter.

Poleward of 30° latitude, heat is still transferred poleward; however, this transfer is primarily by north-south meanders (**planetary waves**) of the jet stream rather than by a direct vertical circulation. Planetary waves are explained in the middle of this Chapter, using the concept of vorticity.

Solved Example

In the free atmosphere (no turbulence), what would be the relative velocity at 30°N for an air parcel that moved northward from the equator?

Solution

Given: $\phi_d = 30°N$, $\phi_s = 0°$. No turbulence.
Find: $U' = ?$ m/s

Use eq. (11.16):

$$U' \approx (7.29\times10^{-5} s^{-1})\cdot(6.357\times10^{6} m)\cdot \left[\frac{1}{\cos(30°)} - \cos(30°)\right] = \underline{\mathbf{134\ m/s}}$$

Check: Units OK. Physics OK.
Discussion: Winds in the real atmosphere are much less than this, where roughly 95% of the reduction is thought to be due to pressure gradient force, and 5% due to turbulent drag. These forces prevent conservation of angular momentum.

VORTICITY

Relative Vorticity

Relative vorticity ζ_r is a measure of the rotation of fluids about a vertical axis relative to the earth's surface. It is defined as positive in the counterclockwise direction. The unit of measurement of vorticity is inverse seconds.

The following two definitions are equivalent:

$$\zeta_r = \frac{\Delta V}{\Delta x} - \frac{\Delta U}{\Delta y} \qquad \bullet(11.17)$$

$$\zeta_r = -\frac{\Delta M}{\Delta n} + \frac{M}{R} \qquad \bullet(11.18)$$

where (U, V) are the (eastward, northward) components of the wind velocity, R is the radius of curvature traveled by a moving air parcel, M is the tangential speed along that circumference in a counterclockwise direction, and n is the direction pointing inward toward the center of curvature.

A physical interpretation for the first equation is sketched in Fig 11.17. If fluid travels along a straight channel, but has shear, then it also has vorticity because a tiny paddle wheel carried by the fluid would be seen to rotate. The second equation is interpreted in Fig 11.18, where fluid following a curved path also has vorticity, so long as radial shear of the tangential velocity does not cancel it.

A special case of the last equation is where the radial shear is just great enough so that winds at different radii sweep out identical angular velocities about the center of curvature. In other words, the fluid rotates as a solid body. For this case, the last equation reduces to

$$\zeta_r = \frac{2M}{R} \qquad \bullet(11.19)$$

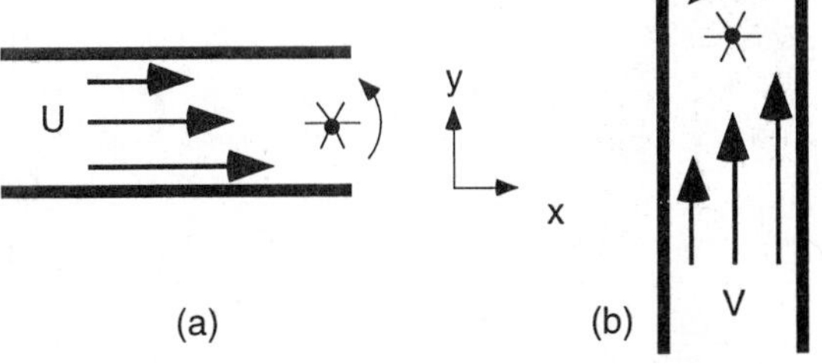

Figure 11.17
Shear-induced relative vorticity. Counter-clockwise turning of the paddle wheels indicates positive vorticity for both examples. (a) $\Delta U/\Delta y$ is negative. (b) $\Delta V/\Delta x$ is positive.

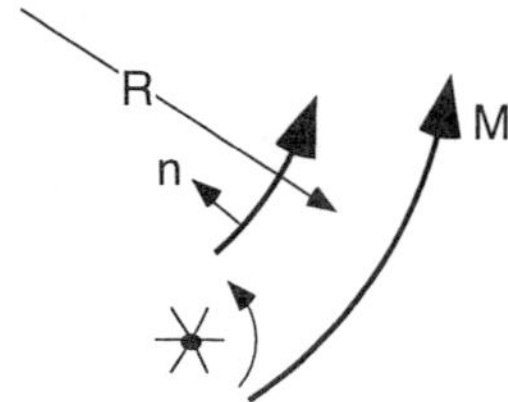

Figure 11.18
Vorticity for flow around a curve, where both the curvature and the shear contribute to positive vorticity in this example (i.e., the paddle wheel rotates counter-clockwise).

> **Solved Example**
> If wind rotates as a solid body about the center of a low pressure system, and the tangential velocity is 10 m/s at radius 300 km, find the relative vorticity.
>
> **Solution**
> Given: M = 10 m/s (cyclonic), R = 300,000 m.
> Find: ζ_r = ? s^{-1}
>
> Use eq. (11.18):
> $$\zeta_r = \frac{2\cdot(10\text{m/s})}{3\times10^5\text{m}} = \underline{\mathbf{6.67x10^{-5}}}\ \text{s}^{-1}$$
>
> **Check**: Units OK. Physics OK.
> **Discussion**: Relative vorticities of synoptic storms are typically about this order of magnitude.

FOCUS • Solid-Body Relative Vorticity

One can derive eq. (11.19) from either eq. (11.17) or eq. (11.18). Consider the sketch at right, where speed M is tangent to the rotating disk at radius R.

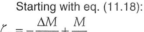

Starting with eq. (11.18):

$$\zeta_r = -\frac{\Delta M}{\Delta n} + \frac{M}{R}$$

Thus:

$$\zeta_r = -\frac{(0-M)}{R} + \frac{M}{R} = \boxed{\frac{2M}{R}}$$

Or, starting with (11.17):

$$\zeta_r = \frac{\Delta V}{\Delta x} - \frac{\Delta U}{\Delta y} = \frac{(M\cos\alpha - 0)}{R\cos\alpha} - \frac{(-M\sin\alpha - 0)}{R\sin\alpha}$$

$$= \zeta_r = \frac{M}{R} + \frac{M}{R} = \boxed{\frac{2M}{R}}$$

Absolute Vorticity

Measured with respect to the "fixed" stars, the total vorticity must include the earth's rotation in addition to the relative vorticity. This sum is called the absolute vorticity ζ_a :

$$\zeta_a = \zeta_r + f_c \qquad \bullet(11.20)$$

where the Coriolis parameter $f_c = 2\Omega\cdot\sin(\phi)$ is a measure of the vorticity of the planet, and where 2Ω = 1.458x10^{-4} s^{-1} .

Potential Vorticity

Potential vorticity ζ_p is defined as the absolute vorticity divided by the depth Δz of the column of air that is rotating:

$$\zeta_p = \frac{\zeta_r + f_c}{\Delta z} = \text{constant} \qquad \bullet(11.21)$$

It has units of (m^{-1}·s^{-1}). In the absence of turbulent drag and heating (latent, radiative, etc.), **potential vorticity is conserved**.

Combining the previous equations yields:

$$\underbrace{\underbrace{-\frac{\Delta M}{\Delta n}}_{shear} + \underbrace{\frac{M}{R}}_{curvature}}_{relative} + \underbrace{f_c}_{planetary} = \underbrace{\zeta_p\cdot\Delta z}_{stretching} \qquad \bullet(11.22)$$

> **Solved Example**
> An 11 km deep layer of air at 45°N latitude has no curvature, but has a shear of –10 m/s across distance 500 km. What is the potential vorticity?
>
> **Solution**
> Assume the shear is in the cyclonic direction.
> Given: Δz = 11000 m, ϕ = 45°N, ΔM = –10 m/s, Δn = 500000 m.
> Find: ζ_p = ? m^{-1}·s^{-1}
>
> Use eq. (11.21):
> $$\zeta_p = \frac{\frac{(10\text{m/s})}{5\times10^5\text{m}} + (1.458\times10^{-4}\text{s}^{-1})\cdot\sin(45°)}{11000\text{m}}$$
> = (2+10.31)x10^{-5} /11000 m = $\underline{\mathbf{1.12x10^{-8}}}$ m^{-1}·s^{-1}
>
> **Check**: Units OK. Physics OK.
> **Discussion**: Planetary vorticity is 5 times greater than the relative vorticity for this case. It cannot be neglected when computing potential vorticity.

where ζ_p is a constant that depends on the initial conditions of the flow. The last term states that if the column of rotating air is stretched vertically, then its relative vorticity must increase or it must move further north where planetary vorticity is greater.

Isentropic Potential Vorticity

Isentropic potential vorticity (IPV) can be defined as

$$\zeta_{IPV} = \zeta_p \cdot \frac{\Delta\theta}{\rho} = \text{constant} \qquad (11.23a)$$

where ζ_p is the potential vorticity measured on an isentropic surface (i.e., a surface connecting points of equal potential temperature θ), and where ρ is air density.

By rewriting the previous equation as

$$\zeta_{IPV} = \frac{\zeta_r + f_c}{\rho} \cdot \left(\frac{\Delta\theta}{\Delta z}\right) \qquad (11.23b)$$

we see that larger IPVs exist where the air is less dense and where the static stability $\Delta\theta/\Delta z$ is greater. For this reason, the IPV is two orders of magnitude greater in the stratosphere than the troposphere.

IPV is measured in potential vorticity units (PVU), defined by $1 \text{ PVU} = 10^{-6}\ \text{K·m}^2\text{·s}^{-1}\text{·kg}^{-1}$. On the average, air in the troposphere has $\zeta_{IPV} < 1.5$ PVU, while in the stratosphere it is greater.

Isentropic potential vorticity is conserved for air moving adiabatically and frictionlessly along an isentropic surface (i.e., a surface of constant potential temperature). Thus, it can be used as a tracer of air. Stratospheric air entrained into the troposphere retains its IPV > 1.5 PVU for a while before losing its identity due to turbulent mixing.

Solved Example

Find the IPV for the previous example, using $\rho = 0.5\ \text{kg/m}^3$ and $\Delta\theta/\Delta z = 3.3\ \text{K/km}$.

Solution

Given: $\zeta_p = 1.12\text{x}10^{-8}\ \text{m}^{-1}\text{·s}^{-1}$, $\rho = 0.5\ \text{kg/m}^3$, $\Delta\theta/\Delta z = 3.3\ \text{K/km}$, $\Delta z = 11$ km

Find: $\zeta_{IPV} = ?$ PVU

Use eq. (11.23b):

$$\zeta_{IPV} = \frac{(1.12\times10^{-8}\text{m}^{-1}\text{s}^{-1})\cdot(11\text{km})}{(0.5\text{kg}\cdot\text{m}^{-3})} \cdot \left(3.3\text{K}\cdot\text{km}^{-1}\right)$$

$= 8.13\text{x}10^{-7}\ \text{K·m}^2\text{·s}^{-1}\text{·kg}^{-1} = \underline{\textbf{0.813 PVU}}$

Check: Units OK. Physics OK.
Discussion: Reasonable value < 1.5 PVU in trop.

MIDLATITUDE TROUGHS AND RIDGES

The jet streams do not follow parallels in zonal flow around the earth to make perfect circles around the north and south poles. Instabilities in the atmosphere cause the jet stream to meander north and south in a wavy pattern as they encircle the globe (Fig. 11.19). These waves are called **planetary waves**, and have a wavelength of roughly 3000 to 4000 km. The number of waves around the globe is called the **zonal wavenumber**, and is typically 7 or 8, although they can range from about 3 to 13.

In Fig 11.19, jet stream winds (shaded) follow the height contours in a general counterclockwise (west to east) circulation as viewed looking down on the North Pole. The jet stream roughly demarks the boundary between the cold polar air from the warmer tropical air, because this temperature difference generates the jet stream winds due to the thermal wind relationship.

Regions of relatively low pressure or low height are called **troughs**. The center of the trough, called the **trough axis**, is indicated with a dashed line. Winds turn cyclonically (counterclockwise in the N. Hemisphere) around troughs; hence, troughs and low-pressure centers (low) are similar.

Ridges of relatively high pressures or heights are between the troughs. **Ridge axes** are indicated with a zig-zag line. Air turns anticyclonically (clockwise

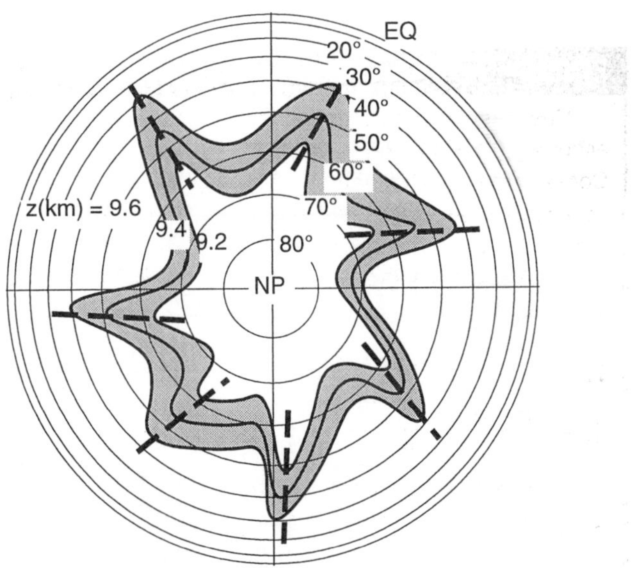

Figure 11.19
Height (z) contours of the 30 kPa pressure surface. View is looking down on the north pole (NP), and circles are latitude lines. Dashed lines indicate troughs. The jet stream is shaded.

in the N. Hemisphere) around ridges; hence, ridges and high-pressure centers (highs) are similar.

BEYOND ALGEBRA • Vorticity of a Wave

Problem: How does relative vorticity ζ_r vary with distance east (x) in a planetary wave?

Solution: Idealize the planetary wave as a sine wave

$$y = A \cdot \sin(2\pi \cdot x / \lambda) \quad \text{(a)}$$

where y is distance north of some reference parallel (such as 45°N), A is north-south amplitude, and λ is east-west wavelength. Assume constant wind speed M following the path of the sine wave. The U and V components can be found from the slope s of the curve. Combining $s \equiv V/U$ & $U^2 + V^2 = M^2$ gives:

$$U = M \cdot (1+s^2)^{-1/2} \quad \text{and} \quad V = M \cdot s \cdot (1+s^2)^{-1/2} \quad \text{(b)}$$

But the slope is just the first derivative of eq. (a)

$$s = \partial y / \partial x = (2\pi A / \lambda) \cdot \cos(2\pi \cdot x / \lambda) \quad \text{(c)}$$

Write eq. (11.17) in terms of partial derivatives:

$$\zeta_r = \partial V / \partial x - \partial U / \partial y \quad \text{(d)}$$

which can be rewritten as:

$$\zeta_r = \partial V / \partial x - (\partial U / \partial x) \cdot (\partial x / \partial y) \quad \text{or}$$

$$\zeta_r = \partial V / \partial x - (\partial U / \partial x) \cdot (1/s) \quad \text{(e)}$$

Plugging eqs. (b) and (c) into (e) gives the answer:

$$\zeta_r = \frac{-2 \cdot M \cdot A \cdot \left(\frac{2\pi}{\lambda}\right)^2 \cdot \sin\left(\frac{2\pi x}{\lambda}\right)}{\left[1 + \left(\frac{2\pi A}{\lambda}\right)^2 \cdot \cos^2\left(\frac{2\pi x}{\lambda}\right)\right]^{3/2}} \quad \text{(f)}$$

This is plotted in Fig a, for an example where λ = 4000 km, A = 1000 km, and M = 50 m/s.

Discussion: $|\zeta_r| = 2M/R$, where R is the radius of curvature for a sine wave, as is given in many calculus textbooks. Notice the narrow, sharp peaks of vorticity. Increased wave amplitude narrows the vort. peaks.

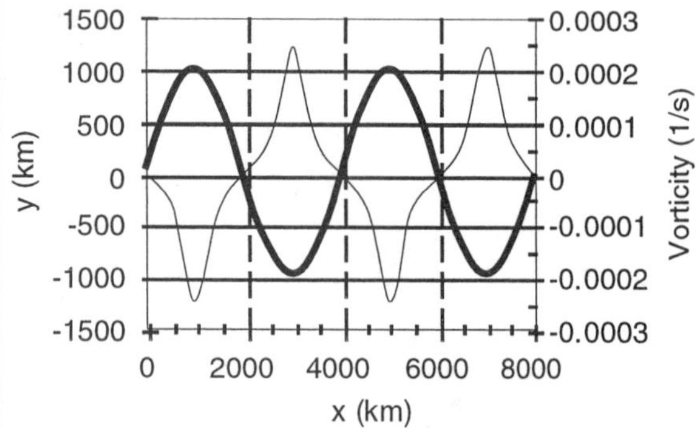

Figure a. *Streamline of an idealized planetary wave (thick), and the corresponding vorticity (thin).*

Two types of instabilities trigger these waves in the general circulation: barotropic instability, and baroclinic instability. Barotropic instability, caused by the earth's rotation, is described next. Baroclinic instability adds the effects of the north-south temperature gradient.

Barotropic Instability & Rossby Waves

Picture the jet stream at midlatitudes, blowing from west to east. If some small disturbance (such as flow over mountains, discussed in Chapter 13) causes the jet to turn slightly northward, then conservation of potential vorticity causes the jet to meander north and south. This meander of the jet stream is called a **Rossby wave** or **planetary wave**.

To understand this process, picture an initially-zonal flow at midlatitudes, such as sketched in Fig 11.20 at point 1. Straight zonal flow has no relative vorticity (no shear or curvature), but there is the planetary vorticity term in eq. (11.22) related to the latitude of the flow. For the special case of a fluid of fixed depth Δz such as the troposphere (Δz = 11 km), the conservation of potential vorticity simplifies to

$$\left[\frac{M}{R} + f_c\right]_{initial} = \left[\frac{M}{R} + f_c\right]_{later} \quad (11.24)$$

where we will focus on the curvature term as a surrogate to the full relative vorticity.

If this flow is perturbed slightly (at point 2) to turn to the north, the air is now moving into higher latitudes where the Coriolis parameter and planetary vorticity are greater. Thus, a negative shear or curvature (negative R) must form in the flow to compensate the increased planetary shear, in order to keep potential vorticity constant. In plain words, the jet turns clockwise (**anticyclonic**) at point 3 until it points southeast.

As it proceeds southward toward its initial latitude (point 4), it has less curvature (i.e., less relative vorticity), but still points southeast. The jet then continues south of its initial latitude to a region where planetary vorticity is less (point 5). To preserve potential vorticity, it develops a **cyclonic** (counterclockwise) curvature and heads back northeast. Thus, initially stable (zonal) flow from point 1 has become wavy, and is said to have become **unstable**.

This Rossby wave requires a variation of Coriolis parameter with latitude to create the instability, which is called **barotropic instability**. Parameter $\beta \equiv \Delta f_c / \Delta y$ gives the rate of change of Coriolis parameter with latitude:

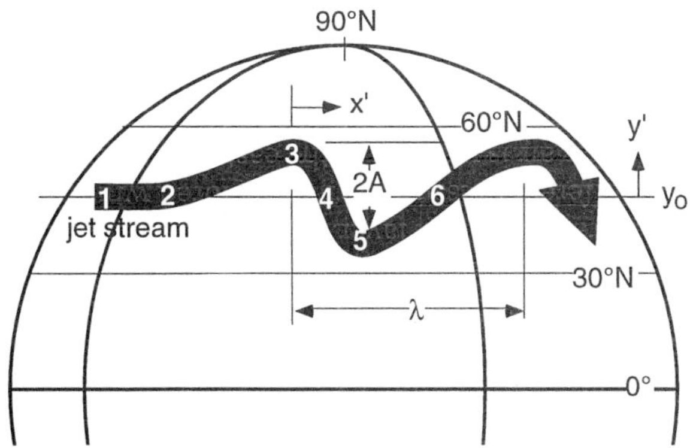

Figure 11.20
Initially zonal flow at point 1, if disturbed at point 2, will develop north-south meanders called Rossby waves.

$$\beta \equiv \frac{\Delta f_c}{\Delta y} = \frac{2 \cdot \Omega}{R_{earth}} \cdot \cos\phi \qquad \bullet(11.25)$$

where R_{earth} =6357 km is the average earth radius. Thus, $2 \cdot \Omega / R_{earth} = 2.29 \times 10^{-11}$ $m^{-1} \cdot s^{-1}$, and β is on the order of (1.5 to 2)$\times 10^{-11}$ $m^{-1} \cdot s^{-1}$.

The path taken by the wave is approximately:

$$y' \approx A \cdot \cos\left[2\pi \cdot \left(\frac{x' - c \cdot t}{\lambda}\right)\right] \qquad (11.26)$$

where y' is the north-south displacement distance from the center latitude Y_o of the wave, x' is the distance east from some arbitrary longitude, c is the phase speed (the speed at which the crest of the wave moves relative to the earth), A is the north-south amplitude of the wave, and λ is the wavelength (see Fig 11.20).

Typical wavelengths are $\lambda \cong 6000$ km, although a wide range of wavelengths can occur. The circumference ($2\pi R_{earth} \cdot \cos\phi$) along a parallel at midlatitudes limits the total number of barotropic waves that can fit around the globe to about 4 to 5. The north-south domain of the wave roughly corresponds to the 30° width of midlatitudes where the jet stream is strongest, giving $A \cong 1665$ km.

These waves propagate relative to the mean zonal wind U_o at **intrinsic phase speed** c_o of about

$$c_o = -\beta \cdot \left(\frac{\lambda}{2\pi}\right)^2 \qquad \bullet(11.27)$$

where the negative sign indicates westward propagation relative to the mean background flow. However, when typical values for β and λ are used in eq. (11.27), the intrinsic phase speed is found to be roughly half the magnitude of the eastward jet-stream speed.

A **phase speed** c relative to the ground is defined as:

$$c = U_o + c_o \qquad \bullet(11.28)$$

which gives the west-to-east movement of the wave crest. For typical values of c_o and background zonal wind speed U_o, the phase speed is positive. Hence, the mean wind pushes the waves toward the east relative to observers on the earth. Eastward moving waves are indeed observed.

BEYOND ALGEBRA • The Beta Plane

One can derive β from the definition for f_c. Starting with eq. (9.10): $f_c = 2\ \Omega \sin\phi$, take the derivative with respect to distance north:

$$\beta \equiv \frac{\partial f_c}{\partial y} = \frac{\partial f_c}{\partial \phi} \cdot \frac{\partial \phi}{\partial y} = \frac{\partial (2\Omega \sin\phi)}{\partial \phi} \cdot \frac{\partial \phi}{\partial y}$$

$$\beta = 2\Omega \cos\phi \cdot \frac{\partial \phi}{\partial y} \qquad (a)$$

As was discussed in an earlier "Beyond Algebra" box, $\partial\phi / \partial y$ is the change of latitude per distance traveled north. The total change in latitude to circumnavigate the earth pole to pole is 2π radians, and the circumference of the earth is $2\pi R$ where R = 6356.766 km is the radius of the earth. Thus:

$$\frac{\partial \phi}{\partial y} = \frac{2\pi}{2\pi \cdot R} = \frac{1}{R} \qquad (b)$$

Plugging eq. (b) into (a) gives the desired answer:

$$\boxed{\beta = \frac{2 \cdot \Omega}{R_{earth}} \cdot \cos\phi} \qquad (11.25)$$

For midlatitude planetary waves confined to a latitude belt such as sketched in Fig 11.19, it is often convenient to assume β = constant. This has the same effect as assuming that a portion of the earth is shaped like a cone (as sketched below in white) rather than a sphere. The surface of the cone is called a **beta plane**, and looks like a lamp shade. Such an idealization allows barotropic effects to be described with slightly simpler math.

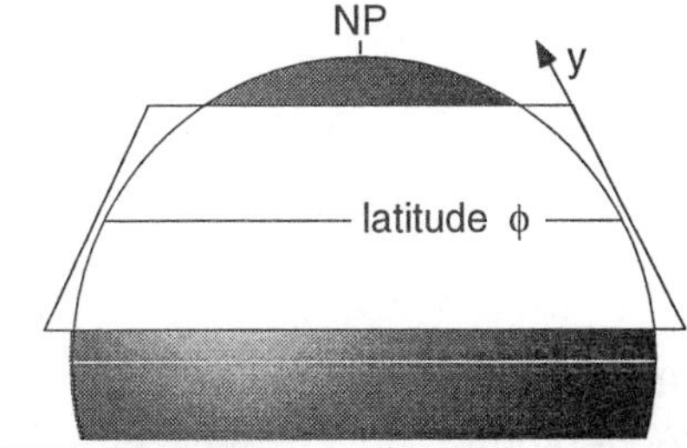

Equation (11.27) is called a **dispersion relation**, because waves of different wavelengths propagate at different phase speeds. Namely, waves that initially coincide would tend to separate or disperse with time.

By combining eqs. (11.27) and (11.28), waves of shorter wavelength (**short waves**) travel faster toward the east than **long waves**. Thus, short waves ride the long wave analogous to a car driving along a hilly road.

Solved Example(§)

A background jet stream of speed 50 m/s meanders with 6000 km wavelength and 1500 km amplitude, centered at 45°N. Plot the path (i.e., the meridional displacement) of the **barotropic** wave between $0 \le x' \le 10000$ km at some initial time and 6 hours later, and find the phase speed.

Solution

Given: $\phi = 45°$, $U_o = 50$ m/s, $\lambda = 6000$ km, $A = 1500$ km, $t = 0$ & 6 h

Find: $\beta = ?$ m^{-1}·s^{-1}, $c = ?$ m/s, $y'(x) = ?$ km

Use eq. (11.25):

$$\beta = \frac{1.458\times10^{-4}\text{s}^{-1}}{6378000\text{m}} \cdot \cos(45°) = \underline{\mathbf{1.62x10}^{-11}\ \text{m}^{-1}\cdot\text{s}^{-1}},$$

Use eq. (11.27):

$$c_o = -(1.62\times10^{-11}\text{m}^{-1}\text{s}^{-1})\left(\frac{6\times10^6\text{m}}{2\pi}\right)^2 = \underline{-14.7}\ \text{m/s}$$

Use eq. (11.28):

$$c = 50 - 14.7\ \text{m/s} = \underline{\mathbf{35.26\ m/s}}$$

Use eq. (11.26) on a spreadsheet for $t = 0$ & 6 h:

$$y' \approx (1500\text{km}) \cdot \cos\left[2\pi \cdot \left(\frac{x' - (35.26\text{m/s})\cdot t}{6\times10^6\text{m}}\right)\right]$$

The results are plotted below:

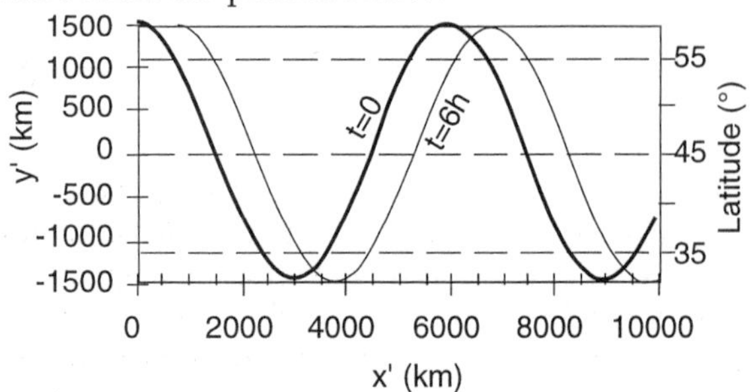

Check: Units OK. Physics OK.

Discussion: The barotropic long wave propagates east about 800 km during the 6 h interval. The jet stream blows along this wavy path at speed 50 m/s.

Solved Example(§)

Same as the previous solved example, except that in addition to the previous long wave, there is also a **barotropic** short wave of 1000 km wavelength with amplitude 300 km.

Solution

Given: Same, plus $\lambda = 1000$ km, $A = 300$ km.

Find: $c = ?$ m/s, $y'(x') = ?$ km

Use eq. (11.27):

$$c_o = -(1.62\times10^{-11}\text{m}^{-1}\text{s}^{-1})\left(\frac{10^6\text{m}}{2\pi}\right)^2 = \underline{\mathbf{-0.41\ m/s}}$$

Use eq. (11.28)

$$c = 50 - 0.41\ \text{m/s} = \underline{\mathbf{49.6\ \ m/s}}$$

Use eq. (11.26):

$$y' \approx (300\text{km}) \cdot \cos\left[2\pi \cdot \left(\frac{x' - (49.6\text{m/s})\cdot t}{1\times10^6\text{m}}\right)\right]$$

The results are plotted below:

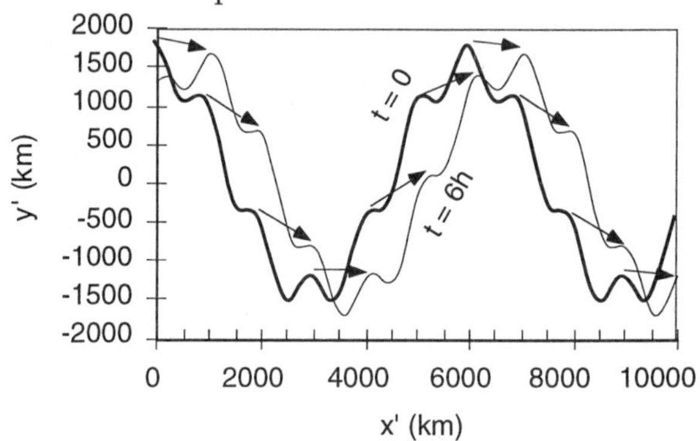

Check: Units OK. Physics OK.

Discussion: The movement of each short wave is indicated with the arrows. Not only do they move with the propagating background long wave, but they also ride the wave toward the east at a speed nearly equal to the wind speed.

Baroclinic Instability & Planetary Waves

Recall from Fig 11.1b that at midlatitudes the cold polar air slides under the warmer tropical air. This causes the air to be statically stable, as can be quantified by a Brunt-Väisälä frequency N_{BV}. The development later in this section extends what we learned of barotropic flows to the more complete **baroclinic** case having both β and N_{BV} effects in an environment with north-south temperature gradient.

First, we take a heuristic approach, and idealize the atmosphere as being a two layer fluid, with a north-south sloping density interface such as idealized in Fig 11.21. This will give us a qualitative picture of baroclinic waves.

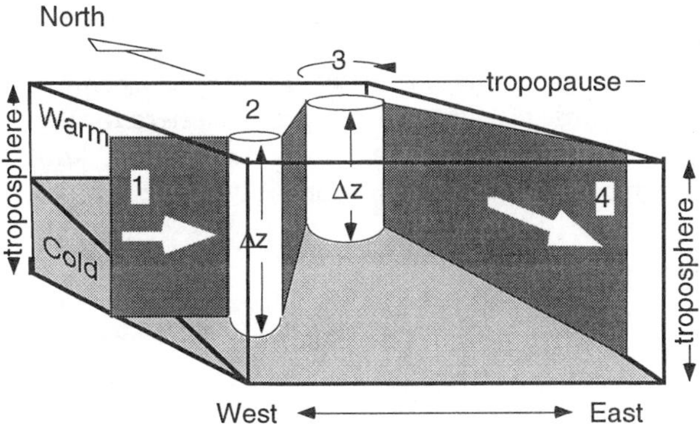

Figure 11.21
Some processes involved in baroclinic instability. Dark gray ribbon represents the jet-stream axis, while white columns indicate the absolute vorticity of the jet.

Qualitative Model

Initially zonal flow of the jet stream (sketched as point 1) has no relative vorticity, but it does have planetary vorticity related to its latitude. The narrow white column in the front left of Fig 11.21 represents air having such absolute vorticity.

If this jet stream is perturbed northward by some outside influence such as a mountain, it rides up on the density surface. The stratosphere is so statically stable that it acts like a lid on the troposphere. Thus, as air meanders northward, it is squeezed between the tropopause and the rising density interface. Namely, Δz shrinks. For this situation, the potential vorticity equation can be written as:

$$\left[\frac{f_c + (M/R)}{\Delta z}\right]_{initial} = \left[\frac{f_c + (M/R)}{\Delta z}\right]_{later} \qquad (11.29)$$

The column depth is less at point 3 than initially at point 2, hence the absolute vorticity at 3 must also be less than at 2, in order for the ratio of absolute vorticity to depth to remain constant. The planetary contribution doesn't help – in fact it is larger at point 3 as discussed in the previous subsection. Hence, the only way to conserve potential vorticity is for the relative vorticity to decrease substantially. As it decreases below its initial value of zero, the jet-stream path curves anticyclonically at point 3.

The jet stream overshoots to the south, and develops cyclonic relative vorticity and turns back to the north. The resulting breakdown of zonal flow into wavy flow is called **baroclinic instability**. The waves look similar to those in Fig 11.20, except with shorter wavelength because now both β and Δz work together to cause the oscillation.

Quantitative Approach

To extend this argument, consider the continuous stratification of Fig 6.11. Between the 290 and 340 K isentropes, for example, the air is statically stable. A column of air bounded at the top bottom and top by these isentropes near the equator shortens as it moves poleward. Thus, we could use the Brunt-Väisälä frequency and the north-south temperature gradient (i.e., the baroclinicity) as a better representation of the physics than the two layer model of Fig 11.21.

Skipping a long derivation, we end up with a north-south displacement of a baroclinic wave that is approximately:

$$y' \approx A \cdot \cos\left(\pi \cdot \frac{z}{Z_T}\right) \cdot \cos\left[2\pi \cdot \left(\frac{x' - c \cdot t}{\lambda}\right)\right] \qquad (11.30)$$

where $Z_T \cong 11$ km is the depth of the troposphere, and A is north-south amplitude. The extra cosine term containing height z means that the north-south wave amplitude first decreases with height from the surface to the middle of the troposphere, and then increases with opposite sign toward the top of the troposphere. Namely, the planetary wave near the tropopause is 180° out of phase compared to that near the ground.

The dispersion relation is:

$$c_o = \frac{-\beta}{\pi^2 \cdot \left[\frac{4}{\lambda^2} + \frac{1}{\lambda_R^2}\right]} \qquad \bullet(11.31)$$

and

$$c = U_o + c_o \qquad \bullet(11.32)$$

Again, intrinsic phase speed c_o is negative with magnitude less than the background jet velocity U_o, allowing the waves to move east relative to the ground.

An "internal" **Rossby radius of deformation λ_R** is defined as:

$$\lambda_R = \frac{N_{BV} \cdot Z_T}{f_c} \qquad \bullet(11.33)$$

where f_c is the Coriolis parameter, and N_{BV} is the Brunt-Väisälä frequency. This radius relates buoyant and inertial forcings. It is on the order of 1300 km.

As for barotropic waves, baroclinic waves have a range of wavelengths that can exist in superposition. However, some wavelengths grow faster than others, and tend to dominate the flow field. For baroclinic waves, this dominant wavelength is on the order of 3000 to 4000 km, and is given by:

$$\lambda \approx 2.38 \cdot \lambda_R \qquad (11.34)$$

FOCUS • Rossby Deformation Radius and Geostrophic Adjustment - Part 1

Definitions:

The spatial distribution of wind speeds and directions is known as the **wind field**. Similarly, the spatial distribution of temperature is called the **temperature field.**

The **mass field** is the spatial distribution of air mass. As discussed in Chapter 1, pressure is a measure of the mass of air above. Thus, the term "mass field" generically means the spatial distribution of pressure (i.e., the **pressure field**) on a constant altitude chart, or of heights (the **height field**) on an isobaric surface.

In baroclinic conditions, the hypsometric equation allows changes to the pressure field to be described by changes in the temperature field. Similarly, the thermal wind relationship relates changes in the wind field to changes in the temperature field.

Geostrophic Adjustment:

In regions of the atmosphere where pressure gradient and Coriolis forces dominate, the atmosphere tends to adjust itself to be in geostrophic balance. This is called **geostrophic adjustment**, and is described quantitatively in the next chapter.

When a disturbance or change is imposed in either the wind field or the pressure field, the other field adjusts to reach a new geostrophic balance. These adjustments are strongest near the disturbance, and gradually weaken with distance. The e-folding distance, beyond which the disturbance is felt only a little, is called the **Rossby radius of deformation,** λ_R.

For example, if winds are forced to increase above their initial geostrophic value at one location, then these winds will redistribute air mass in the horizontal, modifying the pressure field until it reaches a new geostrophic balance with the imposed winds. Thus, **the mass field adjusts to the wind field.**

Alternately, if air mass is added over a particular location, then the modified pressure gradient will force air outward from the disturbance. On the way out, Coriolis force will turn the winds to the right (N. Hem.), until a new geostrophic wind field is achieved. Thus, **the wind field adjusts to the mass field.**

In the real atmosphere, both fields adjust toward a new geostrophic balance. For large scale disturbances ($\lambda > \lambda_R$), most of the adjustment is in the wind field. For small scale disturbances ($\lambda < \lambda_R$), most of the adjustment is in the temperature or pressure fields.

Solved Example(§)

Same as the previous solved example, except in a **baroclinic** case in a standard atmosphere for the one dominant wave. Wave amplitude is 1500 km. Plot results at t = 0 and 6 h for z = 0 and 11 km.

Solution

Given: Same, and use Table 1-4 for std. atmos:

$\Delta T = -56.5-15 = -71.5°C$ across Δz = 11 km.

$T_{avg} = 0.5\cdot(-56.5+15°C) = -20.8°C = 252$ K

Find: λ_R=? km, λ=? km, c=? m/s, $y'(x')$=? km

Use eq. (6.5a):

$$N_{BV} = \sqrt{\frac{(9.8\text{m/s})}{255\text{K}}\left(\frac{-65°\text{K}}{10^4\text{m}} + 0.0098\frac{\text{K}}{\text{m}}\right)} = \underline{0.0113}\ \text{s}^{-1}$$

Use eq. (9.10):

$$f_c = (1.458\times10^{-4}\text{s}^{-1})\cdot\sin(45°) = \underline{1.031\text{x}10^{-4}}\ \text{s}^{-1}$$

Use eq.(11.33):

$$\lambda_R = \frac{(0.0113\text{s}^{-1})\cdot(11\text{km})}{1.03\times10^{-4}\text{s}^{-1}} = \underline{\mathbf{1206\ km}}$$

Use eq. (11.31):

$$\lambda = 2.38\cdot\lambda_R = \underline{\mathbf{2870\ km}}$$

Use eq. (11.31):

$$c_o = \frac{-(1.62\times10^{-11}\text{m}^{-1}\text{s}^{-1})}{\pi^2\cdot\left[\frac{4}{(2870\text{km})^2}+\frac{1}{(1206\text{km})^2}\right]} = \underline{\mathbf{-1.40\ m/s}}$$

Use eq. (11.32)

$$c = 50 - 1.40\ \text{m/s} = \underline{\mathbf{48.6\ \ m/s}}$$

Use eq. (11.30):

$$y' \approx (1500\text{km})\cos\left(\frac{\pi\cdot z}{11\text{km}}\right)\cos\left[2\pi\left(\frac{x' - 48.6\text{m/s}\cdot t}{2870\text{km}}\right)\right]$$

The results are plotted below:

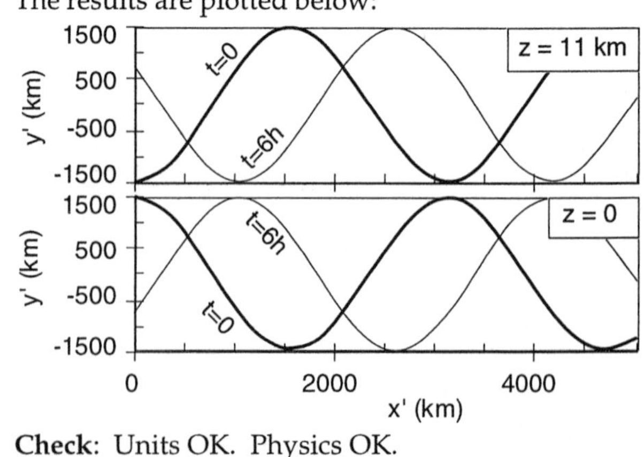

Check: Units OK. Physics OK.

North-south displacement y' is not the only characteristic that is wavy in baroclinic flow. Also wavy are the perturbation velocities (u', v', w'), pressure p', potential temperature θ', and vertical displacement η', where the prime denotes deviation from the mean background state.

To present the wave equations for these other variables, we will simplify the notation by using:

$$a \equiv \pi\cdot z / Z_T \qquad (11.35)$$

and

$$b \equiv 2\pi\cdot(x - c\cdot t)/\lambda \qquad (11.36)$$

Thus:

$$y' = \hat{Y} \cdot \cos(a) \cdot \cos(b)$$

$$\eta' = \hat{\eta} \cdot \sin(a) \cdot \cos(b)$$

$$\theta' = -\hat{\theta} \cdot \sin(a) \cdot \cos(b)$$

$$p' = \hat{P} \cdot \cos(a) \cdot \cos(b) \qquad (11.37)$$

$$u' = \hat{U} \cdot \cos(a) \cdot \cos(b)$$

$$v' = -\hat{V} \cdot \cos(a) \cdot \sin(b)$$

$$w' = -\hat{W} \cdot \sin(a) \cdot \sin(b)$$

Symbols wearing the caret hat (^) represent amplitude of the wave. These amplitudes are defined to be always positive (remember c_o is negative) in the Northern Hemisphere:

$$\hat{Y} = A$$

$$\hat{\eta} = \frac{A \cdot \pi \cdot f_c \cdot (-c_o)}{Z_T} \cdot \frac{1}{N_{BV}^2}$$

$$\hat{\theta} = \frac{A \cdot \pi \cdot f_c \cdot (-c_o)}{Z_T} \cdot \frac{\theta_o}{g}$$

$$\hat{P} = A \cdot \rho_o \cdot f_c \cdot (-c_o) \qquad (11.38)$$

$$\hat{U} = \left[\frac{A \cdot 2\pi \cdot (-c_o)}{\lambda}\right]^2 \cdot \frac{1}{A \cdot f_c}$$

$$\hat{V} = \frac{A \cdot 2\pi \cdot (-c_o)}{\lambda}$$

$$\hat{W} = \frac{A \cdot 2\pi \cdot (-c_o)}{\lambda} \cdot \frac{\pi \cdot (-c_o) \cdot f_c}{Z_T \cdot N_{BV}^2}$$

Each of these amplitudes depends on A, the amplitude of the meridional displacement, which is not fixed but depends on the initial disturbance.

Fig 11.22 illustrates the structure of this baroclinic wave in the Northern Hemisphere. Fig 11.22c corresponds to the surface weather maps used in the previous solved examples. All the variables are interacting to produce the wave. The impact of all these variables on synoptic weather will be discussed in Chapter 13.

Actual variables are found by adding the perturbation variables (eq. 11.37) to the background state. Mean background pressure P_o decreases hydrostatically with height, making the actual pressure at any height be $P = P_o + p'$.

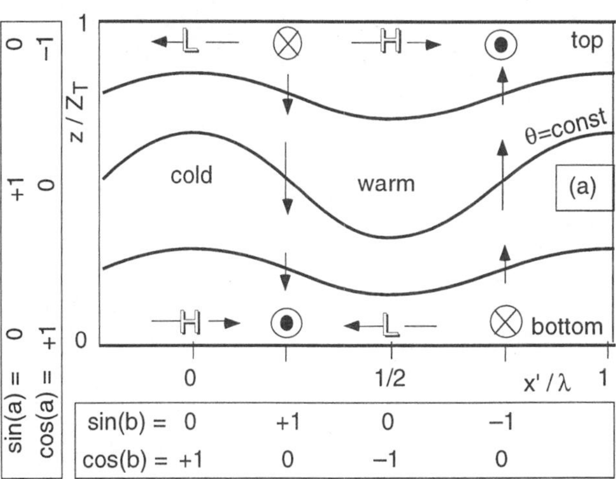

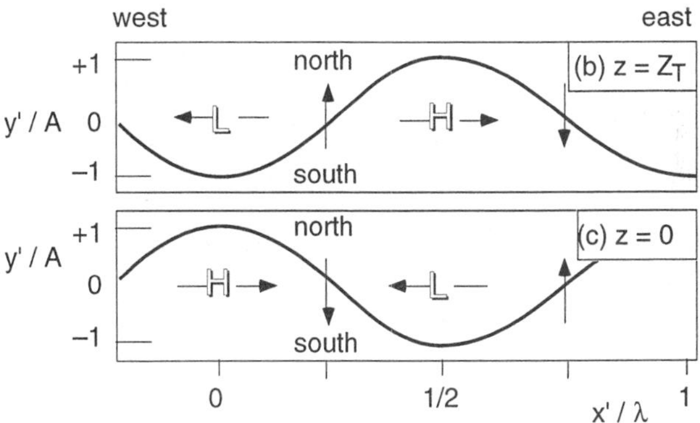

Figure 11.22
Idealized structure of baroclinic wave. (a) Vertical cross section. (b) Weather map at tropopause. (c) Surface weather map. Legend: Arrows indicate wind direction and speed, L and H indicate low and high pressure centers, dot-circle is southward-pointing vector, x-circle is northward pointing vector. Wavy lines on the cross section are isentropes. Also shown are sine and cosine terms from baroclinic wave equations. (after Cushman-Roisin, 1994)

Mean background potential temperature θ_o increases linearly with height (assuming constant N_{BV}), making the actual potential temperature $\theta = \theta_o + \theta'$. Background meridional and vertical winds are assumed to be zero, leaving $V = v'$, and $W = w'$. Background zonal wind U_o is assumed to be geostrophic and constant, leaving $U = U_o + u'$.

Many other factors affect wave formation in the jet stream, including turbulent drag, clouds and latent heating, and nonlinear processes in large-amplitude waves. Also, the more complete solution to baroclinic instability includes waves propagating meridionally as well as zonally (we focused on only zonal propagation here). However, our description above captured most of the important aspects of midlatitude flow that are observed in the real atmosphere.

Solved Example
Same as previous solved example, but find the pressure amplitudes at sea level.

Solution
Given: A = 1500 km, $f_c = 9.47\text{x}10^{-5}\ \text{s}^{-1}$,
c_o = –1.66 m/s, $\rho_o = 1.225\ \text{kg·m}^{-3}$.
Find: $\hat{P}$ = ? kPa

Use eq. (11.38):

$$\hat{P} = (1500\text{km})\left(1.225\frac{\text{kg}}{\text{m}^3}\right)(9.47\times10^{-5}\text{s}^{-1})\left(1.66\frac{\text{m}}{\text{s}}\right)$$
$$= \underline{0.29\ \text{kPa}}$$

Check: Units OK. Physics OK.
Discussion: Between high and low pressure is a distance of half a wavelength, which is 1564 km for this case. The pressure difference between crest and trough of the wave is $2\cdot\hat{P}$ = 0.58 kPa. Thus, the pressure gradient is 0.58/1564 = 0.00037 kPa/km, which is sufficient to drive the winds.

In the real atmosphere, condensation in clouds (which was neglected in eqs. 11.37) contributes to the dynamics to make horizontal pressure gradients even stronger.

Heat Transport

As the jet stream blows along the meandering planetary-wave path (Fig 11.23), it picks up warm air (T' = deviation from mean temperature = positive) and carries it northward (v' = +). Similarly, cold air (T' = negative) is carried southward (v' = –). The average meridional kinematic heat flux is the sum of advective transports divided by the number N of these transports: $F_{y\ waves} = (1/N)\cdot\Sigma(v'\cdot T')$ = (+)·(+) + (–)·(–) = positive. Thus, the north-south meandering jet stream causes a northward heat flux, $F_{y\ waves}$, without requiring a vertical circulation cell.

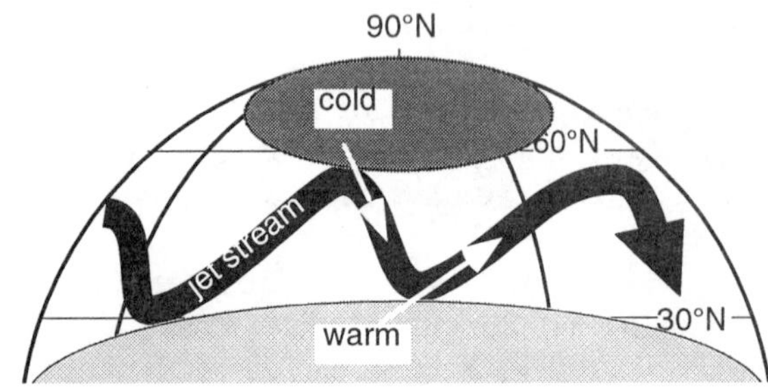

Figure 11.23
Heat transport by planetary waves in midlatitudes.

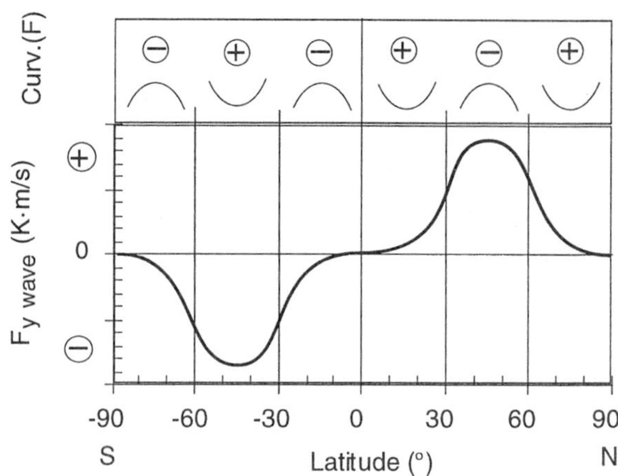

Figure 11.24
Bottom: Sketch of meridional wave flux of heat (positive northward, in kinematic units). Top: Sketch of sign of curvature of the wave flux.

The greatest flux is likely to occur where the waves have the most intense v-component (Fig. 11.23), and also where the north-south temperature gradient is greatest (Fig 11.4). Both of these processes conspire together to create large wave heat fluxes centered in midlatitudes (Fig 11.24 bottom).

Curvature (*Curv*) of a line is defined to be positive when the the line bends concave up (shaped like a bowl). Likewise, curvature is negative for concave down (shaped like a hill). The sign of the curvatures of wave heat flux are indicated in the top of Fig 11.24, which we will use in the next section.

Momentum Transport

Recall from Fig 11.16 that tropical air is faster (u' = positive), and polar air is slower (u' = negative) when moved to 45° latitude, because of angular-momentum conservation. The polar and tropical regions serve as reservoirs of various momentum that can be tapped by the meandering jet stream.

Analogous to heat transport, we find that the north-moving (v' = +) portions of the wave carry fast zonal momentum (u'= +) air northward, and south-moving (v' = –) portions carry slow zonal momentum (u' = –) southward (Fig 11.25). The net meridional transport of zonal momentum $\overline{u'v'}$ is positive, and is largest at the center of midlatitudes. $\overline{u'v'} = (1/N)\cdot\Sigma(v'\cdot u')$ = (+)·(+) + (–)·(–) = positive.

However, the reservoir of tropical momentum is much larger than polar momentum because of the larger circumference of latitude lines there. Hence, tropical momentum can have more influence on mid-latitude air. To account for such unequal influence, a weighting factor $a = \cos(\phi_s)/\cos(\phi_d)$ can be included in the angular momentum expression:

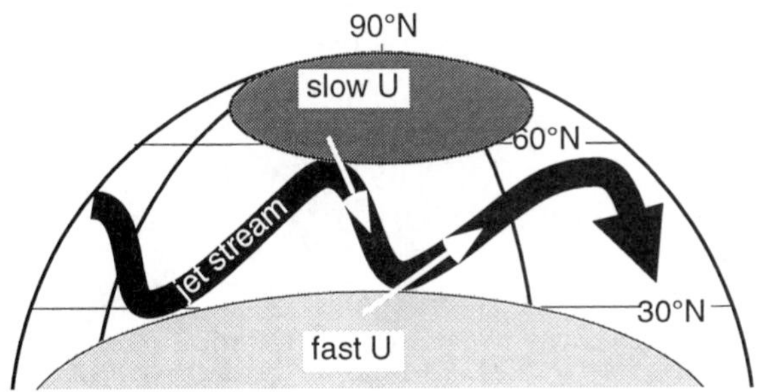

Figure 11.25
Meridional transport of eastward momentum by planetary waves in midlatitudes.

$$a \cdot u' \approx \Omega \cdot R_e \cdot \left[\frac{\cos^2 \phi_s}{\cos \phi_d} - \cos \phi_d \right] \cdot \frac{\cos \phi_s}{\cos \phi_d} \qquad (11.39)$$

Fig 11.26 shows that the weighted zonal velocity is asymmetric about 45° latitude. Namely, from 60° latitude the velocity is –100 m/s, but from a source of 30° latitude the velocity is about +200 m/s. Thus, although the momentum transport is positive everywhere, it decreases to the north. Hence, the meridional gradient *MG* of zonal momentum is negative in the N. Hemisphere mid-latitudes:

$$MG = \Delta \overline{u'v'} \ / \ \Delta y = \text{negative.} \qquad (11.40)$$

Although quantitatively the wind magnitudes from angular-momentum arguments are too large as discussed earlier, the qualitative picture of momentum transport is valid.

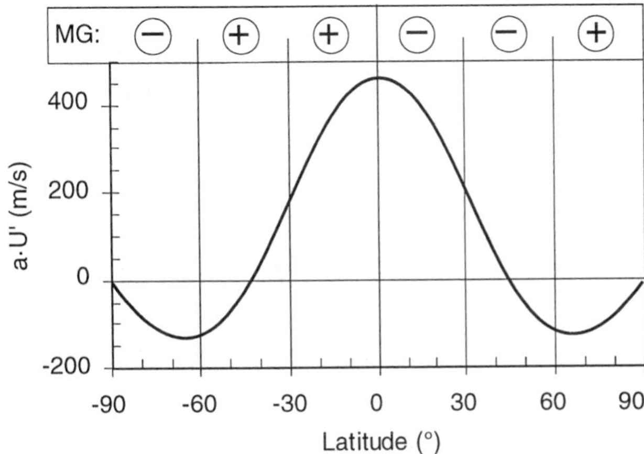

Figure 11.26
Zonal velocity from various source latitudes reaching destination 45°, weighted by the relative amounts of air in the source regions.

THREE-BAND GENERAL CIRCULATION

The preceding sections laid the groundwork to explain why there are three latitude-bands of circulation (Fig 11.27) between the equator and a pole, rather than one big Hadley cell. Namely, on earth we observe **direct** vertical circulation cells in the tropics (0° to 30°) and polar regions (60° to 90°), and an **indirect** cell at midlatitudes (30° to 60°). Direct means circulating in the same sense as the **Hadley cell**, with air rising at low latitudes and sinking at higher latitudes (Fig 11.28).

A measure of vertical cell circulation is:

$$CC = \frac{f_c^{\ 2}}{N_{BV}^2} \frac{\Delta V}{\Delta z} - \frac{\Delta w}{\Delta y} \qquad (11.41)$$

CC is positive for direct cells in the N. Hemisphere, as illustrated in Fig 11.28. In a direct cell, vertical velocity decreases and even changes sign toward the north, making $\Delta w / \Delta y$ negative. Also, northward velocity increases with height, making $\Delta V / \Delta z$ positive for a direct cell. Both of these terms contribute to a positive *CC* in a direct cell. Similarly, *CC* is negative for an indirect cell.

The equations of motion can be combined to yield an equation for the cell circulation *CC*, which is written in abbreviated form as eq. 11.42. Buoyancy associated with a heated surface can drive a vertical circulation, as can momentum transport. But if some of the heating difference is transported by planetary waves, then there will less heat available to drive a direct circulation.

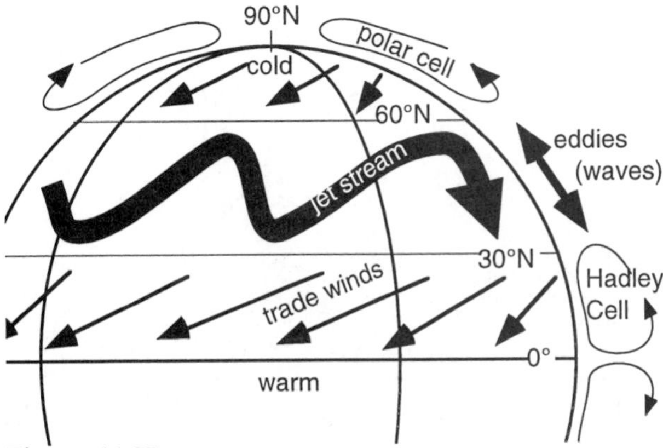

Figure 11.27
Three-band structure of general circulation: 1) vertical Hadley cell in tropics, 2) horizontal planetary waves at midlatitudes, and 3) a weak vertical circulation near the poles.

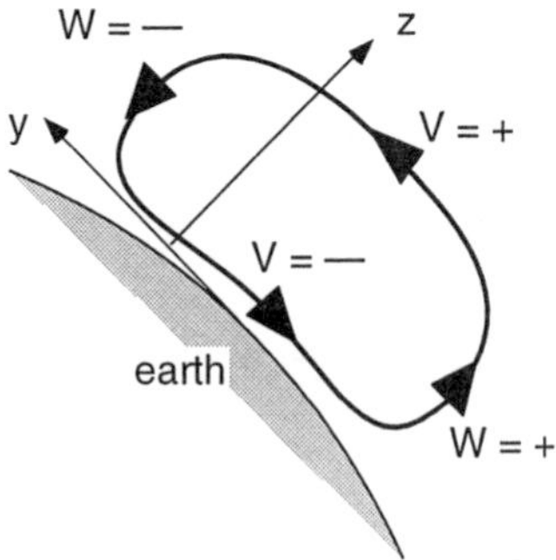

Figure 11.28
Definition of a direct circulation cell in the N. Hemisphere.

Based on the previous discussions of radiative differential heating E_{net}, momentum gradient MG, and heat flux curvature $Curv(F_{y\ wave})$, the contributions of each term to the circulation are listed below the equation. The sign of $\Delta MG/\Delta z$ equals the sign of MG, assuming that northward momentum transport is weakest at the ground, and increases with height because the jet-stream winds increase with height.

$$CC \propto -\frac{\Delta E_{net}}{\Delta y} + Curv(F_{y\ wave}) + \frac{\Delta MG}{\Delta z} \tag{11.42}$$

circulation radiation wave–heat wave–momentum

$CC_{polar} \propto$ positive + positive + positive = **positive**

$CC_{midlat} \propto$ positive + negative + negative = **negative**

$CC_{tropics} \propto$ positive + positive + positive = **positive**

Thus, horizontal planetary waves at midlatitudes are so effective that they reverse the vertical circulation.

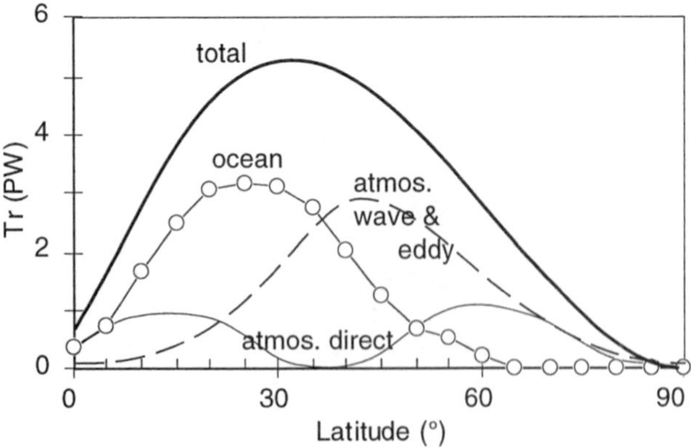

Figure 11.29
Northward transport of heat. (1 PW = 1 petawatt = 10^{15} W). At each latitude, the three thin curves sum to equal the total (thick) curve.

Earlier in this chapter, the differential heating (Fig 11.6) was summed over all latitudes (eq. 11.9) to give the net transport needed by various fluid circulations to compensate the radiative imbalance. The curve that resulted (see the solved problem in that section) is reproduced in Fig 11.29 for just the N. Hemisphere, and is labeled "total".

Observations of the earth/atmosphere/ocean system suggest how the various circulations contribute toward the total circulation (Fig 11.29). As we saw from eq. (11.42), the planetary wave circulation (and its associated high and low pressure eddies) dominate at mid latitudes, where the indirect circulation is called a **Ferrel cell**. Vertically direct circulations dominate elsewhere. The oceans also play a major role.

EKMAN SPIRAL IN THE OCEAN

As the wind blows over the oceans it drags along some of the water. The resulting ocean currents turn under the influence of Coriolis force (Fig 11.30), eventually reaching an equilibrium given by:

$$U = \left[\frac{u_{*water}^{\ 2}}{(K\cdot f_c)^{1/2}}\right]\cdot\left[e^{\gamma\cdot z}\cdot\cos\left(\gamma\cdot z - \frac{\pi}{4}\right)\right] \tag{11.43a}$$

$$V = \left[\frac{u_{*water}^{\ 2}}{(K\cdot f_c)^{1/2}}\right]\cdot\left[e^{\gamma\cdot z}\cdot\sin\left(\gamma\cdot z - \frac{\pi}{4}\right)\right] \tag{11.43b}$$

where the x-axis and U-current direction point in the direction of the surface wind. When these current vectors are plotted vs. depth (z is positive upward), the result is a spiral called the Ekman spiral.

The friction velocity u_{*water} for water is

$$u_{*water}^{\ 2} = \frac{\rho_{air}}{\rho_{water}}\cdot u_{*air}^{\ 2} \tag{11.44}$$

where the friction velocity for air was given in Chapter 4. The depth parameter is

$$\gamma = \sqrt{\frac{f_c}{2\cdot K}} \tag{11.45}$$

where K is the eddy viscosity (Appendix H) in the ocean. The inverse of γ is called the **Ekman layer depth**. Transport of water by Ekman processes is discussed further in the Hurricane chapter.

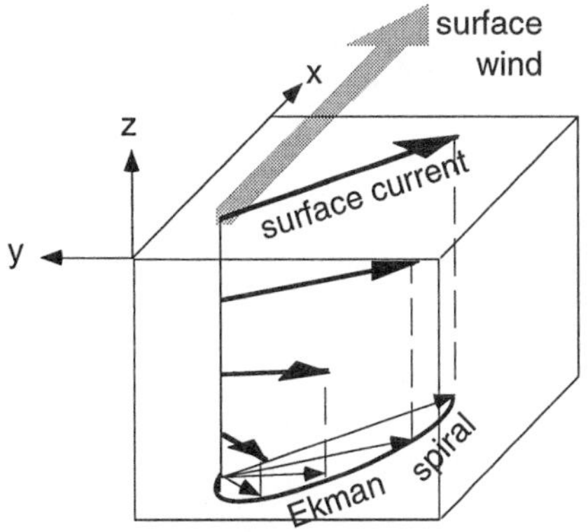

Figure 11.30
Ekman spiral of ocean currents.

Solved Example (§)

Plot the Ekman spiral in the ocean at 45° latitude, using an eddy viscosity of 2×10^{-3} $m^2 \cdot s^{-1}$, and $u_{*water}^2 = 4\times10^{-4}$ m^2/s^2.

Solution

Given: $u_{*water}^2 = 4\times10^{-4}$ m^2/s^2, $K = 0.002$ $m^2 \cdot s^{-1}$, $\phi = 45°$

Find: U & V (m/s) vs. z (m)

First: $f_c = 0.0001031$

Use eq. (11.45):

$\gamma = 0.1605$ m^{-1}

Solve eqs. (11.43) on a spreadsheet, resulting in:

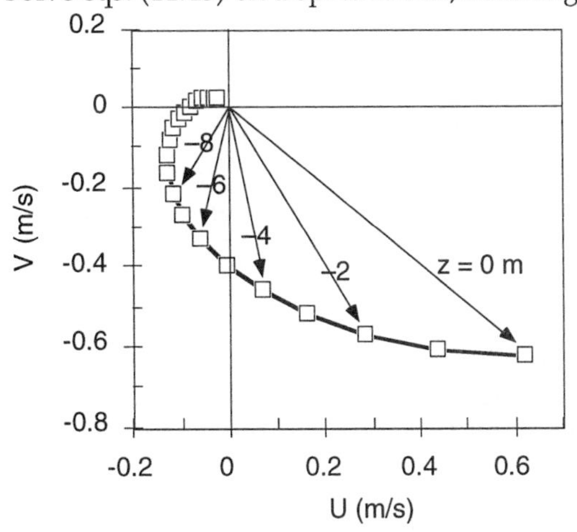

Check: Units OK. Physics OK.

Discussion: The surface current is 45° to the right of the wind direction. Deeper currents decrease in speed and turn to the right due to Coriolis force. The mean transport is in the $-V$ direction.

SUMMARY

Radiation causes net heating of the tropics and cooling at the poles. The global circulation moves the excess heat from the equator to the poles to partially reduce the differential temperature.

The reduction of meridional temperature gradient is not complete, resulting in a jet stream at mid latitudes due to Coriolis force and the thermal wind effect. This jet stream is unstable, and usually meanders north and south in the form of planetary waves, also known as Rossby waves. Vorticity is defined to help explain these waves.

The waves transport so much heat and angular momentum in mid latitudes that a reverse circulation develops there, called the Ferrel cell. Elsewhere, wave and eddy circulations are weaker, allowing direct vertical circulations such as the Hadley cell in the tropics. A weak direct cell also exists in polar regions.

Atmospheric winds drive ocean currents. These currents together with the atmospheric winds transport sufficient heat from the tropics to the poles to completely counteract the radiative imbalance, putting the earth in radiative equilibrium.

Threads

Radiation (Chapt. 2) causes the differential heating that drives the general circulation. Most of this circulation is confined to the troposphere, because of the vertical temperature structure (Chapt. 1). This circulation moves both sensible and latent heat (Chapt. 3).

The jet stream, caused by various dynamical forces (Chapt. 9) meanders north and south creating troughs and ridges. This pattern creates surface high pressure areas and air masses (Chapt. 12), and low pressure areas called cyclones (Chapt. 13). Upward motion in troughs and lows causes moist (Chapt. 5) and dry (Chapt 3) adiabatic cooling (Chapt. 6), which creates clouds (Chapt. 7) and precipitation (Chapt. 8). Downward motion in ridges and highs creates relatively shallow boundary layers (Chapt. 4) that can trap air pollutants (Chapt. 17) and cause air stagnation.

Numerical models (Chapt. 14) of the general circulation are called global climate models (GCMs) when used to forecast climate change (Chapt. 18). The background environment of the general circulation governs the likely areas for hurricane (Chapt. 16) and thunderstorm formation (Chapt. 15). The Ekman spiral process causes sea level to rise ahead of landfalling hurricanes (Chapt. 16), contributing to the destructive storm surge.

EXERCISES

Numerical Problems

N1(§) Plot the idealized temperature and meridional temperature gradient at the following heights (km) above ground.

a. 1 b. 2 c. 4 d. 6
e. 8 f. 10 g. 12 h. 14

N2. Find the radiation in, out and net at latitudes

a. 10° b. 30° c. 45° d. 60°
e. 15° f. 75° g. 5° h. 80°

N3. Find the latitude-band accumulated radiation in and out at latitudes:

a. 10° b. 30° c. 45° d. 60°
e. 15° f. 75° g. 5° h. 80°

N4. Find the latitude-accumulated differential heating at latitudes:

a. 10° b. 30° c. 45° d. 60°
e. 15° f. 75° g. 5° h. 80°

N5. Check to see if the data in Fig 11.6 does give zero net radiation when averaged from pole to pole.

N6. What are the thermal wind vector components at 45°N for a thickness (m) of the 100-85 kPa layer that decreases northward by the amount given below per 1000 km horizontal distance?

a. 50 b. 100 c. 150 d. 200
e. 250 f. 300 g. 350 h. 400

N7. a. Using the idealized surface temperature at the equator and at the pole, find the thermal wind (gradient of geostrophic wind) based on this temperature difference.

b. Same as (a) but at 11 km agl.
c. Same as (a) but at 15 km agl.

N8. Given a Coriolis parameter of 10^{-4} s^{-1}, average temperature of 10°C, and temperature decreasing 10°C toward the east across a distance of 500 km, find the geostrophic wind at the following heights (km) above ground.

a. 1 b. 2 c. 4 d. 6
e. 8 f. 10 g. 12 h. 14

N9(§). Plot the geostrophic (jet) stream speed vs. latitude using eq. (11.14), for these altitudes (km):

a. 1 b. 2 c. 4 d. 6
e. 8 f. 10 g. 12 h. 14

N10. (§) Plot the vertical profile of baroclinic jet-stream speed at latitudes:

a. 20° b. 30° c. 40° d. 60°
e. 15° f. 75° g. 5° h. 80°

N11. Considering conservation of angular momentum, an air parcel from 45° would have what speed relative to earth at destination latitude:

a. 0° b. 30° c. 40° d. 60°
e. 15° f. 75° g. 5° h. 80°

N12. Given the following wind shears (m/s) across 100 km, calculate the relative vorticity.

a. 1 b. 2 c. 3 d. 5
e. 8 f. 10 g. 15 h. 20

N13. Given a wind of 5 m/s rotating clockwise around a high-pressure center, find the relative vorticity at the following radii (km):

a. 50 b. 100 c. 150 d. 200
e. 300 f. 500 g. 800 h. 1000

N14. For the previous two problems, find the absolute vorticity if the flow is located at latitude of 60°N.

N15. If the flow is 11 km thick, find the potential vorticity for the previous problem.

N16(§). Using the shear data from Fig 11.12 for pure zonal flow, plot the following vs. latitude:

a. relative vorticity b. absolute vorticity

N17. Suppose water in your sink is in solid body rotation at one revolution per 5 seconds, and your sink is 1 m in diameter and 0.5 m deep.

a. Find the potential vorticity.

b. If you suddenly pull the stopper and the fluid stretches to depth 1 m in your drain, what is the new relative vorticity and new rotation rate?

N18. If fluid having no relative vorticity at 45°N moves to latitude ___N, find its new relative vorticity.

a. 0° b. 30° c. 40° d. 60°
e. 15° f. 75° g. 5° h. 80°

N19(§). Plot the meridional displacement of a barotropic wave of wavelength 400 km centered at latitude 45°, if imbedded in background flow (m/s) given below. Assume A = 100 km.

a. 10 b. 20 c. 30 d. 50
e. 60 f. 100 g. 150 h. 200

N20. Find the barotropic beta parameter at latitudes

a. 0°	b. 30°	c. 40°	d. 60°
e. 15°	f. 75°	g. 5°	h. 80°

N21. Suppose zonal flow at latitude 45° initially has no relative vorticity, but then flows over a mountain such that its depth changes from 11 km to a depth (km) given below. Find its new relative vorticity. Which way will the wind turn?

a. 5	b. 6	c. 7	d. 8
e. 9	f. 10	g. 12	h. 13

N22. Find the Rossby radius of deformation for an 11 km deep troposphere in the standard atmosphere at latitudes:

a. 20°	b. 30°	c. 40°	d. 45°
e. 50°	f. 55°	g. 60°	h. 70°

N23. Find the dominant wavelength of baroclinic waves for the previous problem.

N24. Find the phase speed for baroclinic waves of the previous problem.

N25(§). Plot the meridional displacement of a baroclinic wave given $\lambda = 4000$ km, $z_T = 11$ km, and $A = 500$ km. The wave is centered at 45° latitude, and is in a standard atmosphere. The intrinsic phase speed (m/s) is:

a. –3	b. –5	c. –7	d. –10
e. –12	f. –15	g. –2	h. –8

N26(§). For the previous problem, find amplitudes of:

a. meridional displacement
b. vertical displacement
c. potential temperature
d. pressure
e. U-wind
f. V-wind
g. W-wind

N27(§). For the previous problem, plot separate vertical cross sections showing perturbation values for those variables.

N28. Measure the circulation of the Hadley cell, if the updraft and downdraft velocities are 0.1 m/s, and the meridional velocities (m/s) are given below. Assume the Hadley cell fills the troposphere between 0° and 30° latitude.

a. 3	b. 5	c. 7	d. 10
e. 12	f. 15	g. 20	h. 25

N29(§). Plot the Ekman spiral at 15°N if the the wind in a neutral boundary layer is blowing from the ENE at 10 m/s, for eddy viscosities (m^2/s) of:

a. 0.001	b. 0.002	c. 0.003	d. 0.004
e. 0.005	f. 0.006	g. 0.007	h. 0.008

Understanding & Critical Evaluation

U1. Express the 5 PW of heat transport typically observed at 30° latitude in other units, such as
a. horsepower b. megatons of TNT
Hint, 1 **Megaton of TNT** $\cong 4.2\text{x}10^{15}$ J.

U2. Although angular momentum conservation is not a good explanation for the jet stream, can it explain the trade winds? Discuss with justification.

U3(§). Combine eqs. (11.1) and (11.3), consider adiabatic cooling with altitude to modify the result, in order to plot the air temperature at 15 km altitude vs. latitude.

U4(§). Combine eqs. (11.1) and (11.4) to reproduce the "out" curve of Fig 11.5.

U5(§). On a spreadsheet, reproduce the results of the solved example calculating E_{in} and E_{out} at all latitudes (taking 5° increments). Use this spreadsheet to calculate and reproduce Fig 11.6 on the differential heating. Also use this to produce the curve of transport Tr vs. latitude.

U6. Derive the equation for jet stream wind speed (11.14). Describe the physical meaning of each term in that equation. Also, describe the limitations of that equation.

U7. a. What is the sign of CC for a direct circulation in the S. Hemisphere?
b. Determine the signs of terms in eq. (11.42) for the S. Hemisphere.

U8. If tangential velocity increases as you cross the wall of a tornado from outside to inside, how does relative vorticity change with radius R from the center of the tornado?

U9. Compare the barotropic and baroclinic relationships for phase speed of planetary waves. Which is fastest (and in what direction) in mid-latitudes?

U10. If the troposphere were isothermal everywhere, find the number of planetary waves that would encircle the earth at 45°N, using baroclinic theory. How does this differ from barotropic theory?

U11. Discuss the differences between barotropic and baroclinic theories and their underlying assumptions for planetary waves.

U12. Using eq. (11.30) for the north-south displacement of a baroclinic wave, discuss (and/or plot), how the location of the crest of the wave changes with altitude z within the troposphere.

U13. How does static stability affect the phase speed of baroclinic planetary waves?

U14. Where with respect to the ridges and troughs in a baroclinic wave would you expect to find the greatest: (a) vertical displacement; (b) vertical velocity; (c) potential temperature perturbation?

U15. Why does the expression for cell circulation contain the Coriolis parameter and Brunt-Väisälä frequency?

U16. Would there be an Ekman spiral in the ocean if there was no Coriolis force? Explain.

Web-Enhanced Questions

W1. Download a northern or southern hemispheric map of the current winds (or pressures or heights from which winds can be inferred), and identify regions of near zonal flow, near meridional flow, extratropical cyclones, and tropical cyclones (if any).

W2. Download an animated loop of geostationary satellite photos (IR or water vapor) for the whole earth disk, and identify regions of near zonal flow, near meridional flow, extratropical cyclones, and tropical cyclones (if any).

W3. Download a still, daytime, visible satellite image for the whole disk from a geostationary satellite, and estimate fractional cloudiness over the different latitude belts. Use this to estimate the incoming solar radiation reaching the ground in each of those belts, and plot the zonally-averaged results vs. latitude.

W4. Download an IR satellite image showing the entire earth (i.e., a whole disk image), and compute the effective radiation temperature averaged around latitude belts. Use this with the Stephan-Boltzmann equation to estimate and plot E_{out} vs. latitude due to IR radiation.

W5. Download rawinsonde data from a string of stations that cross through the center of a low pressure center. Create a vertical slice through the atmosphere similar to Fig 11.3a, and confirm that the point of low pressure on a constant height line corresponds to the point of low height of an isobar.

W6. Download weather maps of temperature for the surface, and for either the 85 kPa or 70 kPa isobaric surfaces. Over your location, or other location specified by your instructor, use the thermal wind relationship to find the change of geostrophic wind with height at the surface, and at the isobaric level you chose above the surface.

W7. Download a weather map for the 100-50 kPa thickness, and calculate the components of the thermal wind vector for that surface. Also, what is the thermal wind magnitude? Draw arrows on the thickness chart representing the thermal wind vectors.

W8. Download data from a string of rawinsonde stations arranged north to south that cross the jet stream. Use the temperature, pressure, and wind speed data to produce vertical cross-section analyses of: (a) temperature; (b) pressure; and (c) wind speed. Discuss how these are related to the dynamics of the jet stream.

W9. Download weather maps of height contours for the 20 or 30 kPa isobaric surface, and draw a line within the region of most closely-spaced contours to indicate the location of the jet stream core. Compare this line to weather maps you can download from various TV and weather networks showing the jet stream location. Comment on both the location and the width of the jet stream.

W10. Download a weather forecast map that shows vorticity. What type of vorticity is it (relative, absolute, potential, isentropic)? What is the relationship between vorticity centers and fronts? What is the relationship between vorticity centers and bad weather (precipitation)?

W11. Download surface weather observations for many sites around your location (or alternately, download a weather map showing wind speed and direction), and calculate the following vorticities at your location: (a) relative; (b) absolute; (c) potential.

W12. Download a map of height contours for either the 50, 30, or 20 kPa surface. Measure the wavelength of a dominant planetary wave near your region, and calculate the theoretical phase speed of both barotropic and baroclinic waves for that wavelength. Compare and discuss.

W13. Download maps at a variety of altitudes and for a variety of fields (e.g., heights, temperature, vertical velocity, etc.), but all valid at the same time, and discuss how they are related to each other using baroclinic theory.

W14. View an animation of a loop of IR or water vapor satellite images for the whole disk of the earth. Identify the three bands of the general circulation, and discuss how they deviate from the idealized picture sketched in this chapter. To hand a result in, print one of the frames from the satellite loop and sketch on it lines that demark the three zones. Speculate on how these zones vary with season.

W15. Download a map of recent ocean surface currents, and compare with a map showing the general circulation winds for the same time. Discuss how the two maps are related.

Synthesis Questions

S1. If the earth did not rotate, how would the steady-state general circulation be different, if at all?

S2. If our current rotating earth with current general circulation suddenly stopped rotating, describe the evolution of the general circulation over time. This is called the **spin-down** problem.

S3. If the earth rotated twice as fast, how would the steady-state general circulation be different, if at all?

S4. If our current rotating earth with current general circulation suddenly started rotating twice as fast, describe the evolution of the general circulation over time. This is a **spin-up** situation.

S5. If radiative heating was uniform between the equator and the poles, how would the steady-state general circulation be different, if at all?

S6. If the equator were snow covered and the poles were not, how would the general circulation change, if at all?

S7. If an ice age caused the polar ice caps to expand to 50° latitude in both hemisphere, describe how general circulation would change, if at all?

S8. Suppose the troposphere were half as deep. How would the general circulation be different, if at all?

S9. What if potential vorticity was not conserved. How would the general circulation be different, if at all?

S10. Suppose the average zonal winds were faster than the phase speed of short waves, but slower than the phase speed for long waves. Describe how the weather and weather forecasting would be different at midlatitudes, if at all?

S11. Suppose that short waves moved faster than long waves, how would the weather at midlatitudes be different, if at all?

S12. Suppose the troposphere was statically unstable everywhere. Could baroclinic waves exist? If so, how would they behave, and how would the weather and general circulation be different, if at all?

S13. If planetary waves did not transport momentum, heat, and moisture meridionally, how would the weather and climate be different, if at all?

S14. If the ocean currents did not transport any heat, how would the atmospheric general circulation and weather be different, if at all?

S15. What if the ocean surface were so slippery that there would be zero wind drag. Would any ocean currents develop? How? Why?

AIR MASSES & FRONTS

CONTENTS

12 Surface weather maps are often drawn to show synoptic features such as high and low-pressure centers, air masses, and fronts (Fig 12.1). These features evolve according to the laws of thermodynamics, dynamics, and continuity.

A high-pressure center, or **high** (H), often contains an **air mass** of well-defined characteristics, such as cold temperatures and low humidity. When different air masses finally move and interact, their mutual border is called a **front**, named by analogy to the battle fronts of World War I.

Fronts are usually associated with low-pressure centers, or **lows** (L). Two fronts per low are most common, although zero to four are observed less frequently. These fronts often rotate counter-clockwise around the low center in the Northern Hemisphere, while the low moves and evolves. Fronts are often the focus of clouds, low pressure, and precipitation. Troughs and dry lines have some, but not all, of the characteristics of fronts.

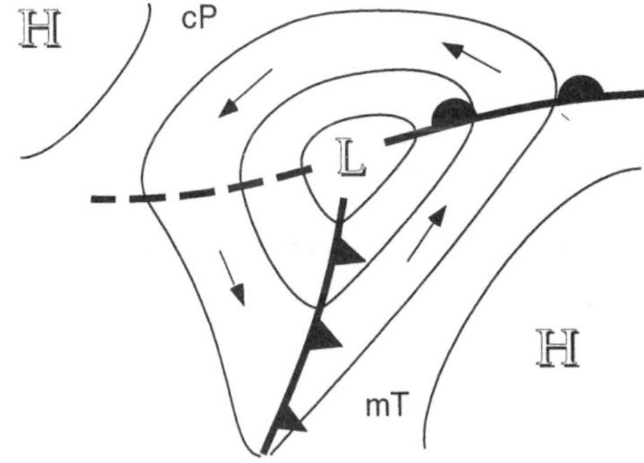

Figure 12.1
Idealized chart showing highs (H), a low (L), isobars (thin lines), and fronts (heavy solid lines) in the N. Hemisphere. Vectors indicate wind. Dashed line is a trough. cP indicates continental polar air; mT indicates maritime tropical air.

ANTICYCLONES OR HIGHS

Characteristics

High-pressure centers, or **highs**, typically have downward motion (**subsidence**) in the mid troposphere, and horizontal spreading of air (**divergence**) near the surface (Fig 12.2a). Highs predominate at 30° and 90° latitude, where the global circulation has downward motion. Transient surface highs also form at mid-latitudes, east of high-pressure ridges in the jet stream, and are an important part of mid-latitude weather variability.

Subsidence impedes cloud development, leading to generally clear skies and fair weather. Winds are also generally light in highs, because gradient-wind dynamics of highs requires weak pressure gradients near the high center (e.g., Fig 9.11 from Chapter 9).

The diverging air near the surface spirals outward due to the weak pressure-gradient force. Coriolis force causes it to rotate clockwise (anticyclonically) around the high-pressure center in the Northern Hemisphere, and opposite in the Southern Hemisphere. For this reason, high-pressure centers are called **anticyclones**.

Downward advection of dry air from the upper troposphere creates dry conditions just above the boundary layer. Subsidence also advects warmer potential temperatures from higher in the troposphere. This strengthens the temperature inversion that caps the boundary layer, and acts to trap pollutants and reduce visibility near the ground.

Subsiding air cannot push through the capping inversion, and therefore does not inject free-atmosphere air directly into the boundary layer. Instead, the whole boundary layer becomes thinner as the top is pushed down by subsidence (Fig 12.2a). This can be partly counteracted by entrainment of free atmosphere air if the boundary layer is turbulent, such as for a convective mixed layer during daytime over land. However, the entrainment rate is controlled by turbulence in the boundary layer (see Chapt. 4), not by subsidence.

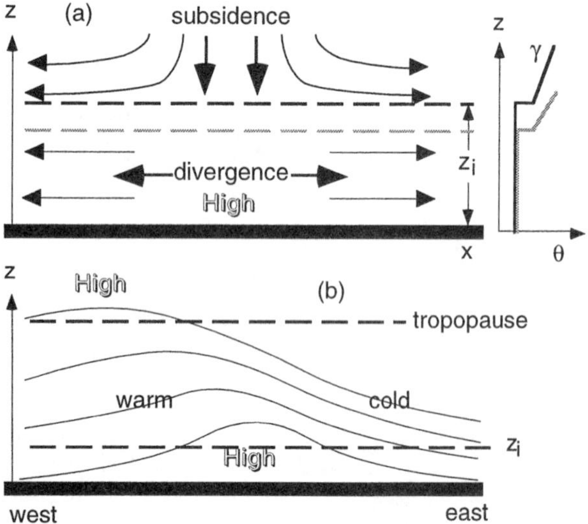

Figure 12.2
(a) Left: vertical circulation above a surface high-pressure center, in the bottom half of the troposphere. Heavy dashed line marks the top of the mixed layer, initially. Gray dashed line shows the top later, assuming no turbulent entrainment into the mixed layer. Right: idealized profile of potential temperature, θ, initially (black line) and later (gray). The mixed layer depth z_i is on the order of 1 km, and the potential- temperature gradient above the mixed layer is represented by γ.
(b) Tilt of high-pressure ridge westward with height, toward the warmer air. Thin lines are height contours of isobaric surfaces.

Vertical Structure

The location difference between surface and upper-tropospheric highs can be explained using gradient wind and thickness concepts.

Because of barotropic and baroclinic instability, the jet stream meanders north and south, creating troughs of low pressure and ridges of high pressure, as discussed in Chapt. 11. **Gradient winds** blow faster around ridges, and slower around troughs, assuming identical pressure gradients. The region east of a ridge and west of a trough has fast-moving air entering from the west, but slower air leaving to the east. Thus, horizontal convergence of air at the top of the troposphere adds more air molecules to the whole tropospheric column, causing a surface high to form east of the upper-level ridge.

West of surface highs, the anticyclonic circulation advects warm air from the equator toward the poles (Fig 12.2b). This heating west of the surface high causes the **thickness** between isobaric surfaces to increase, as explained by the hypsometric equation. Isobaric surfaces near the top of the troposphere are thus lifted to the west of the surface high. These high heights correspond to high pressure aloft; namely, the upper-level ridge is west of the surface high.

The net result is that high-pressure regions tilt westward with increasing height. In Chapt. 13, it will be shown that deepening low-pressure regions also tilt westward with increasing height, at mid-latitudes. Thus, the whole pressure pattern has a consistent phase shift toward the west as altitude increases.

AIR-MASSES

An air mass is a contiguous, widespread body of air that has remained stagnant over a surface for sufficient duration to be modified by the surface (Fig 12.3). Examples are maritime air masses that form with high relative humidity over oceans, or arctic air masses that develop with cold temperatures over polar regions.

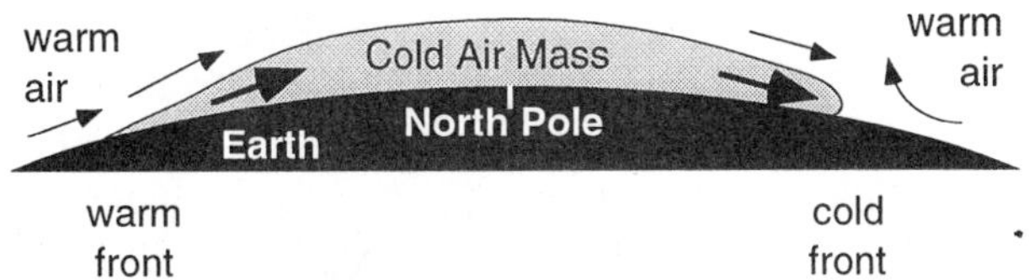

Figure 12.3
Vertical cross section through an arctic air mass.

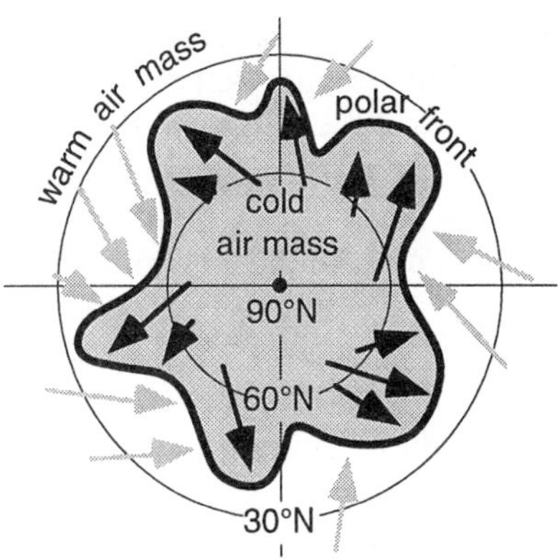

Figure 12.4
Idealized polar front as a boundary of the polar air mass.

Surface highs favor the formation of air masses because the light winds allow long residence times. The semi-permanent highs of the global circulation favor warm air-mass formation at 30° latitude, and cold at 90° latitude.

The boundary between these two global air masses defines the **polar front.** The **Bergen School** of meteorology (see Focus box later in this chapter) idealized this as a continuous front that encircled the poles (Fig 12.4). However, in the real atmosphere only segments of the polar front are observed, marking the boundary between air masses of regional extent.

Air masses are usually classified by their potential temperature and humidity, which by definition are relatively homogeneous within the air mass. Air masses can also be characterized by their visibility, odor, pollen concentration, dust concentration, pollutant concentration, radioactivity, CCN activity, cloudiness, static stability, and turbulence.

Creation

Air masses form as boundary layers within regions of high surface pressure. Because high-pressure centers have light winds, the air masses reside over the surface for a fairly long time before blowing away. During this residence, processes including radiation, conduction, and turbulent transport between the ground and the air cause the air to be modified.

Warm Air Masses

For air over a warmer surface such as the tropical ocean, static instability creates a turbulent air mass consisting of a convective mixed layer (ML). Turbulence is driven by the potential temperature difference $\Delta\theta$ between the surface θ_{sfc} and the air mass θ_{ML}:

$$\Delta\theta = \theta_{sfc} - \theta_{ML} \tag{12.1}$$

A heat flux from the surface into the air is given by eq. (3.22) for light-wind situations, which causes the air mass to warm.

An equilibrium depth z_i is reached when the rate of entrainment of air into the top balances the horizontal **divergence** $\beta = \Delta U / \Delta x + \Delta V / \Delta y$ of air from the sides. The entrainment rate is modulated by the potential-temperature gradient $\gamma = \Delta\theta/\Delta z$ just above the air mass; namely, weaker stability aloft allows greater entrainment rates. When these relationships are used in the heat budget, we can find the warming rate of the air mass and its depth:

$$\Delta\theta \approx \Delta\theta_o \cdot \exp(-t/\tau) \tag{12.2}$$

$$z_i \approx \frac{\Delta\theta}{\tau} \cdot \frac{1}{\beta \cdot \gamma} \tag{12.3}$$

where $\Delta\theta_o$ is the initial value of sea-air temperature difference, and θ_{sfc} is assumed constant.

The e-folding time τ is

$$\tau \approx c \cdot \left(\frac{\theta_{ML\ avg}}{g \cdot \beta \cdot \gamma} \right)^{1/3} \tag{12.4}$$

where c = 140 (dimensionless), and $\theta_{ML\ avg}$ is an average mixed-layer potential temperature (assumed constant, with units of Kelvins). τ is a measure of how long the air must remain over the surface in order to be identified as a **maritime tropical** air mass.

When a cold air mass comes to rest over a warmer surface, it warms rapidly within a deep and vigorously-turbulent mixed layer. Later, the air mass temperature gradually approaches the sea-surface temperature. As the air-sea temperature difference decreases, the ML depth begins to decrease because the decaying turbulence and entrainment cannot counteract the relentless subsidence. Greater divergence causes a shallower ABL that can warm faster. The net result is that creation of this warm air mass is nearly complete after several days.

Solved Example (§)

Air of initial ML potential temperature 10°C comes to rest over a 20°C sea surface. Divergence is 10^{-6} s^{-1}, and assume γ = 3.3 K/km. Find and plot the evolution of air-mass potential temperature and depth, and find the e-folding time.

Solution

Given: $\Delta\theta_o$ = 10°C, γ = 3.3 K/km, β = 10^{-6} s^{-1}.
Find: $\theta_{ML}(t)$ = ? K, $z_i(t)$ = ? m, τ = ? days

Use eq. (12.4) & assume $\theta_{ML\,avg}$ = 15°C = 288 K:

$$\tau \approx 140 \cdot \left(\frac{288\text{K}}{(9.8\text{m}\cdot\text{s}^{-2})(0.00001\text{s}^{-1})(0.0033\text{K/m})} \right)^{1/3}$$

= <u>**1.56 days**</u>

Next, use eqs. (12.1 – 12.3) in a spreadsheet:

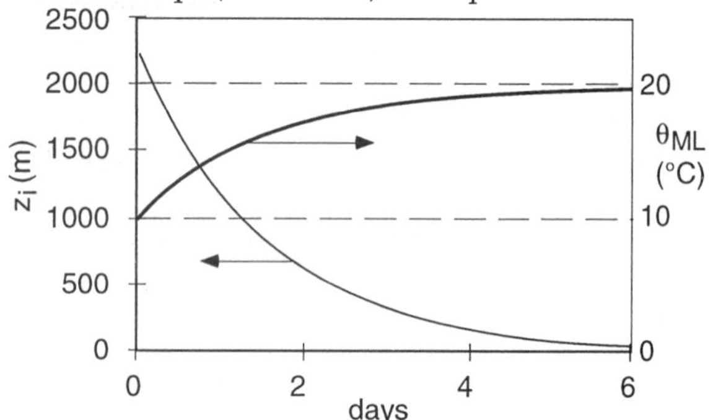

Check: Units OK. Physics OK. Figure reasonable.
Discussion: After about 5 days, the air mass has acquired the same temperature as the underlying surface. During this time, other characteristics would have altered similarly. The mixed layer depth becomes unrealistically small after several days. In the real atmosphere, wind-generated turbulence can maintain a thicker mixed layer.

Cold Air Masses

When air moves over a colder surface such as arctic ice, the bottom of the air first cools by conduction and turbulent transfer with the ground. Turbulence intensity then decreases within the resulting statically-stable boundary layer, reducing the possibility for further conductive cooling.

However, direct radiative cooling of the air, both upward to space and downward to the cold ice surface, causes a temperature change (averaged over a 1 km thick boundary layer) of about 2°C/day. As the air cools below the dew point, water-droplet clouds form. Continued radiative cooling from cloud top causes it to change to an ice cloud, and allows the cloud top to rise as a deeper layer of air becomes chilled (Fig 12.5).

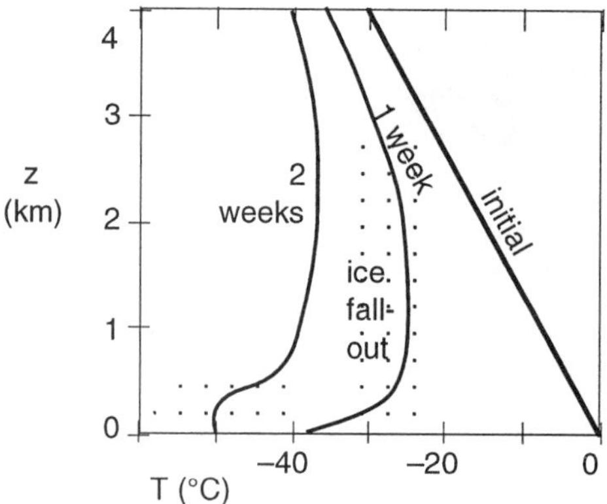

Figure 12.5
Genesis of a continental-polar air mass in the arctic.

The ice-crystals within this cloud are so few and far between, that the weather is sometimes described as cloudless ice-crystal precipitation. This can create some spectacular **halos** and other optical phenomena in sunlight (see Chapt. 19), including sparkling ice crystals known as **diamond dust**. Nevertheless, infrared radiative cooling in this cloudy air is much greater than in clear air, allowing the cooling rate to increase to 3°C/day over a layer as deep as 4 km.

During the two weeks for formation of this **continental-polar** air-mass, most of the ice crystals precipitate out leaving a thinner cloud of 1 km depth. Also, subsidence within the high pressure also reduces the thickness of the cloudy air mass and causes some warming to partially counteract the radiative cooling.

The final result is a nearly-isothermal layer of air that is 3 to 4 km thick, and which has cooled by about 30°C. Final air-mass temperatures are often in the range of –30 to –50 °C, with even colder temperatures near the surface.

While the arctic surface consists of relatively flat sea-ice (except for Greenland), the antarctic has mountains, high ice fields, and significant surface topography (Fig 12.6). As cold air forms by radiation, it can drain downslope as a **katabatic wind**. Steady winds of 10 m/s are common in the interior, with speeds of 50 m/s along some of the steeper slopes.

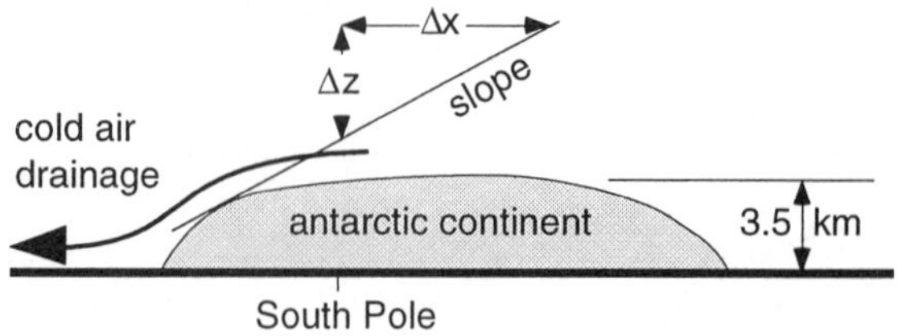

Figure 12.6
Cold katabatic winds draining from Antarctica.

On a surface of slope $\Delta z/\Delta x$, buoyancy force per unit mass translates into a quasi-horizontal slope-force per mass of:

$$\frac{F_{x\ S}}{m} = \frac{g \cdot \Delta\theta}{T_e} \cdot \frac{\Delta z}{\Delta x} \qquad \bullet(12.5a)$$

$$\frac{F_{y\ S}}{m} = \frac{g \cdot \Delta\theta}{T_e} \cdot \frac{\Delta z}{\Delta y} \qquad \bullet(12.5b)$$

where $\Delta\theta$ is the potential-temperature difference between the cold air that is draining and the ambient air above it. The ambient-air absolute temperature is T_e.

The sign of these forces should be such as to accelerate the wind downslope. The equations above work when the magnitude of slope $\Delta z/\Delta x$ is small, because then $\Delta z/\Delta x \cong \sin(\alpha)$, where α is the slope of the topography. With this substitution, one can recognize eq. (12.5) as the slope (buoyancy) term in eq. (10.7).

The katabatic wind speed in the antarctic also depends on turbulent drag force against the ice surface, ambient pressure-gradient force associated with synoptic weather systems, Coriolis force, and turbulent drag caused by mixing of the draining air with the stationary air above it. The Coriolis force cannot be neglected because of the large size of the continent. For example, at an average drainage velocity of 5 m/s, air would need over 2 days to move from the interior to the periphery of the continent, which is a time scale on the same order as the inverse of the Coriolis parameter.

Solved Example

Find the slope force per unit area acting on a katabatic wind of temperature –20°C with ambient air temperature 0°C. Assume a slope of $\Delta z/\Delta x$ = 0.1 .

Solution

Given: $\Delta\theta$ = 20 K, T_e = 273 K, $\Delta z/\Delta x$ = 0.1

Find: $F_{x\ S}/m$ = ? m·s^{-2}

Use eq. (12.5):

$$\frac{F_{x\ S}}{m} = \frac{(9.8\text{m}\cdot\text{s}^{-2})\cdot(20\text{K})}{273\text{K}} \cdot (0.1) = \underline{\mathbf{0.072}}\ \text{m}\cdot\text{s}^{-2}$$

Check: Units OK. Physics OK.

Discussion: This is two orders of magnitude greater than the typical synoptic forces (see Chapter 9). Hence, drainage winds can be strong.

Katabatic drainage removes cold air from the genesis regions, and causes turbulent mixing of the cold air with warmer air aloft. The resulting cold-air mixture is rapidly distributed toward the outside edges of the antarctic continent.

One aspect of the global circulation is a wind that blows around the poles. This is called the **polar vortex**. Katabatic removal of air from over the antarctic reduces the troposphere depth, which enhances the persistence and strength of the polar vortex due to potential vorticity conservation.

Modification

As an air mass moves from its origin, it is modified by the new landscapes under it. For example, a polar air mass will warm and gain moisture as it moves equatorward over warmer vegetated ground. Thus, it gradually loses its original identity.

Heat and moisture transfer at the surface can be described with drag relationships such as eq. (3.21). If we assume for simplicity that wind-induced turbulence creates a well-mixed air-mass of quasi-constant thickness z_i, then the change of air-mass potential temperature θ_{ML} with travel-distance Δx is:

$$\frac{\Delta\theta_{ML}}{\Delta x} \approx \frac{C_H \cdot (\theta_{sfc} - \theta_{ML})}{z_i} \qquad (12.6)$$

where $C_H \cong 0.01$ is the bulk transfer coefficient for heat (see Chapters 3 and 4).

Solved Example

A polar air mass of initial potential temperature –20°C and depth 500 m moves southward over a surface of 0°C. Find the initial rate of temperature change with distance.

Solution

Given: θ_{ML}=–20°C, θ_{sfc} = 0°C, z_i = 500 m

Find: $\Delta\theta_{ML}$ / Δx = ? °C/km

Use eq. (12.6):

$$\frac{\Delta\theta_{ML}}{\Delta x} \approx \frac{(0.01)\cdot[0°\text{C} - (-20°\text{C})]}{(500\text{m})} = \underline{\mathbf{0.4\ °C/km}}$$

Check: Units OK. Physics OK.

Discussion: Neither this answer nor eq. (12.7) depend on wind speed. While quicker speeds give faster position change, the quicker speeds also cause greater heat transfer to/ from the surface (see eq. 2.36). These two effects cancel.

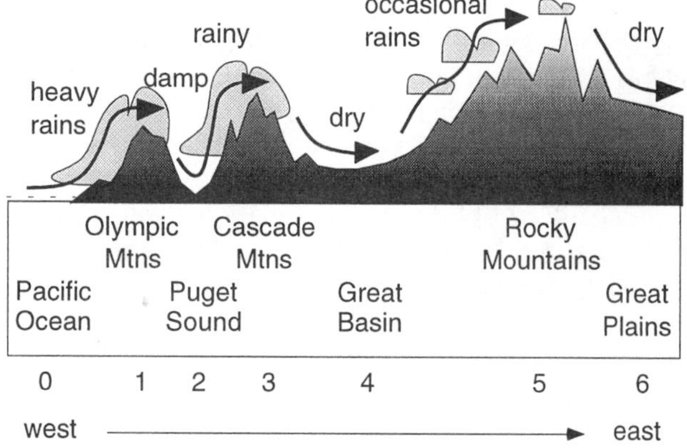

Figure 12.7
Modification of a Pacific air mass by flow over mountains in the northwestern USA. (The numbers identify key points in the solved example.)

If the surface temperature is horizontally homogeneous, then eq. (12.6) can be solved for the air-mass temperature at any distance x from its origin:

$$\theta_{ML} = \theta_{sfc} - (\theta_{sfc} - \theta_{ML\ o}) \cdot \exp\left(-\frac{C_H \cdot x}{z_i}\right) \qquad (12.7)$$

where $\theta_{ML\ o}$ is the initial air mass potential temperature at location $x = 0$.

If an air mass is forced to rise over mountain ranges, the resulting condensation and latent heating will dry and warm the air. For example, an air mass over the Pacific Ocean near the northwestern USA is often classified as maritime polar, because it is relatively cool and humid. As the prevailing westerly winds move this air mass over the Olympic Mountains (a coastal mountain range), the Cascade Mountains, and the Rocky Mountains, there can be substantial precipitation and latent heating (Fig 12.7).

Science Graffito

"It seems odd that scientists, of all people, would question the search for anything thought to be elusive or even impossible to find. Think of the naysayers who used to scoff every time Carl Sagan said there had to be other planets orbiting other suns in other solar systems; now other planets are discovered so often it's hard to keep track of how many there are.

You'd think that the science establishment, having been proven wrong so many times, would adopt a bit more humility. If there's one thing we keep learning, it's that we don't know as much as we think we do." – Editors of *Discover* Magazine, Sept. 99 , **20**, p 57.

Solved Example

A maritime polar air mass has initial conditions of $T = 5°C$ and $RH = 100\%$. Use a thermo diagram to find the temperature and humidity at the following locations: 1 Olympics (elevation ≅ 1000 m), 2 Puget Sound (0 m), 3 Cascades (1500 m), 4 the Great Basin (500 m), 5 Rockies (2000 m), and 6 the western Great Plains (1000 m).

Solution

Given: Elevations from west to east (m) = 0, 1000, 0, 1500, 500, 2000, 1000

Find: T (°C) and RH (%) at the surface

Assume: All condensation precipitates out, no additional heat or moisture transfer from the surface, sea-level pressure is 100 kPa, and we start with air at sea level.

Use Fig 12.7 and thermo diagram Fig 6.12. On the thermo diagram, the air parcels follows the following route: 0 - 1 - 2 - 1 - 3 - 4 - 3 - 5 - 6.

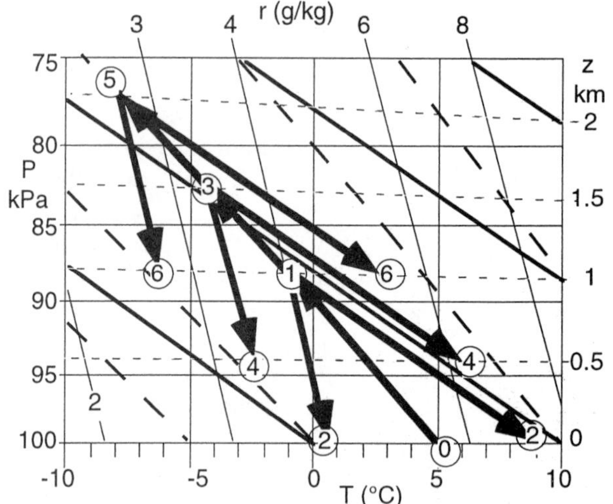

Results, where index = circled numbers:

Index	z(km)	T(°C)	T_d(°C)	RH(%)
0	0	5	5	100
1	1	–1	–1	100
2	0	9	1	55
3	1.5	–4	–4	100
4	0.5	6	–2	54
5	2	–7	–7	100
6	1	2	–6	53

Check: Units OK. Physics OK. Figure OK.

Discussion: The air mass has lost its maritime identity by the time it reaches the Great Plains, and little moisture remains. Humidity for rain in the plains comes from the SE, from the Gulf of Mexico.

SYNOPTIC WEATHER MAPS

Synoptic weather maps give a snapshot of the weather. Fig 12.8a shows an example of a synoptic weather map for the surface, based on surface weather observations. Fig 12.8b shows an upper-air synoptic map valid at the same time, using data from weather balloons, aircraft, satellite, and ground-based remote sensors. By studying both maps, you can get a feeling for the three-dimensional characteristics of the weather.

(a)

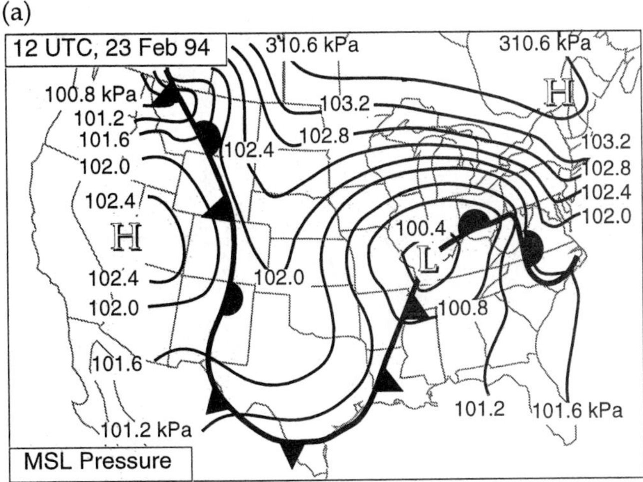

(b)

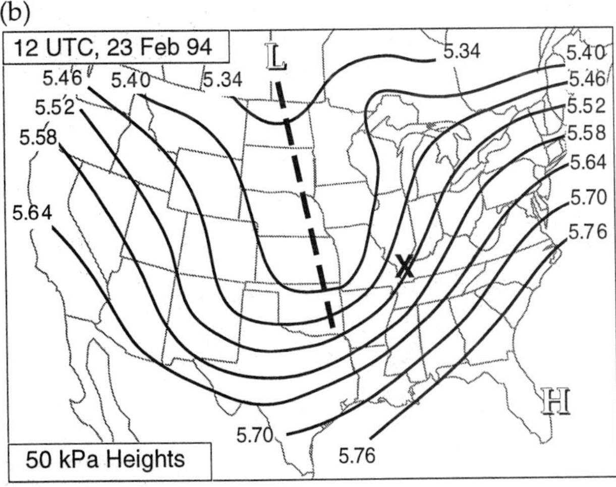

Figure 12.8
Examples of synoptic weather maps showing (a) the pressure pattern at the bottom of the troposphere, and (b) near the middle of the troposphere. The surface weather map (a) shows mean sea-level (MSL) pressure, fronts, and high (H) and low (L) pressure centers. The upper air map (b) shows height (km) contours for the 50 kPa isobaric surface, and also indicates high and low centers, and the trough axis (dashed line). [Based on original analyses by Jon Martin.]

Three steps are needed to create such maps. First, the weather data must be observed and communicated to central locations. Second, the data is quality controlled, where erroneous or suspect values are removed. Third, the data is analyzed, which means it is integrated into a coherent picture of the weather. This last step often involves interpolation to a grid (if it is to be used by computers), or drawing of isopleths and identification of weather features (lows, fronts, etc.) if used by humans. Steps one and three are discussed in more detail next.

Weather Observations

The **World Meteorological Organization (WMO)**, a component of the United Nations, coordinates international weather observing activities, and sets observation standards. Weather observations are taken **synoptically**, which means that they are taken at the same time around the world. By international agreement, the time standard is **Coordinated Universal Time (UTC)**.

The greatest number of manual synoptic surface and upper air observations (using weather balloons) is obtained at 00 UTC, and nearly as many observations are taken at 12 UTC every day. Some countries also take upper air observations at 06 and 18 UTC, as their budgets permit.

Simple weather balloons can be tracked from the ground to estimate winds (**PIBALs** = Pilot Balloons), or can include a radio transmitter to report back temperature, humidity, and pressure (**RAOBs** = Radiosonde Observations), or can use various navigation information such as **GPS** (Global Positioning Satellites) along with the radio transmitter to give winds as well as temperature, humidity, and pressure (**Rawinsondes**). Most of these weather balloons are launched manually.

Many countries also take manual surface weather observations every hour. Weather observations made at airports are coded into a form called **METARS**.

In addition to the manual observations, automatic weather observation systems make more frequent or nearly continuous reports. Examples of automatic surface weather stations are **AWOS** (Automated Weather Observing System), and **ASOS** (Automated Surface Observing Systems) in the USA.

Satellites can radiometrically estimate air temperature to provide remotely-sensed upper air automatic data. Such USA systems include the **TOVS** (TIROS Operational Vertical Sounder, where **TIROS** is Television and Infrared Operational System) on the **NOAA** (National Oceanographic and Atmospheric Administration) civilian polar orbiting satellites, and **SSM/I** (Special Sensor Microwave Imager) system on the **DMSP** (Defense Meteorological Satellite Program) military polar orbiting satellite. Other countries also have weather satellites.

Geostationary satellites can also be used to estimate tropospheric winds by tracking movement of clouds and water vapor patterns. Surface winds over the ocean can be estimated from satellite (such as the **ERS2** European Remote Sensing Satellite) by measuring the scattering of microwaves off of the sea surface. Tropospheric precipitable water can be estimated by satellite from the amount of microwave or IR radiation emitted from the troposphere.

Commercial aircraft provide manual weather observations called Aircraft Reports (**AIREPS**) at specified longitudes as they fly across the oceans. Many commercial aircraft have automatic meteorological reporting equipment such as **ACARS** (Aircraft Communication and Reporting System), **AMDAR** (Aircraft Meteorological Data Relay), and **ASDAR** (Aircraft to Satellite Data Relay).

Remote sensors on the ground include weather radar such as the **NEXRAD** (Weather Surveillance Radar WSR-88D). Microwave wind profilers automatically measure a vertical profile of wind speed and direction. **RASS** (Radio Acoustic Sounding Systems) equipment uses both sound waves and microwaves to measure virtual temperature and winds.

The WMO oversees a **Global Telecommunication System** (GTS) to share this data in near real time with all participating countries. Computers in some of the wealthier countries automatically collect, quality control, organize, and store the vast weather data set.

Figures 12.9 show sources of weather data that were collected by the computers at the European Centre for Medium-Range Weather Forecasts (**ECMWF**) in Reading, England, for a six-hour period centered at 00 UTC on 22 August 1999.

The volume of weather data is immense. There are roughly 350,000 locations (manual stations, automatic sites, and satellite obs) worldwide that report weather observations at 00 UTC. At ECMWF, an estimated 200 megabytes (MB) of weather-observation data are processed and archived every day. Over 2 gigabytes (GB) of weather-forecast results are also produced and archived each day.

The result of all this data is a snapshot of the weather, which can be analyzed on **synoptic weather maps**. Although the word "synoptic" literally means "at the same time", the type of weather that can be seen on such a synoptic map has come to be known as **synoptic-scale** weather (see Table 10-1 in Chapt. 10). The methods used to analyze the weather data to create such maps is discussed next.

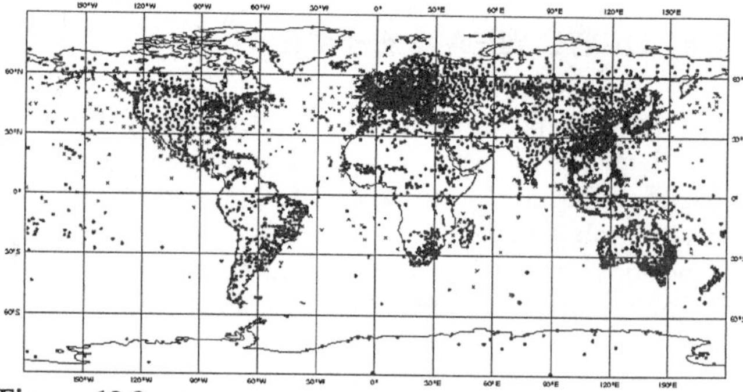

Figure 12.9a
Surface data locations for observations of temperature, humidity, winds, clouds, precipitation, pressure, and visibility collected by ***synoptic weather stations*** *on land and ship. Valid: 00 UTC on 22 Aug 99. Number of observations: 11,170 land + 1,362 ship = 12,532 surface obs. (From the European Centre for Medium Range Weather Forecasts: ECMWF.)*

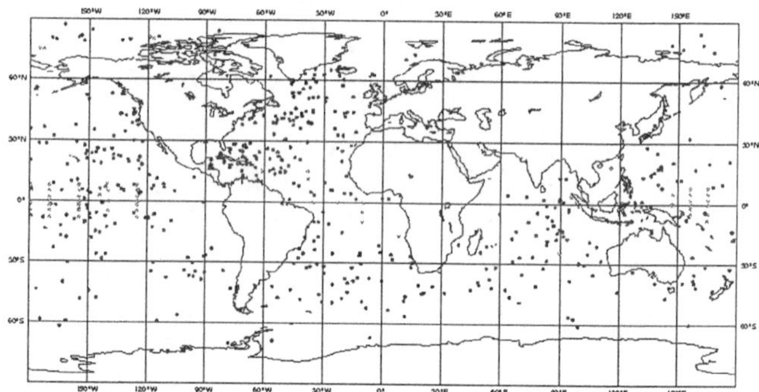

Figure 12.9b
Surface data locations for temperature and winds collected by drifting and moored buoys. Valid: 00 UTC on 22 Aug 99. Number of observations: 1,577 drifters + 112 moored = 1,689 buoys. (From ECMWF.)

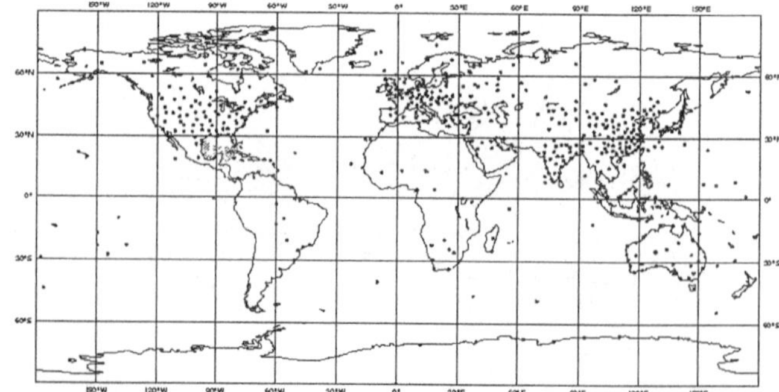

Figure 12.9c
Upper-air sounding locations for temperature, winds, and humidity collected by ***rawinsonde balloons*** *launched from land and ship, and by dropsondes released from aircraft . Valid: 00 UTC on 22 Aug 99. Number of observations: 540 land + 8 ship + 40 dropsondes = 588 soundings. The dropsondes were all in the Gulf of Mexico, to track Hurricane Bret as it approached Texas (From ECMWF.)*

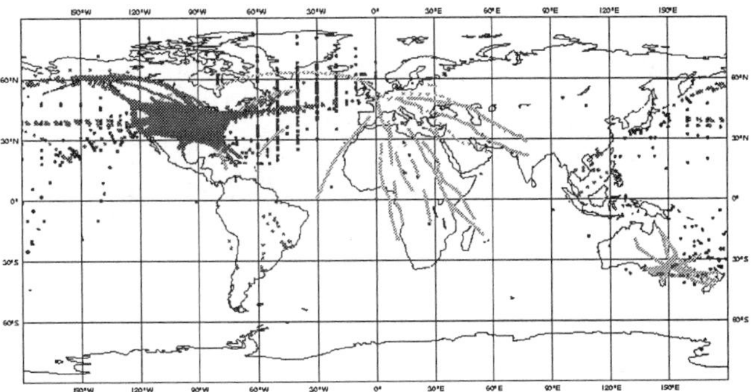

Figure 12.9d
Upper-air data locations for temperature and winds collected by ***commercial aircraft****: AIREP manual reports (black), and AMDAR (light gray) & ACARS (dark gray) automated reports. Valid: 00 UTC on 22 Aug 99. Number of observations: 6,335 AIREP + 2,351 AMDAR + 11,745 ACARS = 20,431 aircraft observations, most at their cruising altitude of 10 to 15 km above sea level. (From ECMWF.)*

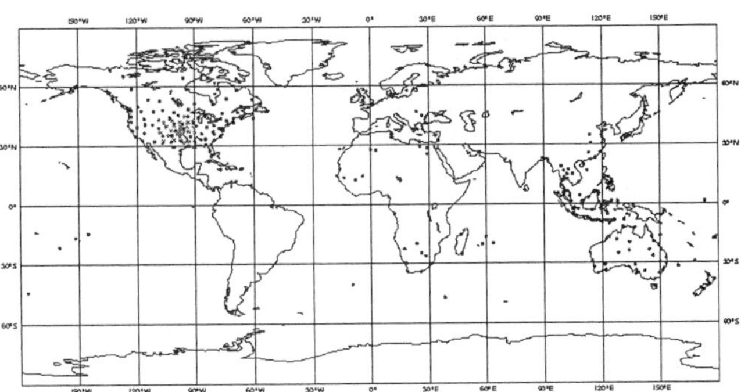

Figure 12.9e
Upper-air data locations for winds collected by: ***pilot balloons*** *and ground-based* ***wind profilers****. Valid: 00 UTC on 22 Aug 99. Number of observations: 260 pilot/rawinsonde balloons + 174 microwave wind profilers = 434 wind soundings. (From ECMWF.)*

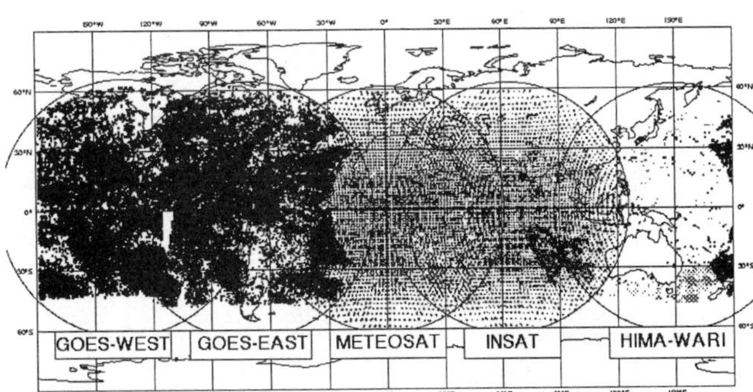

Figure 12.9f
Upper-air data locations for winds collected by ***geostationary satellites*** *from the USA (GOES), Europe (METEOSAT), India (INSAT), and Japan (HIMA-WARI). Based on movement of clouds and water vapor patterns, using IR, water vapor, and visible channels. Valid: 00 UTC on 22 Aug 99. Number of observations: 102,133. (From ECMWF.)*

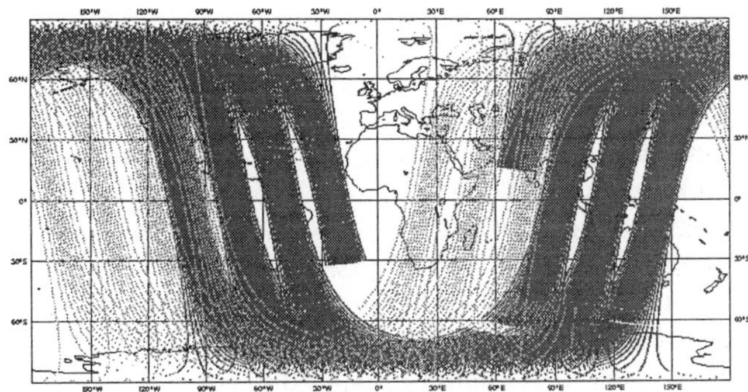

Figure 12.9g
Temperature estimate locations from radiation measurements by ***polar orbiting NOAA satellites*** *in cloudless areas, using TOVS (TIROS Operational Vertical Sounder). Satellites: NOAA 13 (light gray) and NOAA 14 (dark gray) Valid: 00 UTC on 22 Aug 99. Number of observations: 96,668. (From ECMWF.)*

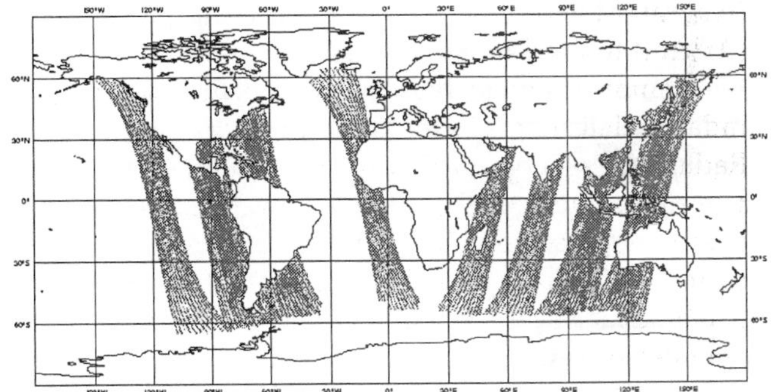

Figure 12.9h
Humidity estimate locations from microwave radiances observed by the SSM/I sensor on ***polar orbiting DMSP*** *satellites. Valid: 00 UTC on 22 Aug 99. Number of observations: 10,356. (From ECMWF.)*

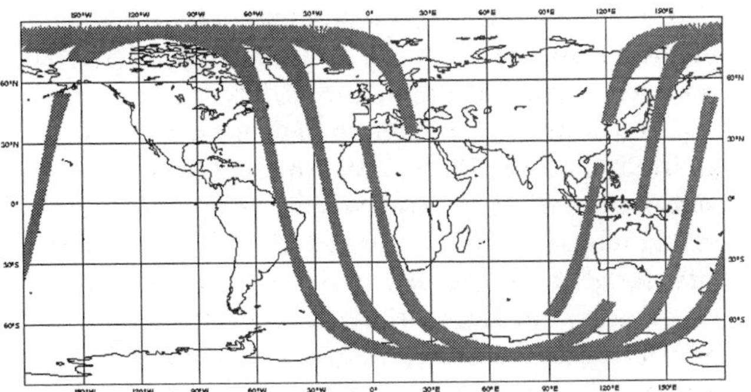

Figure 12.9i
Surface wind estimate locations from microwave scatterometer measurements of the sea surface waves by the ***polar orbiting European ERS2 satellite****. Valid: 00 UTC on 22 Aug 99. Number of observations: 98,553. (From ECMWF.)*

Map Analysis, Plotting & Isoplething

If raw numbers were printed on weather maps, the result would be overwhelming and incomprehensible to most humans. Instead, maps are plotted and simplified by drawing **isopleths** (lines of equal value, see Table 1-5) for certain key variables.

Also, features such as fronts and centers of low and high pressure can be identified. Heuristic models of fronts and lows allow humans to look at these features and anticipate their evolution.

Most weather maps are analyzed by computer. The idealized example below illustrates principles of automated weather-map analysis. While this illustration is for temperature and yields **isotherms**, similar procedures are used for any variable, such as pressure (**isobars**) or humidity (**isohumes**).

First, the synoptic temperature observations are interpolated from the irregular weather-station locations to a grid (Fig 12.10a). Such a grid of numbers is called a **field** of data, and this particular example is a **temperature field**. A discrete temperature field such as stored in a computer array approximates the continuously-varying temperature field of the real atmosphere, and is called an **analysis**.

Next, isotherms are drawn connecting points of equal temperature (Fig 12.10b), according to the following rules:

- isotherms are drawn at regular intervals (such as every 2°C, or every 5°C).
- interpolate where necessary between locations (e.g., the 5°C isotherm must be equidistant between grided observations of 4°C and 6°C.)
- isotherms never cross other isotherms
- isotherms never end in the middle of the map
- each isotherm is labeled, either at the edges of the map (the only places where isotherms can end), or along closed isotherms.
- isotherms have no kinks, except at fronts or jets

Relative maxima and minima are labeled, such as the warm and cold centers in Fig 12.10b.

Frontal zones are identified as regions of tight isotherm packing (Fig 12.10c). Note that any one isotherm need not remain within a frontal zone.

Finally, the front is always drawn on the warm side of the frontal zone (Fig 12.10d), regardless of whether it is a cold, warm, or stationary front. Frontal symbols are drawn on the side of the front toward which the front moves. Semicircles are used when warm air advances, while triangles are used when cold air advances. Frontal movement is determined from successive weather maps at different times, or by the wind direction across the front.

(a)

	west						east
north	4	3	2	2	3	4	5
	16	15	7	5	6	6	7
	20	20	20	20	12	10	9
	23	25	27	30	28	20	13
	25	27	30	33	31	28	26
south	27	28	29	29	29	29	28

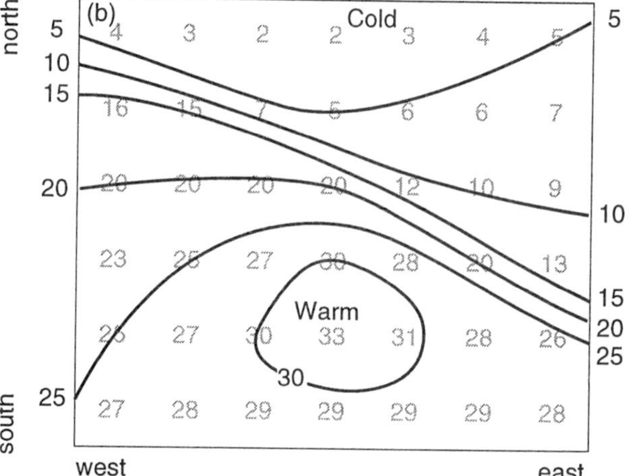

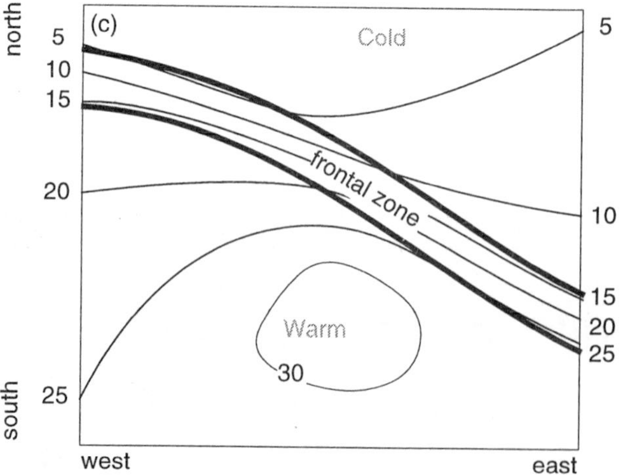

Figure 12.10
Weather map analysis. a) temperature field, with temperature in (°C) plotted on a Cartesian map (abscissa is east-west direction, and ordinate is north-south). b) isotherm analysis. c) frontal zone analysis.

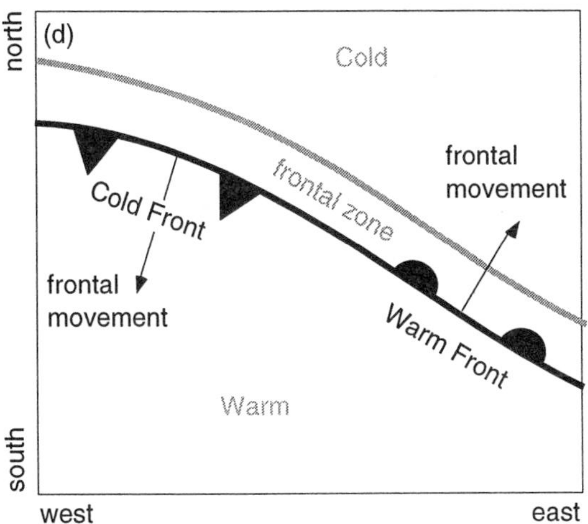

Figure 12.10d
Weather map analysis with frontal symbols.

SURFACE FRONTS

Surface fronts mark the boundaries between air masses at the earth's surface. They usually have the following attributes:

- strong horizontal temperature gradient
- strong horizontal moisture gradient
- strong horizontal wind gradient
- strong vertical shear of the horizontal wind
- relative minimum of pressure
- high vorticity
- confluence
- clouds and precipitation
- high static stability
- kinks in isopleths on weather maps.

FOCUS • The Bergen School of Meteorology

During World War I, Vilhelm Bjerknes, a Norwegian physicist with expertise in radio science and fluid mechanics, was asked in 1918 to form a Geophysical Institute in Bergen, Norway. Cut-off from weather data due to the war, he arranged for a dense network of 60 surface weather stations to be installed. Some of his students were C.-G. Rossby, H. Solberg, T. Bergeron, V.W. Ekman, H.U. Sverdrup, and his son Jacob Bjerknes.

Jacob Bjerknes used the weather station data to identify and classify cold, warm, and occluded fronts. He published his results in 1919, at age 22. The term "front" supposedly came by analogy to the battle fronts during the war. He and Solberg later explained the life cycle of cyclones.

In spite of this long list of attributes, fronts are usually labeled by the **temperature** of the advancing air mass.

Some weather features exhibit only a subset of attributes, and are not labeled as fronts. For example, a **trough** is a line of low pressure, high vorticity, clouds and possible precipitation, wind shift, and confluence. However, it often does not possess the strong horizontal temperature and moisture gradients characteristic of fronts.

Another example of an air-mass boundary that is often not a complete front is the **dry line**. It is discussed later in this chapter.

Horizontal Structure

Cold Fronts (Fig 12.11)

In the central United States, winds ahead of cold fronts typically have a southerly component, and can form strong low-level jets at night and possibly during the day. Warm, humid, hazy air advects from the south.

Sometimes a **squall line** of thunderstorms will form in advance of the front, in the warm air. These can be triggered by wind shear, and by the kinematics (advection) near fronts. They can also be thunderstorms that were initially formed on the cold front, but progressed faster than the front.

Along the front are narrow bands of towering cumuliform clouds with possible thunderstorms and scattered showers. Also along the front the winds are stronger and gusty, and the pressure reaches a relative minimum. Thunderstorm anvils can often spread hundreds of kilometers ahead of the surface front.

Winds shift to a northerly direction behind the front, advecting colder air from the north. This air is often clean with excellent visibilities and clear blue skies during daytime. If sufficient moisture is present, scattered cumulus or broken stratocumulus clouds can form within the cold air mass.

As this air mass consists of cold air advecting over warmer ground, it is statically unstable, convective, and very turbulent. However, at the top of the air mass is a very strong stable layer along the frontal inversion that acts like a lid to the convection. Sometimes over ocean surfaces the warm moist ocean leads to considerable post-frontal deep convection.

The idealized picture presented in Fig 12.11 can differ considerably in the mountains.

Warm Fronts (Fig 12.12)

In the central United States, southeasterly winds ahead of the front bring in cool, humid air from the Atlantic Ocean.

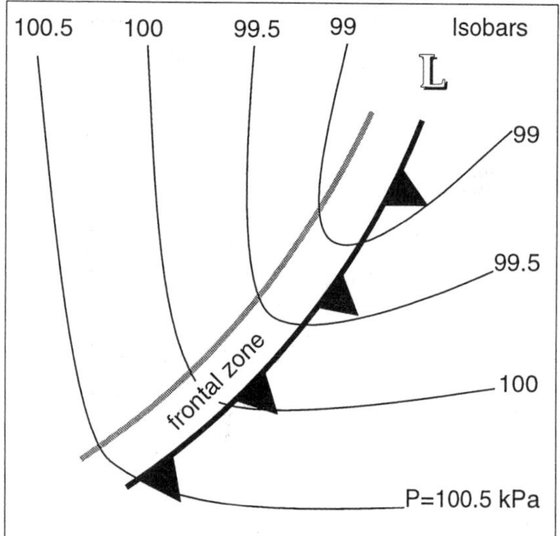

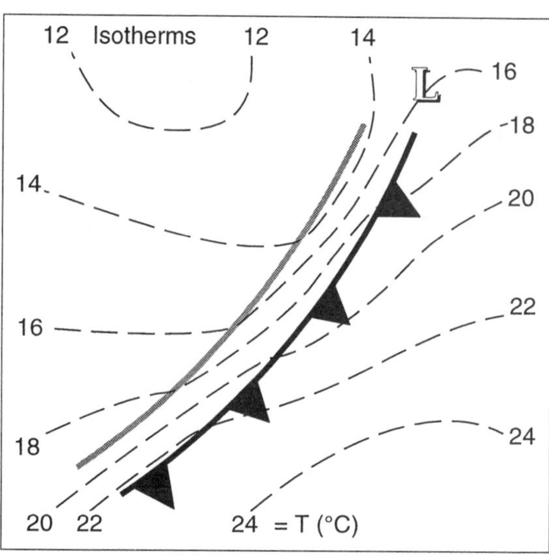

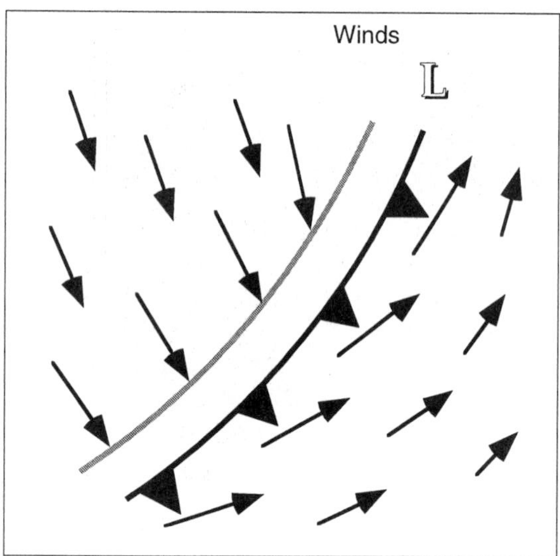

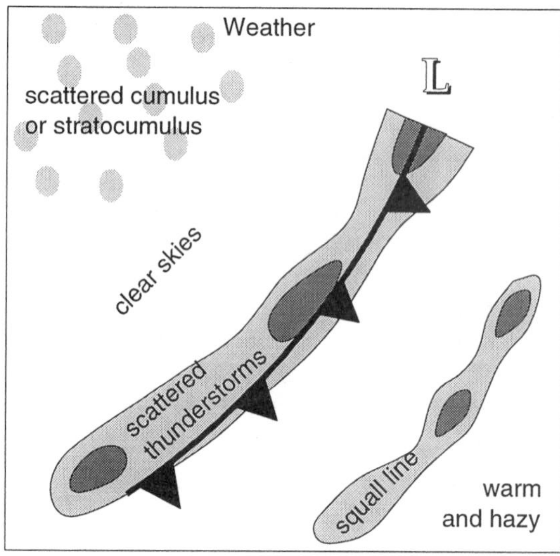

Figure 12.11 (both columns)
Cold front horizontal structure (maps can be overlain).

An extensive deck of stratiform clouds can occur hundreds of kilometers ahead of the surface front. In the cirrostratus clouds at the leading edge of this cloud shield, one can sometimes see halos, sundogs, and other optical phenomena. The cloud shield often wraps around the north side of the low center.

Along the frontal zone can be extensive areas of low clouds and fog, creating hazardous travel conditions. Nimbostratus clouds cause large areas of drizzle and light continuous rain. Moderate rain can form in multiple bands parallel to the front. The pressure reaches a relative minium at the front.

Winds shift to a more southerly direction behind the warm front, advecting in warm, humid, and hazy air. Although heating of air by the surface might not be strong, any clouds and convection that do form can often rise to relatively high altitudes because of weak static stabilities throughout the warm air mass.

Vertical Structure

Suppose that radiosonde observations (**RAOBs**) are used to probe the lower troposphere, providing temperature profiles such as those in Fig 12.13a. To locate fronts by their vertical cross section, one must first convert the temperatures into potential temperatures (Fig 12.13b). Then, lines of equal potential temperature (**isentropes**) can be drawn. Fig 12.13c shows isentropes drawn at 5°C intervals. Often isentropes are labeled in Kelvins.

In the absence of diabatic processes such as latent heating, radiative heating, or turbulent mixing, air parcels follow isentropes when they move adiabatically. For example, consider the $\theta = 35°C$ parcel that is circled above weather station B. Suppose this parcel starts to move westward.

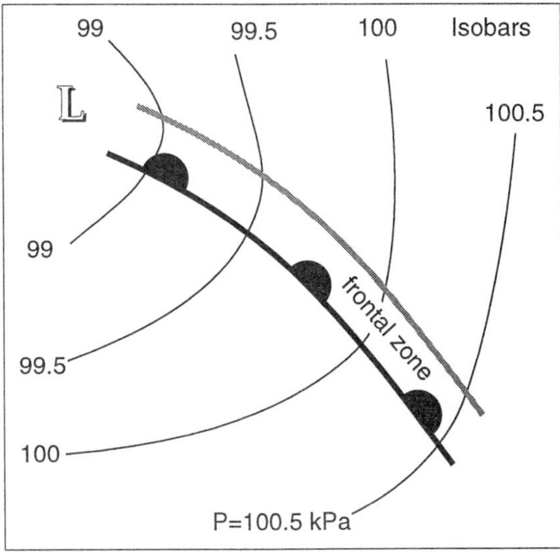

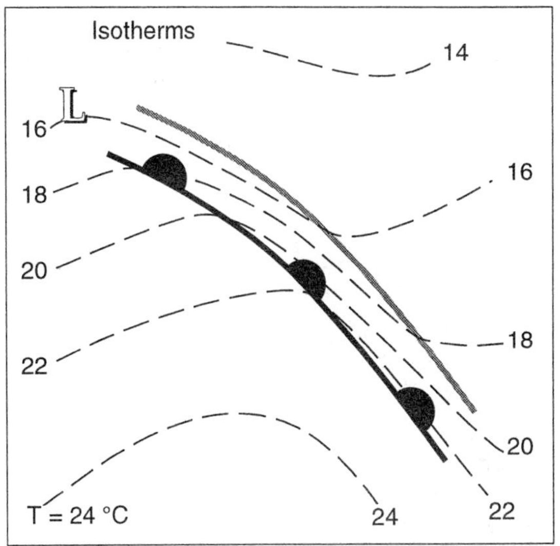

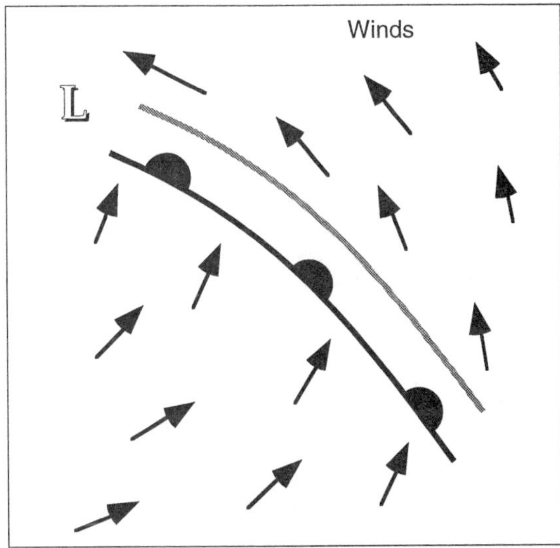

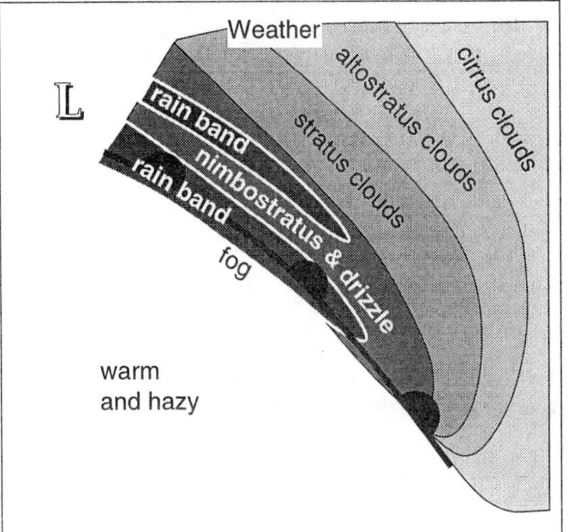

Figure 12.12 (both columns)
Warm front structure (maps can be overlain).

If the parcel were to be either below or above the 35°C isentrope at its new location above point C, buoyant forces would tend to move it vertically to the 35°C isentrope. Such forces happen continuously while the parcel moves, constantly adjusting the altitude of the parcel so it rides on the isentrope.

The net movement is westward and upward along the 35°C isentrope. Similarly, air blowing eastward would move downward along the sloping isentrope. Those air parcels that are forced to rise along isentropic surfaces can form clouds and precipitation, given sufficient moisture.

In three dimensions, one can picture **isentropic surfaces** separating warmer θ aloft from colder θ below. Analysis of the flow along these surfaces provides a clue to the weather associated with the front. Air parcels moving adiabatically must follow the "topography" of the isentropic surface. This is illustrated in Chapt. 13.

The leading boundary of the cold air is called the **surface front** or **frontal zone**. This is the region where isentropes are packed relatively close together (Fig 12.13b & c). The top of the cold air is called the **frontal inversion** (Fig 12.13c). The frontal inversion is also evident at weather stations C and D in Fig 12.13a, where the temperature increases with height. Frontal inversions of warm and cold fronts are gentle and of similar temperature-change.

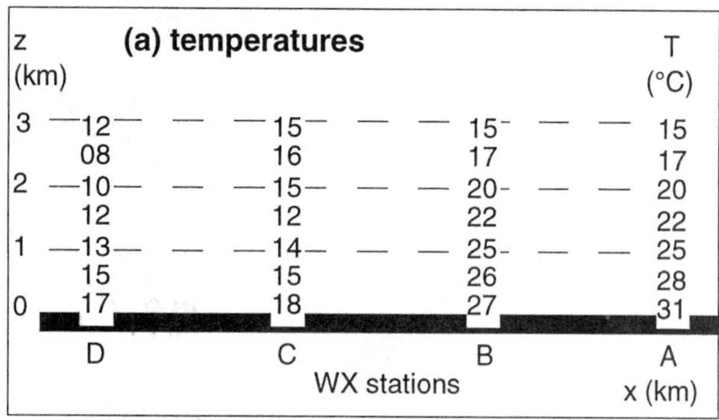

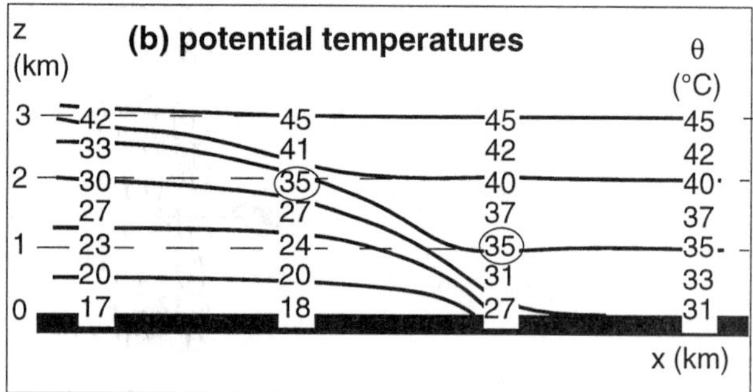

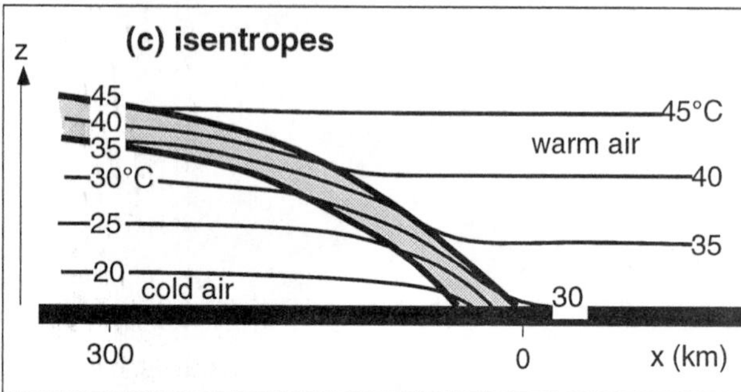

Figure 12.13
Analysis of soundings to locate fronts in a vertical cross section. Frontal zone / frontal inversion is shaded in bottom figure, and is located where the isentropes are tightly packed.

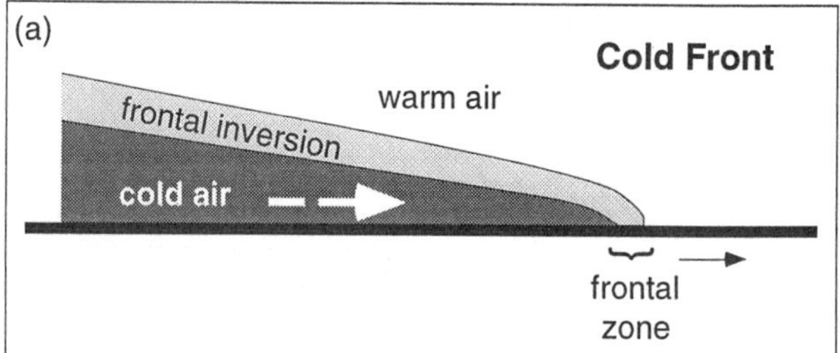

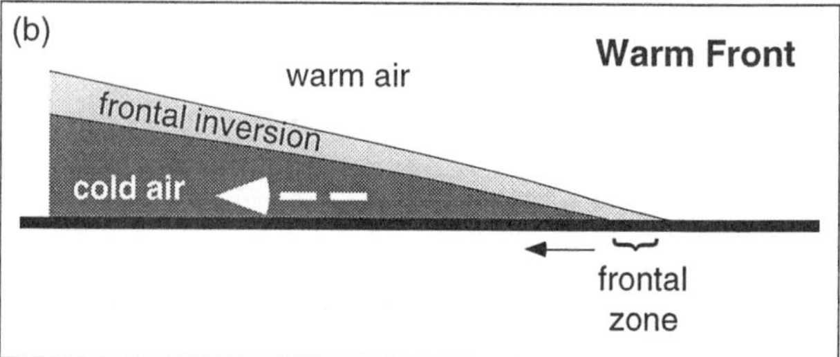

Figure 12.14
Vertical structure of fronts, based on cold air movement.

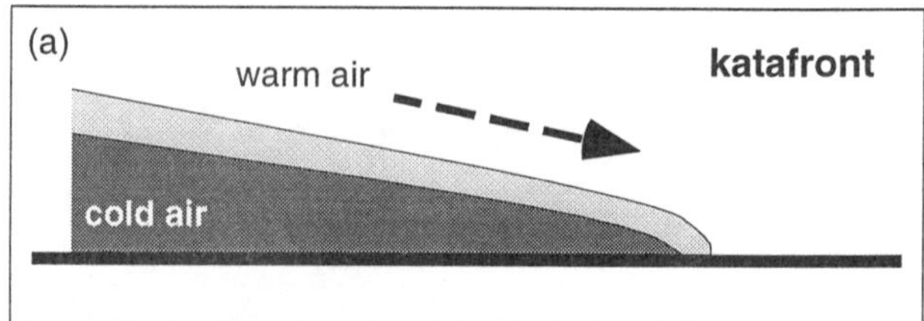

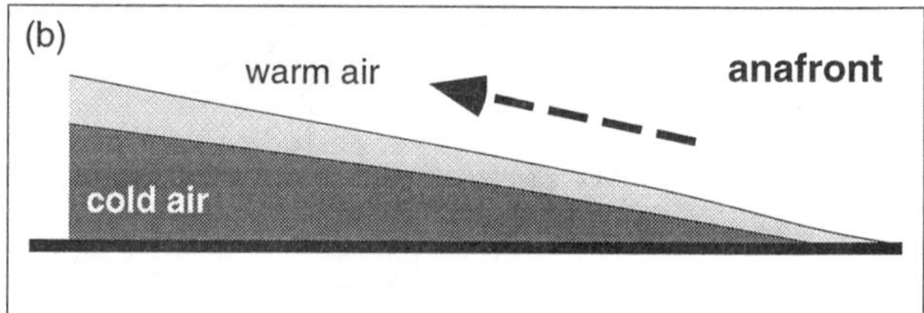

Figure 12.15
Vertical structure of fronts, based on movement of overlying air.

Within about 200 m of the surface, there are appreciable differences in frontal slope. The cold front has a steeper nose (slope ≅ 1:100) than the warm front (slope ≅ 1:300), although wide ranges of slopes have been observed.

Fronts are defined by their temperature structure, although many other quantities change across the front. Advancing cold air at the surface defines the **cold front**, where the front moves toward the warm air mass (Fig 12.14a). Retreating cold air defines the **warm front**, where the front moves toward the cold air mass (Fig 12.14b).

Above the frontal inversion, if the warm air flows down the frontal surface it is called a **katafront**, while warm air flowing up the frontal surface is an **anafront** (Fig 12.15). It is possible to have cold katafronts, cold anafronts, warm katafronts and warm anafronts.

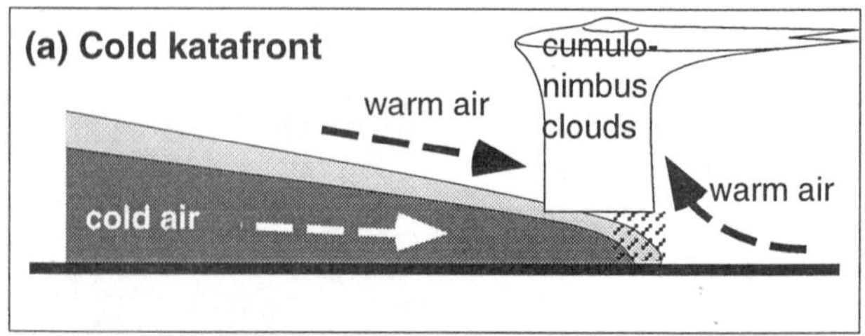

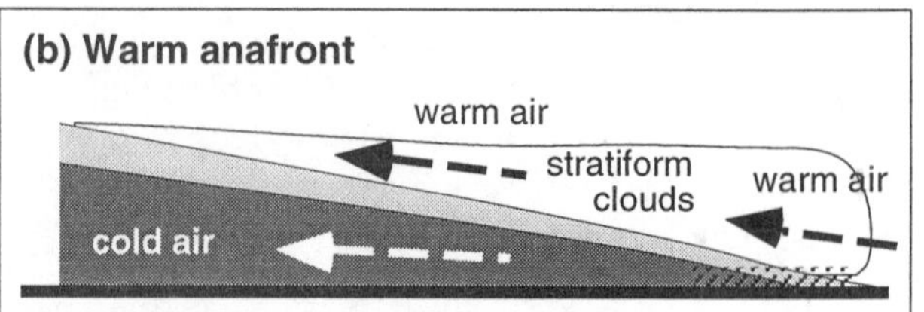

Figure 12.16
Typical fronts in the central USA.

Frequently in the central USA, the cold fronts are katafronts, as sketched in Fig 12.16a. For this situation, warm air is converging on both sides of the frontal zone, forcing the narrow band of cumuliform clouds that is typical along the front. It is also common that warm fronts are anafronts, which leads to a wide region of stratiform clouds caused by the warm air advecting up the isentropic surfaces (Fig 12.16b).

Solved Example

What weather would you expect with a warm katafront?

Solution & Discussion

Cumuliform clouds and showery precipitation would probably be similar to those in Fig 12.16a, except that the bad weather would move in the direction of the warm air at the surface, which is the direction the surface front is moving.

GEOSTROPHIC ADJUSTMENT – PART 2

Picture two air masses initially adjacent (Fig 12.17a). The cold air mass has initial depth H and uniform virtual potential temperature θ_{v1}. The warm air mass has uniform virtual potential temperature $\theta_{v1} + \Delta\theta_v$. The average absolute virtual temperature is $\overline{T_v}$. In the absence of rotation of the coordinate system, one would expect the cold air to spread out completely under the warm air due to buoyancy, reaching a final state that is horizontally homogeneous.

However, on a rotating earth the cold air experiences Coriolis force (to the right in the Northern Hemisphere) as it begins to move southward. Instead of flowing across the whole surface, the cold air spills only distance a before the winds have turned 90°, at which point further spreading stops (Fig 12.17b).

At this quasi-equilibrium, pressure-gradient force associated with the sloping cold-air interface balances Coriolis force, and there is a steady geostrophic wind U_g from east to west. The process of approaching this equilibrium is called **geostrophic adjustment.** Real atmospheres never quite reach this equilibrium.

At equilibrium, the final spillage distance a of the front from its starting location equals the **external Rossby-radius of deformation,** λ_R:

$$a = \lambda_R \equiv \frac{\sqrt{g \cdot H \cdot \Delta\theta_v / \overline{T_v}}}{f_c} \qquad \bullet(12.8)$$

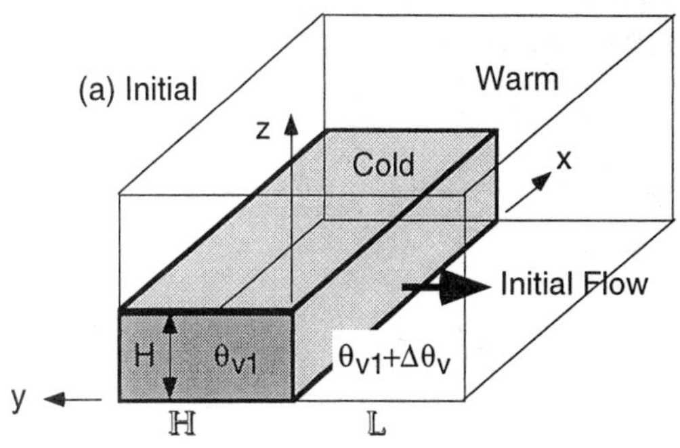

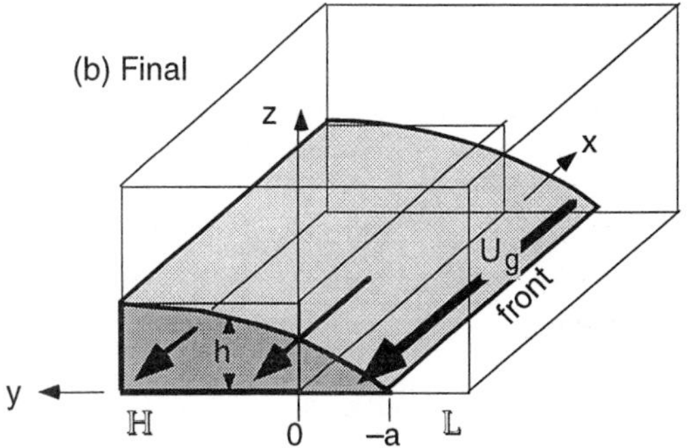

Figure 12.17
Geostrophic adjustment of a cold front. Final state (b) is in dynamic equilibrium, which is never quite attained in the real atmosphere.

where f_c is the Coriolis parameter.

The geostrophic wind U_g in the cold air at the surface is greatest at the front (neglecting friction), and exponentially decreases behind the front:

$$U_g = -\sqrt{g \cdot H \cdot (\Delta\theta_v / \overline{T_v})} \cdot \exp\left(-\frac{y+a}{a}\right) \qquad \bullet(12.9)$$

for $\infty > y \geq -a$. The depth of the cold air h is:

$$h = H \cdot \left[1 - \exp\left(-\frac{y+a}{a}\right)\right] \qquad \bullet(12.10)$$

which smoothly increases to depth H well behind the front (at large y).

Fig 12.17 is highly idealized, having air masses of distinctly different temperatures with a sharp interface in between. For a fluid with a smooth continuous temperature gradient, geostrophic adjustment occurs in a similar fashion, with a final equilibrium state as sketched in Fig 12.18. The top of this diagram represents the top of the troposphere, and the top wind vector represents the jet stream.

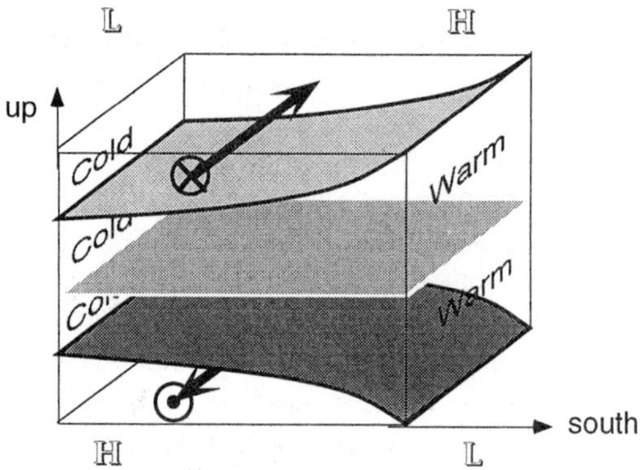

Figure 12.18
Final dynamic state after geostrophic adjustment within an environment containing continuous temperature gradients. Arrows represent geostrophic wind. Shaded areas are isobaric surfaces. H and L indicate high and low pressures. Dot-circle represents the tip of an arrow pointing toward the reader, x-circle represents the tail feathers of an arrow pointing into the page.

This state has high surface pressure under the cold air, and low surface pressure under the warm air. On the cold side, isobaric surfaces are more-closely spaced in height than on the warm side, due to the hypsometric relationship. This results in a pressure reversal aloft, with low pressure (or low heights) above the cold air and high pressure (or high heights) above the warm air.

Horizontal pressure gradients at low and high altitudes create opposite geostrophic winds, as indicated in Fig 12.18. Due to Coriolis force, the air represented in Fig 12.18 is in equilibrium; namely, the cold air does not spread any further.

This behavior of the cold air mass is extremely significant. It means that the planetary-scale flow, which is in approximate geostrophic balance, is unable to complete the job of redistributing the cold air from the poles and warm air from the tropics. Yet some other process must be acting to complete the job, in order to redistribute heat to satisfy the global energy budget (Chapter 11).

That other process is the action of cyclones and Rossby waves. Many small-scale, short-lived cyclones are not in geostrophic balance, and they can act to move the cold air further south, and the warm air further north. These cyclones feed off of the potential energy remaining in the large-scale flow; namely, the energy associated with horizontal temperature gradients. Such gradients have potential energy that can be released when the colder air slides under the warmer air (see Fig 11.1).

BEYOND ALGEBRA • Geostrophic Adjustment

We can verify that the near surface (frictionless) geostrophic wind is consistent with the sloping depth of cold air. The geostrophic wind is related to the horizontal pressure gradient at the surface by:

$$U_g = -\frac{1}{\rho \cdot f_c} \cdot \frac{\Delta P}{\Delta y} \qquad (9.18a)$$

Assume that the pressure at the top of the cold air mass in Fig 12.17 equals the pressure at the same altitude h in the warm air mass. Thus, surface pressures will be different due to only the difference in weight of air below that height.

Going from the top of the sloping cold air mass to the bottom, the pressure increase vertically is given by the hydrostatic eq: $\Delta P_{cold} = -\rho_{cold} \cdot g \cdot h$

A similar equation can be written for the warm air below h. Thus, at the surface, the difference in pressures under the cold and warm air masses are:

$$\Delta P = -g \cdot h \cdot (\rho_{cold} - \rho_{warm})$$

multiplying the RHS by $\overline{\rho} / \overline{\rho}$, where $\overline{\rho}$ is an average density, yields:

$$\Delta P = -g \cdot h \cdot \overline{\rho} \cdot [(\rho_{cold} - \rho_{warm}) / \overline{\rho}]$$

But the ideal gas law can be used to convert from density to virtual temperature, remembering to change the sign because air of greater temperature has less density. Also, $\Delta T_v = \Delta\theta_v$. Thus:

$$\Delta P = -g \cdot h \cdot \overline{\rho} \cdot [(\theta_{v\ warm} - \theta_{v\ cold}) / \overline{T_v}]$$

where this is the pressure change in the negative y direction.

Plugging this into eq. (9.18a) gives:

$$U_g = -\frac{g \cdot (\Delta\theta_v / T_v)}{f_c} \cdot \frac{\Delta h}{\Delta y}$$

which we can write in differential form:

$$\boxed{U_g = -\frac{g \cdot (\Delta\theta_v / T_v)}{f_c} \cdot \frac{\partial h}{\partial y}} \qquad (b)$$

The equilibrium value of h was given by eq. (12.10):

$$h = H \cdot \left[1 - \exp\left(-\frac{y+a}{a}\right)\right] \qquad (12.10)$$

Thus, the derivative is:

$$\frac{\partial h}{\partial y} = \frac{H}{a} \cdot \exp\left(-\frac{y+a}{a}\right)$$

Plugging this into eq. (b) gives:

$$U_g = -\frac{g \cdot (\Delta\theta_v / T_v)}{f_c} \cdot \frac{H}{a} \cdot \exp\left(-\frac{y+a}{a}\right) \qquad (c)$$

But from eq. (12.8) we see that:

$$f_c \cdot a = f_c \cdot \lambda_R = \sqrt{g \cdot H \cdot (\Delta\theta_v / T_v)} \qquad (12.8)$$

Eq. (c) then becomes:

$$\boxed{U_g = -\sqrt{g \cdot H \cdot (\Delta\theta_v / T_v)} \cdot \exp\left(-\frac{y+a}{a}\right)} \qquad (12.9)$$

Thus, the wind is consistent with the sloping height.

Solved Example

A cold air mass of depth 1 km and virtual potential temperature 0°C is imbedded in warm air of virtual potential temperature 20°C. Find the Rossby deformation radius, the maximum geostrophic wind speed, and the equilibrium depth of the cold air mass at $y = 0$. Assume $f_c = 10^{-4}\ s^{-1}$.

Solution

Given: $\overline{T_v} = 280$ K, $\Delta\theta_v = 20$ K, $H = 1$ km, $f_c = 10^{-4}\ s^{-1}$.

Find: $a = ?$ km, $U_g = ?$ m/s at $y = -a$, and $h = ?$ km at $y = 0$.

Use eq. (12.8):

$$a = \frac{\sqrt{(9.8\text{m}\cdot\text{s}^{-2})\cdot(1000\text{m})\cdot(20\text{K})/(280\text{K})}}{10^{-4}\text{s}^{-1}} = \underline{\mathbf{265\ km}}$$

Use eq. (12.9):

$$U_g = -\sqrt{(9.8\text{m}\cdot\text{s}^{-2})(1000\text{m})(20\text{K})/(280\text{K})}\cdot\exp(0)$$
$$= \underline{\mathbf{-26.5\ m/s}}$$

Use eq. (12.10):

$$h = (1\text{km})\cdot[1-\exp(-1)] = \underline{\mathbf{0.63\ km}}$$

Check: Units OK. Physics OK.

Discussion: Frontal-zone widths on the order of 200 km are small compared to lengths(1000s km).

FRONTOGENESIS

Fronts are identified by the large change in temperature across the frontal zone. Hence, the horizontal temperature gradient (temperature change per distance across the front) is one measure of frontal strength. Usually potential temperature is used instead of temperature to simplify the problem when vertical motions are possible.

Physical processes that tend to increase the potential-temperature gradient are called **frontogenetic** — literally they cause the birth or strengthening of the front. Three classes of such processes are kinematic, thermodynamic, and dynamic.

Kinematics

Kinematics refers to motion or advection. This class of processes cannot create potential temperature gradients, but it can strengthen or weaken existing gradients.

From earlier chapters we saw that radiative heating causes north-south temperature gradients between the equator and poles. Also, the general circulation causes meanders in the jet stream which creates transient east-west temperature gradients along troughs and ridges. The standard-atmosphere also has a vertical gradient of potential temperature in the troposphere. Thus, it is fair to assume that temperature gradients exist, which could be strengthened during kinematic frontogenesis.

To illustrate kinematic frontogenesis, consider an initial potential-temperature field with uniform gradients in the x, y, and z directions, as sketched in Fig 12.19. The gradients have the following signs:

$$\frac{\Delta\theta}{\Delta x} = + \qquad \frac{\Delta\theta}{\Delta y} = - \qquad \frac{\Delta\theta}{\Delta z} = + \qquad (12.11)$$

Namely, potential temperature increases toward the east, decreases toward the north, and increases upward. There are no fronts in this picture initially.

We will examine the subset of advections that tend to create a cold front aligned north-south. Define the strength of the front as the potential-temperature gradient across the front:

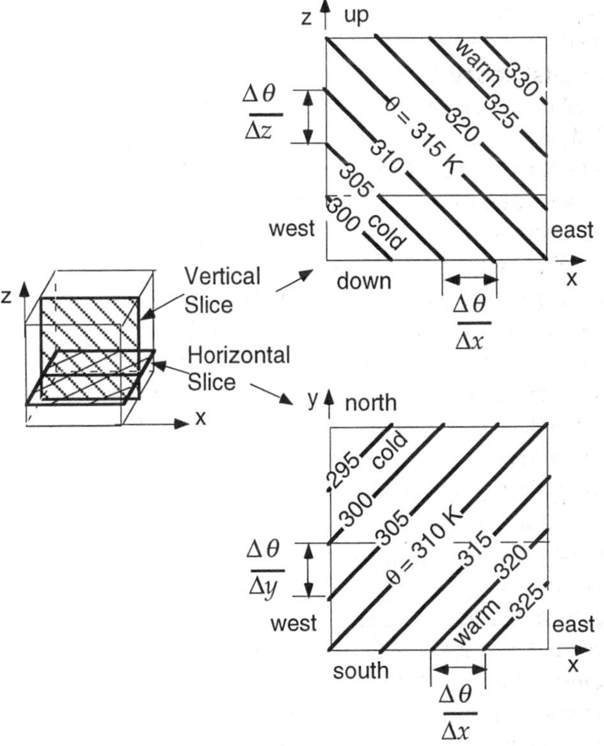

Figure 12.19
Vertical and horizontal slices through a volume of atmosphere, showing initial conditions prior to frontogenesis. The thin horizontal lines show where the planes intersect.

$$\text{Frontal Strength} = FS \equiv \frac{\Delta\theta}{\Delta x} \qquad \bullet(12.12)$$

Change of frontal strength with time due to advection is given by the kinematic frontogenesis eq.:

$$\underbrace{\frac{\Delta(FS)}{\Delta t}}_{\text{Strengthening}} = \underbrace{-\left(\frac{\Delta\theta}{\Delta x}\right)\cdot\left(\frac{\Delta U}{\Delta x}\right)}_{\text{Confluence}} \underbrace{-\left(\frac{\Delta\theta}{\Delta y}\right)\cdot\left(\frac{\Delta V}{\Delta x}\right)}_{\text{Shear}} \underbrace{-\left(\frac{\Delta\theta}{\Delta z}\right)\cdot\left(\frac{\Delta W}{\Delta x}\right)}_{\text{Tilting}} \qquad \bullet(12.13)$$

Confluence

Suppose there is a strong west wind U approaching from the west, but a weaker west wind departing at the east (Fig 12.20). Namely, the air from the west catches-up to air in the east.

For this situation $\Delta U/\Delta x = -$, but $\Delta\theta/\Delta x = +$ from eq. (12.13). Hence, the product of these two terms, when multiplied by the negative sign attached to the confluence term, tend to strengthen the front $[\Delta(FS)/\Delta t = +]$.

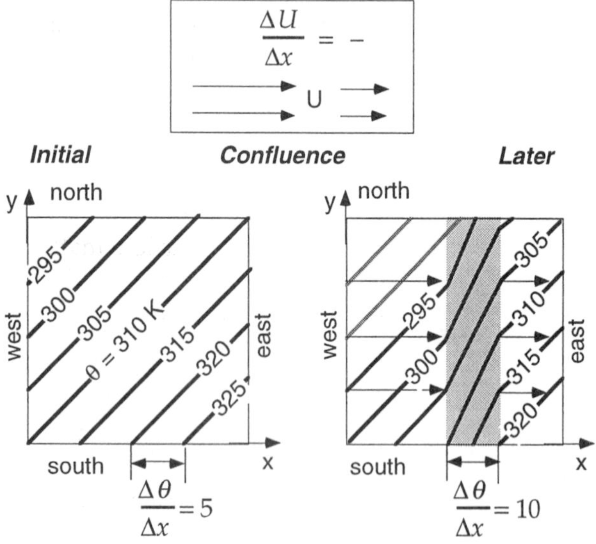

Figure 12.20
Confluence strengthens the frontal zone (shaded), in this case. Arrow tails indicate starting locations for the isotherms.

Shear.

Suppose the wind from the south is stronger on the east side of the domain than the west (Fig 12.21). This is one form of wind shear. As the isentropes on the east advect northward faster than those on the west, the potential temperature gradient is strengthened in between, creating a frontal zone.

While the shear is positive $(\Delta V/\Delta x = +)$, the northward temperature gradient is negative $(\Delta\theta/\Delta y = -)$. Thus the product is positive when the preceding negative sign from the shear term is included. Frontal strengthening occurs for this case.

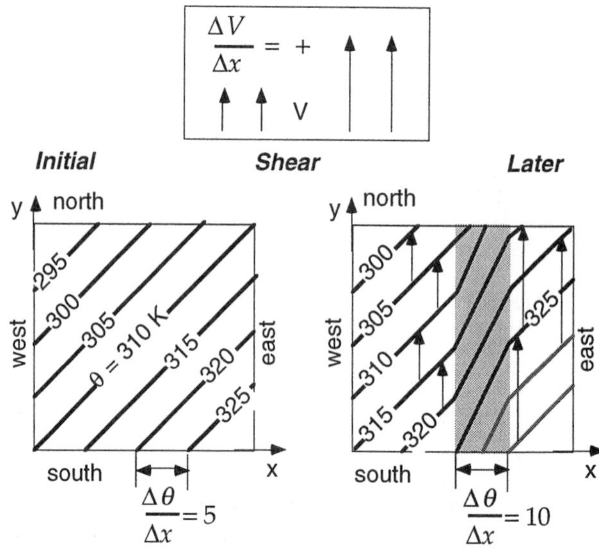

Figure 12.21
Shear strengthens the frontal zone (shaded), in this case. Arrow tails indicate starting locations for the isotherms.

Tilting.

If updrafts are stronger on the cold side of the domain than the warm side, then the vertical potential temperature gradient will be tilted into the horizontal. The result is a strengthened frontal zone (Fig 12.22).

The horizontal gradient of updraft velocity is negative in this example $(\Delta W/\Delta x = -)$, while the vertical potential temperature gradient is positive

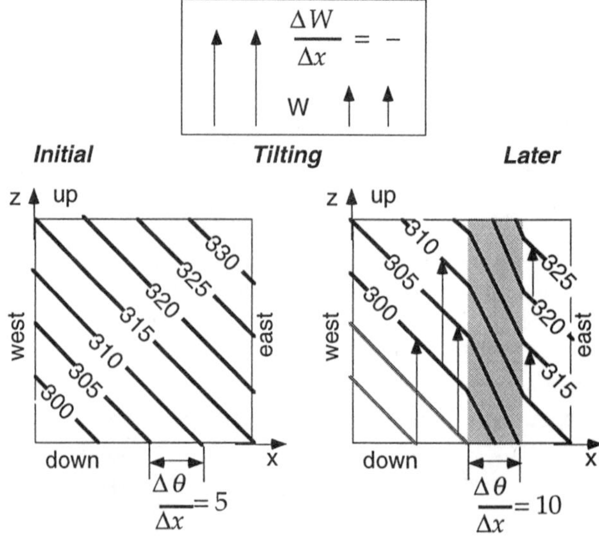

Figure 12.22
Tilting of the vertical temperature gradient into the horizontal strengthens the frontal zone (shaded), in this illustration. However, for most real surface fronts the colder air sinks, causing frontal weakening. Arrow tails indicate starting locations for the isotherms.

($\Delta\theta/\Delta z = +$). The product, when multiplied by the negative sign attached to the tilting term, yields a positive contribution to the strengthening of the front for this case.

While this example was contrived to illustrate frontal strengthening, **for most real fronts the tilting term causes weakening.** Such **frontolysis** is weakest near the surface because vertical motions are smaller there (the wind cannot blow through the ground).

Tilting is important and sometimes dominant for upper-level fronts, described later.

Solved Example

Given an initial environment with $\Delta\theta/\Delta x$ = 0.01°C/km, $\Delta\theta/\Delta y$ = –0.01°C/km, and $\Delta\theta/\Delta z$ = 3.3°C/km. Also suppose that $\Delta U/\Delta x$ = – 0.05 (m/s)/km, $\Delta V/\Delta x$ = 0.05 (m/s)/km, and $\Delta W/\Delta x$ = 0.02 (cm/s)/km. Find the frontogenesis rate.

Solution

Given: (see above)

Find: $\Delta(FS)/\Delta t$ = ? °C·km^{-1}·day^{-1}

Use eq. (12.13):

$$\frac{\Delta(FS)}{\Delta t} = -\left(0.01\frac{°C}{km}\right)\cdot\left(-0.05\frac{m/s}{km}\right) - \left(-0.01\frac{°C}{km}\right)\cdot\left(0.05\frac{m/s}{km}\right) - \left(3.3\frac{°C}{km}\right)\cdot\left(0.0002\frac{m/s}{km}\right)$$

$$= +0.0005 + 0.0005 - 0.00066 \text{ °C·m·s}^{-1}\text{·km}^{-2}$$

$$= \underline{+0.029} \text{ °C·km}^{-1}\text{·day}^{-1}$$

Check. Units OK. Physics OK.

Discussion: Frontal strength $\Delta\theta/\Delta x$ nearly tripled in one day, increasing from 0.01 to 0.029 °C/km.

Thermodynamics

The previous kinematic examples showed **adiabatic** advection (potential temperature as conserved while being blown downwind). However, **diabatic** (non-adiabatic) thermodynamic processes can heat or cool the air at different rates on either side of the domain. These processes include radiative heating/cooling, conduction from the surface, turbulent mixing across the front, and latent heat release/ absorption associated with phase changes of water in clouds.

Define the diabatic warming rate (DW) as:

$$\text{Diabatic Warming Rate} = DW = \frac{\Delta\theta}{\Delta t} \qquad •(12.14)$$

If diabatic heating is greater on the warm side of the front (the east side in this example) than the cold side, then the front will be strengthened:

$$\frac{\Delta(FS)}{\Delta t} = \frac{\Delta(DW)}{\Delta x} \qquad •(12.15)$$

In most real fronts, **turbulent mixing** between the warm and cold sides weakens the front (i.e., causes **frontolysis**). **Conduction** from the surface also contributes to frontolysis. For example, behind a cold front the cold air blows over a usually-warmer surface, which heats the cold air (i.e., air mass modification) and reduces the temperature contrast across the front. Similarly, behind warm fronts the warm air is usually advecting over cooler surfaces.

Over both warm and cold fronts the warm air is often forced to rise. This rising air can cause **condensation** and cloud formation, which strengthens fronts by warming the already-warm air.

Radiative cooling from the tops of stratus clouds reduces the temperature on the warm side of the front, contributing to frontolysis of warm fronts. Radiative cooling from the tops of post-cold frontal stratocumulus clouds can strengthen the front by cooling the already-cold air.

Dynamics

Kinematics and thermodynamics are insufficient to explain observed frontogenesis. While kinematic frontogenesis gives doubling or tripling of frontal strength in a day (see previous solved example), observations show that frontal strength can increase by a factor of 15 during a day. Dynamics can cause this rapid strengthening. Picture an initial state in geostrophic equilibrium, as sketched in Fig 12.23.

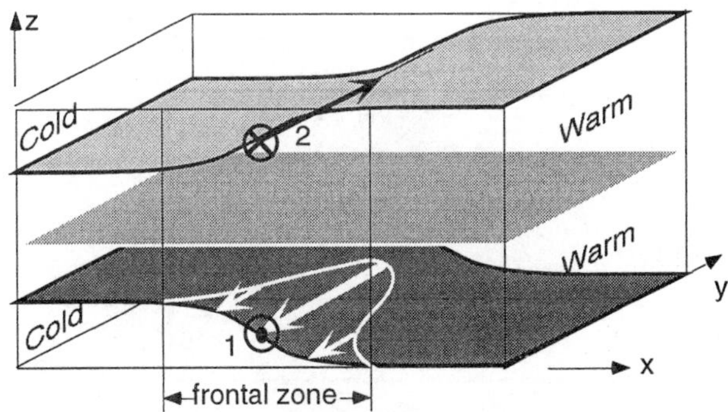

Figure 12.23
Isobaric surfaces near a hypothetical front, before being altered by dynamics. Vectors show equilibrium geostrophic winds, initially (state o). Dot-circle represents the tip of an arrow pointing toward the reader, x-circle represents the tail feathers of an arrow pointing into the page.

This figure shows a special situation where pressure gradients and geostrophic winds exist only midway between the left and right sides. Zero gradients and winds are at the left and right sides. A frontal zone is in the center of this diagram.

Suppose that some external forcing such as kinematic confluence due to a passing Rossby wave causes the front to strengthen a small amount, as sketched in Fig 12.24a. Not only does the potential temperature gradient tighten, but the pressure gradient also increases due to the hypsometric relationship.

The increased pressure gradient implies a different, increased geostrophic wind. However, initially the actual winds are slower, with magnitude equal to the original geostrophic wind speed.

While the actual winds adjust toward the new geostrophic value, they temporarily turn away from the geostrophic direction (Fig 12.25) due to the imbalance between pressure-gradient and Coriolis forces. During this transient state (b), there is a component of wind in the x-direction. This is called **ageostrophic** flow, because there is no geostrophic wind in the x-direction. This flow is sketched in Fig 12.24b.

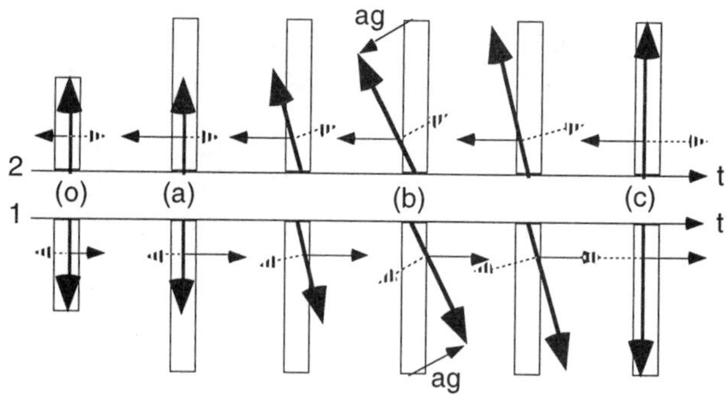

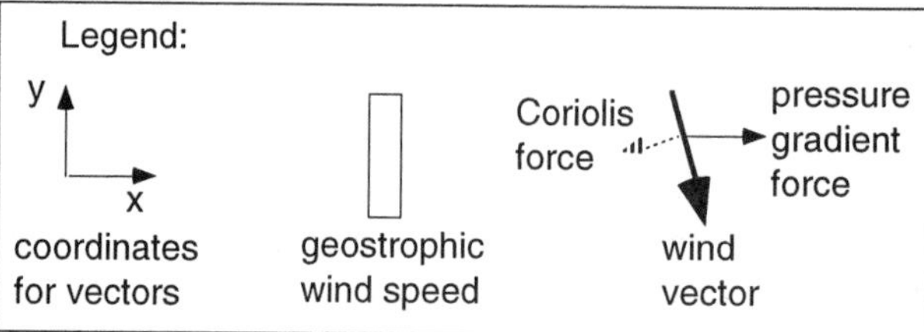

Figure 12.25
Time lines labeled 1 & 2 are for the vectors 1 & 2 in Fig 12.23. Initial state (o) is given by Fig 12.23. Later states (a)-(c) correspond to Figs 12.24(a)-(c). The ageostrophic wind (ag) at time (b) is also indicated. States (o) and (c) are balanced.

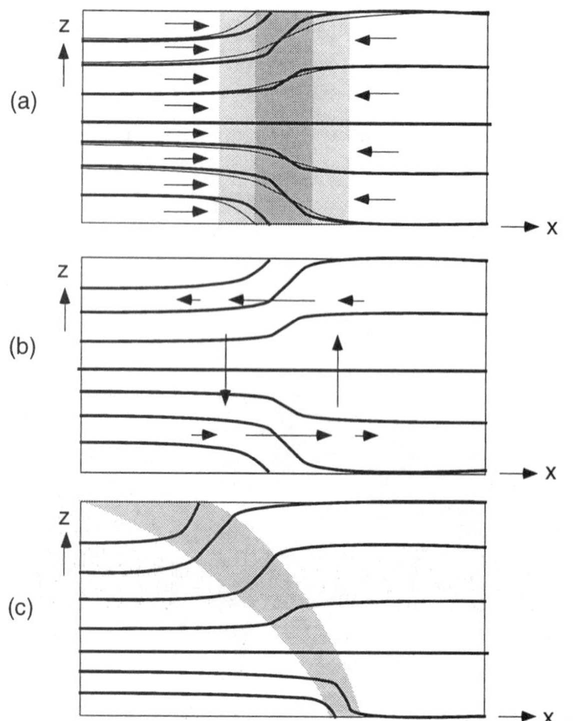

Figure 12.24
Vertical cross section showing dynamic strengthening of a front. (a) Initial state (thin lines) modified by confluence (thick lines). (b) Ageostrophic circulation, called a Sawyer-Eliassen circulation. (c) Final equilibrium state. Frontal zone is shaded, lines are isobars, and arrows are ageostrophic winds.

Because mass is conserved, horizontal convergence and divergence of the U-component of wind cause vertical circulations. The result is a temporary cross-frontal, or **transverse** circulation (Fig 12.24b). It is called a **Sawyer-Eliassen circulation.** The updraft portion of the circulation can drive convection, and cause precipitation.

The winds finally reach their new equilibrium value equal to the geostrophic wind. In this final state, there are no ageostrophic winds, and no cross-frontal circulation. However, during the preceding transient stage, the ageostrophic cross frontal circulation caused extra confluence near the surface. This dynamically-induced confluence adds to the original kinematic confluence to strengthen the surface front. The transverse circulation also tilts the front (Fig 12.24c).

In summary, a large and relatively steady geostrophic wind blows parallel to the front, due to adjustment toward geostrophic equilibrium (Fig 12.17). Superimposed can be a weak, transient, cross-frontal circulation (Fig 12.24b). These two factors are also important for upper-tropospheric fronts, described later.

OCCLUDED FRONTS AND MID-TROPOSPHERIC FRONTS

When three or more air masses come together, such as in an **occluded front**, it is possible for one or more fronts to ride over the top of a colder air mass. This creates lower- or mid-tropospheric fronts that do not touch the surface, and which would not be signaled by temperature changes and wind shifts at the surface. However, such **fronts aloft** can trigger clouds and precipitation observed at the surface.

Occluded fronts occur when cold fronts catch up to warm fronts. What happens depends on the temperature difference between the cold advancing air behind the cold front, and the cold retreating air ahead of the warm front.

Fig 12.26 shows a cold front occlusion, where very cold air catches up to, and under-rides cooler air, forcing the warm air aloft. Most occlusions in interior N. America are of this type, due to the very cold air that advances from Canada in winter.

Observers at the surface would notice stratiform clouds in advance of the front, which would normally signal an approaching warm front. However, instead of a surface warm front, a surface occluded front passes, and the surface temperature decreases like a cold front. The trailing edge of the warm air aloft marks the warm front aloft.

Fig 12.27 shows a warm front occlusion, where cool air catches up to, and over-rides colder air, forcing the warm air aloft. Most occlusions in Europe are of this type, due to the mild cool air that advances from the N. Atlantic and North Sea during winter.

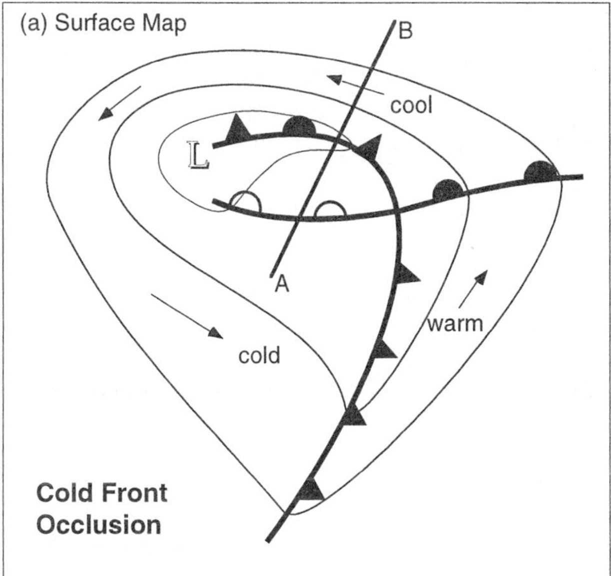

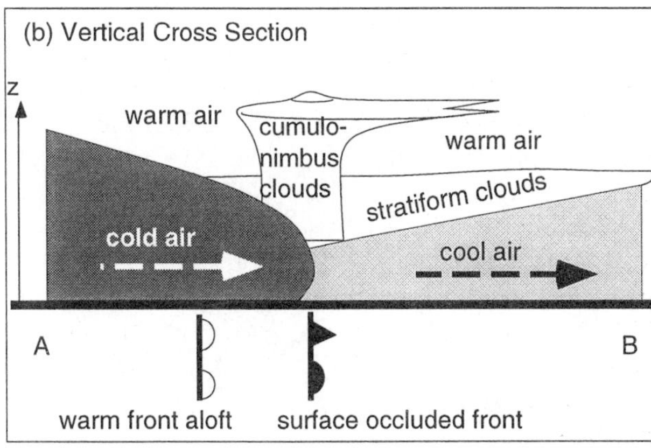

Figure 12.26
Cold front occlusion. (a) Surface map showing position of surface cold front (dark triangles), surface warm front (dark semicircles), surface occluded front (dark triangles and semicircles), and warm front aloft (white semicircles). (b) Vertical cross section along slice A-B from top diagram.

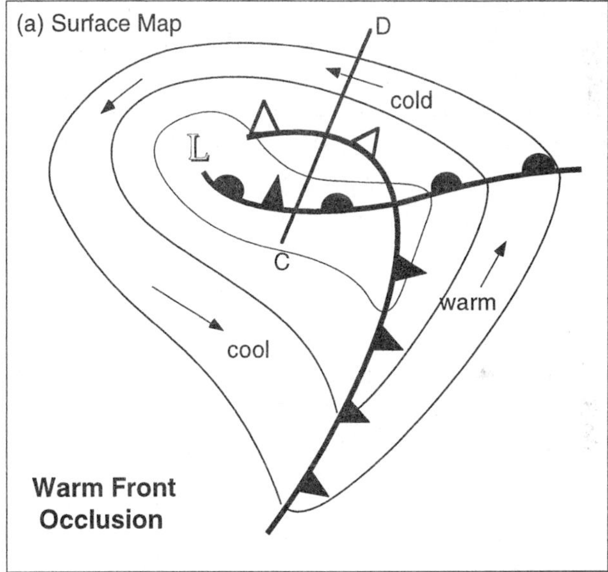

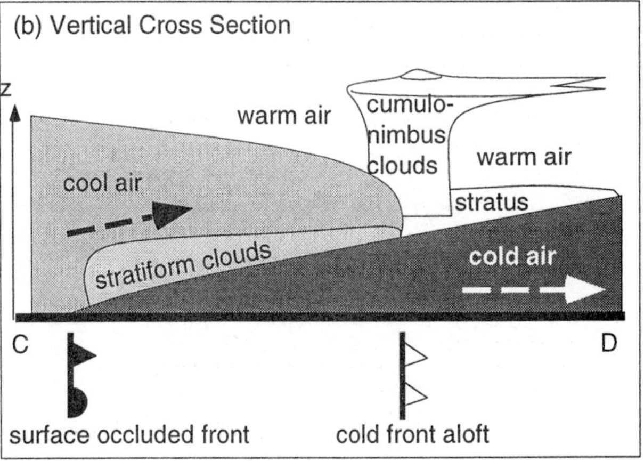

Figure 12.27
Warm front occlusion. (a) Surface map. Symbols are similar to Fig 12.26, except that white triangles denote a cold front aloft. (b) Vertical cross section along slice C-D from top diagram.

Observers at the surface notice stratiform clouds in advance of the front, but with an imbedded shower or thunderstorm. Later, a surface occluded front passes and the surface temperature warms. The leading edge of cool air aloft marks the cold front aloft.

The illustrations above are "text-book" examples of prototypical situations. In real life, more complex maps and cross sections can occur. Sometimes more than three air masses can be drawn together in a low.

UPPER-TROPOSPHERIC FRONTS

Upper-tropospheric fronts are also called **upper-level fronts**, and are sometimes associated with **folds in the tropopause**. A cross section through an idealized upper-level front is sketched in Fig 12.28.

In the lower troposphere, the south-to-north temperature gradient creates the jet stream due to the thermal wind relationship (see Chapter 11). A reversal of the meridional temperature gradient above 10 km in this idealized sketch causes wind velocities to decrease above that altitude. Within the stratosphere, the isentropes are spaced closer together, indicating greater static stability.

South of the core of the jet stream, the tropopause is relatively high. To the north, it is lower. Between these two extremes, the tropopause can wrap around the jet core, and fold back under the jet, as sketched in Fig 12.28. Within this fold the isentropes are tightly packed, indicating an **upper-tropospheric front.** Sometimes the upper-level front connects with a surface front (not shown in Fig 12.28).

Stratospheric air has unique characteristics that allow it to be traced. Relative to the troposphere, stratospheric air has high ozone content, high radioactivity (due to nuclear bomb testing), high static stability, low water-vapor mixing ratio, and high **isentropic potential vorticity**.

The tropopause fold brings air of stratospheric origin down into the troposphere. This causes an injection of ozone and radioactivity into the troposphere.

The heavy line in Fig 12.28 corresponds roughly to the 1.5 PVU (potential vorticity units) isopleth, and is a good indicator of the tropopause. Above this line, PVU values are greater than 1.5 , and increase rapidly with height to values of roughly 10 PVU at the top of Fig 12.28. Below this line, the PVU gradient is weak, with typical values of about 0.5 PVU at midlatitudes. PVU is negative in the Southern Hemisphere, but of similar magnitude.

Sawyer-Eliassen dynamics play a role in formation of tropospheric folds. Picture a tropopause that changes depth north to south, as in Fig 11.11, but does not yet have a fold. If the jet core advects in colder air, then the thickness between the 10 and 20 kPa isobaric surfaces (Fig 11.11b) will decrease as described by the hypsometric equation. This will

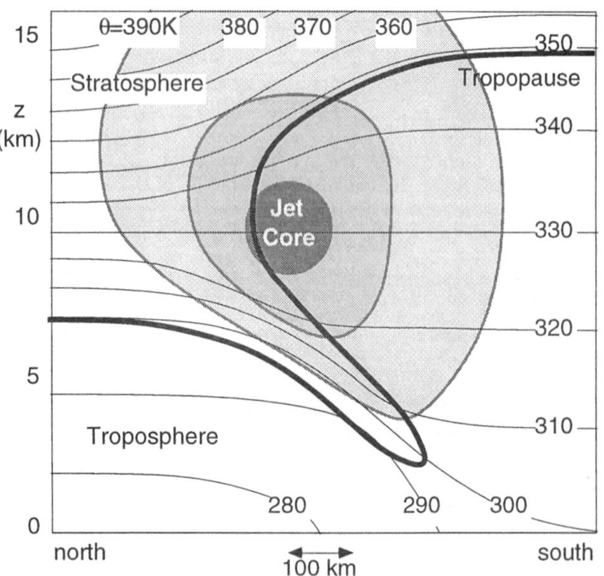

Figure 12.28
Idealized vertical cross-section through the jet stream. Shading indicates west-to-east speeds (into the page): light ≥ 50 m/s, medium ≥ 75 m/s, dark ≥ 100 m/s. Thin lines are isentropes. Thick line is the tropopause.

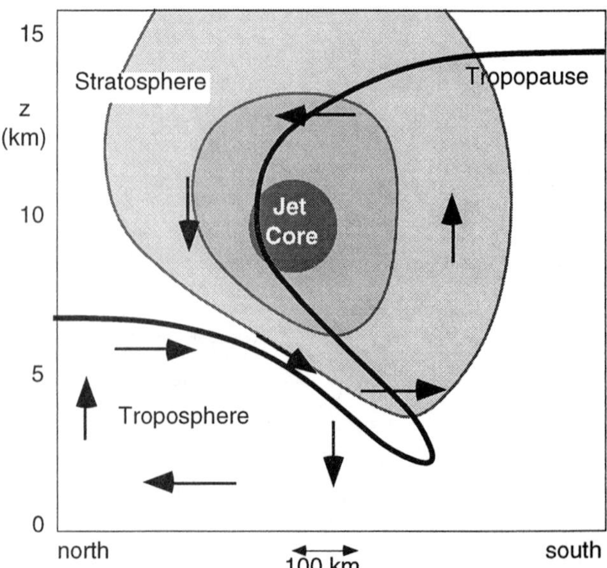

Figure 12.29
Transverse (cross-jet-stream) circulations (arrows) that dynamically form an upper-tropospheric front.

cause greater slope (i.e., greater height gradient) of the 10 kPa surface near the jet core. This temporarily upsets geostrophic balance, causing an ageostrophic circulation above the jet from south to north.

This transient flow continues to develop into a cross jet-steam direct circulation that forms around the jet as sketched in Fig 12.29 (similar to Fig 12.24b). Below the fold is an indirect circulation that also forms due to a complex geostrophic imbalance. Both circulations distort the tropopause to produce the fold, as sketched in Fig 12.29, and strengthen the front due to the kinematic tilting term.

DRY LINES

In western Texas, USA, there often exists a boundary between warm humid air to the east, and warm dry air to the west (Fig 12.30). This boundary is called the **dry line**. Because there is little temperature contrast across it, it cannot be labeled as a warm or cold front. The map symbol for a dry line is like a warm front, except with open semicircles adjacent to each other, pointing toward the moist air.

Moist air on the east side comes from the Gulf of Mexico, while the dry air comes from the semi-arid plateaus of Mexico and the Southwestern USA. Dry lines are observed during roughly 40 to 50% of all days in spring and summer, in that part of the world.

During midday through afternoon, convective clouds are often triggered along the dry lines. Some of these cloud bands grow into organized thunderstorm squall lines that can propagate east from the dry line.

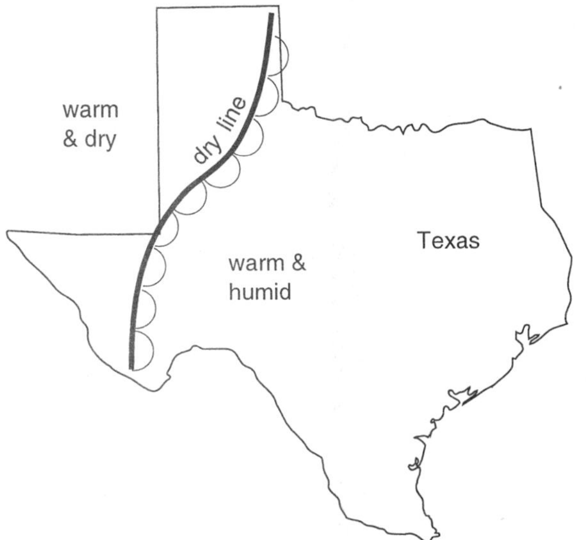

Figure 12.30
Typical location of dry line in the southwestern USA.

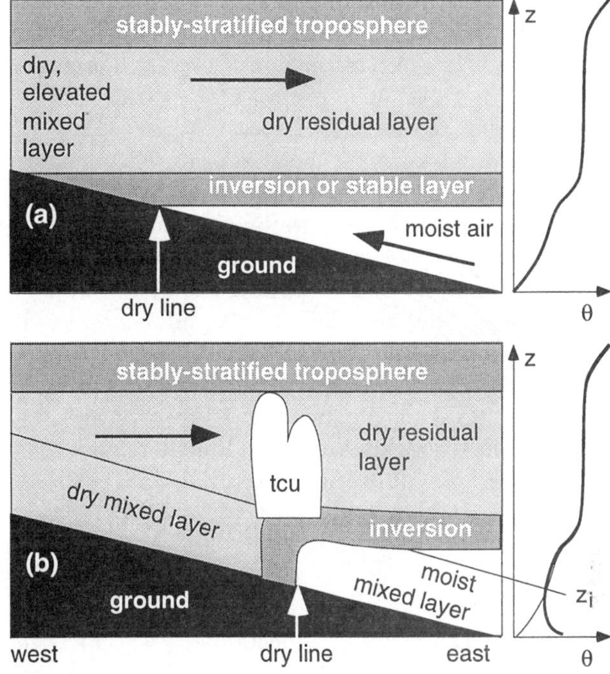

Figure 12.31
Idealized vertical cross section through a dry line (indicated by white arrow). Winds are indicated with black arrows. (a) Early morning. (b) Mid afternoon, with a convective mixed layer of constant-thickness z_i. tcu = towering cumulus clouds.

Dry lines need not be associated with a wind shift, convergence, vorticity, nor with low pressure. Hence, they do not satisfy the definition of a front. However, sometimes dry lines combine with troughs to dynamically contribute to cyclone development.

Dry lines tend to move eastward during the morning, and return westward during evening. This regular diurnal cycle is associated with the daily development of a convective mixed layer, interacting with the sloping terrain.

Over the high plateaus to the west, deep mixed layers can form. Westerly winds advect this dry air eastward, where it overrides cooler, humid air at lower altitudes (Fig 12.31a). Convective turbulence decays in this overrunning air when it loses contact with the heated ground. The result is a non-turbulent residual layer of dry air aloft, with nearly adiabatic lapse rate. Between that elevated residual layer and the more moist air below is a strong temperature inversion.

After sunrise, the warm ground causes a convective mixed-layer to form and grow. For simplicity, we will model a mixed layer of constant depth z_i above local ground level. Because the ground is higher in the west, the top of the mixed layer is able to reach the dry air early in the morning. When this dry air is entrained into the air below, it causes the surface humidity to drop.

Further east, the top of the mixed layer has yet to reach the dry air aloft, so the humidity is still high there. The dry line separates this moist surface air from the dry air to the west.

As the day continues, the mixed layer deepens, allowing locations further east to mix with the dry air aloft (Fig 12.31b). This causes the dry line to move eastward as the day progresses. At night, convective turbulence ceases, a stable boundary layer develops, and prevailing low-altitude easterlies advect moist air back toward the west (Fig 12.31a).

If we assume that initially the inversion is level with respect to sea level, then the progression speed of the dry line eastward $\Delta x/\Delta t$ is directly related to the rate of growth of the mixed layer $\Delta z_i/\Delta t$:

$$\frac{\Delta x}{\Delta t} = \frac{\Delta z_i / \Delta t}{s} \qquad (12.16)$$

where $s = \Delta z/\Delta x$ is the terrain slope.

Suppose the early morning stable boundary layer has a linear profile of potential temperature $\gamma = \Delta\theta/\Delta z$, then the mixed-layer growth proceeds with the square-root of accumulated daytime heating Q_{Ak} (see Chapter 4):

$$z_i = \left(\frac{2 \cdot Q_{Ak}}{\gamma}\right)^{1/2} \qquad (12.17)$$

As a result, the distance Δx that a dry line moves eastward as a function of time is:

$$\Delta x = \frac{1}{s} \cdot \left(\frac{2 \cdot Q_{Ak}}{\gamma}\right)^{1/2} \qquad (12.18)$$

Time is hidden in the cumulative heat term Q_{Ak}.

Solved Example (§)

Suppose the surface heat flux during the day is constant with time, at 0.2 K·m/s in the desert southwest. The early morning sounding gives γ = 10 K/km. The terrain slope is 1:400 . Plot the dry line position with time.

Solution

Given: γ = 10 K/km, $s = 1/400$, $F_H = 0.2$ K·m/s,
Find: $\Delta x(t)$ = ? km.
For constant surface flux, cumulative heating is:

$$Q_{Ak} = F_H \cdot \Delta t$$

where Δt is time since sunrise.

(continued next column)

Solved Example (§) *(continuation)*

Thus, eq. (12.18) can be rewritten as:

$$\Delta x = \frac{1}{s} \cdot \left(\frac{2 \cdot F_H \cdot \Delta t}{\gamma}\right)^{1/2}$$

$$= \frac{400}{1} \cdot \left[\frac{2 \cdot (0.2K \cdot m/s) \cdot \Delta t}{(0.01K/m)}\right]^{1/2}$$

Plotting this with a spreadsheet program:

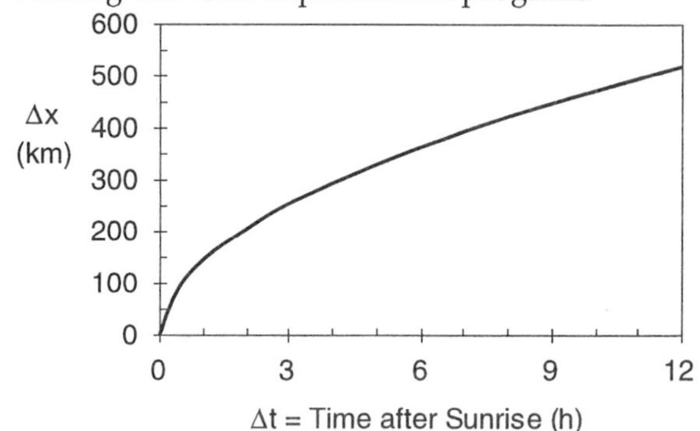

Check: Units OK. Physics OK.
Discussion: Movement eastward is with the square-root of time. Thus, the further east it gets, the slower it goes.

SUMMARY

Surface fronts mark the boundaries between air masses. Changes in wind, temperature, humidity, visibility, and other meteorological variables are frequently found at fronts. Along frontal zones are often low clouds, low pressure, and precipitation. Fronts rotate counterclockwise around the lows in the Northern Hemisphere, as the lows move and evolve.

Fronts strengthen due to advection by the wind (kinematics), external diabatic heating (thermodynamics), and ageostrophic cross-frontal circulations (dynamics). One measure of frontal strength is the temperature change across it.

Air masses can form when air resides over a surface sufficiently long. This usually happens in high-pressure centers of light wind. Air masses over cold surfaces take longer to form than those over warmer surfaces because turbulence is weaker in the statically stable air over the cold surface. As air masses move from their source regions, they are modified by the surfaces over which they flow.

Upper-tropospheric fronts can form as folds in the tropopause. Dry lines form over sloping terrain as a boundary between dry air to the west and moist air to the east; however, dry lines do not usually possess all the characteristics of fronts.

Lows, highs, fronts, and air masses are components of synoptic weather patterns. By plotting synoptic weather maps containing data from many regions in the world, these synoptic features can be identified. Map analysis is routinely performed by computer.

Threads

Mid-latitude highs form because of the meandering jet stream (Chapt. 11). Formation and modification of air masses, and dry-line evolution, are caused by boundary layer (Chapt. 4) processes such as turbulence. Subsidence and divergence in highs are linked by the continuity equation (Chapt. 8). Air mass modification over mountains can be estimated using thermo diagrams (Chapt. 6). Geostrophic adjustment is caused by Coriolis force (Chapt. 8). Isentropes (Chapt. 3) can be used for frontal analysis.

Air masses are drawn into lows (Chapt. 13), which enhances baroclinicity to strengthen the cyclonic circulation. Highs and fronts are predicted numerically (Chapt. 14) during routine weather forecasts. Cold and occluded fronts and dry lines can trigger thunderstorms (Chapt. 15). Air pollution (Chapt. 17) is trapped in highs, causing air stagnation events and air pollution alerts.

EXERCISES

Numerical Problems

N1. For air of average initial mixed-layer temperature 0°C and divergence 10^{-6} s^{-1}, find the e-folding time for warm air-mass formation, given a potential-temperature gradient above the mixed layer of (K/km):

a. 0.5 b. 1 c. 5 d. 0.2
e. 2 f. 4 g. 10 h. 3

N2.(§). If the air of problem (N1) comes to rest over a surface of temperature 25°C, calculate and plot the temperature and depth of the air mass vs. time.

N3.(§) An arctic air mass of initial temperature –30°C is modified as it moves at speed 10 m/s over smooth warmer surface of temperature 0°C. Find and plot the air mass (mixed-layer) temperature vs. downwind distance, for an air-mass thickness (m) of

a. 200 b. 500 c. 1000 d. 2000
e. 100 f. 400 g. 800 h. 1500

N4. Assume the katabatic winds of Antarctica result from a balance between the downslope buoyancy force and turbulent drag force. The surface is smooth and the depth of the katabatic layer is 100 m. The slope, potential temperature difference, and ambient temperature vary according to the table below. Find the katabatic wind speed at the locations in the table.

Location	$\Delta z/\Delta x$	$\Delta\theta$ (K)	T_e (K)
Interior	0.001	40	233
Intermediate	0.005	25	248
Coast	0.01	20	253

N5. Use a thermodynamic diagram. Suppose the wind blows over the Olympic, Cascade, and Rocky Mountains just like in the solved example after Fig 12.7, except that the starting conditions are T = 8°C and RH = 95%.

a. Find the temperature and dew point of the air for the same six locations as in that solved problem.

b. Also, if the air continues to blow eastward of the Rockies and continues to descend to an altitude of 500 m, find the final relative humidity.

N6. Find the external Rossby radius of deformation at 60° latitude for a cold air mass of thickness 500 m and $\Delta\theta$ (°C) of:

a. 10 b. 20 c. 30. d. 8
e. 2 f. 5 g. 15 h. 25

Assume a background potential temperature of 300K.

N7.(§) Find and plot the air-mass depth and geostrophic wind as a function of distance from the front for the cases of the previous exercise. Assume a background potential temperature of 300 K.

N8. Suppose that $\Delta\theta/\Delta x$ = 0.02°C/km, $\Delta\theta/\Delta y$ = 0.01°C/km, and $\Delta\theta/\Delta z$ = 5°C/km. Also suppose that $\Delta U/\Delta x$ = – 0.03 (m/s)/km, $\Delta V/\Delta x$ = 0.05 (m/s)/km, and $\Delta W/\Delta x$ = 0.02 (cm/s)/km. Find the frontogenesis contributions from:

a. confluence b. shear c. tilting
d. and find the strengthening rate.

Understanding & Critical Evaluation

U1. Is there an upper limit on the number of warm fronts and cold fronts that can extend from a low-pressure center? Why?

U2. In high pressure centers, is the boundary layer air compressed to greater density as subsidence pushes down on the top of the mixed layer?

U3. Using the concept of air mass conservation, formulate a relationship between subsidence velocity at the top of the boundary layer (i.e., the speed that the capping inversion is pushed down in the absence of entrainment) to the radial velocity of air within the boundary layer. (Hint: use cylindrical coordinates, and look at the geometry.)

U4. From eqs. (12.1-12.4), determine the relationship for entrainment velocity w_e into the top of the mixed layer. How does this compare to entrainment velocity equations given in Chapt. 4?

U5. Suppose that cold air drains katabatically from the center of Antarctica to the edges. Sketch the streamlines for this air, considering Coriolis force and turbulent drag.

U6. a. Compare and critically discuss eqs. (12.2) and (12.7). Note that for a column of air that moves with wind speed M, then time t for the air mass to move distance x is $t = x/M$.

b. Derive both of these equations using the Eulerian heat budget, and surface heat flux parameterizations for the boundary layer.

c. What assumptions went into these equations, and what are the resulting limitations?

U7. For Fig 12.8, discuss the relationship between the surface low pressure location, and the location of the low-pressure trough at 50 kPa.

U8. If the surface low in Fig 12.8a moves in a direction given by the winds at 50 kPa, predict the track of the surface low.

U9. Other than dry lines, is it possible to have fronts with no temperature change across them? How would such fronts be classified? How would they behave?

U10. Sketch a low with two fronts similar to Fig 12.1, except for the Southern Hemisphere.

U11. Why does the polar front in Fig 12.4 meander north and south?

U12. Suppose that the drainage wind is determined by a balance between quasi-horizontal downslope buoyancy force and turbulent drag force (see eqs. 12.5 and 9.11). Consider wind in only the x-direction. If the Antarctic terrain is smooth with slope of 0.1, find the equilibrium katabatic wind speed if the katabatic air is cooler than the surrounding air by amount (°C):

a. 1	b. 2	c. 5	d. 10
e. 15	f. 20	g. 25	h. 30

U13. Starting with Fig 12.18, suppose that ABOVE the bottom contoured surface the temperature is horizontally uniform. Redraw that diagram but with the top two contoured surfaces sloped appropriately for the temperature state.

U14. For a cold-frontal occlusion (where a cold front catches up to a warm front), speculate and plot a vertical cross section showing isentropes across that front.

U15. Speculate on which is more important for generating ageostrophic cross-frontal circulations: the initial magnitude of the geostrophic wind or the change of geostrophic wind.

U16.(§) Redo the last solved example before the summary, except using eq. (4.2b) for cumulative heating.

U17. By inspection, write a kinematic frontogenesis equation for an east-west aligned front.

U18. Draw figures similar to 12.10 and 12.11, but for an occluded front.

U19. Why should the Rossby radius of deformation depend on the depth of the cold air mass, in a geostrophic adjustment process?

U20. For geostrophic adjustment, the initial outflow of cold air turns due to Coriolis force until it is parallel to the front. Why does it not keep turning and point back into the cold air?

U21. For geostrophic adjustment, what will be the nature of the final winds, if the starting point is a shallow cylinder of cold air 2 km thick and 500 km radius?

U22. Draw isentropes in a vertical cross section through (a) a warm front; (b) an occluded front.

U23. Suppose that an upper-troposphere front (i.e., tropopause fold) was connected with a surface front. Draw a vertical north-south cross section with isentropes for this case.

U24. What types of weather would be expected with upper-tropospheric front that does not have an associated surface front? [Hint, track movement of air parcels as the ride isentropic surfaces.]

U25. What happens in an occluded front where the two cold air masses (the one advancing behind the cold front, and the one retreating ahead of the warm front) have equal temperature?

U26. Would you expect drylines to be possible in parts of the world other than the S.W. USA? If so, where? Justify your arguments.

U27. Both of the following charts correspond to the same weather.

a. Draw isotherms and identify warm and cold centers. Label isotherms every 2°C
b. Draw isobars every 0.2 kPa, and identify high and low pressure centers.
c. Add likely wind vectors to the pressure chart.
d. Identify the frontal zone(s) and draw the frontal boundary on the temperature chart.
e. Use both charts to determine the type of front (cold, warm), and draw the appropriate frontal symbols on the front.
f. Indicate likely regions for clouds, and suggest cloud types in those regions.
g. Indicate likely regions for precipitation.
h. For which hemisphere are these maps?

(i) Temperature (°C):

8	9	11	12	13	14	15
7	9	11	13	14	15	16
8	9	14	18	20	20	18
9	9	16	19	22	24	24
10	11	18	20	22	25	25
12	14	19	20	22	24	25
14	18	19	21	22	23	24
18	19	20	21	22	23	23

(ii) Pressure (kPa). [The first 1 or 2 digits of the pressure are omitted. Thus, 9.5 plotted below = 99.5 kPa, while 0.1 means 100.1 kPa.]

9.5	9.4	9.4	9.5	9.6	9.7	9.8
9.4	9.3	9.2	9.3	9.4	9.6	9.7
9.5	9.2	8.9	9.2	9.3	9.4	9.6
9.5	9.4	9.1	9.4	9.5	9.5	9.6
9.6	9.4	9.2	9.5	9.6	9.7	9.7
9.6	9.4	9.4	9.6	9.7	9.8	9.9
9.6	9.4	9.6	9.7	9.8	9.9	0.0
9.6	9.6	9.7	9.8	9.9	0.0	0.1

Web-Enhanced Questions

W1. Monitor the weather maps on the web every day for a week or more during N. Hemisphere winter, and make a sketch of each low center with the fronts extending (similar to Fig. 12.1). Discuss the variety of arrangements of warm, cold, and occluded fronts that you have observed during that week.

W2. Same as W1, but for the S. Hemisphere during S. Hemisphere winter.

W3. From weather maps showing vertical velocity (or showing ω), find typical values of that vertical velocity near the center of highs. (If ω is used instead of w, it is acceptable to leave the units in mb/s). Compare this vertical velocity to the radial velocity of air diverging away from the high center in the boundary layer.

W4. For a wintertime situation, access a N. (or S.) Hemisphere surface weather map from the web, or access a series of weather maps from different agencies around the world in order to get information for the whole Hemisphere. Draw the location of the polar front around the globe.

W5. Use weather maps on the web to find a location where air has moved over a region and then becomes stationary. Monitor the development of a new air mass (or equivalently, modification of the old air mass) with time at this location. Look at temperature and humidity.

W6. From a time-sequence of high-resolution visible satellite photos (perhaps from polar orbiting satellites), determine the motion of icebergs in the S. Hemisphere just offshore from Antarctica. Recalling the relationship between ocean current direction and surface wind direction (Ekman spiral), and knowing that 90% of the iceberg is under water, estimate the wind directions over the ocean near those icebergs. [Hint: Coriolis force acts to the left in the S. Hemisphere.]

W7. For a weather situation of relatively zonal flow from west to east over western N. America, access upper air soundings for weather stations in a line more-or-less along the wind direction. Show how the sounding evolves as the air flows over each major mountain range. How does this relate to air mass modification?

W8. Access from the web weather maps for sea level, and for 50 kPa, over the same location and for the same time. Compare the locations of the low pressure centers or troughs at those two altitudes. Discuss how each level might affect the other.

W9. Access simple weather maps from the web that plot values of pressure or temperature at the weather stations, but which do not have the isopleths drawn. Print these, and then draw your own isobars or isotherms. If you can do both isobars and isotherms for a given time over the same region, then identify the frontal zone, and determine if the front is warm, cold, or occluded. Plot these features on your analyzed maps. Identify highs and lows.

W10. Access from the web surface weather maps showing plotted station symbols, along with the frontal analysis. Compare surface temperature, wind, and pressure along a line of weather stations that crosses through the frontal zone. How do the observations compare with the idealized frontal model?

W11. Access upper air soundings for a line of RAOB stations across a front. Use this data to draw vertical cross sections of:

a. pressure or height b. temperature
c. humidity d. potential temperature

W12. Download upper soundings from a station under or near the jet stream. Use the data to see if there is a tropopause fold. [Hint: assuming that you don't have measurements of radioactivity or isentropic potential vorticity, use mixing ratio or potential temperature as a tracer of stratospheric air.]

W13. Access a sequence of surface weather maps from the web that show the movement of fronts. Do you see any fronts that are labeled as stationary fronts, but which are moving? Do you see any fronts labeled as warm or cold fronts, but which are not moving? Are there any fronts that move backwards compared to the symbology labeling the front (i.e., are the fronts moving opposite to the direction that the triangles or semicircles point)? [Hint: often fronts are designated by how they move relative to the low center. If a cold front, for example, is advancing cyclonically around a low, but the low is moving toward the west (i.e., backwards in mid latitudes), then relative to people on the ground, the cold front is retreating.]

W14. Search the web for surface weather maps that indicate a dry line in the S.W. USA. If one exists, then search the web for upper-air soundings just east and just west of the dry line. Plot the resulting soundings of potential temperature and humidity, and discuss.

Synthesis Questions

S1. Does the geostrophic adjustment process affect the propagation distance of cold fronts in the real atmosphere, for cold fronts that are imbedded in the cyclonic circulation around a low-pressure center? If this indeed happens, how could you detect it?

S2. What if there was no inversion at the top of the boundary layer? Redraw Fig 12.2 for this situation.

S3. What if boundary-layer processes were so slow that air masses took 5 times longer to form compared to present air mass formation of about 3 to 5 days? How would the weather and general circulation be different, if at all?

S4. Suppose that air masses formed within the whole depth of the troposphere, instead of forming only within the boundary layer. How would the weather and general circulation be different than now, if at all?

S5. Suppose that the sun caused radiative cooling of earth, while IR radiation from space caused warming of earth. How would the weather and climate be different, if at all?

S6. What if precipitation did not occur as air flows over mountain ranges. Thus, mountains could not modify air masses by this process. How would weather and climate be different from now, if at all?

S7. a. Suppose that all the weather observations over land were accurate, and all the ones over oceans had large errors. At midlatitudes where weather moves from west to east, discuss how forecast skill would vary from coast to coast across a continent such as N. America.

b. How would forecast skill be different if observations over oceans were accurate, and over land were inaccurate?

S8. Suppose that the width of frontal zones were infinitesimally small. How would weather be different, if at all?

S9. What if turbulence were always so intense that frontal zones were usually 1000 km in width. How would weather be different, if at all?

S10. Suppose a major volcanic eruption injected a thick layer of ash and sulfate aerosols at altitude 20 km. Would the altitude of the tropopause adjust to become equal to this height? Discuss.

S11. Suppose the slope of the ground in a dry line situation was not as sketched in Fig 12.31. Instead, suppose the ground was bowl shaped, with high plateaus on both the east and west ends. Would drylines exist? If so, what would be their characteristics, and how would they evolve?

CYCLONES

CONTENTS

13 **Cyclones** are synoptic-scale regions of **low sea-level pressure** with **upward vertical velocity** and **cyclonically** (counterclockwise in the N. Hemisphere) rotating winds. The upward vertical velocity supports clouds, precipitation, and sometimes thunderstorms. Cyclones are often called **lows** for short. The term **storm system** is sometimes used by broadcast meteorologists to indicate the nasty weather that often accompanies cyclones.

Cyclones can cause floods, blizzards, strong winds and wind chills, and hazardous travel conditions over thousands of square kilometers. They are the regions of bad weather that come from the west every three to seven days to locations in midlatitudes. One such storm is described next, as motivation for understanding the dynamics of cyclones. These dynamics are used at the end of the chapter to re-examine the storm.

CASE STUDY OF A CYCLONE OVER NORTH AMERICA

Overview and Storm Track

On 20 February 1994, an upper-level trough forms east of the Rocky Mountains, in western Texas. Divergence in the upper-tropospheric jet stream between this trough and a ridge east of Texas causes a broad region of upward motion over the southern plains. This **upper level disturbance** forms a cloud shield and precipitation. As upward motion continues during the next two days, a surface low develops over western Texas.

As the cyclone tracks northeast toward Illinois (Fig 13.1) during 22 to 23 Feb, it intensifies and forms well-developed cold and warm fronts. A squall line of intense thunderstorms with hail and weak tornadoes is triggered in the Southern Plains along and ahead of the cold front. North of the low, heavy snow falls in the upper plains and the midwest.

The main low quickly tracks further northeast, dropping heavy snow across the Ohio valley. It occludes and begins to weaken, but still produces heavy snow and freezing rain in New England.

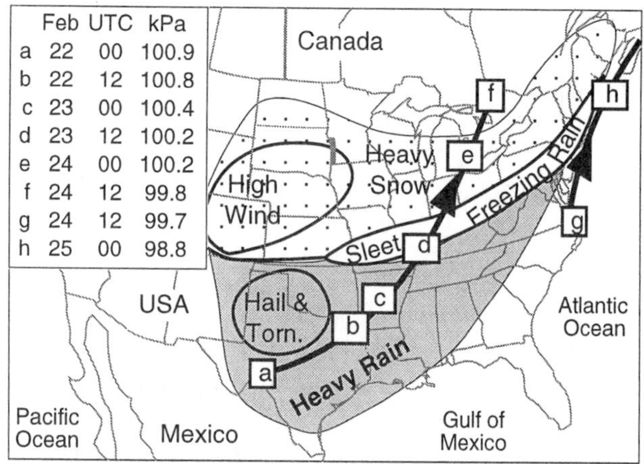

Figure 13.1
Track of a cyclone, its weather, and central pressure evolution during February 1994.

Meanwhile, a secondary cyclone forms off the Carolina coast, caused in part by support from the upper-level flow. This storm rapidly intensifies over the warm Gulf-stream, and during 24 Feb the first cyclone merges/shifts into it. The combined cyclone causes blizzard conditions from Washington D.C. to Prince Edward Island, Canada, in the subsequent three days. A third low has also formed in Alberta, Canada, east of the Rockies. It quickly sweeps across the northern plains bringing strong winds, which whip up the previously fallen snow creating whiteout conditions and deep snow drifts.

In summary, the original cyclone is born, completes its full life cycle, and dies within just a few days while causing havoc across a large portion of N. America east of the Rockies. Weather maps showing this cyclone are given at the end of this Chapter.

Storm Data

• **20 Feb 94** – Heavy rains (15 cm) cause flooding in south Texas.

• **21 Feb** – Thunderstorms and heavy rain continue in north Texas and southern Oklahoma, with 2 cm diameter hail, an F0 tornado, and straight-line winds gusting to 25 m/s. Heavy snow begins in the Colorado Rockies.

• **22 Feb** – Thunderstorms continue in Texas and Oklahoma, with more hail (most 2 cm diameter, but some 5 cm), lightning starts some fires, and heavy rains causing flash floods that wash out highways and close bridges.

Snow ends in the mountains of Colorado, leaving up to 38 cm of new snow on the ground. Snow and heavy snow fall all day, leaving 20 cm accumulation in Kansas, and 25 cm in Nebraska, causing schools to close early. Heavy snow starts in South Dakota, Iowa, N. Missouri, and Illinois.

Snow mixed with freezing rain begins in southern Missouri, leaving an ice glaze 3.5 cm thick. Snow starts in Ohio, dropping 15 cm, followed by a little freezing rain and gusts to 12 m/s.

• **23 Feb** – Thunderstorms continue in Texas with gusts up to 22 m/s, rolling a mobile home. An F1 tornado damages barns, mobile homes, and an aircraft hangar, and hail 4.5 cm diameter falls in Texas.

Winds of 38 m/s are reported in ski areas near and east of the continental divide in the Colorado Rockies.

Heavy snow continues in Nebraska (35 cm), South Dakota (30 cm), Kansas (26 cm), Missouri (25 cm), and Illinois (23 cm at O'Hare airport in Chicago, with strong winds and drifting snow). Heavy snow starts in North Dakota.

Thundersnow in Iowa accumulates at a rate of 9 cm/h near Des Moines and Waterloo. A band of convective snow showers leaves up to 28 cm in a 3 h period as it sweeps across N.E. Iowa. Winds of 17 m/s reduce visibilities to 0.1 km, cause drifts up to 1.5 m, resulting in power outages and many closed roads. A dry air mass advects into extreme S.E. Iowa.

Thunderstorm straight-line winds and an F0 tornado damage numerous buildings in Alabama and Ohio. Heavy rains cause flash floods in Tennessee, washing out bridges and cars, and requiring evacuation of 5 families.

Heavy wet snow spreads through more of the midwest, starting in SE. Minnesota (23 cm) and S. Wisconsin (40 cm, with 18 m/s winds, whiteout conditions in blowing snow, numerous accidents, schools closed). In Michigan, 33 cm of new snow with winds gusting to 22 m/s cause near blizzard conditions (schools closed in 12 counties).

Heavy snow and freezing rain spreads into New England, including Maine, New Hampshire, Vermont, Massachusetts, New York, and New Jersey.

• **24 Feb** – Mountain downslope winds increase in Colorado, with gusts up to 48 m/s at the National Center for Atmospheric Research in Boulder. A roof blows off a community college in Lakewood, CO, causing evacuation of 1500 people.

Blizzard conditions with heavy snow occurs in North Dakota (23 cm new snow, with gusts to 20 m/s). There are near blizzard conditions in South Dakota as snowfall ends there.

Michigan reports 34 cm total snow, drifts up to 1 m, and a 25-car pileup including a fully-loaded gasoline tanker on interstate-96, resulting in one death. Heavy snow ends in Wisconsin.

Heavy rains and flash floods end in Tennessee.

Heavy snow and ice storms continue in New England, enhanced by the new cyclone moving up the East Coast. Many states are hit: Maine (38 cm snow in NW., but more freezing rain in SE.), New Hampshire (28 cm), Vermont (45 cm snow, with sleet and freezing rain mixed in S.), Massachusetts (20 cm snow, with sleet and freezing rain along E. coast), and New York (55 cm snow, with ice storms near coast and downed power lines).

• **25 Feb** – Strong winds cause a blizzard in N. Illinois, with an additional 15 cm new snow, gusts to 17 m/s, whiteout conditions, dangerous wind chill, and schools closed. Heavy snow in Indiana leaves 25 cm in places.

An "Alberta Clipper" cyclone sweeps across the Plains States, causing a ground blizzard (gusts to 22 m/s with 10 cm new snow), dangerous wind chills, and a convective snow shower of 85 dBZ observed by a WSR-88D radar. Blowing of old snow causes drifts up to 3 m. Strong gusts occur in Kansas and North Dakota. Snow starts again in Wisconsin, with 25 cm new snow, with additional lake-effect snow near Great Lakes. Heavy snow falls in Ohio.

Heavy snow changes to sleet and freezing rain, and ends in New Hampshire, Vermont, & New York.

• **26 Feb** – Blizzard ends in N. Illinois with peak gusts 30 m/s, closing 4 interstate highways and leaving hundreds of stranded cars. Similar conditions in N. Indiana, with blowing and drifting snow, whiteout conditions, gusts to 25 m/s, and drifts to 1 m, leave 1400 stranded motorists. A snow emergency declared in 8 counties. In N. Ohio is 15 cm of new snow, gusts to 20 m/s, and convective snowfall rates of 5 cm/h. Heavy snow ends in S. New York, leaving up to 23 cm of new snow.

• **To be continued**– To understand how such a storm can develop, we examine cyclone dynamics next. We will return to the case study at the end of this chapter.

CYCLOGENESIS

Cyclogenesis is the birth and growth of cyclones. Such intensification can be defined by the:

- sea-level pressure decrease,
- upward-motion increase, and
- vorticity increase.

These characteristics are not independent; for example, upward motion can reduce surface pressure, which draws in air that begins to rotate due to Coriolis force. However, we can gain insight into the workings of the storm by examining the dynamics and thermodynamics that govern each of these characteristics. Each of these will be explored in detail in sections of this chapter, after first examining how mountains can trigger cyclone formation.

The following conditions have been found to favor <u>rapid</u> cyclogenesis:

- strong baroclinicity – a large horizontal temperature gradient
- weak static stability – temperature decreasing with height faster than the tropospheric standard atmosphere
- mid or high-latitude location – earth's contribution to vorticity increases toward the poles
- large moisture input – latent heating due to cloud condensation adds energy and reduces static stability
- large-amplitude wave in the jet stream – a trough to the west and ridge to the east of the surface low enhance horizontal divergence aloft, which strengthens updrafts
- terrain elevation decrease toward east – cyclogenesis to the lee of mountains

A cyclone that develops extremely fast is called a **cyclone bomb**, and the process is called **explosive cyclogenesis**. To be classified as a bomb, the central pressure of a cyclone must decrease at a rate of 0.1 kPa per hour for at least 24 hours. Explosive cyclogenesis often occurs when cold air moves over the northern edge of a very warm ocean current, such as over the Gulf Stream off the N.E. USA. In these regions, strong evaporation and heat transfer from the sea surface enhances many of the factors listed above for rapid cyclogenesis. Cyclone bombs can cause sudden major winter storms, with high seas and freezing rain that are hazards to shipping and road travel.

The opposite of cyclogenesis is **cyclolysis**. This is literally death of a cyclone. Most cyclones go through a cycle of formation, growth, and death, over a period of about a week at midlatitudes. However, lifetimes less than a day and greater than two weeks have been observed.

If cyclones were to exist forever while they are blown around the world by the jet stream, then the weather would be much easier to forecast. However, real cyclones form and evolve while they move from west to east, which is why weather forecasting is such a challenge. While numerical weather forecast models do an increasingly good job at describing this complex evolution, the best human forecasters use their knowledge of cyclone dynamics, together with practical experience gained on the job, to refine these numerical forecasts. Cyclone dynamics are discussed next.

LEE CYCLOGENESIS

Rossby showed in 1939 that evolution of the planetary waves bearing his name can be described by conservation of absolute vorticity, as was discussed in Chapt. 11. One of the many ways to create Rossby waves is by topographic modification of the flow, as is described next.

Stationary Planetary Waves

When air flows across a mountain range such as the Rocky Mountains, planetary-scale Rossby waves are triggered in the atmosphere (Fig 13.2b). These Rossby waves are stationary relative to the ground. Otherwise, they behave like the barotropic and baroclinic waves in the general circulation, discussed in Chapter 11. The north-south (meridional) component of wave-amplitude often decays to the east of the mountains, as boundary-layer turbulence and deep convection cause damping.

At the crest of the wave, the air is turning anticyclonically (clockwise in the Northern Hemisphere). Hence, over the western portion of the mountains (location C in Fig 13.2b), one would generally expect to find a predominance of highs. Further east, at the trough, the air curves cyclonically.

Thus, to the lee of (i.e., downwind of) the mountains, cyclones are favored. The process of cyclone generation and enhancement is called **lee cyclogenesis**.

(a)

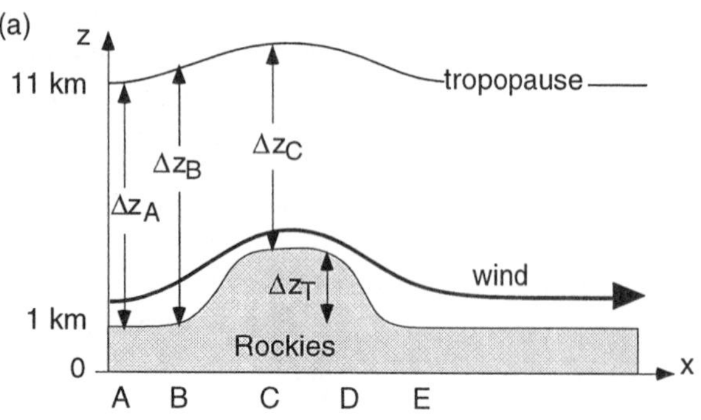

(b)

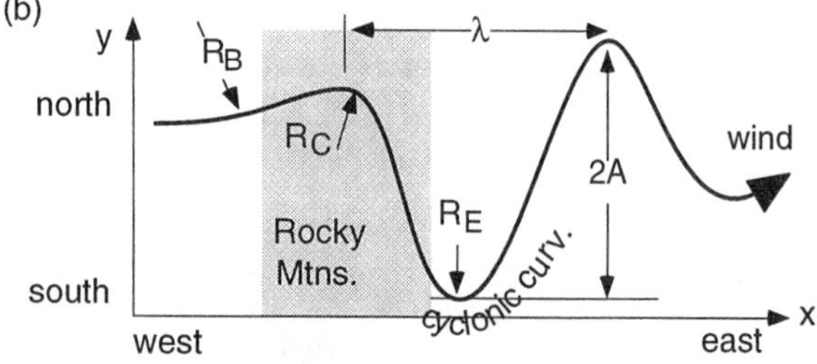

Figure 13.2
Cyclogenesis to the lee of the mountains. $\Delta z_C < \Delta z_A < \Delta z_B$. (a) Vertical cross section. (b) Map of jet-stream flow.

The wavelength λ of these waves is given approximately by

$$\lambda \approx 2 \cdot \pi \cdot \left[\frac{M}{\beta}\right]^{1/2} \qquad (13.1)$$

where M is the average wind speed. The rate of increase β of the Coriolis parameter f_c with distance north is:

$$\frac{\Delta f_c}{\Delta y} \equiv \beta = \frac{2 \cdot \Omega}{R_{earth}} \cdot \cos\phi \qquad \bullet(13.2)$$

where $2 \cdot \Omega = 1.458\text{x}10^{-4}\ \text{s}^{-1}$. β is on the order of 1.5 to $2\text{x}10^{-11}\ \text{m}^{-1} \cdot \text{s}^{-1}$ at mid latitudes.

The north-south amplitude A depends on the height of the mountains Δz_T relative to the upstream depth of the troposphere Δz_A:

$$A \approx \frac{f_c}{\beta} \cdot \frac{\Delta z_T}{\Delta z_A} \qquad (13.3a)$$

However, the ratio $f_c / \beta \equiv R_{earth} \cdot \tan(\phi)$, where R_{earth} = 6357 km is the average earth radius, and ϕ is latitude. At midlatitudes where $\tan(\phi) \cong 1$, the amplitude is

$$A \approx \frac{\Delta z_T}{\Delta z_A} \cdot R_{earth} \qquad (13.3b)$$

Solved Example

For wind blowing 16.2 m/s over a 1 km-high mountain range at 45°N, find the wavelength and amplitude of the planetary wave. Assume the tropospheric depth is 10 km upstream of the mtns.

Solution

Given: ϕ = 45°N, M = 16.2 m/s, Δz_T= 1 km, Δz_A= 10 km.

Find: λ = ? km, A = ? km

First, find β at 45°N, using eq. (13.2):

$$\beta = (2.294 \times 10^{-11} \text{m}^{-1} \cdot \text{s}^{-1}) \cdot \cos(45°) = 1.62\text{x}10^{-11}\ \text{m}^{-1} \cdot \text{s}^{-1}$$

Use eq. (13.1):

$$\lambda \approx 2 \cdot \pi \cdot \left[\frac{16.2 \text{m} \cdot \text{s}^{-1}}{1.62 \times 10^{-11} \text{m}^{-1} \cdot \text{s}^{-1}}\right]^{1/2} = \underline{6283 \text{ km}}$$

Use eq. (13.3b):

$$A \approx \frac{1\text{km}}{10\text{km}} \cdot (6378\text{km}) = \underline{638 \text{ km}}$$

Check: Units OK. Physics OK.
Discussion: The perimeter of the 45° meridian is $2 \cdot \pi \cdot R_{earth} \cdot \cos(45°)$ = 28,243 km. Thus, about 4.5 wavelengths fit round the earth – it is indeed a planetary-scale wave. Also, $2 \cdot A \cong 11°$ latitude.

The north-south distance Δy between crest and trough is just $\Delta y = 2 \cdot A$.

Thus, wavelength depends on wind speed, but not mountain height. Wave amplitude depends on mountain height, but not wind speed.

Conservation of Potential Vorticity

Another way to examine these waves is by their vorticity. Neglecting shear and assuming constant wind-speed M for simplicity, eq. (11.22) becomes:

$$\zeta_p = \frac{(M/R) + f_c}{\Delta z} = \text{constant} \qquad \bullet(13.4)$$

which states that potential vorticity is conserved.

Suppose air at point A (Fig 13.2b), west of the Rockies, blows directly from west to east, and initially has no curvature ($M/R = 0$). Thus, eq. (13.4) reduces to

$$\zeta_p = \frac{f_{cA}}{\Delta z_A} \qquad (13.5a)$$

This initial condition sets the constant value of the potential vorticity for this example.

As the air approaches the mountain, some of the near-surface air is blocked by the ridge. Thus, the higher air is forced to rise over this blocked air, even before reaching the mountain (Fig 13.2a). This region of influence extends west from the mountains a distance roughly equal to the internal Rossby radius of deformation λ_R (see eq. 11.33). Within this region, the air column stretches from Δz_A to Δz_B, causing a cyclonic curvature (location B in Fig 13.2a).

Then, when the air flows over the mountains, the previous two equations combine to give the anticyclonic radius of curvature over the west portion of the mountains (at location C):

$$R_C = \frac{-M}{f_{cA} \cdot (\Delta z_T / \Delta z_A)} \qquad (13.5b)$$

where $\Delta z_T = \Delta z_A - \Delta z_B$ is the relative terrain height above the surrounding land (Fig 13.2a).

As the air moves south, the Coriolis parameter decreases, thereby requiring the radius of anticyclonic curvature to increase. Eventually, the flow is straight, blowing toward the southeast (location D). As it overshoots south, and as the air column stretches as it leaves the mountains, the curvature must become positive (cyclonic). The radius of curvature R_E is similar in magnitude to R_C, but of opposite sign. This is the region where cyclone development is favored.

Solved Example

Suppose a noncurving wind blows from the west at 20 m/s (assume constant), at latitude 40°N. Initially, the tropospheric depth is 10 km. Then, the air flows over the Rocky Mountains, where you can assume a relative terrain depth of 1 km above the neighboring plains. Find the potential vorticity, and the radius of curvature over the mountains.

Solution

Given: $\phi = 40°N$, $M = 20$ m/s, $R_{initial} = \infty$,
$\Delta z_A = 10$ km, $\Delta z_T = 1$ km.
Find: $\zeta_p = ?\ m^{-1} \cdot s^{-1}$, $R_C = ?$ km

Assume: shear can be neglected.
First, find the Coriolis parameter using eq. (9.8):
$f_{cA} = 1.46 \times 10^{-4} \cdot \sin(40°) = 9.38 \times 10^{-5}\ s^{-1}$
Use eq. (13.5a):

$$\zeta_p = \frac{9.38 \times 10^{-5}\,s^{-1}}{10\text{km}} = \underline{\mathbf{9.38x10^{-9}}\ m^{-1}\,s^{-1}}$$

Use eq. (13.5b) to find anticyc. radius of curv.:

$$R_C = \frac{-(20\text{m/s})}{(9.38 \times 10^{-5} s^{-1}) \cdot (1\text{km}/10\text{km})} = \underline{\mathbf{-2132\ km}}$$

Check: Units OK. Physics OK.
Discussion: We expect $R_E \cong -R_C$, namely about 2000 km. Thus, mountains trigger long waves, and cyclones are favored east of the Rockies.

Equatorward Propagation Along the Lee Side

Picture a cyclone that already exists at the lee side of the mountain range (Fig 13.3). The bottom of this cyclone rests on a sloping surface (slope $\alpha = \Delta z / \Delta x$). Circulation around the low will cause downslope flow on the south side, and upslope on the north.

Rotation about the low will cause vertical columns of air along the southern side of the cyclone to stretch. Those on the north side shrink. For any potential vorticity ζ_p of the cyclone, stretching intensifies the relative vorticity on the south side, according to:

$$\Delta\zeta_r = 2 \cdot R \cdot \alpha \cdot \zeta_p \qquad (13.6)$$

A similar change of opposite sign occurs on the north side. This causes the low to propagate southward along the eastern side of the mountains.

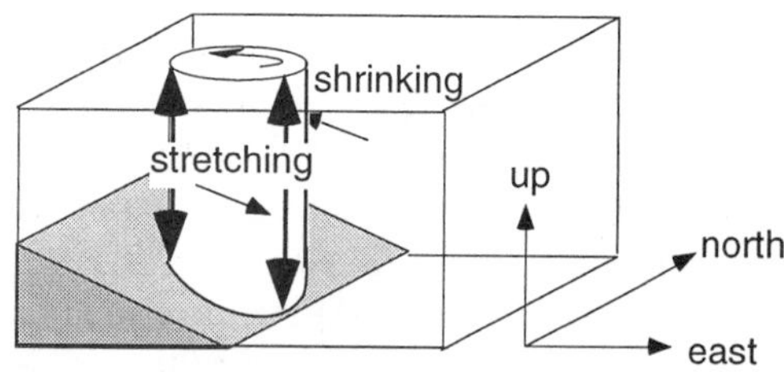

Figure 13.3
Cyclone on on the eastern slope of a mountain.

Another way to examine the same process is by the adiabatic warming that occurs at the bottom of the column (south side) as it descends down the mountain slope. No warming occurs at the top of the column, because air is not flowing downward there.

This differential heating reduces the overall static stability of the troposphere; namely, $\Delta\theta/\Delta z$ decreases. A decrease of static stability must be accompanied by an increase of relative vorticity, in order to maintain constant isentropic potential vorticity ζ_{IPV} (eq. 11.23b). Thus, vorticity intensifies on the south side of the cyclone, and decreases on the north side. The low tends to propagate south along the eastern side of the mountains.

Solved Example

A cyclone of radius 500 km has potential vorticity of $2\text{x}10^{-8}$ m^{-1}·s^{-1}. It rests on a mountain slope of 1/1000. Find the increase of relative vorticity along the south side of the cyclone.

Solution

Given: R= 500 km, ζ_p = $2\text{x}10^{-8}$ m^{-1}·s^{-1}, α= 0.001
Find: $\Delta\zeta_r$ = ? s^{-1} .

Assumption: Neglect changes of latitude.

Use eq. (13.6):

$$\Delta\zeta_r = 2\cdot(500{,}000\text{m})\cdot(0.001)\cdot(2\times10^{-8}\text{s}^{-1}\cdot\text{m}^{-1})$$
$$= \underline{2\text{x}10^{-5}}\ \text{s}^{-1}\ .$$

Check: Units OK. Physics OK.
Discussion: This intensification on the cyclone fringe is easily as large as the average relative vorticity within center of the cyclone, resulting in southward cyclone movement.

FOCUS • Maximum Advection on WX Maps

One trick to locating the region of maximum advection is to find the region of smallest area between crossing isopleths on a weather map, where one set of isopleths must define a wind.

For example, consider temperature advection by the geostrophic wind. Temperature advection will occur only if the winds blow across the isotherms at some nonzero angle. Stronger temperature gradient with stronger wind component perpendicular to that gradient gives stronger temperature advection.

But stronger geostrophic winds are found where the isobars are closer together. Stronger temperature gradients are found where the isotherms are closer together. In order for the winds to cross the isotherms, the isobars must cross the isotherms. Thus, the greatest temperature advection is where the tightest isobar packing crosses the tightest isotherm packing. At such locations, the area bounded between neighboring isotherms and isobars is smallest.

This is illustrated in the surface weather map below, where the smallest area is shaded to mark the maximum temperature advection. There is a jet of strong geostrophic winds (tight isobar spacing) running from NW. to SE. There is also a front with strong temperature gradient (tight isotherm spacing) from NE. to SW. However, the place where the jet and temperature gradient <u>together</u> are strongest is the shaded area.

This approach works for other variables too. If isopleths of vorticity and height contours are plotted on an upper-air chart, then the smallest area between crossing isopleths indicates the region of maximum **vorticity advection**. For advection by the **thermal wind**, packing of **thickness contours** is used instead of packing between isobars or height contours.

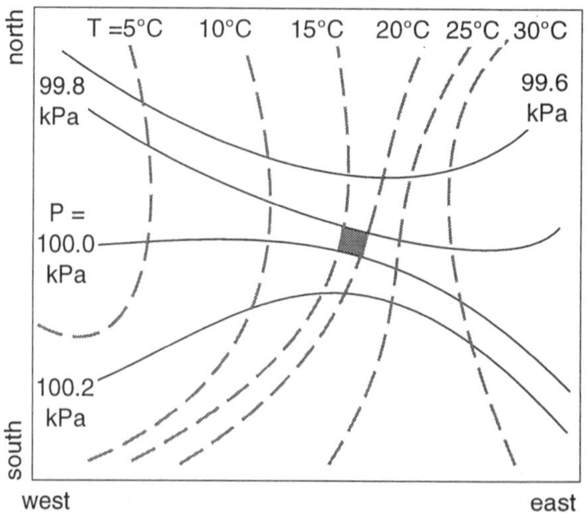

Fig. a. *Thin solid lines are isobars. Thick gray dashed lines are isotherms. Shaded region marks the greatest temperature advection.*

CYCLONE SPIN-UP

Even in the absence of slope or lee effects, baroclinic instability (see Chapter 11) can cause cyclogenesis. Vorticity increase at low altitudes is one measure of cyclone strength.

Vorticity Tendency

A forecast equation for relative vorticity is:

$$\underbrace{\frac{\Delta \zeta_r}{\Delta t}}_{spin-up} = \underbrace{-U\frac{\Delta \zeta_r}{\Delta x} - V\frac{\Delta \zeta_r}{\Delta y}}_{horizontal\ advection} \underbrace{-V\frac{\Delta f_c}{\Delta y}}_{beta} \underbrace{+f_c\frac{\Delta W}{\Delta z}}_{stretching}$$

$$\underbrace{-W\frac{\Delta \zeta_r}{\Delta z}}_{vert.advect.} \underbrace{+\zeta_r\frac{\Delta W}{\Delta z}}_{stretching} \underbrace{+\frac{\Delta U}{\Delta z}\cdot\frac{\Delta W}{\Delta y} - \frac{\Delta V}{\Delta z}\cdot\frac{\Delta W}{\Delta x}}_{tilting} \underbrace{-C_D\frac{M}{z_i}\zeta_r}_{turb.\ drag} \qquad \bullet(13.7)$$

The first term gives the **spin-up** rate of a cyclone, because it indicates the increase of vorticity. This term is also called the **tendency** term, because it shows how vorticity tends to change with time. A positive tendency corresponds to a strengthening cyclone.

Fig 13.4 shows a physical interpretation of the other terms. Advection (Fig 13.4a) changes vorticity by blowing-in air with different vorticity from some other location. **Positive vorticity advection (PVA)** occurs when greater vorticity is blown into a region.

The opposite is **negative vorticity advection (NVA)**. Advection can work in both horizontal directions and in the vertical direction, although just one horizontal direction is sketched in Fig 13.4a.

The **beta effect** is named because $\Delta f_c / \Delta y \equiv \beta$ (see eq. 13.2). The beta effect is similar to the advection affect, except that winds from the north blow-in vorticity associated with the earth's rotation (Fig 13.4b).

Stretching has already been discussed in Chapter 11. As shown in Fig 13.4c, a short vortex tube will stretch if the top rises faster than the base. This difference in rise is given by $\Delta W/\Delta z$. When such a tube stretches in the vertical, mass conservation requires horizontal convergence, which shrinks the tube radius. Because of this convergence/divergence effect, this term is also known as the **divergence term**. The two stretching terms account for the earth's rotation and the relative vorticity.

The **tilting term** shows how horizontal vorticity can be tilted into vertical vorticity. For example, Fig 13.4d shows that there is vorticity associated with the vertical shear of horizontal wind near the ground. The vortex tubes are shown between the U wind vectors. If these tubes were then tilted by a vertical velocity that increases toward the north, then the horizontal vortex tubes would become more vertical, thereby contributing to the vertical vorticity.

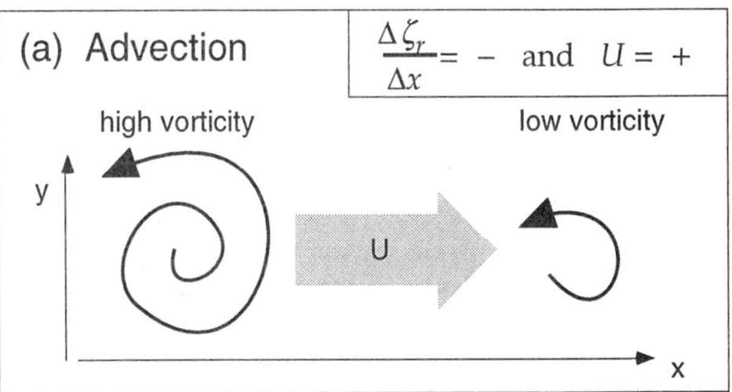

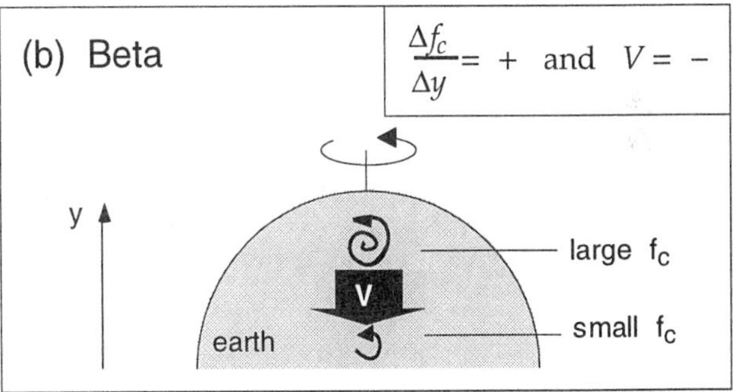

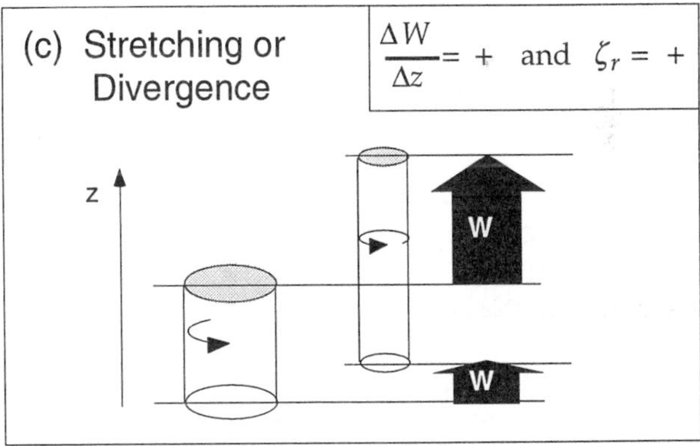

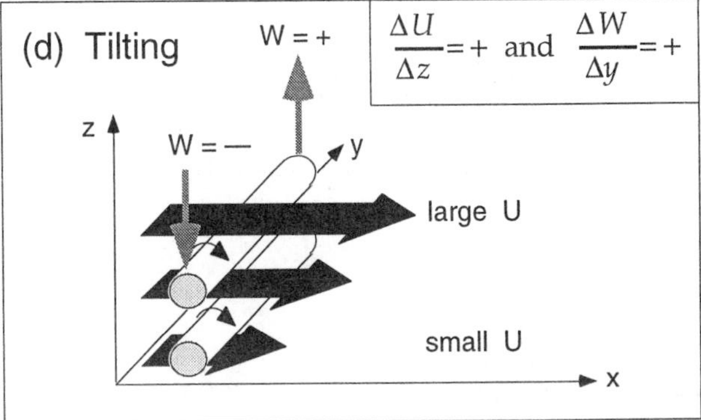

Figure 13.4
Physical interpretation of terms in the vertical vorticity equation (continued next column).

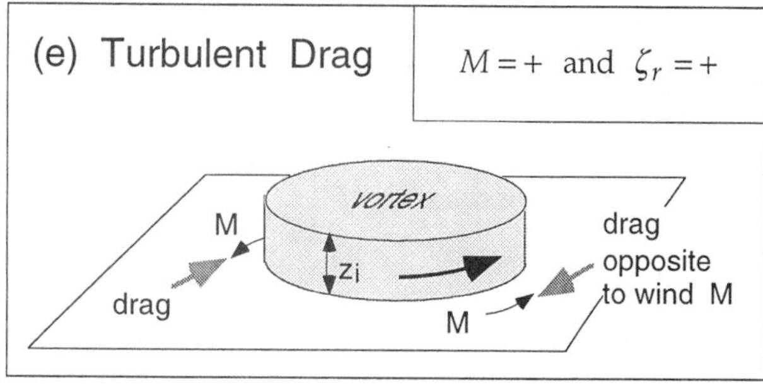

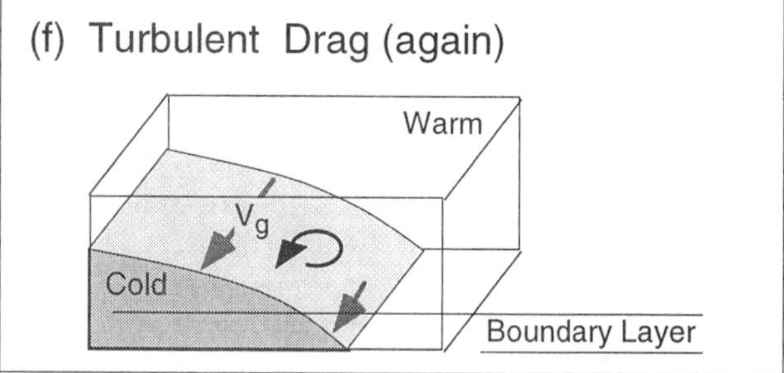

Figure 13.4 *(continuation)*
Physical interpretation of terms in the vertical vorticity equation

Finally, the **turbulent-drag term** has two effects. One effect is to reduce rotation regardless of direction, which causes **spin-down** (Fig 13.4e). This occurs in the turbulent boundary layer (of depth z_i) due to frictional drag against the ground. In the absence of continued spin-up processes, the cyclone will exponentially spin-down and die.

The other turbulent-drag effect causes spin up, although it is not apparent from eq. (13.7). Recall from Fig 12.17 (geostrophic adjustment) that there is a geostrophic wind along sloping isentropic surfaces. If such a surface intersects the boundary layer, then turbulent drag will slow the winds near the ground, but not above the boundary layer (Fig 13.4f). The resulting shear produces cyclonic relative vorticity.

The net spin-up or spin-down depends on the sum of all these terms.

Quasi-Geostrophic Approximation

Above the boundary layer, and away from fronts, jets, and thunderstorms, the terms in the second line of the vorticity equation are smaller than those in the first line, and can be neglected. Also, for synoptic scale, extratropical weather systems, the winds are almost geostrophic (**quasi-geostrophic**).

These weather phenomena are simpler to analyze than thunderstorms and tropical hurricanes, and can be well approximated by a set of equations (quasi-geostrophic vorticity and omega equations) that are less complicated than the full set of **primitive equations** of motion (Newton's second law, the first law of thermodynamics, continuity, and ideal gas law).

As a result of the simplifications above, the vorticity forecast equation simplifies to the following **quasi-geostrophic vorticity equation**:

•(13.8)

$$\underbrace{\frac{\Delta\zeta_g}{\Delta t}}_{spin-up} = \underbrace{-U_g\frac{\Delta\zeta_g}{\Delta x} - V_g\frac{\Delta\zeta_g}{\Delta y}}_{horizontal\ advection} \underbrace{-V_g\frac{\Delta f_c}{\Delta y}}_{beta} \underbrace{+f_c\frac{\Delta W}{\Delta z}}_{stretching}$$

where the relative **geostrophic vorticity** ζ_g is defined similar to eq. (11.17), except using geostrophic winds U_g and V_g:

$$\zeta_g = \frac{\Delta V_g}{\Delta x} - \frac{\Delta U_g}{\Delta y} \qquad \bullet(13.9a)$$

For solid body rotation, eq. (11.19) becomes:

$$\zeta_g = \frac{2 \cdot G}{R} \qquad \bullet(13.9b)$$

where G is the geostrophic wind speed and R is the radius of curvature.

The prefix "quasi-" is used for the following reasons. If the winds were perfectly geostrophic or gradient, then they would be parallel to the isobars. Such winds never cross the isobars, and could not cause convergence into the low. With no convergence there would be no vertical velocity.

However, we know from observations that vertical motions do exist and are important for causing clouds and precipitation in cyclones. Thus, the last term in the quasi-geostrophic vorticity equation includes W, a wind that is not geostrophic. When such an **ageostrophic** vertical velocity is included in an equation that otherwise is totally geostrophic, the equation is said to be **quasi-geostrophic**, meaning partly geostrophic. The quasi-geostrophic approximation will also be used later in this chapter to estimate vertical velocity in cyclones.

Within a quasi-geostrophic system, the vorticity and temperature fields are closely coupled together, due to the dual constraints of geostrophic and hydrostatic balance. This implies close coupling between the wind and mass fields, as was discussed in Chapts. 11 and 12 in the sections on geostrophic adjustment. While such close coupling is not observed for every weather system, it is a reasonable approximation for synoptic-scale, extratropical systems, as previously mentioned.

Solved Example

Suppose an initial flow field has no geostrophic vorticity, but a straight north to south geostrophic wind blows at 10 m/s at latitude 45°. Also, the top of a 1 km thick column of air rises at 0.01 m/s, while its base rises at 0.008 m/s. Find the rate of geostrophic-vorticity spin-up.

Solution

Given: $V = -10$ m/s, $\phi = 45°$, $W_{top} = 0.01$ m/s, $W_{bottom} = 0.008$ m/s, $\Delta z = 1$ km.

Find: $\Delta\zeta_g/\Delta t = ?\ s^{-2}$

First, use eq. (9.10):

$$f_c = (1.458\times10^{-4}\,s^{-1})\cdot\sin 45° = 0.000103\ s^{-1}$$

Next, use eq. (13.2):

$$\frac{\Delta f_c}{\Delta y} \equiv \beta = \frac{1.458\times10^{-4}\,s^{-1}}{6.357\times10^{6}\,m}\cdot\cos 45° = 1.62\text{x}10^{-11}\ m^{-1}\cdot s^{-1}$$

Use the definition of a gradient (see Appendix A):

$$\frac{\Delta W}{\Delta z} = \frac{W_{top} - W_{bottom}}{z_{top} - z_{bottom}} = \frac{(0.01-0.008)\text{m/s}}{(1000-0)\text{m}} = 2\text{x}10^{-6}\ s^{-1}$$

Finally, use eq. (13.8):

$$\underbrace{\frac{\Delta\zeta_g}{\Delta t}}_{spin-up} = \underbrace{-(-10\text{m/s})\cdot(1.62\times10^{-11}\text{m}^{-1}\cdot\text{s}^{-1})}_{beta} + \underbrace{(0.000103\text{s}^{-1})\cdot(2\times10^{-6}\,\text{s}^{-1})}_{stretching}$$

$$=(1.62\text{x}10^{-10} + 2.06\text{x}10^{-10})\ s^{-2} = \underline{\mathbf{3.68x10^{-10}}}\ s^{-2}$$

Check: Units OK. Physics OK.

Discussion: Even without any initial geostrophic vorticity, the rotation of the earth can spin-up the flow if the wind blows appropriately.

Application to Idealized Weather Patterns

An idealized weather pattern is shown in Fig 13.5. Every feature in the figure is on the 50 kPa isobaric surface (i.e., in the mid troposphere), except the 𝕃 which indicates the location of the surface low center. All three components of the geostrophic vorticity equation can be studied.

Geostrophic and gradient winds are parallel to the height contours. The trough axis is a region of cyclonic (counterclockwise) curvature of the wind, which yields a large positive value of geostrophic vorticity. At the ridge there is negative (clockwise) relative vorticity. Thus, the **advection** term is positive and contributes to spin-up of the cyclone because the wind is blowing higher positive vorticity into the area of the surface low.

For any fixed pressure gradient, the gradient winds are slower than geostrophic when curving cyclonically ("slow around lows"), and faster than geostrophic for anticyclonic curvature, as sketched with the heavy wind arrows in the Figure. Examine the 50 kPa flow immediately above the surface low. Air is departing faster than entering. This imbalance (divergence) draws air up from below. Hence, W increases from near zero at the ground to some positive updraft speed at 50 kPa. This **stretching** helps to spin-up the cyclone.

The **beta** term, however, contributes to spin-down because air from lower latitudes (with smaller Coriolis parameter) is blowing toward the location of the surface cyclone. This effect is small when the wave amplitude is small. The sum of all three terms in the quasigeostrophic vorticity equation is often positive, providing a net spin-up and intensification of the cyclone.

In real cyclones, contours are often more closely spaced in troughs, causing relative maxima in jet stream winds called jet streaks. Vertical motions associated with horizontal divergence in jet streaks are discussed later in this chapter. These motions violate the assumption that air mass is conserved along an "isobaric channel". Rossby also pointed out in 1940 that the gradient wind balance is not valid for non-stationary motion. Thus, the model of Fig 13.5 has weaknesses that limit its applicability.

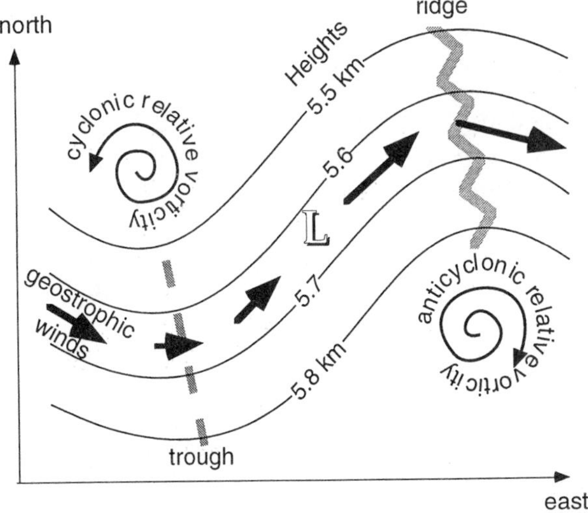

Figure 13.5
An idealized 50 kPa chart with equally-spaced height contours, as introduced by J. Bjerknes in 1937. The location of the surface low 𝕃 is indicated.

UPWARD MOTION

Upward vertical motion is another measure of cyclone vigor and cloud formation. In height z coordinates, the vertical velocity is defined as $W \equiv \Delta z/\Delta t$, where t is time. An analogous vertical velocity, **omega** (ω), can be defined in pressure coordinates:

$$\omega \equiv \frac{\Delta P}{\Delta t} \qquad \bullet(13.10)$$

Because pressure decreases with height, ω is negative for updrafts, while W is positive. Using the hydrostatic equation, the relationship between ω and W is:

$$\omega = -\rho \cdot g \cdot W \qquad \bullet(13.11)$$

where ρ is air density and $g = 9.8\ \text{m·s}^{-2}$ is gravitational acceleration. For example, at $P = 50$ kPa where standard atmospheric density is $\rho \cong 0.69\ \text{kg·m}^{-3}$, an updraft of $W = 1$ m/s gives $\omega = -6.8$ Pa/s. Either W or ω can be used to represent vertical motion. Most computer forecasts use ω.

We will use two approaches to investigate vertical motion: the continuity equation and the omega equation. The continuity approach shows how jet-stream winds diverging horizontally at the top of the troposphere can cause vertical motion in the mid troposphere. The omega equation utilizes quasi-geostrophic dynamics within the bottom half of the troposphere to diagnose vertical motion from the vorticity and thermal wind. Both approaches give complementary insight.

Continuity Effects

Vertical motion in a cyclone is often driven by changes of horizontal wind speed of the upper-tropospheric jet stream. Mass continuity requires that horizontal divergence be compensated by vertical convergence, and vice versa. Namely, air leaving a region horizontally must be balanced by replacement air entering the same region vertically, so as not to leave a vacuum (Fig 13.6). The jet stream is near the base of the stratosphere, where strong static stability aloft impedes vertical motion above the jet. Thus, most of the compensating vertical motion occurs in the troposphere, beneath the level of the jet.

For the column of air sketched in Fig 13.6, the incompressible continuity equation in two dimensions becomes

$$W_{mid} = \frac{\Delta M}{\Delta s} \cdot \Delta z \qquad (13.12)$$

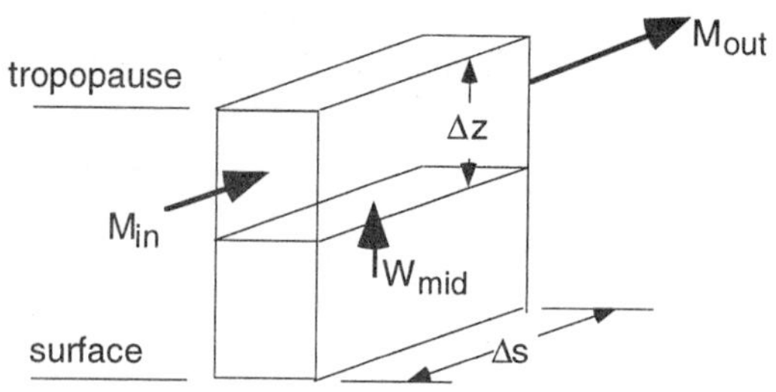

Figure 13.6
A column of tropospheric air, showing that mid-level vertical motion W is related to upper level divergence $M_{out} - M_{in}$, where M is wind speed.

where $\Delta M = M_{out} - M_{in}$ is the change of horizontal wind speed, Δs is horizontal distance between inflow and outflow, $\Delta M/\Delta s$ is the upper level **horizontal divergence**, Δz is the upper portion of the troposphere (between the 50 kPa surface and the tropopause), and W_{mid} is the vertical velocity across the 50 kPa surface. We will assume for simplicity that all of the horizontal divergence aloft can be represented by the winds in one direction, as sketched.

Two mechanisms that cause jet-stream divergence are examined next: curvature of the jet stream, and jet streaks. The first is a large-scale feature in approximate geostrophic balance. The second is a smaller-scale feature that causes an ageostrophic (non-geostrophic) divergent flow.

Jet-Stream Curvature

Recall that air blows slower around troughs than around ridges, due to centrifugal force (Chapter 9). Consider a column of air situated east of an upper-level trough. The gradient wind speed horizontally entering the column is less than that leaving (Fig 13.7). This results in a wind speed difference of ΔM across a distance $\Delta s = d$.

Assume the jet stream oscillates north-south as a sine wave. If the north-south distance between crest and trough is Δy (= 2·amplitude of wave), and the east-west wavelength is λ, then the distance d between crest and trough is:

$$\Delta s = d = \left[(\lambda/2)^2 + \Delta y^2\right]^{1/2} \qquad (13.13)$$

Gradient wind speeds around anticyclones and cyclones were discussed in Chapter 9. The speed difference between outflow (anticyclonic) and inflow (cyclonic) is

$$\Delta M = 0.5 \cdot f_c \cdot R \cdot \left[2 - \sqrt{1 - \frac{4 \cdot G}{f_c \cdot R}} - \sqrt{1 + \frac{4 \cdot G}{f_c \cdot R}}\right] \qquad (13.14)$$

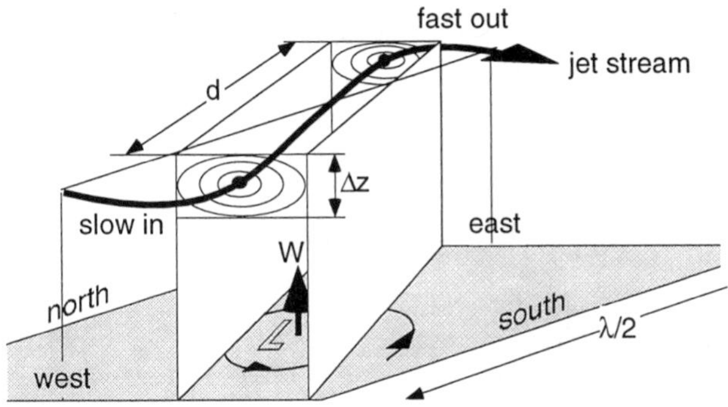

Figure 13.7
Factors affecting total mass within a column of air. The column is between the two vertical planes.

where G is geostrophic wind speed. The radius of curvature R at crests and troughs of sine waves is:

$$R = \frac{1}{2\pi^2} \cdot \frac{\lambda^2}{\Delta y} \tag{13.15}$$

The effect of upper-level divergence due to jet-stream curvature can be found by combining the equations above:

$$W_{mid} = \frac{\frac{f_c \cdot \Delta z \cdot \lambda^2}{4\pi^2 \cdot \Delta y}\left[2 - \sqrt{1 - \frac{8\pi^2 G \cdot \Delta y}{f_c \cdot \lambda^2}} - \sqrt{1 + \frac{8\pi^2 G \cdot \Delta y}{f_c \cdot \lambda^2}}\right]}{[(\lambda/2)^2 + \Delta y^2]^{1/2}} \tag{13.16}$$

Thus, knowing the wavelength, amplitude, and geostrophic wind speed, it is possible to estimate the vertical motion.

Solved Example

Given a jet stream thickness of 5 km, density 0.364 kg·m^{-3}, wavelength 6000 km, amplitude 500 km, with average geostrophic wind speed of 30 m/s. Find the radius of curvature at the crest, distance d along the flow, in vs. outflow wind-speed difference, and mid-level vertical velocity.

Solution

Given: ρ_{top} = 0.364 kg·m^{-3}, G = 30 m/s,
Δz = 5 km, λ = 6000 km,
Δy = 2·500 km = 1000 km.
Find: R = ? km, d = ? km, ΔM= ? m/s, W_{mid}= ? m/s
Assume: f_c = 0.0001 s^{-1}

(continued next column)

Solved Example *(continuation)*

Use eq. (13.15):

$$R = \frac{1}{2\pi^2} \cdot \frac{(6000\text{km})^2}{(1000\text{km})} = \underline{\mathbf{1824\ km}}$$

From Chapter 9 use:

$$\frac{G}{f_c \cdot R} = Ro_c = \frac{30\text{m/s}}{(0.0001\text{s}^{-1}) \cdot (1824\text{km})} = \underline{\mathbf{0.16}}$$

Use eq. (13.14):

$$\Delta M = \frac{10\text{m/s}}{2(0.164)} \cdot \left[2 - \sqrt{1-4(0.164)} - \sqrt{1+4(0.164)}\right]$$
$$= \underline{\mathbf{11.6\ m/s}}$$

Use eq. (13.13):

$$d = [(3000\text{km})^2 + (1000\text{km})^2]^{1/2} = \underline{\mathbf{3162\ km}}$$

Use eq. (13.12):

$$W_{mid} = \left(\frac{11.6\text{m/s}}{3162\text{km}}\right) \cdot 5\text{km} = \underline{\mathbf{0.018\ m/s}}$$

Check: Units OK. Physics OK.
Discussion: A vertical velocity of a couple cm/s averaged over the whole cyclone is typical, although local updrafts in clouds can be faster.

Jet Streaks

A jet streak is a relative maximum of wind speed within the jet stream. Picture a jet streak as sketched in Fig 13.8a, blowing from west to east. This forms in the upper-tropospheric jet stream in regions where isobars or height contours are tightly packed (i.e., closely spaced). Wind speed reaches a maximum at the center of the streak.

Wind cannot only advect temperature or humidity, but it can also advect itself. As a result, the wind maximum advects itself eastward. An outcome is that the jet maximum finds itself in a region where the isobars are further apart. Namely, the actual wind speed in the exit region of the streak is faster than the theoretical geostrophic wind. This is a situation that causes an ageostrophic wind to the south.

To quantify this effect, start with the full equations of motion from Chapter 9 (eqs. 9.15). Consider only the equation for U wind, assume approximate steady state (no time change), and neglect boundary-layer drag. The result is:

$$0 = -U\frac{\Delta U}{\Delta x} - f_c\left(V - V_g\right) \tag{13.17}$$

We can define an **ageostrophic wind** V_{ag} as the difference between the actual wind and the geostrophic wind:

$$V_{ag} \equiv V - V_g \qquad \bullet(13.18a)$$

Similarly:

$$U_{ag} \equiv U - U_g \qquad \bullet(13.18b)$$

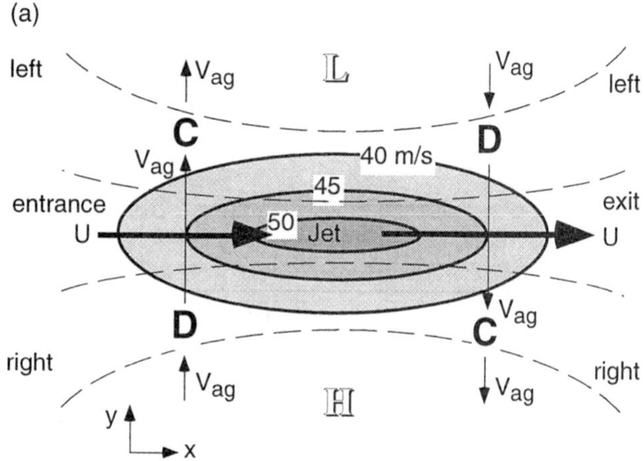

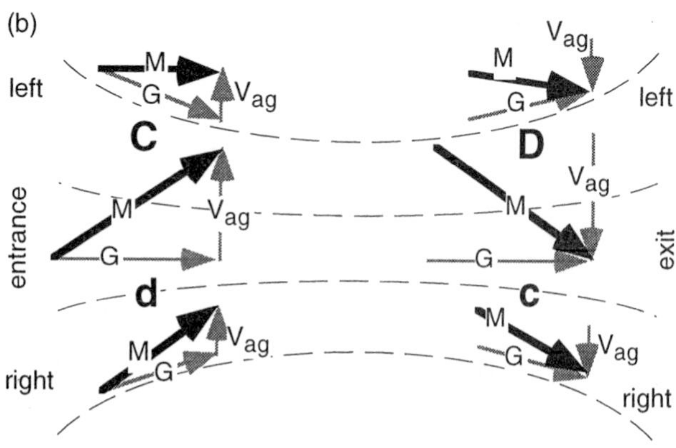

Figure 13.8
(a) Jet streak, showing isobars or height contours (dashed lines), isotachs (shaded contours), jet stream wind and jet axis (heavy arrows), ageostrophic components V_{ag} of wind (thin arrows), and regions of horizontal convergence C and divergence D.
(b) Gray vectors represent the geostrophic wind G parallel to the isobars, and the ageostrophic winds as in part (a). The vector sum of the gray vectors is the total wind speed M, shown by the heavy black vectors. Upper-case C and D represent larger convergence and divergence than lower-case c and d.

With this definition, the equation of motion becomes

$$V_{ag} = \frac{U}{f_c} \frac{\Delta U}{\Delta x} \qquad \bullet(13.19a)$$

For a north-south aligned jet axis, the corresponding ageostrophic wind would be

$$U_{ag} = -\frac{V}{f_c} \frac{\Delta V}{\Delta y} \qquad \bullet(13.19b)$$

In the **exit region** of the jet, U is positive but $\Delta U/\Delta x$ is negative (wind decrease eastward). Thus, eq. (13.19a) says that a negative V_{ag} ageostrophic wind develops (i.e., blows from north to south). In the **entrance region** the U wind increases in the x direction, causing a positive V_{ag} (i.e., blows from south to north, as sketched in Fig 13.8a).

The magnitude of ageostrophic wind is largest near the jet axis, it decreases in magnitude further off to the side of the jet. As a result, there is horizontal divergence of the ageostrophic wind in the left exit region of the jet, and convergence in the right exit region. A similar but opposite convergence-divergence pair occurs in the entrance region. These four regions are called the **four quadrants** of the jet streak.

The total wind speed M is the vector sum of the geostrophic components G parallel to the isobars, and the ageostrophic components V_{ag} as indicated in Fig 13.8b. Focusing on the total wind speed (thick black vectors in Fig 13.8b), one can see strong convergence in the left entrance region (i.e., the vectors come towards each other at a large angle). Similarly, there is large divergence in the left exit region.

In the right entrance and exit regions, however, the total wind vectors are almost parallel to the wind vectors along the center axis of the jet. Thus, divergence and convergence are weaker there.

Fig 13.9 shows the vertical and horizontal circulations induced by the ageostrophic motion near the exit region of the jet streak. Such a flow is called a **secondary circulation**. The vertical components of this circulation, which are caused by the convergence and divergence regions aloft, also induce a pair of horizontal divergence and convergence at the surface.

The sense (direction) of this circulation below the **exit region** of the jet is **indirect** (opposite to the sense of rotation of a Hadley cell). The sense below the **entrance region** is **direct**.

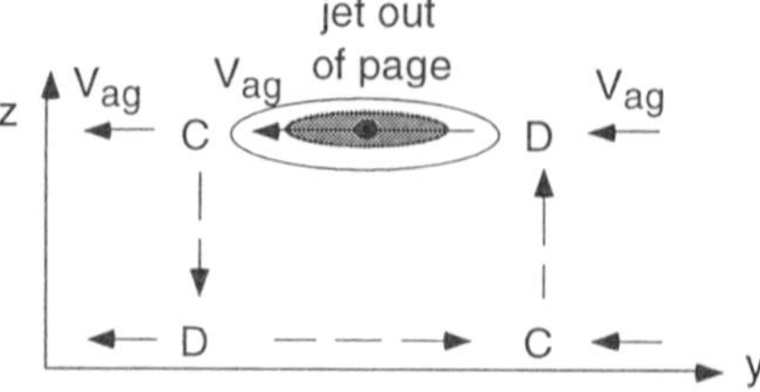

Figure 13.9
Vertical cross-section through the exit region of a jet (shaded), where the jet points out of the page. The V_{ag} ageostrophic wind (thin solid) is driven by jet-streak dynamics, while the dashed circulation is driven by mass continuity. Regions of horizontal convergence C and divergence D are indicated.

Regions of divergence aloft remove mass from the column of air, thereby lowering sea-level pressure. Such sea-level pressure drop and associated upward motion cause cyclogenesis. Hence, cyclones are favored below the left exit region and right entrance region of the jet streak.

To find the mid-level vertical velocity below the left exit quadrant, let $\Delta s = \Delta y$ be the north-south width of the jet streak. Assume the ageostrophic wind V_{ag} smoothly approaches zero within a distance Δy to the side of the jet. Using eq. (13.19a) in eq. (13.12), the upward motion is

$$W_{mid} = \left| \frac{U \cdot \Delta U}{f_c \cdot \Delta x} \right| \cdot \frac{\Delta z}{\Delta y} \qquad (13.20)$$

Solved Example

Suppose a jet streak of width 1000 km has a core wind speed of 50 m/s that decreases eastward to 25 m/s across a distance of 1000 km. If the jet region is 5 km thick, find the mid-level vertical velocity. Assume $f_c = 10^{-4}\ s^{-1}$.

Solution

Given: U = 50 m/s, ΔU = 25 – 50 = –25 m/s, $\Delta x = \Delta y$ = 1000 km, Δz = 5 km.

Find: W_{mid} = ? m/s

Use eq. (13.20):

$$W_{mid} = \frac{(50\text{m/s}) \cdot (25\text{m/s})}{(10^{-4}\text{s}^{-1}) \cdot (10^6\text{m})} \cdot \frac{(5\text{km})}{(1000\text{km})} = \underline{0.0625\ \text{m/s}}$$

Check: Units OK. Physics OK.

Discussion: This is a larger average vertical velocity over a much smaller horizontal area than for the previous solved example. Such a vertical velocity can significantly intensify a cyclone.

Omega Equation

The **omega equation** is the name of a diagnostic equation used to find vertical motion (or omega, in pressure units). We will use a form of this equation developed by K. Trenberth, based on quasi-geostrophic dynamics and thermodynamics.

The full omega equation is one of the nastier-looking equations in meteorology (see the Beyond Algebra box at the end of this Chapter). To simplify it, we focus on one part of the full equation, apply it to the bottom half of the troposphere (the layer between 100 to 50 kPa isobaric surfaces), and convert the result from ω to W.

The approximate omega equation is thus:

$$W_{mid} \cong \frac{-2 \cdot \Delta z}{f_c} \left[U_{TH} \frac{\overline{\Delta \zeta_g}}{\Delta x} + V_{TH} \frac{\overline{\Delta \zeta_g}}{\Delta y} + V_{TH} \frac{\beta}{2} \right] \qquad \bullet(13.21a)$$

where W_{mid} is the vertical velocity in the mid-troposphere (at P = 50 kPa), Δz is the 100 to 50 kPa thickness, U_{TH} and V_{TH} are the thermal wind components for the 100 to 50 kPa layer, f_c is Coriolis parameter, β is the change of Coriolis parameter with y (see eq. 13.2), ζ_g is the geostrophic vorticity, and the overbar represents an average over the whole depth of the layer. An equivalent form is:

$$W_{mid} \cong \frac{-2 \cdot \Delta z}{f_c} \left[M_{TH} \frac{\overline{\Delta(\zeta_g + (f_c/2))}}{\Delta s} \right] \qquad \bullet(13.21b)$$

where s is distance along the thermal wind direction, and M_{TH} is the thermal wind speed.

Regardless of the form, the terms in square brackets represent the advection of vorticity by the thermal wind, where vorticity consists of the geostrophic relative vorticity plus a part of the vorticity due to the earth's rotation. The geostrophic vorticity at the 70 kPa isobaric surface is often used to approximate to the average geostrophic vorticity over the whole 100 to 50 kPa layer.

A physical interpretation of the omega equation is that greater upward velocity occurs where there is greater advection of cyclonic (positive) geostrophic vorticity by the thermal wind. Greater upward velocity favors clouds and heavier precipitation. Also, by moving air upward from the surface, it reduces the pressure under it, causing the surface low to move toward that location and deepen.

Weather maps can be used to determine the location and magnitude of the maximum upward motion. The idealized map of Fig 13.10a shows the height (z) contours of the 50 kPa isobaric surface, along with the trough axis. Also shown is the location of the surface low and fronts.

At the surface, the greatest vorticity is often near the low center. At 50 kPa, it is often near the trough axis. At 70 kPa, the vorticity maximum (often called the **vort max** for short) is usually between those two locations. In Fig 13.10a, the darker shading corresponds to regions of greater cyclonic vorticity at 70 kPa.

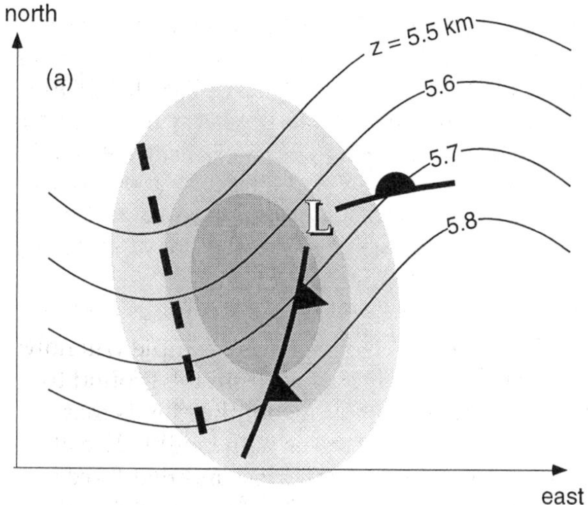

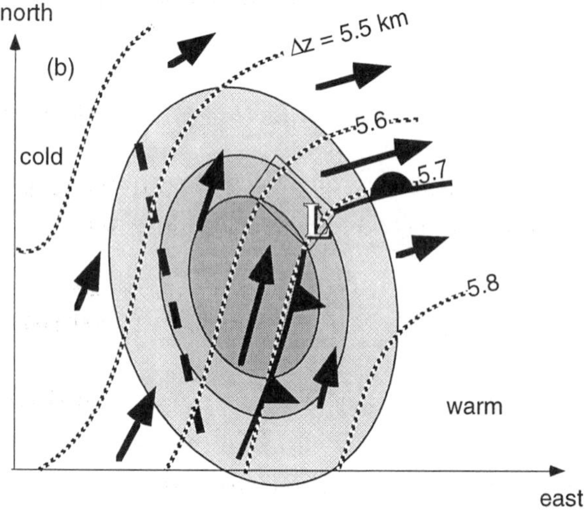

Figure 13.10
(a) Weather at three different pressure heights: (1) 50 kPa heights (thin solid lines) and trough axis (thick dashed line); (2) surface low pressure center (L) and fronts; (3) 70 kPa vorticity (shaded).
(b) Trough axis, surface low and fronts, and vorticity shading are identical to Fig (a). Added are: 100 to 50 kPa thickness (dotted lines), thermal wind vectors (arrows), and region of maximum vorticity advection by the thermal wind (thin rectangular box). It is within this box that the omega equation gives the greatest updraft speed.

Fig 13.10b shows the thickness (Δz) of the layer of air between the 100 and 50 kPa isobaric surfaces. Thickness lines are often nearly parallel to surface fronts, with the tightest packing on the cold side of the fronts. Recall that thermal wind is parallel to the thickness lines, with cold air to the left, and with the greatest velocity where the thickness lines are most tightly packed. Thermal wind direction is represented by the arrows in Fig 13.10b, with longer arrows denoting stronger speed.

Recall from the Focus box earlier in this chapter that advection is greatest where the area between crossing isopleths is smallest. This rule also works for advection by the thermal wind. The dotted lines represent the isopleths that drive the thermal wind. The thin black lines around the shaded areas are isopleths of vorticity. The smallest area between these crossing isopleths indicates the greatest vorticity advection by the thermal wind, and is outlined by a rectangular box. For this particular example, the greatest updraft would be expected within this box.

Be careful when you identify the smallest area. In Fig 13.10b, there is another area equally as small further south-south-west from the low center. However, the cyclonic vorticity is being advected away from this region rather than toward it. Hence, this is a region of <u>negative</u> vorticity advection by the thermal wind, which would imply <u>downward</u> vertical velocity.

Solved Example

The 100 to 50 kPa thickness is 5 km and $f_c = 10^{-4}$ s^{-1}. A west to east thermal wind of 20 m/s blows through a region where the avg. cyclonic vorticity decreases by 10^{-4} s^{-1} toward the east across a distance of 500 km. Use the omega eq. to find mid tropospheric upward velocity.

Solution

Given: $U_{TH} = 20$ m/s, $V_{TH} = 0$, $\Delta z = 5$ km, $\Delta\zeta = -10^{-4}$ s^{-1}, $\Delta x = 500$ km, $f_c = 10^{-4}$ s^{-1}.
Find: $W_{mid} = ?$ m/s

Use eq. (13.21a):

$$W_{mid} \cong \frac{-2 \cdot (5000\text{m})}{(10^{-4}\text{s}^{-1})}\left[(20\text{m/s})\frac{(-10^{-4}\text{s}^{-1})}{(5\times10^{5}\text{m})} + 0 + 0\right]$$

$= \mathbf{\underline{0.4\ m/s}}$

Check: Units OK. Physics OK.
Discussion: At this updraft speed, an air parcel would take 7.6 h to travel from the ground to the tropopause.

SEA-LEVEL-PRESSURE TENDENCY

Sea-level pressure is another measure of cyclone intensity. When pressure drops in an intensifying low center, the low is said to **deepen**. In other words, the low becomes lower. When sea-level pressure increases, the low **fills** (i.e., the low weakens). The pressure change with time is called the **pressure tendency**. The amount of air mass in the column of atmosphere above the low center determines the surface pressure in a hydrostatic environment.

Mass Budget

We will approach this subject heuristically. Picture a small cylinder filled with air, as shown in Fig 13.11a. Near the middle of the cylinder is a weightless, frictionless piston. Above and below the piston, the air pressures are identical, and the air densities are identical.

Suppose that some of the air is withdrawn from the top of the cylinder, as shown in Fig 13.11b. Pressure in the top of the cylinder will decrease, which will cause the piston to rise until the bottom pressure equals the top pressure. At that point, the densities above and below the piston are also identical. Similarly, air mass could have been added to the top of the cylinder, causing the pressure to increase and piston to move down.

Thus, we can use the vertical motion of the piston as a surrogate measure of net mass flow and pressure change. Namely, upward motion indicates decreases in mass and pressure, while downward motion indicates increases.

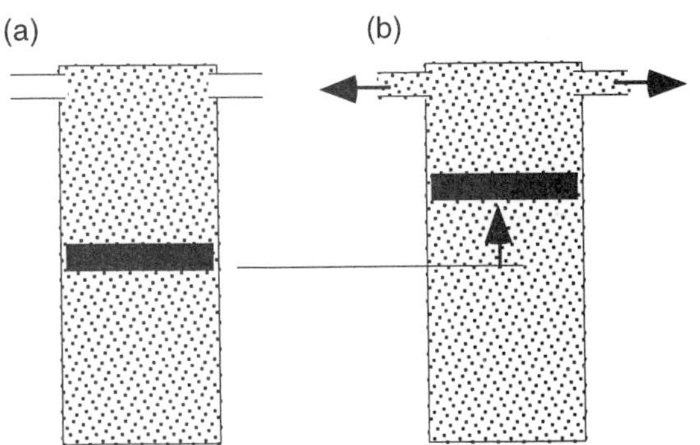

Figure 13.11
(a) Column of air in a closed cylinder, with a piston (black) in the middle. (b) Piston location changes after some air is withdrawn from the top. Assume that no air leaks past the piston.

That seems as a simple enough rule. However, picture what would happen if air were withdrawn or added to the bottom of the cylinder. Upward piston motion would correspond to an increase in mass and pressure, not a decrease as before. Thus, we can use vertical piston movement as a surrogate measure of pressure change only if we change the sign of the result depending on whether mass is added at the top or bottom of the column.

We will extend this reasoning to the atmosphere, where the cylinder of the previous example will now be visualized as a column of air from the ground to the top of the atmosphere. A complication is that atmospheric density decreases with height. Recall that pressure is force per unit area, and that force is mass times acceleration. Sea-level pressure P_S is a measure of the total **mass m of air in a column** above the surface

$$P_S = \frac{g}{A} \cdot m \qquad \bullet(13.22)$$

where $g = 9.8\ \mathrm{m \cdot s^{-2}}$ is gravitational acceleration, and A is the horizontal cross-section area under the column.

Changes in sea-level pressure with time t are caused by changes in total air mass above the surface:

$$\frac{\Delta P_S}{\Delta t} = \frac{g}{A} \cdot \frac{\Delta m}{\Delta t} \qquad \bullet(13.23)$$

Recall from the definition of density that

$$m = \rho \cdot Volume = \rho \cdot A \cdot z \qquad (13.24)$$

where z is the height of the volume. Define the change of height with time as a surrogate velocity $W_{surrogate}$ analogous to the piston movement.

Thus, if the mass flow in or out of the column occurs at a height z where the air density $\rho(z)$ is known, then the equations above can be combined to give the pressure tendency:

$$\frac{\Delta P_S}{\Delta t} = \pm g \cdot \rho(z) \cdot W_{surrogate}(z) \qquad (13.25)$$

The proper sign for the right-hand-side of the equation must be chosen depending on the cause of the surrogate vertical motion (i.e., whether it is driven at the top or bottom of the troposphere). There can be more than one mechanism adding or subtracting mass to a column of air, so we will generalize the right-hand-side of eq. (13.25) to be a sum of terms.

Four mechanisms will be included here in a simplified model for pressure tendency:

- Upper-level divergence
- Boundary-layer pumping
- Advection
- Diabatic heating

Divergence aloft has just been described in the previous section as a mass removal process. Namely, sea-level pressure can drop if the jet stream or a jet streak removes mass aloft. That resulting W_{mid} will be used as one of the forcing terms in the net pressure tendency equation, to be given later in this section.

Boundary-layer pumping was discussed in Chapter 9. Sometimes lows deepen so rapidly, that the accelerating boundary-layer winds blow directly from high to low pressure as an ageostrophic non-equilibrium flow. Later, if the low persists for a sufficiently-long duration, the winds can turn cyclonically and behave as a classical equilibrium boundary-layer flow, with a vertical velocity as was given in Chapter 9. This W_{BL} to will be used later in the net pressure tendency equation.

Advection describes the movement of cyclones by the steering-level winds. Namely, surface pressure can drop at a fixed point on the ground if the horizontal wind blows in an air column having less total mass than the column of air that is blowing out. Analogous to advection terms in other equations earlier in this book (e.g., heat, moisture, and momentum budget eqs.), the advection term will be approximated as $-M_c \cdot (\Delta P_s/\Delta s)$, where M_c is the speed of movement of the whole column of air along path s, and $\Delta P_s/\Delta s$ is the horizontal surface pressure gradient along that path.

Diabatic (i.e., non adiabatic) **heating** due to condensation is discussed in detail in the next section. Other diabatic effects such as direct radiative heating can also affect the mass budget, but are not discussed here.

Diabatic Heating

When water vapor condenses in clouds, latent heat is released. This heat warms the air in a column, causing it to expand (Fig 13.12). According to the hypsometric equation, horizontal pressure gradients develop which push the air out of the column. Hence, there is divergence of mass from the column, which lowers the sea-level pressure.

In an overly simplified point-of-view, the heated air expands out of the top of the column and overflows. The amount of overflow out of the top in this simple view equals the amount of mass diverged from the sides in a more realistic view. None-the-less, vertical velocity W at the top of the column is a surrogate measure of the mass loss.

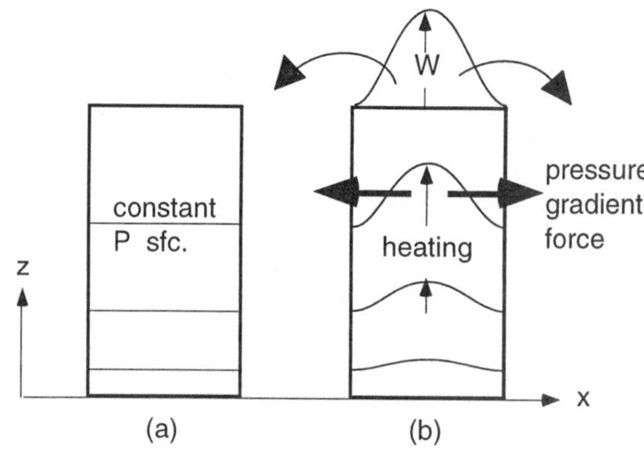

Figure 13.12
(a) Column of air before heating. (b) Column of air after heating, where the air expands and overflows out of the column.

If all of the condensation falls as precipitation, then the latent-heating rate is related to the rainfall rate RR:

$$\frac{\Delta T_v}{\Delta t} = \frac{1}{\Delta z} \cdot \frac{L_v}{C_p} \cdot \frac{\rho_{liq}}{\rho_{air}} \cdot RR \qquad (13.26)$$

where T_v is the virtual temperature in the column, t is time, L_v is the latent heat of vaporization, C_p is the specific heat at constant pressure for air, ρ_{liq} is the density of liquid water (1025 kg/m^3), ρ_{air} is the density of air, Δz is the column depth, and where RR has units of mm/h.

The net effect on sea-level pressure can be found by combining the hypsometric equation and the equation above:

$$\frac{\Delta P_s}{\Delta t} = -\frac{g}{\overline{T_v}} \cdot \frac{L_v}{C_p} \cdot \rho_{liq} \cdot RR \qquad \bullet(13.27)$$

where g =9.8 m·s^{-2} is gravitational acceleration, and $\overline{T_v}$ is the average virtual temperature in the column (≅ 300 K). Because RR is the net precipitation reaching the ground, the equation above represents the net or average effect of latent heating over the whole cyclone depth.

Although this equation depends on absolute temperature, which can vary, to first order we can approximate it as

$$\frac{\Delta P_s}{\Delta t} \approx -b \cdot RR \qquad \bullet(13.28)$$

where $b = -\frac{g}{\overline{T_v}} \cdot \frac{L_v}{C_p} \cdot \rho_{liq} \cong 0.084$ kPa/mm$_{rain}$.

Solved Example

Radar measures a medium intensity rain of 10 mm/h. Find the diabatic heating contribution to sea-level pressure tendency.

Solution

Given: $RR = 10$ mm/h
Find: $\Delta P_s/\Delta t = ?$ kPa/h

Use eq. (13.28):

$$\Delta P_s/\Delta t = (0.084 \text{ kPa/mm}_{rain})\cdot(10 \text{ mm/h}) = \underline{\mathbf{0.84 \text{ kPa/h}}}$$

Check: Units OK. Physics OK.
Discussion: A column of air could not rain at this rate for more than a few hours, due to the maximum precipitable water available in the column of air. If additional water vapor is advected into the column to cause additional rain, the water vapor is usually imbedded in air of appropriate temperature to partially compensate the otherwise-excessive pressure fall.

Net Pressure Tendency

Combining all of the mechanisms just described using the idealized geometry of Fig 13.6 yields the following form for the net **sea-level pressure tendency equation**:

$$\underbrace{\frac{\Delta P_s}{\Delta t}}_{\text{pressure tendency}} = \underbrace{-M_c \cdot \frac{\Delta P_s}{\Delta s}}_{\text{horizontal advection}} + \underbrace{g \cdot \rho_{BL} \cdot W_{BL}}_{\text{boundary-layer pumping (in)}} - \underbrace{g \cdot \rho_{mid} \cdot W_{mid}}_{\text{upper-level horiz. divergence (out)}} - \underbrace{b \cdot RR}_{\text{latent heating}} \qquad \bullet(13.29)$$

where ρ_{BL} and ρ_{mid} are the average air densities at the top of the boundary layer and midpoint of the column, respectively, and the speed of movement of the column is M_c measured positive in the direction of movement, along horizontal path s. A negative value of $\Delta P_s/\Delta t$ corresponds to pressure drop and cyclone intensification.

Imagine a scenario where baroclinic instability in the jet stream causes a wave to move over some location on the earth that initially lacks any strong highs or lows. If the upper-level trough is west of this location, and ridge east, then divergence aloft can begin to suck air out of the column (time A in Fig 13.13). Similarly, the presence of a jet streak at the right location could initiate upper-level divergence.

Solved Example

Suppose upper level divergence has an effective vertical velocity of 0.03 m/s over a cyclone, while boundary layer pumping causes a vertical velocity of 0.01 m/s. The rainfall rate is 5 mm/h. Sea-level pressure increases 1 kPa across a distance of 200 km to the west of the cyclone, and there is a steering wind from the west at 20 m/s. Find the sea-level pressure tendency.

Solution

Given: $W_{mid} = 0.03$ m/s , $W_{BL} = 0.01$ m/s,
$\Delta P/\Delta s = -1$ kPa/200 km, $RR = 5$ mm/h
Find: $\Delta P_s/\Delta t = ?$ kPa/h
Assume: $\rho_{BL} = 1.112 \text{ kg·m}^{-3}$, $\rho_{mid} = 0.5 \text{ kg·m}^{-3}$

Use eq. (13.29): $\Delta P_s/\Delta t =$

$$-\left(20\frac{\text{m}}{\text{s}}\right)\left(\frac{-1\text{kPa}}{2\times10^5\text{m}}\right)+\left(9.8\frac{\text{m}}{\text{s}^2}\right)\left(1.112\frac{\text{kg}}{\text{m}^3}\right)\left(0.01\frac{\text{m}}{\text{s}}\right)$$

$$-\left(9.8\frac{\text{m}}{\text{s}^2}\right)\left(0.5\frac{\text{kg}}{\text{m}^3}\right)\left(0.03\frac{\text{m}}{\text{s}}\right)-\left(0.084\frac{\text{kPa}}{\text{mm}}\right)\left(5\frac{\text{mm}}{\text{h}}\right)$$

$\Delta P_s/\Delta t = 0.0001 + 0.00011 - 0.00015 - 0.00012$ kPa/s
advect + B.L. – jet diver – diabatic
$= -\ 0.00006$ kPa/s = **$\underline{-\ 0.22 \text{ kPa/h}}$**

Check: Units OK. Physics OK.
Discussion: The cyclone would continue to deepen, given this scenario. Advection is quite important. Don't forget to convert from Pa to kPa in the boundary-layer (BL) and jet-divergence terms. Also convert the rainfall term to kPa/s from kPa/h.

As sea-level pressure falls, pressure gradient force begins to spin-up the circulation around the low. Initially the circulation is weak, resulting in weak boundary-layer pumping. At this stage the upper level divergence is continuing to withdraw air from the column faster than boundary-layer pumping fills it, allowing the low to intensify (time B).

The rising air in the cyclone causes clouds to form, with the resulting precipitation. Latent heating warms the air, which contributes to the loss of mass and decrease of sea-level pressure.

As the low matures, it wraps the cold and warm air masses further around the low center, thereby modifying the temperature gradients and thermal winds. The low center begins to lag behind the region of maximum divergence aloft. The surface low and associated air masses also interact with and modify the jet-stream aloft.

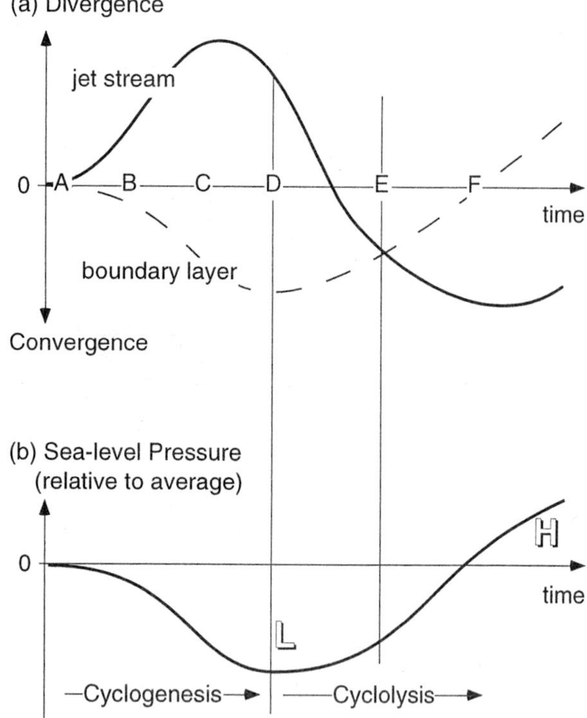

Figure 13.13
(a) Divergence contributions from the jet stream and boundary layer, and (b) their modification of sea-level pressure during cyclone evolution.

After the time of maximum divergence (time C), withdrawal of air aloft decreases while input of air from the boundary layer increases, eventually reaching the point (time D) where input and output match. The low ceases to deepen at this point.

Later as the low occludes, it lags so far behind the upper-air wave that jet-stream divergence ceases over the column. Meanwhile the still-spinning surface low continues to pump boundary-layer air into the column. With the upper-level trough east of the surface low, convergence aloft also pumps air into the column from above. Thus, input of mass is now occurring at both the top and bottom (time E), causing the sea-level pressure to rise. The cyclone rapidly fills and spins down as pressure gradients weaken.

Finally at (time F), the low has filled completely, and circulations disappear. During its lifetime, the low has done its job of moving warm air poleward and cold air equatorward. Analogous to LeChatelier's principle, it has partially undone the baroclinic instability that produced it in the first place.

SELF DEVELOPMENT OF CYCLONES

Up to this point in the discussion, cyclogenesis has been treated as a response to various imposed forcings. However, there are also some positive feedbacks that allow the cyclone to enhance its own intensification. This is often called **self development**.

Condensation

As discussed in the quasigeostrophic vorticity subsection, divergence of the upper-level winds east of the trough (Fig 13.14) causes a broad region of upward motion there. Rising air forms clouds and possibly precipitation if sufficient moisture is present. Such a cloud region is sometimes called an **upper-level disturbance** by broadcast meteorologists, because the bad weather is not yet associated with any strong surface low.

Latent heating of the air due to condensation enhances buoyancy and increases upward motion. The resulting **stretching** enhances spin-up of the vorticity, and it withdraws some of the air away from the surface, leaving lower pressure.

Diabatic heating also increases the average temperature of the air column, which pushes the 50 kPa pressure surface upward (i.e., increasing its height), according to the hypsometric relationship. This builds or strengthens a ridge west of the initial ridge axis.

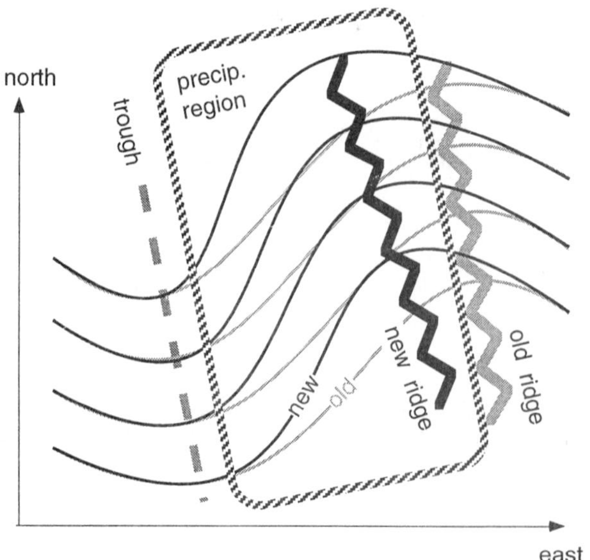

Figure 13.14
50 kPa chart showing cloud and precipitation region in the upper troposphere, causing latent heating and a westward shift of the ridge axis.

The result is a shortening of the wavelength between trough and ridge in the 50 kPa flow, causing tighter turning of the winds and greater vorticity (Fig 13.14). Vorticity advection also increases.

As the low strengthens, there can be more precipitation and more latent heating. This positive feedback shifts the ridge further west, which enhances the vorticity and the vorticity advection. The net result is rapid strengthening of the cyclone.

Temperature Advection

Cyclone intensification can also occur when warm air exists slightly west from the ridge axis, as sketched in Fig 13.15. For this situation, warm air advects into the region just west of the ridge, causing ridge heights to increase. Also cold air advects into the trough, causing heights to fall there. The net result is intensification of the wave amplitude (Fig 13.16). Stronger wave amplitude can cause stronger surface lows due to enhanced upper-level divergence.

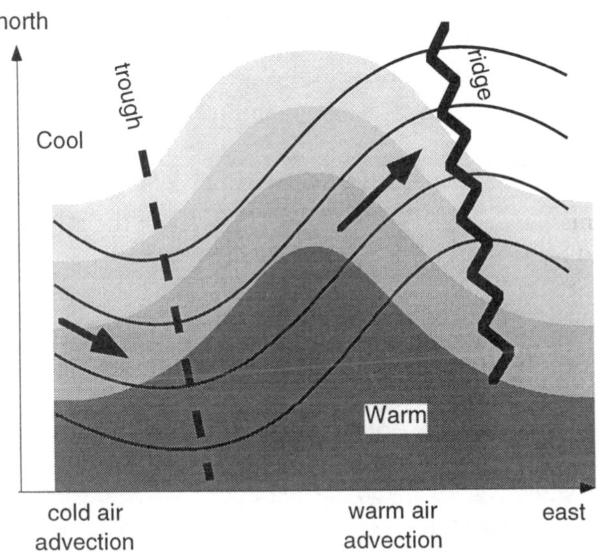

Figure 13.15
50 kPa chart showing a temperature field (shaded) that is 1/4 wavelength west of the wave in the height contours (solid lines).

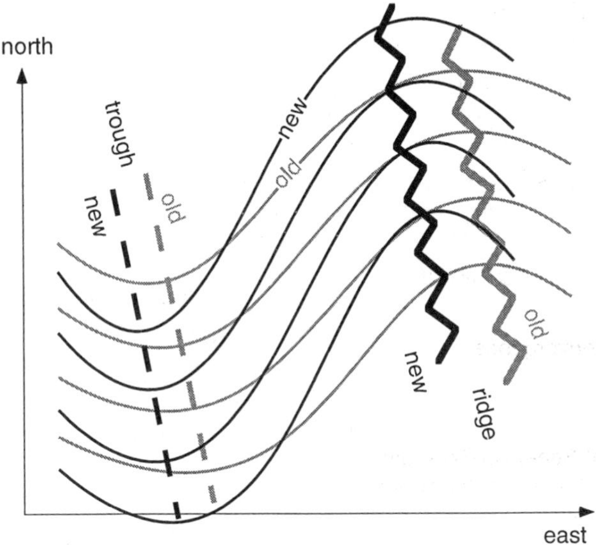

Figure 13.16
50 kPa chart showing the westward shift and intensification of north-south wave amplitude caused by differential temperature advection.

Science Graffito & On Doing Science

"There are no certainties in science. Scientists do not assert that they know what is immutably 'true' — they are committed to searching for new, temporary theories to explain, as best they can, phenomena." Science is "a process for proposing and refining theoretical explanations about the world that are subject to further testing and refinement." – U.S. Supreme Court.

ISO-SURFACES AND THEIR UTILITY

Before we return to the case-study cyclone, it is helpful to understand the types of weather maps that are used to examine cyclone dynamics.

Lows are three-dimensional beasts that evolve in time. To examine their inner workings we usually slice them in the horizontal, vertical, or along special curved surfaces called iso-surfaces. In this way we can study a two-dimensional dissection of the beast, which often reveals insights into their physics. Different iso-surfaces have different advantages. See Table 1-5 of Chapter 1 for a comprehensive list of iso-surface names.

Height

The only constant height surface used regularly is the surface at **mean sea level (MSL)**. Weather maps analyzed on this surface are somewhat fictitious, because sea-level is underground virtually everywhere except at the oceans. Nevertheless, this surface is used to represent the weather where people live. It also represents the bottom boundary on the atmosphere.

A frequently plotted map is MSL pressure and surface fronts.

Pressure

Isobaric (equal pressure) surfaces must be identified by their pressure. Commonly-used isobaric charts include the: 95 kPa surface, 85 kPa surface, 70 kPa surface, 50 kPa surface, 30 kPa surface, and 25 kPa surface. Because atmospheric pressure monotonically decreases with height, we know that the 25 kPa surface is at higher altitude than the 50 kPa surface, and so forth.

In mountainous regions such as the Rocky mountains, the 95 kPa and 85 kPa surfaces do not exist because they would be below ground. These surfaces are not usually plotted in that region (unless data is extrapolated to a hypothetical isobaric surface below ground).

On an isobaric chart, we can plot various weather elements (anything except pressure). For example, we can plot temperatures observed on the 50 kPa surface. We could draw lines of equal temperature (isotherms) along this isobaric surface. The meteorological jargon for such an analysis is a chart of "50 kPa isotherms".

An analysis of heights-above-sea-level along a 50 kPa isobaric surface is called a chart of "50 kPa heights". Similarly, we could look at "85 kPa isohumes", "25 kPa isotachs", "50 kPa absolute vorticity", or any other combination of our choosing.

Although confusing at first, sometimes two different weather maps are plotted on the same screen. This is done to illustrate physical relationships between both fields of data. An example is a chart of 50 kPa heights and isotherms. Such a chart tells us about temperature advection by the wind.

Sometimes the two superimposed charts are not on the same atmospheric surface, such as sea-level pressure (a constant height chart) and 95 kPa isotherms (a constant pressure chart). By using perhaps solid lines for isobars and dashed lines for isotherms, the result is hopefully not too cluttered.

Thickness

Often a map of height-thickness between pressures of 100 kPa and 50 kPa is analyzed. Greater thickness corresponds to warmer temperature, according to the hypsometric equation. In meteorological jargon, this map would be called the "100-50 kPa thickness" chart.

Horizontal temperature gradients cause horizontal thickness gradients. The **thermal wind** flows parallel to thickness lines, with low thickness (cold air) to the left of the thermal wind vector in the Northern Hemisphere. Recall that the thermal wind indicates how the geostrophic wind changes with height. We have seen that thermal wind advection of geostrophic vorticity is important for cyclogenesis.

Potential Temperature

An **isentropic** surface connects points of equal potential temperature. Such a surface also curves up and down. For example, the global average surface temperature decreases from equator to pole, and the average potential temperature increases with altitude in the troposphere. These two effects cause isentropic surfaces to slope downward toward the equator, on the average. However, any snapshot of an isentropic surface can differ from the average, particularly near fronts.

Isentropic surfaces are identified by their temperature, usually in Kelvins. An example is the 300 K isentropic surface, which is close to the surface near the equator, but rises to 4 to 8 km near the poles. The 350 K surface is at a higher altitude, mostly in the stratosphere. Again, we can plot other weather elements on such a surface, such as 300 K heights or isohumes.

Isentropic surfaces cannot cross other isentropic surfaces. They can also intersect the ground.

Isentropic surfaces are used because air parcels tend to follow them when blown by the wind, under adiabatic conditions (see Fig 12.13 in the previous chapter). Diabatic processes such as radiation, turbulence, and condensation can warm or cool the parcel, and cause it to change (jump) to a different isentropic surface.

Fig 13.17 sketches an idealized isentropic surface corresponding to the 280 K isentrope of Fig 6.11. Fig 13.17a is a 3-D representation, where there is a cold dome of air in the north-east. This could be the cold air behind a cold front. The 280 K surface slopes upward to the north-east. The dashed lines on this figure represent height contours of the isentropic surface above mean sea level.

A vertical view of the same scene is shown in Fig 13.17b. This is an **isentropic chart**. Height contours are shown again as the dashed lines. Other variables could also be shown on the isentropic surface, such as temperature, humidity, wind, vorticity, etc.

This particular example shows wind vectors on the isentropic surface. Because air parcels tend to follow isentropic surfaces as they advect, the wind vector in the upper-right portion of Fig 13.17b implies that air parcels are rising as they blow toward the higher height contours. From such rising, we would expect adiabatic cooling and clouds to form in that region.

At the lower left side of this figure, air is blowing down the isentropic surface toward lower altitudes, in this hypothetical circulation. Descending air parcels warm adiabatically, and become cloud free.

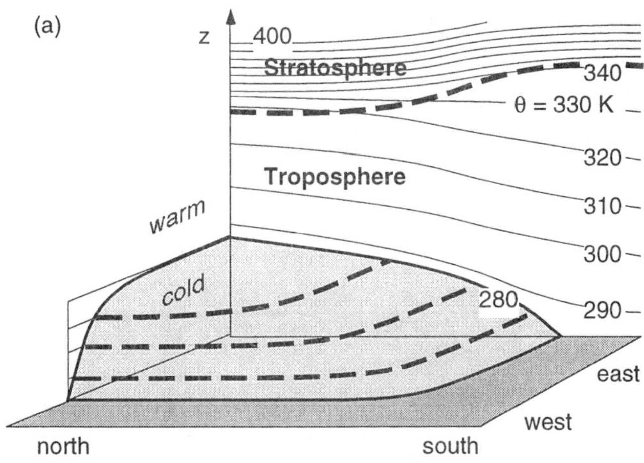

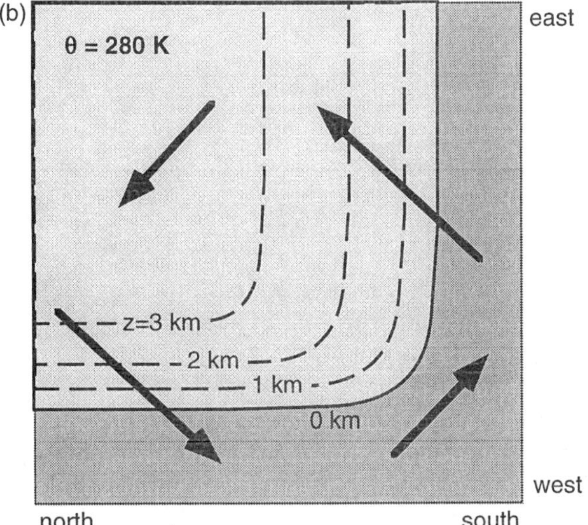

Figure 13.17
Isentropic surface. (a) Three-dimensional sketch of the 280 K isentropic surface, corresponding to Fig 6.11. (b) Vertical view of the same scene. This is an isentropic chart. Dashed lines are height contours of the 280 K isentropic surface, and vectors are wind directions along that surface.

Potential Vorticity (PVU)

The **1.5 PVU** isentropic-potential-vorticity surface can be plotted as a marker for the tropopause. It often is found at high altitude (over 15 km) near the equator, and lower altitude (6 km) near the poles. These surfaces can fold back on themselves, such as near fronts. Vertical cross sections of PVU can show intrusions of stratospheric air into the troposphere near fronts – the **tropopause folding** that was discussed in Chapter 12. An example of such a fold is shown in Fig 13.23 for the synoptic-storm case study in the next section.

CASE STUDY OF A CYCLONE (CONTINUED)

Now we return to the cyclone described at the start of this chapter. Recall that Fig 13.1 showed the storm track and weather over several days of the lifetime of the cyclone.

Here, we look at a snapshot of the cyclone at 12 UTC on 23 Feb 1994, which is near the time of peak intensity of the storm. First, a stack of charts at different isobaric surfaces are presented, each valid at the same time. Then a thickness chart, and a vertical cross section are presented. Taken together, these weather maps give us an understanding of the 3-D structure of the storm.

Surface Charts

Although the cyclone central pressure is not very low in an absolute sense (Fig 13.18 top), it is quite low relative to the neighboring intense highs. The strong pressure gradient between high and low is sufficient to drive very strong winds (refer to the storm data at the start of this chapter).

There is quite a large surface temperature change across the frontal zone (Fig 13.18 middle). Cold air ($\theta \le 0°C$) is advecting from the north behind the low, while warm air ($\theta \cong 20°C$) is advecting from the south ahead of the low.

The cold air (see the 0°C isentrope) is dammed-up against the Rockies, causing the western part of the front to become stationary. The Appalachian mountains also slow the northward march of the warm front, resulting in some frontal kinks east of the low center.

North of the low in the cold air (Fig 13.18 bottom), there is a broad area of light to moderate snow that the radar cannot detect. Imbedded are regions of moderate to heavy snow that the radar can see. Closer to the low, and just east of it where temperatures are near 0°C, sleet and freezing rain is falling.

Well in advance of the cold front in the southeast USA is a squall line of intense thunderstorms. These storms produce heavy rain, hail, and some small tornadoes. The squall line is feeding on latent heat, because of the high humidities near the Gulf of Mexico.

Surface Charts

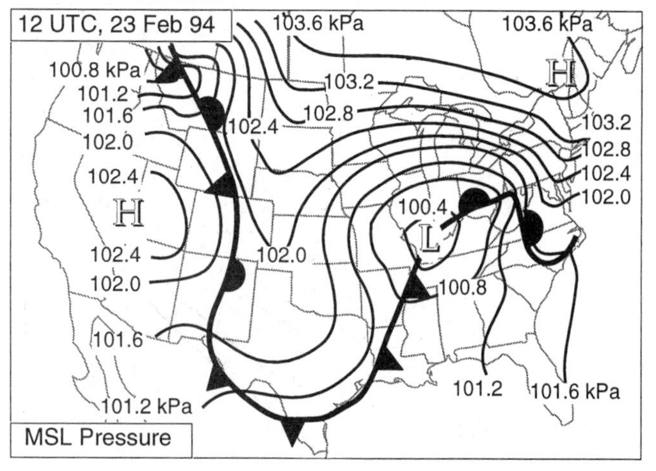

85 kPa Charts

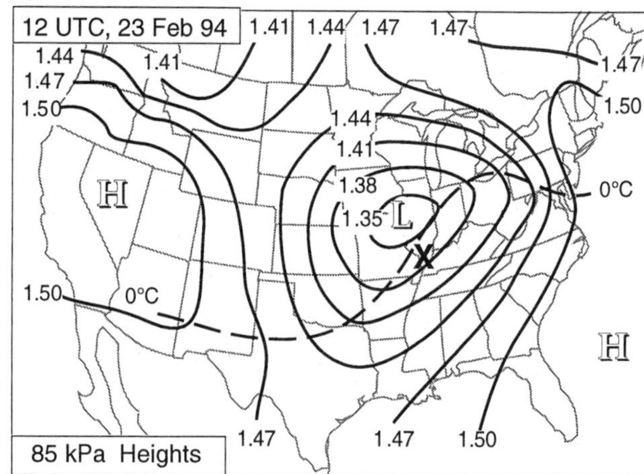

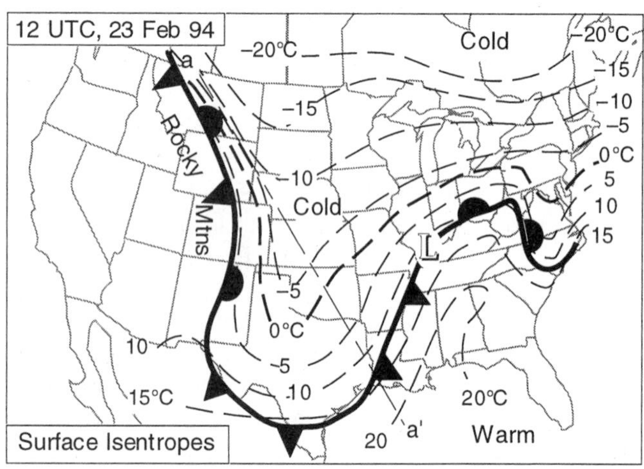

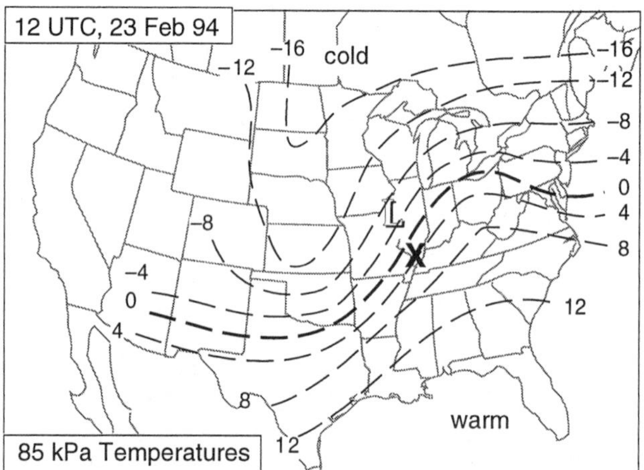

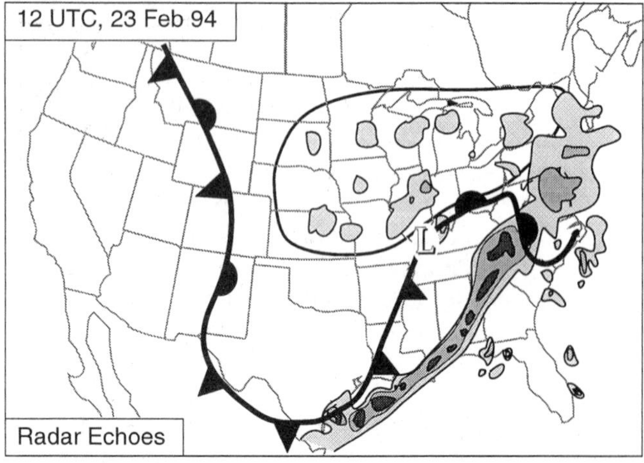

Figure 13.18

Surface weather maps. (Top) Mean sea level pressure (kPa), and surface fronts. (Middle) Surface isentropes (°C), and fronts. (Bottom) Radar echoes of precipitation. Levels: 1 - light precip. (light gray); 3 - heavy precip. (medium gray); 5 - intense precip. (dark). Heavy line encircles snow. [from Jon Martin.]

Figure 13.19

85 kPa weather maps. (Top) geopotential height (km). (Bottom) temperature (°C). The 0°C isotherm is thick dashed. X marks the surface low. [Courtesy of Jon Martin.]

Pressure level 85 kPa corresponds to about 1.5 km above ground. At this height, the low center is over western Illinois (Fig 13.19 top). This is northwest of the surface low. Because winds at this pressure level follow the height contours, we expect northerly winds in the Central Plains (Nebraska and Kansas), and southerly winds to the east of the low center.

From Texas to Pennsylvania is a strong baroclinic zone, which is evident by the close spacing between isotherms (Fig 13.19 bottom). There is strong cold-air advection in Oklahoma, Texas, Arkansas, and Missouri. This is marked by the height contours crossing the isotherms. Warm air advection occurs east of the low center.

70 kPa Charts

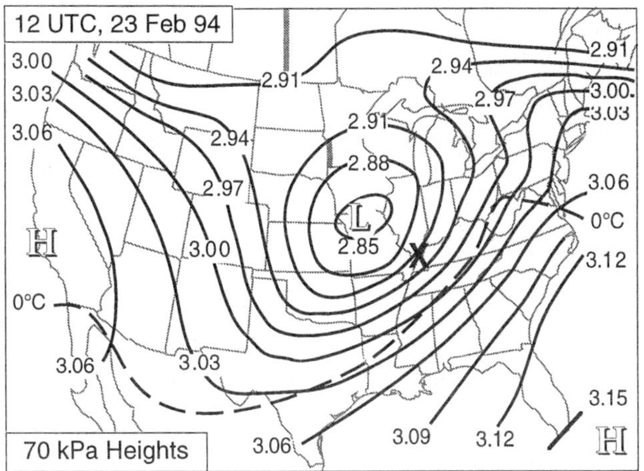

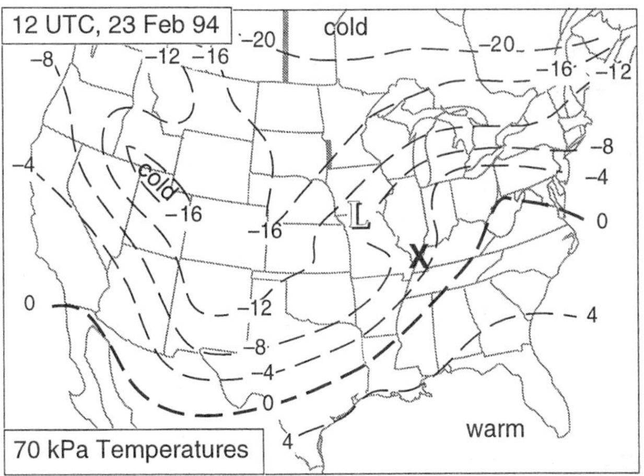

Figure 13.20
70 kPa weather maps. (Top) Geopotential heights (km). (Bottom) Temperature (°C). The 0°C isotherm is thick dashed. X marks surface low. [Courtesy of Jon Martin.]

Pressure level 70 kPa is roughly 3 km above ground. At this height, the low is centered over northern Missouri — significantly northwest of the surface low (Fig 13.20 top). The tightly packed height contours from Arkansas through West Virginia suggest a strong southwesterly wind in that region.

Cold-air advection is occurring in Oklahoma, Texas, and New Mexico (Fig 13.20 bottom). Subsidence and clearing skies are expected in regions of cold-air advection aloft. Warm-air advection is associated with ascending air, clouds, and precipitation, such as over and east of the Great Lakes.

50 kPa Charts

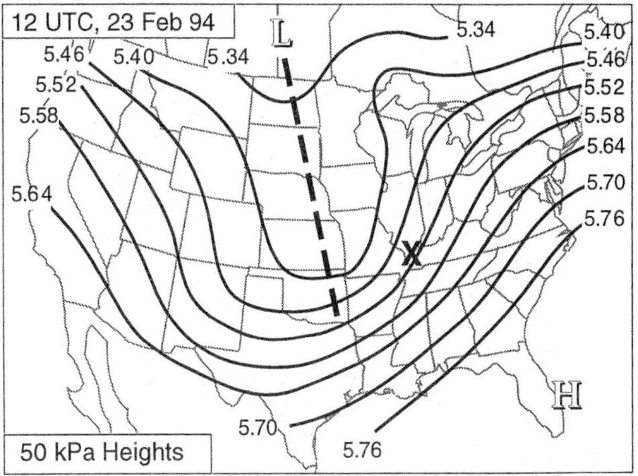

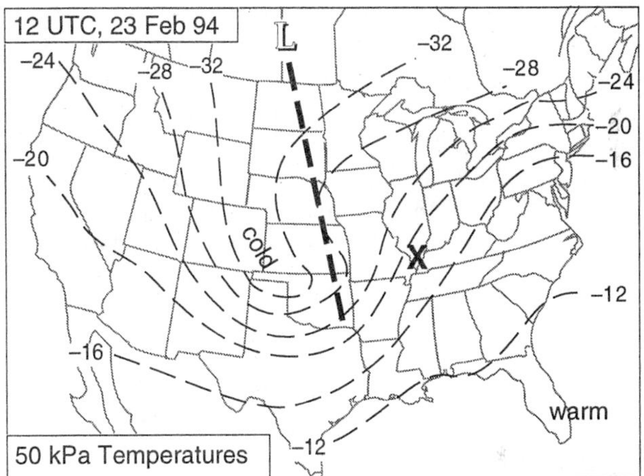

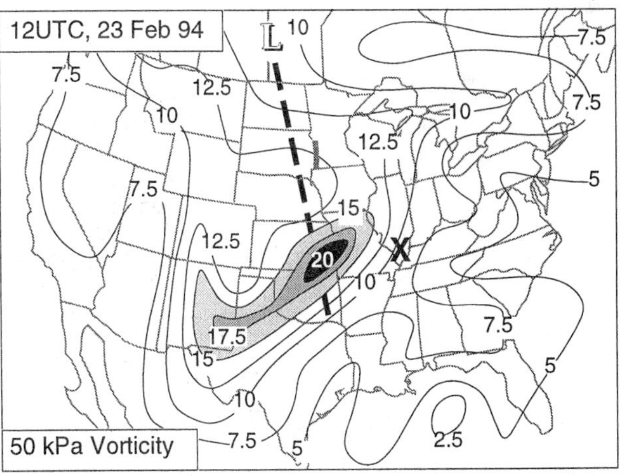

Figure 13.21
50 kPa weather maps. (Top) Geopotential heights (km). (Mid.) Temperature (°C). (Bot.) Absolute vorticity (10^{-5} s^{-1}). Thick dashed line is trough axis. X marks sfc. low. [from Martin.]

Instead of a closed low center as in the charts for lower altitudes, the 50 kPa chart shows a **trough** of low pressure (heavy dashed line in Fig 13.21 top) through the central plains. This trough axis is again further to the west, compared to the surface low. The trough is a portion of the north-south wave in the flow, such as discussed in Chapter 11.

The region of cold air (Fig 13.21 middle) is west of the trough, which looks similar to the idealized Fig 13.15. Hence, we expect self-development to occur, with strengthening of the wave amplitude.

Fig 13.21 (bottom) shows absolute vorticity on the 50 kPa pressure surface. The vorticity maximum (**vort max**) is often at, or just behind, the low center. Winds advect the vort max (positive vorticity advection) ahead, thereby supporting propagation of the low. This vorticity map is from a 24 h forecast of the U.S. National Weather Service (NWS) Nested Grid Model (NGM), valid at 12 UTC on 23 Feb 94.

25 kPa Charts

The 25 kPa pressure level is roughly at 10 km above ground. On this chart (Fig 13.22), the trough over central North America is quite evident, with the trough axis at roughly 100°W longitude. From this N. Hemispheric perspective, we see that the trough is part of the jet stream that meanders between about 30° and 50°N latitude around the North Pole.

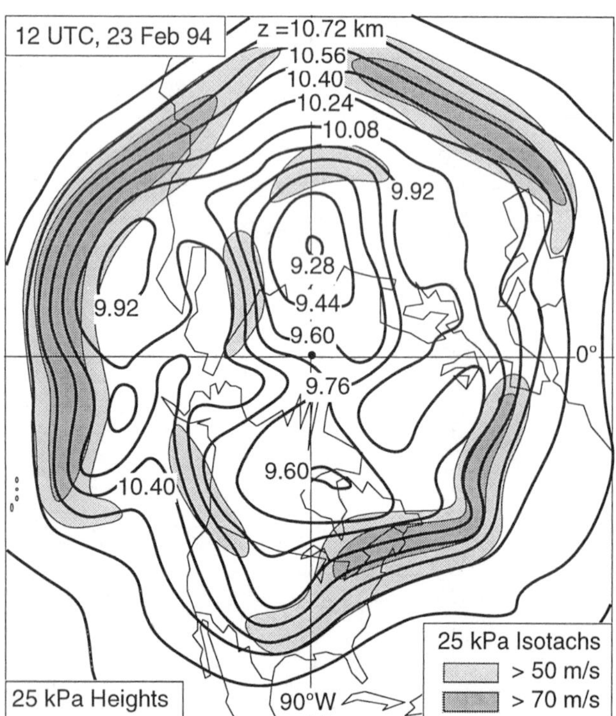

Figure 13.22
25 kPa heights (dark lines) and isotachs (shaded). [ECMWF]

A **jet streak** is evident from the Great Lakes extending east over Nova Scotia, Canada. Isotachs are contoured and shaded for 50 and 70 m/s. The right entrance region to the jet streak is over Lake Ontario. Horizontal divergence aloft favors cyclone development at the surface, in this region. Such development indeed occurred, as indicated by the storm track of Fig 13.1.

Vertical Cross Section

Fig 13.23 shows a vertical cross section from northwest to southeast, across North America (along section line **a – a'** in Fig 13.18b). Plotted are isentropes (equal potential temperature), and a single contour of isentropic potential vorticity.

The cold air mass is evident in Kansas (to the right of LBF in this plot). The surface front is just northwest of Lake Charles, Louisiana (LCH), as indicated by the tight packing of the isentropes near the surface.

The 1.5 PVU line marks the tropopause. Above this line, the isentropes are more closely spaced, indicating the greater static stability in the stratosphere.

There is a **tropopause fold** between North Platte and Lake Charles, as indicated by the 1.5 PVU line dipping down from the stratosphere. An upper-level front is evident in this region, based on the close spacing between isentropes.

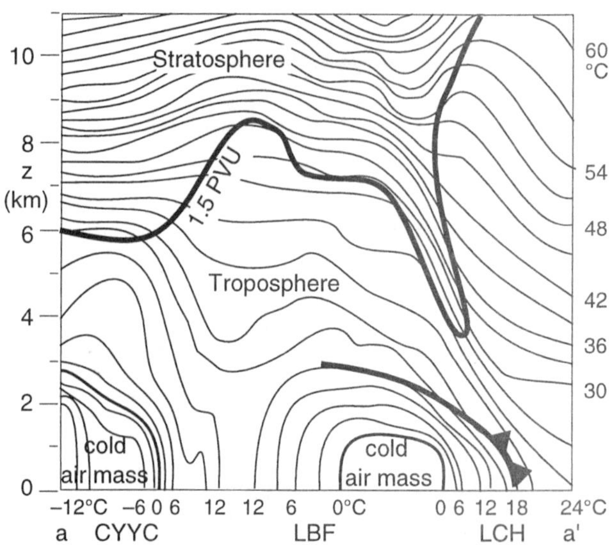

Figure 13.23
Vertical cross section from northwest (CYYC = Calgary, Alberta) through the central USA (LBF = North Platte, Nebraska) to the southeast (LCH = Lake Charles, Louisiana). Thin lines are isentropes (°C). Thick line in top half of fig is the 1.5 potential vorticity unit (PVU) contour. Cold-frontal symbols indicate the surface front and frontal inversion top. Valid 00 UTC 24 Feb 94. [This is 12 h later than the other figs.]

100–50 kPa Thickness Chart

The cold air pushing into the central plains of the USA from Canada has small 100–50 kPa thickness (Fig 13.24), as expected from the hypsometric equation. A thermal wind parallel to the thickness lines would create a jet stream that dips south toward Texas and then NE. toward Ohio, as is indeed observed.

(a)

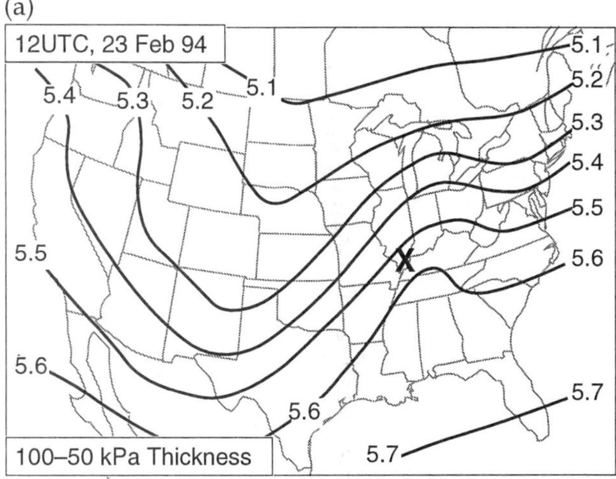

(b)

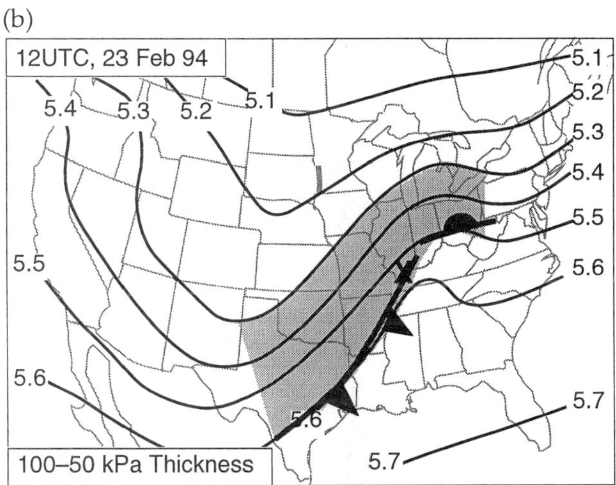

Figure 13.24
(a) 100–50 kPa thickness (km), from a 24 h forecast of the NWS NGM, valid at 12 UTC on 23 Feb 94. X marks sfc. low.
(b) Regions of tight packing of the thickness contours (shaded) are helpful for identifying frontal locations, in non-mountainous regions. Fronts would be near the warm (thick) side of this zone.

SUMMARY

Cyclones are regions of low pressure, cyclonic vorticity, and upward motion. These lows can cause clouds, precipitation and strong winds. Cyclogenesis is the birth or growth of cyclones, while cyclolysis is their decline and death.

Mountain ranges can disturb the planetary flow, causing north-south waves in the otherwise west-to-east jet stream. The troughs of such waves contain cyclonic curvature, and usually exist just east of the mountains. This tends to enhance cyclogenesis to the lee of the Rockies, such as in Colorado and Alberta.

Vorticity is one measure of cyclone intensity. Vorticity changes with terrain height, advection, stretching, tilting, turbulent drag, and with latitude.

Upward motion is another measure of cyclone intensity. It is often driven by jet streaks, and by curvature in the jet stream. Greater updrafts are possible in environments with weaker static stability, and with closely-packed and tightly-curved isobars on sea-level weather maps. Advection of geostrophic vorticity by the thermal wind also contributes to upward motion, as described by the omega equation.

Divergence aloft tends to reduce the sea-level pressure, and lows can deepen due to latent heating associated with precipitation. This is partially counteracted by inflow of air due to boundary-layer pumping. Once a surface low exists, it can enhance itself via temperature advection and latent-heat release in clouds.

A variety of weather maps can be used to study synoptic weather features such as cyclones, such as for the 23 Feb 94 case-study shown in this chapter.

Threads

Almost every chapter preceding this one contributed to the understanding of cyclones. We utilized concepts of temperature, density, pressure, and the standard atmosphere from Chapt. 1. Cyclones form as nature's way to redistribute heat caused by the radiative imbalance (Chapt. 2) between the equator and poles. This radiative imbalance causes baroclinicity primarily in the boundary layer (Chapt. 4) via the heat budget (Chapt. 3).

The bad weather of cyclones includes clouds (Chapts. 5 - 7), precipitation (Chapt. 8), and strong winds (Chapt. 9). Cyclones are triggered by waves in the jet stream of the general circulation (Chapt. 11), and help to transport heat, momentum, and moisture at midlatitudes. The convergence at cyclones draws together different air masses (Chapt. 12), thereby forming fronts and enhancing baroclinicity.

Cyclones can be predicted numerically by solving the primitive equations (Chapt. 14). Thunderstorms and tornadoes are triggered by the frontal weather near cyclones. Extratropical cyclones are **cold core systems**, which differ from warm core tropical cyclones (i.e., hurricanes, Chapt. 16). Polluted air (Chapt. 17) converges near cyclones, but pollutants can be washed out by rain. Changes in climate (Chapt. 18) might alter the tracks or intensities of cyclones.

BEYOND ALGEBRA • The Omega Equation

Full Omega Equation

The omega equation describes vertical motion in pressure coordinates. One form of the quasi-geostrophic omega equation is:

$$\left\{\nabla_p^2 + \frac{f_o^2}{\sigma}\frac{\partial^2}{\partial p^2}\right\}\omega = \frac{-f_o}{\sigma}\cdot\frac{\partial}{\partial p}\left[-\vec{V}_g \bullet \vec{\nabla}_p\left(\zeta_g + f_c\right)\right]$$

$$-\frac{\Re}{\sigma \cdot p}\cdot\nabla_p^2\left[-\vec{V}_g \bullet \vec{\nabla}_p T\right]$$

where f_o is a reference Coriolis parameter f_c at the center of a beta plane, σ is a measure of static stability, V_g is a vector geostrophic wind, $\Re$ is the ideal gas law constant, p is pressure, T is temperature, ζ_g is geostrophic vorticity, and • means vector dot product.

$\vec{\nabla}_p(\ ) = \partial(\ )/\partial x|_p + \partial(\ )/\partial y|_p$ is the "**del**" operator, which gives quasi-horizontal derivatives along an isobaric surface. Another operator is the **Laplacian**:

$$\nabla^2{}_p(\ ) = \partial^2(\ )/\partial x^2\Big|_p + \partial^2(\ )/\partial y^2\Big|_p .$$

Although the omega equation looks particularly complicated and is often shown to frighten unsuspecting people, it turns out to be virtually useless. The result of this equation is a small difference between very large terms on the RHS that often nearly cancel each other, and which can have large error.

Trenberth Omega Equation

Trenberth developed a more useful form that avoids the small difference between large terms:

$$\left\{\nabla_p^2 + \frac{f_o^2}{\sigma}\frac{\partial^2}{\partial p^2}\right\}\omega = \frac{2f_o}{\sigma}\cdot\left[\frac{\partial \vec{V}_g}{\partial p} \bullet \vec{\nabla}_p\left(\zeta_g + (f_c/2)\right)\right]$$

For the omega subsection of this chapter, we focus on the vertical (pressure) derivative on the LHS, and ignore the Laplacian. This leaves:

$$\frac{f_o^2}{\sigma}\frac{\partial^2 \omega}{\partial p^2} = \frac{2f_o}{\sigma}\cdot\left[\frac{\partial \vec{V}_g}{\partial p} \bullet \vec{\nabla}_p\left(\zeta_g + (f_c/2)\right)\right]$$

Upon integrating over pressure from p=100 to 50 kPa:

$$\frac{\partial \omega}{\partial p} = \frac{-2}{f_o}\cdot\left[\vec{V}_{TH} \bullet \vec{\nabla}_p\left(\zeta_g + (f_c/2)\right)\right]$$

where the definition of thermal wind V_{TH} is used, along with the mean value theorem for the last term.

The hydrostatic eq. is used to convert the LHS: $\partial\omega/\partial p = \partial W/\partial z$. The whole eq. is then integrated over height, with $W = W_{mid}$ at $z = \Delta z$ (= 100 - 50 kPa thickness) and $W = 0$ at $z = 0$. This gives W_{mid} =

$$\frac{-2\cdot\Delta z}{f_c}\left[U_{TH}\frac{\overline{\Delta\left(\zeta_g + (f_c/2)\right)}}{\Delta x} + V_{TH}\frac{\overline{\Delta\left(\zeta_g + (f_c/2)\right)}}{\Delta y}\right]$$

But f_c varies with y, not x. The result is eq. (13.21a).

EXERCISES

Numerical Problems

N1. What direction do cyclones spin in the Southern Hemisphere? Why?

N2(§). Plot a curve of β vs. latitude, for both the N. and S. Hemispheres.

N3. For latitude 45°N, calculate the typical wavelength of stationary planetary waves to the lee of the mountains, for a wind speed (m/s) of:

a. 10 b. 20 c. 30 d. 40
e. 50 f. 60 g. 70 h. 80

N4. For latitude 45°N, calculate the typical amplitude of stationary planetary waves to the lee of the mountains, assuming a tropospheric depth of 11 km and mountain heights (km) of:

a. 0.5 b. 1 c. 1.5 d. 2
e. 2.5 f. 3 g. 3.5 h. 4

N5. Winds rotate cyclonically at speed 20 m/s with radius of curvature 1000 km. Find the potential vorticity of air of thickness 11 km at latitude

a. 30° b. 35° c. 40° d. 45
e. 50° f. 55° g. 60° h. 65°

N6. Straight west-to-east flow of 30 m/s is initially 11 km deep at latitude 50°. It flows over mountains of height (km) given below. Calculate the potential vorticity and radius of curvature in the wave crest and trough.

a. 0.5 b. 1 c. 1.5 d. 2
e. 2.5 f. 3 g. 3.5 h. 4

N7. For a potential vorticity of 10^{-8} $m^{-1}\cdot s^{-1}$ in a cyclone of radius 300 km, find the change of relative vorticity if the cyclone is over a mountain slope of

a. 0.0001 b. 0.0002 c. 0.0005 d. 0.001
e. 0.002 f. 0.005 g. 0.01 h. 0.02

N8. Suppose the U-component geostrophic wind increases 5 m/s per 100 km of distance northward. There is no other shear. Find the advection of geostrophic vorticity for a constant V-component of geostrophic wind (m/s) of

a. –10 b. 5 c. –8 d. 4
e. 10 f. –5 g. 8 h. –4

N9. The top of a 2 km thick layer of air rises 0.01 m/s and its base rises 0.005 m/s. A south-to-north wind blows at 5 m/s. Find the rate of spin up of geostrophic vorticity at latitude:

a. 30°	b. 35°	c. 40°	d. 45
e. 50°	f. 55°	g. 60°	h. 65°

N10. At a height where the density is 1 kg/m^3, find omega for vertical velocities (m/s) of

a. 0.02	b. 0.05	c. 0.1	d. 0.2
e. 0.5	f. 1	g. 2	h. 5

N11. Given a vertical velocity of 0.1 m/s, find the corresponding value of ω at the following altitudes (km), assuming a standard atmosphere:

a. 0	b. 0.2	c. 0.5	d. 1
e. 2	f. 5	g. 10	h. 20

N12(§). Plot the radius of curvature at the crest of a wave in the jet stream, as a function of wavelength, for wave amplitudes (km) of

a. 200	b. 400	c. 600	d. 800
e. 1000	f. 1200	g. 1500	h. 2000

N13. Calculate the radius of curvature at the trough in the 25 kPa height contour lines at 12 UTC on 23 Feb 94.

N14. Calculate the speed difference between out- and in-flowing jet streams at 60° latitude for a geostrophic wind of 10 m/s and a radius (km) of curvature of

a. 300	b. 400	c. 600	d. 800
e. 1000	f. 1200	g. 1500	h. 2000

N15. Calculate the contribution to mid-tropospheric vertical velocity for the previous problem, for d = 4000 km and a vertical thickness of the jet stream of 5 km.

N16. Calculate the contribution to sea-level pressure tendency for the previous problem, if the density is 0.3 kg/m^3.

N17. Find the north-south component of ageostrophic wind in the exit region of a jet streak at 50° latitude if the average geostrophic velocity is 40 m/s and the zonal wind gradient is –20 m/s across a distance of (km)

a. 100	b. 200	c. 400	d. 600
e. 800	f. 1000	g. 1200	h. 1500

N18. Find the jet-streak contribution to mid-tropospheric vertical velocity for the previous problem, if the thickness of the jet is 5 km and the width of the jet is: (i) 400 km (ii) 800 km

N19. Find the contribution to surface pressure tendency for the previous problem.

N20. The 100 to 50 kPa thickness is 5.4 km and f_c = 10^{-4} s^{-1}. A west to east thermal wind of 25 m/s blows through a region where the average cyclonic vorticity decreases by 10^{-4} s^{-1} toward the east across a distance (km) given below. Use the omega eq. to find the mid-tropospheric upward velocity.

a. 100	b. 200	c. 300	d. 400
e. 500	f. 700	g. 1000	h. 2000

N21. How much mass is in a column of air 10 km in radius if the sea-level pressure (kPa) is

a. 102	b. 100	c. 98	d. 96
e. 94	f. 92	g. 90	h. 88

N22. What is the change of temperature with time for a rain rate (mm/h) of:

a. 1	b. 2	c. 5	d. 10
e. 20	f. 40	g. 60	h. 80

N23. What is the diabatic heating contribution to surface pressure tendency for the rain rates of the previous problem?

N24. Find the boundary-layer pumping contribution to sea-level pressure tendency for density 1 kg/m^3 and vertical velocity (m/s) of:

a. 0.001	b. 0.002	c. 0.005	d. 0.01
e. 0.02	f. 0.05	g. 0.1	h. 0.2

Understanding & Critical Evaluation

U1. Suppose straight west-to-east flow at latitude 60° and initial depth 11 km encounters a semi-infinite plateau of terrain height 1 km. Describe mathematically the nature of the flow over the plateau.

U2. The tilting effect of the vorticity equation has two terms, but only one term was sketched in Fig 13.2d. Sketch the other term.

U3. Fig 13.5 examines flow east of a trough. Draw a similar sketch, except east of a ridge, and indicate the physical effects causing vorticity change. Also, indicate if the result is spin-up or spin-down.

U4. How does the radius of curvature of a planetary wave vary with latitude, everything else being equal?

U5. Sketch a figure similar to Fig 13.13a, except for the temperature wave located 1/4 wavelength to the right instead of to the left of the height field. How would that alter the flow (sketch your result similar to Fig 13.13b)?

U6. Using the case-study weather map for 25 kPa heights and isotachs, calculate the jet streak and curvature contributions to mid-tropospheric vertical velocity and to sea-level pressure tendency over the Ohio Valley, valid at 12 UTC on 23 Feb 94.

U7. Using the case-study weather maps, calculate the terms in the vorticity equation at the center of the surface low.

U8. Can cyclones form if there is no baroclinicity (e.g., no north-south temperature gradient in the air)? Why?

U9. Can the vorticity, vertical velocity, and sea-level pressure tendency equations be used to study or predict cyclolysis? Describe.

U10. If cyclones often form to the lee of mountain ranges, do cyclones often weaken near the upwind side of mountain ranges? Why? How can you verify your answer?

U11. What is the wavelength of stationary planetary waves near the equator?

U12. Consider Fig 13.2. Instead of the Rockies being as illustrated in that figure, suppose that all the land to the east of the Rockies was a plateau at the same height as the top of the Rockies. Would the triggering of cyclones, and the wavelength of planetary waves be different? Why?

U13. In the winter, the tropopause is often lower and the wind speeds faster than during summer. How would the triggering of stationary planetary waves be different, if at all?

U14. Is there poleward propagation of cyclones at the upwind slopes of mountain ranges? Describe.

U15. a. For the set of four maps of horizontal structure of a cold front (in Chapt. 12), use the "area between crossing isopleths" technique to locate the region of maximum temperature advection.

b. Same, but for the warm front maps.

U16. Suppose that the only two terms in the vorticity tendency equation were the spin-up term and the turbulent drag term. If there was some initial vorticity, describe how the vorticity would evolve with time.

U17. Is it possible that there is a tilting term that can take vorticity about a vertical axis, and tilt it into horizontal vorticity. Devise a term or terms that describe this process. Could this process happen in nature?

U18. In Figs. 13.4, no diagram was shown for vertical advection. Draw a sketch to illustrate this process.

U19. a. Summarize the quasi-geostrophic approximation.

b. What are the limitations of the quasi-geostrophic vorticity equation?

U20. Recall the limitation on the max wind speed and pressure gradient near high-pressure centers, but not near lows. How does that affect eq. (13.14)?

U21. Derive eq. (13.13). What limitations apply to this equation?

U22. Derive eq. (13.14).

U23. For eq. (13.16), how does W_{mid} vary with:

a. G b. λ c. Δy d. latitude?

U24. In Fig 13.8b, why are some divergence and convergence regions indicated as being stronger than others? Discuss.

U25. For Fig 13.9, what causes the portion of the secondary circulation near the ground? (Remember that Newton's law states that forces are needed to drive winds.)

U26. For a jet streak axis that is aligned from SW to NE, how could you use eqs. (13.19) for this situation?

U27. Recall that the thermal wind is really the change of geostrophic wind with height, which is proportional to the horizontal temperature gradient. Rewrite the omega equation in terms of:

a. the vertical gradient of geostrophic wind

b. the horizontal temperature gradient.

U28. Show that eq. (13.21b) is identical to (13.21a). What assumptions are needed to convert between the two equations, if any?

U29. What if there was no baroclinicity (no horizontal temperature gradient). How would that affect the omega eq.?

U30. What are the limitations on the heuristic arguments of Fig 13.11, where a vertical velocity was used as a surrogate for terms in the mass budget?

U31. Create a relationship between surface pressure tendency and dBZ values of reflectivity observed by weather radar.

U32. Given that Doppler radar can measure both precipitation rate and radial components of velocity, what quantitative information can be used from it to compute intensification of cyclones?

U33. Is it possible for the latent heating term in the net pressure tendency equation to increase, rather than decrease, the central pressure of a cyclone? Describe the conditions necessary for this to happen.

U34. Would anticyclones have any self development processes analogous to those for cyclones? Describe.

U35. For the case study cyclone in this chapter, describe the terms in, and evolution of:

a. vorticity tendency b. vertical motion
c. surface pressure tendency

U36. Knowing the acceleration rate for a parcel of air at sea level (given Newton's second law and assuming that you know the horizontal pressure gradient), is it possible for air to accelerate so slowly compared to the deepening of a cyclone bomb that the wind speed is not in equilibrium? Discuss using order-of-magnitude numbers.

U37. Why are the surface temperatures in Fig 13.23 much warmer than the temperatures shown along a – a' in Fig 13.18 middle?

Web-Enhanced Questions

W1. For an exciting storm system that affected you, develop a case study of weather maps from the web and discuss cyclone evolution based on the physical processes reviewed in this chapter.

W2. Search the web for sources of storm damage data ("Storm Data").

W3. For an intense cyclone, download a set weather maps and upper air maps for a snapshot of the weather. Use this to develop a 3-D image in your mind of the storm, and its dynamics.

W4. Search the web for maps showing typical or historical tracks of extra-tropical cyclones. What are the names of some of the favored cyclone formation sites (such as the Alberta Clipper)?

W5. Track an intense cyclone as it approaches the west coast of the N. America from the Pacific. Download a sequence of maps showing how the storm changes as it passed over the various mountain ranges in the western part of the continent.

W6. For an exciting storm that caused significant damage, download data from the web to create a summary map of its track and damage, similar to Fig 13.1.

W6. Search the web for case studies of cyclone bombs, or explosive cyclogenesis. Discuss the evolution of these storms.

W7. From web maps of the jet stream (upper air charts at 20, 25, or 30 kPa), measure the wavelength of the various waves. Also measure the wave amplitude. How do they vary with latitude? With season?

W8. From web maps of the jet stream (upper air charts at 20, 25, or 30 kPa), estimate the vorticity from the radius of curvature and the wind speed.

W9. Download upper air maps of temperature and height contours for a level such as 85 or 70 kPa. Superimpose these two fields if they are not already on the same map. Use the technique described in the "Advection" focus box to locate the region of maximum temperature advection.

W10. Same as the previous problem, but for a level such as 50 kPa where vorticity and height maps are available to locate the region of maximum vorticity advection. Locate both the regions of max positive and negative vorticity advection by the geostrophic wind. Which region would most likely favor cyclone development?

W11. Download a map of 70 kPa vorticity, and of 100 - 50 kPa thickness contours. From the thickness, determine the thermal wind near a cyclone. Use the result to find the region of max vorticity advection by the thermal wind, which according to the omega equation would favor upward motion and strengthening of cyclones.

W12. Search the web for an isentropic weather map for the current weather. If you cannot find one, then search the web for a map from a previous case study, or from a figure in a paper about isentropic analysis. From this map, discuss the elevation variation of the isentropic surface, state how high it is within the troposphere, and discuss any other info such as if winds are likely to blow up or down the isentropic surface in various regions.

W13. Locate a jet streak on a current weather map, or a past map if necessary. Indicate where on the map you would expect cyclones to develop or strengthen. Given the change of wind velocity at different locations near the entrance or exit of the jet streak, plug those numbers into the formula to estimate upward motion associated with jet streak divergence.

W14. Study the spacing of height contours at a jet stream level (20, 25, or 30 kPa) as the jet stream flows around neighboring ridges and troughs, for a map you download from the web. Does the spacing between contours remain relatively constant around both the ridge and trough? If not, given the changes in spacing, and given what you know about gradient wind speeds (e.g., slow around lows), discuss the variation of actual wind speed around the ridge and trough.

W15. Search the web for radar observations in a cyclone having heavy precipitation. From the dBZ values, estimate the rainfall rate (see an earlier chapter for the relationship). Then, from the rainfall rate, estimate the latent heating contribution to the surface pressure tendency. Also, see if there is any self-development of the cyclone associated with this latent heating.

W16. Search the web for the site of a major weather map provider (unisys, cola, intellicast, etc.), government weather service (NWS, AES), university forecasting project, military forecast service (US Navy NOGAPS), numerical forecast center (NCEP, ECMWF, CMC), or research organization that produces lots of weather maps. In addition to traditional maps on various isosurfaces (e.g., height, pressure, thickness, potential temperature, etc.), what other weird, interesting, and useful maps can you find?

Synthesis Questions

S1. Suppose there were no major north-south mountain ranges in N. America. How would the weather and climate be different, if at all?

S2. What if average temperature decreased toward the east over N. America, rather than toward the north. How would this change in the direction of baroclinicity change the weather and climate, if at all?

S3. Suppose that instead of the Rockies, there was a trench in the land as wide as the Rockies are wide, and as deep as the Rockies are high. How would the weather and climate be different, if at all?

S4. What if the earth rotated twice as fast. How would the triggering of cyclones and the nature of stationary planetary waves be different, if at all?

S5. Suppose there is a major mountain range running east-west across the center of N. America, and no north-south ranges. How would weather and climate be different, if at all?

S6. Most numerical weather forecast models (to be described in the next chapter) use the primitive equations (the conservation equations for heat, moisture, and momentum, along with the continuity eq. and ideal gas law) to make their forecasts.

a. Is it possible to use the quasi-geostrophic vorticity and omega equations to make a forecast instead?

b. How would you get the wind components and temperature from such a forecast?

c. Is such a forecast system simpler or more complicated to solve than the primitive equations?

S7. Suppose the jet stream is at its current altitude, and meanders north and south, and creates horizontal divergence and convergence as it does now. But what if the static stability in the stratosphere equaled that in the troposphere. What change would there be to jet stream forcing of surface pressure systems, if any? Discuss.

S8. Suppose that extra-tropical cyclones did not exist. How would the weather and climate be different, if at all?

S9. Suppose that the troposphere were as strongly stable as the stratosphere. How would the weather and climate be different, if at all?

S10. Suppose that earth was a larger planet that had a very hot core and surface, similar to Jupiter. If the heat from the solid earth to the atmosphere was the dominant source of heat that balanced net longwave radiative cooling from the atmosphere, how would the weather and climate on Earth be different, if at all?

NUMERICAL WEATHER PREDICTION

CONTENTS

14 Most weather forecasts are made by computer. Computers can keep track of the myriad of complex nonlinear interactions between winds, temperature, and moisture at thousands of locations and altitudes around the world — an impossible task for humans. Also, data observation, collection, analysis, and dissemination are increasingly automated.

Fig 14.1 gives an example of an automated forecast for the case-study cyclone that was discussed in Chapter 13. Output directly from the computer, this meteogram is easier for non-meteorologists to interpret than weather maps. Automated text and voice-synthesis forecasts are also produced. But to produce such forecasts, the equations describing the atmosphere first must be solved.

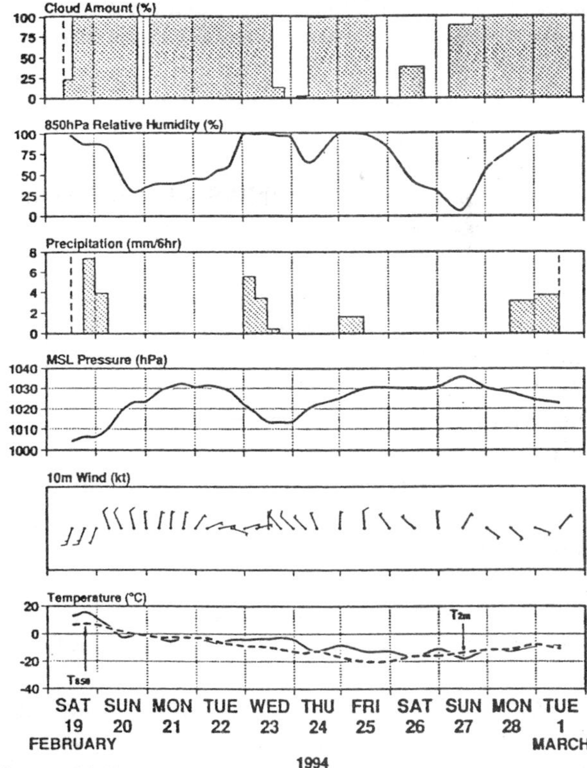

Figure 14.1
Ten-day weather forecast for Des Moines, Iowa, plotted as a ***meteogram*** *(time series), based on initial conditions observed at 12 UTC on 19 Feb 94. (a) Cloud amount (%), (b) relative humidity (%) at 85 kPa, (c) precipitation (mm/6 hours), (d) mean sea-level (MSL) pressure x10 (kPa), (e) wind at 10 m (full barb = 10 kt), (f) temperature (°C) at heights 2 m (solid) and 85 kPa (dashed). Produced by ECMWF.*

SOLUTIONS TO ATMOSPHERIC EQUATIONS

Weather forecasts are made by solving the **equations of motion** for the atmosphere. These are the non-linear partial-differential equations of dynamics, thermodynamics, mass continuity, and moisture conservation that have been introduced (in simplified form) in earlier chapters.

Unfortunately, no one has yet succeeded in solving the full governing equations analytically. An **analytical solution** is itself an algebraic equation or number that can be applied at every location in the atmosphere. For example, the equation $y^2 + 2xy = 8x^2$ has an analytical solution $y = 2x$, which allows us to find y at any location x.

Approximate Solutions

To get around this problem, three alternatives are used. One is to find an exact analytical solution to a simplified (approximate) version of the governing equations. A second is to conceive a simplified physical model, for which exact equations can be solved. The third is to find an approximate **numerical solution** to the full governing equations.

An atmospheric example of the first method is the geostrophic wind, which is an exact solution to a highly simplified equation of motion. This is the case of steady-state (equilibrium) winds above the boundary layer where friction can be neglected, and for regions where the isobars are nearly straight.

Early **numerical weather prediction** (NWP) efforts used the second method. For example, Rossby derived a simplified set of equations for the early computers by modeling the atmosphere as if it were a layer of water surrounding the earth.

Modern NWP uses the third method. Here, the full, partial-differential equations (also known as the **primitive equations**) are solved using finite-difference approximations, but only at discrete locations called **grid points**. Usually these grid points are at regularly-spaced intervals around the world, rather than at each city or town.

Routine operational weather forecasts have been produced by computer since 1975. Numerical forecast skill exceeded human forecast skill soon after 1975, and we have been dependent on the computerized forecast products ever since.

Parameterizations and Models

A **parameterization** is an approximation to an unknown term by one or more known terms or factors. Details and rules of parameterization are described in an "On Doing Science" box in Appendix H. In meteorology, some physical processes are not known well enough to provide exact physical laws, but the net effect of that process can be observed and parameterized. For other situations, the exact physical processes are known, but are so complicated or computationally unwieldy that a simpler parameterization provides an answer that is good enough.

For example, one might be able to use a simple parameterization for net rainfall rate, instead of following the evolution of billions of individual cloud droplets and ice crystals as they grow and collide. This is called a **cloud microphysics parameterization**. Another example is turbulence, where one can parameterize the net vertical heat flux transported by turbulence, instead of calculating the motion and heat transport of each swirl and eddy. This is called a **turbulence parameterization.**

Also used in NWP are parameterizations for other physical phenomena: **radiation**, **boundary-layer**, **mountain-wave drag**, **vegetation**, **surface effects**, and many more. All of these parameterizations are generically called **physical parameterizations**, or **physics** for short.

Similarly, one can design different numerical algorithms to approximate by algebra the various partial differential equations of motion. Some might be more accurate, while others might be more computationally efficient. These are generically called **numerical parameterizations**, or **numerics** for short.

Because parameterizations are only approximations, there is not a single, correct parameterization. Different scientists might propose different parameterizations for the same phenomenon. The different algorithms perform better for different situations.

The computer code that incorporates one particular numerical parameterization, and certain particular physics parameterizations, is called a **numerical model**. People developing these extremely large sets of computer code are called **modelers**. It typically takes teams of modelers (meteorologists, physicists, and computer scientists) several years to develop a new numerical model.

Different weather forecast centers develop different numerical models, containing different physics and numerics. These models are given names and acronyms, such as the GEM model (Global Environmental Multiscale). The different models usually give slightly different forecasts.

FOCUS • History of NWP (part 1)

The first equations of fluid mechanics were formulated by Leonhard Euler in 1755, using the differential calculus invented by Isaac Newton in 1665, Gottfried Wilhelm Leibniz in 1675, and using partial derivatives devised by Jean le Rond d'Alembert in 1746.

Terms for molecular viscosity were added by Claude-Louis Navier in 1827 and George Stokes in 1845. The equations describing fluid motion are often called the **Navier-Stokes equations**. These primitive equations for fluid mechanics were refined by Herman von Helmholtz in 1888.

About a decade later Vilhelm Bjerknes in Norway suggested that these same equations could be used for the atmosphere. He was a very strong proponent of using physics, rather than empirical rules, for making weather forecasts.

In 1922, Lewis Fry Richardson in England published a book describing the first experimental numerical weather forecast — which he made by solving the primitive equations with mechanical desk calculators. His book was very highly regarded and well received, as one of the first works that combined the many physical theories in a thorough, interactive way.

It took him 6 weeks to make a 6 h forecast. Unfortunately, his forecast of surface pressure was off by an order of magnitude compared to the real weather. Because of the great care that Richardson took in producing these forecasts, most of his peers concluded that NWP was not feasible. This discouraged further work on NWP until two decades later.

John von Neumann, a physicist at Princeton University's Institute for Advanced Studies, and Vladimir Zworykin, an electronics scientist at RCA's Princeton Laboratories and key inventor of television, proposed in 1945 to initiate NWP as a way to demonstrate the potential of the recently-invented electronic computers. Their goal was to simulate the general circulation as a first step toward climate modification. During the first few years they couldn't agree how to approach the problem.

Von Neumann brought together a group of theoretical meteorologists, including Carl-Gustav Rossby, Arnt Eliassen, Jule Charney, and George Platzman. Eventually, they realized the necessity to simplify the full primitive equations in order to focus their limited computer power on the long waves of the general circulation.

The first electronic computer, the ENIAC, filled a large room at Princeton, and used vacuum tubes that generated tremendous heat and frequently burned out. Its limited capacity precluded solution of the full primitive equations, a predicament that was happily accepted, knowing Richardson's failure with the primitive eqs.

Charney and von Neumann developed a simple one-layer barotropic model (conservation of absolute

(*continued next column*)

FOCUS • History of NWP (part 2)

vorticity) that Charney presented at NWP conferences in 1948 and 1949.

The research team had many hurdles: translating the differential equations into discrete form, writing the code in machine language (FORTRAN and C++ had not yet been invented), deciding how large a forecast domain was necessary, and doing many feasibility calculations by hand (using slide rules and mechanical calculators).

Their first ENIAC forecasts were made in March-April 1950, for three case studies over North America. The results were quite promising. Soon thereafter, Bert Bolin, Joseph Smagorinsky and Norman Phillips joined the research team, and they made several more numerical forecasts.

Meanwhile, Rossby returned to his native Sweden to begin a NWP effort there. Together with Bolin (back from Princeton) and Phillips, they developed a barotropic model based on a simplified physical picture of the atmosphere. A Swedish electronic computer called BESK became operational in 1953, and at the time was the most powerful computer in the world.

With support from the Swedish Air Force, the first routine operational numerical forecasts were started in Sweden in December 1954. These were 24, 48, and eventually 72-hour forecasts of 50 kPa heights, not of surface conditions. This phase of operational forecasts ended in May 1955, by which time 60 to 70 forecasts had been produced with good skill.

In 1953 in the USA, IBM announced its specifications for a new computer. The US Weather Bureau, USAF Air Weather Service, and Navy WX Service formed a Joint Numerical Weather Prediction Unit (JNWPU) in 1954 to use this computer, with director George Cressman.

This group decided to focus on operational weather forecasts using the quasi-geostrophic baroclinic model that had been derived in 1948 by Charney and Eliassen. Computer power limited the model to three layers, eventually placed at 40, 70, and 90 kPa. In 1955 the IBM 701 computer was introduced, and by May the JNWPU began its first routine operational numerical weather forecasts for North America. Also, Hans Panofsky proposed a scheme for automated analysis of weather data.

Meanwhile, Phillips and Smagorinsky were using NWP in a research mode to learn more about how the general circulation works. However, after von Neumann left Princeton in 1955 (and died in 1956), the NWP group began to fall apart. The US military services ceased their support, Charney and Phillips left for MIT, and Rossby died in 1957.

The numerical weather forecasts were often of poor quality, and were not appreciated by the experienced human forecasters. During the first two years of

(*continued next column*)

FOCUS • History of NWP (part 3)

operation of the JNWPU, the numerical forecasts were largely ignored.

In 1958, the National Meteorological Center (NMC) was organized to perform the operational numerical forecasting for the National Weather Service. The USAF organized a Global Weather Central in Omaha, Nebraska, to handle their numerical forecasting, and the Navy's numerical modeling moved to Monterey, California.

At first, the numerical forecasts were initialized using hand-analyzed data, where the observations were manually interpolated to the model grid points. By the mid 1950's, a variety of automated schemes were employed. Eventually, an **objective analysis** method was adopted of fitting new observations to previous forecast fields.

The first baroclinic (multi-layer) models used in 1955 and 1956 failed because they were poorly calibrated and awkward to use. Modelers then retreated to the barotropic model. Because this model had only one layer, it left the computer underutilized. To take advantage of the extra computer power, modelers expanded the barotropic domain to the hemisphere, but were disappointed to discover deteriorated forecast quality.

These barotropic models did not include topography (due to computer limitations), and the resulting long-waves in the general circulation were quite unrealistic in their westward movement. It was found that the forecast quality improved by fudging the longest waves (eliminating them from the forecast, and then re-inserting them later). This was a very disappointing time in NWP.

Eventually, Cressman and Phillips found a solution by altering the characteristics of the air above the modeled troposphere. The resulting 50 kPa barotropic forecasts were of such high quality that ten years elapsed before baroclinic models could beat them.

Finally, in 1963 the new computers were powerful enough that a six-layer primitive equation (PE) model could be implemented. Forecast scores dramatically improved, finally making NWP a useful forecast tool.

Since then, improvements in forecast quality have been closely tied to growth in computer power (memory, speed, & file storage, see Fig 14.2). Added to the models were more layers, a finer mesh of grid points in the horizontal, an increase to a fully global domain (in 1966), topography, and landscape characteristics including snow and ice coverage. Also improved were parameterizations of physical processes such as radiation, clouds, precipitation, and turbulence.

The NMC regional (limited-domain) models evolved into a limited-area fine-mesh (LFM) model in 1976, which was phased out in 1994. A nested-grid model (NGM) was introduced in 1990 to make 2-day forecasts. The NGM used a grid-point representation of the

(*continued next column*)

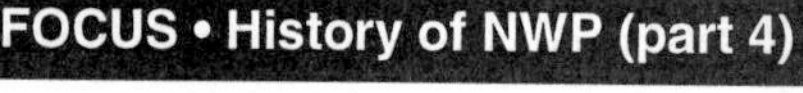

FOCUS • History of NWP (part 4)

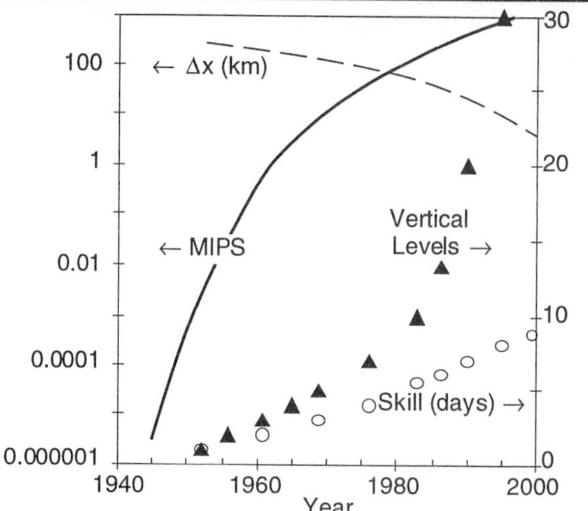

Figure 14.2
Improvement in horizontal grid resolution (Δx), number of model levels, and duration of a skillful 50 kPa forecast as a function of computer power (MIPS = millions of instructions per second). These data are smoothed.

hydrostatic primitive equations. It had 16 layers, and horizontal resolutions of 80 and 160 km on inner (N. America) and outer (N. Hemisphere) grids.

An "eta" hydrostatic, grid-point model was introduced by NMC in the early 1990's. It had greater vertical resolution (38 layers) to handle topography better. It made a 2-day forecast over 1/3 of the N. Hem. By 1999 it had 32 km horiz. resolution and 45 layers.

Spectral representations of horizontal fields replaced grid points in the NMC global models at about 1980. By the late 1990s, these models evolved into a hydrostatic Global Spectral Model (GSM), with 42 layers, and 80 km horizontal resolution. The GSM was used for 5-day aviation (AVN) forecasts, and for medium-range forecasts (MRF) out to 15 days, using a time step of 9 minutes, and taking 14 min runtime/day on a Cray YMP.

In the mid 1990's, U.S. military numerical forecasts were consolidated in Monterey at the Fleet Numerical Meteorological and Oceanographic Center (FNMOC). NMC was reorganized into the National Centers for Environmental Prediction (NCEP), with their Environmental Modeling Center still near Washington, DC. They introduced a MesoEta model, which by 1999 had 10 km horizontal grid spacing, and 60 layers. The Canadian Meteorological Center (CMC) developed a Global Environmental Multiscale (GEM) model in the late 1990s, with variable grid resolution as fine as 15 km.

Meanwhile, in the mid 1970s the European Community established a European Centre for Medium-Range Weather Forecasts (ECMWF) in Reading,

(*continued next column*)

FOCUS • History of NWP (part 5)

England. Their first medium-range forecast in 1979 was based on a grid-point model. In the 1990s, their 10-day forecasts were produced with spectral models.

Starting in 1992, ECMWF began running an Ensemble Prediction System (EPS). By the end of 1999, 50 alternative analyses were constructed every day from the one operational analysis. These perturbed analyses differed slightly at those locations where small analysis errors could grow rapidly. The 50 analyses were used as starting points for 50 different ten-day forecasts, run on a slightly coarser grid mesh. The spread of the resulting 50 forecasts were used as a measure of forecast confidence, and provided probability information.

By late 1999, the ECMWF numerical forecasts had one of the highest verification skills in the world. Their model had a horizontal resolution equivalent to about 60 km, with 60 layers in the vertical. This works out to 8.3 million grid points in the model, at which tens of meteorological variables were computed every 20 minutes. This ten-day forecast took approximately one and a half hours on a Fujitsu VPP700 computer utilizing 24 processors, and involved approx. 10^{14} computations. [*For a thorough history of NWP, see F. Nebeker, 1995: "Calculating the Weather". Academic Press. 255 pp.*]

ON DOING SCIENCE • Mathematics

"To have to translate one's verbal statement into mathematical formulae compels one to scrutinize the ideas therein expressed. Next the possession of formulae makes it much easier to deduce the consequences. In this way absurd implications, which might have passed unnoticed in a verbal statement, are brought clearly into view & stimulate one to amend the formula.

Mathematical expressions have, however, their special tendencies to pervert thought: the definiteness may be spurious, existing in the equation but not in the phenomena to be described; and the brevity may be due to the omission of the more important things, simply because they cannot be mathematized. Against these faults we must constantly be on our guard."

– L.F. Richardson, 1919

GRID POINTS

To numerically forecast the weather at every point in the atmosphere would require an infinitely-large computer. That might never happen. Instead, we make numerical forecasts for a finite number of regularly-spaced locations called **grid points**.

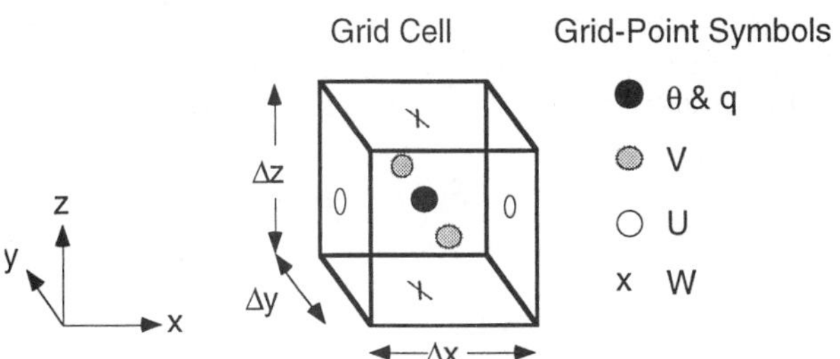

Figure 14.3
Example of a grid cell with staggered grid points.

Each grid point holds the average value for a volume of surrounding air. This volume is called a **grid cell** or **grid volume**. The size of the grid cell in the three Cartesian directions is Δx, Δy, and Δz (see Fig 14.3). Typical values are $\Delta x = \Delta y$ = tens to hundreds of kilometers, while Δz = tens to hundreds of meters. Thus, many grid cells are needed to fill the atmosphere around the earth.

Often, grid points are arranged in a **staggered** arrangement around the cell, with different variables being represented by grid points at different locations (Fig 14.3). One common arrangement is to represent thermodynamic variables such as potential temperature θ, specific humidity q, liquid water content, cloudiness, etc. in the center of the grid cell. Vertical velocities are at the top and bottom faces of the grid cell, to indicate vertical flow (W) across those cell boundaries. Similarly, U and V are at the other faces to describe flow across the other cell boundaries (Fig 14.3).

Cells are identified by a set of indices (i, j, k) that indicate their (x, y, z) positions within the domain. Fig 14.4 shows a two-dimensional example. By using these indices as subscripts, we can specify any variable at any location. For example, θ_{32} is the potential temperature in the center of the shaded grid cell, at x location $i = 3$, and z location $k = 2$.

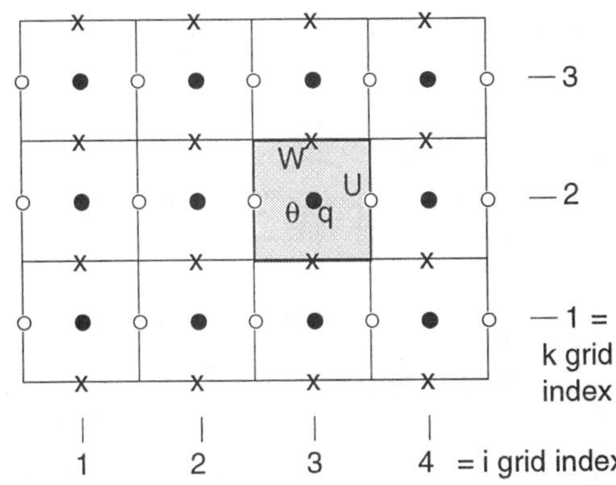

Figure 14.4
Arrangement and indexing of grid cells within a 2-D domain. Each variable shown in the shaded cell has indices $i = 3$, $k = 2$.

FINITE-DIFFERENCE EQUATIONS

The physical equations of motion, which are essentially smooth analytical functions, must be discretized to work at these grid points. As an example, consider the physics of **temperature advection** in the x-direction, as given by a combination of eqs. 3.13c & 3.18:

$$\frac{\Delta\theta}{\Delta t} = -U\frac{\Delta\theta}{\Delta x} \qquad (14.1)$$

To evaluate the right hand side of this equation at the shaded grid cell in Fig 14.4, first we must get the temperature gradient across each side of the grid cell. For the left side of the cell, use the temperature difference between the two neighboring cells and multiply the result by the U_{22} velocity between those cells:

$$U\frac{\Delta\theta}{\Delta x}\bigg|_{left\ \ side} = U_{22}\left[\frac{\theta_{32}-\theta_{22}}{\Delta x}\right]$$

Similarly, for the right side of the cell:

$$U\frac{\Delta\theta}{\Delta x}\bigg|_{right\ \ side} = U_{32}\left[\frac{\theta_{42}-\theta_{32}}{\Delta x}\right]$$

By averaging the advection at the left and right sides, we can find the advection where we want it, at the center:

$$(14.2)$$

$$U\frac{\Delta\theta}{\Delta x}\bigg|_{center} = \frac{1}{2}\left\{U_{32}\left[\frac{\theta_{42}-\theta_{32}}{\Delta x}\right]+U_{22}\left[\frac{\theta_{32}-\theta_{22}}{\Delta x}\right]\right\}$$
$$= \frac{1}{2\cdot\Delta x}\left\{U_{32}[\theta_{42}-\theta_{32}]+U_{22}[\theta_{32}-\theta_{22}]\right\}$$

All of the terms in the equation above are associated with the same time t.

The left side of eq. (14.1) can also be rearranged into a forecast equation by expanding the gradients. One form of the expansion is:

$$\frac{\Delta\theta}{\Delta t} \approx \frac{\theta(t+\Delta t)-\theta(t-\Delta t)}{2\cdot\Delta t}$$

When combined with eq. (14.2) to give the potential temperature at some future time as a function of the potential temperatures and winds at earlier times:

$$\theta_{32}(t+\Delta t) = \theta_{32}(t-\Delta t) - 2\Delta t\cdot\left[U(t)\cdot\frac{\Delta\theta(t)}{\Delta x}\bigg|_{center}\right] \qquad (14.3)$$

Typical time-step durations Δt are on the order of a few minutes.

The equation above is a form of the **leapfrog scheme**. It gets its name because the forecast starts from the previous time step (t–Δt) and leaps over the present step (t) to make a forecast for the future (t+Δt). Although it leaps over the present step, it utilizes the present conditions to determine the future conditions. Fig 14.5 shows a sketch of this scheme.

The two leapfrog solutions (one starting at t–Δt and the other starting at t, illustrated above and below the time line in Fig 14.5) sometimes diverge from each other, and need to be occasionally averaged together to yield a consistent forecast. Without such averaging the solution would become unstable, and would numerically blow up (see next section). There are many other numerical solutions that work better than the leapfrog method; the only reason for showing the leapfrog solution here is that it allows explicit time differencing to be illustrated with relatively simple algebra.

By combining eqs. (14.2 & 3), we get our desired forecast equation:

$$(14.4)$$

$$\theta_{32}(t+\Delta t) = \theta_{32}(t-\Delta t) -$$
$$\frac{\Delta t}{\Delta x}\cdot\left\{U_{32}\cdot[\theta_{42}-\theta_{32}]+U_{22}\cdot[\theta_{32}-\theta_{22}]\right\}_t$$

where the subscript t at the very right indicates that all of the terms inside the curly brackets are evaluated at time t.

In general, for any grid cell at location ik, the forecast equation for temperature (considering only advection in the <u>x-direction</u>) is:

$$(14.5)$$

$$\theta_{ik}(t+\Delta t) = \theta_{ik}(t-\Delta t) -$$
$$\frac{\Delta t}{\Delta x}\cdot\left\{U_{ik}\cdot[\theta_{(i+1)k}-\theta_{ik}]+U_{(i-1)k}\cdot[\theta_{ik}-\theta_{(i-1)k}]\right\}_t$$

Thus, to forecast the temperature at one location, we need the temperatures at three locations and two times, and the winds at two different locations and one time. The arrangement of locations and times needed to forecast one aspect of physics for any grid point is called a **stencil**.

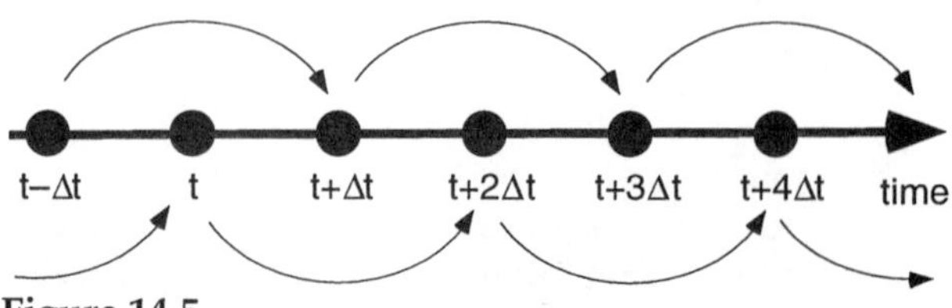

Figure 14.5
Time line illustrating the "leapfrog" time-differencing scheme.

When advection in the other two directions is also considered, as well as radiative, turbulent and latent heating of the air, the resulting finite-difference equation for temperature can be quite long and messy. However, the result is trivial for the computer to compute because the original partial differential equations have been converted to algebra, assuming no bugs were introduced while programming.

Although solving the equation is trivial, it does take a finite time. Considering that this time must be spent for each grid point in the domain, and that the computation must be repeated for a succession of short time steps in order to reach forecast durations of several days, the total computer time accumulates.

Solved Example

At $t = 0$, $\theta_{32} = 10°C$. At $t = 15$ min, $\theta_{22} = 11°C$, $\theta_{32} = 10.1°C$, $\theta_{42} = 9°C$, $U_{22} = U_{32} = 5$ m/s. Numerically forecast the potential temperature θ_{32} at $t = 30$ min, assuming $\Delta x = 50$ km.

Solution

Given: $\Delta t = 15$ min, $\Delta x = 50$ km,
$U = 5$ m/s $= 0.3$ km/min
Find: $\theta_{32} = ?°C$ at $t = 30$ min.

Use eq. (14.4):

$$\theta_{32}(t+\Delta t) = 10°C - \frac{(15\,\text{min})}{(50\text{km})} \cdot \left(0.3\frac{\text{km}}{\text{min}}\right)\{[9-10.1]+[10.1-11]°C\}$$

$= 10°C + 0.18°C$ = <u>**10.18°C**</u>

Check: Units OK. Physics OK.
Discussion: The wind is blowing warmer air into the grid point of interest, thereby causing warming. The potential temperature evolution every 15 minutes is forecast to be: 10, 10.1, and 10.18 °C.

NUMERICAL STABILITY

Not considering coding bugs, computer viruses, or user errors, there are four potential causes of errors in numerical weather forecasting: **round-off error**, **truncation error**, **numerical instability**, and **dynamical instability**. Dynamical instability will be discussed later in the section on chaos.

Round-off error exists because computers represent numbers by a limited number of binary bits (e.g., 32, 64, or 80 bits). As a result, some real decimal numbers can be only approximately represented in the computer. For example, a 32-bit computer can resolve real numbers that are different from each other by about $3\text{x}10^{-8}$ or greater. Any finer differences are missed.

The slight error between decimal and binary representations of a number can accumulate, or can cause unexpected outcomes of conditional tests. Most modern computers use many bits to represent numbers, so with proper care in programming, this is usually not a problem.

Truncation error: When an analytical variable such as potential temperature is represented at one grid point as a function of its values at other grid points, the result is an infinite sum of terms, each of greater power of Δx or Δt. This is a **Taylor series**. The most important terms in the series are the first ones — the ones of lowest power (said to be of lowest order). However, the higher-order terms do slightly improve the accuracy. For practical reasons, the numerical forecast can consider only the first few terms from the Taylor series. Such a series is said to be truncated; i.e., higher-order terms are neglected.

One first order scheme is called the **Euler method**, but is rarely used because the truncation errors are relatively large and can lead to numerical instability. The **leapfrog** scheme of the previous section is a second order scheme. An even better scheme that is accurate to fourth order or higher is called the **Runge-Kutta** method, and is described in books on numerical methods.

Numerical instability results in forecasts that **blow up**. Namely, the numerical solution rapidly diverges from the true solution and can approach unrealistic values (±∞) that can have incorrect sign. Truncation error is one cause of numerical instability.

Numerical instability can also occur if the wind speeds are large, the grid size is small, and the time step is too large. For example, eq. (14.5) models advection by using temperature in <u>neighboring</u> grid cells. But what happens if the wind speed is so strong that temperature from a more distant location in the real atmosphere (beyond the neighboring cell) can arrive during the time step Δt? Such a physical situation is not accounted for in the numerical approximation of eq.(14.5). This can create numerical errors that amplify, causing the model to blow up.

Such errors can be minimized by taking a small enough time step. The specific requirement for stability of advection processes in one dimension is

$$\Delta t \le \frac{\Delta x}{|U|} \qquad \bullet(14.6)$$

with similar requirements in the y and z directions. This is known as the **Courant-Friedrichs-Lewy (CFL)**

stability criterion, or the **Courant condition.** When modelers use finer mesh grids with smaller Δx values, they must also reduce Δt to preserve numerical stability. The combined effect greatly increases model run time on the computer. For example, if Δx and Δy are reduced by half, then so must Δt, thereby requiring 8 times as many computations.

For advection, one way to avoid the time step limitation above is to use a **semi-Lagrangian** method. This scheme uses the wind at each grid point to calculate a **backward trajectory**. The backward trajectory indicates the source location for air blowing into the grid cell of interest. This source location need not be adjacent to the grid-cell of interest. By carrying the values of meteorological variables from the source to the destination during the time step, advection can be successfully modeled.

For other physical processes such as diffusion and wave propagation, there are other requirements for numerical stability. To preserve overall stability in the model, one must satisfy the most stringent condition; that is, the one requiring the smallest time step. At ECMWF for example, their numerical methods allow a maximum time step of about 15 min.

Solved Example

To numerically forecast a hurricane with winds of 50 m/s, a fine-mesh model might be used with Δx = 5 km. What maximum time increment is allowed for a leapfrog version of advection?

Solution

Given: U = 50 m/s, Δx = 5 km

Find: Δt = ? min.

Use eq. (14.6): Δt = 5000 m / (50 m/s) = <u>**100 s**</u>

Check: Units OK. Physics OK.

Discussion: This 1.67 minute time increment is quite small. To make a 2-day forecast would require 1728 time steps.

Suppose a numerical domain of 500 km x 500 km in the horizontal, and 15 km in the vertical (with Δz = 500 m) is used. This means that there are about 300,000 grid points within the domain. Suppose 10 variables (winds, temperature, cloud droplets, etc.) are forecast at each grid point for each time step. This would require solving 10 x 300,000 x 1728 = $5.184\text{x}10^9$ equations.

However, each equation consists of many terms (such as advection, diffusion, condensation, etc.), and the computation stencil for each term requires many calculations (such as illustrated in eq. 14.5). One can easily see how the number of calculations needed to make a hurricane forecast is tremendous, and requires a powerful computer.

THE NUMERICAL-FORECAST PROCESS

Weather forecasting is an **initial value problem**. As shown in eq. (14.5), one must know the initial conditions (at times t and $t - \Delta t$) on the right hand side in order to forecast the temperature at later time ($t+\Delta t$). Thus, to make forecasts of real weather, one must start with observations of real weather.

Weather observation platforms and instruments were already discussed in Chapt. 12. Weather data from these instruments are communicated to central locations. Government forecast centers use this data to make the forecasts.

There are three phases of this forecast process. First is pre-processing, where weather observations from irregular locations and times around the world are transformed into a regular grid of initial conditions. Second is the actual computerized weather forecast, where the finite-difference approximation of the equations of motion are solved. Finally, post-processing is performed to refine and correct the forecasts, and to produce additional secondary products tailored for specific customers.

Balanced Mass and Flow Fields

Over the past few decades it was learned by hard experience that numerical models give bad forecasts if they are initialized with the raw observed data. One reason is that the observation network has large gaps, such as over the oceans and in much of the Southern Hemisphere. Also, while there are many observations at the surface, there are fewer observations aloft. Observations can also contain errors, and local phenomena such as drainage winds can bias the observations.

The net effect of such gaps, errors, and inconsistencies is that the numerical representation of this initial condition is **imbalanced**. By imbalanced, we mean that the observed winds disagree with the theoretical winds, where theoretical winds such as the geostrophic wind are based on temperature and pressure fields.

Balanced and imbalanced flows can be illustrated with a pond of water. Suppose the true water-level is everywhere level, and the water currents are zero (Fig 14.6a). This flow system is **balanced**, because with a level pond surface we indeed expect no currents. Let this state represent the "true" initial condition of the pond.

Consider what happens to a numerical model of the pond if observation errors are incorporated into the initial conditions. Suppose that the water level in

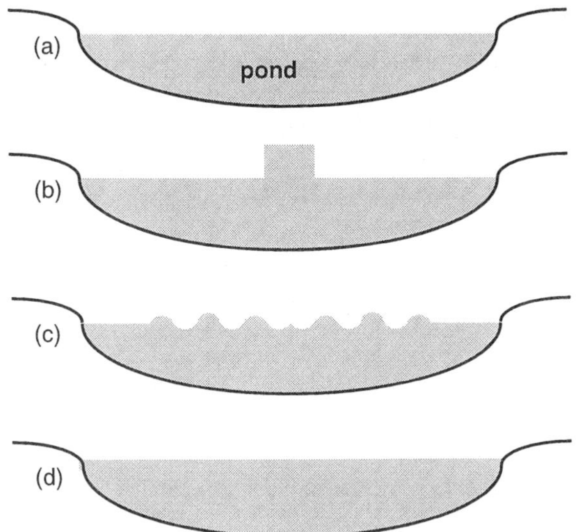

Figure 14.6
Demonstration of imbalanced initial conditions. (a) True initial state of a pond of water. (b) Modeled initial state including an erroneous observation of water depth in the center of the pond. (c) Wave generation as the modeled pond adjusts itself toward a balanced state. (d) Final balanced state with slightly higher water everywhere.

the center of the pond is erroneously "observed" to be 1 m higher than the level everywhere else (Fig 14.6b). This **mass field** (i.e., the distribution of water mass in the pond) is **not balanced** with the **flow field** (i.e., the motions or circulations within the pond, which for this example are assumed to be zero).

Because the pond model uses the full equations of motion, the model response is similar to the response of a real pond. Namely, excess water at the center quickly sinks, generating waves that quickly propagate across the pond (Fig 14.6c). This is how the flow tries to adjust to the imbalance.

Water mass eventually reaches a new balanced state of slightly deeper water everywhere, with no motions (Fig 14.6d). The transient waves and currents are an artifact of the poor initial conditions, and are not representative of the true flow in the pond. Hence, the forecast results are not to be trusted during the first several minutes of the forecast period while the model is adjusting itself to a balanced state.

Numerical weather forecasts of the atmosphere have the same problem, but on a longer time scale than a pond. Namely, the first few hours of a weather forecast are relatively useless while the model adjusts to imbalances in the initial conditions. During this startup period, simulated atmospheric waves are bouncing around in the model, both vertically and horizontally.

After the first 18 h or so of forecast, the dynamics are fairly well balanced, and give essentially the same forecast as if the fields were balanced from the start.

However, spurious waves in the model might also cause unjustified rejection of good data during data assimilation (see next subsection).

Also, the erroneous waves can generate erroneous clouds that cause erroneous precipitation, etc. The net result could be an unrealistic loss of water from the model that could reduce the chance of future cloud formation and precipitation. Change of water content is just one of many **irreversible processes** that can permanently harm the forecast.

In summary, initialization problems cause a transient period of poor forecast quality, and can permanently degrade longer-term forecast skill or cause rejection of good data. Hence, methods to reduce startup imbalances, such as described next, are highly desirable.

Data Assimilation and Analysis

The technique of incorporating observations into the model's initial conditions is called **data assimilation**. Most assimilation techniques capitalize on the tendency of models to create a balanced state during their forecasts.

One can utilize the balanced state from a previous forecast as a **first guess** of the initial conditions for a new forecast. When new weather observations are incorporated with the first guess, the result is called a weather **analysis**. Although the analysis represents current or recent-past weather (not a forecast), the analyzed field is usually **not** exactly equal to the raw observations because the analysis has been smoothed and partially balanced.

First, an automated initial screening of the raw data is performed. During this quality control phase, some observations are rejected because they are unphysical (e.g., negative humidities), or they disagree with most of the surrounding observations. In locations of the world where the observation network is especially dense, neighboring observations are averaged together to make a smaller number of more-accurate observations.

When incorporating the remaining weather observations into the analysis, the raw data from various sources are not treated equally. Some sources have greater likelihood of errors, and are weighted less than those observations of higher quality. Also, observations made slightly too early or too late, or made at a different altitude, are weighted less.

Let σ_g be the standard deviation associated with the first guess. Let σ_o be the standard deviation of the raw observations from a sensor such as a rawinsonde (Table 14-1). Larger σ indicates larger errors.

An **objective optimum interpolation** analysis weights the first guess Z_g and the observation Z_o by

Table 14-1. Typical errors of weather observations (from ECMWF Meteor. Bulletin M3.2, 1994)

Sensor Type	σ_o
Wind errors in the lower troposphere	(m/s)
Surface stations and ship obs	3 to 4
Drifting buoy	5 to 6
Rawinsonde, pilot bal., wind prof.	2 to 3
Aircraft and satellite	3
Pressure errors	(kPa)
Surface weather stations	0.1
Ship and drifting buoy	0.2
S. Hemisphere manual analysis	0.4
Geopotential height errors	(m)
Surface weather stations	7
Ship and drifting buoy	14
S. Hemisphere manual analysis	32
Rawinsonde	13 to 26

their respective errors to produce an analysis field Z_a:

$$Z_a = Z_g + (Z_o - Z_g) \cdot \frac{\sigma_g^{\,2}}{\sigma_g^{\,2} + \sigma_o^{\,2}} \tag{14.7}$$

where Z represents geopotential height of the 50 kPa surface in this example. The word "objective" means performed by computer. If the observation has larger errors than the first guess, then the analysis weights the observation less and the first-guess more.

Optimum interpolation is not perfect, leaving some imbalances that cause atmospheric gravity waves to form in the subsequent forecast. A **normal-mode initialization** modifies the analysis further by removing the characteristics that might excite gravity waves. Another scheme, called **variational analysis,** attempts to match secondary characteristics of the analysis field to observations by minimizing a **statistical cost function**. For example, the radiance that would be expected from the analyzed temperatures is compared to radiance measured by satellite, allowing corrections to be made to the temperature analysis as appropriate.

To illustrate the initialization process, suppose a forecast was started using initial conditions at midnight UTC, and that a 6-hour forecast was produced, valid at 6 UTC. This 6 UTC forecast could serve as the first guess for new initial conditions, into which the new 6 UTC weather observations could be incorporated. The resulting 6 UTC analysis could then be used to start the next forecast run. The process could then be repeated for successive 6-hour forecasts.

Solved Example

A drifting buoy observes a wind of 10 m/s, while the first guess for the same location gives an 8 m/s wind with 2 m/s likely error. Find the analysis wind speed.

Solution

Given: M_o = 10 m/s, σ_o = 6 m/s from Table 14-1.
M_g = 8 m/s, σ_g = 2 m/s
Find: M_a = ? m/s

Use eq. (14.7), except using M in place of Z:

$$M_a = (8\text{m/s}) + (10 - 8\text{m/s}) \cdot \frac{(2\text{m/s})^2}{(2\text{m/s})^2 + (6\text{m/s})^2}$$

= 8 m/s + (2 m/s)·(4/40) = **8.2 m/s**

Check: Units OK. Physics OK.
Discussion: Because the drifting buoy has such a large error, it is given very little weight in producing the analysis. If it had been given equal weight as the first guess, then the average of the two would have been 9 m/s. It might seem disconcerting to devalue a real observation compared to the artificial value of the first guess, but it is necessary to avoid startup problems.

FOCUS • The Pacific Data Void

One hazard of data assimilation is that the resulting analysis does not represent truth, because the analysis includes a previous forecast as a first guess. If the previous forecast was wrong, then the subsequent analysis is poor.

Even worse are situations where there is little or no observation data. For data-sparse regions, the first-guess from the previous forecast dominates the "analysis". This means that future forecasts start from old forecasts, not from observations. Forecast errors tend to accumulate and amplify, causing very poor forecast skill further downstream.

One such region is over the N.E. Pacific Ocean. From Fig 12.9c, there are no RAOBs in that region to provide data at the dynamically important mid-tropospheric altitudes. Ships and buoys provide some surface data, and aircraft and satellites provide data near the tropopause, but there is a sparsity of data in the middle. This is known as the **Pacific data void.**

Poor forecast skill is indeed observed downstream of this data void, in British Columbia, Canada, and Washington and Oregon, USA. Their weather forecast problem is exacerbated by the complex terrain of mountains and shoreline.

Forecasts

Unfortunately, it takes time to communicate to numerical forecast centers the vast quantities of weather observations obtained around the world (see Chapter 12). At ECMWF, for example, they wait about 8 hours after the official observation time before they have received sufficient observations to produce the analysis and start the forecast.

By the time they start the forecast, the analysis is already about half a day old. Hence, their forecast needs to advance about half a day before it catches up to "present". This wasted initial forecast period is not lamented, because startup problems associated with the still-slightly-imbalanced initial conditions yield preliminary results that should be discarded anyway.

The computer continues advancing the forecast by taking small time steps, as illustrated in Fig 14.5. As the forecast reaches key times, such as 1 day, 2 days, etc. out to 10 days or so, the forecast fields are saved on a disk for display and post-processing.

Recall that weather consists of the superposition of many different scales of motion, from small turbulent eddies to large planetary-scale Rossby waves. Unfortunately, the forecast quality of the smaller scales deteriorates much more rapidly than that for the larger scales. For example, cloud forecasts might be good out to 2 to 12 hours, frontal forecasts might be good out to 12 to 36 hours, while the Rossby wave forecasts might be useful out to several days. Fig 14.7 indicates the ranges of horizontal scales over which the forecast is reasonably accurate.

This can be deceiving when you look at a weather forecast, because all scales are superimposed on the weather map regardless of the forecast duration. Thus, when studying a 5 day forecast, you should try to ignore all the small features on the weather map such as thunderstorms or frontal positions. Even though they exist on the forecast map, they are probably wrong. Only the positions of the major ridges and troughs in the jet stream might possess any forecast skill at this forecast duration. Maps in the next section illustrate such deterioration of small scales.

> **FOCUS • Electricity & Operational Reliability**
>
> Most of the large computing centers have elaborate power back-up systems to prevent damage to their computers in the event of electrical power failure. At ECMWF, for example, two 800 kW rotary UPS (Uninterruptable Power Supply) systems are continuously on-line to maintain the supply to the computer facility. Meanwhile, petrol-powered generators automatically start and switch on line within 20 s to provide an additional 1600 kW standby power generation. These systems provide sufficient power for the computers and their large cooling/air-conditioning systems.

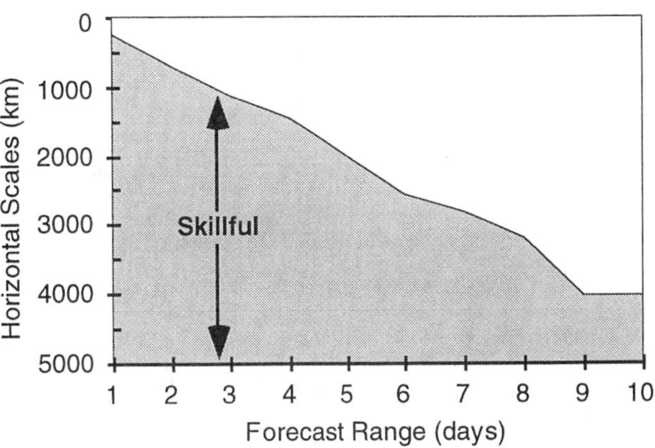

Figure 14.7
Range of horizontal scales having reasonable forecast skill (shaded) for various forecast durations. (from ECMWF)

Case Study: 22-25 Feb 1994 – N. America

Figures 14.8 show the weather valid at 00 UTC on 24 February 1994, which is the same case study as used in the previous two chapters. Fig 14.8a gives the verifying **analysis**; namely, a smoothed fit to the actual weather measured at 00 UTC on 24 Feb 1994.

Figs 14.8b-e give the ECMWF weather **forecasts** valid for the same time, but initialized 1.5, 3.5, 5.5, and 7.5 days earlier. For example, Fig 14.8b was initialized with weather observations from 12 UTC on 22 Feb 94, and the resulting 1.5 day forecast valid at 00 UTC on 24 Feb 94 is shown in the figure. Fig 14.8c was initialized from 12 UTC on 20 Feb 94, and the resulting 3.5 day forecast is shown in the figure. Thus, each succeeding figure is the result of a longer-range forecast, which started with earlier observations, but ended at the same time.

Solid isobars are MSL pressure in mb (1000 mb = 100 kPa), plotted every 5 mb. Dashed isotherms are 85 kPa temperatures, plotted every 2.5°C, with the 0°C line bold. The map domain covers eastern North America. It is centered on Lake Ontario, and extends from southern Texas at the lower left, to Great Slave Lake in the Northwest Territories of Canada at the upper left, to just south of Greenland at the upper right, to near Puerto Rico at the lower right.

These figures demonstrate the inconsistency of forecasts started at different times with different initial conditions. Such inconsistency is inherent in all forecasts, and illustrates the limits of forecastability. The analysis (Fig 14.8a) shows a low centered near Detroit, Michigan, with a cold front extending southwest toward Arkansas. The 1.5-day forecast (Fig 14.8b) is reasonably close, but the 3.5-day forecast (Fig 14.8c) shows the low too far south and the cold front too far west. The 5.5 and 7.5-day forecasts (Fig 14.8d&e) show improper locations for the fronts and lows, but the larger scales are OK.

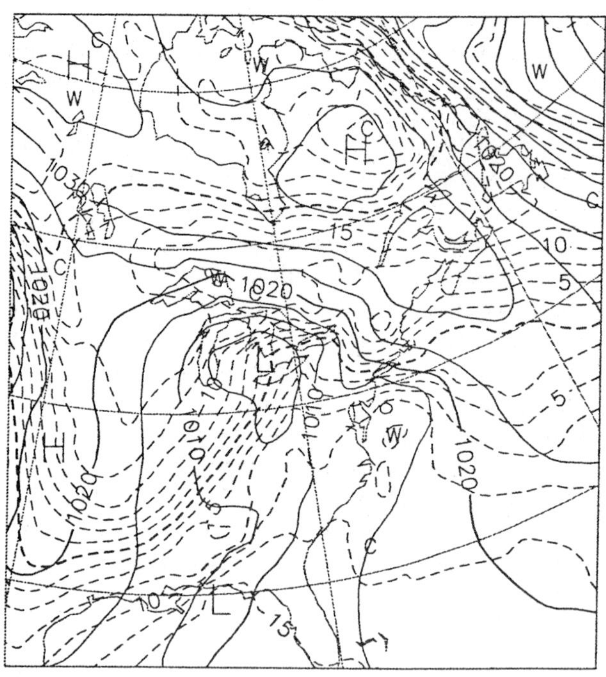

Figure 14.8a
Analysis, valid 00 UTC 24 Feb 94. (Courtesy of ECMWF.)

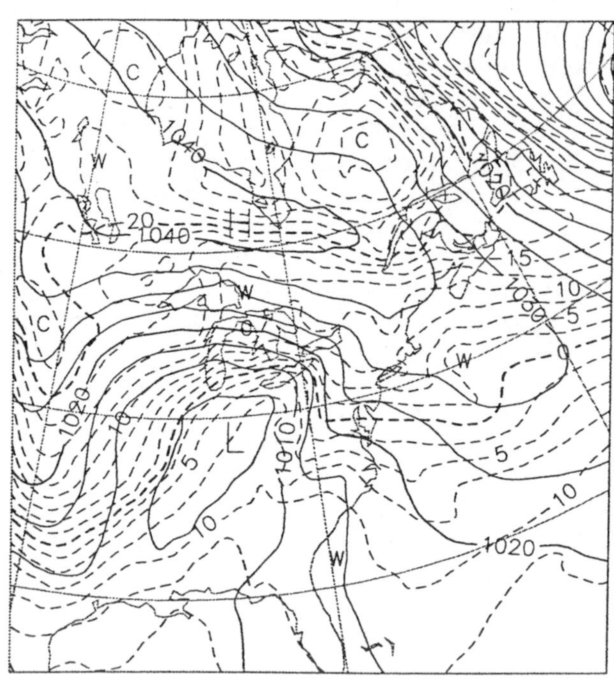

Figure 14.8c
3.5-day forecast, valid 00 UTC 24 Feb 94, started from initialization data at 12 UTC on 20 Feb 94. (From ECMWF.)

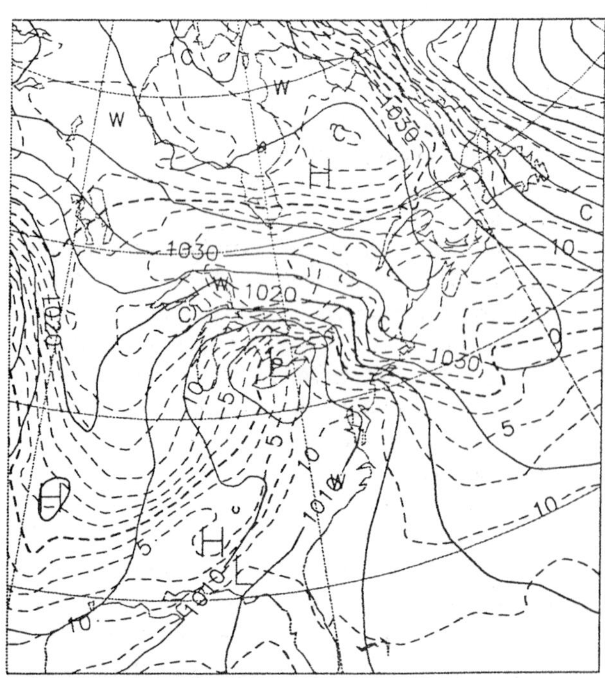

Figure 14.8b
1.5-day forecast, valid 00 UTC 24 Feb 94, started from initialization data at 12 UTC on 22 Feb 94. (From ECMWF.)

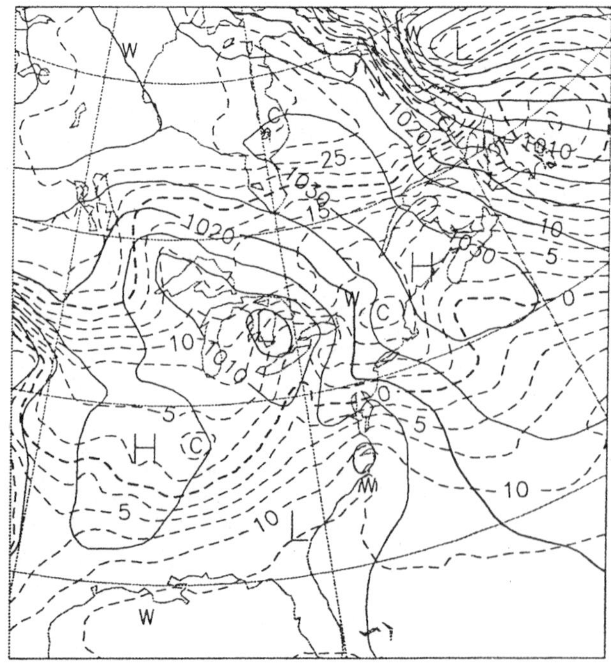

Figure 14.8d
5.5-day forecast, valid 00 UTC 24 Feb 94, started from initialization data at 12 UTC on 18 Feb 94. (From ECMWF.)

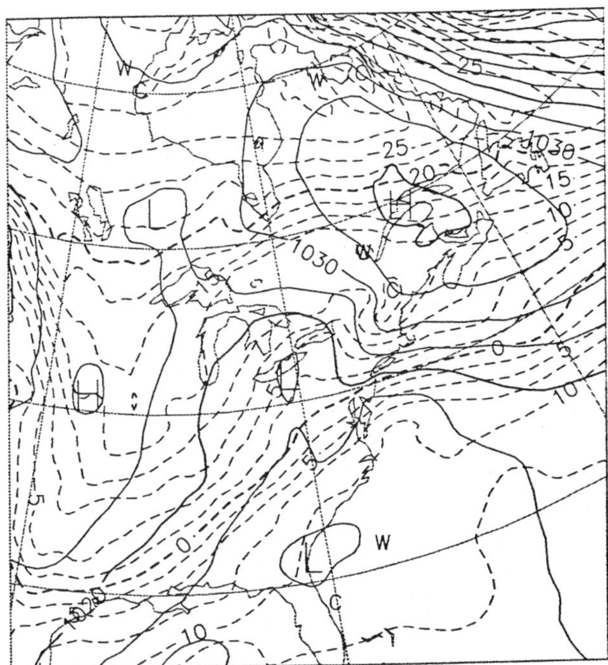

Figure 14.8e
7.5-day forecast, valid 00 UTC 24 Feb 94, started from initialization data at 12 UTC on 16 Feb 94. (From ECMWF.)

Post-processing

After the dynamical computer model has completed its forecast, additional **post-processing** computations can be made with the saved output. Fundamental saved fields consist of winds, temperature, and specific humidity. The secondary, post processed fields are products for human forecasters, for the general public, or for specific industries such as agriculture.

Secondary thermodynamic variables include: potential temperature, virtual potential temperature, liquid-water or equivalent potential temperature, wet-bulb temperature, near-surface (z = 2 m) temperature, surface skin temperature, surface heat fluxes, surface albedo, wind chill temperature, static stability, short and long-wave radiation, and various storm-potential indices.

Secondary moisture variables include: relative humidity, cloudiness (altitudes and coverage), precipitation type and amount, visibility, near-surface dew-point (z = 2 m), soil wetness, and snowfall.

Secondary dynamic variables include streamlines, trajectories, absolute vorticity, potential vorticity, isentropic potential vorticity, vorticity advection, Richardson number, CAPE, dynamic stability, near-surface winds (z = 10 m), surface stress, surface roughness, mean-sea-level pressure, and turbulence.

While many of the above variables are computed at central numerical-computing facilities, additional computations can be made by separate organizations. Local forecast offices of the National Weather Service can tailor the numerical guidance to produce local forecasts of maximum and minimum temperature, precipitation, cloudiness, and storm and flood warning for the neighboring counties.

Consulting firms, broadcast companies, and airlines, for example, acquire the fundamental and secondary fields via data networks such as the internet. From these fields they compute products such as computerized flight plans for aircraft, crop indices and threats such as frost, hours of sunshine, and heating or cooling-degree days for utility companies.

Universities also acquire the primary and secondary output fields, to use for teaching and research. Some of the applications result in weather maps that are put back on the internet and served on the world-wide web (www).

Forecast Refinement

Automated forecasts can often be improved by tailoring the results to specific locales. For example, towns might be located in valleys or near coastlines. These are landscape features that can modify the local weather, but which are not captured by the coarse mesh numerical model. A number of automated statistical techniques can be applied as post-processing to tune the model output toward the climatologically-expected local weather.

Two classical statistical methods are the **Perfect Prog Method** (**PPM**) and **Model Output Statistics** (**MOS**). Both methods use a best-fit statistical regression to relate input fields (**predictors**) to different output fields (**predictands**). An example of a predictand is surface visibility, while predictors for it might include relative humidity, wind speed, and precipitation. The PPM method uses *observations* for the predictors to determine regression coefficients, while MOS uses *model forecast* fields. Once the coefficients are known, both methods use the model forecast fields as the predictors.

Best fit regressions are found using multi-year sets of predictors and predictands. The parameters of the resulting best fit regression equations are held constant during their subsequent usage.

The PPM method has the advantage that it does not depend on the particular forecast model, and can be used immediately after changing the forecast model. The PPM produces best predictand values only when the model produces perfect predictor forecasts, which is rare.

The MOS advantage is that any systematic model errors can be compensated by the statistical regression. A disadvantage of MOS is that a multi-year set of model output must first be collected and statistically fit, before the resulting regression can be used for future forecasts. Both MOS and PPM have a disadvantage that the statistical parameters are fixed.

A newer alternative method is the **Kalman filter**, which continually refines the statistical parameters after each use. This adaptive method has the advantage of MOS, in that it uses model output for the predictors. It learns from its mistakes, and can automatically and quickly retune itself after any changes in the fundamental numerical model. The Kalman filter is becoming increasingly popular.

Solved Example

Given the following simplified MOS regression:

$$T_{\min} = -295 + 0.4 \cdot T_{15} + 0.25 \cdot \Delta Z + 0.6 \cdot T_d$$

for daily minimum temperature (K) in winter at Madison, Wisconsin, where T_{15} = observed sfc. temperature (K) at 15 UTC, ΔZ = model fcst. 100-85 kPa thickness (m), and T_d = model fcst. dew point (K). Predict T_{min} given: T_{15} = 273 K, ΔZ = 1,200 m, and T_d = 260 K.

Solution

Given: T_{15} = 273 K, ΔZ = 1,200 m, T_d = 260 K.
Find: T_{min} = ? K

$$T_{\min} = -295 + 0.4 \cdot (273) + 0.25 \cdot (1200) + 0.6 \cdot (260)$$
$$= 270.2 \cong \underline{-3°C}.$$

Check: Units OK. Physics OK.
Discussion: Chilly, but typical for Madison.

FORECAST QUALITY

Accuracy vs. Skill

There are many different ways to measure the quality of a forecast. One of the least useful is forecast **accuracy**. For example, in Vancouver, Canada, the skies are cloudy 327 days each year on the average. If I forecast clouds every day of the year, then my accuracy (= number of correct forecasts / total number of forecasts) will be 327/365 = 90% on the average. Although this accuracy is quite high, it shows no skill. To be skillful, I must beat climatology to successfully forecast which days will be sunny.

Skill measures forecast improvement above the climatic average. On some days the forecast is better than others, so these measures of skill are usually averaged over a long time (months to years) and over a large area (such as all the grid points in the USA, Canada, Europe, or the world).

Verification Scores

Verification is the process of determining the quality of a forecast. Quality can be measured in different ways, based on various statistical definitions. First, we must define the terms. Let:

A = initial analysis (based on observations)
V = verifying analysis (based on later obs.)
F = forecast
C = climatological conditions
n = number of grid points being averaged

An **anomaly** is defined as the difference from climatology at any instant in time. For example

$F - C$ = predicted anomaly
$A - C$ = persistence anomaly
$V - C$ = verifying anomaly

A **tendency** is the change with time:

$F - A$ = predicted tendency
$V - A$ = verifying tendency

An **error** is the difference from the observations (i.e., from the verifying analysis):

$F - V$ = forecast error
$A - V$ = persistence error

The first error is used to measure forecast accuracy. Note that the persistence error is the negative of the verifying tendency.

As defined earlier in this book, the overbar represents an average. In this case, the average is over a number n of grid points:

$$\overline{X} = \frac{1}{n} \sum_{k=1}^{n} X_k \qquad \bullet(14.8)$$

where k is an arbitrary grid-point index, and X represents any variable.

The simplest quality statistic is the **mean error**.

$$\overline{(F - V)} = \text{mean forecast error} \qquad \bullet(14.9)$$

$$\overline{(A - V)} = \text{mean persistence error} \qquad \bullet(14.10)$$

Positive errors at some grid points can cancel out negative errors at other grid points, giving a false impression of overall error. However, this statistic can indicate a mean **bias** (i.e., a mean difference) between the forecast and verification data.

A more-useful way to quantify error is by the root-mean-square (**RMS**), because errors at individual grid points contribute to the RMS error regardless of their sign:

$$\sqrt{\overline{(F-V)^2}} = \text{RMS forecast error} \qquad \bullet(14.11)$$

$$\sqrt{\overline{(A-V)^2}} = \text{RMS persistence error} \qquad \bullet(14.12)$$

This statistic not only includes contributions from each individual grid point, but it also includes any mean bias error.

RMS error increases with forecast duration. It is also greater in winter than summer, because summer usually has more quiescent weather. Fig 14.9 shows the maximum (winter) and minimum (summer) RMS errors of the 50 kPa heights in the Northern Hemisphere, based on ECMWF forecasts.

Anomaly correlations indicate whether the forecast (or persistence) is varying from climatology in the same direction as the observations. For example, if the forecast is for warmer-than-normal temperatures and the verification confirms that warmer-than-normal temperatures were observed, then there is a positive correlation between the forecast and the weather. At other grid points where the forecast is poor, there might be a negative correlation. When averaged over all grid points, one hopes that there are more positive than negative correlations, giving a net positive correlation.

By dividing the correlations by the standard deviations of forecast and verification anomalies, the result is normalized into an **anomaly correlation coefficient.** This coefficient varies between 1 for a perfect forecast, to 0 for an awful forecast. (Actually, for a really awful forecast the correlation can reach a minimum of –1, which indicates that the forecast is opposite to the weather. Namely, the model forecasts warmer-than-average when colder actually occurs, and vice-versa.)

The definitions of these correlation coefs. are:

anomaly correlation for the forecast =

$$\frac{\overline{\left[(F-C)-\overline{(F-C)}\right]\cdot\left[(V-C)-\overline{(V-C)}\right]}}{\sqrt{\overline{\left[(F-C)-\overline{(F-C)}\right]^2}\cdot\overline{\left[(V-C)-\overline{(V-C)}\right]^2}}} \qquad (14.13)$$

anomaly correlation for persistence =

$$\frac{\overline{\left[(A-C)-\overline{(A-C)}\right]\cdot\left[(V-C)-\overline{(V-C)}\right]}}{\sqrt{\overline{\left[(A-C)-\overline{(A-C)}\right]^2}\cdot\overline{\left[(V-C)-\overline{(V-C)}\right]^2}}} \qquad (14.14)$$

A **persistence forecast** is one where you assume the weather will not change from the initial conditions. This type of forecast also requires no skill, even though it is accurate over a day or two. Fig 14.10 compares persistence and NWP-forecast anomaly correlations produced at ECMWF for a year. One measure of forecast skill is the vertical separation between the forecast and persistence curves in Fig 14.10.

We see that ECMWF beats persistence over the full 10 days of forecast. Also, using 60% correlation

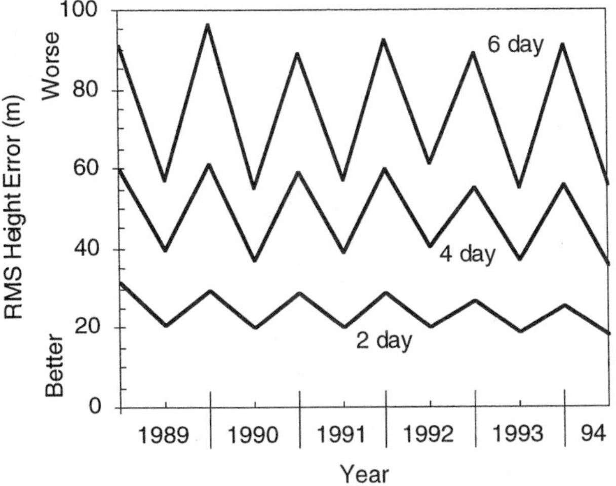

Figure 14.9
Maximum (winter) and minimum (summer) RMS error of 50 kPa heights over the Northern Hemisphere, for ECMWF forecast durations of 2, 4, and 6 days.

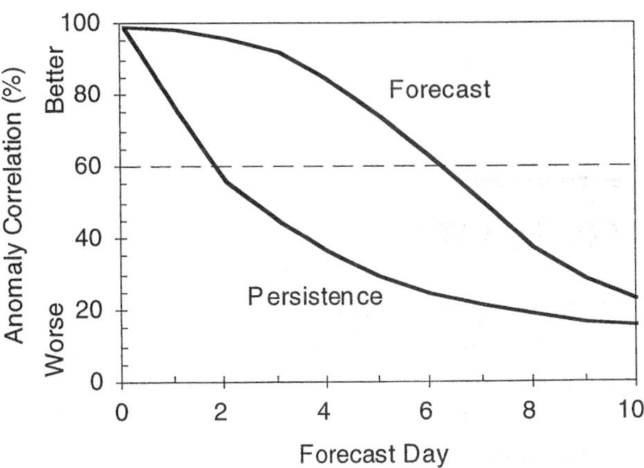

Figure 14.10
Forecast skill of 50 kPa heights over Europe, based on the 1200 UTC values from every day of the year 16 Aug 93 to 15 Aug 94. A persistence forecast is one where the weather is assumed not to change with time from the initial conditions. A value of 100% is a perfect forecast, while a value of 0 is no better than climatology. (Provided by ECMWF)

as an arbitrary measure of quality, we see that a quality persistence forecast extends out to only about 2 days, while a quality ECMWF forecast is obtained out to a bit over 6 days, for this particular year.

Similar statistics can be defined for forecast *accuracy* by eliminating the –*C* in the above two equations. These are called **absolute correlations.**

Another measure of forecast skill is the **S1 score,** which verifies horizontal gradients of height or pressure. It is often applied to 50 kPa heights, and to mean sea level (MSL) pressure. Recall that horizontal pressure gradients are related to the geostrophic wind, and are related to temperature gradients via the hypsometric equation. Thus, this one statistic includes a lot of physics.

Define horizontal gradients between neighboring grid points in the forecast height field as:

$$\Delta F_x = F_{i,j} - F_{i+1,j} \qquad \text{(east-west gradient)}$$

$$\Delta F_y = F_{i,j} - F_{i,j+1} \qquad \text{(north-south gradient)}$$

where *i* is the east-west grid-point index, and *j* is the north-south index. For the verifying analyses:

$$\Delta V_x = V_{i,j} - V_{i+1,j} \qquad \text{(east-west gradient)}$$

$$\Delta V_y = V_{i,j} - V_{i,j+1} \qquad \text{(north-south gradient)}$$

With these definitions, the **S1 score** is:

$$S1 = 100 \cdot \frac{\overline{\Delta F_x - \Delta V_x} + \overline{\Delta F_y - \Delta V_y}}{\overline{\max(\Delta F_x , \Delta V_x)} + \overline{\max(\Delta F_y , \Delta V_y)}} \qquad (14.15)$$

Lower values of *S1* correspond to better forecasts, because *S1* represents a type of error.

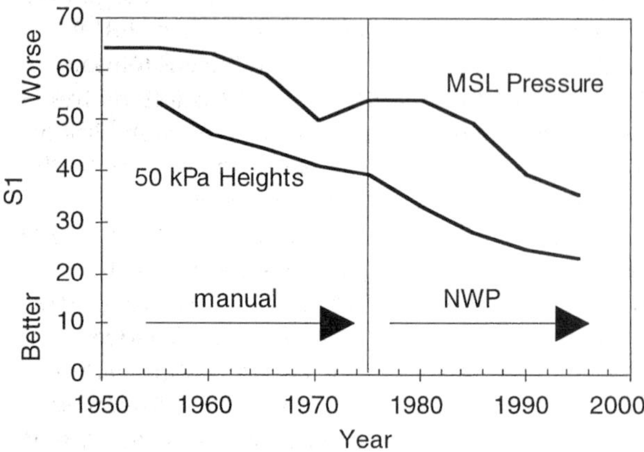

Figure 14.11
S1 skill scores (smoothed) for NMC forecasts of summer mean sea level (MSL) pressure, and for 50 kPa heights. The statistical domain includes most of North America and portions of the neighboring oceans. Forecast skill is gradually improving.

The *S1* errors associated with the NMC forecasts is plotted in Fig 14.11. Note that in the mid 1970s when automated forecasts were introduced, the *S1* score became temporarily worse than that based on the earlier manual forecasts, for MSL pressure forecasts. Since that time, both the MSL pressure and 50 kPa height scores have continued to improve.

Solved Example (§)

Given the following synthetic analysis (*A*), forecast (*F*), verification (*V*), and climate (*C*) fields of 50 kPa height (km). Each field represents a weather map (north at top, S bottom, E right).

Analysis:

5.3	5.3	5.3	5.4
5.4	5.3	5.4	5.5
5.5	5.4	5.5	5.6
5.6	5.5	5.6	5.7
5.7	5.6	5.7	5.7

Forecast:

5.5	5.2	5.2	5.3
5.6	5.4	5.3	5.4
5.6	5.5	5.4	5.5
5.7	5.6	5.5	5.6
5.7	5.7	5.6	5.6

Verification:

5.4	5.3	5.3	5.3
5.5	5.4	5.3	5.4
5.5	5.5	5.4	5.5
5.6	5.6	5.5	5.6
5.6	5.7	5.6	5.7

Climate:

5.4	5.4	5.4	5.4
5.4	5.4	5.4	5.4
5.5	5.5	5.5	5.5
5.6	5.6	5.6	5.6
5.7	5.7	5.7	5.7

Find the mean forecast and persistence errors, the RMS errors and anomaly correlations for the forecast and persistence, and the S1 score.

Solution

Use eq. (14.9): mean fcst error = 0.01 km = **10 m**
Use eq. (14.10): mean persistence error = **15 m**
Use eq. (14.11): RMS forecast error = **63 m**
Use eq. (14.12): RMS persistence error = **87 m**
Use eq. (14.13): fcst. anomaly correl = **81.3%**
Use eq. (14.14): persist. anomaly correl = **7.7%**
Use eq. (14.15): S1 score = **31.4**

Check: Units OK. Physics OK.
Discussion: Contour the analysis field, and see a wave with ridge & trough. The verification shows the wave moving east, but the forecast amplifies it too much.

NONLINEAR DYNAMICS AND CHAOS

Predictability

Recall that NWP is an initial value problem, where these initial values are based on observed weather conditions. Unfortunately, the observations include instrumentation and sampling errors. We have already examined how such errors cause startup problems. How do these errors affect the long-range predictability?

Ed Lorenz at the Massachusetts Institute of Technology suggested that the equations of motion (which are **nonlinear** because they contain products of variables) are very sensitive to initial conditions. Such sensitivity means that substantially different weather forecasts can result from slightly different initial conditions.

This is a sad state of affairs. Our initial conditions will always have small errors, and thus our forecasts will always become inaccurate with time. Thus, there is a limit to the predictability of weather.

A simple physical illustration of **sensitive dependence to initial conditions** is a toy balloon. Inflate one with air and then let it go to fly around the room. Repeat the experiment, being careful to inflate the balloon the same amount and to point it in the same direction. You probably know from experience that the path and final destination of the balloon will differ greatly from flight to flight. In spite of how simple a toy balloon seems, the dynamical equations describing its flight are extremely sensitive to initial conditions, making predictions of flight path virtually impossible.

Lorenz Strange Attractor

Another illustration of sensitive dependence to initial conditions was suggested by Lorenz. Suppose we examine 2-D convection within a tank of water, where the bottom of the tank is heated (Fig 14.12). The vertical temperature gradient from bottom to top drives a circulation of the water, with warm fluid trying to rise. The circulation can modify the temperature distribution within the tank.

A very specialized, highly-simplified set of equations that approximates this flow is:

$$\frac{\Delta C}{\Delta t} = \sigma \cdot (L - C)$$

$$\frac{\Delta L}{\Delta t} = r \cdot C - L - C \cdot M \qquad (14.16)$$

$$\frac{\Delta M}{\Delta t} = C \cdot L - b \cdot M$$

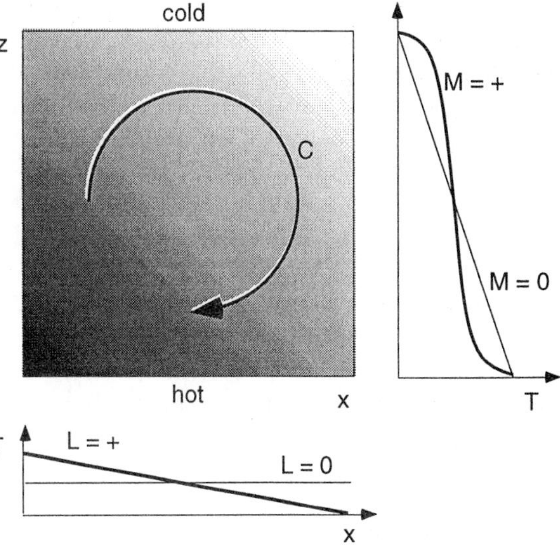

Figure 14.12
Tank of fluid (shaded), showing circulation C. The vertical M and horizontal L distributions of temperature are also shown.

where C gives the circulation (positive for clockwise, and greater magnitude for a more vigorous circulation), L gives the left-right distribution of temperature (positive for warm water on the left), and M indicates the amount of vertical mixing (0 for a linear temperature gradient, and positive when temperature is more uniformly mixed within the middle of the tank). Each of these variables is dimensionless.

Fig 14.13 shows forecasts of C and M vs. time, made with parameter values:

$\sigma = 10.0$, $b = 8/3$, and $r = 28$

and initial conditions:

$C(0) = 13.0$, $L(0) = 8.1$, and $M(0) = 45$.

Note that all three variables were forecast together, even though L was not plotted to reduce clutter. From Fig 14.13 it is apparent that the circulation changes direction chaotically, as indicated by the change of sign of C. Also, the amount of mixing in the interior of the tank increases and decreases, as indicated by chaotic fluctuations of M.

When one dependent variable is plotted against another, the result is a **phase-space** plot of the solution. Because the Lorenz equations have three dependent variables, the phase space is three-dimensional. Fig 14.14 shows a two-dimensional view of the solution, which looks like a butterfly.

This solution exhibits several important characteristics that are similar to the real atmosphere. First, it is irregular or chaotic, meaning that it is impossible

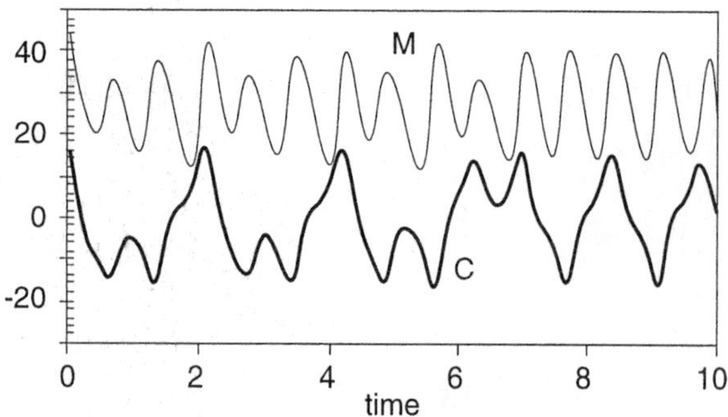

Figure 14.13
Time evolution of circulation C and mixing M.

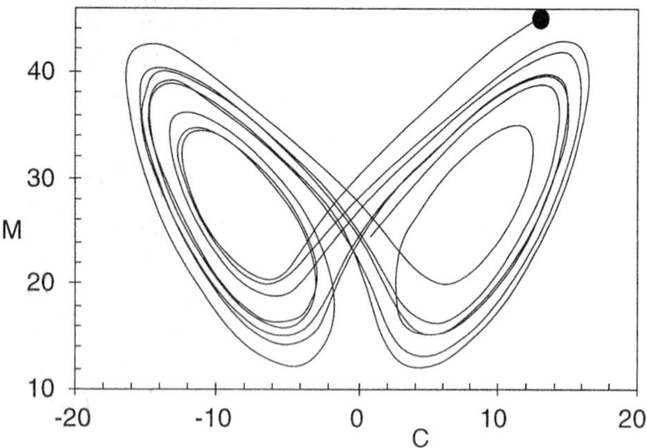

Figure 14.14
"Butterfly" showing evolution of the solution to the Lorenz equations in phase space. Solid dot indicates initial condition.

to guess the solution in the future. Second, the solution is bounded within a finite domain:

$$-20 < C < 20, \quad -30 < L < 30, \quad \text{and } 0 \le M < 50.$$

which implies that the solution will always remain physically reasonable. Third, the solution (M vs. C) appears to flip back and forth between two favored regions, (i.e., the separate wings of the butterfly). These wings tend to attract the solution toward them, but in a rather strange way.

Hence, they are called **strange attractors**. Fourth, the exact solution is very dependent on the initial conditions, as illustrated in the solved examples next. Yet, the eventual solution remains attracted to the same butterfly.

Science Graffito

"Computations indicate that a perfect model should produce three-day forecasts ... which are generally good; one-week forecasts ... which are occasionally good; and two-week forecasts ... which, although not very good, may contain some useful information" – E. Lorenz, 1993: *The Essence of Chaos.* Univ. of Washington Press, Seattle. 227 pp.

Solved Example (§)

Solve the Lorenz equations for the parameters and initial conditions listed previously in this chapter. Use a dimensionless time step of $\Delta t = 0.01$, and forecast from $t = 0$ to $t = 10$.

Solution

Given: $C(0) = 13.0$, $L(0) = 8.1$, and $M(0) = 45$,
and $\sigma = 10.0$, $b = 8/3$, and $r = 28$.
Find: $C(t) = ?, L(t) = ?, M(t) = ?$

First, rewrite eqs. (14.16) in the form of a forecast:

$$C(t+\Delta t) = C(t) + \Delta t \cdot [\sigma \cdot (L(t) - C(t))]$$

$$L(t+\Delta T) = L(t) + \Delta t \cdot [r \cdot C(t) - L(t) - C(t) \cdot M(t)]$$

$$M(t+\Delta T) = M(t) + \Delta t \cdot [C(t) \cdot L(t) - b \cdot M(t)]$$

As an example, for the 1st step:

$$C(0.01) = 13.0 + 0.01 \cdot [10.0 \cdot (8.1 - 13.0)] = 12.51$$

Next, set this up on a spreadsheet, a portion of which is reproduced below.

t	C	L	M
0.00	13.00	8.1	45.00
0.01	12.51	5.809	44.85
0.02	11.84	3.643	44.38
0.03	11.02	1.666	43.63
0.04	10.08	-0.07	42.65
0.05	9.069	-1.55	41.51
0.06	8.007	-2.76	40.26
0.07	6.931	-3.71	38.96
0.08	5.866	-4.44	37.67
0.09	4.836	-4.96	36.4
0.10	3.856	-5.32	35.19

Note that your answers might be different than these, due to different round-off errors and mathematical libraries on the spreadsheets.

Plots. These answers are already plotted in Figs 14.13 - 14.14.

Check: Units dimensionless. Physics OK.
Discussion: Note that the forecast equations above use the Euler time-differencing scheme, which is the least accurate. Nevertheless, it illustrates the Lorenz attractor.

Solved Example (§)

Repeat the previous solved example, but for a slightly different initial condition: $M(0) = 44$.

Solution

Given: $C(0) = 13.0$, $L(0) = 8.1$, and $M(0) = 44$, and $\sigma = 10.0$, $b = 8/3$, $r = 28$.

Find: $C(t) = ?$, $L(t) = ?$, $M(t) = ?$

As in the previous solved example.

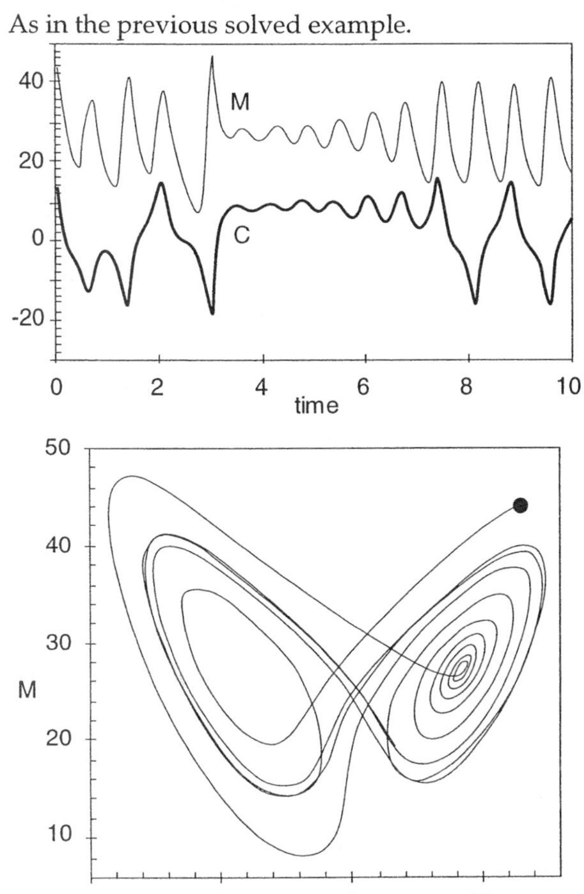

Check: Units OK. Physics OK.

Discussion: These are quite different from Figs 14.13 & 14, demonstrating sensitive dependence to initial conditions.

Ensemble Forecasts

To estimate the dependence of weather forecasts to initial conditions, some numerical forecast centers such as ECMWF repeatedly forecast the same time period, but from slightly different initial conditions. This yields **ensemble forecasts**. Fig 14.15 shows 10-day ensemble forecasts of 85 kPa temperature, precipitation, and 50 kPa heights at Des Moines, Iowa, for the same case-study period as discussed in the previous chapters.

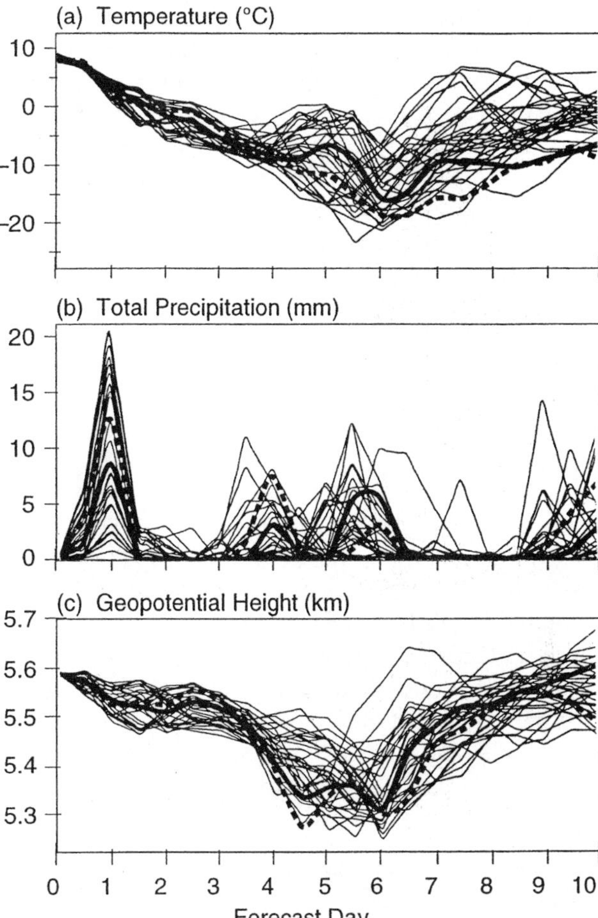

Figure 14.15

Ten-day ensemble forecasts for Des Moines, Iowa, starting from 19 Feb 1994. (a) Temperature (°C) at 85 kPa. (b) Total precipitation (mm). (c) Geopotential height (km) of the 50 kPa surface. (Courtesy of ECMWF).

The heavy dashed line in these figures shows the single forecast run made with the high-resolution model, starting from the "best" initial conditions. This is the "official" forecast produced by ECMWF. The thick solid line starts from the same initial conditions, but is made with a coarser grid resolution to save forecast time. Note that the forecast changes substantially when the grid resolution changes.

The thin lines are forecasts from slightly different initial conditions. To save computer time, these multiple forecasts are also made with the lower grid-resolution. The forecasts of temperature and geopotential height start out quite close, and diverge slowly during the first 3.5 days. At that time (roughly when the cyclone reaches Des Moines late 22 Feb 94), the solution rapidly diverges, signaling a sudden loss in forecast skill that is never regained.

By looking at the spread of the resulting forecasts from each other, we can estimate the probability for error of the "official" forecast. Unfortunately, there is still no way of knowing which of the many forecasts will be closest to reality. Alternately, forecasts of the risks (chances) of hazardous weather are possible.

By averaging all the ensemble members together, one can find an **ensemble average** forecast that is usually more skillful than any individual member. This is one of the strengths of ensemble forecasting.

After several days into the forecast (Fig 14.15), the ensemble forecasts seem chaotic. Yet this **chaotic** solution is bounded within a finite region. This is related to the concept of a "strange attractor," such as studied by Lorenz.

Studies of chaos often focus on the **eventual** state of the solution, at times far from the initial condition. At these long times, the dynamics have forgotten the initial state. Although this eventual state is somewhat useless as a weather forecast, it can provide some insight into the range of possible climatic conditions that are allowed by those dynamical equations.

It also indicates what weather conditions are **not** likely to occur, which could be valuable information for some customers. For example, if none of the ensemble forecasts give temperatures below freezing on one particular day, then a categorical "no freeze" forecast could be made with high confidence.

ON DOING SCIENCE • Scientific Revolutions

In the late 1950s and early 1960s, Ed Lorenz was making numerical forecasts of convection to determine if statistical forecasts were better than NWP forecasts using the nonlinear dynamical equations. One day he re-ran a numerical forecast, but entered slightly different initial data. He got a strikingly different answer.

He was curious about this effect, and allowed himself to become sidetracked from his original investigations. This lead to his description of chaos, discovery of a strange attractor, and realization of the sensitive dependence of some equations to initial conditions.

He published his results in 1963 ("Deterministic nonperiodic flow". *J. Atmos. Sci.*, **20**, p 130-141). From 1963 to about 1975, this groundbreaking paper was rarely cited by other scientists — a clue that it was not yet accepted by his colleagues. However, between 1975 to 1980, researchers were becoming increasingly aware of his work. From 1980 to present, this paper has been cited on the order of 100 times per year.

About a decade and a half elapsed before this new theory gained wide acceptance, which is typical of many paradigm shifts. Namely, it takes about one human generation for a scientific revolution to mature, because typically the older scientists (who hold the power) are not willing to make the shift. The scientific revolution occurs when this group retires and younger scientists (with the newer ideas) take their place.

USEFUL INDICATORS ON WEATHER MAPS PRODUCED BY NWP

We switch gears now from numerical and dynamical considerations, to practical ones. One outcome of a numerical weather prediction is a set of weather maps that sometimes do not show everything that forecasters need. Human forecasters have developed special tricks for interpreting numerical forecast maps, a few of which are described here.

Certain weather boundaries and events are often delineated by key values of isopleths or other features on maps. A sample of these are given below:

• **Cloud Likelihood.** The 70% relative humidity isohumes on 70 kPa forecast charts often enclose areas likely to have clouds.

• **Precipitation Likelihood.** The 90 % relative humidity isohumes on 70 kPa forecast charts often enclose areas where precipitation is likely.

• **Forecast Confidence.** If the members of an ensemble forecast have only small spread, then whatever analysis problems there might be on this day, they will have negligible effect on the accuracy of the forecast. However, on days with greater spread, forecasters should not believe the details of the forecast, although the general evolution might be OK. The ensemble range gives a reliable estimate about possible alternative or extreme scenarios, or at least what type of weather is very unlikely.

• **Rain vs. Snow.** The 5.4 km thickness contour on a 100-50 kPa chart often separates rain (where thicknesses are greater) from snow (where thicknesses are less), although the threshold contour varies regionally.

An alternate indicator is the –5°C isotherm on the 85 kPa chart. Regions on the cold side of this isotherm will likely have snow. This latter method is a bit more reliable than the thickness method.

• **Cyclone Development.** Regions of rising air usually favor cyclone development, and often contain the associated clouds and precipitation.

Air often rises downstream (east) of the trough axis on a 50 kPa chart. In fact, the ridge axis further east often marks the northeast boundary of clouds and/or precipitation in the N. Hemisphere.

For another clue to cyclone development, picture isopleths of vorticity plotted on a 50 kPa chart, and locate any vorticity maxima. Through any vorticity maximum (**vort max**), superimpose contours of 100-50 kPa thickness. Recall that the thermal wind blows parallel to these thickness contours.

Downwind (following thickness contours) of the vort max is often a region of upward motion and cyclone development. Upwind is downward motion (**subsidence**) and fair weather.

Another indicator of cyclone development is regions of positive vorticity advection (PVA) on weather maps.

• **Cyclone Movement.** Low centers often move in the same direction as the direction of the sea-level isobars in the **warm sector** of the storm (i.e., in the warm air mass). The cyclones are actually steered by the winds aloft, which happen to be parallel to the warm-sector surface winds because of the weak vertical shear within the warm sector.

• **Cyclone Death.** Surface low centers tend to occlude and die as they move from a position south of the 30 kPa jet stream to a position north.

• **Occluded Fronts.** Precipitation associated with occluded fronts often aligns itself along a 100–50 kPa thickness ridge, to the east of the low-pressure center at the surface.

ELEMENTS OF A GOOD WEATHER BRIEFING

Whether getting a weather briefing or giving one, you can usually understand the weather better by examining weather maps in an organized sequence. Normally, the reason for getting a briefing is to learn the forecast at one specific location or route. However, past and present weather should be reviewed first to get a sense of the evolution of the weather.

While everyone has their own style, below is an organization that works well for me.

1) Weather maps should be briefed in the following order: **past, present, future**
   The following steps apply to each time phase:
2) First look at a global or hemispheric map showing the jet stream or 50 kPa contours or other large-scale pattern, to get the **big picture**.
3) Next, introduce maps of special fields such as vorticity or vertical motion, or any other maps that can be used to **explain the physics**.
4) Then, discuss the surface weather map, starting from the downstream edge and progressing upstream. At mid-latitudes, this is from **east toward west**.
5) Discuss the **features on the surface weather maps** (highs, lows, fronts, and their clouds and precipitation).
6) Indicate **record, extreme, or unusual events**.
7) **Focus** on your own locale, or other points of interest such as travel routes or destinations. Tailor this discussion to the particular user. Also suggest safety precautions for any storms in the forecast.
8) During the forecast phase, give some indication of **reliability** or likely success of the forecast, and suggest **alternative forecasts** that might occur.
9) Encourage **questions** and interaction, and indicate where additional information can be found.

SUMMARY

By approximating the equations of motion with finite-difference equations, computers can solve the equations to make weather forecasts at grid points around the world. The forecast is made by starting with initial weather conditions, and then taking many small time steps into the future.

Errors in the initial conditions can cause startup problems in the forecast, due to erroneous imbalances between mass and flow within the model. These errors can be reduced by using a data assimilation scheme that uses a first guess based on a previous forecast, and then incorporates the new observations weighted by their accuracy.

To avoid numerical instability, small time steps are required during the forecast. Even so, there are limits to predictability. Larger-scale weather features can be successfully predicted out to about a week, while the smaller features can be predicted out to a day or so. Forecast quality is best measured by various skill scores, which indicate the improvement over the climatological average conditions.

Another limit to predictability is the sensitive dependence of the equations of motion to initial conditions. Any errors in the initial conditions (which always exist) cause the forecasts to diverge from reality, eventually approaching chaotic states where forecast skill is poor.

Nevertheless, useful forecast skill exists out to several days. The resulting weather maps can be interpreted to give the local weather, and can be tailored to the needs of specific customers. At present, most of the weather forecasts are made by computer. Weather observations and forecast dissemination is becoming increasingly automated, with little need for human intervention.

Threads

Vilhelm Bjerknes (during 1900 to 1920) and Lewis F. Richardson(during 1910 to 1930) were early proponents of using physics to forecast weather. Based on their work, most meteorologists consider 7 equations to be the minimum necessary to forecast the weather. These partial differential equations, when approximated by algebraic equations via finite difference methods, form the basis for numerical weather prediction.

The 7 equations, currently known as the **primitive equations**, consist of:

- 3 forecast equations, one for each wind direction, based on Newton's second law (Chapt. 9);
- 1 forecast equation for air density, from the continuity equation (Chapt. 9);
- 1 forecast equation for temperature, from heat conservation as expressed by the first law of thermodynamics (Chapt. 3);
- 1 diagnostic equation for pressure, based on the ideal gas law (Chapt. 1);
- 1 forecast equation for humidity, based on moisture conservation (Chapt. 5).

Initial conditions come from weather observations (Chapt. 12). Boundary conditions and external forcings include solar and IR radiation (Chapt. 2), and surface drag (Chapt. 4).

The resulting numerical forecasts give just numbers at an array of grid points. However, those numbers can be analyzed and graphed to reveal cyclones (Chapt. 13), air masses and fronts (Chapt. 12), hurricanes and thunderstorms (Chapts. 15 & 16), clouds and precipitation (Chapts. 7 & 8), vertical atmospheric structure and stability (Chapts. 1 & 6).

Forecast grids covering the whole globe can forecast the general circulation (Chapt. 11), while grids with very fine grid spacing can forecast local winds (Chapt. 10). When the forecast is extended over decades and centuries, the results can hint at possible climate change (Chapt. 18).

Science Graffito

"The scheme is complicated because the atmosphere is complicated." – L.F. Richardson, 1922: *Weather Prediction by Numerical Process*. Cambridge University Press.

EXERCISES

Numerical Problems

N1. At $t = 0$, $\theta_{32} = 15°C$. At $t = 10$ min, $\theta_{22} = 15°C$, $\theta_{32} = 16°C$, $\theta_{42} = 17°C$, $U_{22} = 10$ m/s, and $U_{32} = 9$ m/s.

a. Numerically forecast θ_{32} at $t = 20$ min, assuming $\Delta x = 30$ km.

b. Is there any indication of separation of the two leapfrog solutions?

N2. For a model using a grid spacing of $\Delta x = 100$ km and a time step of 30 min, is the model numerically unstable if the maximum wind speed (m/s) is:

a. 1 b. 2 c. 5 d. 10
e. 15 f. 20 g. 50 h. 100

N3. Suppose the first guess pressure in an optimum interpolation is 100 kPa, with an error of 0.2 kPa. Find the analysis pressure if an observation of $P = 102$ kPa was observed by:

a. surface weather station b. ship
c. Southern Hemisphere manual analysis

N4. Using Fig 14.7, estimate at what forecast range (days) do we loose the ability to forecast:

a. tornadoes b. hurricanes
c. fronts d. cyclones
e. Rossby waves f. thunderstorms
g. Boras h. lenticular clouds

N5. Using the MOS regression from the solved example in this chapter, calculate the predictand if each of the predictors based on forecast-model output increased by

a. 1% b. 2%. c. 3% d. 4%
e. 5% f. 6%. g. 7% h. 8%

N6. Using Fig 14.9, has there been any improvement in RMS height errors between 1989 to 1994? Discuss.

N7.(§) Given the following fields of 50 kPa height (km). Find the:

a. mean forecast error
b. mean persistence error
c. RMS forecast error
d. RMS persistence error
e. forecast anomaly correlation
f. persistence anomaly correlation
g. S1 score
h. Draw height contours by hand for each field, to show locations of ridges and troughs.

Analysis:

5.2	5.3	5.4	5.3
5.3	5.4	5.5	5.4
5.4	5.5	5.6	5.5
5.5	5.6	5.7	5.6
5.6	5.7	5.8	5.7

Forecast:

5.3	5.4	5.5	5.4
5.5	5.4	5.5	5.6
5.6	5.6	5.6	5.6
5.8	5.7	5.6	5.7
5.9	5.8	5.7	5.8

Verification:

5.3	5.3	5.3	5.4
5.4	5.3	5.4	5.5
5.5	5.4	5.5	5.5
5.7	5.5	5.6	5.6
5.8	5.7	5.6	5.6

Climate:

5.4	5.4	5.4	5.4
5.4	5.4	5.4	5.4
5.5	5.5	5.5	5.5
5.6	5.6	5.6	5.6
5.7	5.7	5.7	5.7

Understanding & Critical Evaluation

U1. Use the meteogram of Fig 14.1.
- a. After 20 Feb, when does the low pass closest to Des Moines, Iowa?
- b. During which days does it rain, and which does it snow?
- c. During which days is there cold-air advection?
- d. Based on the wind direction, guess whether the low center passes north or south of Des Moines.
- e. After 20 Feb, when does the cold front pass Des Moines?
- f. What is the total amount of precipitation that fell during the midweek storm?
- g. How does this forecast, which was initialized with data from 19 Feb, compare with the actual observations (refer to previous chapter)?

U2. Speculate on the capability of weather forecasting if digital computers had not been invented.

U3. Write a finite-difference equation similar to eq. (14.2) for the shaded grid cell of Fig 14.4, but for:
- a. vertical advection
- b. advection in the y-direction

U4. Draw the stencil of grid points used for computing horizontal advection in eq. (14.2). Namely, in a diagram similar to Fig 14.4, show only those grid points involved in the computation.

U5. For the case study forecast of Fig 14.8, first photocopy the figures. Then, on each map
- a. Draw the likely location for fronts.
- b. Indicate the locations of low centers
- c. Comment on the forecast accuracy for fronts, cyclones, and the large-scale flow for this case.

U6. Suppose you are making weather forecasts for Pittsburgh, Pennsylvania, which is close to the intersection of the 40°N parallel and 80°W meridian, shown by the intersection of latitude and longitude lines in Figs 14.8 just south of Lake Erie. During the 7.5 days prior to 00 UTC 24 Feb 94, your temperature forecasts for 00 UTC 24 Feb would likely change as you received newer updated forecast maps.

What is your temperature forecast for 00 UTC 24 Feb, if you made it ___ days in advance from the ECMWF forecast charts of Fig 14.8?
- a. 7.5 b. 5.5 c. 3.5 d. 1.5
- e. and which forecast was closest to the actual analysis?

U7. For Fig 14.9, speculate why the RMS height errors are greater in winter than summer.

U8. Fit an exponential curve to the persistence data of Fig 14.10. What is the e-folding time?

U9.(§) For the Lorenz equations, with the same parameters and initial conditions as used in this chapter, reproduce the results similar to the first solved example, except for all 1000 time steps. Also:
- a. Plot L and C on the same graph vs. time.
- b. Plot M vs. L c. Plot L vs C.

U10.(§) Suppose the Lorenz equations were modified by assuming that $C = L$. For the second two Lorenz equations, replace every C with L, and recalculate for the first 1000 time steps.
- a. Plot L and M vs. time on the same graph.
- b. Plot M vs. L.

Note that the solution converges to a steady-state solution. On the graph of M vs. L, this is called a **fixed point**. This fixed point is an **attractor**, but not a strange attractor.
- c. Describe what type of physical circulation is associated with this solution.

U11. Use the ensemble forecast for Des Moines in Fig 14.15.

a. What 85 kPa temperature forecast, and with what reliability, would you make for forecast day: 1, 3, 5, 7, and 9 ?

b. Which temperature ranges would you be confident to forecast would NOT occur, for day: 1, 3, 5, 7, and 9 ?

c. In spite of the forecast uncertainty, are you confident about the general trends in temperature?

d. Could you confidently forecast when rain is most likely? If so, how much rain would you predict?

U12. Using the forecast charts of Fig 14.8, would you forecast rain or snow for Chicago (SW tip of Lake Michigan), and how would that forecast have changed during the preceding days?

U13. What type of weather would you expect if the 70 kPa relative humidity is 95% and the 100–50 kPa thickness is 5.0 km?

U14. In what order are weather maps presented in the weather briefing given by your favorite local TV meteorologist? What are the advantages and disadvantages of this approach compared to the order suggested in this chapter?

U15. a. Using the grid of Fig. 14.4, and eq. (14.2), circle each grid point that was involved in the calculation of potential temperature in the center of the shaded grid cell. Next circle the points that would have also be used if north-south advection (i.e., in the *z* or *k* direction) were also included. The set of points you circled gives the stencil for that forecast.

b. Instead of the staggered grid of Fig 14.4, suppose that all of the variables (temperature, winds, humidity, etc.) were represented at the same grid points, only in the center of each cell. Rewrite eq. (14.2) for this case, remembering that differences and averages between two grid points are valid halfway between those points, and that before two variables can be multiplied, they must be valid at exactly the same location (regardless of whether that location is on or between grid points).

U16. If the atmosphere is balanced, and if observations of the atmosphere are perfectly accurate, why would numerical models of the atmosphere start out imbalanced?

U17. Experiment with the Lorenz equations on a spreadsheet. Over what range of values of the parameters σ, b, and r, do the solutions still exhibit chaotic solutions similar to that shown in Fig 14.14?

U18. A pendulum swings with a regular oscillation.

a. Plot the position of the pendulum vs. time.

b. Plot the velocity of the pendulum vs. time.

c. Plot the position vs. velocity. This is a called a **phase diagram** according to chaos theory. How does it differ from the phase diagram (i.e., the butterfly) of the Lorenz strange attractor?

Web-Enhanced Questions

W1. Search the web for info about each of the following operational weather forecast centers. Describe the full title, location, computers that they use, and models that they run. Also answer any special questions indicated below for these forecast centers.

a. CMC (and list the branches of CMC).

b. NCEP (and list the centers that make up NCEP)

c. ECMWF

d. FNMOC

W2. Search the web for the government forecast centers in Germany, Japan, China, Australia, or any other country specified by your instructor.

W3. Based on web searches, for each of the numerical models listed below:

a. Define the full title

b. Indicate at which center or university it is run

c. Find the max forecast duration for each run

d. Find the domain (i.e., world, N. Hem. N. America, Canada, Oklahoma, etc.)

GEM
ETA
MESOETA
MRF
AVN
NGM
NOGAPS
COAMPS
ECMWF
MC2
UW-NMS
MM5
RAMS
ARPS
WRF

W4. Search the web for models in addition to those listed in the previous exercises, that are being run operationally. Describe the basic characteristics of these models.

W5. Search the web for a discussion of MOS. What is it, and why is it useful to forecasting?

W6. Find on the web different forecast models that produced precipitation forecasts for Vancouver, Canada (or other city specified by your instructor). Do this for as many models as possible that are valid at the same time and place. Specify the date/time for your discussion. Try to pick an interesting day when precipitation is starting or ending, or a storm is passing. Compare the forecasts from the different models, and if possible search the web for observation data of precipitation against which to validate the forecasts.

W7. At which web sites can you find forecast sea states (e.g., wave height, etc.)?

W8. Based on results of a web search, discuss different ways that ensemble forecasts can be presented via images and graphs.

W9. What types of daily forecasts are currently being made by a university (not a government operational center) closest to your location?

W10. What are the broad categories of observation data that are used to create the analyses (the starting point for all forecasts). Hint, see the ECMWF data coverage web site, or similar sites from NCEP or the Japanese forecast agency.

W11. Search the web for verification scores for the national weather forecast center that forecasts for your location. How have the scores changed by season, by by year? How do the anamoly correlation scores vary with forecast day, compared to the results from ECMWF shown in this chapter?

W12. Find a web site that shows plots of the Lorenz "butterfly", similar to Fig 14.14. Even better, search the web for a 3-D animation, showing how the solution chaotically shifts from wing to wing.

W13. Search the web for other equations that have different strange attractors. Discuss how the equations and attractors differ from those of Lorenz.

W14. Examine from the web the forecast maps that are produced by various forecast centers. Instead of looking at the quality of the forecasts, look at the quality of the weather map images that are served on the web. Which forecast centers produce the maps that are most attractive? Which are easiest to understand? Which are most useful?

W15. Search the web for a meteogram of the weather forecast for your town (or for a town near you, or a town specified by the instructor). What are the advantages and disadvantages of using meteograms to present weather forecasts, rather than weather maps?

Synthesis Questions

S1. Learn what an analog computer is, and how it differs from a digital computer. If automated weather forecasts were made with analog rather than digital computers, how would forecasts be different, if at all?

S2. a. Suppose that there were no weather observations in the western half of N. America. How would the forecast quality over Washington, DC, and Ottawa, Canada, be different, if at all? Given that national legislators live in those cities, speculate on the changes that they would require of the national weather services in the USA and Canada in order to improve the forecasts.

b. Extending the discussion from part (a), suppose that weather observations are back to normal in N. America, but the seats of government were moved to Seattle and Vancouver. Given what you know about the Pacific data void, speculate on the changes that they would require of the national weather services in the USA and Canada in order to improve the forecasts.

S3. How many grid points are needed to forecast over the whole world with roughly 1 m grid spacing? When do you anticipate computer power will have the capability to do such a forecast? What, if any, are the advantages to such a forecast?

S4. Design a grid arrangement different from that in Fig 14.4, but which is more efficient (i.e., involves fewer calculations) or utilizes a smaller stencil.

S5. a. Suppose one person developed and ran a NWP model that gave daily forecasts with twice the skill as those produced by any other NWP model run operationally around the world. What power and wealth could that person accumulate, and how would they do it? What would be the consequences, and who would suffer?

b. Same question as part (a), but for one country rather than one person.

S6. Look up the Runge-Kutta finite difference method in a book on numerical methods. Write eq. (14.1) in finite difference form using this method. How does the result differ from that shown in eqs. (14.2) to (14.5)? Can the Runge-Kutta method be implemented on a computer spreadsheet program? Try it.

S7. Suppose that there was not a CFL numerical stability criterion that restricted the time step that can be used for NWP. How would NWP be different, if at all? Even without a numerical stability criterion, would there be any other restrictions on the time step? If so, discuss.

S8. Speculate on the ability of national forecast centers to make timely weather forecasts if a computer hacker destroyed the internet and other world-wide data networks.

S9. If greater spread of ensemble members in an ensemble forecast means greater uncertainty, then is greater spread desirable or undesirable in an ensemble forecast?

S10. Which would likely give more-accurate forecasts: a categorical model with very fine grid spacing, or an average of ensemble runs where each ensemble member has coarse grid spacing? Why?

Science Graffiti

ON SENSITIVE DEPENDENCE TO INITIAL CONDITIONS

"Does the flap of a butterfly's wings in Brazil set off a tornado in Texas?" – E. Lorenz, 1972.

"Can a man sneezing in China cause a snow storm in New York? – Steward, 1941: *Storm*.

"Did the death of a prehistoric butterfly change the outcome of a US presidential election?" – Bradbury, 1980

S11. Finite-difference equations are approximations to the full, differential equations that describe the real atmosphere. However, such finite-difference equations can also be thought of as exact representations of a numerical atmosphere that behaves according to different physics. How is this numerical atmosphere different from the real atmosphere? How would physical laws differ for this numerical atmosphere, if at all?

S12. Suppose that electricity did not exist. How would you make numerical weather forecasts? Also, how would you disseminate the results to customers?

S13. How good must a numerical forecast be, to be good enough? Discuss.

Science Graffito

"Consumers...have often been satisfied with forecasts that meteorologists...have judged to be of little or no value." – Frederik Nebeker, 1995.

THUNDERSTORMS

CONTENTS

15 Thunderstorms are among the most violent and difficult-to-predict weather elements. Dangerous aspects include:

- vigorous updrafts and turbulence,
- downbursts and gust fronts,
- mesocyclones and tornadoes,
- lightning and thunder, and
- hail and intense precipitation.

Yet, thunderstorms can be studied. They can be probed with radar and aircraft, and simulated in the laboratory and by computer. They form in the air, and must obey the same laws of fluid mechanics and thermodynamics as the rest of the atmosphere.

Thunderstorms are also quite beautiful and majestic. Aesthetics and science merge as thunderstorms to provide endless fascination and enjoyment to those who study them.

CONVECTIVE CONDITIONS

Two conditions are needed to form deep moist convection such as a thunderstorm: **conditional instability** and a **trigger mechanism**. Conditional instability is a special state of the ambient environment prior to storm formation. The trigger mechanism is an external influence that starts the storm. Once initiated, a thunderstorm can often maintain itself.

Conditional Instability

Picture an environmental sounding plotted as the heavy solid line in Fig 15.1. The corresponding surface temperature and dew point are indicated by the two black dots. Such a situation is typical of pre-storm environments found in early afternoon on days when thunderstorms are likely.

During the day, the sun heats the ground and evaporates water into the air. Air near the surface becomes statically unstable, allowing air parcels from the surface-layer to rise. Temperature in such a rising parcel decreases dry adiabatically (thin diagonal solid line in Fig 15.1), and the dew point follows an isohume (thin dotted line).

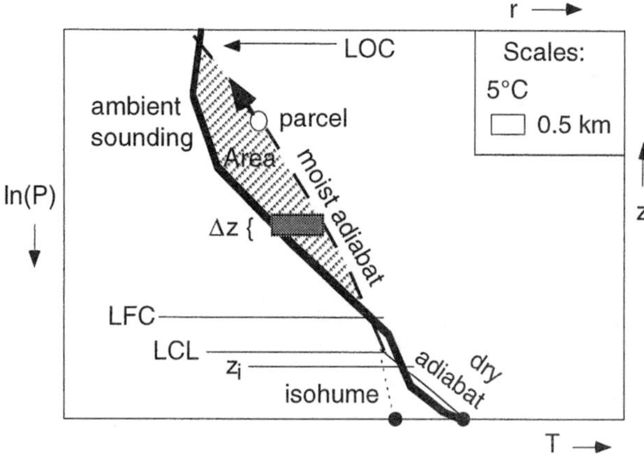

Figure 15.1
Thermodynamic diagram showing prestorm environment.

For this example, the rising parcel hits a stable layer in the environmental sounding, at height z_i. Heat and moisture trapped in the boundary layer below z_i can thus accumulate during the day, and serve as the fuel for future thunderstorms. Without such a lid at z_i, the sensible and latent heat would be continuously released as small clouds, preventing sufficient energy build-up to drive strong thunderstorms.

Next, suppose some external trigger mechanism forces the air to rise past the boundary-layer top z_i. As the air is pushed above the **lifting condensation level** (LCL), clouds form and latent heat is added to the air. The temperature of the rising cloudy air follows a moist adiabat (heavy dashed line).

If the ambient sounding is conducive to thunderstorm development, as is sketched in Fig 15.1, and if the trigger mechanism pushes the air sufficiently high, then the cloudy air parcel will reach an altitude where it is again warmer than the environment. This altitude is called the **level of free convection** (LFC). Above this point, the air parcel continues to rise and accelerate under its own buoyancy, while still following the moist adiabat.

Eventually, the rising parcel reaches an altitude where it becomes colder than the environment. This altitude is the **limit of convection** (LOC), and marks cloud-top. Often, this top is near the tropopause. In the region between the LFC and the LOC, the environment is said to be **conditionally unstable**. This means that the environment is statically stable with respect to dry motions (and is stable in its initial state), but is unstable with respect to saturated motions such as the cloudy air parcel that is rising through it.

Trigger Mechanisms

Ambient conditions such as described above are necessary, but insufficient to create a thunderstorm. Also needed is a mechanism to **trigger** the storm by forcing the boundary-layer air to rise past the low-level stable layer. Any external process that forces the boundary-layer air to rise through this stable lid can be the trigger.

Triggers include:

boundaries between air masses:

- fronts,
- dry lines,
- sea-breeze fronts,
- gust fronts from other thunderstorms,

and other triggers:

- atmospheric buoyancy waves,
- mountains, and
- localized regions of excess surface heating.

For example, an advancing cold front can drive under the ambient boundary-layer air, forcing it upward through the stable layer at z_i. Similarly, horizontal winds hitting a mountain slope are forced upward.

CAPE

Violent updrafts in thunderstorms indicate that a tremendous amount of energy is being released. A measure of the amount of energy available to create such motion is the **convective available potential energy** (CAPE). The concept of CAPE can be illustrated in a thermodynamic diagram (Fig 15.1).

Between the LFC and the LOC, the air parcel is warmer than the environment, and there is a positive buoyant force tending to accelerate the parcel upward. Recall from Chapter 6 that the buoyant force per unit mass is $F/m = (T_{vp} - T_{ve}) \cdot (|g| / T_{ve})$, where $|g|$ is the magnitude of gravitational acceleration (9.8 m/s^2), T_{vp} is the parcel virtual temperature, and T_{ve} is environmental virtual temperature at the same altitude as the parcel.

Energy (work) is force times distance. Thus, the incremental buoyant energy per unit mass ($\Delta E/m$) associated with the rise of an air parcel across some small increment of distance Δz is $\Delta E/m = (F/m) \cdot \Delta z$. For now, use $T_v \cong T$, which gives:

$$\frac{\Delta E}{m} \cong \frac{|g|}{T_e}(T_p - T_e) \cdot \Delta z \qquad (15.1)$$

This increment of energy is proportional to the area inside the rectangle plotted in Fig 15.1. The width of the shaded rectangle is the parcel–environment temperature difference, and the height is Δz.

The total energy is proportional to the total shaded area in Fig 15.1, where this shaded area can be approximated by a sum of rectangles. Thus, the total convective available potential energy is:

$$CAPE = \frac{|g|}{T_e} \cdot Area \qquad (15.2a)$$

or

$$CAPE = \sum \frac{|g|}{T_e}(T_p - T_e) \cdot \Delta z \qquad \bullet(15.2b)$$

CAPE is used by meteorologists to estimate the possible intensity of thunderstorms.

Define the kinetic energy per unit mass for updrafts as $KE/m = 0.5 \cdot w^2$, where w is the updraft speed. If all of the potential energy were converted into kinetic energy, then KE = CAPE. If there were no frictional drag, then the updraft velocity at the LOC would be:

$$w = [2 \cdot CAPE]^{1/2} \qquad (15.3)$$

The inertia of such large updraft velocities can cause the rising cloud-top to overshoot above the LOC. This is called **penetrative convection**, and is observed in satellite pictures as a turret of cloud that temporarily overshoots into the lower stratosphere from the top of the thunderstorm anvil.

Solved Example

Using the scales in Fig 15.1, find the CAPE and the theoretical updraft velocity at the LOC.

Solution

Given: Fig 15.1

Find: CAPE = ? m^2/s^2, w = ? m/s

On Fig 15.1, pave the shaded area with tiles, each the same size as the "scale". Try to make the total area of the tiles equal to the total shaded area:

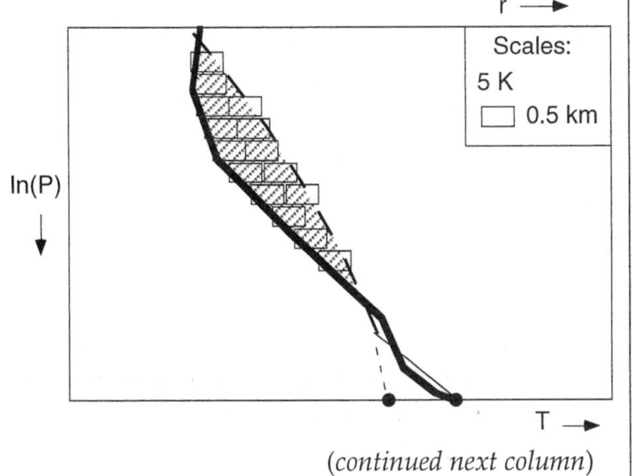

(*continued next column*)

Solved Example (*continuation*)

For temperature differences, 1°C = 1 K. Thus, each tile has an area of 5 K x 500 m = 2500 K·m, and there are 15 tiles. The total area is *Area* = 15 x 2500 K·m = 37500 K·m.

Use eq. (15.2a):

$$CAPE = [(9.8\text{m}\cdot\text{s}^{-1})\cdot(37500\text{K}\cdot\text{m})/(300\text{K})]$$
$$= \underline{\mathbf{1225}}\ \text{m}^2/\text{s}^2$$

Use eq. (15.3):

$$w = \left[2\cdot(1225\text{m}^2\cdot\text{s}^{-2})\right]^{1/2} = \underline{\mathbf{49.5\ m/s}}$$

Check: Units OK. Physics OK. Figure OK.
Discussion: This is quite a violent updraft, which is why aircraft avoid flying through T-storms. Both the CAPE and the resulting vertical velocity are unrealistically large in this example, which was the result of my exaggerating the difference between environmental sounding and moist lapse rate in order to better illustrate the shaded area.

SEVERE-THUNDERSTORM ENVIRONMENT

Thunderstorms come in a wide range of intensities. The most intense storms are produced with an environment similar to that sketched in Fig 15.2. Key aspects of this environment are summarized below.

- The **dew-point temperature** in the boundary layer (the bottom layer in Fig 15.2) is a fairly good indicator of the fuel supply for the storm. Larger dew points imply a warmer, moister boundary layer, and favor more-intense thunderstorms.

  High dew points are possible only if both the air temperature and relative humidity are high — a condition that implies copious amounts of sensible and latent heat to fuel the storm. In the central United States, dew-point temperatures of 20°C or greater are sometimes observed, and are often associated with violent, juicy storms.
- **Cold air aloft** increases the conditional instability. It corresponds to an environmental sounding that remains well to the left of the rising parcels (see Fig 15.1), and thereby contributes to larger values of CAPE.
- **Dry air** at mid levels contributes to downburst intensity, described in the next section.

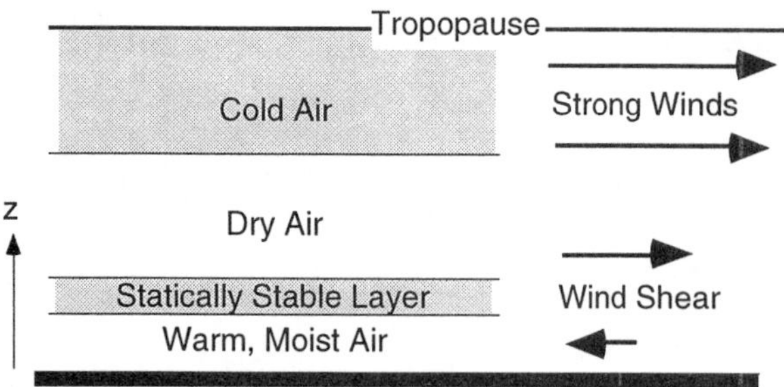

Figure 15.2
An environment conducive to severe thunderstorms.

• As previously mentioned, the **statically-stable layer** in Fig 15.2 acts as a **lid** or cap on the boundary layer, initially trapping the storm's fuel supply near the ground. This allows fuel levels to accumulate and build to the point where its eventual release can feed a violent storm.

• **Strong winds aloft** favor tornadic thunderstorms. As these winds are disturbed by the storm, vorticity can develop that causes the whole thunderstorm to spin. This is called a **mesocyclone.** Stretching of the vorticity due to updrafts can increase vorticity to the point where tornadoes form within the cloud. As the tornado stretches vertically, eventually its bottom appears below cloud base: first as a rotating isolated-lowering of the cloud base called the **wall cloud**, next as a **funnel cloud** (a tornado that does not reach the ground), and finally as a **tornado** (when it reaches the ground).

• Low altitude **wind shear** in the pre-storm environment favors long-lasting thunderstorms. Without any shear, the whole environment including the thunderstorm and the boundary layer would move *en masse.* This would result in a storm that consumes the limited fuel supply immediately under it, and dies after about 15 minutes. With shear, the boundary layer winds move relative to the storm, and can continuously feed fuel into a longer-lasting storm. Low-altitude shear can also generate mesocyclones.

DOWNBURSTS AND GUST FRONTS

Three factors can create or enhance downdrafts. One is a **vertical pressure gradient** that differs from the background hydrostatic pressure profile. Another is **precipitation drag** associated with falling rain drops or ice crystals. The third is **evaporative cooling** of falling rain within initially-drier air.

Precipitation Drag

As rain falls, it tends to drag some air along with it. If the rain is falling at its terminal velocity (see Chapter 8) then the pull of gravity on the rain drops is balanced by air drag. Because drag acts between air and rain, it not only retards the drop velocity but it enhances the air downdraft velocity.

The presence of liquid or solid water in a cloudy, rainy air parcel adds weight to the parcel — an effect called **liquid-water loading.** It has the same effect as if the parcel were colder and more dense. This effect can be quantified by the virtual temperature T_v (see Chapter 1, eq. 1.14), which is defined to include a term for liquid- or solid-water mixing ratio, r_L.

The buoyant force of rain-laden air is found by using the virtual temperature in equation (6.2a):

$$\frac{F}{m}=\frac{T_{v\ parcel}-T_{v\ environ}}{T_{v\ environ}}\cdot|g| \tag{15.4}$$

If the downdraft air falls a distance of Δz with no change in its buoyant force per mass, then by equating kinetic and potential energy, the downdraft speed at the bottom of the fall is

$$w=-\sqrt{2\cdot(F/m)\cdot\Delta z} \tag{15.5}$$

Solved Example

10 g/kg of liquid water exists as rain drops in saturated air of temperature 10°C and pressure 80 kPa. Find the: (a) virtual temperature, (b) buoyancy force per mass associated with just the liquid-water loading, and (c) the downdraft velocity at the bottom of a 1 km fall, associated with that liquid-water loading.

Solution

Given: r_L = 10 g/kg, T = 10°C
Find: T_v = ? °C, F/m = ? m/s^2.

(a) First, find the saturation vapor pressure using Table 5-1: e_s = 1.233 at T = 10°C.
Use eq. (5.3):
r_s = 0.622·(1.233kPa)/(80–1.233kPa)=9.74 g/kg
Use eq. (1.14):
$T_v = T\cdot[1+0.61\cdot r_s - r_L]$
= (283K)·[1 + 0.61·(0.00974)] – (283K)·[0.01]
= 284.68 – 2.83 K = 281.85 K = <u>8.85°C</u>

(b) Use eq. (15.4):

$$\frac{F}{m}=\frac{-2.83\text{K}}{284.68\text{K}}\cdot\left|9.8\frac{\text{m}}{\text{s}^2}\right| = \underline{-0.097}\ \text{m/s}^2 .$$

(continued next column)

Solved Example *(continuation)*

(c) Use an eq. similar to (15.5), except using the liquid-water loading F/m.

$$w = -\sqrt{2 \cdot (F/m) \cdot \Delta z}$$
$$= -\sqrt{2 \cdot (-0.097\text{m}/\text{s}^2) \cdot (-1000\text{m})}$$
$$= \underline{-13.9 \text{ m/s}}$$

Check: Units OK. Physics OK.
Discussion: Although the water vapor in the air adds buoyancy equivalent to a temperature increase of 1.68°C, the liquid water loading decreases buoyancy equivalent to a temperature decrease of 2.83°C. The net effect is that this saturated, liquid-water laden air acts heavier (colder, T = 8.85°C) than dry air at the same temperature (10°C).

Evaporative Cooling

In the downdraft region of a thunderstorm, the rain-filled air of the storm can mix with neighboring environmental air that is drier. The resulting mixture of air is not saturated, and allows evaporation of water from the raindrops. This evaporation cools the air due to absorption of latent heat, and contributes to its negative buoyancy.

The temperature change associated with evaporation of r_L grams of liquid water per kilogram of air is:

$$\Delta T = -\left(\frac{L_v}{C_p}\right) \cdot r_L \qquad (15.6)$$

where (L_v/C_p) = 2.5 K·kg$_{air}$·$(g_{water})^{-1}$.

Evaporative cooling of falling rain is often a much larger effect than the liquid water loading. In regions such as the western Great Plains of the United States (e.g., near Denver), the environmental air is often so dry that evaporative cooling causes dangerous downdrafts called **downbursts** (Fig 15.3). Hazardous downbursts can occur even under virga, where precipitation evaporates before reaching the ground. The smaller-diameter, but intense downbursts, are called **microbursts**.

Downbursts of 0.5 to 5 km in diameter have been observed. For extreme cases, downdraft speeds of nearly 10 m/s have been observed 100 m above ground. This is particularly hazardous to landing and departing aircraft, because this vertical velocity can sometimes exceed aircraft climb rate. Doppler radars can detect some of the downbursts and give early warning to pilots.

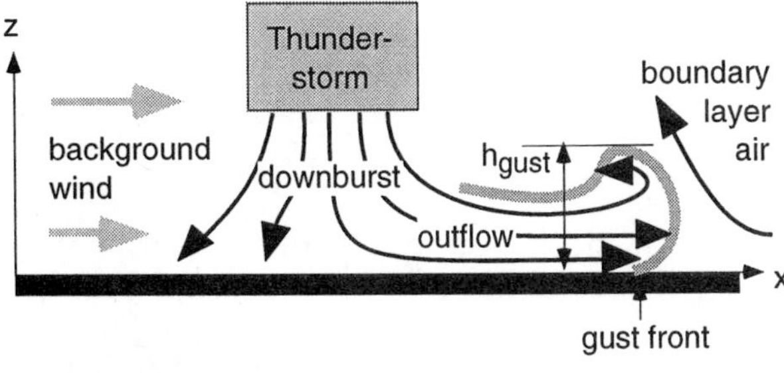

Figure 15.3
Side view of a downburst, showing straight-line outflow winds, and the gust front.

Solved Example
If all of the liquid water of the previous solved example evaporates, find the temperature decrease of the air.

Solution
Given: r_L = 10 g/kg
Find: ΔT = ? K

Use eq. (15.6):

$$\Delta T = -\left(2.5K \cdot \frac{kg_{air}}{g_{water}}\right) \cdot \left(10 \frac{g_{water}}{kg_{air}}\right) = \underline{-25 \text{ K}}$$

Check: Units OK. Physics OK.
Discussion: The temperature change associated with evaporation of rain is nearly ten times that associated with rainfall drag (see previous solved example). That is why dry air at mid-levels in the environment contributes to dangerous storms with severe turbulence.

Both this and the previous example neglected turbulent drag of the downburst air against the environmental air. This effect can greatly reduce the actual downburst speed compared to the idealized calculations above.

Gust Fronts

When downdraft or downburst air reaches the ground, it spreads out (Fig 15.3). The leading edge of the spreading air is the **gust front** (Fig 15.4). Gust fronts can be 100 to 1000 m deep, 5 to 100 kilometers wide, and last for 2 to 20 minutes. Gust fronts typically spread with speeds of 5 to 15 m/s.

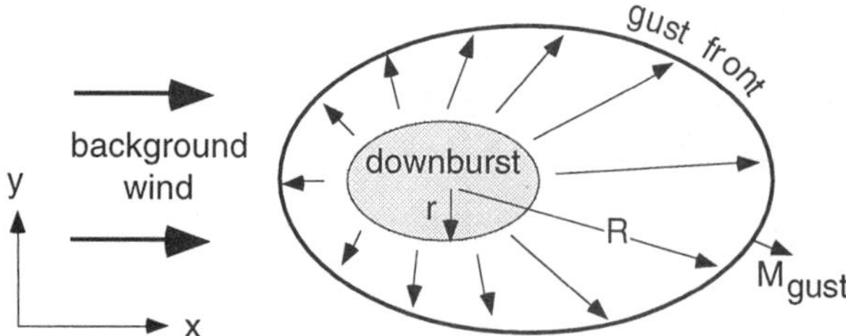

Figure 15.4
Top view of a downburst, straight-line outflow winds (thin arrows), and gust front.

FOCUS • Downbursts & Flight Operations

In 1975 a jet airliner crashed killing 113 people at JFK airport in New York, while attempting to land during a downburst event. Similar crashes prompted a major research program to understand and forecast downbursts. One outcome is a recommended procedure for pilots to follow, which is counter to their normal reactions based on past flight experience.

Picture an aircraft flying through a downburst while approaching to land, as sketched below. First the aircraft experiences strong headwinds (1) , then the downburst (2) , and finally strong tail winds (3). Pilots are now trained to safely respond to this situation in a way that is counter-intuitive.

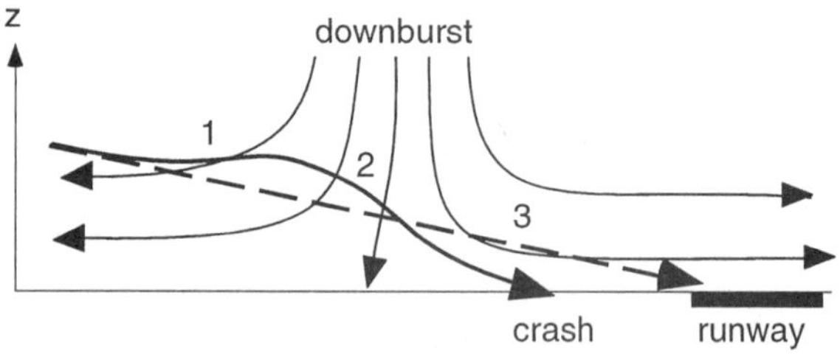

The dashed line shows the desired glideslope, while the solid heavy line shows the flight path of the aircraft that had crashed at JFK. Initially, the aircraft was on the desired glideslope.

Upon reaching point 1, strong headwinds blowing over the aircraft wings increased the lift, causing the aircraft to rise above the glide slope. The pilot's intuitive reaction was to reduce engine power and lower the nose of the aircraft, to descend back to the glideslope. Unfortunately, the aircraft was in a nose-low and reduced-power configuration as it reached point 2, opposite to what was needed to counteract the downburst. As the aircraft flew toward point 3, strong tailwinds reduced the airflow over the wings, reducing the lift and causing the aircraft to sink until it crashed.

Since that time, downburst and wind-shear sensors have been (and are continuing to be) deployed at airports to provide warnings to pilots. Also, pilots are now trained to maintain extra airspeed if they fly through an unavoidable downburst event.

Behind the gust front, winds of 30 m/s are possible. These are called **straight-line winds,** to distinguish them from tornadic winds. Nevertheless, these straight-line winds can cause significant damage — blowing down trees and overturning mobile homes. When these gust fronts move over dry soil, the result is a **dust storm** or **sand storm,** called a **haboob**.

Because the gust front formed from downdraft air that was chilled by evaporative cooling, a temperature decrease of 1 to 3 °C is often noticed during gust front passage. The leading edge of this cold air acts like a miniature cold front as it progresses away from the downburst. It hugs the surface because of its cold temperature, and plows under the warmer boundary-layer air.

As the boundary-layer air is forced to rise over the advancing gust front (Fig 15.3), it can reach its LCL and trigger new thunderstorms. Thus, a parent thunderstorm with gust front can spawn offspring storms in a process that is called **propagation**. The result is a sequence of storms that move across the country, rather than a single long-lived storm.

The speed of advance of the gust front, as well as the gust-front depth, are partly controlled by the continuity equation. Namely, the vertical supply rate of air by the downburst must balance the horizontal removal rate by expansion of outflow air away from the downburst. This can be combined with an empirical expression for depth of the outflow.

The result for gust-front advancement speed M_{gust} is

$$M_{gust} = \left[\frac{0.2 \cdot w_d \cdot r^2 \cdot g \cdot \Delta T_v}{R \cdot T_v}\right]^{1/3} \tag{15.7}$$

where w_d is the downburst velocity, r is the radius of the downburst, R is the distance of the gust front from the downburst center, ΔT_v is the virtual temperature difference between the environment and the cold outflow air, and T_v is the environmental virtual temperature. The gust front depth h_{gust} is:

$$h_{gust} = 0.85 \cdot \left[\left(\frac{w_d \cdot r^2}{R}\right)^2 \cdot \frac{T_v}{g \cdot |\Delta T_v|}\right]^{1/3} \tag{15.8}$$

Both equations show that the depth and speed of the gust front decreases as the distance R of the front increases away from the downburst. Also, colder outflow increases the outflow velocity, but decreases its depth.

Solved Example

Given a downburst speed of 5 m/s within an area of radius 0.5 km. If the gust front is 2 km away from the downburst, and is 2°C colder than the environment air of 300 K, find the depth and advance rate of the gust front.

Solution

Given: $w_d = -5$ m/s, $r = 0.5$ km, $R = 2$ km, $\Delta T_v = -2$ K, $T_v = 300$ K.

Find: $M_{gust} = ?$ m/s, $h_{gust} = ?$ m

Use eq. (15.7): $M_{gust} =$

$$= \left[\frac{0.2 \cdot (-5\text{m/s}) \cdot (500\text{m})^2 \cdot (9.8\text{m}\cdot\text{s}^{-2})(-2\text{K})}{(2000\text{m}) \cdot (300\text{K})} \right]^{1/3}$$

= **2.0 m/s**

Use eq. (15.8): $h_{gust} =$

$$= 0.85 \cdot \left[\left(\frac{(-5\text{m/s}) \cdot (500\text{m})^2}{(2000\text{m})} \right)^2 \cdot \frac{(300\text{K})}{(9.8\text{m}\cdot\text{s}^{-2})(2\text{K})} \right]^{1/3}$$

= **181 m**

Check: Units OK. Physics OK.

Discussion: As the gust front advances, it entrains and mixes with the boundary layer air. This dilutes the cold outflow, warms it toward the boundary-layer temperature, and retards its gust advancement velocity.

LIGHTNING AND THUNDER

Lightning

Lightning occurs when the voltage difference is large enough to ionize the air and make it conductive. This difference, called the breakdown potential, is $B = 3\text{x}10^9$ V/km for dry air (where V is volts). Thus, by measuring the length Δz of any electrical spark, including lightning, one can estimate the voltage difference $\Delta V_{lightning}$ that caused the spark:

$$\Delta V_{lightning} = B \cdot \Delta z \qquad \bullet(15.9)$$

Over 29 million lightning flashes were detected during 1998 by the U.S. National Lightning Detection Network (NLDN). The median peak current was 20 kA (where A is Amperes), for negative currents. High electrical currents in lightning are less likely than low currents. Based on measurements of lightning surges in electrical power lines, the probability P that lightning current will exceed value I (kA) is well approximated by a log-normal distribution:

Solved Example

What voltage difference is necessary to create a lightning bolt that reaches between the ground and a cloud that is 2 km high?

Solution

Given: $\Delta z = 2$ km, $B = 3\text{x}10^9$ V/km

Find: $\Delta V_{lightning} = ?$ Volts

Use eq. (15.9):

$\Delta V_{lightning} = (3\text{x}10^9 \text{ V/km})\cdot(2 \text{ km}) =$ **6x10^9** V

Check: Units OK. Physics OK.

Discussion: Six billion volts is more than enough to cause cardiac arrest, so it is wise to avoid being struck by lightning. High-voltage electrical transmission lines are often about $3.5\text{x}10^5$ V.

$$P = \exp\left\{ -0.5 \cdot \left[\frac{\ln((I - I_0)/I_1)}{s_I} \right]^2 \right\} \qquad (15.10)$$

for $I \geq I_o + I_1$, where $I_o = 2$ kA, $I_1 = 3.5$ kA, and $s_I = 1.5$. As plotted in Fig 15.5, there is a 50% chance that a lightning current will exceed about 20 kA.

A standardized wave shape is used by electrical engineers to model the wave shape of the lightning surge in electrical power lines. It is given by

$$e = e_o \cdot a \cdot \left[\exp\left(-\frac{t}{T_1} \right) - \exp\left(-\frac{t}{T_2} \right) \right] \qquad (15.11)$$

where e and e_o can apply to current or voltage. The nominal constants are: $T_1 = 70$ µs, $T_2 = 0.15$ µs, and $a = 1.016$. The surge, plotted in Fig 15.6 reaches a peak in 1 µs, and decreases by half in 50 µs. The actual time constants vary from case to case.

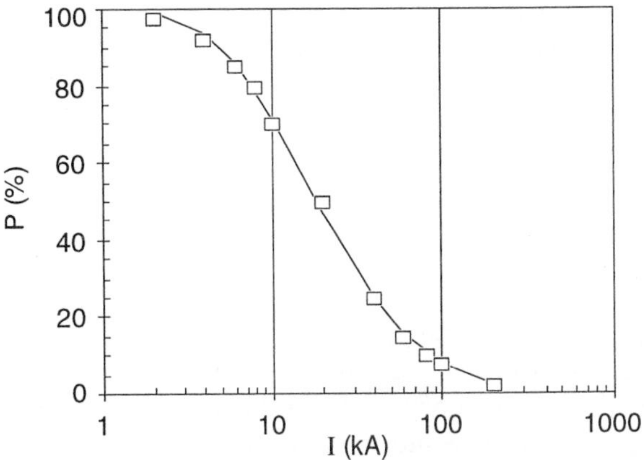

Figure 15.5
Probability P that lightning current will exceed I in an electrical power line. Data points (observations), curve (15.10).

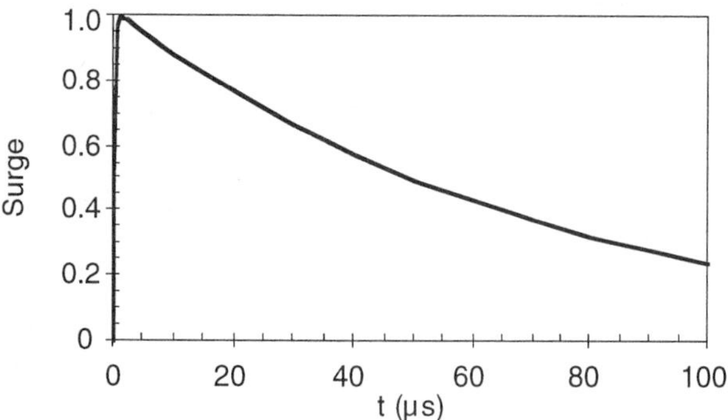

Figure 15.6
Lightning surge wave shape in an electrical power line.

The sudden increase and decrease in flow of electricity through the lightning path in air generates electromagnetic radiation much like a broadcast antenna. These radio signals are known as **sferics**, and can be heard on AM radio bands. Networks of receiving stations on the ground can measure the time and amplitude of the sferics.

Solved Example

What is the probability that a lightning strike will cause a current of 80 kA or greater in a power line? How soon after the strike will the current be reduced to 8 kA?

Solution

Given: I = 80 kA, $I_{later} / I_{initial} = 8/80 = 0.1$
Find: P = ? %, t = ? μs

Use eq. (15.10):

$$P = \exp\left\{-0.5 \cdot \left[\frac{\ln((80-2\text{kA})/3.5\text{kA})}{1.5}\right]^2\right\} = \underline{12\%}$$

Use eq. (15.11):

$$0.1 = 1.016 \cdot \left[\exp\left(-\frac{t}{70\mu\text{s}}\right) - \exp\left(-\frac{t}{0.15\mu\text{s}}\right)\right]$$

For long times, the last exponential is negligible.

$$t \approx -70\mu\text{s} \cdot \ln\left(\frac{0.1}{1.016}\right) = \underline{\textbf{162.3 μs}}$$

Check: Units OK. Physics OK. Agrees with Figs.
Discussion: During the short but intense electrical surge, insulation can be destroyed, power transformers can be melted, and circuit breakers can open. This results in power surges or interruptions that reach homes and businesses, causing comp#ters to go berserk, file$ to be cor@upte#, and #@!*.&%$@#¶£X$#... &@% .. . .

Triangulation can be used to locate the strikes, and received amplitude can be used to estimate lightning stroke strength. This forms the basis of ground-based **lightning detection networks**. Satellite-based lightning detection is also being refined.

Some networks receive **very low frequency (VLF)** radio waves (10 kHz frequency; 30 km wavelength), where the domain between the ground and the ionosphere is a waveguide for these waves. If the ground station can measure the peak electric field E (V/m) associated with the VLF wave passage, and if distance D (m) to the lightning can be determined by triangulation, then the peak current I (A) flowing in the return stroke can be estimated from:

$$I = \frac{-2\pi \cdot \varepsilon_o \cdot c^2}{v_L} \cdot E \cdot D \tag{15.12}$$

where $c = 3.00986 \times 10^8$ m/s is the speed of light, $\varepsilon_o = 8.854 \times 10^{-12}$ A·s·(V·m)$^{-1}$ is the permittivity of free space, and v_L = 1.0 to 2.2 x 10^8 m/s is the velocity of the current through the return-stroke path.

Solved Example

What electric field would be measured at a lightning detection station 200 km from a lightning stroke of intensity 20 kA?

Solution

Given: I = 20 kA, D = 200 km
Find: E = ? V/m
Assume: $v_L = 2.0 \times 10^8$ m/s

Solve eq. (15.12) for E:

$$E = \frac{I}{D} \cdot \frac{-v_L}{2\pi \cdot \varepsilon_o \cdot c^2} = \frac{(20\text{kA})}{(200\text{km})} \cdot \left\{\frac{-2\times10^8 \text{m/s}}{2\pi \cdot \left[8.854\times10^{-12}\text{A}\cdot\text{s}/(\text{V}\cdot\text{m})\right] \cdot (3\times10^8\text{m/s})^2}\right\}$$

$$= \underline{\textbf{-4 V/m}}$$

Check: Units OK. Physics OK.
Discussion: Lightning detection stations can also determine the polarity of the lightning, and the waveform. Cloud-to-ground lightning has a different waveform and polarity than cloud-to-cloud lightning. Thus, algorithms processing signals from the network of stations can also determine lightning type, charge (positive or negative) as well as intensity and location.

Often, lightning flashes consist of one to ten quick return strokes in the same ionized path, with a typical time interval between strokes of 50 to 300 ms. 91% of the return strokes have **negative charges** to ground, with the remaining 9% **positive** to ground.

When lightning strikes a metal-skinned vehicle such as a car or aircraft, the electricity flows on the outside of the metal skin, which acts like a Faraday cage. Thus if you are inside the vehicle, there is little direct hazard to health from the lightning, other than the surprise of the loud thunder and the temporary blindness by the flash. However, while the electricity flows around the metal skin, it can induce secondary currents within the wires of the car or plane, which can destroy electronic equipment and avionics.

Thunder

When the electricity travels through the ionized air path, it rapidly heats the air to incandescence, which we see as the flash. This hot air expands and causes a shock wave that changes to a sound wave. We hear the sound wave as thunder as it spreads from the bolt.

Sound travels much slower than the speed of light. The sound velocity s relative to the ground depends mostly on absolute temperature T and wind speed M, according to:

$$s = s_o \cdot \left(\frac{T}{T_o}\right)^{1/2} + M \cdot \cos(\phi) \qquad \bullet(15.13)$$

where ϕ is the angle between wind direction and sound-propagation direction, s_o = 343.15 m/s, and T_o = 293 K. At 20°C in calm air, the speed of sound relative to the ground equals s_o. Sound travels faster in warmer air, and when blown by a tail wind.

To estimate the distance to the lightning in kilometers, count the number of seconds between seeing the lightning and hearing the thunder, then divide that time by 3 . This algorithm assumes typical temperatures and calm winds,

When light waves or sound waves propagate from one medium into another, they **refract** (bend). The amount of bending depends on the **index of refraction**, n, which is proportional to the ratio of wave propagation speeds in the different media. This relationship is called **Snell's Law**, and is discussed in more detail in Chapt. 19.

Sound waves refract as they propagate through regions of varying temperature or wind. For calm conditions, the index of refraction is:

$$n = \sqrt{T_o / T} \qquad (15.14)$$

Solved Example

A thunderstorm approaching from the northwest creates a visible lightning stroke. There is a 5 m/s wind from the southeast blowing into the storm, and the temperature is 30°C. If you hear thunder 10 s after seeing the lightning, how far away was the stroke?

Solution

Given: T = 30°C = 303 K, M = 5 m/s, ϕ = 180°.
Find: x = ? km

Use eq. (15.13):

$$s = (343.15\text{m/s}) \cdot \left(\frac{303\text{K}}{293\text{K}}\right)^{1/2} + (5\text{m/s}) \cdot \cos(180°)$$

$= 349 - 5 = 344$ m/s

$x = s \cdot t$ = (344 m/s)·(10 s) = <u>**3.44 km**</u>

Check: Units OK. Physics OK.
Discussion: If the wind had been blowing in the opposite direction, the distance estimate would have been 3.54 km. Given typical errors in timing the thunder after the lightning, we see that the winds play only a small role.

where T_o = 293 K. Snell's law can be written as

$$n \cdot \cos(\alpha) = const \qquad (15.15a)$$

or

$$\cos\alpha_2 = \sqrt{T_2 / T_1} \cdot \cos\alpha_1 \qquad \bullet(15.15b)$$

where α is the elevation angle of the ray path of the sound, and subscripts 1 and 2 denote before and after refraction across a sharp change of temperature.

For gradual changes in temperature, the ray path equation is:

$$\Delta\alpha = \frac{\gamma}{2 \cdot T} \cdot \Delta x \qquad (15.16)$$

where T is the absolute temperature at ground level, $\gamma = -\Delta T/\Delta z$ is the lapse rate (assumed constant with height), and $\Delta\alpha$ is the angle change (radians) of a ray that moves a small increment of distance Δx horizontally. Normally the absolute temperature changes by only a small percentage with height, making $\Delta\alpha$ nearly constant.

To solve eq (15.16) iteratively, start with a known ray angle α at a known height z and known x location. Next, take a step in the horizontal by solving the following equations sequentially:

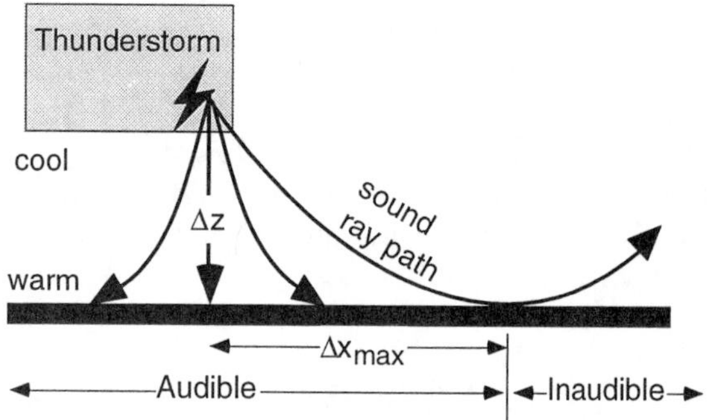

Figure 15.7
Thunder ray paths, and audibility at ground level in calm winds.

$$\begin{aligned} x_{new} &= x_{old} + \Delta x \\ \alpha_{new} &= \alpha_{old} + \Delta\alpha \\ \Delta z &= \Delta x \cdot \tan(\alpha) \\ z_{new} &= z_{old} + \Delta z \end{aligned} \tag{15.17}$$

Repeat for the next increment of Δx. A plot of each pair of z_{new} vs x_{new} is the ray path.

Sound rays bend toward the colder air. In the atmosphere, temperature usually decreases smoothly with height. Thus, a ray of thunder traveling downward from a cloud will tend to curve upward (Fig 15.7). Those sound rays that are pointed more vertically downward will reach the ground near the storm, allowing you to hear the thunder.

Other rays leaving the cloud at shallower angles will be refracted away from the ground, preventing persons on the ground from hearing it. The greatest distance Δx_{max} away from the storm at which you can still hear thunder (neglecting wind effects) is

$$\Delta x_{\max} \approx 2 \cdot \sqrt{\frac{T \cdot \Delta z}{\gamma}} \tag{15.18}$$

where Δz is the height of sound origin.

When wind is included, the ray path equation is a bit more complicated:

$$\frac{n(\cos\alpha)\left[1 - m^2 n^2 \sin^2\alpha\right] - mn^2 \sin^2\alpha}{1 + mn(\cos\alpha)\left[1 - m^2 n^2 \sin^2\alpha\right]^{1/2} - m^2 n^2 \sin^2\alpha} = const \tag{15.19}$$

where $m = M/s_o$ is the **Mach number** of the wind.

Solved Example (§)

For the previous solved example (except assume calm winds), if the lapse rate is adiabatic and the lightning altitude was 2 km, find the maximum distance of thunder audibility. Plot the path of this sound ray.

Solution

Given: T = 303 K, γ= 9.8 K/km, Δz = 2 km
Find: Δx_{max} = ? km, and plot z vs. x

Use eq. (15.18):

$$\Delta x_{\max} \approx 2 \cdot \sqrt{\frac{(303\text{K}) \cdot (2\text{km})}{(9.8\text{K/km})}} = \underline{\textbf{15.7 km}}$$

Use eq. (15.16):

$$\Delta\alpha = \frac{(9.8\text{K/km})}{2 \cdot (303\text{K})} \cdot (1\text{km}) = 0.01617 \text{ radians}$$

Next, use eqs. (15.17) in a spreadsheet, and plot backward from the ground, where initially: x = 0 km, z = 0 km, and α = 0 radians. This backwards trajectory will allow us to follow the sound ray path back to the lightning that caused it.

x (km)	alpha (rad)	Δz (km)	z (km)
0	0	0	0
1	0.01617162	0.01617303	0.01617303
2	0.03234323	0.03235452	0.04852754
3	0.04851485	0.04855295	0.09708049
4	etc.		

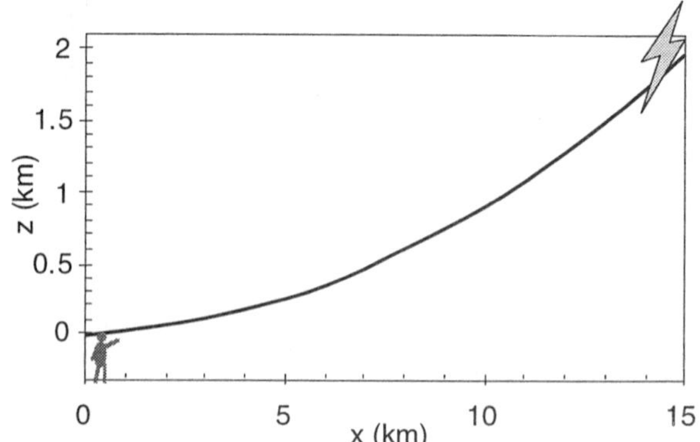

Check: Units OK. Physics OK.
Discussion: The ray path reaches 2 km altitude at a lateral distance slightly greater than 15 km, which agrees with x_{max}. This ray, which is tangent to the ground, happens to leave the cloud at an elevation angle of about 14°.

TORNADOES

Tornadoes form when rotation exists on a broad scale in thunderstorms. Rotation on the scale of the whole thunderstorm updraft is called a **mesocyclone**, and is approximately 10 to 15 km in diameter for the more-severe, long-lasting thunderstorms called **supercell** storms. Tornadic circulations are much smaller — about 10 to 1000 m in diameter.

Origin of Rotation

As of the writing of this book, there is still debate on the origin of rotation in thunderstorms, and how that rotation intensifies at low altitudes to form tornadoes. Apparently, the existence of mid-altitude vorticity is insufficient by itself to explain tornadoes. Also, there is some evidence that different mechanisms might trigger different tornadoes.

One theory assumes that there is weak cyclonic vorticity that already exists in the boundary layer of small synoptic-scale or mesoscale cyclones (**mesocyclone**). This vorticity is spun-up by vortex stretching as the boundary-layer air rises in convective updrafts that form in, or move over, the mesocyclone.

Another theory is that there is a cascade of vertical vorticity from larger to smaller-scale vortices. For example, weak vorticity from a large, mid-level mesocyclone could cascade to a low-altitude mesocyclone. From there, it cascades to an intensifying tornado vortex aloft, and finally to a tornado that touches the ground.

A third theory is that vertical shear of the horizontal wind in the boundary layer is responsible, by creating horizontal vorticity (rotation about a horizontal axis). This vorticity is later tilted into vertical vorticity (rotation about a vertical axis) by convective updrafts. By itself, this vorticity is too weak to explain the intense tornadoes that can form. Thus, stretching of the weak vertical vortex by convective updrafts is still needed to increase its vorticity.

A fourth, newer theory is that horizontal vorticity is generated baroclinicly by the rain-cooled, downdraft air that hits the ground and spreads out horizontally under the storm. Shear between this cool outflow air and the warmer ambient air that it is under-running creates horizontal vorticity. Recent observations show that there are many mesoscale frontal boundaries under thunderstorms that can generate horizontal vorticity, and that these vorticity sources are not always under the mesocyclone that is visible by radar. The horizontal vorticity is then tilted into vertical, either by the downdraft, the gust front, or other mesoscale boundary. Once in the vertical, the updraft stretches the vortex to spin-up the vorticity.

We will focus on the third theory here, which until recently was the accepted tornadogenesis paradigm. It utilizes horizontal wind vectors as can be plotted on a hodograph.

A **hodograph** indicates the horizontal wind direction and speed over a range of heights. For example, Fig 15.8 shows an idealized hodograph in the ambient environment outside of a thunderstorm. The dashed lines will be discussed first, where these are with respect to the ground, with wind components aligned to the east (U') and the north (V').

The wind vector for each height is drawn as an arrow, with the tail of every arrow at the origin. A line that connects the tips of the arrows represents the change of wind speed and direction with height. Usually, just this line is plotted on the hodograph without the arrows.

In many environments for severe storms, the wind **veers** with height; namely, it turns clockwise, such as illustrated in Fig 15.8. This figure is typical, in that the low altitude winds are from the southeast, while the winds veer with increasing height to become more westerly. The thunderstorm itself has significant movement, and is indicated with the black dot in that figure. Usually, this storm-movement dot is on the concave side of the hodograph curve.

Define a new coordinate system, the origin of which moves with the storm. Wind vectors can then be drawn (black arrows) that are relative to the storm,

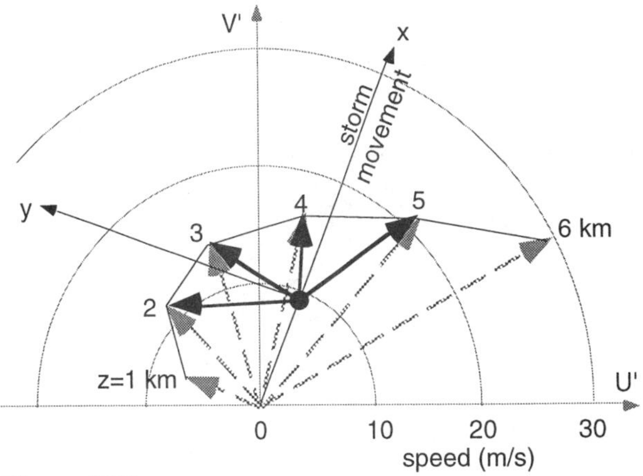

Figure 15.8
Hodograph of ambient wind near the thunderstorm (thin solid line). Gray dashed lines are wind vectors relative to the ground. Thick black arrows are wind vectors relative to the storm movement (indicated by the black dot).

rather than relative to the ground. For example, we see that the wind at altitude 2 km is to the left of the y-axis, while the 3 km wind is slightly to the right.

These two wind vectors are replotted in Fig 15.9, again relative to the storm movement (the black dot). The change of wind vector with height is like a vertical shear of the U-wind, as plotted at the far end of the rectangle. Recall from Chapter 11 that this shear defines a vorticity. You can think of this vorticity as a rotating vortex tube. Because the axis of rotation of the vorticity is aligned with the average wind direction, V, in the horizontal, this vorticity is called **streamwise vorticity**.

In the absence of a thunderstorm, this vortex tube lays flat on the ground. However, thunderstorm convection (updrafts) can distort the vortex tube, as plotted in Fig 15.9. In this situation, the wind V is forced to bend upward to become W, and the vortex tube bends with it. This process tilts the horizontal vorticity into vertical vorticity, ζ.

The region of rotating updraft air in a thunderstorm is the **mesocyclone** (lightly shaded in Fig 15.9). Mesocyclones are precursors to tornadoes. In summary, these mesocyclones can form from vertical shear of the horizontal wind, which is quite common in the environment surrounding and prior to thunderstorms.

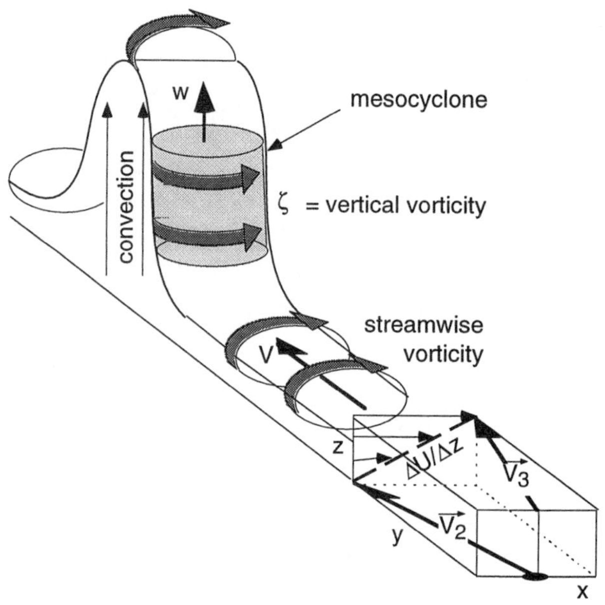

Figure 15.9
Vertical shear $\Delta U/\Delta z$ of the horizontal wind (between V_2 and V_3) creates streamwise vorticity. When the mean wind V (relative to storm movement) is bent upward by rising convective motions of the developing thunderstorm, V changes into W, and streamwise vorticity is tilted into vertical vorticity ζ. This rotating updraft (shaded) is the mesocyclone.

Solved Example

Plot a hodograph for the following winds:

z (km)	U (m/s)	V (m/s)
1	-3	3
2	0	6
3	5	5
4	10	2

Solution

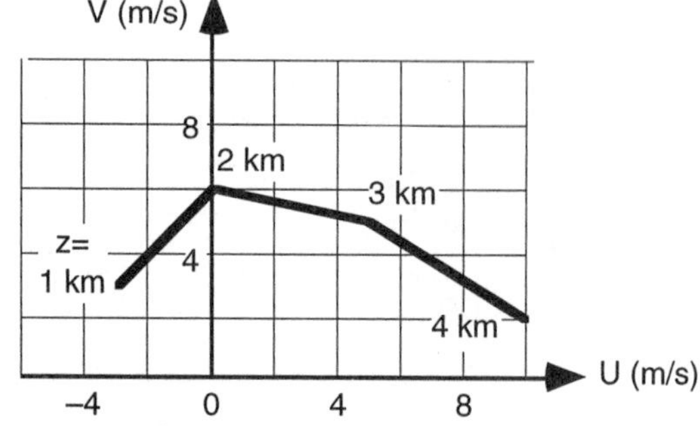

Discussion: This exhibits the vertical shear similar to that in Fig 15.8, suggesting that it could be tilted into the vertical, and then stretched by the updraft to produce intense tornadic vorticity.

Mesocyclones and Helicity

In the mesocyclone, air is both rotating and rising (Fig 15.10a) resulting in spiral or helical motion (Fig 15.10b).

Helicity H is a measure of the corkscrew nature of the winds. It is a scalar, defined as:

$$H = U \cdot \left[\frac{\Delta W}{\Delta y} - \frac{\Delta V}{\Delta z}\right] + V \cdot \left[-\frac{\Delta W}{\Delta x} + \frac{\Delta U}{\Delta z}\right] + W \cdot \left[\frac{\Delta V}{\Delta x} - \frac{\Delta U}{\Delta y}\right] \tag{15.20}$$

For environmental air possessing only vertical shear of the horizontal winds, eq. (15.20) reduces to

$$H \approx V \cdot \frac{\Delta U}{\Delta z} - U \cdot \frac{\Delta V}{\Delta z} \tag{15.21a}$$

This is the portion of helicity associated with **streamwise vorticity**, and is easily calculated from the information in a hodograph.

After convection tilts the vortex tube to form the **mesocyclone**, most of the helicity is moved into the last term of eq. (15.20):

$$H = W \cdot \left[\frac{\Delta V}{\Delta x} - \frac{\Delta U}{\Delta y}\right] = W \cdot \zeta \qquad \bullet(15.21b)$$

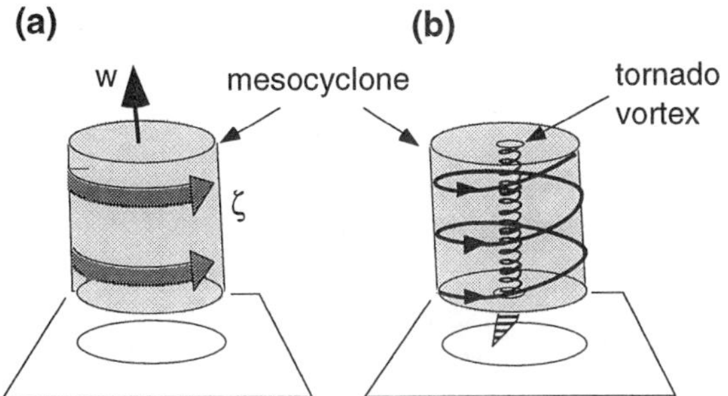

Figure 15.10
(a) Updraft and rotation in a mesocyclone combine to give (b) helical flow. The tornado vortex has a much smaller radius than the mesocyclone.

Larger values of the background helicity (eq. 15.21a) can lead to larger values of mesocyclone vorticity (eq. 15.21b). This causes some thunderstorms to become supercell storms, which are more likely to form tornadoes, hail, and strong straight-line winds. Hence, meteorologists calculate helicity from hodographs as a tool to forecast tornadic storms.

Solved Example
(a) Find the helicity associated with these winds:

z (km)	U (m/s)	V (m/s)
1	0	6
2	4	8

(b) If helicity is preserved when streamwise vorticity is tilted into a mesocyclone of updraft 10 m/s, find the vertical vorticity of the mesocyclone.

Solution
Given: (a) see above (b) $W = 10$ m/s
Find: (a) $H = ?$ m·s^{-2} (b) $\zeta = ?$ s^{-1}

(a) Use eq. (15.21a) with winds averaged between the two heights where needed:

$$H \approx (7\text{m/s})\cdot\frac{(4-0\text{m/s})}{(2-1\text{km})} - (2\text{m/s})\cdot\frac{(8-6\text{m/s})}{(2-1\text{km})}$$

$= \underline{\mathbf{0.024}}$ m·s^{-2}

(b) Use eq. (15.21b):

$$(0.024\text{m}\cdot\text{s}^{-2}) = (10\text{m/s})\cdot\zeta$$

$\zeta = \underline{\mathbf{0.0024}}$ s^{-1}

Check: Units OK. Physics OK.
Discussion: This vorticity is roughly 100 times the vorticity of a synoptic-scale cyclone. Time-lapse movies of mesocyclone thunderstorms show visible rotation.

Tornadoes and Swirl Ratio

Once a mesocyclone has formed, it is still necessary to concentrate the rotation to form a tornado (Fig 15.10b). This concentration is perhaps formed by the vertical stretching of the vortex in the updraft, similar to the water vortices that form in a bath tub when the drain is opened.

The characteristics of the resulting tornado, however, can vary depending on the dimensionless **swirl ratio**:

$$S = \frac{M_{tan}}{W} \qquad \bullet(15.22)$$

where M_{tan} is the tangential component of velocity around the perimeter of the updraft region, and W is the average updraft velocity. For a cylindrical mesocyclone, this can be rewritten as

$$S = \frac{R_o \cdot M_{tan}}{2 \cdot z_i \cdot M_{rad}} \qquad \bullet(15.23)$$

where R_o is the mesocyclone radius, z_i is the boundary-layer depth, and M_{rad} is the radial (inflow) component of velocity. Typically, $R_o \cong 2 \cdot z_i$; thus, $S \cong M_{tan} / M_{rad}$.

For small swirl ratios (0.1 to 0.5), tornadoes simulated in laboratories have a single, well-defined funnel (Fig 15.11a) that looks somewhat smooth (laminar). The tornado core has low pressure, updraft, and has a core radius R_1 of 5% to 25% of the mesocyclone updraft radius, R_o.

At larger swirl ratios, a turbulent downdraft begins to form in the top part of the tornado core, which has a much larger diameter than the laminar bottom (Fig 15.11b). The bottom of the tornado still has a laminar updraft core. The updraft and downdraft portions of the core meet at a vertical stagnation point called the **breakdown bubble**.

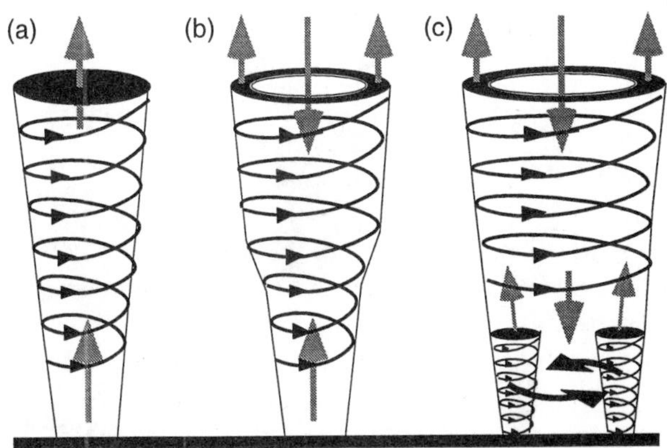

Figure 15.11
Tornadoes with swirl ratios (S) of: (a) $S = 0.3$; (b) $S = 0.8$; and (c) $S = 2.0$.

As the swirl ratio approaches 1, the turbulent downdraft extends downward, with the breakdown bubble eventually reaching the ground. As the swirl ratio continues to increase, the tornado looks quite turbulent and has a large diameter, with a downdraft core surrounded by a helix of rotating, rising air.

For even larger swirl ratio, interlocking **multiple vortices** form and chase each other around the perimeter of the large tornado downdraft (Fig 15.11c). The process of changing from a single laminar funnel to a turbulent, multiple-vortex tornado, is called **tornado breakdown**.

> **Solved Example**
> What is the swirl ratio for a mesocyclone with radial velocity 2 m/s, tangential velocity 10 m/s at radius 2000 m in a boundary layer 1 km deep?
>
> **Solution**
> Given: M_{tan} = 10 m/s, M_{rad} = 2 m/s, R_o = 2000 m, z_i = 1 km
> Find: S = ? (dimensionless)
>
> Use eq. (15.23):
> $$S = \frac{(2000\text{m})\cdot(10\text{m / s})}{2\cdot(1000\text{m})\cdot(2\text{m / s})} = \underline{5.0}$$
>
> **Check**: Units OK. Physics OK.
> **Discussion**: Creates a multiple vortex tornado.

Core Pressure Deficit

Centrifugal force is large for winds rotating around a tornado. However, the horizontal size of tornadoes is so small that Coriolis force is negligible. Hence, the equations of motion for the tangential velocity M_{tan} around a tornado describe a **cyclostrophic balance** between pressure gradient force and centrifugal force (Chapter 9, eq. 9.33), which in radial coordinates is:

$$0 = \underbrace{-\frac{1}{\rho}\frac{\Delta P}{\Delta R}}_{\textit{pressure gradient}} + \underbrace{\frac{M_{\tan}{}^2}{\Delta R}}_{\textit{centri-fugal}} \qquad (15.24)$$

where $\Delta R = R_o$ is the radius of the tornado, and ΔP is the pressure deficit between the tornado core and the surrounding environment.

Solving for the pressure deficit gives:

$$\Delta P \approx \rho \cdot M_{\tan}{}^2 \qquad \bullet(15.25)$$

which is a special form of **Bernoulli's equation**.

Physically, this equation states that low pressure in the tornado core tends to suck air into it, but is balanced by centrifugal force that tends to throw air out from the center of rotation. The net result is no radial flow across the tornado perimeter, which acts like the walls of a vacuum-cleaner hose.

At the bottom of the tornado, friction near the ground reduces the wind speed, preventing centrifugal force from balancing pressure gradient force. As a result, the low pressure inside the tornado sucks air into it, analogous to air being sucked into the end of a vacuum-cleaner hose.

> **Solved Example**
> What core pressure deficit is expected in a tornado having tangential winds of 100 m/s?
>
> **Solution**
> Given: M = 100 m/s.
> Find: ΔP = ? kPa
>
> Use eq. (15.25), and assume: $\rho = 1\ \text{kg/m}^3$:
> $$\Delta P \approx (1\text{kg}\cdot\text{m}^{-3})\cdot(100\text{m/s})^2 = 10^4\ \text{kg·m}^{-1}\text{·s}^{-2} = \underline{\mathbf{10\ kPa}}$$
>
> **Check**: Units OK. Physics OK.
> **Discussion**: This is quite a large pressure deficit — roughly 10% of the sea-level pressure. However, most tornadoes are not this violent. Typical tangential winds of 60 m/s or less would correspond to core pressure deficits of 3.6 kPa.

Tornado Intensity Scales

Two tornado intensity scales are used to classify the severity of tornadoes. One is the **TORRO** (T) scale, developed by the Tornado and Storm Research Organization in Europe. The other is the **Fujita** (F) scale, developed in the USA by Ted Fujita, and adopted by the US. National Weather Service.

The TORRO scale is based on tangential wind speed M_{tan}. The lower wind-speed bound for each T range (such as for T3 or T5) is defined approximately by:

$$M_{\tan} = a\cdot(T+4)^{1.5} \qquad (15.26a)$$

where T is a dimensionless TORRO value, and a = 2.365 m/s. It is related to the Beaufort (see Chapt. 16) wind scale (B) by:

$$B = 2\cdot(T+4) \qquad (15.26b)$$

Table 15-1. TORRO tornado intensity.

Scale	Speed (m/s)	Tornado Description
T0	17 – 24	Light
T1	25 – 32	Mild
T2	33 – 41	Moderate
T3	42 – 51	Strong
T4	52 – 61	Severe
T5	62 – 72	Intense
T6	73 – 83	Moderately Devastating
T7	84 – 95	Strongly Devastating
T8	96 – 107	Severely Devastating
T9	108 – 120	Intensely Devastating
T10	121 – 134	Super

The TORRO scale ranges from T0 for a very weak tornado through T10 for a super tornado (Table 15-1), and even larger T values are possible with stronger winds. T values do not depend on damage caused, although stronger winds usually cause more damage.

The Fujita scale is based on damage caused by the tornado. It ranges from F0 for slight damage, to F12 for damage caused by wind speeds of Mach 1. The lower wind speed M_{tan} bound for each F range (such as F2 or F3) is defined by:

$$M_{\tan} = a \cdot (2F + 4)^{1.5} \tag{15.27}$$

where a = 2.25 m/s.

Tornadoes of F5 intensity and greater all cause nearly total destruction of most buildings and trees, so it is virtually impossible to identify tornadoes of F6 or greater based on damage surveys. Thus, the normal range of reported Fujita intensity is F0 to F5 (Table 15-2).

Table 15-2. Fujita tornado intensity.

Scale	Speed (m/s)	Damage Classification	Relative Frequency
F0	18 – 32	Light or Gale	29%
F1	33 – 50	Moderate	40%
F2	51 – 70	Significant	24%
F3	71 – 92	Severe	6%
F4	93 – 116	Devastating	2%
F5	117 – 142	Incredible	< 1%

DOPPLER RADAR

Doppler weather radars can not only measure the intensity of rainfall by the amount of reflected microwave energy, they can also measure the speed of the air toward or away from the radar. This latter feature allows the radar to detect motions within the thunderstorm that might be associated with tornadoes.

To do this, the radar measures the frequency shift $\Delta\nu$ of the received signal compared to the transmitted frequency, ν. When the microwaves from the radar bounce off of droplet-laden air that is moving away from the radar, the frequency becomes slightly lower. Conversely, higher frequencies are found for air moving toward the radar.

The magnitude of frequency shift is given by the Doppler equation:

$$\Delta\nu = \frac{2 \cdot M_r}{\lambda} \tag{•15.28}$$

where M_r is the radial velocity of the air relative to the radar, and λ is the wavelength of the microwaves. Many Doppler radars transmit wavelengths of λ = 10 cm. The wavelength and frequency are related by the speed of light c = 3x10^8 m/s according to:

$$\nu = c \ / \ \lambda \ . \tag{•15.29}$$

Solved Example

What Doppler shift is expect for a radial velocity of 100 m/s tangent to a tornado?

Solution

Given: M_r = 100 m/s, λ = 0.1 m
Find: $\Delta\nu$ = ? s^{-1}

Use eq. (15.28): $\Delta\nu = \frac{2 \cdot (100\text{m/s})}{(0.1\text{m})}$ = <u>**2000**</u> s^{-1}

Check: Units OK. Physics OK.
Discussion: Note that the original transmitted frequency is given by eq. (15.29): $\nu = c / \lambda$ = (3x10^8 m/s) / (0.1 m) = 3x10^9 s^{-1} . Thus, the frequency shift is less than one part per million compared to the original frequency. For this reason, Doppler radars need very sensitive detectors, which are quite expensive.

SUMMARY

Once triggered, thunderstorms (**cumulonimbus** clouds) can often sustain themselves within an unstable environment with strong wind shear. Their fuel is mostly the latent heat released during condensation of humid air drawn aloft from the boundary layer. Their ability to grow depends on the conditional instability of the environment, as measured by the convective available potential energy (CAPE).

Wind shear in the ambient environment can be converted into rotating mesocyclones within the storm updraft. From these mesocyclones comes the most violent weather, including tornadoes, hail, and heavy rains. The rotation within thunderstorms can be studied with Doppler radar, and can be analyzed using helicity and swirl ratio. Tornado intensity can be classified with TORRO and Fujita scales.

Downbursts are formed by precipitation loading and evaporation. Downbursts change into gust fronts upon hitting the ground. The resulting straight-line winds are particularly hazardous to mobile homes, while the downbursts are a flight hazard.

The formation of lightning is not completely understood, but the characteristics of lightning can be measured. Thunder propagates as a sound wave, and can be reflected and refracted as other wave phenomena.

Threads

Thunderstorms form in air that has conditionally unstable static stability (Chapt. 6), with tops rising to the tropopause (Chapt. 1). Solar radiation (Chapt. 2) during the day adds heat (Chapt. 3) to the boundary layer (Chapt. 4), and evaporates water (Chapt. 5) from the surface. This warm, humid boundary layer air serves as the fuel for thunderstorms, which can be triggered by fronts from cyclones (Chapt. 13), and by drylines (Chapt 12). Once formed, thunderstorms are steered by mid- and upper-tropospheric winds that make up the jet stream (Chapt. 11) in the general circulation.

Condensation of water vapor in the buoyant, rising air makes the cumulonimbus cloud (Chapt. 7), and the cold-cloud ice process in the top of the cloud contributes to rapid precipitation formation (Chapt. 8). Thunderstorm rain drops have fairly large diameter, which makes them ideal for contributing to rainbows (Chapt. 19). Weather radars can measure both rainfall intensity (Chapt. 8) and Doppler velocity.

The vertical and horizontal winds are driven by pressure-gradient and buoyancy forces (Chapt. 9), and rotating tornadoes are also influenced by centrifugal force. Boundary layer wind shear (Chapt. 4) creates vorticity (Chapt. 11) that can be tilted into the vertical vorticity of mesocyclones. The strong winds hitting buildings creates a damaging dynamic pressure (Chapt. 10) that can be estimated with Bernoulli's equations. Bernoulli's equation also relates tornado core pressure to tangential wind speed.

With fine-enough grid spacing, thunderstorm clouds and winds can be predicted numerically (Chapt. 14) by computer. Hurricane (Chapt. 16) spiral bands consist of lines of thunderstorms. Thunderstorms can transport air pollutants (Chapt. 17) vertically from the boundary layer and deposit them higher in the troposphere and lower stratosphere, and thunderstorm rain can scrub out some of the pollutants. Changes in global patterns of aerosol concentrations might alter thunderstorm formation. Climate change (Chapt. 18) might alter the atmospheric stability, thereby changing the frequency of thunderstorms.

Science Graffito

Mobile homes and trailers are so vulnerable to thunderstorm and tornadic winds that, to many non-scientists, those homes seem to attract tornadoes. This has lead disaster preparedness officers in some towns to suggest putting mobile home parks outside of town, to serve as "tornado bait" to lure tornadoes away from town.

EXERCISES

Numerical Problems

N1. Given the following sounding. Find the CAPE and the updraft velocity assuming no friction. Hint, use a thermo diagram.

z (km)	T (°C)	U (m/s)	V (m/s)
10	–45	50	3
8	–40	30	5
6	–30	10	5
4	–10	5	7
2	+10	0	7
1.5	5	–3	5
0.5	15	–5	3
0	25	–1	1

Use a surface dew-point temperature (°C) of:

a. 25 b. 20 c. 16 d. 12
e. 10 f. 8 g. 6 h. 4

N2. Find the total virtual temperature, the buoyancy force per unit mass, and the downdraft velocity at the bottom of a 2 km fall, given:

	r_{liq} (g/kg)	T (°C)	P (kPa)
a.	10	10	80
b.	5	10	80
c.	5	20	90
d.	1	2	70
e.	10	15	80
f.	5	15	80
g.	2	20	90
h.	1	5	70

N3. If instead of falling, all the liquid water of the previous exercise evaporates, find the final air temperature.

N4. A downburst of radius 0.6 km and speed 8 m/s descends through an environment of 30°C. Find the gust front advancement speed and depth at a distance of 5 km from the downburst center, given a temperature deficit (°C) of

a. 1 b. 2 c. 3 d. 4
e. 5 f. 6 g. 7 h. 8

N5(§). For the previous problem, plot curves of advancement speed and depth versus gust-front distance from the downburst center.

N6. What voltage difference is needed to cause lightning of length (km)

a. 1 b. 2 c. 3 d. 4
e. 5 f. 6 g. 7 h. 0.5

N7. What is the probability of power lines being struck by lightning that generates a current (kA) in excess of:

a. 5.5 b. 7 c. 10 d. 50
e. 100 f. 200 g. 500 h. 1000

N8. When, after a lightning strike, will the surge in a power line decay to the following fraction of its peak value:

a. 0.5 b. 0.2 c. 0.1 d. 0.05
a. 0.02 b. 0.01 c. 0.005 d. 0.002

N9. How intense is the lightning if a lightning detection receiving station 300 km away detects an electric field strength of (V/m):

a. –1 b. –2 c. –3 d. –4
e. –5 f. –10 g. –20 h. –0.1

N10(§). For calm conditions, draw a curve of range vs. time between seeing lightning and hearing thunder. Draw additional curves for head and tail winds (m/s) of:

a. 1 b. 2 c. 5 d. 10
e. 15 f. 20 g. 25 h. 30

N11. What is the maximum distance you can expect to hear thunder from lightning 5 km high in calm conditions with average temperature 300 K, if the lapse rate (°C/km) is:

a. 9.8 b. 8 c. 6.5 d. 5
e. 4 f. 2 g. 1 h. –2

N12(§). Plot a hodograph for the winds of exercise N1.

N13§). Calculate the helicity between every pair of levels in the sounding of exercise N1.

N14. In a synoptic-scale cyclone, we might expect vorticities of 10^{-5} s^{-1} and average updraft velocities of 0.1 m/s. Find the helicity.

N15. Find the Doppler shift of 10 cm microwaves for radial velocities (m/s) of:

a. –100 b. –75 c. –50 d. –20
e. 100 f. 75 g. 50 h. 20

Understanding & Critical Evaluation

U1. Look at a thermodynamic diagram, and comment about how the amount of conditional instability varies with general surface temperature. Namely, what amount of instability might be expected in the tropics vs. in the arctic?

U2(§). For a lightning stroke 2 km above ground in a calm adiabatic environment of average temperature 300 K, plot thunder ray paths leaving downward from the lightning stroke at elevation angles from –90° to 90°, every 10°.

U3. In a synoptic-scale anticyclone, is the helicity positive or negative?

U4. Suppose the swirl ratio is 1 for a tornado of radius 300 m in a boundary layer 1 km deep. Find the radial velocity and core pressure deficit for each tornado intensity of the

a. Fujita scale.

b. TORRO scale.

U5. On a thermodynamic diagram, plot a conditionally unstable sounding, and compare it to soundings that are absolutely statically stable and statically unstable.

U6. a. Can thunderstorms fully utilize all the CAPE?

b. What are limitations on use of the CAPE equations?

c. Why might vertical velocity in thunderstorms be not as great as that given by eq. (15.3)?

U7. Devise an equation for CAPE that utilizes virtual potential temperature differences, which is a better measure of buoyancy.

U8. Suppose that a downburst of cold air descends and warms adiabatically. As this air parcel descends, it eventually reaches an altitude where it is no longer colder than the environment. Neglecting air drag, calculate the final velocity at that neutrally buoyant altitude, assuming that buoyancy force decreased linearly as the parcel descended.

U9. What is the name of the temperature to which raindrops cool when they are falling through unsaturated air?

U10. How does the shape of the lightning surge curve change with changes of parameters T_1 and T_2?

U11. How do gust front depth and advancement speed vary with:

a. virtual temperature of the outflow air?

b. downburst velocity?

Plot the variation of gust front depth and advancement speed with distance R of the gust front from the center of the downburst.

U12. a. What is the probability that lightning current will be exactly 10 kA?

b. What is the probability that lightning current will be between 9 and 11 kPa?

U13. Is it possible for the speed of sound relative to the ground to be zero? Under what conditions would this occur? Are these conditions likely?

U14. If there was no refraction of sound waves in air, then what is the maximum distance from lightning that one could hear thunder?

U15. Given the typical hodograph of ambient winds near a thunderstorm, is it possible for meso-anticyclones to develop in addition to, or instead of, mesocyclones? Discuss.

U16. Is it possible for anticyclones to have positive helicity? Discuss.

U17. The denominator of the swirl ratio can be interpreted physically as the volume of air flowing in toward the tornado from the ambient environment. What is a physical interpretation of the numerator?

U18. Show how the relationship between core pressure deficit and tangential velocity around a tornado can be found from Bernoulli's equation. What assumptions are needed, and what are the resulting limitations?

U19. What is the maximum vertical velocity of large falling rain drops relative to the ground, knowing that air can be dragged along with the drops as a downburst?

U20. If there were no drag of rain drops against air, could there still be downbursts?

U21. Can the TORRO scale be used to classify the intensity of winds triggered by meteor strikes and volcanic eruptions? Discuss.

U22. Can thunderstorms form over Antarctica? Why?

Web-Enhanced Questions

Search the web for tutorials on thunderstorm processes to answer the following questions. Although it is dangerous to list web sites because they are so short-lived, the Univ. of Illinois has (as of the writing of this book) a thunderstorm tutorial as part of their ww2010 project:
http://ww2010.atmos.uiuc.edu/(Gh)/guides/mtr/svr/home.rxml

W1. Thunderstorm hazards:
a. List 6 of the hazards of thunderstorms.
b. Which hazard causes the most deaths in N.America?
c. Which hazard causes the most monetary losses?

W2. Microburst behavior:
a. What is the difference between a **macroburst** and a microburst?
b. What are typical and maximum winds observed?
c. List the 3 stages of a microburst, and indicate which one causes the greatest surface gusts.
d. What is the typical duration of a microburst?
e. How are downbursts hazardous to pilots, and what can pilots do about it?

W3. Storm types:
a. Name 2 types of multicell thunderstorms.
b. What is the typical lifetime of a single-cell thunderstorm?
c. What is another name for a single-cell thunderstorm?

W4. a. Rank the following storms from least to greatest danger:
(i) multicell cluster (ii) single cell
(iii) supercell (iv) multicell line.
b. Next, rank these same storms from worst to best predictability.

W5. Supercell thunderstorms:
a. What makes a supercell storm different from a single-cell storm?
b. Where are supercell storms found relative to the squall line?
c. List 3 variations or types of supercell storms.
d. Where in N. America are supercell storms typically found?
e. What makes supercell storms unique?

W6. List the evolution stages as a multicell storm evolves into a supercell storm.

W7. Define the following:
a. **gustnado** b. **rainfoot** c. **dustfoot**

For the next questions, also refer to a Storm Spotter Glossary on the web. As of the writing of this book, the following glossary was available from the NWS:
http://www.nssl.noaa.gov/~nws/branick2d.html
(Also, the figures at the end of this glossary are particularly good.)

W8. What do the following acronyms mean?
a. LEWP b. WER c. BWER d. RFD.

W9. List 4 components that storm spotters look for, to gauge the type of thunderstorm and its development.

W10. For each combination of strong/weak updraft/downdraft, list the type of storm or hazard associated with it.

W11. What is the Lemon technique?

W12. Do environments with wind shear cause unorganized or organized storms? Discuss.

W13. What clues can you use to discriminate between gust fronts and wall clouds?

W14. What is the difference between scud clouds and virga?

W15. a. Under which part of the thunderstorm are wall clouds expected?
b. What do rotating wall clouds indicate?

W16. What happens to a supercell during cold-air undercutting?

W17. a. What is a mesocyclone?
b. Draw a map of its attributes.
c. Describe the evolution of a tornadic mesocyclone.

W18. Why can severe supercell storms move to the right of the steering-level winds?

W19. What is a cyclic storm?

W20. Search the web for upper air soundings for RAOB sites near current severe weather. Find a site that includes textual information listing various stability and storm indices, along with the plotted sounding. What info is in this text that is relevant to severe storms?

W21. From what web site can you get current weather watches and warnings for thunderstorms and tornadoes, for:

a. USA b. Canada c. Europe
d. Japan e. China f. Australia
g. Other country specified by your instructor?

W22. Tornado intensity scales:

a. Search for web sites that describe the TORRO tornado intensity scale. (http://www.torro.org.uk/)

b. Make a table comparing the TORRO scale for tornado classification with the Fujita intensity scale.

c. What percentage of tornadoes are in the strong to violent (F4-F5) category, and what percentage of all tornado deaths do they cause?

W23. Search the web for tornado chaser sites. In particular, look for information from the National Severe Storms Lab (NSSL). List 5 storm chasing tips that you consider most important.

W24. Search the web for information that lists the main medical/physiological effects to a human that is struck by lightning.

W25. From the web, find at what altitude in supercells that tornadoes first form.

W26. Tornado Alley:

a. Where is "tornado alley"?

b. Why are there so many tornadoes in tornado alley?

W27. Search the web for info on the general trend in tornado deaths during the 1900s, and how they compare to the trend in tornado reports.

W28. Search the web to find the probability that any one house will be struck by a tornado during 1 year.

W29. Lightning detection.

a. Search the web for information on lightning detection networks. (Many such networks are private, and provide the lightning strike maps only to paying customers. However, they might have public web pages with propaganda that describe their networks.)

b. Search the web for info on satellite platforms for detecting lightning.

W30. Search the web for information and images of thunderstorm **sprites** (electrical discharges visible as fleeting red glows in the mesosphere 30 to 90 km above ground, over thunderstorm anvils). What are typical numbers of sprites observed above thunderstorms?

Synthesis Questions

S1. Suppose that atmospheric boundary layers did not have a capping inversion. Discuss changes to thunderstorm intensity and frequency, if any.

S2. If there were no thunderstorms on earth, how would the global general circulation be different, if at all?

S3. What if thunderstorms could penetrate through both the troposphere and stratosphere. How would average vertical atmospheric structure be different, if at all?

S4. Suppose that the "cold cloud" process for creating large cloud hydrometeors did not exist. Would thunderstorms exist? If so, what would be their characteristics?

S5. What if there were no mountain ranges on earth. How would thunderstorm frequency, location, and characteristics be different, if at all? How would it change the local weather at your location, if at all?

S6. Suppose that latent heat of condensation cools the air instead of warms it. What would be the characteristics of thunderstorms?

S9. Which types of aircraft would likely have greater difficulty with downbursts: large, heavy commercial jet aircraft; or small, light, general-aviation planes?

S10. Discuss 5 ways to harness the power of lightning.

S11. Under what atmospheric conditions could sound waves from thunder be focused toward one point on the ground. If those conditions could be made to occur, what would happen at that point? Could this be used as a weapon?

S12. Is there an upper limit to the tangential speed of tornadoes? If so, what is it, and why? Discuss the dynamics that would be involved.

S13. Could two Doppler radars scanning the same domain from different locations be used to estimate the 3-D wind field? If so, what assumptions would be needed, and what would be the limitations?

HURRICANES

CONTENTS

16 Hurricanes (Fig 16.1) form over tropical oceans where the sea-surface temperature is at least 26°C (Fig 16.2) throughout a depth of 60 m or more. At the equator there is no Coriolis force; hence there is no rotation available to be concentrated into hurricanes. Thus, hurricanes are most likely to form between 10° to 30° latitude, during autumn when ocean temperatures are greatest.

At the center of the storm is a cloud-free eye of relatively low surface pressure, warm temperature, and subsiding air. Around the eye is a conical eye wall of extremely intense thunderstorms that extend into the base of the stratosphere – about 15 km high. Bands of thunderstorms spiral out from the eye wall.

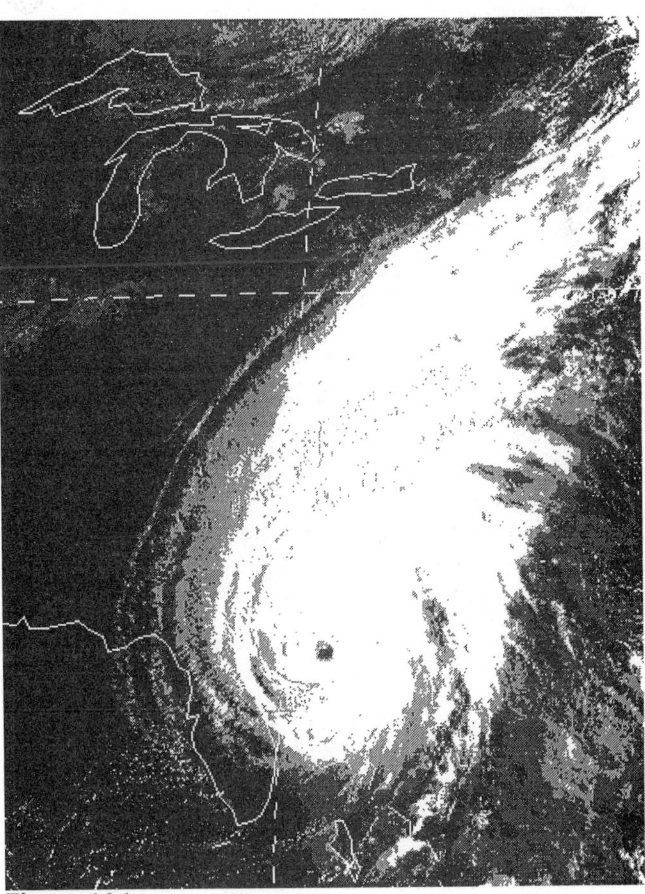

Figure 16.1
Visible satellite image of category 4 hurricane Floyd, at 12 UTC on 15 Sep 1999, in the Atlantic Ocean just east of Florida, USA.

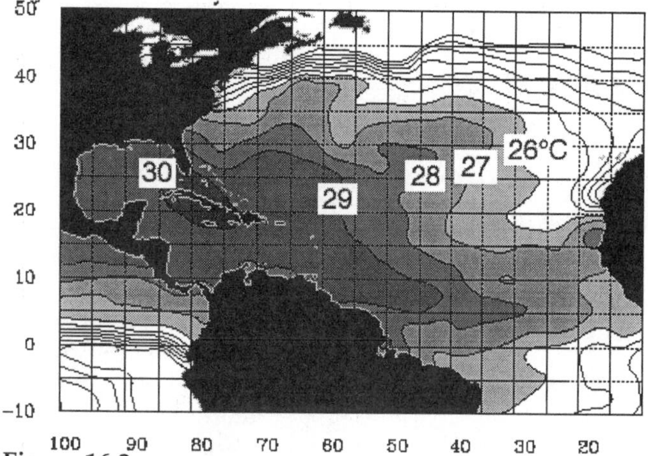

Figure 16.2
Sea surface temperature (°C) in the Atlantic Ocean averaged over the week of 5 to 11 Sep 1999, just prior to hurricane Floyd, a category 4 hurricane. Shaded regions are warmer than 26°C, the isotherm increment is 1°C, and the isotherm range is 20 to 30°C. [Adapted from an image from the National Hurricane Center.]

A cross section through an idealized hurricane is shown in Fig 16.3. Boundary-layer air is drawn toward the eye, and picks up significant amounts of water vapor due to evaporation as it blows over the increasingly rough ocean. Water vapor is the main source of energy for the storm. The cyclonically-spiraling air never reaches the eye, but rises in the eye wall, producing extremely heavy rain.

The rising air hits the stratosphere where it spreads out, creating the anticyclonic outflow that is observed by satellite. The outflow eventually subsides back toward the sea surface.

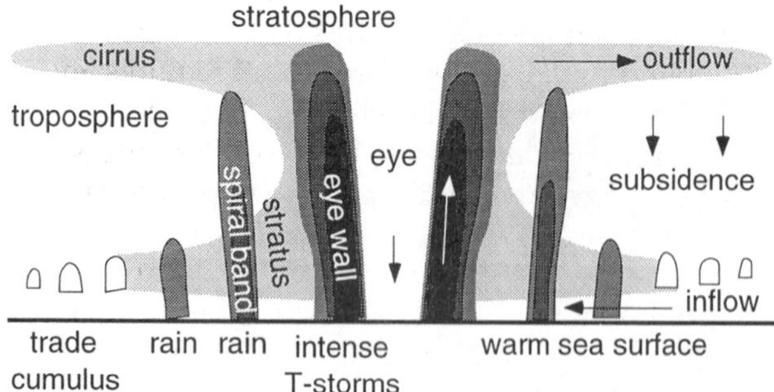

Figure 16.3
Idealized anatomy of a hurricane. Darker shading indicates more intense radar echoes in thunderstorms. Depth ≅ 15 km. Width ≅ 1500 km. Trade cumulus are cumulus clouds in the trade-wind region of the general circulation.

DYNAMICS

Initial Spin Up

While tornadoes form by tilting of the horizontal wind shear, such is not the case for hurricanes. For hurricanes, rotation is associated with convergence into a low center. During this convergence, angular momentum associated with the earth's rotation is concentrated into angular momentum associated with hurricane winds. Hurricanes are also called **tropical cyclones** because of their cyclonic rotation.

The absolute **angular momentum** AAM is:

$$AAM = M_{\tan} \cdot R + 0.5 \cdot f_c \cdot R^2 \qquad \bullet(16.1)$$

where M_{tan} is the tangential velocity at radius R from the hurricane center, and f_c is the Coriolis parameter. At latitudes typical of hurricanes (20°), $f_c \cong 0.00005 \text{ s}^{-1}$. The adjective "absolute" means that this quantity includes both the relative angular momentum of the hurricane as well as the background angular momentum associated with the earth's rotation.

If the hurricane were to experience no friction against the surface, then angular momentum would be conserved while air converges. Thus, tangential velocity of the hurricane winds at some small radius R_{final} can be created even if the air at some larger initial radius R_{init} possesses no rotation:

$$M_{\tan} = \frac{f_c}{2} \cdot \left(\frac{R_{init}^2 - R_{final}^2}{R_{final}} \right) \qquad (16.2)$$

In real hurricanes, friction cannot be neglected. Hence, actual tangential winds are less than those calculated with eq. (16.2).

Insipient hurricanes are easily destroyed by vertical shear of the ambient horizontal wind. For a hurricane to successfully form, background winds within about 4° latitude of the storm must have less than 8 m/s shear over the depth of the troposphere.

Solved Example

In initially rotationless air at radius 400 km from a tropical-cyclone center, find its tangential velocity after it converges to a radius of 100 km. The cyclone is centered at 10° latitude.

(continued next column)

Solved Example *(continuation)*

Solution

Given: $\phi = 10°$, $R_{init} = 400$ km, $R_{final} = 100$ km

Find: $M_{tan} = ?$ m/s

First, find the Coriolis parameter:

$f_c = 2 \cdot \Omega \cdot \sin\phi = (1.458 \times 10^{-4} s^{-1}) \cdot \sin(10°)$

$= 0.0000253 \text{ s}^{-1}$

Use eq. (16.2):

$$M_{\tan} = \frac{(0.0000253\text{s}^{-1})}{2} \cdot \left(\frac{(400\text{km})^2 - (100\text{km})^2}{100\text{km}} \right)$$

$= \underline{\mathbf{19\ m/s}}$

Check: Units OK. Physics OK.

Discussion: This velocity is already 60% of the 32 m/s required in order for the tropical storm to be reclassified as a hurricane. At the equator, there is no Coriolis force, hence there are no hurricanes.

Subsequent Development

As the hurricane winds continue to spiral in toward the cyclone center, centrifugal force gains importance. The **gradient wind** equations (9.25) can be reframed in cylindrical coordinates to be

$$\frac{1}{\rho} \cdot \frac{\Delta P}{\Delta R} = f_c \cdot M_{\tan} + \frac{M_{\tan}^2}{R} \tag{16.3}$$

where $\Delta P/\Delta R$ is the radial pressure gradient, and ρ is air density. The last term represents centrifugal force.

As the winds continue to spiral-in and rotationally accelerate, Coriolis force can be neglected compared to the growing centrifugal force. Inward-directed pressure-gradient force is balanced by outward-directed centrifugal force. As was introduced in Chapt. 9, the resulting **cyclostrophic wind**, M_{cs} is:

$$M_{cs} = M_{\tan} = \sqrt{\frac{R}{\rho} \cdot \frac{\Delta P}{\Delta R}} \tag{16.4}$$

This equation is approximately valid within about 100 km of the center of rotation, and includes the eye and surrounding eye wall.

Solved Example

Find the cyclostrophic wind at a radius of 50 km. The pressure gradient is 1 kPa/20 km. Assume the density is 1 kg/m^3.

Solution

Given: $R = 50$ km, $\Delta P = 1$ kPa, $\Delta R = 20$ km,

$\rho = 1$ kg/m^3.

Find: $M_{cs} = ?$ m/s

Use eq. (16.4):

$$M_{cs} = \sqrt{\frac{(50\text{km})}{(1\text{kg} \cdot \text{m}^{-3})} \cdot \frac{(1000\text{Pa})}{(20\text{km})}} = \underline{\mathbf{50\ m/s}}$$

Check: Units OK. Physics OK.

Discussion: Well above the 32 m/s needed to be classified as a hurricane.

THERMODYNAMICS

Warm Core

The center of hurricanes is warmer than the surrounding air, due to latent heat release by the organized convection. While the eye might be only 0 to 2°C warmer near the surface, it can be 10°C warmer at 12 km altitude. Hence, a hurricane is a **warm core** system. Such a situation is expected because winds and temperature are coupled via the thermal-wind relationship.

Consider first the radial component of horizontal winds. We know that air converges toward the eye in the bottom of the troposphere, but diverges from the eye near the top. To converge in the boundary layer requires an eye of lower pressure $P_{B\ eye}$ than the surroundings $P_{B\ \infty}$. To diverge aloft requires an eye of higher pressure in the eye $P_{T\ eye}$ than the surroundings $P_{T\ \infty}$ at the same altitude. Subscripts B and T denote bottom and top of the troposphere.

Hence the pressure in the eye must decrease with height more slowly than the surrounding vertical pressure gradient (Fig 16.4). According to the hypsometric equation, this occurs if the core is warmer than the surroundings.

Suppose a hurricane is approximately $z_{max} \cong 15$ km deep, has a temperature in the eye averaged over the whole hurricane depth of $\overline{T}_{eye} = 273$ K, and has ambient surface pressure distant from the storm of $P_{B\ \infty} = 101.3$ kPa. The pressure difference at the top of the hurricane ($\Delta P_T = P_{T\ \infty} - P_{T\ eye}$) is approximately related to the pressure difference at the bottom ($\Delta P_B = P_{B\ \infty} - P_{B\ eye}$) by:

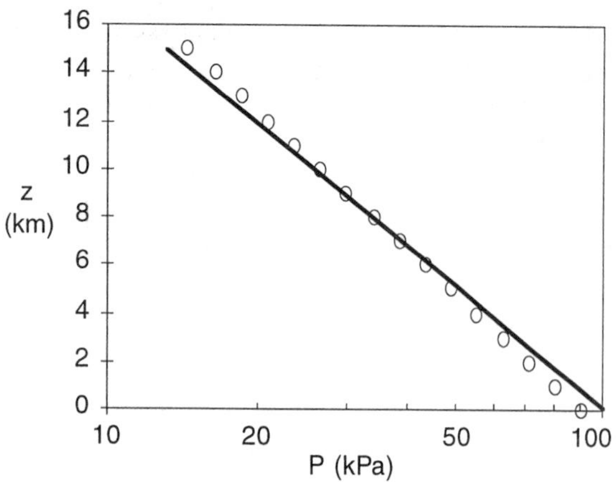

Figure 16.4
Pressure decrease with height in a warm hurricane core (circles) and in a cooler surrounding environment (line). Near the surface, the eye has lower pressure, but aloft higher pressure than the surroundings.

$$\Delta P_T \approx a \cdot \Delta P_B - b \cdot \Delta T \qquad \bullet(16.5)$$

where $a \cong 0.15$ (dimensionless), $b \cong 0.7$ kPa/K, and $\Delta T = T_{eye} - T_\infty$ is the temperature difference averaged over the depth of the troposphere. Note that ΔP_T is negative.

Next, consider the tangential winds. These winds spiral cyclonically around the eye at low altitudes, but spiral anticyclonically near the top of the troposphere. Hence, the tangential velocity must decrease with altitude and eventually change sign. The ideal gas law can be used with eq. (16.3) to show how tangential winds M_{tan} vary with altitude z:

Solved Example

The surface pressure in the eye of a hurricane is 90 kPa, while the surrounding pressure is 100 kPa. If the core is 10 K warmer than the surroundings, find the pressure difference at the top of the hurricane between the eye and surroundings.

Solution

Given: $\Delta P_B = 100 - 90 = 10$ kPa, $\Delta T = 10$ K
Find: $\Delta P_T = ?$ kPa

Use eq. (16.5):
$\Delta P_T \approx (0.15)\cdot(10\text{kPa}) - (0.7\text{kPa/K})\cdot(10\text{K})$
$= \mathbf{-5.5\ kPa}$

Check: Units OK. Physics OK.
Discussion: This answer is negative, meaning that the eye has higher pressure than the surroundings, aloft. This pressure reversal drives the spiraling outflow aloft.

BEYOND ALGEBRA • Pressure Reversal

Derivation of eq. (16.5)

Use the hypsometric equation to relate the pressure at the top of the hurricane eye to the pressure at the bottom of the eye. Do the same for the surroundings. Use those equations to find the pressure difference at the top $\Delta P_T = P_{B\,\infty} \cdot \exp[-g \cdot z_{max} / (\Re \cdot \overline{T}_\infty)] - P_{B\,eye} \cdot \exp[-g \cdot z_{max} / (\Re \cdot \overline{T}_{eye})]$.

Then, using $P_{B\,eye} \equiv \Delta P_B - P_{B\,\infty}$, collecting the exponential terms that are multiplied by $P_{B\infty}$, and finally using a first-order series approximation for those exponentials, one gets eq. (16.5), where

$a = \exp[-g \cdot z_{max} / (\Re \cdot \overline{T}_{eye})] \cong 0.15$,

$b = -(g \cdot z_{max} \cdot P_{B\infty}) / (\Re \cdot \overline{T}_{eye} \cdot \overline{T}_\infty) \cong 0.7$ kPa/K,

$g = 9.8$ m·s^{-2} is gravitational acceleration, and $\Re = 287.04$ m^2·s^{-2}·K^{-1} is the gas constant for dry air.

$$\left(\frac{2 \cdot M_{tan}}{R} + f_c\right) \cdot \frac{\Delta M_{tan}}{\Delta z} = \frac{g}{\overline{T}} \cdot \frac{\Delta T}{\Delta R} \qquad \bullet(16.6)$$

where T is absolute temperature, and the overbar denotes an average over depth. Because the tangential winds decrease with height, this equation says that the temperature must decrease with distance R from the eye. Hence, the hurricane has a warm core.

Solved Example

Suppose at a radius of 40 km the tangential velocity decreases from 50 m/s at the surface to 10 m/s at 10 km altitude. Find the radial temperature gradient at a latitude where the Coriolis parameter is 0.00005 s^{-1}. Assume $g/T = 0.0333$ m·s^{-2}·K^{-1}.

Solution

Given: $R = 40$ km, $M_{tan} = 50$ m/s at $z = 0$,
$M_{tan} = 10$ m/s at $z = 10$ km, $f_c = 0.00005$ s^{-1},
$g/T = 0.0333$ m·s^{-2}·K^{-1}
Find: $\Delta T/\Delta R = ?$ K/km

Use eq. (16.6):

$$\frac{g}{\overline{T}} \cdot \frac{\Delta T}{\Delta R} = \left(\frac{2\cdot(30\text{m/s})}{40{,}000\text{m}} + 0.00005\text{s}^{-1}\right) \cdot \frac{(10-50\text{m/s})}{(10{,}000\text{m})}$$

$= -6.2\text{x}10^{-6}$ s^{-2}.

$$\frac{\Delta T}{\Delta R} = \frac{-6.2\times10^{-6}\text{s}^{-2}}{0.0333\text{m}\cdot\text{s}^{-2}\cdot\text{K}^{-1}} = -1.86\text{x}10^{-4}\ \text{K/m}$$

$= \mathbf{-0.19\ K/km}$.

Check: Units OK. Physics OK.
Discussion: From the center of the eye, the temperature decreases about 7.4 K at a radius of 40 km, for this example. Indeed, the core is warm.

FOCUS • Warm vs. Cold Core Cyclones

For a cyclone to survive and intensify, air must be constantly withdrawn from the top. This removal of air mass from the cyclone center (**core**) counteracts the inflow of boundary-layer air at the bottom, which always happens due to surface drag. The net result is low surface pressure in the core, which drives the winds.

Warm-core cyclones (e.g., hurricanes) and **cold-core cyclones** (e.g., extratropical lows) differ in the way they cause horizontal divergence to remove air from the top of the cyclones [see Fig a, parts (a) and (b)].

Hurricanes are vertically stacked, with the eye of the hurricane near the top of the tropopause almost directly above the eye near the surface (Fig a.a). Intense latent heating in the hurricane warms the whole depth of the troposphere near the core, causing high pressure aloft because warm layers of air have greater thickness than cold. This high aloft causes air to diverge horizontally at the top of the hurricane, which is why visible and IR satellite loops show anticyclonic flow in the cirrus and other high clouds that spiral away from the hurricane.

Extratropical lows are not vertically stacked, but have low pressure that tilts westward with increasing height (Fig a.b). As the surface circulation around the cyclone advects in cold, polar, boundary-layer air on the west side of the cyclone, the small thicknesses in that sector cause pressure to decrease more rapidly with height. The net result is an upper-level trough west of, and a ridge east of, the surface low. A jet stream meandering through this trough-ridge system would cause horizontal divergence, as was shown in Figs 13.6 and 13.7.

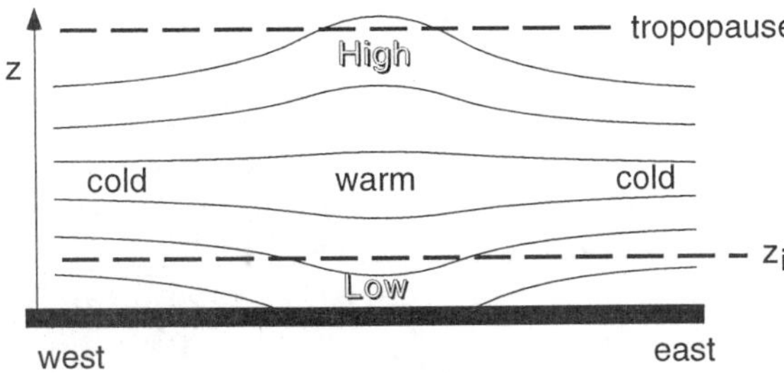

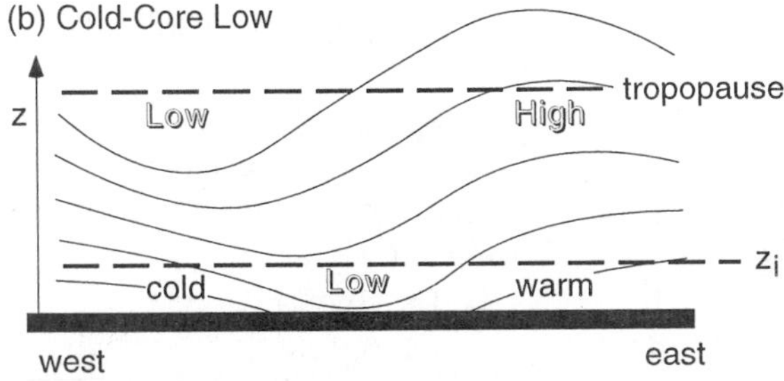

Figure a.
Vertical cross section of (a) warm, and (b) cold core cyclones. Thin lines are isobars, and z_i is ABL top.

BEYOND ALGEBRA • Warm Core Winds

Derivation of eq. (16.6)

There are 3 steps: (1) scale analysis; (2) differentiation of eq. (16.3) with respect to height *z*, and (3) simplification of the pressure term.

(1) Scale Analysis

Differentiate the ideal gas law $P = \rho \cdot \Re \cdot T$ with respect to *z*, and use the chain rule of calculus:

$$\frac{\partial P}{\partial z} = \rho \Re \frac{\partial T}{\partial z} + \Re T \frac{\partial \rho}{\partial z}$$

Then divide this by the ideal gas law:

$$\frac{1}{P}\frac{\partial P}{\partial z} = \frac{1}{T}\frac{\partial T}{\partial z} + \frac{1}{\rho}\frac{\partial \rho}{\partial z}$$

Between z = 0 to 20 km, typical variations of the other variables are: P = 101 to 5.5 kPa, T = 288 to 216 K, and ρ = 1.23 to 0.088 kg/m^3. Thus, the temperature term in the eq. above varies only 1/4 as much as the other two terms. Based on this scale analysis, we can neglect the temperature term, which leaves:

$$\frac{1}{P}\frac{\partial P}{\partial z} \cong \frac{1}{\rho}\frac{\partial \rho}{\partial z} \qquad \text{(a)}$$

(2) Differentiate eq. (16.3) with respect to z:

$$\frac{\partial}{\partial z}\left[\frac{1}{\rho}\frac{\partial P}{\partial R}\right] = f_c \cdot \frac{\partial M_{tan}}{\partial z} + \frac{2M_{tan}}{R} \cdot \frac{\partial M_{tan}}{\partial z}$$

Upon switching the left and right sides:

$$\left[\frac{2M_{tan}}{R} + f_c\right] \cdot \frac{\partial M_{tan}}{\partial z} = \frac{\partial}{\partial z}\left[\frac{1}{\rho}\frac{\partial P}{\partial R}\right] \qquad \text{(b)}$$

(3) Simplify the pressure term:

First, use the chain rule:

$$\frac{\partial}{\partial z}\left[\frac{1}{\rho}\frac{\partial P}{\partial R}\right] = \frac{1}{\rho} \cdot \frac{\partial}{\partial z}\left[\frac{\partial P}{\partial R}\right] + \frac{\partial P}{\partial R} \cdot \frac{\partial}{\partial z}\left[\rho^{-1}\right]$$

R and *z* are independent, allowing the order of differentiation to be reversed in the first term on the RHS:

$$\frac{\partial}{\partial z}\left[\frac{1}{\rho}\frac{\partial P}{\partial R}\right] = \frac{1}{\rho} \cdot \frac{\partial}{\partial R}\left[\frac{\partial P}{\partial z}\right] - \frac{1}{\rho^2}\frac{\partial P}{\partial R} \cdot \frac{\partial \rho}{\partial z}$$

Use the hydrostatic eq. $\partial P / \partial z = -\rho \cdot g$ in the first term on the RHS:

$$\frac{\partial}{\partial z}\left[\frac{1}{\rho}\frac{\partial P}{\partial R}\right] = \frac{1}{\rho} \cdot \frac{\partial}{\partial R}\left[-\rho \cdot g\right] - \frac{1}{\rho}\frac{\partial P}{\partial R} \cdot \frac{1}{\rho}\frac{\partial \rho}{\partial z}$$

But *g* is constant. Also, substitute (a) in the last term:

$$\frac{\partial}{\partial z}\left[\frac{1}{\rho}\frac{\partial P}{\partial R}\right] = -\frac{g}{\rho} \cdot \frac{\partial \rho}{\partial R} - \frac{1}{\rho}\frac{\partial P}{\partial R} \cdot \frac{1}{P}\frac{\partial P}{\partial z}$$

(*continued next column*)

BEYOND ALGEBRA • Warm Core Winds 2

Derivation of eq. (16.6) *(continuation)*

Substitute the ideal gas law in the first term on the RHS:

$$\frac{\partial}{\partial z}\left[\frac{1}{\rho}\frac{\partial P}{\partial R}\right] = -\frac{g}{\rho\cdot\Re}\cdot\frac{\partial(P\cdot T^{-1})}{\partial R} - \frac{1}{P}\frac{\partial P}{\partial R}\cdot\frac{1}{\rho}\frac{\partial P}{\partial z}$$

Use the chain rule on the first term on the right, and substitute the hydrostatic eq. in the last term:

$$\frac{\partial}{\partial z}\left[\frac{1}{\rho}\frac{\partial P}{\partial R}\right] = -\frac{P\cdot g}{\rho\cdot\Re}\cdot\frac{\partial(T^{-1})}{\partial R} - \frac{g}{\rho\cdot\Re\cdot T}\cdot\frac{\partial P}{\partial R} + \frac{g}{P}\frac{\partial P}{\partial R}$$

Substitute the ideal gas law in the 2nd term on the right:

$$\frac{\partial}{\partial z}\left[\frac{1}{\rho}\frac{\partial P}{\partial R}\right] = \frac{P\cdot g}{\rho\cdot\Re\cdot T^2}\cdot\frac{\partial T}{\partial R} - \frac{g}{P}\cdot\frac{\partial P}{\partial R} + \frac{g}{P}\frac{\partial P}{\partial R}$$

But the last two terms cancel. Using the ideal gas law in the remaining term leaves:

$$\frac{\partial}{\partial z}\left[\frac{1}{\rho}\frac{\partial P}{\partial R}\right] = \frac{g}{T}\cdot\frac{\partial T}{\partial R} \qquad \text{(c)}$$

(4) Completion:

Finally, equate (b) and (c):

$$\left[\frac{2M_{\tan}}{R} + f_c\right]\cdot\frac{\partial M_{\tan}}{\partial z} = \frac{g}{T}\cdot\frac{\partial T}{\partial R} \qquad (16.6)$$

which is the desired answer, when converted from derivatives to finite differences.

Solved Example

Suppose air in the eye of a hurricane has the following thermodynamic state: $P = 70$ kPa, $r = 1$ g/kg, $T = 15°C$. Find the entropy.

Solution

Given: $P = 70$ kPa, $r = 1$ g/kg, $T = 288$ K.
Find: $s = ?$ J·kg^{-1}·K^{-1} .

Use eq. (16.7):

$$s = \left(1004\frac{J}{kg_{air}\cdot K}\right)\cdot\ln\left(\frac{288K}{273K}\right) + \left(2500\frac{J}{g_{water}}\right)\cdot\left(1\frac{g_{water}}{kg_{air}}\right)\frac{1}{288K} - \left(287\frac{J}{kg_{air}\cdot K}\right)\cdot\ln\left(\frac{70kPa}{100kPa}\right)$$

$s =$ **165** J·kg^{-1}·K^{-1}

Check: Units OK. Physics OK.
Discussion: The actual value of entropy is meaningless, because of the arbitrary constants T_o and P_o. However, the difference between two entropies is meaningful, because the arbitrary constants cancel out.

Carnot Cycle

Hurricanes are analogous to **Carnot heat engines** in that they convert thermal energy into mechanical energy. One measure of the energy involved is the **total entropy** s per unit mass of air:

$$s = C_p\cdot\ln\left(\frac{T}{T_o}\right) + \frac{L_v\cdot r}{T} - \Re\cdot\ln\left(\frac{P}{P_o}\right) \qquad \bullet(16.7)$$

where $C_p = 1004$ J·kg^{-1}·K^{-1} is the specific heat of air at constant pressure, T is absolute temperature, $L_v =$ 2500 J/$g_{\text{water vapor}}$ is the latent heat of vaporization, r is mixing ratio, $\Re$ = 287 J·kg^{-1}·K^{-1} is the gas-law constant, P is pressure. $T_o = 273$ K and $P_o = 100$ kPa are arbitrary reference values.

A thermo diagram is used to illustrate hurricane thermodynamics (Fig 16.5). Because of limitations in range and accuracy of thermo diagrams, you might find slightly different answers than those given here in Table 16-1. For simplicity, assume constant sea surface temperature of $T_{SST} = 28°C$.

As an initial condition at Point 1 in Fig 16.5, consider relatively warm ($T = 28°C$) dry air ($r \cong 0$, $T_d = -70°C$) in the boundary layer ($z = 0$, $P = 100$ kPa), but outside of the hurricane. Using eq. (16.7), the initial entropy is $s_1 \cong 98$ J·kg^{-1}·K^{-1}.

In the first part of the hurricane's Carnot cycle, air in the boundary layer spirals in from Point 1 toward the eye wall of the hurricane (Point 2) isothermally ($T = 28°C$, because of heat transfer with the sea surface) at constant height ($z = 0$ = sea level). Pressure decreases to $P = 90$ kPa as the air approaches the low-pressure eye. Evaporation from the sea surface increases the mixing ratio to saturation ($r \cong 28$ g/kg, $T_d \cong 28°C$), thereby causing entropy to increase to $s_2 \cong 361$ J·kg^{-1}·K^{-1}. This evaporation is the major source of energy for the storm.

From Point 2, air rises moist adiabatically to Point 3 in the thunderstorms of the eye wall. The moist-adiabatic process conserves entropy ($s_3 = 366$ J·kg^{-1}·K^{-1}, within the accuracy of the thermo diagram, thus $s_2 \cong s_3$). During this process, temperature drops to $T \cong -18°C$ and mixing ratio decreases to about $r \cong 3.7$ g/kg. However, the decrease of pressure ($P = 90$ to 25 kPa) in this rising air compensates to maintain nearly constant entropy.

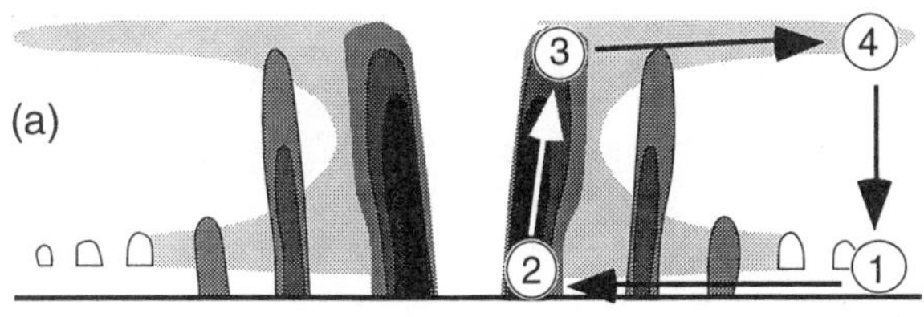

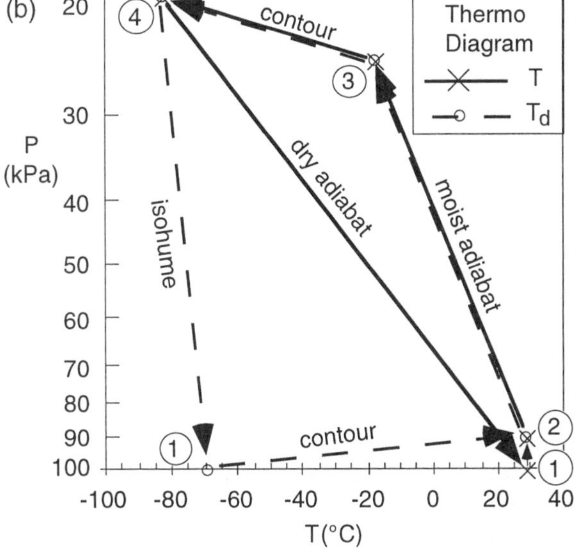

Figure 16.5
(a) Circulation of air through the hurricane. (b)Thermodynamic diagram showing hurricane processes. Solid lines show the temperature changes, dashed lines show dew point temperature changes.

Once the cloudy air reaches the top of the troposphere at Point 3, it spirals outward to Point 4 at roughly constant altitude. (Sorry, the thermo diagram of Fig 6.2 does not go high enough to simulate a real hurricane of 15 km depth, so we will use $z \cong 10$ km here.) The divergence of air is driven by a pressure gradient of $P = 25$ kPa in the eye to 20 kPa outside the hurricane.

During this high-altitude outflow from Points 3 to 4, air rapidly loses heat due to infrared radiation, causing its temperature to decrease from $T = T_d =$ –18°C to – 83°C. The air remains saturated, and mixing ratio decreases from $r \cong 3.7$ to near 0. The cooling also converts more water vapor into precipitation. Entropy drops to 98 J·kg^{-1}·K^{-1}.

Finally, the air subsides dry adiabatically from Points 4 to 1, with no change of mixing ratio ($r \cong 0$, $T_d \cong -70$°C). Temperature increases adiabatically to $T = 28$°C, due to compression as the air descends into higher pressure ($P = 100$ kPa). This dry adiabatic process also preserves entropy, and thus is called an **isentropic** process (and dry adiabats are also known as **isentropes**). The final state of the air is identical to the initial state, at Point 1 in Fig 16.5.

Table 16-1. Example of thermodynamic states within a hurricane, corresponding to points (Pt) in Fig 16.5. Reference: $T_o = 273$ K, $P_o = 100$ kPa.

Pt	P (kPa)	T (°C)	T_d (°C)	r (g/kg)	s [J/(K·kg)]
1	100	28	–70	≅ 0	98
2	90	28	28	28	361
3	25	–18	–18	3.7	366
4	20	–83	–83	≅ 0	98

This Carnot process is a closed cycle; namely, the air can recirculate through the hurricane. However, during this cycle, entropy is gained near the sea surface where the temperature is warm, while it is lost near cloud top where temperatures are much colder.

The gain of entropy at one temperature and loss at a different temperature allows the Carnot engine to produce mechanical energy ME according to

$$ME = (T_B - T_{T\ avg}) \cdot (s_{eyewall} - s_\infty)_B \qquad \bullet(16.8)$$

where subscripts B and T denote bottom and top of the troposphere, *eyewall* denotes boundary layer air under the eye wall, and ∞ denotes the ambient conditions at large distances from the hurricane (e.g.: $P_\infty \cong 101.3$ kPa). This mechanical energy drives the hurricane-force winds, ocean waves, atmospheric waves, and mixing of both the atmospheric and ocean against buoyant forces.

If all of the mechanical energy were consumed trying to maintain the hurricane force winds against the frictional drag in the boundary layer (an unrealistic assumption), then the hurricane could support the following maximum pressure ratio at the surface:

$$\ln\left(\frac{P_\infty}{P_{eye}}\right)_B = \frac{ME}{T_B \cdot \Re} \qquad (16.9)$$

FOCUS • The Power of Hurricanes

One measure of hurricane power is the rate of dissipation of kinetic energy by wind drag against the sea surface. This is proportional to the ABL wind speed cubed, times the sea-surface area over which that speed is valid, summed over all areas under the hurricane. K. A. Emanuel (1999, *Weather*, **54**, 107-108) estimates dissipation rates of 3 x 10^{12} W for a typical Atlantic hurricane (with max winds of 50 m/s at radius 30 km), and 3 x 10^{13} W for a Pacific typhoon (with max winds of 80 m/s at radius 50 km).

Solved Example

Find the mechanical energy and minimum possible eye pressure that can be supported by the hurricane of Table 16-1.

Solution

Given: $s_{eyewall} = 361$ J·kg^{-1}·K^{-1}, $s_\infty = 98$ J·kg^{-1}·K^{-1}

$T_B = 28°C$, $T_{T\,avg} \cong 0.5·(-18-83) = -50.5°C$.

Find: ME = ? kJ/kg, P_{eye} = ? kPa

Use eq. (16.8): $ME =$

$$(28+50.5)K \cdot (361-98)J \cdot kg^{-1} \cdot K^{-1} = \underline{\textbf{20.6 kJ/kg}}$$

Use eq. (16.9):

$$\ln\left(\frac{P_\infty}{P_{eye}}\right)_B = \frac{(20{,}600\text{J} \cdot \text{kg}^{-1})}{(301\text{K}) \cdot (287\text{J} \cdot \text{kg}^{-1} \cdot \text{K}^{-1})} = 0.238$$

$$P_{eye} = (101.3\text{kPa})/\exp(0.238) = \underline{\textbf{79.8 kPa.}}$$

Check: Units OK. Physics OK.
Discussion. This eye pressure is lower than the actual eye pressure of 90 kPa. The difference is related to the *ME* of winds and waves.

A HURRICANE MODEL

Although hurricanes are quite complex and not fully understood, we can build an idealized model that mimics some of the features of real hurricanes.

Pressure Distribution

Eye pressures of 95 to 99 kPa at sea level are common in hurricanes, although a pressure as low as P_{eye} = 87 kPa has been measured. One measure of hurricane strength is the pressure difference ΔP_{max} between the eye and the surrounding ambient environment P_∞ = 101.3 kPa:

$$\Delta P_{\max} = P_\infty - P_{eye} \tag{16.10}$$

The surface pressure distribution across a hurricane can be approximated by:

$$\frac{\Delta P}{\Delta P_{\max}} = \begin{cases} \dfrac{1}{5} \cdot \left(\dfrac{R}{R_o}\right)^4 & \text{for } R \le R_o \\ \left[1 - \dfrac{4}{5} \cdot \dfrac{R_o}{R}\right] & \text{for } R > R_o \end{cases} \tag{16.11}$$

where $\Delta P = P(R) - P_{eye}$, and R is the radial distance from the center of the eye. This pressure distribution

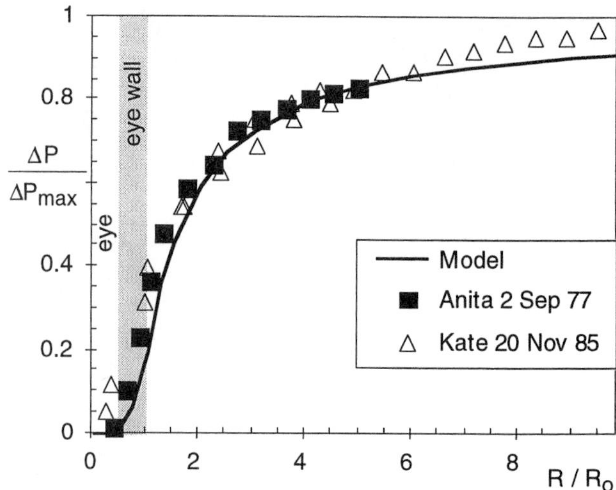

Figure 16.6
Surface pressure distribution across a hurricane. Model is eq. (16.11).

is plotted in Fig 16.6, with data points from two hurricanes.

R_o is the critical radius where the maximum tangential winds are found. In the hurricane model presented here, R_o is twice the radius of the eye. Eyes range from 4 to 100 km in radius, with average values of 15 to 30 km radius. Thus, one anticipates average values of critical radius of: $30 < R_o < 60$ km, with an observed range of $8 < R_o < 200$ km.

Tangential Velocity

To be classified as a hurricane, the sustained winds (averaged over 1-minute) must be 32 m/s or greater. While most anemometers are unreliable at extreme wind speeds, maximum hurricane winds have been reported in the 75 to 95 m/s range.

As sea-level pressure in the eye decreases, maximum tangential winds M_{max} around the eye wall increase (Fig 16.7).

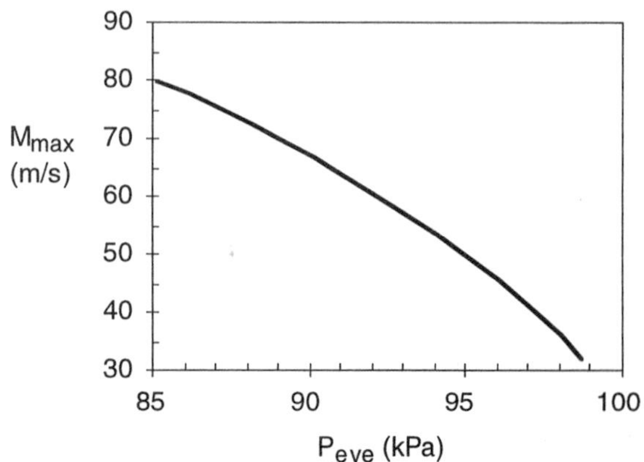

Figure 16.7
Maximum sustained winds around a hurricane increase as sea-level pressure in the eye decreases.

An empirical approximation for this relationship, based on Bernoulli's equation, is :

$$M_{max} = a \cdot (\Delta P_{max})^{1/2} \qquad \bullet(16.12)$$

where $a = 20$ (m/s)·kPa$^{-1/2}$.

If winds are assumed to be cyclostrophic, then the previous approximation for pressure distribution (eq. 16.11) can be used to give a distribution of tangential velocity M_{tan} in the boundary layer:

$$\frac{M_{tan}}{M_{max}} = \begin{cases} (R/R_o)^2 & \text{for } R \le R_o \\ (R_o/R)^{1/2} & \text{for } R > R_o \end{cases} \qquad (16.13)$$

where the maximum velocity occurs at critical radius R_o. This is plotted in Fig 16.8, with data points from a few hurricanes.

For the hurricanes plotted in Fig 16.8, the critical radius of maximum velocity was in the range of R_o = 20 to 30 km. This is a rough definition of the outside edge of the **eye wall** for these hurricanes, within which the heaviest precipitation falls. The maximum velocity for these storms was M_{max} = 45 to 65 m/s.

Winds in Fig 16.8 are relative to the eye. However, the whole hurricane including the eye is often moving. Hurricane **translation** speeds can be M_t = 5 m/s as they drift westward in the tropics, with extreme speeds of 25 m/s as they later move poleward. Typical translation speeds of the hurricane as it moves over the ocean are 10–15 m/s.

The total wind speed relative to the surface is the vector sum of the translation speed and the rotation speed. On the right quadrant of the storm relative to its direction of movement in the Northern Hemisphere, the translation speed adds to the rotation speed, while on the left it subtracts.

Solved Example
What max winds are expected if the hurricane eye has surface pressure of 95 kPa?

Solution
Given: $P_{B\ eye}$ = 95 kPa, Assume $P_{B\ \infty}$ = 101.3 kPa
Find: M_{max} = ? m/s.
Assume translation speed can be neglected.

Use eq. (16.12): $M_{max} = 20 \cdot (101.3 - 95)^{1/2}$ = **50 m/s**

Check: Units OK. Physics OK.
Discussion: This is a level 3 hurricane on the Siffir-Simpson scale.

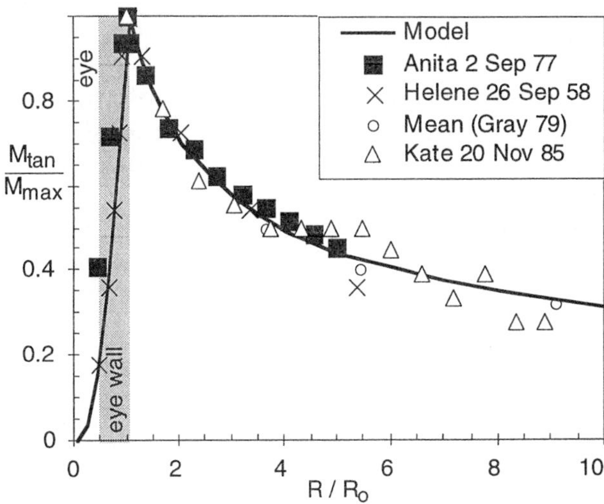

Figure 16.8
Tangential winds near the surface at various radii around hurricanes. Model is eq. (16.13).

Total speed relative to the surface determines ocean wave and surge generation. Thus, the right quadrant of the storm near the eye wall is most dangerous. Also, tornadoes are likely there.

FOCUS • Saffir-Simpson Hurricane Scale

In the early 1970s, consulting engineer Herbert Saffir and US National Hurricane Center director Robert Simpson developed this scale to give public-safety officials an estimate of hurricane wind and storm-surge damage potential.

Table 16–2. *HC* = hurricane category, TD = tropical depression (not a hurricane), TS = tropical storm (not a hurricane), P_{eye} = sea-level pressure in the eye (kPa), M_{max} = maximum wind speed (m/s), *S* = approximate storm surge height (m).

HC	P_{eye} (kPa)	M_{max} (m/s)	*S* (m)
TD	–	< 17	
TS	–	17 – 32	
1	≥ 98.0	33 – 42	1.2 – 1.6
2	97.9 – 96.5	43 – 49	1.7 – 2.5
3	96.4 – 94.5	50 – 57	2.6 – 3.9
4	94.4– 92.0	58 – 69	4.0 – 5.5
5	< 92.0	> 70	> 5.5

Solved Example (§)

Replot the speed model of eqs. (16.13), except including effects of hurricane movement of 10 m/s. Assume the maximum tangential speed relative to the eye is 50 m/s, and the max winds occur at critical radius of 25 km. Use a coordinate system with the x-axis aligned with the hurricane movement (**translation**) direction, and plot the resulting wind speed U as a function of distance y perpendicular to the translation direction.

Solution

Given: $M_t = 10$ m/s, $M_{max} = 50$ m/s, $R_o = 25$ km
Plot: U vs. y

Use eqs. (16.13), and let $U = (M_t \pm M_{tan})$

Use a coordinate system aligned with the eye movement. Thus, we must add tangential winds on the right of the eye to the translation speed of 10 m/s. Subtract tangential winds on the left from the translation speed.

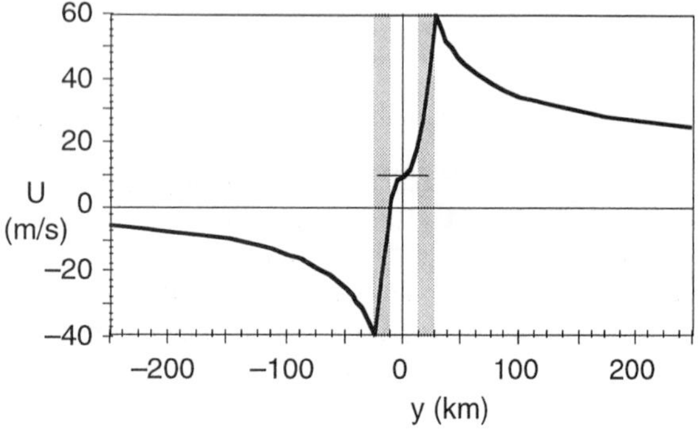

Check: Units OK. Physics OK.
Discussion: Peak winds are 60 m/s to the right of the eye, and 40 m/s to the left. The eye-wall is highlighted in this figure. Translation speed of the eye is indicated by the dash in the center of the fig.

Radial Velocity

For an idealized hurricane, air mass is trapped in the boundary layer as it converges horizontally toward the eye wall. Horizontal continuity in cylindrical coordinates requires:

$$M_{rad} \cdot R = \text{constant} \tag{16.14}$$

where M_{rad} is the radial velocity, negative for inflow. Thus, as R decreases toward R_o, the magnitude of inflow must increase. Inside of R_o, thunderstorm convection removes air mass vertically, implying that horizontal continuity is no longer satisfied.

As wind velocities increase inward toward the eye wall from outside (see previous subsections), wave height and surface roughness also increase. The resulting turbulent drag against the ocean surface tends to couple the radial and tangential velocities, which we can approximate by $M_{rad} \propto M_{tan}^2$. Drag-induced inflow such as this eventually converges and forces ascent via the **boundary-layer pumping** process (Chapt. 9).

The following equations utilize the concepts above, and are consistent with the tangential velocity in the previous subsection:

$$\frac{M_{rad}}{M_{max}} = \begin{cases} -\dfrac{R}{R_o} \cdot \left[\dfrac{1}{5}\left(\dfrac{R}{R_o}\right)^3 + \dfrac{1}{2} \dfrac{W_s}{M_{max}} \dfrac{R_o}{z_i} \right] & \text{for } R \le R_o \\ -\dfrac{R_o}{R} \cdot \left[\dfrac{1}{5} + \dfrac{1}{2} \cdot \dfrac{W_s}{M_{max}} \cdot \dfrac{R_o}{z_i} \right] & \text{for } R > R_o \end{cases} \tag{16.15}$$

where W_s is negative, and represents the average subsidence velocity in the eye. Namely, the horizontal area of the eye, times W_s, gives the total kinematic mass flow downward in the eye. The boundary-layer depth is z_i, and M_{max} is still the maximum tangential velocity.

As an example, Fig 16.9 shows a plot of the equations above, using $z_i = 1$ km, $R_o = 25$ km, $M_{max} = 50$ m/s, and $W_s = -0.2$ m/s . In the eye, subsidence causes air to weakly diverge (positive M_{rad}) toward the eye wall. Inside the eye wall, the radial velocity rapidly changes to inflow (negative M_{rad}), reaching an extreme value of –7.5 m/s for this example. Outside of the eye wall, the radial velocity smoothly decreases as required by horizontal mass continuity.

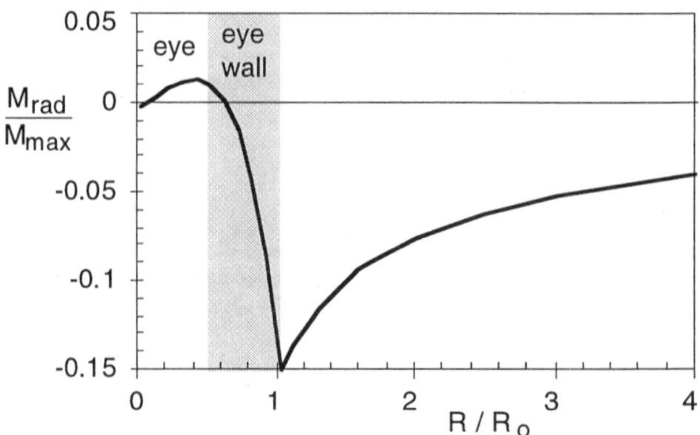

Figure 16.9
Radial winds in a hurricane boundary layer. (Negative values indicate motion inward, converging toward the center.)

Vertical Velocity

At radii less than R_o, the converging air rapidly piles up, and rises out of the boundary layer as thunderstorm convection within the eye wall. The vertical velocity out of the top of the boundary layer, as found from mass continuity, is

$$\frac{W}{M_{max}} = \begin{cases} \left[\frac{z_i}{R_o} \left(\frac{R}{R_o} \right)^3 + \frac{W_s}{M_{max}} \right] & \text{for } R < R_o \\ 0 & \text{for } R > R_o \end{cases} \tag{16.16}$$

For simplicity, we are neglecting the upward motion that occurs in the spiral rain bands at $R > R_o$.

As before, W_s is negative for subsidence. Although subsidence acts only inside the eye for real hurricanes, the relationship above applies it every where inside of R_o for simplicity. Within the eye wall, the upward motion overpowers the subsidence, so our simplification is of little consequence.

Using the same values as for the previous figure, the vertical velocity is plotted in Fig 16.10. The maximum upward velocity is 1.8 m/s in this case, which represents an average around the eye wall. Updrafts in individual thunderstorms can be much faster.

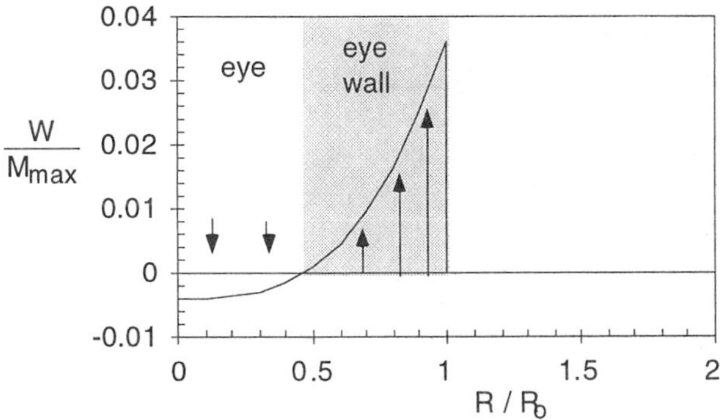

Figure 16.10
Vertical velocity at various radii around hurricanes, out of the top of the boundary layer. $W_s = -0.2$ m/s in the eye.

Temperature

Suppose that pressure difference between the eye and surroundings at the top of the hurricane is equal and opposite to that at the bottom. From eq. (16.5) the temperature T averaged over the hurricane depth at any radius R is found from:

$$\Delta T(R) = c \cdot \left[\Delta P_{max} - \Delta P(R) \right] \tag{16.17}$$

where $c = 1.64$ K/kPa, the pressure difference at the bottom is $\Delta P = P(R) - P_{eye}$, and the temperature

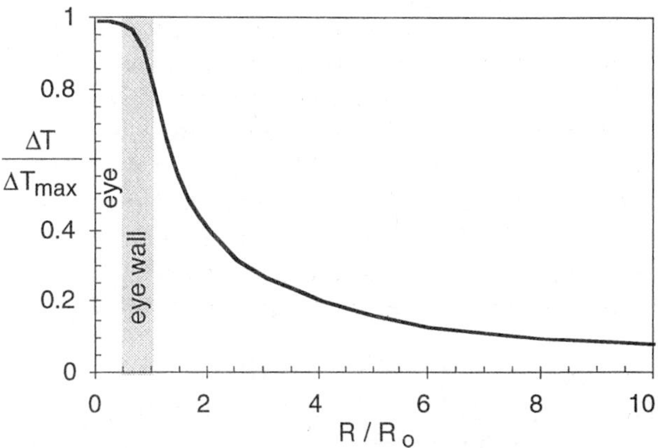

Figure 16.11
Temperature distribution, averaged over a 15 km thick hurricane, showing the warm core.

difference averaged over the whole hurricane depth is $\Delta T(R) = T_{eye} - T(R)$. When used with eq. (16.11), the result is:

$$\frac{\Delta T}{\Delta T_{max}} = \begin{cases} 1 - \frac{1}{5} \cdot \left(\frac{R}{R_o} \right)^4 & \text{for } R \le R_o \\ \frac{4}{5} \cdot \frac{R_o}{R} & \text{for } R > R_o \end{cases} \tag{16.18}$$

where $\Delta T_{max} = T_{eye} - T_{\infty} = c \cdot \Delta P_{max}$, and $c =$ 1.64 K/kPa. This is plotted in Fig 16.11.

Science Graffito

"The tempest arose and wearied me so that I knew not where to turn; ...[my] eyes never beheld the seas so high, angry and covered by foam. The wind not only prevented our progress, but offered no opportunity to run behind any headland for shelter; hence we were forced to keep out in this bloody ocean, seething like a pot on a hot fire. Never did the sky look more terrible; for one whole day and night it blazed like a furnace, and the lightning broke forth with such violence that each time I wondered if it had carried off my spars and sails; the flashes came with such fury and frightfulness that we all thought this ship would be blasted. All this time the water never ceased to fall from the sky; I don't say it rained, because it was like ...[a] deluge. The people were so worn out that they longed for death to end their dreadful suffering."

– Christopher Columbus, 1502.

Composite Picture

A coherent picture of hurricane structure can be presented by combining all of the idealized models described above. The result is sketched in Fig 16.12. For real hurricanes, sharp cusps in the velocity distribution would not occur because of vigorous turbulent mixing in the regions of strong shear.

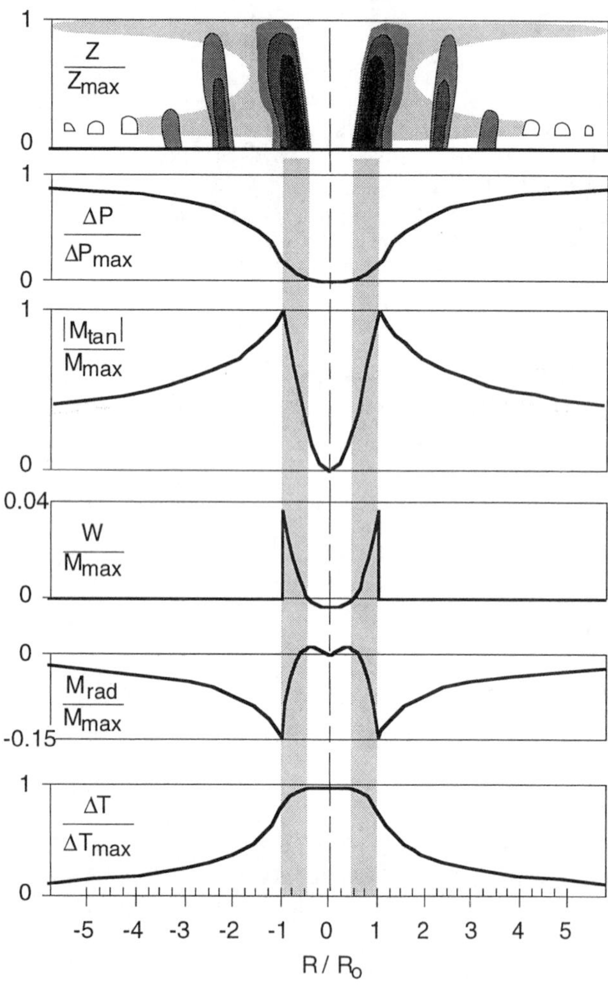

Figure 16.12
Composite hurricane structure based on the idealized model. Pressure differences are at sea level. All horizontal velocities are in the boundary layer, while vertical velocity is across the top of the boundary layer. Temperature differences are averaged over the whole hurricane depth.

Solved Example

A hurricane of critical radius R_o = 25 km has a central pressure of 90 kPa. Find the wind components, vertically-averaged temperature excess, and pressure at radius 40 km from the center. Assume W_s = –0.2 m/s, and z_i = 1 km.

Solution

Given: $P_{B\,eye}$ = 90 kPa, R_o = 25 km, R = 40 km, W_s = –0.2 m/s, and z_i = 1 km.
Find: P = ? kPa, T = ? °C, M_{tan} = ? m/s, M_{rad} = ? m/s, W = ? m/s
Assume $P_{B\,\infty}$ = 101.3 kPa
Figure: Similar to Fig 16.12.

Note that $R > R_o$. First, find maximum values:

$$\Delta P_{max} = (101.3 - 90\text{kPa}) = 11.3 \text{ kPa}$$

Use eq. (16.12):

$$M_{max} = [20(\text{m/s})\cdot\text{kPa}^{-1/2}]\cdot(11.3\text{kPa})^{1/2} = 67 \text{ m/s}$$

$$\Delta T_{max} = c\cdot\Delta P_{max} = (1.64 \text{ K/kPa})\cdot(11.3\text{kPa}) = 18.5°\text{C}$$

Use eq. (16.11):

$$\Delta P = (11.3\text{kPa})\cdot\left[1 - \frac{4}{5}\cdot\frac{25\text{km}}{40\text{km}}\right] = 5.65 \text{ kPa}$$

$P = P_{eye} + \Delta P = 90 + 5.65 =$ **95.65 kPa**.

Use eq. (16.18):

$$\Delta T = (18.5°\text{C})\cdot\frac{4}{5}\cdot\frac{(25\text{km})}{(40\text{km})} = \mathbf{9.25°C}$$

averaged over the whole hurricane depth.

Use eq. (16.13):

$$M_{\tan} = (67\text{m/s})\cdot\sqrt{\frac{25\text{km}}{40\text{km}}} = \mathbf{53\ m/s}$$

Use eq. (16.15): M_{rad} =

$$-(67\text{m/s})\cdot\frac{(25\text{km})}{(40\text{km})}\cdot\left[\frac{1}{5} + \frac{1}{2}\cdot\frac{(-0.2\text{m/s})}{(67\text{m/s})}\cdot\frac{(25\text{km})}{(1\text{km})}\right]$$

M_{rad} = **–6.8 m/s**

Use eq. (16.16): W = **0 m/s**

Check: Units OK. Physics OK.
Discussion: Based on the Siffir-Simpson hurricane damage scale, this hurricane is borderline between level 4 and 5, and thus is very intense.

STORM SURGE

Much of the hurricane-caused damage results from inundation of coastal areas by high seas. The high seas are caused by the reduced atmospheric pressure in the eye, and by wind blowing the water against the coast to form a large propagating surge called a Kelvin wave. In addition, high tides and high surface-waves can exacerbate the damage.

Atmospheric Pressure Head

In the eye of the hurricane, atmospheric surface pressure is lower than ambient. Hence the force per area pushing on the top of the water is less. This allows the water to rise in the eye until the additional head (weight of fluid above) of water compensates for the reduced head of air (Fig 16.13).

The amount of rise Δz of water in the eye is

$$\Delta z = \frac{\Delta P_{\max}}{\rho_{liq} \cdot g} \qquad \bullet(16.19)$$

where g is gravitational acceleration = 9.8 m/s^2, ρ_{liq} = 1000 kg/m^3 is the density of liquid water, and ΔP_{max} is the atmospheric surface pressure difference between the eye and the undisturbed environment.

To good approximation, this is

$$\Delta z = a \cdot \Delta P_{\max} \qquad (16.20)$$

where a = 0.1 m/kPa. Thus, in a strong hurricane with eye pressure of 90 kPa (causing $\Delta P_{max} \cong$ 10 kPa), the sea level would rise 1 m.

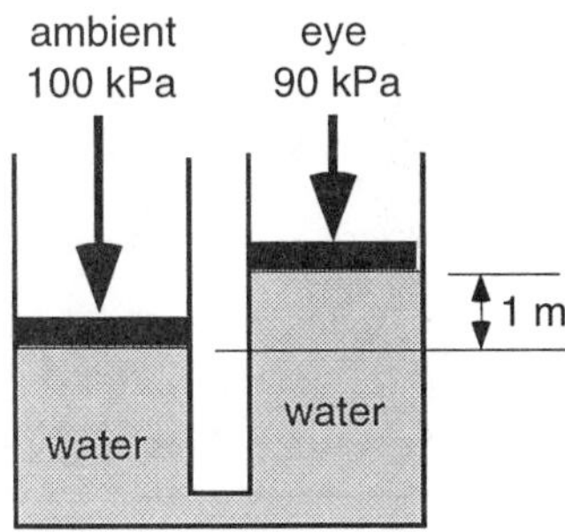

Figure 16.13
Reduced atmospheric pressure in the eye allows sea-level to rise.

Ekman Transport

Recall from Chapter 11 that ocean currents are generated by wind drag on the sea surface. The Ekman spiral describes how the current direction and speed varies with depth. The net **Ekman transport**, accumulated over all depths, is exactly perpendicular to the surface wind direction.

In the Northern Hemisphere, this net transport of water is to the right of the wind, and has magnitude:

$$\frac{Vol}{\Delta t \cdot \Delta y} = \frac{\rho_{air}}{\rho_{water}} \cdot \frac{C_D \cdot V^2}{f_c} \qquad (16.21)$$

where Vol is the volume of water transported during time interval Δt, Δy is a unit of length of coastline parallel to the mean wind, V is the wind speed near the surface (actually at 10 m above the surface), C_D is the drag coefficient, f_c is the Coriolis parameter, ρ_{air} = 1.225 kg·m^{-3} is the air density at sea level, and ρ_{water} = 1000 kg·m^{-3} is water density.

As a hurricane approaches the eastern coast of continents, the winds along the front edge of the hurricane are parallel to the coast, from north to south in the Northern Hemisphere. Hence, there is net Ekman transport of water directly toward shore, where it begins to pile up and make a storm surge (Fig 16.14).

If the hurricane were to hover just offshore for sufficient time to allow a steady-state condition to develop, then the Ekman transport toward the shore would be balanced by downslope sloshing of the surge.

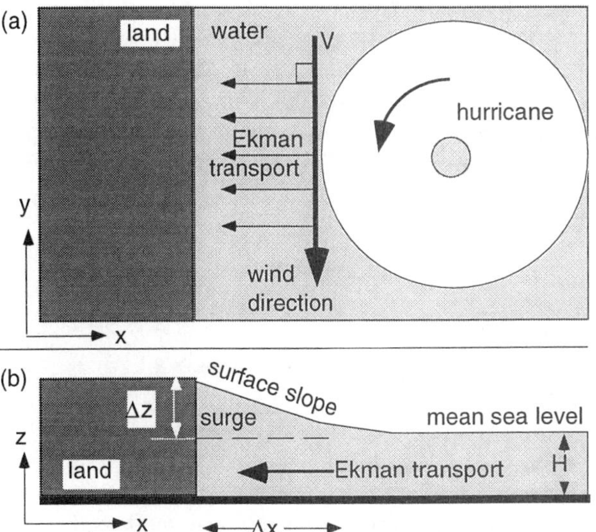

Figure 16.14
Onshore Ekman transport prior to hurricane landfall, in the Northern Hemisphere. (a) Top view. (b) View from south.

The surge slope $\Delta z/\Delta x$ for that hypothetical equilibrium is:

$$\frac{\Delta z}{\Delta x} \approx \frac{\rho_{air}}{\rho_{water}} \cdot \frac{C_D \cdot V^2}{g \cdot H} \qquad (16.22)$$

where H is the unperturbed ocean depth (e.g., 50 m) near the coast.

Solved Example

For hurricane-force winds of 40 m/s, over a continental shelf portion of ocean of depth 50 m, find the volume transport rate and equilibrium surge slope. Assume $C_D = 0.01$.

Solution

Given: $\rho_{air} = 1.225$ kg·m^{-3} , $\rho_{water} = 1000$ kg·m^{-3}
$M = 40$ m/s, $H = 50$ m, $C_D = 0.01$
Find: $Vol/(\Delta t \cdot \Delta y) = ?$ m^2/s, and $\Delta z/\Delta x = ?$
Assume: $f_c = 0.00005$ s^{-1}

Use eq. (16.21):

$$\frac{Vol}{\Delta t \cdot \Delta y} = \frac{(1.225\text{kg/m}^3)}{(1000\text{kg/m}^3)} \cdot \frac{(0.01)\cdot(40\text{m/s})^2}{(0.00005\text{s}^{-1})}$$
$$= \underline{392}\ \text{m}^2/\text{s}$$

Use eq. (16.22):

$$\frac{\Delta z}{\Delta x} \approx \frac{(1.225\text{kg/m}^3)}{(1000\text{kg/m}^3)} \cdot \frac{(0.01)\cdot(40\text{m/s})^2}{(9.8\text{m/s}^2)\cdot(50\text{m})} = \underline{\mathbf{0.00004}}$$

Check: Units OK. Physics OK.
Discussion: The flow of water is tremendous. As it starts to pile up, the gradient of sea-level begins to drive water away from the surge, so it does not continue growing. The slope corresponds to 4 cm rise per km distance toward the shore. Over tens to hundreds of kilometers, the rise along the coast can be significant.

Kelvin Wave

Because the hurricane has finite size, Ekman transport is localized to the region immediately in front of the hurricane. Thus, water is piled higher between the hurricane and the coast than it is further north or south along the coast. When viewed from the East, the surge appears as a long-wavelength wave, called a **Kelvin wave** (Fig 16.15).

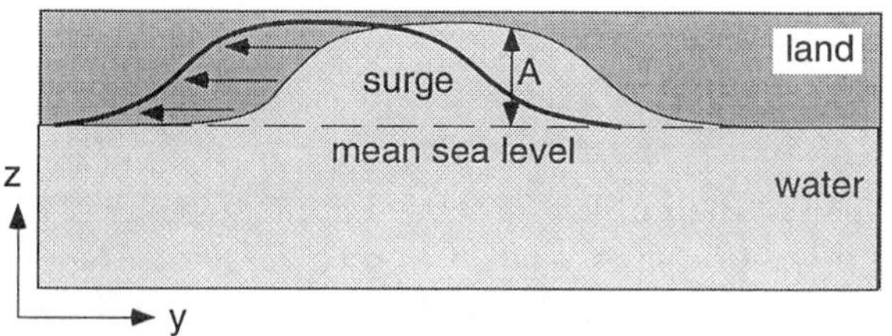

Figure 16.15
The surge, viewed from the east, is a Kelvin wave that propagates south.

The propagation speed of the wave, called the **phase speed** c, is

$$c = \sqrt{g \cdot H} \qquad \bullet(16.23)$$

which is also known as the **shallow-water wave speed**. Typical phase speeds are 15 to 30 m/s. These waves always travel with the coast to their right in the Northern Hemisphere, so they propagate southward along the east coast of continents, and northward along west coasts.

As the wave propagates south along the East Coast, it will hug the coast and inundate the shore immediately south of the hurricane. Meanwhile, Ekman transport continues to build the surge in the original location. The net result is a continuous surge of high water along the shore that is closest to, and south of, the hurricane. Typical surge depths can be 2 to 10 m at the coast, with extreme values of 13 – 20 m.

Should the hurricane move southward at a speed nearly equal to the Kelvin wave speed, then Ekman pumping would continue to reinforce and build the surge, causing the amplitude of the wave A to grow according to:

$$\frac{\Delta A}{\Delta t} \approx \frac{\rho_{air}}{\rho_{water}} \cdot \frac{C_D \cdot V^2}{\sqrt{g \cdot H}} \qquad (16.24)$$

where the amplitude A is measured as maximum height of the surge above mean sea level.

Solved Example

Using info from the previous solved example, find the Kelvin wave phase speed, and the growth rate if the hurricane follows the wave southward.

Solution

Given: (same as previous example)
Find: $c = ?$ m/s, $\Delta A/\Delta t = ?$ m/s

Use eq. (16.23):

$$c = \sqrt{(9.8\text{m/s}^2)\cdot(50\text{m})} = \underline{\mathbf{22.1\ m/s}}$$

(continued next column)

Solved Example *(continuation)*

Use eq. (16.24):

$$\frac{\Delta A}{\Delta t} \approx \frac{(1.225\text{kg/m}^3)}{(1000\text{kg/m}^3)} \cdot \frac{(0.01)\cdot(40\text{m/s})^2}{\sqrt{(9.8\text{m/s}^2)\cdot(50\text{m})}}$$

$= \underline{\mathbf{0.00089\ m/s}}$

Check: Units OK. Physics OK.

Discussion: Luckily, hurricanes usually turn northward. If hurricanes were to translate southward with speed 22.1 m/s, matching the Kelvin wave speed, then an exceptionally dangerous situation would develop with amplification of the surge by over 3 m/hour.

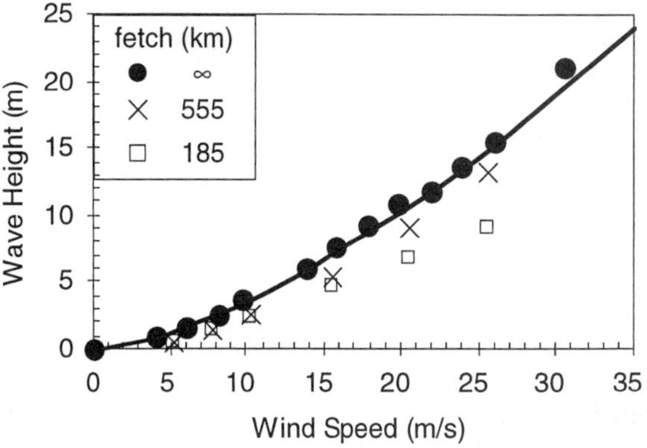

Figure 16.16
Wave height of surface wind-generated waves. Solid line is eq. (16.25) for unlimited fetch and duration. Data points are ocean observations with different fetch.

SURFACE WIND-WAVES

Waves are generated on the sea surface by action of the winds. Greater winds acting over longer distances (called **fetch**) for greater time durations can excite higher waves. High waves caused by hurricane-force winds are not only a hazard to shipping, but can batter structures and homes along the coast.

Four coastal hazards of a hurricane are:

- wave scour of the beach under structures,
- wave battering of structures,
- surge flooding, and
- wind damage.

The first two hazards exist only right on the coast, in the beach area. Also, the surge rapidly diminishes by 10 to 15 km inland from the coast. Tornadoes, lightning, and rain can cause additional problems even at larger distances from the coast.

For wind speeds M up to hurricane force, the maximum-possible wave height (for unlimited fetch and duration) can be estimated from:

$$h = h_2 \cdot \left(\frac{M}{M_2}\right)^{3/2} \qquad (16.25)$$

where $h_2 = 4$ m and $M_2 = 10$ m/s. Wave heights are plotted in Fig 16.16.

As winds increase beyond hurricane force, the wave tops become partially chopped off by the winds. Thus, wave height does not continue to increase according to eq. (16.25). For extreme winds of 70 m/s, the sea surface is somewhat flat, but poorly defined because of the mixture of spray, foam, and chaotic seas that appear greenish white during daytime.

Wavelengths of the wind-waves also increase with wind speed. Average wavelengths λ can be approximated by:

$$\lambda = \lambda_2 \left(\frac{M}{M_2}\right)^{1.8} \qquad (16.26)$$

where $\lambda_2 = 35$ m, and $M_2 = 10$ m/s. Wavelengths are plotted in Fig 16.17.

The longest wavelength waves are called **swell**, and can propagate large distances, such as across whole oceans. Hence, a hurricane in the middle tropical Atlantic can cause large surf in Florida well before the storm reaches the coast.

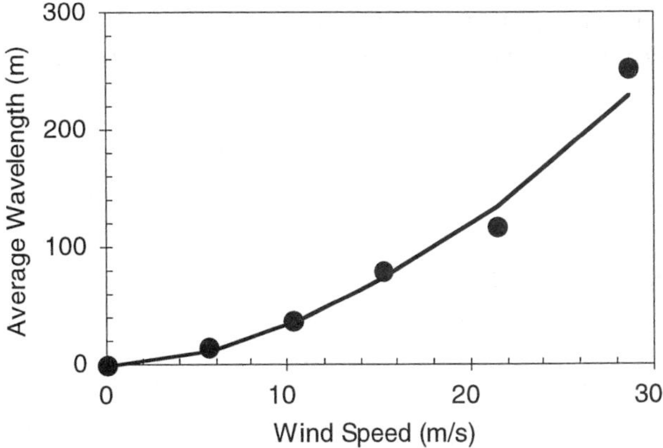

Figure 16.17
Wave length of wind waves. Solid line is eq. (16.26).

FOCUS • Beaufort Wind Scale

In 1805, Admiral Beaufort of the British Navy devised a system to estimate and report wind speeds based on the amount of canvas sail that a full-rigged frigate could carry. It was updated in 1874, as listed below. For a modern table of corresponding wind effects on land, see Ahrens "Meteorology Today".

Table 16–3. Legend: *B* = Beaufort number; *D* = modern description; *M* = wind speed in knots (1 m/s ≅2 knots), S1 = speed through smooth water of a well-conditioned man-of-war with all sail set, and clean full; S2 = sails that a well-conditioned man-of-war could just carry in chase, full and by; S3 = sails that a well-conditioned man-of-war could scarcely bear.

B	*D*	*M* (kt)	Deep Sea Criteria
0	Calm	0 – 1	Becalmed
1	Light Air	1 – 3	Just sufficient to give steerageway
2	Slight Breeze	4 – 6	S1 = 1 – 2 knots
3	Gentle Breeze	7 – 10	S1 = 3 – 4 knots
4	Moderate Breeze	11 – 16	S1 = 5 – 6 knots
5	Fresh Breeze	17 – 21	S2 = Royals, etc.
6	Strong Breeze	22 – 27	S2 = Topgallant sails
7	High Wind	28 – 33	S2 = Topsails, jib, etc.
8	Gale	34 – 40	S2 = Reefed upper topsails and courses
9	Strong Gale	41 – 49	S2 = Lower topsails and courses
10	Whole Gale	48 – 55	S3 = lower main topsail and reefed foresail
11	Storm	56 – 65	S3 = storm staysails
12	Hurricane	> 65	S3 = no canvas

Solved Example

Find the maximum possible wave height (assuming unlimited fetch) and wavelength for hurricane force winds of 35 m/s.

Solution

Given: $M = 50$ m/s
Find: $h = ?$ m, $\lambda = ?$ m

Use eq. (16.25):

$$h = (4\text{m}) \cdot \left(\frac{35\text{m/s}}{10\text{m/s}} \right)^{3/2} = \underline{\mathbf{26.2\ m}}$$

Use eq. (16.26):

$$\lambda = (35\text{m}) \cdot \left(\frac{35\text{m/s}}{10\text{m/s}} \right)^{1.8} = \underline{\mathbf{334\ m}}$$

Check: Units OK. Physics OK. Agrees with Figs.
Discussion: Wavelengths are much longer than wave heights. Thus, wave slopes are small — less that 1/10. Only when these waves reach shore does wave slope grow until the waves break as surf.

SUMMARY

Hurricanes are born over tropical oceans, and die when they leave regions of warm sea-surface temperature. Evaporation from the warm ocean into the windy boundary layer increases the energy in the storm, which ultimately drives its circulation similar to a Carnot-cycle heat engine. Hurricanes die over land not due to the extra drag caused by buildings and trees, but due mostly to the lack of strong evaporation of water from the surface.

Because hurricanes are born in the trade-wind regions of the general circulation, they are blown westward by the trade winds. Many eventually reach the eastern shores of continents where the general circulation turns them poleward. While near the shore, they can cause damage due to storm-surge flooding, wind-wave battering, beach erosion, wind damage, heavy rain, tornadoes, and lightning. The surge is caused by Ekman transport of water toward shore.

Hurricanes are tropical cyclones. They have low-pressure centers, called eyes, and rotation is cyclonic (counterclockwise in the Northern Hemisphere) near the surface. The hurricane core is warm, which causes high pressure to form in the eye near the top of the storm. This high pressure drives diverging, anticyclonic winds out of the hurricane.

Updrafts are strongest in the eye wall of thunderstorms encircling the clear eye. Rotation is initially gathered from the absolute angular momentum associated with the earth's rotation. As the storm develops and gains speed, centrifugal force dominates over Coriolis force within about 100 km of the eye, causing winds that are nearly cyclostrophic in the bottom third of the troposphere. Simple analytical models can be built to mimic the velocities, temperature, and pressure across a hurricane.

Threads

Energy for hurricanes comes ultimately from the sun (Chapt. 2), which heats the sea surface (Chapt. 3). Evaporation of sea water into the boundary layer (Chapt. 4) increases the humidity (Chapt. 5). Latent heat release during condensation drives the atmosphere like a heat engine. This thermodynamic cycle can be described using a thermodynamic diagram (Chapt. 6).

Some of this energy is realized as motions (Chapt. 9) that drive the strong winds around the eye of the tropical cyclone. These warm-core cyclones are different from cold-core extratropical cyclones (Chapt. 13). The relationship between the warm core temperature, the low pressure at the surface, and the high pressure aloft is described by the hypsometric equation (Chapt. 1) and thermal wind effects (Chapt. 9). Radial velocity is governed partly by boundary layer turbulence (Appendix H, Chapt. 4) and surface drag, and vertical velocity can be described using the continuity equation (Chapt. 9).

Once formed in the subtropics, hurricanes move toward the west in the trade winds of the general circulation (Chapt. 11). Surrounding the hurricane eye is an eye wall of clouds (Chapt. 7), heavy precipitation (Chapt. 8), and strong thunderstorms (Chapt. 15). **Hurricane frequency** of occurrence is modulated by the climate (Chapt. 18), and is less frequent during **El Niño** events.

EXERCISES

Numerical Problems

N1. At 10° latitude, find the absolute angular momentum associated with the following radii and tangential velocities:

	R (km)	M_{tan} (m/s)
a.	50	50
b.	100	30
c.	200	20
d.	500	5
e.	1000	0
f.	30	85
g.	75	40
h.	300	10

N2. If there is no rotation in the air at initial radius 500 km and latitude 10°, find the tangential velocity at radii (km):

a. 450 b. 400 c. 350 d. 300
e. 250 f. 200 g. 150 h. 100

N3. Assume $\rho = 1$ kg/m^3. Find the value of cyclostrophic wind for:

	R (km)	$\Delta P/\Delta R$ (kPa/100 km)
a.	100	5
b.	75	8
c.	50	10
d.	25	15
e.	100	10
f.	75	10
g.	50	20
h.	25	25

N4. For the previous problem, find the value of gradient wind, for latitude of 20°.

N5. At sea level, the pressure in the eye is 93 kPa and that outside is 100 kPa. Find the corresponding pressure difference at the top of the hurricane, assuming that the core (averaged over the hurricane depth) is warmer than surroundings by (°C):

a. 5 b. 2 c. 3 d. 4
e. 1 f. 7 g. 10 h. 15

N6. At radius 50 km the tangential velocity decreases from 35 m/s at the surface to 10 m/s at altitude (km)

a. 2 b. 4 c. 6 d. 8
e. 10 f. 12 g. 14 h. 16

Find the radial temperature gradient in the hurricane. The latitude = 10°, and average temperature = 0°C.

N7. Find the total entropy for:

	P (kPa)	T (°C)	r (g/kg)
a.	100	26	22
b.	100	26	0.9
c.	90	26	24
d.	80	26	0.5
e.	100	30	25
f.	100	30	2.0
g.	90	30	28
h.	20	–36	0.2

N8. On a thermo diagram of Chapter 6, plot the data points from Table 16-1.

N9. Starting with saturated air at sea-level pressure of 90 kPa in the eye wall with temperature of 26°C, calculate (by equation or by thermo diagram) the state of that air parcel as it moves to:

a. 20 kPa moist adiabatically, and thence to

b. a point where the potential temperature is the same as that at 100 kPa at 26°C, but at the same height as in part (a). Thence to

c. 100 kPa dry adiabatically and conserving humidity. Thence to

d. Back to the initial state.

e to h: Same as a to d, but with initial T = 30°C.

N10. Given the data from Table 16-1, what would be the mechanical energy available if the average temperature at the top of the hurricane were

a. –18 b. –25 c. –35 d. –45
e. –55 f. –65 g. –75 h. –83

N11. For the previous problem, find the minimum possible eye pressure that could be supported.

N12. Use P_∞ = 100 kPa at the surface. What maximum tangential velocity is expected for an eye pressure (kPa) of:

a. 86 b. 88 c. 90 d. 92
e. 94 f. 96 g. 98 h. 100

N13. For the previous problem, what are the peak velocity values to the right and left of the storm track, if the hurricane translates with speed (m/s):

(i) 10 (ii) 20

N14. For radius (km) of:

a. 5 b. 10 c. 15 d. 20
e. 25 f. 30 g. 50 h. 100

find the hurricane-model values of pressure, temperature, and wind components, given a pressure in the eye of 95 kPa, critical radius of R_o = 20 km, and W_s = –0.2 m/s. Assume the vertically-averaged temperature in the eye is 0°C.

N15. (§) For the previous problem, plot the radial profiles of those variables between radii of 0 to 200 km.

N16. Use P_∞ = 100 kPa at the surface. Find the rise of sea level in the eye of a hurricane with central pressure (kPa) of:

a. 86 b. 88 c. 90 d. 92
e. 94 f. 96 g. 98 h. 100

N17. Find the Ekman transport rate and surge slope if winds (m/s) of:

a. 10 b. 20 c. 30 d. 40
e. 50 f. 60 g. 70 h. 80

in advance of a hurricane are blowing parallel to the shore, over an ocean of depth 50 m. Use C_D = 0.005 and assume a latitude of 30°.

N18. For the previous problem, find the growth rate of the Kelvin wave if the hurricane tracks south parallel to shore at the same speed as the wave.

N19. What is the Kelvin wave speed in an ocean of depth (m):

a. 200 b. 150 c. 100 d. 80
e. 60 f. 40 g. 20 h. 10

N20. Find the wind-wave height and wavelength expected for wind speeds (m/s) of:

a. 10 b. 15 c. 20 d. 25
e. 30 f. 40 g. 50 h. 60

N21. For the previous problem, give the:

(i) Beaufort wind category
(ii) Saffir-Simpson hurricane category

Understanding & Critical Evaluation

U1.(§) Using relationships from Chapter 1, plot an environmental pressure profile vs. height across the troposphere assuming an average temperature of 273 K. Assume the environmental sea-level pressure is 100 kPa. On the same graph, plot the pressure profile for a warm hurricane core of average temperature (K): a. 280 b. 290 c. 300
assuming a sea-level pressure of 95 kPa.

U2. Suppose that the pressure-difference magnitude between the eye and surroundings at the hurricane top is only half that at the surface. How would that change, if at all, the temperature model for the hurricane? Assume the sea-level pressure distribution is unaltered.

U3. For the hurricane model given in this chapter, describe how the pressure, tangential velocity, radial velocity, vertical velocity, and temperature distribution are consistent with each other, based on dynamic and thermodynamic relationships. If they are not consistent, quantify the source and magnitude of the discrepancy, and discuss the implications and limitations. Consider the idealizations of the figure below.

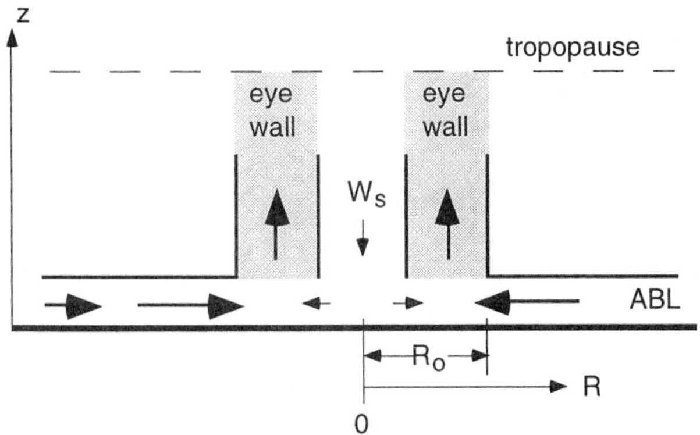

Figure 16.18
Idealized hurricane.

Hints: a. Use the cyclostrophic relationship to show that tangential velocity is consistent with the pressure distribution.

b. Use mass conservation for inflowing air trapped within the boundary layer to show how radial velocity should change with R, for $R > R_o$.

c. Assuming wave drag causes radial velocity to be proportional to tangential velocity squared, show that the radial velocity and tangential velocity equations are consistent. For $R < R_o$, use the following alternate relationship based on observations in hurricanes: $M_{tan}/M_{max} = (R/R_o)^2$.

d. Rising air entering the bottom of the eye wall comes from two sources, the radial inflow in the boundary layer from $R > R_o$, and from the subsiding air in the eye, which hits the ground and is forced to diverge horizontally in order to conserve mass. Combine these two sources of air to compute the average updraft velocity within the eye wall.

e. Use mass continuity in cylindrical coordinates

$$\frac{1}{R}\frac{\Delta(R \cdot M_{rad})}{\Delta R} = -\frac{\Delta W}{\Delta z}$$

to derive vertical velocity from radial velocity, for $R < R_o$. (Note, for $R > R_o$, it was already assumed in part (a) that air is trapped in the ABL, so there is zero vertical velocity there.)

f. Use the hypsometric relationship, along with the simplifications described in the temperature subsection of the hurricane model, to relate the radial temperature distribution (averaged over the whole hurricane depth) to the pressure distribution.

U4. Derive eq. (16.5), using the hints given in the paragraph after that eq. Hint: use the first-order series expansion of: $\exp(-c) \cong 1 - c$.

U5. An article by Willoughby and Black (1996: Hurricane Andrew in Florida: dynamics of a disaster, *Bull. Amer. Meteor. Soc.*, **77**, 543-549) shows tangential wind speed vs. radial distance.

a. For their Figs. 3b and 3d, compare their observations with the hurricane model in this chapter.

b. For their Figs. 3e - 3g, determine translation speed of the hurricane and how it varied with time as the storm struck Florida.

U6. a. What is the relationship between angular momentum and vorticity?

b. Re-express eq. (16.1) as a function of vorticity.

U7. In Fig. 16.5b, why does the T_d line from Point 1 to Point 2 follow a contour that slopes upward to the right, even though the air parcel in Fig 16.5a is moving horizontally, staying near sea level?

U8. a. Re-express eq. (16.7) in terms of potential temperature.

b. Discuss the relationship between entropy and potential temperature.

U9. Fig. 16.12 for a storm surge caused by Ekman transport is for a hurricane just off the east coast of a continent. Instead, suppose there was a hurricane-force cyclone just off the west coast of a continent at mid-latitudes.

a. Would there still be a storm surge caused by Ekman transport?

b. Which way would the resulting Kelvin wave move (north or south) along the west coast?

U10. Devise a mathematical relationship between the Saffir-Simpson hurricane scale, and the Beaufort wind scale.

Web-Enhanced Questions

For the following questions, search the web for Hurricane Frequently Asked Questions (Hint: http://www.aoml.noaa.gov/hrd/tcfaq/tcfaqHED.html) and for hurricane tutorials (Hint: http://covis.atmos.uiuc.edu/guide/hurricanes/html/menu.html)

W1. Search for satellite images and movies of different hurricanes. Discuss similarities and differences.

W2. What averaging period is used when measuring "max sustained surface winds"?

W3. What is a "Cape Verde" hurricane?

W4. What wind speeds are required for "super-typhoon" classification?

W5. Easterly waves:
- a. What causes the African easterly jet?
- b. What are the typical wavelength and period of easterly waves?
- c. How many waves are typically generated each year?

W6. What is a sub-tropical cyclone?

W7. What are the differences between mid-latitude cyclones and tropical cyclones?

W8. Which quadrant of a land-falling hurricane has conditions that favor tornado formation?

W9. Define these acronyms:
a. CDO b. TUTT c. CLIPER

W10. Search for web sites that show hurricane tracks for
- a. the current hurricane season.
- b. past hurricane seasons.

W11. What is the range of typical diameters of the eye?

W12. What names will be used for the first 3 tropical cyclones in the Atlantic next year?

W13. Why are some hurricane names "retired"?

W14. Which hurricane myth is your favorite?

W15. What are "concentric eyewall circles", and in what situations do they form?

W16. What is the highest category hurricane that an anchored mobile home could survive without major damage?

W17. Intense (category 3-5) hurricanes account for what percentage of tropical cyclone landfalls, but cause what percent of the damage in N. America?

W18. How many times more destructive is a category 4 hurricane compared to category 1, considering "potential damage"?

W19. What is the depth of the greatest storm surge recorded for a tropical cyclone?

W20. On the average:

a. Which region has the most hurricanes/typhoons per year, Atlantic or NE Pacific?

b. What is the total global average number of hurricanes/typhoons per year?

W21. Are hurricanes more or less likely during an El Niño year? Why?

W22. How many models are available for forecasting hurricane track?

W23. Which 3 months are the peak hurricane season in the Atlantic?

W24. Give 2 reasons why hurricanes/typhoons are unlikely to hit the W. Coast of N. America.

W25. What max wind speeds (horizontal and vertical) at flight level have "hurricane hunters" observed, and in what part of the storm are they found?

W26. Where (web site) can you get hurricane preparedness info?

W27. What is Dr. Bill Gray's latest forecast for Hurricane activity in the Atlantic? Is it forecast to be greater or less than average? [Hint: look for a web site at Colorado State University (...colostate.edu)].

W28. What type of "hurricane potential" information is available on the web? (Hint: http://grads.iges.org/ pix/hurpot.html)

W29. a. At what web sites can you get current sea-surface temperature maps?

b. Also, can you find any animations showing how sea-surface temperature has varied over a year, or over several months?

Synthesis Questions

S1. What if the earth rotated twice as fast. Describe changes to hurricane characteristics, if any.

S2. Suppose that global warming caused the sea surface temperature to exceed 26°C from the equator to 60° latitude in late summer and early autumn. How would hurricane characteristics change, if at all?

S3. What if the average number of hurricanes tripled. How would the momentum, heat, and moisture transport by hurricanes change the global circulation, if at all?

S4. Some science fiction novels describe "supercanes" with supersonic wind speeds. Are these physically possible? Describe the dynamics and thermodynamics necessary to support such a storm in steady state, or use the same physics to show why they are not possible.

S5. Suppose the tropical tropopause was at 8 km altitude, instead of roughly 16 km altitude. How would hurricane characteristics change, if at all?

S6. What if the earth's climate was such that the tropics were cold and the poles were hot, with sea surface temperature greater than 26°C reaching from the poles to 60° latitude. Describe changes to hurricane characteristics, if any.

S7. Suppose that the sea surface was perfectly smooth, regardless of the wind speed. How would hurricane characteristics change, if at all?

S8. What if Coriolis force was equal in magnitude to the pressure-gradient and centrifugal force terms, even for air in the hurricane eye wall rotating around the eye. Describe how hurricane force winds would differ, if at all.

S9. Is it possible to have a hurricane without a warm core? Be aware that in the real atmosphere, there are hurricane force cyclones a couple times a year over the northern Pacific Ocean, during winter.

S10. Suppose that the thermodynamics of hurricanes were such that air parcels, upon reaching the top of the eye wall clouds, do not loose any heat by IR cooling as they horizontally diverge away from the top of the hurricane. How would the Carnot cycle change, if at all, and how would that affect hurricane intensity?

S11. Is it possible for a hurricane to move from the northern to the southern hemisphere? If so, discuss the processes that would make it happen.

Science Graffito

In 1989, category 5 hurricane Hugo moved directly over the US. Virgin Islands in the Caribbean Sea. When the hurricane reached the island of St. Croix, it temporarily stopped its westward translation, allowing the intense eye wall to blast the island with violent winds for hours. The following is an eye witness account.

"It had been many years since St. Croix was in the path of a major storm. Hurricane Hugo reached into the Lesser Antilles with a deliberate vengeance. St. Croix was somewhat prepared. Many hundreds of people had moved into schools and churches to take refuge. But no one was ready for what happened next. By 1800 hours winds were a steady 50 kts with gusts up to 70 kts from the northwest. I was on the top floor of the wooden Rectory at the St. Patrick's Church in Frederiksted with my husband and 8 month old son."

"By 2000 hours it was apparent that our comfortable room with a view was not going to provide a safe haven. The electricity had been out for some time and a very big gust from the north blew the air conditioner out of the window, landing at the foot of our bed. We evacuated with only one diaper change and bottle of baby juice, leaving behind the playpen, high chair, and bundles of accessories brought from home. We followed Fr. Mike down the wooden staircase. Drafts were everywhere and glass doors exploded just as we passed on our way to Fr. O'Connor's main living quarters on the first floor, where the walls were made of thick coral blocks."

"We settled in again in spite of the persistent crashing and banging against the heavy wooden shutters. We had to shout to hear each other across the room and our ears were popping. In the bathroom, the plumbing sounded like a raging sea. The water in the toilet bowl sloshed around and vibrated. Mercifully the baby slept."

"Soon the thick concrete walls and floors were vibrating accompanied by a hum that turned into the "freight train howl". The banging intensified and persisted for the next 4 hours. By 0100 hours we were tense and sweaty and wondering if it would ever end and if there was anything left outside. Fr. O'Connor was praying and feared that many people must be dead. He got up to open a closet door and a wall of water flowed into the bedroom. At that point we moved to the dining room with a group of 8 other people trapped in the rectory and waited. "

(continued next column)

Science Graffito

(continuation)

There was concern that the rest of the roof would go and it was decided we would make a run for the schoolhouse made of 2-foot thick concrete walls. I held the baby in my arms and with flimsy flip-flops, just about skated across the cement courtyard dodging flying branches and sheets of galvanized aluminum. The window was opened for us as a big, old mahogany tree blocked the door.

Shortly after, the eye was over us. The thick wooden shutters were flung open and about 100 people outside climbed in the window. The housing project nearby had been stripped of its north and east walls. The eye remained over us for 2 hours then the wind started up with the same intensity coming from the southwest. Only now the room was packed. Strangers were sharing the same mattresses. People slept in desks and chairs made for elementary children and it was hot. Toddlers and infants wailed.

There was no generator, only an occasional flashlight could be seen. Fears of surges were on everyone's' minds. We were only 200 meters from the west shoreline. Hugo had slowed down its eastward track to 4 mph and the eye passed straight through the middle of the 23-mile long island of St. Croix. It seems like the storm was in a fixed permanent position.

When dawn broke, the winds still howled. By 0800 it was safe to open the windows and the landscape made me burst into tears. There was not a leaf left on a tree, there was not a tree left standing, just tangled branches lying sideways everywhere and not one blade of green grass. The wind had burned the ground and turned everything brown. The gray skies, light rain, and brown landscape persisted for several weeks.

There were only 2 deaths reported but within weeks several dozen people died from heart attacks, strokes, electrocutions and other accidents associated with reconstruction. The majority of the island residents functioned without power for 3 to 6 months, using generators or candle power and gas stoves.

– Susan Krueger Allick, 1999

AIR POLLUTION DISPERSION

CONTENTS

17 Every living thing pollutes. Life is a chemical reaction, where input chemicals such as food and oxygen are converted into growth or motion. The reaction products are waste or pollution.

The only way to totally eliminate pollution is to eliminate life — not a particularly appealing option. However, a system of world-wide population control could stem the increase of pollution, allowing residents of our planet to enjoy a high quality of life.

Is pollution bad? From an anthropocentric point of view, we might say "yes". To do so, however, would deny our dependence on pollution. In the earth's original atmosphere, there was very little oxygen. Oxygen is believed to have formed as pollution from plant life. Without this pollutant, animals such as humans would likely not exist now.

However, it is reasonable to worry about other **chemicals that threaten our quality of life**. We call such chemicals **pollutants**, regardless of whether they form naturally, or **anthropogenically** (man-made). Many of the natural sources are weak emissions from large area sources, such as forests or swamps. Anthropogenic sources are often concentrated at points, such as at the top of smoke stacks (Fig 17.1). Such high concentrations are particularly hazardous, and have received significant study.

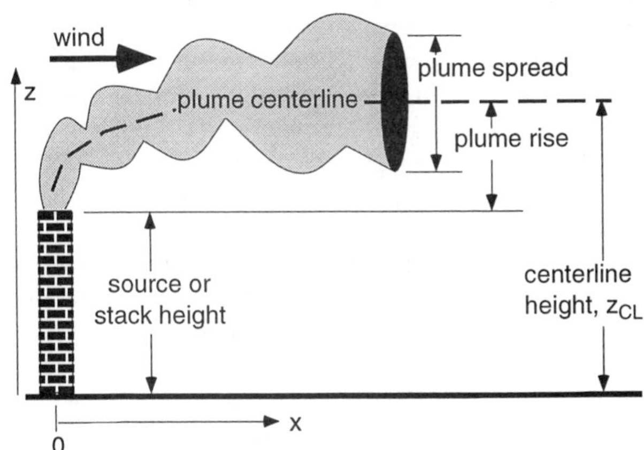

Figure 17.1
Pollutant plume characteristics.

DISPERSION FACTORS

Dispersion is the name given to the spread and movement of pollutants. Pollution dispersion depends on

- wind speed and direction
- plume rise, and
- atmospheric turbulence.

Pollutants disperse with time by mixing with the surrounding cleaner air, resulting in an increasingly dilute mixture within a plume of increasing size.

Wind and **turbulence** are characteristics of the ambient atmosphere, as were described in earlier chapters. While emissions out of the top of the stack often have strong internal turbulence, this quickly decays, leaving the ambient flow to do the majority of the dispersing.

The direction that the effluent travels is controlled by the synoptic weather and general circulation. Destinations of known emissions can be found using a **forward trajectory** along the mean wind, while source locations of polluted air can be found from a **backward trajectory**.

The stream of polluted air downwind of a smoke stack is called a **smoke plume**. If the plume is buoyant, or if there is a large effluent velocity out of the top of the smoke stack, the center of the plume can rise above the initial emission height. This is called **plume rise**.

The word "plume" in air pollution work means a long, slender, nearly-horizontal region of polluted air. However, the word "plume" in boundary-layer studies refers to the relatively wide, nearly vertical updraft portion of buoyant air that is convectively overturning. Because smoke plumes emitted into the boundary layer can be dispersed by convective plumes, one must take great care to not confuse the two usages of the word "plume".

The goal of calculating dispersion is to predict or diagnose the pollutant concentration at some point distant from the source. **Concentration** c is often measured as a mass per unit volume, such as mg/m^3, or $\mu g/m^3$. It can also be measured as volume ratio of pollutant gas to clean air, such as parts per million (**ppm**) or parts per billion (**ppb**).

A **source - receptor** framework is used to relate emission factors to predicted downwind concentration values. We can examine pollutants emitted at a known rate from a **point source** such as a **smoke stack**. We then follow the pollutants as they are blown downwind and mix with the surrounding air. Eventually, the mixture reaches a receptor such as a sensor, person, plant, animal or structure, where we can determine the expected concentration.

In this chapter, we will assume that the mean wind is known, based on either weather observations, or on forecasts. We will focus on the plume rise and dispersion of the pollutants, which allows us to determine the concentration of pollutants downwind of a known source.

AIR QUALITY STANDARDS

To prevent or reduce health problems associated with air pollutants, many countries set air quality standards. These standards prescribe the maximum average concentration levels allowed by law. Failure to satisfy these standards can result in fines, penalties, and increased government regulation.

In the USA, the standards are called **National Ambient Air Quality Standards** (**NAAQS**). In Canada, they are called National Air Quality Objectives. In the Great Britain, they are called National Air Quality Strategy Standards and Objectives. Other countries have similar names for such goals. Table 17–1 lists standards for a few countries.

In theory, these concentrations are not to be exceeded anywhere at ground level at any time. In practice, meteorological events sometimes occur, such as light winds and shallow ABLs during anticyclonic conditions, that trap pollutants near the ground and cause concentration values to become undesirably large.

Also, temporary failures of air pollution control measures at the source can cause excessive amounts of pollutants to be emitted. Regulations in some of the countries allow for a small number of concentration **exceedences** without penalty.

To avoid expensive errors during the design of new factories, smelters, or power plants, **air pollution modeling** is performed to determine the likely pollution concentration based on expected emission rates. Usually, the greatest concentrations happen near the source of pollutants. The procedures presented in this chapter illustrate how concentrations at receptors can be calculated from known emission and weather conditions.

By comparing the predicted concentrations against the air quality standards of Table 17–1, engineers can modify the factory design as needed to ensure compliance with the law. Such modifications can include building taller smoke stacks, removing the pollutant from the stack effluent, changing fuels or raw materials, or utilizing different manufacturing or chemical processes.

Table 17–1. Air quality concentration (μg/m^3) standards for the USA (US), Canada (CAN), and Great Britain (UK) for some of the commonly-regulated chemicals. Concentrations represent averages over the time periods listed. For Canada, listed are the "max. acceptable" levels.

Avg.Time	US	CAN	UK
Sulfur Dioxide (SO_2)			
1 yr	80	60	
1 day	365	300	
3 h	1300		
1 h		900	
15 min			260
Nitrogen Dioxide (NO_2)			
1 yr	100	100	40
1 day		200	
1 h		400	280
Carbon Monoxide (CO)			
8 h	10,000	15,000	11,300
1 h	40,000	35,000	
Ozone (O_3)			
1 yr		30	
1 day		50	
8 h	157		100
1 h	235	160	
Fine Particulates, diameter < 10 μm **(PM-10)**			
1 yr	50	70	
1 day	150	120	50
Fine Particulates, diameter < 2.5 μm **(PM-2.5)**			
1 yr	15		
1 day	65		
Lead (Pb)			
1 yr			0.5
3 mo	1.5		

TURBULENCE STATISTICS

For air pollutants emitted from a point source such as the top of a smoke stack, mean wind speed and turbulence both affect the pollutant concentration measured downwind at ground level. The mean wind causes pollutant **transport**. Namely it blows or advects the pollutants from the source to locations downwind. However, while the plume is advecting, turbulent gusts acts to spread, or **disperse**, the pollutants as they mix with the surrounding air. Hence, we need to study both mean and turbulent characteristics of wind in order to predict downwind pollution concentrations.

Review of Basic Definitions

Recall from Chapter 4 that variables such as velocity components, temperature, and humidity can be split into mean and turbulent parts. For example:

$$M = \overline{M} + M' \qquad (17.1)$$

where M is instantaneous speed in this example, $\overline{M}$ is the mean wind speed (usually averaged over time or horizontal distance), and M' is the instantaneous deviation from the mean value.

The mean wind speed at any height z is

$$\overline{M}(z) = \frac{1}{N}\sum_{i=1}^{N} M_i(z) \qquad (17.2)$$

where M_i is the wind speed measured at some time or horizontal location index, i, and N is the total number of observation times or locations.

However, smoke plumes can spread in the vertical direction. Recall from Chapt. 4 that the ABL wind speed often varies with height. Hence, the wind speed that affects the pollutant plume must be defined as an average speed over the vertical thickness of the plume.

If the wind speeds at different, equally spaced layers, between the bottom and the top of a smoke plume are known, and if k is the index of any layer, then the average over height is:

$$\overline{\overline{M}} = \frac{1}{K}\sum_{k=1}^{K} \overline{M}(z_k) \qquad (17.3)$$

where the sum is over only those layers spanned by the plume. K is the total number of layers in the plume.

This works for nearly horizontal plumes that have known vertical thickness. For the remainder of this chapter, we will use just one overbar (or sometimes no overbar) to represent an average over both time (index i), and vertical plume depth (index k).

The coordinate system is often chosen so that the x-axis is aligned with the mean wind direction, averaged over the whole smoke plume. Thus,

$$\overline{M} \equiv \overline{U} \qquad (17.4)$$

There is no lateral (crosswind) mean wind ($\overline{V} \approx 0$) in this coordinate system. The mean vertical velocity is quite small, and can usually be neglected ($\overline{W} \approx 0$, except near hills) compared to plume dispersion rates. However, u', v', and w' can be non-zero, and are all important.

Recall from Chapt. 4 that **variance** $\sigma_A{}^2$ of any quantity A is defined as

$$\sigma_A{}^2 = \frac{1}{N}\sum_{k=1}^{N}(A_k - \overline{A})^2 = \frac{1}{N}\sum_{k=1}^{N}(a'^{\,2}) = \overline{a'^{\,2}} \qquad (17.5)$$

Standard deviation is

$$\sigma_A = (\sigma_A{}^2)^{1/2}. \qquad (17.6)$$

Chapt. 4 gives estimates of velocity standard deviations.

Isotropy (again)

Recall from Chapter 4 that turbulence is said to be **isotropic** when:

$$\sigma_u{}^2 = \sigma_v{}^2 = \sigma_w{}^2 \qquad (17.7)$$

As will be shown later, the rate of smoke dispersion depends on the velocity variance. Thus, if turbulence is isotropic, then a smoke puff would tend to expand isotropically, as a sphere; namely, it would expand equally in all directions.

There are many situations where turbulence is **anisotropic** (not isotropic). During the daytime over bare land, rising thermals create stronger vertical motions than horizontal. Hence, a smoke puff would disperse more in the vertical. At night, vertical motions are very weak, while horizontal motions can be larger. This causes smoke puffs to **fan** out horizontally at night, and for other stable cases.

Similar effects operate on smoke plumes formed from continuous emissions. For this situation, only the vertical and lateral velocity variances are relevant. Fig 17.2 illustrates isotropy and anisotropy.

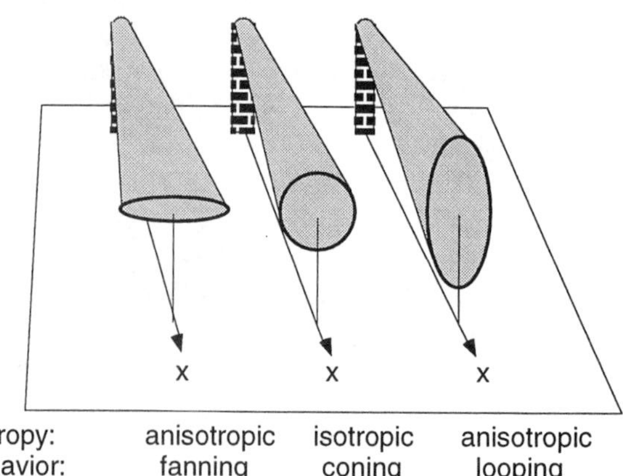

Figure 17.2
Isotropic and anisotropic dispersion of smoke plumes.

Solved Example (§)

Given an x-axis aligned with the mean wind U=10 m/s, and the y-axis aligned in the crosswind direction, V. Listed at right are measurements of the V-component of wind.

a. Find the V mean wind speed and standard deviation.

b. If the vertical standard deviation is 1 m/s, is the flow isotropic?

t (h)	V (m/s)
0.1	2
0.2	–1
0.3	1
0.4	1
0.5	–3
0.6	–2
0.7	0
0.8	2
0.9	–1
1.0	1

Solution

Given: Velocities: σ_w = 1 m/s, U = 10 m/s
Find: $\overline{V}$ = ? m/s, σ_v = ? m/s, isotropy = ?
Assume V wind is constant with height.

Use eq. (17.2), except for V instead of M:

$$\overline{V}(z) = \frac{1}{N}\sum_{i=1}^{N} V_i(z) = \frac{1}{10}(0) = \underline{\mathbf{0\ m/s}}$$

Use eq. (17.5), but for V: $\sigma_v{}^2 = \frac{1}{n}\sum_{i=1}^{n}(V_i - \overline{V})^2$

$= (0.1)\cdot(4+1+1+1+9+4+0+4+1+1) = 2.6\ \mathrm{m^2/s^2}$

Use eq. (17.6)

$$\sigma_v = \sqrt{2.6\mathrm{m}^2\cdot\mathrm{s}^{-2}} = \underline{\mathbf{1.61\ m/s}}$$

Use eq. (17.7): $\sigma_v > \sigma_w$, therefore **Anisotropic**.

Check: Units OK. Physics OK.
Discussion: Dispersion looks like Fig 17.2 stable.

Pasquill-Gifford (PG) Turbulence Types

During weak advection, the nature of convection and turbulence are controlled by the wind speed, incoming solar radiation (**insolation**), cloud shading, and time of day or night. Pasquill and Gifford (PG) suggested a practical way to estimate the nature of convection, based on these forcings.

They used the letters "A" through "F" to denote different turbulence types, as sketched in Fig 17.3 (reproduced from Fig 4.24 in Chapter 4). "A" denotes free convection in statically unstable conditions. "D" is forced convection in statically neutral conditions. Type "F" is for statically stable turbulence. Type "G" was added later to indicate meandering, wavy plumes in otherwise-nonturbulent flow. PG turbulence types can be estimated using Tables 17-2.

Early methods for determining pollutant dispersion utilized a different plume spread equation for each Pasquill-Gifford type. One drawback is that there are only 7 discrete categories (A – G); hence, calculated plume spread would suddenly jump when the PG category changed in response to changing atmospheric conditions.

Newer air pollution models do not use the PG categories, but use the fundamental meteorological conditions (such as shear and buoyant TKE generation, or values of velocity variances that are continuous functions of wind shear and surface heating), which vary smoothly as atmospheric conditions change.

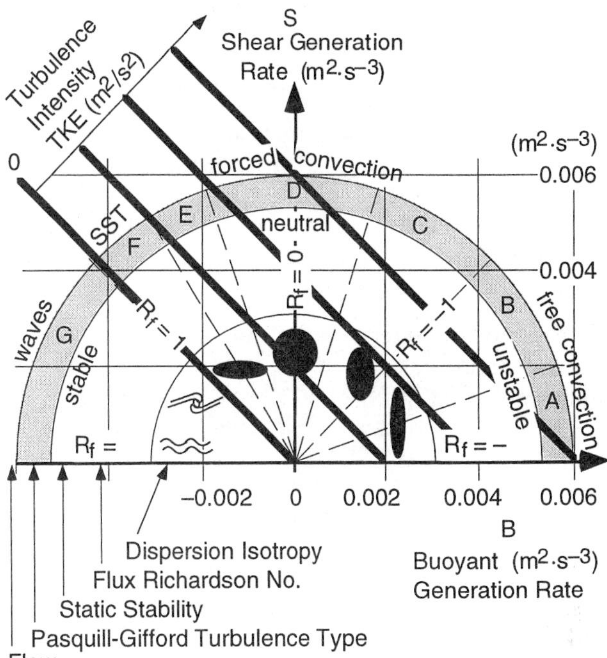

Figure 17.3
Rate of generation of TKE by buoyancy (abscissa) and shear (ordinate). Shape and rates of plume dispersion (dark spots or waves). Dashed lines separate sectors of different Pasquill-Gifford turbulence type. Isopleths of TKE intensity (dark diagonal lines). R_f is flux Richardson number (see Chapt. 4). SST is stably-stratified turbulence. (See Chapt. 4 for turbulence details.)

Table 17-2a. Pasquill-Gifford turbulence types, for **Daytime**.

M (m/s)	Insolation: Strong	Moderate	Slight
< 2	A	A to B	B
2 to 3	A to B	B	C
3 to 4	B	B to C	C
4 to 6	C	C to D	D
> 6	C	D	D

Table 17-2b. Pasquill-Gifford turbulence types, for **Nighttime**.

M (m/s)	Cloud Coverage: ≥ 4 / 8 low cloud or thin overcast	≤ 3 / 8
< 2	G	G
2 to 3	E	F
3 to 4	D	E
4 to 6	D	D
> 6	D	D

Solved Example

Determine the PG turbulence type during night with 25% cloud cover, and winds of 5 m/s.

Solution

Given: $M = 5$ m/s, clouds = 2/8 .
Find: PG = ?

Use Table 17–2b. PG = "D"

Check: Units OK. Physics OK.
Discussion: As wind speeds increase, the PG category approaches "D" (statically neutral), for both day and night conditions. This corresponds to forced convection.

Science Graffito

"If I had only one day left to live, I would live it in my statistics class — it would seem so much longer."

– Anonymous. [from C.C Gaither (Ed.), 1996: *Statistically Speaking: A Dictionary of Quotations*, Inst. of Physics Pub., 420 pp].

DISPERSION STATISTICS

Snapshot vs. Average

Snapshots of smoke plumes are similar to what you see with your eye. The plumes have fairly-well defined edges, but each plume wiggles up and down, left and right (Fig 17.4a). The concentration through such an instantaneous smoke plume can be quite variable, so a hypothetical profile is sketched in Fig 17.4a.

A time exposure of a smoke-stack plume might appear as sketched in Fig 17.4b. When averaged over a time interval such as an hour, most of the pollutant is found near the centerline of the plume. Average concentration decreases smoothly with distance away from the centerline. The resulting profile of concentration is often bell shaped, or **Gaussian**. Air quality standards in most countries are based on averages, as was listed in Table 17–1.

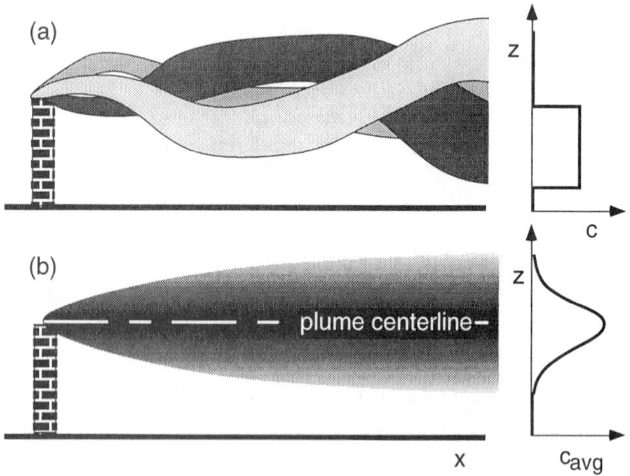

Figure 17.4
(a) Snapshots of instantaneous smoke plumes at different times, also showing a concentration c profile for the dark-shaded plume. (b) Average over many plumes, with the average concentration, c_{avg}, profile shown at right.

Center of Mass

The **plume centerline** height z_{CL} can be defined as the location of the **center of mass** of pollutants. In other words, it is the average height of the pollutants. Suppose you measure the concentration c_k of pollutants at a range of equally-spaced heights z_k through the smoke plume. The center of mass is a weighted average defined by:

$$z_{CL} = \bar{z} = \sum_{k=1}^{K} c_k \cdot z_k \Bigg/ \sum_{k=1}^{K} c_k \qquad \bullet(17.8)$$

where K is the total number of heights, and k is the height index, and the overbar denotes a mean.

For passive tracers, the plume centerline is at the same height as the top of the stack. For buoyant plumes in hot exhaust gases, the centerline rises above the top of the stack. Plume rise can also occur if the exit velocity out of the stack is large enough to blow the pollutant upward.

A similar center of mass can be found for the crosswind (lateral) location, assuming measurements are made at equal intervals across the plume. Passive tracers blow downwind. Thus, the center of mass of a smoke plume, when viewed from above such as from a satellite, follows a mean wind trajectory from the stack location (see the discussion of streamlines, streaklines, and trajectories in Chapt 10).

Standard Deviation – Sigma

For time-average plumes such as in Fig 17.4b, the plume edges are not easy to locate. They are poorly defined because the bell curve gradually approaches zero concentration with increasing distance from the centerline. Thus, we cannot use edges to measure plume spread (depth or width).

Instead, the standard deviation σ_z of pollutant location is used as a measure of plume spread, where standard deviation is the square root of the variance $\sigma_z{}^2$. The vertical-position variance must be weighted by the pollution concentration as shown here:

$$\sigma_z{}^2 = \frac{\sum_{k=1}^{K} c_k \cdot (z_k - \bar{z})^2}{\sum_{k=1}^{K} c_k} \qquad \bullet(17.9)$$

where $\bar{z}$ is the average height found from the previous equation.

A similar equation can be defined for lateral standard deviation: σ_y. The vertical and lateral dispersion need not be equal, because the dispersive nature of turbulence is not the same in the vertical and horizontal when turbulence is anisotropic.

When the plume is compact, the standard deviation and variance are small. These statistics increase as the plume spreads. Hence we expect sigma to increase with distance downwind of the stack. Such spread will be quantified in the next section.

Gaussian Curve

The Gaussian or "normal" curve is bell shaped, and is given in one-dimension by:

$$c(z) = \frac{Q_1}{\sigma_z \sqrt{2\pi}} \cdot \exp\left\{ -0.5 \left[\frac{z - \bar{z}}{\sigma_z} \right]^2 \right\} \qquad \bullet(17.10)$$

where $c(z)$ is the one-dimensional concentration (g/m) at any height z, and Q_1 (g) is the total amount of pollutant emitted.

This curve is symmetric about the mean location, and has tails that asymptotically approach zero as z approaches infinity (Fig 17.5). The area under the curve is equal to Q_1, which physically means that pollutants are conserved. The **inflection points** in the curve (points where the curve changes from concave down to concave up) occur at exactly one σ_z from the mean. Between $\pm 2 \cdot \sigma_z$ are 95% of the pollutants; hence, the Gaussian curve exhibits strong central tendency.

Eq. (17.10) has three parameters: Q_1, $\bar{z}$, and σ_z. These parameters can be estimated from measurements of concentration at equally-spaced heights through the plume, in order to find the best-fit Gaussian curve. The last two parameters are found with eqs. (17.8) and (17.9). The first parameter is found from:

$$Q_1 = \Delta z \cdot \sum_{k=1}^{K} c_k \qquad \bullet(17.11)$$

where Δz is the height interval between neighboring measurements.

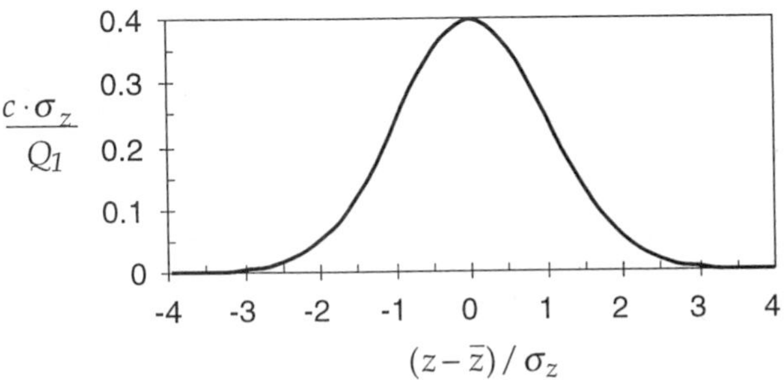

Figure 17.5
Gaussian curve. Each unit along the abscissa corresponds to one standard deviation away from the center.

Nominal Plume Edge

For practical purposes, the edge of a Gaussian plume is defined as the location where the concentration is 10% of the centerline concentration. This gives a plume spread (e.g., depth from top edge to bottom edge) of

$$d = 4.3 \cdot \sigma_z \qquad \bullet(17.12)$$

Solved Example (§)

Given the following concentration measurements. Find the plume centerline height, standard deviation of height, and nominal plume depth. Plot the best-fit curve through these data points.

z (km)	c(µg/m^3)
2	0
1.8	1
1.6	3
1.4	5
1.2	7
1	6
0.8	2
0.6	1
0.4	0
0.2	0
0	0

Solution

Given: Δz = 0.2 km, with concentrations above

Find: $\bar{z}$ = ? km, σ_z = ? km, Q_1 = ? km·µg/m^3, d = ? km, and plot $c(z)$ = ? µg/m^3

Use eq. (17.8) to find the plume centerline height:

$$\bar{z} = \frac{30.2\text{km} \cdot \mu\text{g} \cdot \text{m}^{-3}}{25\mu\text{g} \cdot \text{m}^{-3}} = \underline{\mathbf{1.208\ km}}$$

Use eq. (17.9):

$$\sigma_z^2 = \frac{1.9575\text{km}^2 \cdot \mu\text{g} \cdot \text{m}^{-3}}{25\mu\text{g} \cdot \text{m}^{-3}} = 0.0783\ \text{km}^2$$

$$\sigma_z = \sqrt{0.0783\text{km}^2} = \underline{\mathbf{0.28\ km}}$$

Use eq. (17.12):

$$d = 4.3 \cdot (0.28\text{km}) = \underline{\mathbf{1.2\ km}}$$

Use eq. (17.11):

$$Q_1 = (0.2\text{km}) \cdot 25\mu\text{g} \cdot \text{m}^{-3} = 5.0\ \text{km} \cdot \mu\text{g/m}^3$$

Use eq. (17.10) to plot the best-fit curve:

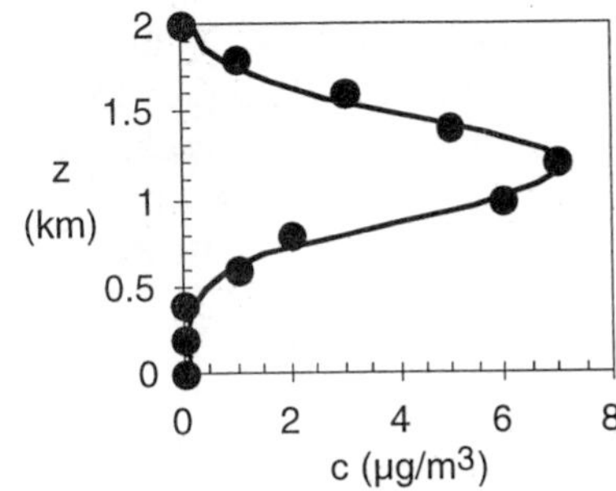

Check: Units OK. Physics OK. Sketch OK.
Discussion: The curve is a good fit to the data.

TAYLOR'S STATISTICAL THEORY

Statistical theory explains how plume dispersion statistics depend on turbulence statistics and downwind distance.

Passive Conservative Tracers

Many pollutants consist of gases or very fine particles. They passively ride along with the wind, and trace out the air motion. Hence, the rate of dispersion depends solely on the air motion (wind and turbulence) and not on the nature of the pollutant. These are called **passive tracers**. If they also do not decay, react, fall out, or stick to the ground, then they are also called **conservative tracers**, because all pollutant mass emitted into the air is conserved.

Some pollutants are not passive or conservative. Dark soot particles can adsorb sunlight to heat the air, but otherwise they might not be lost from the air. Thus, they are active because they alter turbulence by adding buoyancy, but are conservative. Radioactive pollutants are both nonconservative and active, due to radioactive decay and heating.

For passive conservative tracers, the amount of dispersion (σ_y or σ_z) depends not only on the intensity of turbulence (σ_v or σ_w , see Chapt. 4), but on the distribution of that turbulence among eddies of different sizes. For a plume of given spread, eddies as large as the plume diameter cause much greater dispersion than smaller-size eddies. Thus, dispersion rate increases with time or downwind distance, as shown below.

Dispersion Equation

G.I. Taylor theoretically examined an individual passive tracer particle as it moved about by the wind. Such an approach is **Lagrangian**, as discussed in Chapter 3. By averaging over many such particles within a smoke cloud, he derived a **statistical theory** for turbulence.

One approximation to his result is

$$\sigma_y^2 = 2 \cdot \sigma_v^2 \cdot t_L^2 \cdot \left[\frac{x}{M \cdot t_L} - 1 + \exp\left(-\frac{x}{M \cdot t_L} \right) \right] \qquad \bullet(17.13a)$$

$$\sigma_z^2 = 2 \cdot \sigma_w^2 \cdot t_L^2 \cdot \left[\frac{x}{M \cdot t_L} - 1 + \exp\left(-\frac{x}{M \cdot t_L} \right) \right] \qquad \bullet(17.13b)$$

where x is distance downwind from the source, M is wind speed, and t_L is the Lagrangian time scale.

Thus, the spread σ_y and σ_z of passive tracers increases with turbulence intensity (σ_v or σ_w) and with downwind distance x.

The **Lagrangian time scale** is a measure of how quickly a variable becomes uncorrelated with itself. For very small-scale atmospheric eddies, this time scale is only about 15 seconds. For convective thermals, it is on the order of 15 minutes. For the synoptic-scale high and low pressure systems, the Lagrangian time scale is on the order of a couple days. We will use a value of 1 minute for dispersion in the boundary layer.

Dispersion Near and Far from the Source

Close to the source (at small times after the start of dispersion), eq. (17.13a) reduces to

$$\sigma_y \approx \sigma_v \cdot \left(\frac{x}{M} \right) \qquad \bullet(17.14)$$

while far from the source it can be approximated by:

$$\sigma_y = \sigma_v \cdot \left(2 \cdot t_L \cdot \frac{x}{M} \right)^{1/2} \qquad \bullet(17.15)$$

There are similar equations for σ_z.

Thus, we expect plumes to initially spread linearly with distance near to the source, but change to square-root with distance further downwind.

Solved Example (§)

Plot vertical and horizontal plume spread σ_z and σ_y vs. distance, using a Lagrangian time scale of 1 minute and wind speed of 10 m/s at height 100 m in a neutral boundary layer of depth 500 m. There is a rough surface of mixed crops.

a) Plot on both linear and log-log graphs.

b) Also plot the short and long-distance limits of σ_y on the log-log graph.

Solution

Given: $z = 100$ m, $z_o = 0.25$ m from Table 4-1,
$M = 10$ m/s, $t_L = 60$ s, $h = 500$ m

Find: σ_z and σ_y (m) vs. x (km).

Use eq. (4.13): $u_* = \dfrac{0.4 \cdot (10\text{m/s})}{\ln(100\text{m} / 0.25\text{m})} = 0.668$ m/s

Use eq. (4.25b): $\sigma_v =$
$1.6 \cdot (0.668\text{m/s}) \cdot [1 - 0.5 \cdot (100\text{m} / 500\text{m})] = 0.96$ m/s

Similarly, use eq. (4.25c): $\sigma_w = 0.75$ m/s

(continued next column)

Solved Example (§) *(continuation)*

Use eqs. (17.13a & b) to produce the graphs:

A linear graph is shown above, and log-log below.

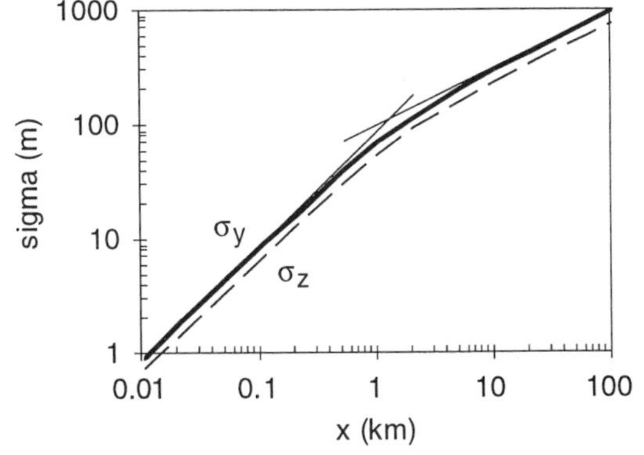

Eqs. (17.14) & (17.15) are plotted as the thin solid lines.

Check: Units OK. Physics OK. Sketch OK.
Discussion: The cross over between short and long time limits is at $x \cong 2 \cdot M \cdot t_L = 1.2$ km.

DISPERSION IN NEUTRAL & STABLE BOUNDARY LAYERS

To calculate pollutant concentration at the surface, one needs to know both the height of the plume centerline, and the spread of pollutants about that centerline. **Plume rise** is the name given to the first issue. **Dispersion** (from Taylor's statistical theory) is the second. When they are both used in an expression for the average spatial distribution of pollutants, pollution concentrations can be calculated.

Plume Rise

Ground-level concentration generally decreases as plume-centerline height increases. Hence, plume rise above the physical stack top is often desirable. The centerline of plumes can rise above the stack top due to the initial momentum associated with exit velocity out of the top of the stack, and due to buoyancy if the effluent is hot.

Neutral Boundary Layers

Statically neutral situations are found in the residual layer (not touching the ground) during light winds at night. They are also found throughout the bottom of the boundary layer (touching the ground) on windy, overcast days or nights.

The height z_{CL} of the plume centerline above the ground in neutral boundary layers is:

$$z_{CL} = z_s + \left[a \cdot l_m^2 \cdot x + b \cdot l_b \cdot x^2 \right]^{1/3} \qquad \bullet(17.16)$$

where $a = 8.3$, $b = 4.2$, x is distance downwind of the stack, and z_s is the physical stack height. This equation shows that the plume centerline keeps rising as distance from the stack increases. It ignores the capping inversion at the ABL top, which would eventually act like a lid on plume rise and upward spread.

A **momentum length scale**, l_m, is defined as:

$$l_m \approx \frac{W_o \cdot R_o}{M} \qquad \bullet(17.17)$$

where R_o is the stack-top radius, W_o is stack-top exit velocity of the effluent, and M is the ambient wind speed at stack top. l_m can be interpreted as a ratio of vertical emitted momentum to horizontal wind momentum.

A **buoyancy length scale**, l_b, is defined as:

$$l_b \approx \frac{W_o \cdot R_o^2 \cdot g}{M^3} \cdot \frac{\Delta\theta}{\theta_a} \qquad \bullet(17.18)$$

where $g = 9.8$ m/s^2 is gravitational acceleration, $\Delta\theta = \theta_p - \theta_a$ is the temperature excess of the effluent, θ_p is the initial stack gas potential temperature at stack top, and θ_a is the ambient potential temperature at stack top. l_B can be interpreted as a ratio of vertical buoyancy power to horizontal power of the ambient wind.

Science Graffito

"The solution to pollution is dilution." – Anonymous.

This aphorism was accepted as common sense during the 1800s and 1900s. By building taller smoke stacks, more pollutants could be emitted, because the pollutants would mix with the surrounding clean air and become dilute by the time they reached the surface.

However, by 2000, society realized that there were also global implications to emitting more pollutants. Issues included greenhouse gases, climate change, and stratospheric ozone destruction. Thus, government regulations changed to include total emission limits.

Solved Example (§)

At stack top, the effluent has velocity 20 m/s, temperature 200°C, and emission rate 250 g/s of SO_2. The stack is 75 m high, and has a radius of 2 m at the top. At stack top, the ambient wind is 5 m/s, and ambient potential temperature is 20°C.

For a neutral boundary layer, plot plume centerline height vs. downwind distance.

Solution

Given: $W_o = 20$ m/s, $Q = 250$ g/s, $z_s = 75$ m,
$\theta_p = 473$ K + (9.8 K/km)·(0.075 km) = 474 K,
$\theta_a = 293$ K, $M = 5$ m/s.

Find: $z(x) = ?$ m.

Use eq. (17.17):

$$l_m \approx \frac{(20\text{m/s})\cdot(2\text{m})}{(5\text{m/s})} = 8\text{ m}$$

Use eq. (17.18):

$$l_b \approx \frac{(20\text{m/s})\cdot(2\text{m})^2\cdot(9.8\text{m}\cdot\text{s}^{-2})}{(5\text{m/s})^3}\cdot\frac{(474-293\text{K})}{293\text{K}}$$

= 3.87 m

Use eq. (17.16):

$$z_{CL} = (75\text{m}) + \left[8.3\cdot(8\text{m})^2\cdot x + 4.2\cdot(3.87\text{m})\cdot x^2\right]^{1/3}$$

This is shown as the solid line on the plot below:

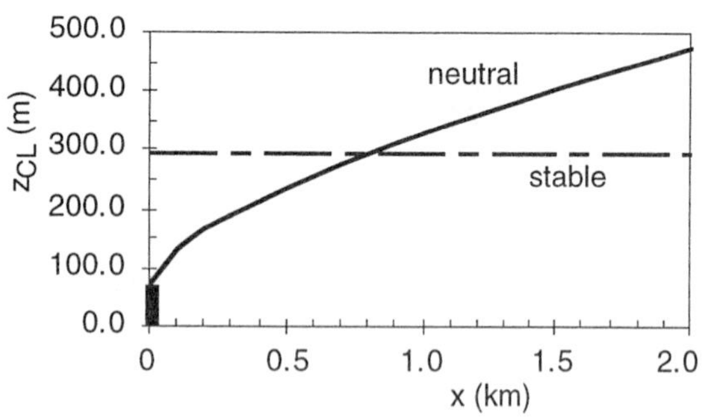

Check: Units OK. Physics OK. Sketch OK.

Discussion: The equation for plume rise in neutral conditions gives a plume rise that continues to increase with downwind distance. However, real plumes usually hit an elevated inversion, and stop rising. This is a good reason for using a thermo diagram to locate the stable layers aloft.

Stable Boundary Layers

In statically stable situations, the ambient potential temperature increases with height. This limits the plume-rise centerline to a final equilibrium height $z_{CL\ eq}$ above the ground:

$$z_{CL\ eq} = z_s + 2.6\cdot\left(\frac{l_b\cdot M^2}{N_{BV}^2}\right)^{1/3} \qquad \bullet(17.19)$$

where the Brunt-Väisälä frequency N_{BV} is used as a measure of static stability (see eq. 6.5 in Chapter 6).

Solved Example

At stack top, the effluent has velocity 20 m/s, temperature 200°C, and emission rate 250 g/s of SO_2. The stack is 75 m high, and has a radius of 2 m at the top. At stack top, the ambient wind is 5 m/s, and ambient potential temperature is 20°C.

For a stable boundary layer with $\Delta\theta_a/\Delta z$ = 5°C/km, find the final equilibrium height.

Solution

Given: $W_o = 20$ m/s, $Q = 250$ g/s, $z_s = 75$ m,
$\theta_p = 473$ K + (9.8 K/km)·(0.075 km) = 474 K,
$\theta_a = 293$ K, $M = 5$ m/s. $\Delta\theta_a/\Delta z = 5$°C/km

Find: $\Delta z_{CL\ eq} = ?$ m

This problem is identical to the previous solved example, except for the ambient static stability. First, use eq. (6.5b) from Chapter 6:

$$N_{BV}^2 = \frac{(9.8\text{m}\cdot\text{s}^{-2})}{(293\text{K})}\cdot\frac{5\text{K}}{1000\text{m}} = 1.67\text{x}10^{-4}\ \text{s}^{-2}$$

Use eq. (17.19):

$$z_{CL\ eq} = (75\text{m}) + 2.6\cdot\left(\frac{(3.87\text{m})\cdot(5\text{m/s})^2}{1.67\times10^{-4}\text{s}^{-2}}\right)^{1/3}$$

= 75 m + 216.7 m = **291.75 m**

Compute on a spreadsheet, and plot z_{CL} vs. x. This is plotted in the previous solved example as the dashed line.

Check: Units OK. Physics OK. Sketch OK.

Discussion: Because of the static stability, the plume reaches a final, equilibrium height. However, it does not reach this height instantly. Instead, it approaches it a bit slower than the neutral plume rise curve plotted in the previous solved example.

Gaussian Concentration Distribution

For neutral and stable boundary layers (PG types C through F), the sizes of turbulent eddies are relatively small compared to the depth of the boundary layer. This simplifies the problem by allowing turbulent dispersion to be modeled analogous to molecular diffusion. For this situation, the average concentration distribution about the plume centerline is well approximated by a 3-D Gaussian bell curve:

•(17.20)

$$c = \frac{Q}{2\pi \sigma_y \sigma_z M} \cdot \exp\left[-0.5 \cdot \left(\frac{y}{\sigma_y}\right)^2\right] \cdot \left\{\exp\left[-0.5 \cdot \left(\frac{z - z_{CL}}{\sigma_z}\right)^2\right] + \exp\left[-0.5 \cdot \left(\frac{z + z_{CL}}{\sigma_z}\right)^2\right]\right\}$$

where Q is the source emission rate of pollutant (g/s), σ_y and σ_z are the plume-spread standard deviations in the crosswind and vertical, y is lateral (crosswind) distance of the receptor from the plume centerline, z is vertical distance of the receptor above ground, z_{CL} is the height of the plume centerline above the ground, and M is average ambient wind speed at the plume centerline height.

For receptors at the ground ($z = 0$), eq. (17.20) reduces to:

•(17.21)

$$c = \frac{Q}{\pi \sigma_y \sigma_z M} \cdot \exp\left[-0.5 \cdot \left(\frac{y}{\sigma_y}\right)^2\right] \cdot \exp\left[-0.5 \cdot \left(\frac{z_{CL}}{\sigma_z}\right)^2\right]$$

The above two equations assume that the ground is flat, and that any pollutants that hit the ground are "reflected" back into the air. Also, they do not work for dispersion in statically unstable mixed layers.

To use these equations, the turbulent velocity variances are first found from the equations in Chapter 4. Next, plume spread is found from Taylor's statistical theory (eqs. 17.13). Plume centerline heights are then found from the previous subsection. Finally, they are all used in eqs. (17.20) or (17.21) to find the concentration at a receptor.

Recall that Taylor's statistical theory states that the plume spread increases with downwind distance. Thus, σ_y, σ_z, and z_{CL} are functions of x, which makes concentration c a strong function of x, in spite of the fact that x does not appear explicitly in the two equations above.

Solved Example (§)

Given a "surface" wind speed of 10 m/s at 10 m above ground, neutral static stability, boundary layer depth 800 m, surface roughness length 0.1 m, emission rate of 300 g/s of passive, non-buoyant SO_2, wind speed of 20 m/s at plume centerline height, and Lagrangian time scale 1 minute.

Plot isopleths of concentration at the ground for plume centerline heights of: (a) 100m, (b) 200m

Solution

Given: M = 10 m/s at z = 10 m, z_o = 0.1 m,
M = 20 m/s at z = 100 m = z_{CL}, neutral,
Q = 300 g/s of SO_2, t_L = 60 s, h = 800 m
Find: c (µg/m^3) vs. x (km) and y (km), at z = 0.
Assume z_{CL} is constant.

Use eq. (4.13):
$u_* = 0.4 \cdot (10 \text{ m/s})/\ln(10 \text{ m}/0.1 \text{ m}) = 0.869$ m/s

(a) Use eq. (4.25b) & (4.25c):
$\sigma_v = 1.6 \cdot (0.869\text{m/s}) \cdot [1 - 0.5(100/800)] = 1.3$ m/s
$\sigma_w = 1.25 \cdot (0.869\text{m/s}) \cdot [1 - 0.5(100/800)] = 1.02$ m/s

Use eq. (17.13a & b) to get σ_y and σ_z vs. x.
Then use eq. (17.21) to find c at each x and y:

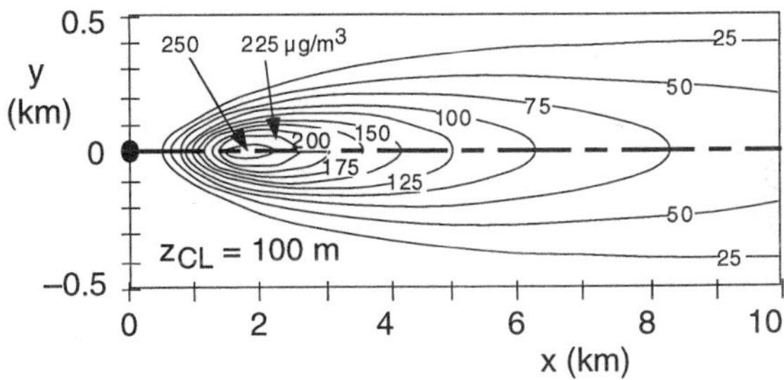

(b) Similarly, for a higher plume centerline:

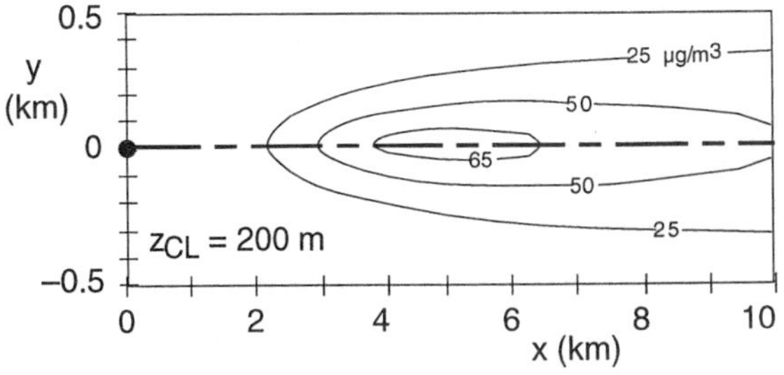

Check: Units OK. Physics OK.
Discussion: The pattern of concentration at the ground is called a **footprint**. Higher plume centerlines cause lower concentrations at the ground. That is why engineers design tall smoke stacks, and try to enhance buoyant plume rise.

Faster wind speeds also cause more dilution.

BEYOND ALGEBRA • Diffusion Equation

The Gaussian concentration distribution is a solution to the diffusion equation, as is shown here.

For a conservative, passive tracer, the budget equation says that concentration c in a volume will increase with time t if greater tracer flux F_C enters a volume than leaves. In one dimension (z), this is:

$$\frac{dc}{dt} = -\frac{\partial F_c}{\partial z} \qquad \text{(a)}$$

If turbulence consists of only small eddies, then turbulent tracer flux flows down the mean tracer gradient:

$$F_c = -K\frac{\partial c}{\partial z} \qquad \text{(b)}$$

where K, the eddy diffusivity, is analogous to a molecular diffusivity (see K-Theory in Appendix H), and F_C is in kinematic units (concentration times velocity).

Plugging eq. (b) into (a), and assuming constant K, gives the 1-D **diffusion equation**:

$$\boxed{\frac{dc}{dt} = K\frac{\partial^2 c}{\partial z^2}} \qquad \text{(c)}$$

This parabolic differential equation can be solved with initial conditions (IC) and boundary conditions (BC). Suppose a smoke puff of mass Q grams of tracer is released in the middle of a vertical pipe that is otherwise filled with clean air at time $t = 0$. Define the vertical coordinate system so that $z = 0$ at the initial puff height. Dispersion up and down the pipe is one-dimensional.

IC: $c = 0$ at $t = 0$ everywhere except at $z = 0$.

BC1: $\int c \; dz = Q$, at all t, where integration is $-\infty$ to ∞

BC2: c approaches 0 as z approaches $\pm\infty$, at all t.

The solution is:

$$c = \frac{Q}{(4\pi Kt)^{1/2}}\exp\left(\frac{-z^2}{4Kt}\right) \qquad \text{(d)}$$

You can confirm that this is a solution by plugging it into eq. (c), and checking that the LHS equals the RHS. It also satisfies all the initial and boundary conditions.

Comparing eq. (d) with eq. (17.10), we can identify the standard deviation of height as

$$\sigma_z = \sqrt{2Kt} \qquad \text{(e)}$$

which says that tracer spread increases with the square root of time, and greater eddy-diffusivity causes faster spread rate. Thus, the solution is Gaussian:

$$\boxed{c = \frac{Q}{\sqrt{2\pi}\cdot\sigma_z}\exp\left[-\frac{1}{2}\left(\frac{z}{\sigma_z}\right)^2\right]} \qquad \text{(17.10)}$$

Finally, using **Taylor's hypothesis** that $t = x/M$, we can compare eq. (e) with the σ_z version of eq. (17.15), and conclude that:

$$K = \sigma_w{}^2 \cdot t_L \qquad \text{(f)}$$

which shows how K increases with turbulence intensity.

DISPERSION IN UNSTABLE BOUNDARY LAYERS (CONVECTIVE MIXED LAYERS)

During conditions of light winds over an underlying warmer surface (PG types A & B), the boundary layer is **statically unstable** and in a state of **free convection**. Turbulence consists of thermals of warm air that rise from the surface to the top of the mixed layer. These vigorous updrafts are surrounded by broader areas of weaker downdraft. The presence of such large turbulent structures and their asymmetry causes dispersion behavior that differs from the usual Gaussian plume dispersion.

As smoke is emitted from a point source such as the top of a smoke stack, some of the emissions are by chance emitted into the updrafts of passing thermals, and some into downdrafts. Thus, the smoke appears to **loop** up and down, as viewed in a snapshot. However, when averaged over many thermals, the smoke disperses in a unique way that can be described deterministically. This description works only if variables are normalized by free-convection scales.

The first step is to get the meteorological conditions such as wind speed, ABL depth, and surface heat flux. These are then used to define the ABL convective scales such as the Deardorff velocity w_*. Source emission height, and downwind receptor distance are then normalized by the convective scales to make dimensionless distance variables.

Next, the dimensionless (normalized) variables are used to calculate the plume centerline height and vertical dispersion distance. These are then used as a first guess in a Gaussian equation for cross-wind-integrated concentration distribution, which is a function of height in the ABL. By dividing each distribution by the sum over all distributions, a corrected cross-wind-integrated concentration can be found that has the desirable characteristic of conserving pollutant mass.

Finally, the lateral dispersion distance is estimated. It is used with the cross-wind-integrated concentration to determine the dimensionless Gaussian concentration at any lateral distance from the plume centerline. Finally, the dimensionless concentration can be converted into a dimensional concentration using the meteorological scaling variables.

Although this procedure is complex, it is necessary, because non-local dispersion by large convective circulations in the unstable boundary layer works completely differently than the small-eddy dispersion in neutral and stable ABLs. The whole procedure can be solved on a spreadsheet, which was used to produce Figs 17.7 and 17.8.

Relevant Variables

Physical Variables:

c = concentration of pollutant (g/m^3)

c_y = **crosswind-integrated concentration** (g/m^2), which is the total amount of pollutant within a long-thin box that is 1 m^2 on each end, and which extends laterally across the plume at any height z and downwind location x (see Fig 17.6)

Q = emission rate of pollutant (g/s)

x = distance of a receptor downwind of the stack (m)

z = height of a receptor above ground (m)

z_{CL} = height of the plume centerline (center of mass) above the ground (m)

z_s = stack-top height = source height (m)

σ_y = lateral standard deviation of pollutant (m)

σ_z = vertical standard deviation of pollutant (m)

σ_{zc} = vertical standard deviation of crosswind-integrated concentration of pollutant (m)

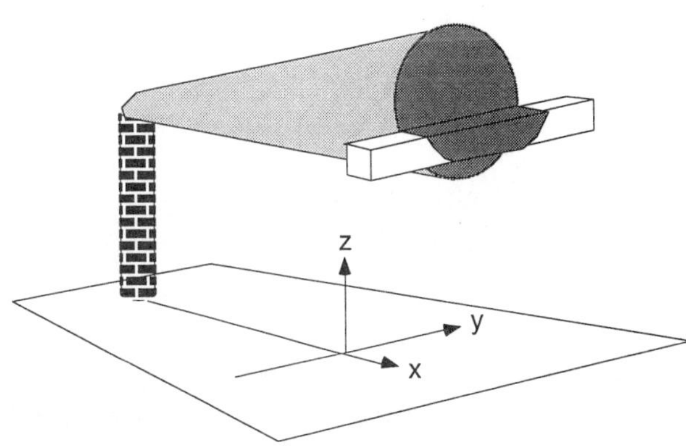

Figure 17.6
Crosswind integrated concentration c_y is the total amount of pollutants in a thin conceptual box (1 m^2 at the end) that extends crosswind (y-direction) across the smoke plume. This concentration is a function of only x and z.

Mixed-Layer Scaling Variables:

F_H = effective surface kinematic heat flux (K·m/s), see eqs. (3.22)

M = mean wind speed (m/s)

$$w_* = \left[\frac{|g| \cdot z_i \cdot F_H}{T_v}\right]^{1/3} \cong 0.02 \cdot w_B \qquad (17.22)$$

= **Deardorff velocity** (m/s)

z_i = depth of the convective mixed layer (m)

Dimensionless Scales:

These are usually denoted by uppercase symbols (except for M and Q, which have dimensions).

$$C = \frac{c \cdot z_i^2 \cdot M}{Q} = \text{dimensionless concentration} \qquad \bullet(17.23)$$

$$C_y = \frac{c_y \cdot z_i \cdot M}{Q} = \text{dimensionless crosswind-integrated concentration} \qquad (17.24)$$

$$X = \frac{x \cdot w_*}{z_i \cdot M} = \text{dimensionless downwind distance of receptor from source} \qquad \bullet(17.25)$$

$Y = y / z_i$ = dimensionless crosswind (lateral) distance of receptor from centerline (17.26)

$Z = z / z_i$ = dimensionless receptor height (17.27)

$Z_{CL} = z_{CL} / z_i$ = dimensionless plume centerline height (17.28)

$Z_s = z_s / z_i$ = dimensionless source height (17.29)

$\sigma_{yd} = \sigma_y / z_i$ = dimensionless lateral standard deviation (17.30)

$\sigma_{zdc} = \sigma_{zc} / z_i$ = dimensionless vertical standard deviation of crosswind-integrated concentration (17.31)

As stated in more detail earlier, to find the pollutant concentration downwind of a source during convective conditions, three steps are used: (1) Find the plume centerline height. (2) Find the crosswind integrated concentration at the desired x and z location. (3) Find the actual concentration at the desired y location.

Plume Centerline

For neutrally-buoyant emissions, the dimensionless height of the center of mass (plume centerline Z_{CL}) varies with dimensionless distance downwind X:

$$Z_{CL} \approx 0.5 + \frac{0.5}{1+0.5\cdot X^2}\cdot \cos\left[2\pi\frac{X}{\lambda}+\cos^{-1}(2\cdot Z_s-1)\right] \qquad \bullet(17.32)$$

where Z_s is the dimensionless source height, and the dimensionless wavelength parameter is $\lambda = 4$.

The centerline tends to move down from elevated sources, which can cause high concentrations at ground level (see Fig 17.7). Then further downwind, they rise a bit higher than half the mixed-layer depth, before reaching a final height at $0.5 \cdot z_i$. For buoyant plumes, the initial downward movement of the centerline is much weaker, or does not occur at all.

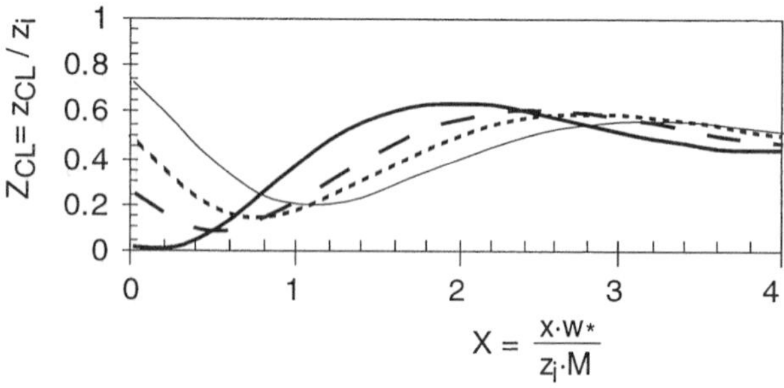

Figure 17.7
Height of the averaged pollutant centerline z_{CL} with downwind distance x, normalized by mixed-layer scales. Dimensionless source heights are $Z_s = z_s / z_i = 0.025$ (thick solid line); 0.25 (dashed); 0.5 (dotted); and 0.75 (thin solid). The plume is neutrally buoyant.

Crosswind Integrated Concentration

The following algorithm provides a quick approximation for the crosswind integrated concentration. Find a first guess dimensionless C_y' as a function of dimensionless height Z using a Gaussian approach for vertical dispersion:

$$C_y' = \exp\left[-0.5\cdot\left(\frac{Z-Z_{CL}}{\sigma_{zdc}'}\right)^2\right] \qquad \bullet(17.33)$$

where the prime denotes a first guess, and where the vertical dispersion distance is:

$$\sigma_{zdc}' = a\cdot X \qquad (17.34)$$

with $a \cong 0.25$. This calculation is done at K equally-spaced heights between the ground to the top of the mixed layer.

Next, find the average over all heights $0 \le Z \le 1$:

$$\overline{C_y'} = \frac{1}{K}\sum_{k=1}^{K} C_y' \qquad (17.35)$$

where index k corresponds to height z. Finally, calculate the revised estimate for dimensionless crosswind integrated concentration at any height:

$$C_y = \frac{C_y'}{\overline{C_y'}} \qquad \bullet(17.36)$$

Examples are plotted in Fig 17.8 for various source heights.

Concentration

The final step is to assume that lateral dispersion is Gaussian, according to:

$$C = \frac{C_y}{(2\pi)^{1/2}\cdot\sigma_{yd}}\exp\left[-0.5\cdot\left(\frac{Y}{\sigma_{yd}}\right)^2\right] \qquad \bullet(17.37)$$

The dimensionless standard deviation of lateral dispersion distance from an elevated source is

$$\sigma_{yd} \approx b\cdot X \qquad (17.38)$$

where $b \cong 0.5$.

At large downwind distances (i.e., at $X \ge 4$), the dimensionless crosswind integrated concentration always approaches $C_y \to 1.0$, at all heights. Also, directly beneath the plume centerline, $Y = 0$.

ON DOING SCIENCE • The Citizen Scientist

Scientists and engineers have at least the same responsibilities to society as do other citizens. Like our fellow citizens, we ultimately decide the short-term balance between environmental quality and material wealth, by the goods that we buy and by the government leaders we elect. Be informed. Take a stand. Vote.

Perhaps we have more responsibility, because we can also calculate the long-term consequences of our actions. We have the ability to evaluate various options and build the needed technology . Take action. Discover the facts. Design solutions.

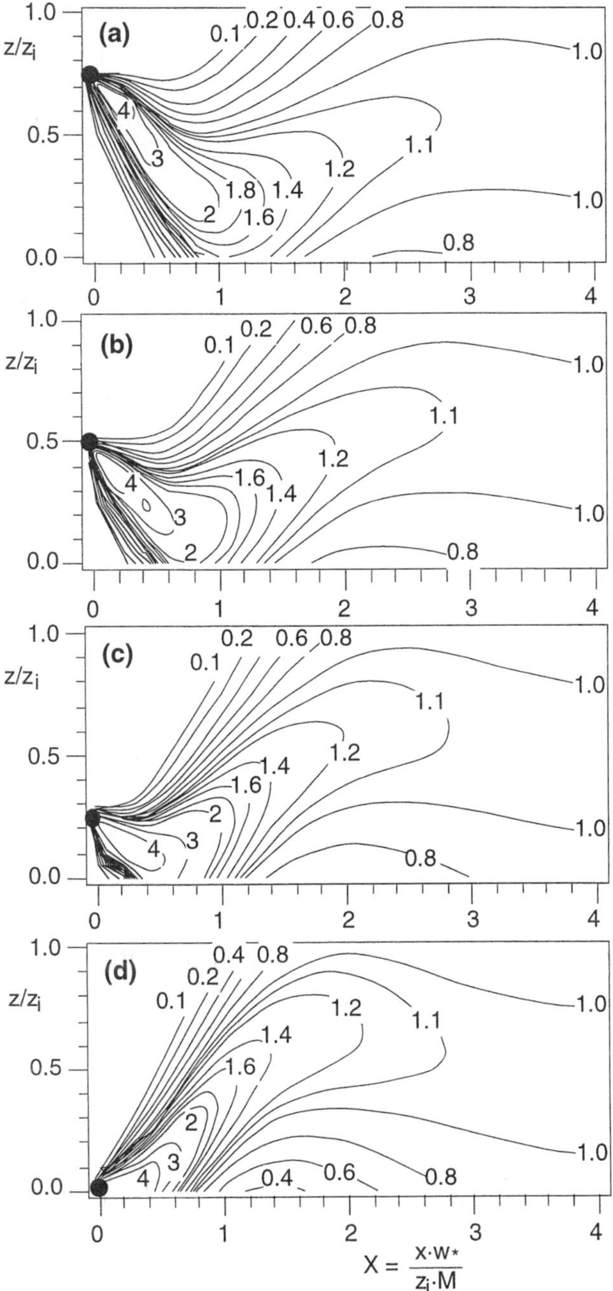

Figure 17.8
Isopleths of dimensionless crosswind-integrated concentration $C_y = c_y \cdot z_i \cdot M / Q$ in a convective mixed layer, where c_y is crosswind integrated concentration, z_i is depth of the mixed layer, M is mean wind speed, and Q is emission rate of pollutants. Source heights are $Z_s = z_s/z_i$ = (a) 0.75, (b) 0.5, (c) 0.25, (d) 0.025, and are plotted as the large black dot.

Solved Example

Source emissions of 200 g/s of SO_2 occur at height 150 m. The environment is statically unstable, with a Deardorff convective velocity of 1 m/s, a mixed layer depth of 600 m, and a mean wind speed of 4 m/s.

Find the concentration at the ground 3 km downwind from the source, directly beneath the plume centerline.

Solution

Given: Q = 200 g/s, z_s = 150 m, z_i = 600 m, M = 4 m/s, w_* = 1 m/s

Find: c (μg/m³) at x = 3 km, $y = z = 0$.

Use eq. (17.25):

$$X = \frac{(3000\text{m})\cdot(1\text{m/s})}{(600\text{m})\cdot(4\text{m/s})} = 1.25$$

Use eq. (17.29): $Z_s = (150\text{m})/(600\text{m}) = 0.25$

From Fig 17.8c, read $C_y \cong 0.9$ at X = 1.25, Z = 0.

Use eq. (17.38): $\sigma_{yd} \approx 0.5\cdot(1.25) = 0.625$

Use eq. (17.37) with $Y = 0$:

$$C = \frac{0.9}{\sqrt{2\pi}\cdot 0.625} = 0.574$$

Finally, use eq. (17.23) rearranging it first to solve for concentration in physical units:

$$c = \frac{C\cdot Q}{z_i^2\cdot M} = \frac{(0.574)\cdot(200\text{g/s})}{(600\text{m})^2\cdot(4\text{m/s})} = \underline{79.8\ \mu\text{g/m}^3}$$

Check: Units OK. Physics OK.
Discussion: We were lucky that the dimensionless source height was 0.25, which allowed us to use Fig 17.8c. For other source heights not included in that figure, we would have to create new figures using equations (17.33) through (17.36).

Science Graffito

"The service we render to others is really the rent we pay for our room on the earth." – Sir Wilfred Grenfell.

SUMMARY

Pollutants emitted from a smoke stack will blow downwind and disperse by turbulent mixing with ambient air. By designing a stack of sufficient height, pollutants at ground level become sufficiently dilute as to not exceed local environmental air-quality standards. Additional buoyant plume rise above the physical stack top can further reduce ground-level concentrations.

Air-quality standards do not consider instantaneous samples of pollutant concentration. Instead, they are based on averages over time. For such averages, statistical descriptions of dispersion must be used, including the center of mass (plume centerline) and the standard deviation of location (proportional to plume spread).

For emissions in the boundary layer, the amount of dispersion depends on the type of turbulence. This relationship is described by Taylor's statistical theory.

During daytime conditions of free convection, thermals cause a peculiar form of dispersion that often brings high concentrations of pollutants close to the ground. At night, turbulence is suppressed in the vertical, causing little dispersion. As pollutants remain aloft for this case, there is often little hazard at ground level. Turbulent dispersion is quite anisotropic for these convective and stable cases.

In neutral conditions of overcast and strong winds, turbulence is more isotropic. Smoke plumes disperse at roughly equal rates in the vertical and lateral directions, and are well described by Gaussian formulae.

Various classification schemes have been designed to help determine the appropriate turbulence and dispersion characteristics. These range from the detailed examination of the production of turbulence kinetic energy, through examination of soundings plotted on thermo diagrams, to look-up tables such as those suggested by Pasquill and Gifford.

Finally, although we used the words "smoke" and "smoke stack" in this chapter, most emissions in North America are sufficiently clean that particulate matter is not visible. This clean-up has been expensive, but commendable.

Threads

Because most pollutants are emitted from near the surface, and most receptors are at the surface, the mean transport and turbulent dispersion of pollutants are primarily controlled by boundary layer characteristics (Chapt. 4). The nature of the turbulence depends on the radiatively-driven (Chapt. 2) heating (Chapt. 3), and the dynamic forces and winds (Chapts. 9 & 10). Many emissions, particularly from fuel combustion, contain water vapor (Chapt. 5) as a combustion product, which can condense and form billowy white emission plumes.

Once emitted, the dispersion depends partly on the static stability (Chapt. 6) of the ambient atmosphere. Cumulus clouds (Chapt. 7) can withdraw pollutants out of the top of the ABL, and precipitation (Chapt. 8) can scrub out some pollutants. The transport and dispersion of pollutants, and their interaction with weather (Chapts. 12 and 13) and clouds, can be included as part of numerical weather forecasts (Chapt. 14).

However, not all pollutants are immediately scrubbed out of the smoke plume, and can be carried by the general circulation (Chapt. 11) to the whole globe, which could affect climate change (Chapt. 18). As pollution concentrations increase, the visibility and color of the air is affected (Chapt. 19).

EXERCISES

Numerical Problems

N1. Given the following wind measurements.
a. Find the variance in each direction.
b. Determine if the flow is isotropic.
c. Speculate on the shape of smoke plumes.

t (min)	U (m/s)	V (m/s)	W (m/s)
1	8	1	0
2	11	2	–1
3	12	0	1
4	7	–3	1
5	12	0	–1

N2. Determine the Pasquill-Gifford turbulence type
a. Strong sunshine, clear skies, winds 1 m/s
b. Thick overcast, winds 10 m/s, night
c. Clear skies, winds 2.5 m/s, night
d. Noon, thin overcast, winds 3 m/s.
e. Cold air advection 2 m/s over a warm lake.
f. Sunset, heavy overcast, calm.
g. Sunrise, calm, clear.
h. Strong sunshine, clear skies, winds 10 m/s.

N3. If ABL turbulence is such that the TKE production rates are $B = 0.005$ and $S = 0.003$ $m^2 \cdot s^{-3}$:
a. Specify the nature of flow/convection
b. Estimate the Pasquill-Gifford turbulence type.
c. Determine the dispersion isotropy
d. Estimate the flux Richardson number (Chapt.4)
e. Is turbulence strong or weak?

N4. Given the following pollutant concentrations measured with height.

a. Find the height of center of mass.
b. Find the vertical height variance.
c. Find the vertical height standard deviation.
d. Find the total amount of pollutant emitted.
e. Find the nominal plume spread (depth)

z (km)	c ($\mu g/m^3$)
1.5	0
1.4	0
1.3	5
1.2	25
1.1	20
1.0	45
0.9	55
0.8	40
0.7	30
0.6	10

N5.(§) For the previous problem, find the best-fit Gaussian curve through the data, and plot the data and curve on the same graph.

N6. Given lateral and vertical velocity variances of 1.0 and 0.5 m^2/s^2, respectively. Find the variance of plume spread in the lateral and vertical, at distance 3 km downwind of a source in a wind of speed 5 m/s. Use a Lagrangian time scale of:

a. 15 s	b. 30 s	c. 1 min	d. 2 min
e. 5 min	f. 10 min	g. 15 min	h. 20 min

N7.(§) For a Lagrangian time scale of 2 minutes and wind speed of 10 m/s, plot the standard deviation of vertical plume spread vs. downwind distance for a vertical velocity variance (m^2/s^2) of:

a. 0.1	b. 0.2	c. 0.5	d. 1.0
e. 2	f. 3	g. 5	h. 10

N8.(§) For the previous problem, plot the results if
(i) only the near-source equation
(ii) only the far source equation
is used over the whole range of distances.

N9. Plot the following soundings (see next column) on the boundary-layer $\theta - z$ thermo diagram from Chapt. 6. Determine the boundary-layer structure, including location and thickness of components of the boundary layer. Speculate whether it is daytime or nighttime, and whether it is winter or summer. For daytime situations, calculate the mixed-layer depth. This depth controls pollution concentration (shallow depths are associated with **air-pollution episodes**, and during calm winds to **air stagnation events**).

z (m)	a. T (°C)	b. T (°C)
2500	–11	8
2000	–10	10
1700	–8	8
1500	–10	10
1000	–5	15
500	0	18
100	4	18
0	7	15

N10. For the ambient soundings of the previous example, assume that a smoke stack of height 100 m emits effluent of temperature 6°C with mixing ratio 3 g/kg. (Hint, assume the smoke is an air parcel, and use a thermo diagram.)

a. How high would the plume rise, assuming no dilution with the environment?
b. Would steam condense in the plume?

N11. Given the following emission parameters:

	W_o (m/s)	R_o (m)	$\Delta\theta$ (K)
a.	5	3	200
b.	30	1	50
c.	20	2	100
d.	2	2	50
e.	5	1	50
f.	30	2	100
g.	20	3	50
h.	2	4	20

Find the momentum and buoyant length scales. Assume $g/\theta_a \cong 0.0333\ m \cdot s^{-2} \cdot K^{-1}$, and M = 5 m/s for all cases.

N12.(§) For the previous problem, plot the plume centerline height vs. distance if the physical stack height is 100 m and the atmosphere is statically neutral.

N13. For buoyant length scale of 5 m, physical stack height 10 m, environmental temperature 10°C, and wind speed 2 m/s, find the equilibrium plume centerline height in a statically stable boundary layer, given ambient potential temperature gradients of $\Delta\theta/\Delta z$ (K/km):

a. 1	b. 2	c. 4	d. 7
e. 10	f. 12	g. 15	h. 20

N14. Given $\sigma_y = \sigma_z$ = 300 m, z_{CL} = 500 m, z = 200 m, Q = 100 g/s, M = 10 m/s. For a neutral boundary layer, find the concentration at y (km) =

a. 0	b. 0.1	c. 0.2	d. 0.5
e. 1	f. 2	g. 5	h. 10

N15(§). Plot the concentration footprint at the surface, downwind of a stack with z_{CL} = 50 m, σ_v = 1 m/s, σ_w = 0.5 m/s, M = 2 m/s, Lagrangian time scale = 1 minutes, Q = 400 g/s of SO_2, in a stable boundary layer.

N16. Calculate the dimensionless downwind distance, given a mixed layer depth of 2 km, wind speed 3 m/s, and surface heat flux of 0.15 K·m/s. Assume $g/T_v \equiv 0.0333$ m·s^{-2}·K^{-1}. The actual distance x (km) is:

a. 0.2	b. 0.5	c. 1	d. 2
e. 5	f. 10	g. 20	h. 50

N17. If w_* = 1 m/s, mixed layer depth is 1 km, wind speed is 5 m/s, Q = 100 g/s, find the

a. dimensionless downwind distance at x = 2 km
b. dimensionless concentration if c = 100 µg/m^3
c. dimensionless crosswind integrated concentration if c_y = 1 mg/m^2

N18.(§) For a convective mixed layer, plot dimensionless plume centerline height with dimensionless downwind distance, for dimensionless source heights of:

a. 0	b. 0.01	c. 0.02	d. 0.05
e. 0.1	f. 0.2	g. 0.5	h. 1.0

N19.(§) For the previous problem, plot isopleths of dimensionless crosswind integrated concentration, similar to Fig 17.8, for convective mixed layers.

N20. Source emissions of 300 g/s of SO_2 occur at height 200 m. The environment is statically unstable, with a Deardorff convective velocity of 1 m/s, and a mean wind speed of 5 m/s.

Find the concentration at the ground at distances 1, 2, 3, and 4 km downwind from the source, directly beneath the plume centerline. Assume the mixed layer depth (m) is:

a. 400	b. 600	c. 800	d. 1000
e. 1200	f. 1500	g. 2000	h. 2500

(Hint: Interpolate between figures if needed, or derive your own figures.)

Understanding & Critical Evaluation

U1. Why does a "nominal" plume edge need to be defined? Why cannot the Gaussian distribution be used, with the definition that plume edge happens where the concentration becomes zero. Discuss, and support your arguments with results from the Gaussian distribution equation.

U2. The Lagrangian time scale is different for different size eddies. In nature, there is a superposition of turbulent eddies acting simultaneously. Describe the dispersion of a smoke plume under the influence of such a spectrum of turbulent eddies.

U3. Fig 17.3 shows how dispersion isotropy can change as the relative magnitudes of the shear and buoyancy TKE production terms change. Also, the total amount of spread increases as the TKE intensity increases. Discuss how the shape and spread of smoke plumes vary in different parts of that figure.

U4. Eq. (17.8) gives the center of mass (i.e., plume centerline height) in the vertical direction. Create a similar equation for plume center of mass in the horizontal, using a cylindrical coordinate system centered on the emission point.

U5. Compare the two equations for variance: (17.5) and (17.9). Why is the one weighted by pollution concentration, and the other not?

U6. For the Gaussian curve (eq. 17.10), set Q_1 = 100 g/m, and $\bar{z} = 0$. Plot on graph paper the curve for σ_z (m) =

a. 100	b. 200	c. 300	d. 400

Compare the areas under each curve, and discuss the significance of the result.

U7. While Taylor's statistical theory equations give plume spread as a function of downwind distance, x, these equations are also complex functions of the Lagrangian time scale t_L. For a fixed value of downwind distance, plot curves of the variation of plume spread (eq. 17.13) as a function of t_L. Discuss the meaning of the result.

U8. a. Derive eqs. (17.14) and (17.15) for near-source and far-source dispersion from Taylor's statistical theory equations (17.13).

b. Why do the near and far source dispersion equations appear as straight lines in a log-log graph (see the solved example after eq. (17.15)?

U9. For plume rise in statically neutral conditions, write a simplified version of the plume rise equation (17.16) for the special case of:

a. momentum only b. buoyancy only

Also, what are the limitations and range of applicability of the full equation and the simplified equations?

U10. For plume rise in statically stable conditions, the amount of rise depends on the Brunt-Väisälä frequency. As the static stability becomes weaker, the Brunt-Väisälä frequency changes, and so changes the plume centerline height. In the limit of extremely weak static stability, compare this plume rise equation with the plume rise equation for statically neutral conditions. Also, discuss the limitations of each of the equations.

U11. In eq. (17.20), the "reflected" part of the Gaussian concentration equation was created by pretending that there is an imaginary source of emissions an equal distance underground as the true source is above ground. Otherwise, the real and imaginary sources are at the same horizontal location and have the same emission rate.

In eq. (17.20), identify which term is the "reflection" term, and show why it works as if there were emissions from below ground.

U12. In the solved example in the Gaussian Concentration Distribution subsection, the concentration footprints at ground level have a maximum value neither right at the stack, nor do concentrations monotonically increase with increasing distances from the stack. Why? Also, why are the two figures in that solved example different?

U13. Show that eq. (17.20) reduces to eq. (17.21) for receptors at the ground.

U14. For Gaussian concentration eq. (17.21), how does concentration vary with:

a. σ_y b. σ_z c. M

U15. Give a physical interpretation of crosswind integrated concentration, using a different approach than was used in Fig 17.6.

U16. For plume rise and pollution concentration in a statically unstable boundary layer, what is the reason for, or advantage of, using dimensionless variables?

U17. If the Deardorff velocity increases, how does the dispersion of pollutants in an unstable boundary layer change?

U18. In Fig 17.8, at large distances downwind from the source, all of the figures show the dimensionless concentration approaching a value of 1.0. Why, and what is the significance or justification for such behavior?

Web-Enhanced Questions

W1. Search the web for the government agency of your country that regulates air pollution. In the USA, it is the Environmental Protection Agency (EPA). Find the current air pollution standards for the chemicals listed in Table 17–1.

W2. Search the web for an air quality report for your local region (such as town, city, state, or province). Determine how air quality has changed during the past decade or two.

W3. Search the web for a site that gives current air pollution readings (or an **air quality index**) for your region. In some cities, this pollution reading is updated every several minutes, or every hour. If that is the case, see how the pollution reading varies with hour during a typical workday.

W4. Search the web for information on health effects of different exposures to different pollutants.

W5. Air pollution models are computer codes that use equations similar to the ones in this chapter, to predict air pollution concentration. Search the web for a list of names of a few of the popular air pollution models endorsed by your country or region.

W6. Search the web for inventories of emission rates for pollutants in your regions. What are the biggest polluters?

W7. Search the web for an explanation of **emissions trading**. Discuss why such a policy is or is not good for industry, government, and people.

W9. Search the web for information on **acid rain**.

W10. Search the web for information on **forest death** (**waldsterben**) caused by pollution or acid rain.

W11. Search the web for instruments that can measure concentration of the chemicals listed in Table 17–1.

W12. Search the web for "web-cam" cameras that show a view of a major city, and discuss how the visibility during fair weather changes during the daily cycle on a workday.

W13. Search the web for information of plume rise and/or concentration predictions for complex (mountainous) terrain.

W14. Search the web for information to help you discuss the relationship between "good" ozone in the stratosphere and mesosphere, vs. "bad" ozone in the boundary layer.

W15. For some of the major industry in your area, search the web for information on control technologies that can, or have, helped to reduce pollution emissions.

W16. Search the web for satellite photos of emissions from major sources, such as a large industrial complex, smelter, volcano, or a power plant. Use the highest-resolution photographs to look at lateral plume dispersion, and compare with the dispersion equations in this chapter.

W17. Search the web for information on forward or backward trajectories, as used in air pollution. One example is the Chernobyl nuclear accident, where radioactivity measurements in Scandinavia were used with a back trajectory to suggest that the source of the radioactivity was in the former Soviet Union.

W18. Search the web for information on chemical reactions of air pollutants in the atmosphere.

W19. Search the web for satellite photos and other information on **urban plumes**.

Synthesis Questions

S1. Suppose that there was not a diurnal cycle, but that the atmospheric temperature profile was steady, and equal to the standard atmosphere. How would local and global dispersion of pollutants from tall smoke stacks be different, if at all?

S2. In the present atmosphere, larger-size turbulent eddies often have more energy than smaller size one. What if the energy distribution were reversed, with the vigor of mixing increasing as eddy sizes <u>de</u>crease. How would that change local dispersion, if at all?

S3. What if tracers were not passive, but had a special magnetic attraction only to each other. Describe how dispersion would change, if at all.

S4. What if a plume that is rising in a statically neutral environment has buoyancy from both the initial temperature of the effluent out of the top of the stack, and also from additional heat gained while it was dispersing. A real example was the black smoke plumes from the oil well fires during the Gulf War.

Sunlight was strongly absorbed by the black soot and unburned petroleum in the smoke, causing solar warming of the black smoke plume. Describe any resulting changes to plume rise.

S4. Suppose that smoke stacks produced smoke rings, instead of smoke plumes. How would dispersion be different, if at all?

S5. When pollutants are removed from exhaust gas before the gas is emitted from the top of a smoke stack, those pollutants don't magically disappear. Instead, they are converted into water pollution (to be dumped into a stream or ocean), or solid waste (to be buried in a dump or landfill). Which is better? Why?

S6. Propose methods whereby life on earth could produce zero pollution. Defend your proposals.

S7. What if the same emission rate of pollutions occurs on a fair-weather day with light winds, and an overcast rainy day with stronger winds. Compare the dispersion and pollution concentrations at the surface for those situations. Which leads to the least concentration at the surface, locally? Which is better globally?

S8. Suppose that all atmospheric turbulence was extremely anisotropic, such that there was zero dispersion in the vertical , but normal dispersion in the horizontal.

a. How would that affect pollution concentrations at the surface, for emissions from tall smoke stacks?

b. How would it affect climate, if at all?

S9. What if ambient wind speed was exactly zero. Discuss the behavior of emission plumes, and how the resulting plume rise and concentration equations would need to be modified.

S10. What if pollutants that were emitted into the atmosphere were never lost or removed from the atmosphere. Discuss how the weather and climate would be different, if at all?

S11. If there were no pollutants in the atmosphere (and hence no cloud and ice nuclei), discuss how the weather and climate would be different, if at all.

S12. Divide the current global pollutant emissions by the global population, to get the net emissions per person. Given the present rate of population increase, discuss how pollution emissions will change over the next century, and how it will affect the quality of life on earth, if at all.

CLIMATE CHANGE

CONTENTS

Science Graffito

"The most important fact about Spaceship Earth: an instruction book didn't come with it." – R. Buckminster Fuller.

18 Earth's average surface temperature has been surprisingly steady over millennia. Factors that control the temperature are solar output, earth orbital characteristics, ocean and atmosphere circulations, plate tectonics, clouds, ice and snow, natural and anthropogenic gases, particulates and aerosols including volcanic emissions, and life. The myriad of interactions between these processes and conditions is what makes studies of climate change so challenging.

Some of these factors tend to reduce climate change, and are classified as **negative feedbacks.** Others amplify change, and are **positive feedbacks.** The steadiness of our temperature suggests that negative feedbacks dominate.

Heuristic models are presented below to isolate and study certain physical processes. These are oversimplifications of nature, yet they serve to illustrate some climatic controls and responses.

Better experiments on climate change are possible with a special class of large-scale, long-duration numerical weather forecast models known as **global climate models** (**GCM**s). These models are quite complex. From GCM climate simulations, we have improved our understanding of atmospheric physics, feedback processes, and climate sensitivities. However, due to the large number of approximations in most GCMs, forecasts of actual climate change are still somewhat uncertain.

RADIATIVE EQUILIBRIUM

The earth is heated by the sun, and is cooled by longwave (IR) radiation to space (Fig 18.1). From historical records we know that the earth has had nearly constant absolute temperature (less than 4% change over the past 100 million years), suggesting a balance between radiation inflow and outflow.

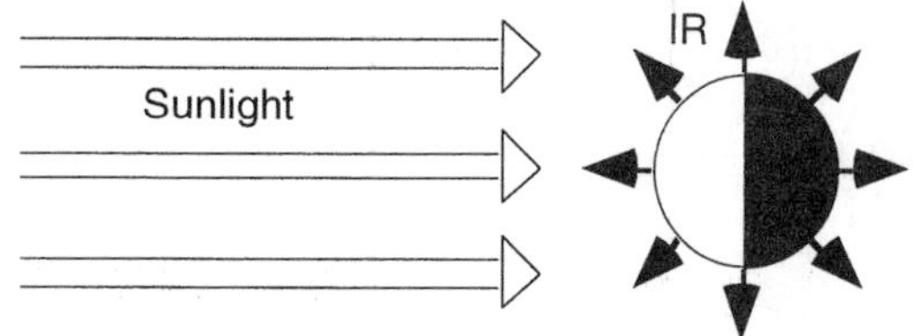

Figure 18.1
Solar input (white arrows) and infrared output (dark arrows) from earth.

Incoming radiation is the solar constant, minus reflected sunlight, times the interception area:

$$Radiation\ \ In = (1-A)\cdot S\cdot \pi R_{earth}^2 \qquad \bullet(18.1)$$

where A = 0.3 is global albedo, S = 1368 W·m^{-2} is the annual average solar constant, and R_{earth} is the earth radius. The area of a disk rather than the surface area of a sphere is used because the area of intercepted solar radiation is equivalent to the area of the shadow cast behind the earth.

Assume the earth is a black body in the infrared, and that it emits IR radiation from its whole surface area (continents and oceans). Multiplying the Stefan-Boltzmann law (eq. 2.18), which gives the emissions per unit area, times surface area of the earth (assumed spherical here) gives:

$$Radiation\ \ Out = \sigma_{SB}\cdot T_e^4\cdot 4\pi R_{earth}^2 \qquad \bullet(18.2)$$

where σ_{SB} = 5.67x10^{-8} W·m^{-2}·K^{-4} is the Stefan-Boltzmann constant, and T_e is called the **effective radiation emission temperature**.

The real earth is quite heterogeneous, with oceans, continents, forests, clouds and ice caps of differing temperatures. Each of these regions emits different amounts of radiation. Also, some radiation is emitted from the surface, some from various heights in the atmosphere, and some from clouds. The effective radiation temperature is defined so that the total emissions from the actual earth-atmosphere system equal the theoretical emissions from a uniform earth of temperature T_e.

We can equate incoming and outgoing radiation, and solve for T_e:

$$T_e = \left[\frac{(1-A)\cdot S}{4\cdot\sigma_{SB}}\right]^{1/4} \qquad \bullet(18.3)$$

$$\cong 255\ \text{K} = -\ 18°\text{C}$$

Fig 18.2 shows the variation of effective temperature with solar irradiance, assuming constant albedo.

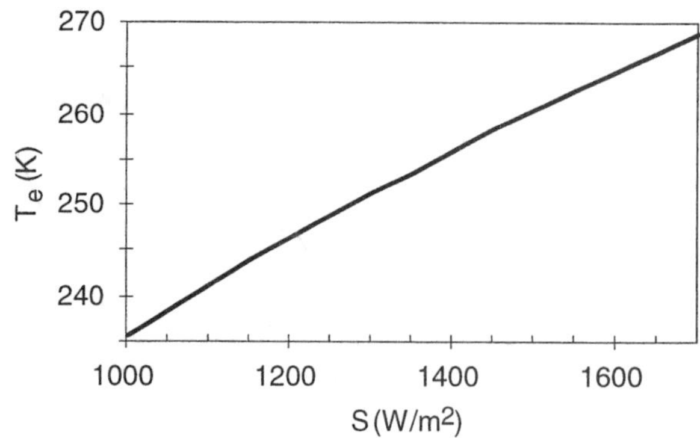

Figure 18.2
Variation of earth effective radiation temperature T_e with solar irradiance, S assuming an albedo of 0.3 .

This effective temperature is too cold compared to the observed surface temperature of 15°C. Hence, we infer that this simple model is missing some important physics.

While the annual average irradiance is about S = 1367 W·m^{-2} , this value varies from about 1366 when there are few sunspots to 1368 when there are many. We have been using this latter value in previous examples in this book. Sunspot activity has an 11 year cycle. These irradiance changes would cause an effective temperature change of about 0.05°C, which is essentially undetectable.

This radiation balance is a **negative feedback**. Namely, increases of incoming solar radiation are mostly compensated by increases in outgoing IR, yielding small temperature changes. The earth system attempts to remain in a stable equilibrium.

Solved Example
Find the radiative equilibrium temperature of the earth-atmosphere system.

Solution:
Given: S = 1368 W·m^{-2} solar constant
A = 0.3 global albedo (reflectivity)
Find: T_e = ? K radiative equilibrium temp.

Use eq. (18.3):

$$T_e = \left[\frac{(1-0.3)\cdot(1368\ \text{W}\cdot\text{m}^{-2})}{4\cdot(5.67\times10^{-8}\ \text{W}\cdot\text{m}^{-2}\cdot\text{K}^{-4})}\right]^{1/4}$$

T_e = 254.9 K ≅ 255 K = **– 18°C**

Check: Units OK. Physics OK.
Discussion: The effective radiation temperature does not depend on the radius of the planet.

Because the radiative feedback dominates over all others, we can be assured that the earth's climate is relatively stable, as indeed the historical records demonstrate. Fifty million years ago, the average surface temperature was about 10°C warmer than now. During the past ten thousand years, the climate has oscillated about ±1°C about the recent 15°C average. Although these temperature changes are relatively small compared to the earth's absolute temperature, they can cause dramatic changes in sea level and glaciation.

GREENHOUSE EFFECT

We can increase the sophistication of the previous model by adding a layer of atmosphere (see Fig 18.3) that is opaque to infrared radiation, but transparent to visible. Sunlight still heats the earth's surface. Radiation from the surface heats the opaque atmosphere. Radiation from the top of the atmosphere is lost to space, while radiation from the bottom heats the earth.

The incoming radiation from the sun must balance the outgoing radiation from the atmosphere, in order for the earth-atmosphere system to remain in radiative equilibrium. Hence

$$T_A = T_e = 255 \ K \tag{18.4}$$

as before, where T_e is the effective emission temperature of the whole earth-atmosphere system, and T_A is the temperature of the atmosphere. For this case, the two temperatures are equal, because the atmosphere is opaque.

Other energy balances can be made for the atmosphere separately, and for the earth's surface separately, with output from one system being the input to the other. The opaque atmosphere is assumed to emit as much IR radiation upward to space, as downward toward the earth. The amount emitted in each direction from the atmosphere is given by the Stefan-Boltzmann law as $\sigma_{SB} \cdot T_A{}^4$. Incoming IR radiation from the earth's surface is $\sigma_{SB} \cdot T_s{}^4$. If the atmosphere is at **steady state** (i.e., no temperature change with time), then incoming IR radiation equals outgoing:

$$\sigma_{SB} \cdot T_s^4 = 2 \cdot \sigma_{SB} \cdot T_e^4 \tag{18.5a}$$

where T_s is the earth surface temperature.

The sum of atmospheric and solar radiation reaching the earth's surface is exactly twice that of the solar radiation alone. This extra heating from the atmosphere causes the earth to warm until it emits exactly twice the radiation as before. This allows the earth's surface temperature to be estimated by:

$$T_s = 2^{1/4} \cdot T_e \qquad \bullet(18.5b)$$

$$\cong 303 \text{ K} = 30°\text{C}$$

Warming of the earth's surface due to radiation emitted from the atmosphere is known as the **Greenhouse effect**.

This temperature is too warm compared to the observed average surface temperature of 15°C, from which infer that we are still missing some important physics.

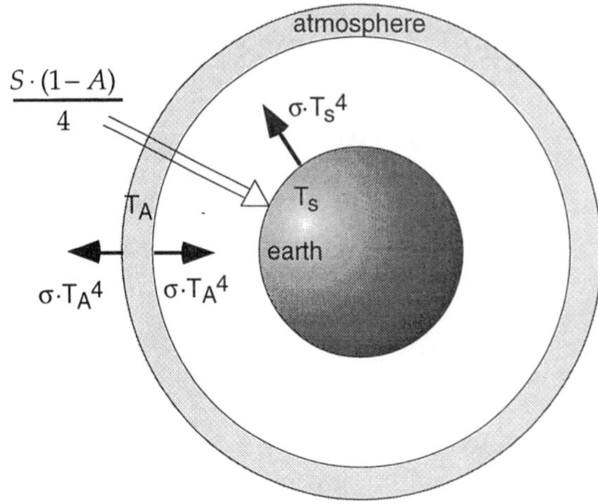

Figure 18.3
Greenhouse effect with an idealized one-layer atmosphere. Distance between earth and atmosphere is exaggerated. In fact, we assume that the radius of the earth approximately equals the radius at the top of the atmosphere.

Solved Example
For the previous solved example, find the earth surface temperature under a one-layer atmosphere.

Solution
Given: $T_e = 255$ K
Find: $T_s = ?$ K

Use eq. (18.5b):
$T_s = 1.189 \cdot (255\text{K}) = 303 \text{ K} = \underline{30°\text{C}}$

Check: Units OK. Physics OK.
Discussion: Although the radiation reaching the earth's surface doubles, the surface temperature increases by a smaller amount.

ATMOSPHERIC WINDOW

Between 8 and 11μm wavelengths, the atmosphere is semi-transparent to infrared (IR) radiation. This is the so-called **atmospheric window**. Suppose this window allows 10% of the IR radiation from the earth to be lost directly to space, without first being absorbed by the atmosphere. Using Kirchoff's law from Chapt. 2, this gives an absorptivity = emissivity (e) = 90% for this hypothetical atmosphere.

As before, the incoming fluxes must balance the outgoing, for an equilibrium atmosphere. The same holds for the earth's surface. Also, as before, the flux output from one system is often an input to the other.

For a one-layer atmosphere with a window as sketched in Fig 18.4, the equilibrium energy balance (input = output) of the atmosphere is:

$$e \cdot \sigma_{SB} \cdot T_s^{\ 4} = 2 \cdot e \cdot \sigma_{SB} \cdot T_A^{\ 4} \qquad (18.6)$$

For the earth's surface, the corresponding balance is:

$$\sigma_{SB} \cdot T_e^{\ 4} + e \cdot \sigma_{SB} \cdot T_A^{\ 4} = \sigma_{SB} \cdot T_s^{\ 4} \qquad (18.7)$$

Solving these for the atmosphere and surface temperatures yields:

$$T_A^4 = \frac{T_e^4}{2-e} \qquad \bullet(18.8)$$

$$T_s^4 = \frac{2 \cdot T_e^4}{2-e} \qquad \bullet(18.9a)$$

Thus:

$$T_s \cong \ 296 \text{ K} = 23°\text{C} \qquad (18.9b)$$

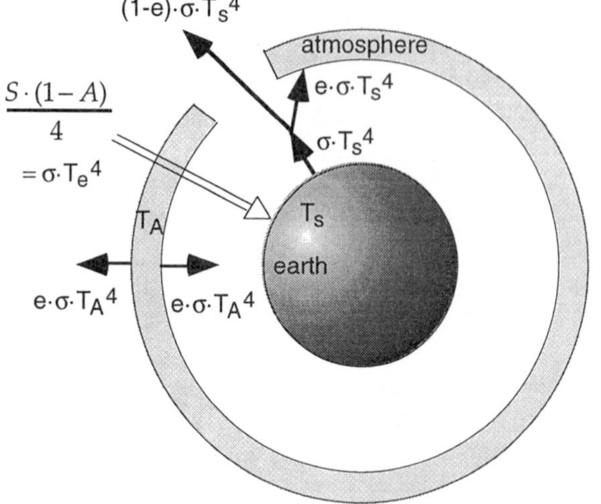

Figure 18.4
Radiation balance in a hypothetical one-layer atmosphere with an IR window.

This surface temperature is a bit cooler and closer to realistic than for a windowless atmosphere. The implications of this conclusion are tremendous. Namely, if **anthropogenic greenhouse gases** carbon dioxide (CO_2), nitrous oxide (N_2O), methane (CH_4) and CFC-12 ($C\,Cl_2\,F_2$) accumulate in the atmosphere, they can close the atmospheric window by absorbing more of the IR radiation emitted from earth. As a result, surface temperatures would probably warm toward those of the previous solved example. This is the threat of **global warming**.

Finally, be assured that Fig 18.4 is just a cartoon designed to illustrate that some IR radiation from earth can escape to space through a translucent atmosphere. A spatial hole in the atmosphere does not really exist.

Solved Example

For the previous solved example, assume an atmospheric emissivity of 0.9 , and find the surface temperature.

Solution

Given: T_e = 255 K, e = 0.9
Find: T_s = ? K

Use eq. (18.9a):

$$T_s = \left(\frac{2}{2-0.9}\right)^{1/4} \cdot (255\text{K}) = 296 \text{ K} = \underline{23°\text{C}}$$

Check: Units OK. Physics OK.
Discussion: This model is a bit more sophisticated than the previous one, and gives a slightly better result (closer to the observed temperature of 15°C).

WATER-VAPOR AND CLOUD FEEDBACKS

Cloudless Atmosphere

One **positive feedback** involves water vapor. Warmer surface temperature cause more evaporation from the oceans. Water vapor is a **natural greenhouse gas**. Increased water vapor in the air tends to close the atmospheric window, which traps more radiation and further increases the surface temperature.

Although this is a positive feedback, it does not cause a run-away climate change of forever-increasing temperatures. The reason is that once the atmospheric window is closed, any increase in water vapor will have no additional effect on this feedback. For a closed window, the negative feedback of radiative equilibrium still dominates, and a new equilibrium is reached.

As a mathematical contrivance to illustrate this feedback, suppose that absorptivity and emissivity e of the atmosphere increase with humidity, and can ultimately be related to surface temperature T_s by:

$$e = e_w + \frac{T_s - T_{sw}}{T_r} \qquad (18.10)$$

where $e_w = 0.9$ is the equilibrium emissivity from the previous section corresponding to an atmospheric window, $T_{sw} = 296$ K is the equilibrium surface temperature for that same case, and $T_r = 45$ K is an arbitrary reference temperature, chosen to show the exponential growth of this positive feedback. Note that eq. (18.10) has no physical basis.

Equations (18.9 & 18.10) are a closed set. Starting with initial conditions of $T_s = T_{sw}$ in eq. (18.10), we disturb the equilibrium at time step 7 by adding 0.1 K to the surface temperature. For each subsequent time step we recompute e and T_s, subject to the restriction that $e \le 1$. Albedo is held constant at 0.3, and the solar constant is 1368 $W{\cdot}m^{-2}$.

This hypothetical scenario is sketched in Fig 18.5. The disturbance grows exponentially after the initial disturbance, until the window is closed. At that point, the new equilibrium state is that of the closed-window greenhouse described previously ($T_s = 303$ K, $e = 1$). This new temperature (30°C) would still allow life to thrive on earth.

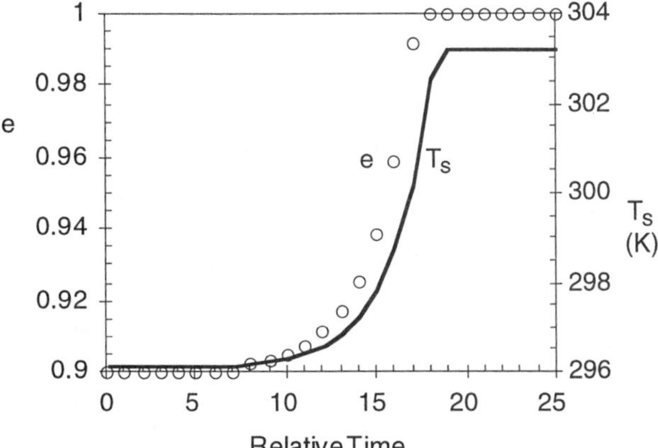

Figure 18.5
Variation of emissivity (e) and surface temperature (Ts) under a water-vapor positive feedback. The initial equilibrium conditions corresponding to an open window are disturbed at time step 7 by the addition of 0.1 K to the surface temperature.

Cloudy Atmosphere

The previous analysis neglected clouds. As water vapor and temperature increase in the atmosphere, so might cloud formation. Averaged over the whole globe, clouds cause a net cooling. The reflection of sunlight from cloud top more than offsets the warming caused by absorption and re-emission of terrestrial radiation.

Increases of moisture in the previous example, might cause increases in cloudiness, which increases the global albedo A. This might cause the final equilibrium temperature to be not as warm as shown in Fig 18.5.

ICE–ALBEDO–TEMPERATURE FEEDBACK

Another positive feedback involves the formation of ice and snow. As surface temperature cools, there will be less snow melt. As ice covers more of the oceans, and as snow covers increasingly large regions of land and ice, more and more sunlight will be reflected. This, in turn, enhances the cooling and increases snow coverage.

Analogous to the previous illustration, we will contrive a model for the temperature influence on albedo:

$$A = A_w + \frac{T_{sw} - T_s}{T_r} \qquad (18.11)$$

subject to the constraint that $A \le 0.75$, which is a typical albedo for snow. Let A_w be the present-day global albedo of 0.3.

This equation, together with eqs. (18.3 & 18.9), provide a closed set that can be solved for surface temperature and albedo. Initial conditions are $A = A_w = 0.3$, and $T_s = T_{sw} = 296$ K. The emissivity is held constant at 0.9 in this example, and an arbitrary reference temperature of $T_r = 65$ K is used. The outcome is shown in Fig 18.6.

As before, run-away global cooling is not possible, because once the globe is completely snow covered, any additional snow depth will not alter the albedo. The final temperature for this example is 229 K, which is quite chilly. Nevertheless, we are back to the negative feedback of radiative equilibrium.

As before, these simplistic arguments ignore many of the other feedbacks and processes such as **dynamical feedback**, **evaporative-cooling feedback**, **lowering of sea level** as more water is captured in glaciers, etc.

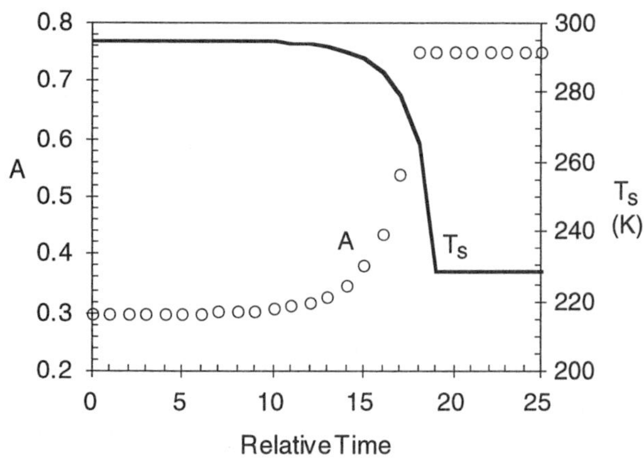

Figure 18.6
Illustration of ice-albedo feedback. Initially at equilibrium for an atmospheric window, the equilibrium is disturbed at time step 7 by subtracting 0.1 K from the surface temperature.

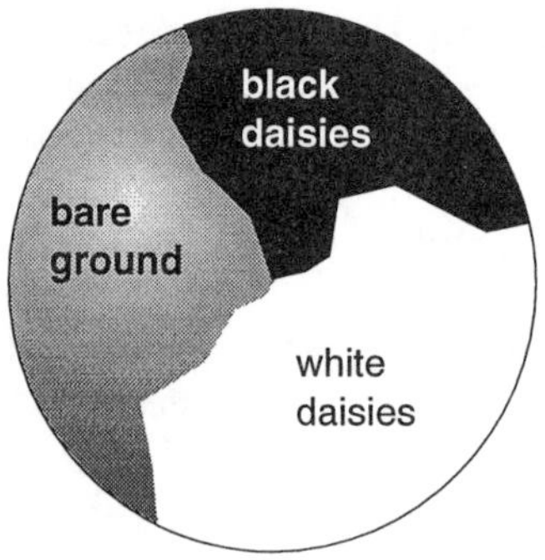

Figure 18.7
Coverage of daisies on Daisyworld.

DAISYWORLD

Does life on earth merely respond and adapt to changes of climate? Lovelock suggested that life controls the climate in a way that favors continuance of life (**homeostasis**). This theory is called the **gaia** hypothesis, after the Greek word for "mother earth".

To demonstrate such "biological homeostasis of the global environment", Lovelock and colleagues presented a simple hypothetical model called **Daisyworld**. On daisyworld, only two types of plants exist: black daisies and white daisies. These daisies grow and die depending on the temperature. However, as their populations increase, they alter the global albedo due to their color, and thus alter the global radiation temperature (Fig 18.7).

Physics

A highly simplified and parameterized set of equations were suggested by Lovelock to illustrate biological homeostasis. We will present the equations in an order and with modifications that aid spreadsheet computations.

First, there are two constants. One is the Stefan-Boltzmann constant $\sigma = 5.67\text{x}10^{-8}$ W·m^{-2}·K^{-4}. The other is the solar constant, S = 1368 W/m^2. A solar luminosity parameter, $L \cong 1$, provides a convenient way to experiment with increases or decreases in solar output. From these quantities, a solar forcing ratio is defined as

$$q = \frac{L \cdot S}{4 \cdot \sigma} \tag{18.12}$$

Parameters are:
$A_W = 0.75$ albedo of white daisies
$A_B = 0.25$ albedo of black daisies
$A_G = 0.5$ albedo of bare ground.
$D = 0.3$ daisy death rate (equal for black and white).
$Tr = 0.6$ horizontal transport parameter (varies between 0 for no transport, to 1 for total mixing in the horizontal). When $Tr = 0$, the temperature in a daisy patch is controlled locally by the albedo of that patch. For $Tr = 1$, the average global albedo determines the local temperature.

The areas of the globe covered by daisies are:
C_W = fractional area covered by white daisies
C_B = fractional area covered by black daisies.
Our initial conditions will be no daisies: $C_W = C_B = 0$. Also

$$C_G = 1 - C_B - C_W \tag{18.13}$$

is the fractional area of bare ground.

The planetary albedo is an average of the albedoes of the daisies, weighted by their abundance:

$$A = C_G \cdot A_G + C_B \cdot A_B + C_W \cdot A_W \tag{18.14}$$

The effective radiation temperature (Kelvins) is:

$$T_e^{\,4} = q \cdot (1 - A) \tag{18.15}$$

and the surface temperature under a one-layer windowless atmosphere is

$$T_s^{\,4} = 2 \cdot T_e^{\,4} \tag{18.16}$$

Surface temperatures of individual daisy patches are:

$$T_B^{\,4} = (1 - Tr) \cdot q \cdot (A - A_B) + T_s^{\,4} \tag{18.17a}$$

and

$$T_W^{\,4} = (1 - Tr) \cdot q \cdot (A - A_W) + T_s^{\,4} \tag{18.17b}$$

for black and white daisies, respectively.

These daisies grow fastest at an optimum temperature of $T_o = 22.5°C = 295.5$ K. Growth rate is less for warmer or colder temperature, and growth ceases altogether at temperatures colder than 5°C or warmer than 40°C. The growth rate β is:

$$\beta_B = 1 - b \cdot (T_o - T_B)^2 \qquad (18.18a)$$

$$\beta_W = 1 - b \cdot (T_o - T_W)^2 \qquad (18.18b)$$

where $b = 0.003265$ K^{-2}, and subject to the constraint that $\beta \geq 0$.

This biological homeostasis model is a prognostic model. It forecasts how daisy coverage changes with time as populations of black and white daisies grow and interact. The population equation for daisies is

$$C_{B\ new} = C_B + \Delta t \cdot C_B \cdot (C_G \cdot \beta_B - D) \qquad (18.19a)$$

$$C_{W\ new} = C_W + \Delta t \cdot C_W \cdot (C_G \cdot \beta_W - D) \qquad (18.19b)$$

which gives the new values after a time step of Δt. We use $\Delta t = 1$. The daisy coverages are constrained such that $C_{B\ new} \geq C_s$ and $C_{W\ new} \geq C_s$, where $C_s = 0.01$ is the seed coverage. Without such seeding of the planet, no daisies could grow.

These new coverages are then used back in eq. (18.13), and the computation process is repeated for many time steps. After sufficient steps, the model approaches a stable equilibrium.

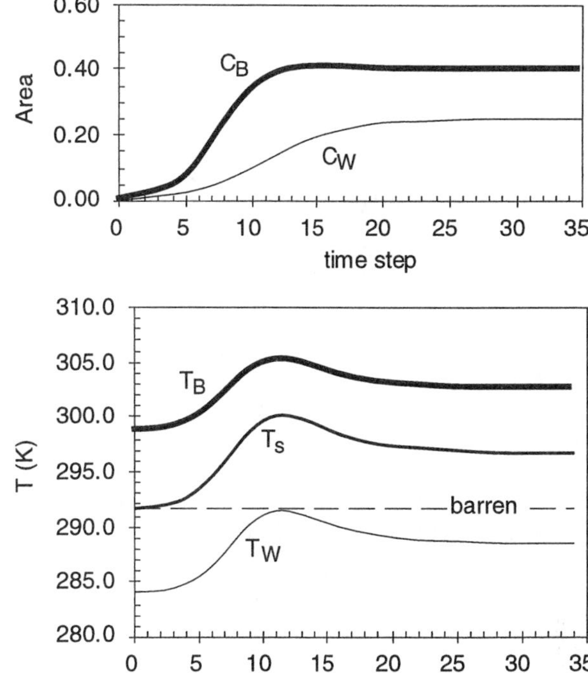

Figure 18.8
Sample Daisyworld forecast for $L = 1.2$.

Equilibrium and Homeostasis

For $L = 1.2$ from the solved example, daisy populations reach equilibrium values of $C_B = 0.39$ and $C_W = 0.25$ (Fig 18.8). The equilibrium surface temperature is $T_s = 296.75$ K, which is warmer than the corresponding temperature of a barren planet of 291.65 K. This latter quantity is found using $A = A_G$ in eqs. (18.15 & 18.16).

By rerunning the model for various values of luminosity, we can discover the equilibrium states that are possible. These are plotted in Fig 18.9. **Homeostasis** is evident in that the fecund planet maintains a nearly-steady surface temperature conducive to life, for a wide range of luminosities. It does this by producing predominantly black daisies for low luminosity, in order to maximize absorption of the solar energy. Alternately, white daisies dominate for high luminosity to reflect the excess solar energy and maintain temperate conditions.

Although this conceptual model is too simplistic to describe the real earth, it serves to make us think about new possibilities.

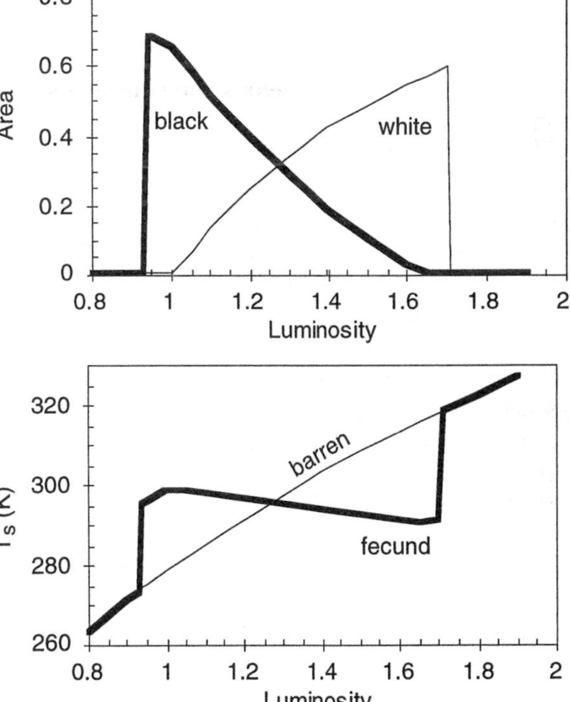

Figure 18.9
Equilibrium conditions in daisyworld for various solar luminosities. Top: areal coverage of daisies. Bottom, surface temperature.

Solved Example(§)

Use the parameter values of the previous page, along with $L = 1.2$, to make a Daisyworld forecast.

Solution

Given: $L = 1.2$, $Tr = 0.6$

Find: $C_B = ?$, $C_W = ?$, $T_s = ?$ K, $T_B = ?$ K, $T_W = ?$ K

First use eq. (18.12):

$$q = \frac{1.2\cdot(1368\text{W}\cdot\text{m}^{-2})}{4\cdot(5.67\times10^{-8}\text{W}\cdot\text{m}^{-2}\cdot\text{K}^{-4})} = 7.24\text{x}10^{9}\ \text{K}^4$$

Initialize with $C_B = C_W = 0$.

Iterate, solving eqs. (18.13) - (18.19) on a spreadsheet (where book width forced me to wrap columns 9 and 10 under columns 2 and 3):

t	C_B $beta_B$	C_W $beta_W$	C_G	A	T_s (K)	T_B (K)	T_w (K)
0	0.00 0.97	0.00 0.57	1.00	0.50	291.6	298.7	284.1
1	0.01 0.97	0.01 0.57	0.98	0.50	291.6	298.7	284.1
2	0.02 0.96	0.01 0.58	0.97	0.50	291.8	298.8	284.2
3	0.03 0.96	0.02 0.60	0.96	0.50	292.0	299.0	284.4
4	0.04 0.95	0.02 0.63	0.94	0.49	292.5	299.3	284.8
5	0.07 0.94	0.03 0.67	0.90	0.49	293.2	299.9	285.4
. . . etc. . .							
98	0.39 0.84	0.25 0.84	0.36	0.46	296.7	302.5	288.5

Check: Units OK. Physics OK.

Discussion: Because of the "seed" coverage, values of C_W and C_B of 0.01 actually correspond to no daisies.

At steady state, the growth parameters should be equal: $\beta_B = \beta_W$. This is a good check to see if steady state has been obtained.

The temperature of a barren planet happens to equal the value of T_s from the first iteration (before any daisies). Thus, $T_{s\ barren} = 291.6$ K.

These answers are plotted in Fig 18.8.

GCMs

Global climate models (**GCMs**) are special types of numerical weather prediction (NWP) models (see Chapter 14). A regular grid of nodes covers the globe, and simulations are made at each node for each time step by numerically solving sets of finite-difference equations.

Some climate feedback processes are explicitly included via the dynamics, thermodynamics, and physics in the model. Other feedback processes must be approximated and parameterized. The net result is a model that includes many but not all of the complex interactions that exist in the real atmosphere.

To simulate the climate, the run duration must be long enough to model years to decades to centuries. For these models to run on present-day computers, sacrifices must be made in grid resolution to reduce the run time. **Coarse horizontal resolutions** of about 1° latitude are used on the whole globe, and only a limited number of vertical grid layers are used.

The coarse resolution is a drawback in that much of the important physics (boundary layers, turbulence, radiation, cloud physics) is **subgrid scale**, and cannot be resolved. For example, clouds of diameter less than about 2° of latitude ($\cong$ 222 km) would be missed. Instead, these subgrid physical processes must be parameterized (approximated).

Parameterizations are only approximations to nature. GCM models are so highly parameterized that their forecast quality is dubious.

However, the nature of interactions between processes can be studied fairly successfully using GCMs. For example, two hypothetical climate forecasts could be made: one with mountains and another without. By comparing the two forecasts, the **sensitivity** of climate to topography can be documented. Similarly, the sensitivities of climate to a host of other processes can also be examined, by making appropriate model runs.

All numerical models need initial (ICs) and boundary (BCs) conditions. While short-term weather forecast models rely heavily on initial conditions, climate models are virtually independent of initial conditions, because of the long forecast duration.

Conversely, boundary conditions are critical for climate models, and less important for short-term forecast models. In order for climate forecasts to not drift from the true climate, the models need very accurate parameterizations of the heat, moisture, and momentum budgets at the earth's surface. These parameterizations must be validated against observations of the real climate, before we are confident to use them to study hypothetical climates.

Operational numerical weather forecast models attempt to accurately forecast actual synoptic features, such as the accurate location and intensity of cyclones. Such forecast quality rapidly deteriorates with forecast duration, as shown in Chapter 14.

GCM modelers, however, do not even attempt to forecast the precise nature of individual storms. Instead, they want the overall statistics of cyclone intensity and location to agree with observed statistics. They want the net effect of many storms to be accurate, even though the individual storms in the forecast match nothing in the observations. Attributes of these models are compared in Table 18-1.

Before leaving this section on GCMs, note that there is a class of operational forecast models called **medium-range weather forecast models**. These models often have the best attributes of both the GCM and the short-range operational models.

Table 18-1. Comparison of global climate models (GCMs) to operational numerical weather forecast models (OFMs).

Attribute	GCM	OFM
duration	years	days
grid resolution	coarse	fine
ICs	less important	important
BCs	important	less important
parameterized	heavily	less
fcst. quality	statistical	deterministic

SENSITIVITY

Definition

We will use globally-averaged surface temperature T_s as a measure of the climate. How sensitive is the climate to external influences? By discovering the most important influences, we can avoid inadvertent alteration of surface temperature in the real atmosphere, and can spend extra care in parameterizing those influences in GCMs.

Suppose that initially the climate is in equilibrium, and that the global radiation budget is balanced:

$$0 = S_1 - L_1$$

where $S_1 = \mathit{Radiation\ In} = (1-A)\cdot S\cdot \pi R_{earth}^2$ and $L_1 = \mathit{Radiation\ Out} = \sigma_{SB}\cdot T_e^4\cdot 4\pi R_{earth}^2$. Both S_1 and L_1 have units of $\mathrm{W/m^2}$.

Then some external influence causes an additional input of energy ΔN ($\mathrm{W/m^2}$).

$$0 \neq S_1 - L_1 + \Delta N$$

To reach a new equilibrium, the surface temperature might change, altering S and L to new values (units are still $\mathrm{W/m^2}$):

$$0 = S_2 - L_2 + \Delta N$$

The difference between new and initial budgets is:

$$0 = \Delta S - \Delta L + \Delta N$$

But the changes of S and L are associated with changes in surface temperature:

$$\Delta S = \frac{\Delta S}{\Delta T_s}\cdot \Delta T_s$$

and

$$\Delta L = \frac{\Delta L}{\Delta T_s}\cdot \Delta T_s$$

Substituting these into the previous equation, and rearranging, gives:

$$\frac{\Delta T_s}{\Delta N} = \left[\frac{\Delta L}{\Delta T_s} - \frac{\Delta S}{\Delta T_s}\right]^{-1} \qquad (18.20)$$

The ratio of temperature change (the climate response) to external forcing is called the **climate sensitivity factor** λ:

$$\lambda \equiv \frac{\Delta T_s}{\Delta N} \qquad (18.21a)$$

and has units of $\mathrm{K / (W\cdot m^{-2})}$. Different values of λ can be found for different processes affecting climate. Smaller values of λ indicate less sensitive response and minimal surface temperature change.

Magnitudes

Different researchers have suggested different values of the climate sensitivity factor (Table 18-2).

Table 18-2. Approximate climate sensitivity factors.

Process	λ [K / (W·m^{-2})]
Stefan Boltzmann IR ($L \propto T^4$)	0.27
Satellite-observed IR ($L \propto T$)	0.55
Water vapor increase with T	0.5
Water vapor & cloud	0.75
Ice albedo	0.27 to 2.7
Evaporation from tropical sea	0.3
Biological (Daisyworld)	–0.3

Usage

Equation (18.21) can be rearranged to solve for the temperature change expected from various forcings.

$$\Delta T_s = \lambda \cdot \Delta N \qquad \bullet(18.21b)$$

Doubling of carbon dioxide from 300 ppm (parts per million by volume) to 600 ppm in the atmosphere is expected to increase the greenhouse effect by absorbing more IR radiation. This will trap more IR radiation in the earth-atmosphere, causing a net energy increase of $\Delta N = 4\ W/m^2$. Using eq. (18.21b), the anticipated **direct** climatic response due to only the Stefan-Boltzmann process is global warming:

$$\Delta T_s = (0.27\text{K} \cdot \text{W}^{-1} \cdot \text{m}^2) \cdot (4\text{W/m}^2) = \underline{1°\text{C}} \qquad (18.22)$$

However, this direct warming would be amplified by the **water-vapor feedback** process. Using the water-vapor sensitivity factor of 0.5, we would expect <u>2°C</u> global warming. When the combined effects of increased water vapor and increased clouds are considered, GCM simulations suggest a net sensitivity factor is 0.75.

Thus, the $\Delta N = 4\ W/m^2$ associated with CO_2 doubling is anticipated to cause roughly <u>3°C</u> total global warming, when the direct and **indirect** processes are included. There is much uncertainty in this number, however, with different GCMs suggesting global warming ranging from 1.7°C to 5.4°C.

What happens when other processes alter cloud cover? A **10% increase in cloudiness** could reflect sufficient sunlight to reduce the energy budget by $\Delta N = -4\ W/m^2$, even considering IR trapping below clouds. Using eq. (18.21b), the anticipated climatic response is global cooling of about <u>– 3°C</u>.

Mt. Pinatubo erupted in June 1991 producing about 30 million tons of **stratospheric aerosols** including sulfuric acid. Over parts of the world, net radiation was reduced by about $\Delta N = -4\ W/m^2$. However, when averaged over the whole globe, including regions not affected by Mt. Pinatubo, the global cooling is estimated to be <u>– 0.25 to – 0.5°C</u>.

If the **sunspot number** causes the solar constant to increase about $\Delta N = 2\ W/m^2$, then global warming is expected to be <u>0.54°C</u>.

There are many other factors that affect global temperature, but which are not yet included in GCMs. More research is needed before global climate predictions can be made with confidence.

Solved Example

Over the past 150k years, it is believed that orbital variations could have altered solar irradiance by 50 W/m^2 or more, at certain latitudes. What climatic response would be expected?

Solution

Given: $\Delta S = \Delta N = 50\ W/m^2$

Find: $\Delta T_s = ?\ °C$

Use eq. (18.21b) with $\lambda = 0.27$ for the Stefan-Boltzmann factor:

$$\Delta T_s = (0.27\text{K} \cdot \text{W}^{-1} \cdot \text{m}^2) \cdot (50\text{W} \cdot \text{m}^{-2}) = \underline{13.5°\text{C}}$$

Check: Units OK. Physics OK.

Discussion: This estimate is in much closer agreement with data than if the satellite derived sensitivity parameter ($\lambda = 0.55$) had been used.

SUMMARY

Radiative equilibrium between absorbed solar radiation and emitted infrared radiation provide a strong negative feedback to keep the earth's temperature stable. Re-radiation by the atmosphere back to earth is the greenhouse effect, and causes surface temperatures to be warmer than otherwise.

Increased sophistication in heuristic climate models is possible by including aerosols, clouds, emissivities, horizontal atmospheric and oceanic circulations, vertical mixing, snow cover, and vegetation. The most sophisticated models are global climate models (GCMs), that employ numerical weather forecasting techniques over long time scales.

While water-vapor and ice-albedo processes cause positive feedback, runaway climate change does not occur because of the dominant negative feedback of radiative equilibrium. The sensitivity of global-average temperature change to various forcings can be estimated using a climate sensitivity factor. This factor can be used to estimate the temperature-change impact of forcings such as carbon dioxide doubling, earth orbital changes, cloudiness increase, and volcanic influences.

Life forms might have evolved in a manner to help maintain a global climate conducive to plant growth. Such homeostasis is described in the gaia hypothesis, and is heuristically modeled with Daisyworld.

Threads

This chapter is the culmination of all previous chapters. Much of the climate change issue relates to radiation (Chapt. 2) within the setting of our existing atmospheric structure (Chapt. 1). The radiative input alters the heat budget (Chapt. 3), which controls clouds (Chapt. 7) and precipitation (Chapt. 8) via the dependence of saturation humidity on temperature (Chapt. 5).

It also alters the global circulation (Chapt. 11), which can affect the frequency, intensity, and track of cyclones (Chapt. 13) and other storms (Chapts. 15 and 16).

Much of the concern about climate change is related to pollutants (Chapt. 17) emitted into the atmosphere by people in the boundary layer (Chapt. 4). Venting of greenhouse gases out of the boundary layer, and longevity of these tracers in the upper atmosphere depends, in part, on the vertical stability of the atmosphere (Chapt. 6).

Global climate models (GCMs) are forms of numerical weather prediction models (Chapt. 14). The sensitivities of the GCMs to various processes is determined by all the physics described earlier in this book, and allows you to judge the accuracy of climate change predictions from these models.

On Doing Science • Ethics & Data Abuse

In the race to reach scientific conclusions, and with the natural human desire to be successful, scientists sometimes inappropriately manipulate their data. The following are considered to be unethical in science:

"• **Massaging** – performing extensive transformations... to make inconclusive data appear to be conclusive;

• **Extrapolating** – developing curves [or proposing theories] based on too few data points, or predicting future trends based on unsupported assumptions about the degree of variability measured;

• **Smoothing** [or **trimming**] – discarding data points too far removed from expected or mean values;

• **Slanting** [or **cooking**]– deliberately emphasizing and selecting certain trends in the data, ignoring or discarding others which do not fit the desired or preconceived pattern;

• **Fudging** [or **forging**]– creating data points to augment incomplete data sets or observations; and

• **Manufacturing** – creating entire data sets *de novo*, without benefit of experimentation or observation."

[from C.J. Sindermann, 1982: *Winning the Games Scientists Play*. Plenum. 290 pp.; and Sigma Xi, 1984: *Honor in Science*. Sigma Xi Pubs., 41 pp.]

EXERCISES

Numerical Problems

N1. Find the earth effective radiation temperature for albedoes of:

a. 0.1	b. 0.2	c. 0.3	d. 0.4
e. 0.5	f. 0.6	g. 0.7	h. 0.8

N2. Calculate the effective radiation temperature for the following planets.

Planet	Albedo
a. Mercury	0.06
b. Venus	0.78
c. Mars	0.17
d. Jupiter	0.45

N3. For a global albedo of 0.3, find the earth effective radiation temperature for solar constant values (W/m^2) of:

a. 1300	b. 1325	c. 1350	d. 1375
e. 1400	f. 1425	g. 1450	h. 1475

N4. For an idealized one-layer atmosphere that is opaque to all IR radiation, find the surface temperature for:
(i) exercise N1
(ii) exercise N3

N5. For an atmosphere with a window in the IR, find the surface temperature under a one-layer atmosphere with emissivity:
a. 0.50 b. 0.55 c. 0.60 d. 0.65
e. 0.70 f. 0.75 g. 0.80 h. 0.85
Assume an effective emission temperature of 255 K.

N6(§). Recompute Fig 18.5, but for a solar constant (W/m^2) of:
a. 1300 b. 1325 c. 1350 d. 1375
e. 1400 f. 1425 g. 1450 h. 1475

N7(§). Recompute Fig 18.6, but for an atmospheric emissivity of:
a. 0.70 b. 0.75 c. 0.80 d. 0.85
e. 0.90 f. 0.92 g. 0.95 h. 0.99

Understanding & Critical Evaluation

U1. If the atmosphere can be idealized as having two layers, each opaque to IR, extend the results from the Greenhouse section to find an equation for surface temperature as a function of effective radiation temperature.

U2. After a major (hypothetical) volcanic eruption, assume that the ash in the atmosphere quickly spreads evenly around the globe. This dirty air absorbs 25% of any sunlight that shines through the atmosphere. For simplicity, assume that IR absorption/emission is not affected by the ash, and that the albedo of the earth's surface has not changed by ash deposits. Using a simplified climate model with a one-layer atmosphere with no atmospheric window, what is the new temperature of the earth's surface (averaged over the whole globe)?

U3. If water-vapor feedback tends to cause a sudden climatic change to warmer temperatures, and ice-albedo feedback tends to cause a sudden change in the opposite direction, why is our climate so stable at its present value?

U4. How would the surface temperature change from clear sky conditions to conditions under:
a. overcast stratocumulus clouds (low, warm, thick, opaque to both visible and IR)
b. thin overcast cirrostratus (transparent to visible, opaque to IR, high, cold)
c. both (a) and (b)

U5.(§) For a daisyworld
a. Redo the calculation from the body of this chapter.
b. What happens to the temperature over black and white daisies if the transport parameter changes to 0 or 1?
c What happens if you start with initial conditions of 100% coverage by white or black daisies?
d. What happens if you take a time step increment of 2, 4, or 8?
e. What parameter values prevent homeostasis from occurring (i.e., eliminate the nearly constant temperature conditions on a fecund daisyworld?
f. Is it possible for homeostasis to occur on earth, given the large fraction of area covered by oceans?

U6. If every thing else is constant in a global-climate version of a numerical weather prediction model, how would run time change if grid resolution is made twice as coarse?

U7. In year 1815, a massive **volcanic eruption on Tambora** sent an estimated 200 megatons of aerosols into the stratosphere. Extrapolate from the Mt. Pinatubo data given in the body to estimate the magnitude and sign of climate change. Would this temperature change be sufficient to explain the "**year without summer**" in 1816, when crops failed in Europe and USA?

U8. For the last 2 million years the global climate is believed to have been 1.5 to 3 degrees colder than now. For the 100 million years before that, the climate ranged from 2 to 5 degrees warmer than now. If our climate is presently warming, does that mean we are approaching or diverging from "normal"?

U9. Starting with a single layer atmosphere with no window. Suppose that initially the earth-atmosphere system is in equilibrium. Then, suddenly, the sun becomes hotter and the insolation doubles. Describe the fluxes and energy budgets while the earth and atmosphere adjust toward their new equilibrium.

U10. Three feedbacks were discussed in this chapter: water-vapor/cloud, ice-albedo, and biological (daisy-world) as if they were independent of each other. How might they interact, and what would be the resulting change in the equilibrium conditions on earth?

U11. Refer back to the numerical weather prediction chapter, and add more information to Table 18–1 to compare GCMs with operational numerical weather forecast models.

U12. What are the limitations on the use of sensitivity factors, as in eq. (18.21b)?

U13. In the Daisyworld Fig 18.9, the daisies are able to control the climate only over a limited range of luminosities. What changes to the growth characteristics of the daisies would allow control of the climate over a wider range of luminosities (and thus would allow greater chance of survival of an extreme luminosity change event)? Check your theory by modifying the growth equations accordingly, and performing the calculation on a spreadsheet or other program.

Web-Enhanced Questions

W1. Where you live, look up on the web the daily, monthly, and yearly range of temperatures. How do these temperature ranges compare with climatic variations associated with CO_2 doubling?

W2. Use the web to find an up-to-date definition of
a. El Niño b. La Niña c. El Viejo

W3. What is meant by an "anomaly map" (for example, of sea surface temperature (SST))? Find examples of such maps on the web.

W4. Search the web for images of yearly, or longer period, average cloud cover. Discuss how the result is related to radiation emission, reflection, and absorption in different parts of the world.

W5. Use the web to help define the southern Oscillation index (SOI).

W6. Use the web to find what is meant by teleconnections.

W7. During the 1997 – 1998 El Niño, how much warmer than average was the warmest SST in the eastern tropical Pacific Ocean, based on information you can get from the web?

W8. Use the web to learn the relationship between trade winds, upwelling, and El Niño.

W9. Search the web to find the worldwide weather consequences of the 1982-83 El Niño.

W10. Use the web to learn about the US government agency NCDC. How might it be useful to you? What similar agencies exist in other countries?

W11. Search the web for animations of SST variations. Also, how are they associated with rainfall shifts, cloud coverage changes, etc.?

W12. Search the web for information of consequences of global change on ecology and human health. Also, what is "Environmental Distress Syndrome"?

W13. Search the web to find a list of the pandemics during this past millennium. Which, if any, were related to climatic conditions?

W14. Search the web to identify the research and governmental groups making climate change predictions. Also, what methods are they using to do this (physical models, statistical methods, other)?

W15. Search the web for information on the current value of the solar constant, and its variations over the past several years.

W16. Search the web for information on anthropogenically produced greenhouse gases in the atmosphere, and how their concentrations have changed during the past several decades. Don't just limit your research to CO_2, such as can be obtained from the Muana Loa site.

W17. Search the web for a precise graph showing the atmospheric windows in the IR portion of the spectrum.

Synthesis Questions

S1. What if the earth did not rotate, the atmosphere did not move, and there was no conduction from the warm side to the cold side of the earth. Create an energy balance for a single-layer atmosphere with no atmospheric window.

S2. Same as the previous problem, but with an atmospheric window.

S3. Suppose that the atmosphere could be described by two layers, each without a window. Find the energy balances for each window, for the earth's surface, and for the whole earth-atmosphere system. Also find the earth surface temperature.

S4. Same as the previous exercise, except each atmospheric layer has a window in the same wavelength band.

S5. Same as the previous exercises, except that the atmospheric windows of the two layers are in different wavelength bands.

S6. Same as exercise S3, except with three atmospheric layers. Based on the result, extrapolate to a large number of layers.

S7. What if Earth had an internal heat source that created as much excess heat as the insolation at the top of the atmosphere. Describe the resulting energy balances of all relevant systems, and find the final equilibrium temperatures of the earth's surface and the atmosphere (assume single layer with no window).

S8. Suppose that transparent windows in the atmosphere opened over the north and south pole, similar to the ozone hole. Given that these spatial windows are over only the colder parts of the earth's surface, calculate the resulting energy balances for all relevant systems, and compute the average earth surface and atmosphere temperatures. State all assumptions.

S9. What if solar radiation from the sun ionized the sunny side of the atmosphere resulting in complete closure of the IR atmospheric window there. But assume that the shady side remains un-ionized, and a window can exist. Calculate the resulting energy budgets, and the equilibrium temperatures of the earth and the atmosphere. State all assumptions.

S10. Re-derive the Daisyworld equations, but for a situation where daisies always cover the whole earth. The relative coverages of black and white daisies can vary, but there is never any bare ground. How does the result differ, both in its equilibrium and its sensitivity to disturbances, from the Daisyworld described in this chapter?

S11. Suppose that global warming causes increased thunderstorm activity. Discuss all the possible feedback mechanisms and sensitivities, and whether the net result is positive or negative feedback.

S12. Same as the previous exercises, except for hurricanes.

S13. For both the water-vapor/cloud feedback, and the ice-albedo feedback, it was shown that the positive feedbacks do not cause temperatures that approach absolute zero or infinity. Instead, the earth-atmosphere would reach a new, steady equilibrium.

a. Given the temperatures associated with those new equilibria, could humans survive on Earth? Why? Is there any evidence in past history to support your claim?

b. Once in that new equilibrium, how difficult would it be to return to our present equilibrium? Would there also be a positive feedback to return to our present climate?

c. Suppose that the earth-atmosphere system frequently (every century) flip-flopped between the two equilibrium states. What would be the long term (millennium) average climate for this situation?

S14. The sensitivity of various processes in GCMs can be computed, as was listed in Table 18–2. However, what if there is an important process that is not known yet, but which has a sensitivity similar to the largest ones in Table 18–2. Devise an overall model accuracy measure that includes the possibility that important physical processes have been omitted from the GCM.

This has happened frequently in the evolution of GCMs. An example are sulfates in the atmosphere, which were found in the late 1900s to be produced by microscopic life in the ocean. The resulting atmospheric sulfate aerosols can reflect some solar radiation, thereby causing global cooling. This could be a negative feedback mechanism that offsets global warming, but which was accidently left out. How many other mechanisms that we don't know about are still missing? How much do you believe the temperature change forecasts associated with doubling of CO_2?

S15. Would global warming alter the general circulation? How?

OPTICS

CONTENTS

19 Light can be considered as photon particles or electromagnetic waves, either of which travel along paths called **rays**. The interaction of these light rays with hexagonal ice crystals or spherical rain drops creates the myriad of optical phenomena that adorn the sky.

To first order, light rays travel in straight lines within a uniform transparent medium such as air or water, but can **reflect** (bounce back) or **refract** (bend) at an interface between two media. Refraction can also occur where a single medium such as air smoothly changes its density. **Scattering** and **diffraction** of light create other optical phenomena.

The beauty of nature and the utility of physics come together in the explanation of atmospheric optical phenomena. Liquid water drops refract and reflect sunlight to create rainbows. Solid ice crystals refract sunlight to create halos, sun dogs, and various arcs, while reflections from ice crystals create subsuns, parhelic circles and sun pillars.

RAY GEOMETRY

When a **monochromatic** (single color) light ray reaches an interface between two media such as air and water, a portion of the incident light from the air can be reflected back into the air, some can be refracted as it enters the water (Fig 19.1), and some can be absorbed and changed into heat (not sketched). Similar processes occur across an air-ice interface.

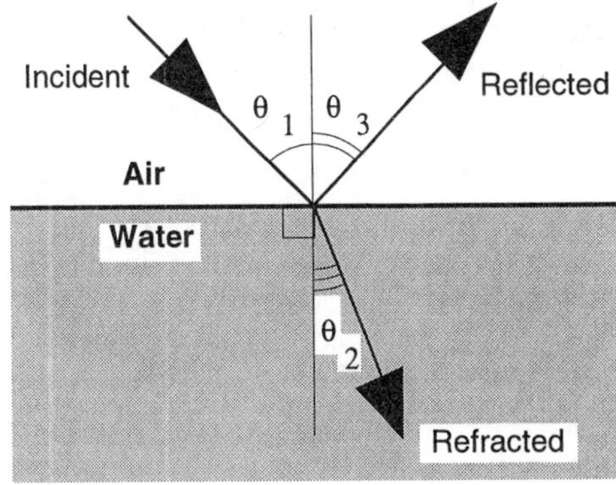

Figure 19.1
Geometric optics at an air-water interface.

Reflection

The angle of the reflected ray always equals the angle of the incident ray, measured with respect to a line normal (perpendicular) to the interface:

$$\theta_1 = \theta_3 \qquad \bullet(19.1)$$

The reflected angle does not depend on color.

Refraction

Snell's law gives the relationship between the incident angle θ_1 and refracted angle θ_2:

$$\frac{\sin\theta_1}{\sin\theta_2} = \frac{c_1}{c_2} = \frac{n_2}{n_1} \qquad \bullet(19.2)$$

where c_i is the speed of light through medium i. Incident and refracted rays are always in the same plane. This plane includes the line that is normal to the surface. It is the plane that gives the smallest angle between the incident ray and the surface.

Refractive index n_i for medium i relative to a vacuum is defined as:

$$n_i \equiv \frac{c_o}{c_i} \qquad \bullet(19.3)$$

where $c_o = 3 \times 10^5$ km/s is the speed of light in a vacuum. The ratio of refractive indices is sometimes defined as

$$\mu_{12} \equiv \frac{n_1}{n_2} \qquad \bullet(19.4)$$

where subscripts 1 and 2 refer to the media containing the incident ray and refracted rays, respectively. Different colors and different media have different refractive indices, as indicated in Table 19-1.

Table 19-1. Refractive index relative to a vacuum.

wavelength, λ		index of refraction, n		
µm	color	air (at 15°C & 101.3kPa)	ice	water (at 15°C)
0.7	red	1.0002753	1.307	1.329
0.6	orange-yel	1.0002763	1.310	1.333
0.5	green-blue	1.0002781	1.314	1.339
0.4	violet	1.0002817	1.320	1.345

The index for liquid water decreases by about 0.002 as the temperature increases from 15°C to 35°C. *Red light is bent less than violet light* as it passes through an interface. Thus, refraction causes white light to be split into a **spectrum** of colors. This phenomenon is called **dispersion**.

FOCUS • Newton and Colors

To confirm his laws of motion, Isaac Newton wanted to view the motions of the planets. He built his own telescopes for this purpose. However, the images he observed through his lenses were blurry. For example, images of stars were spread into a streak of colors.

After experimenting with different lenses, he concluded that neither the glass nor the construction was flawed. He realized that there must be some unknown physics causing this optical phenomenon. Like many great scientists, he allowed himself to get side-tracked to study this phenomenon in more detail.

One of his experiments was to obtain a triangular prism, and to allow sunlight to pass through it. He observed that the white sunlight is composed of a spectrum of colors: red, orange, yellow, green, blue, indigo, and violet. Newton must have had a unique sense of color, because most people cannot discriminate between indigo and violet in the spectrum.

Newton concluded that the refraction of light through a lens inevitably causes color **dispersion**. Thus, a pinpoint of white starlight would be spread into a smear of colors. His solution to the telescope problem was to design a telescope without glass lenses. Instead he invented a reflection telescope using curved mirrors, because reflection does not cause separation of light into colors.

Sometimes it is easier to work with x and y components of the incident ray, where the x-axis might be aligned with the axis of a columnar ice crystal, for example, and the y-axis might be on the crystal surface (Fig 19.2). The relationship between the component angles and the incident angle is:

$$\tan^2\theta_1 = \tan^2\alpha_1 + \tan^2\beta_1 \qquad (19.5)$$

This relationship also applies to refracted angles, and will be used extensively later in this chapter to discuss ice-crystal optics.

Solved Example

Rays of red and violet light in air strike a water surface, both with incident angle of 60°. What is the angle of refraction for each color?

Solution

Given: $\theta_1 = 60°$

Find: $\theta_2 = ?°$ for red and violet.

First, solve eq. (19.2) for the refraction angle:

$\theta_2 = \arcsin[(n_1/n_2)\cdot\sin(\theta_1)]$

and use refractive indices from Table 19-1.

For red: $\theta_2 = \arcsin[(1.0002753/1.329)\cdot\sin(60°)]$

θ_2= **40.68°**

Similarly for violet: θ_2= **40.10°**

Check: Units OK. Physics OK.

Discussion: Had there been no bending, then both answers would have been 60°. Angles closer to 60° for this example correspond to less bending. The answers above confirm the statement that red light is bent less than violet. The amount of bending is fairly large, about 20°.

Component angles α_1 and β_1 of the incident ray do *not* individually obey Snell's law (eq. 19.2). Nevertheless, Snell's law can be reformulated in terms of components as follows:

$$c_{\alpha\beta} = \frac{\mu_{12}^2}{1+(1-\mu_{12}^2)\cdot\left\{\tan^2\alpha_1 + \tan^2\beta_1\right\}} \qquad (19.6)$$

and

$$\tan^2\alpha_2 = c_{\alpha\beta}\cdot\tan^2\alpha_1 \qquad (19.7a)$$

$$\tan^2\beta_2 = c_{\alpha\beta}\cdot\tan^2\beta_1 \qquad (19.7b)$$

where α_2 and β_2 are the components of the refracted ray (analogous to α_1 and β_1).

These equations are abbreviated as

$$\alpha_2 = S_\alpha(\alpha_1,\beta_{1,},\mu_{12}) = \arctan\left[\left(c_{\alpha\beta}\cdot\tan^2\alpha_1\right)^{1/2}\right]$$

$$\bullet(19.8)$$

$$\beta_2 = S_\beta(\alpha_1,\beta_{1,},\mu_{12}) = \arctan\left[\left(c_{\alpha\beta}\cdot\tan^2\beta_1\right)^{1/2}\right]$$

where S represents Snell's law for components.

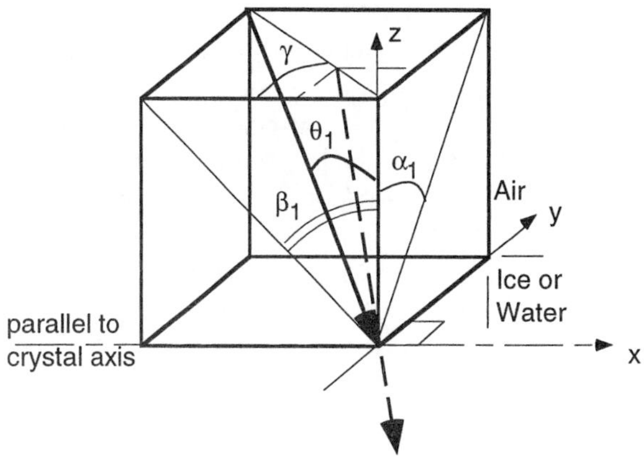

Figure 19.2

Components of refraction geometry. Incident ray is heavy solid arrow. Refracted ray is heavy dashed arrow. The tail of the refracted ray is extended to show that it lies in the same plane as the incident ray.

Solved Example

A ray of red light in air strikes a water surface, with incidence angle components of 45° and 54.74°. What are the corresponding component angles of the refracted ray?

Solution

Given: $\alpha_1 = 45°$, $\beta_1 = 54.74°$.

Find: $\alpha_2 = ?°$, $\beta_2 = ?°$

From Table 19-1 for red light:

$\mu_{12}^2 = (1.0002753/1.329)^2 = 0.5665$

Next, solve eq. (19.6):

$$c_{\alpha\beta} = \frac{0.5665}{1+(1-0.5665)\cdot(1+2)} = 0.24625$$

Then use eq. (19.7):

$\tan^2\alpha_2 = 0.24625\cdot[\tan^2 45°] = 0.24625$

$\tan^2\beta_2 = 0.24625\cdot[\tan^2 54.74°] = 0.49266$

Thus: $\alpha_2 = \arctan[(0.24625)^{0.5}]$ = **26.39°**

β_2 = **35.065°**

Check: Units OK. Physics OK.

Discussion: Using eq. (19.5) with the incident angle components of 45° and 54.74°, we find that the incident ray angle is $\theta_1 = \arctan[(1+2)^{0.5}] = 60°$. This is the same as the previous solved example. Using eq. (19.5) on the answers above, we find $\theta_2 = \arctan[\,(0.2462+0.49266)^{0.5}] = 40.68°$. This is also the same as the previous solved example, which verifies eqs. (19.6) & (19.7).

Huygens' Principle

Huygens suggested that every point along a wave front acts like a generator of new spherical secondary wavelets. Wave-front position after some time interval is located at the tangent to all of the new wavelets. Thus, when a portion of a wave front encounters a medium with a slower light velocity, then that portion of the wave slows, causing the whole wave front to turn into the medium (see Fig 19.3).

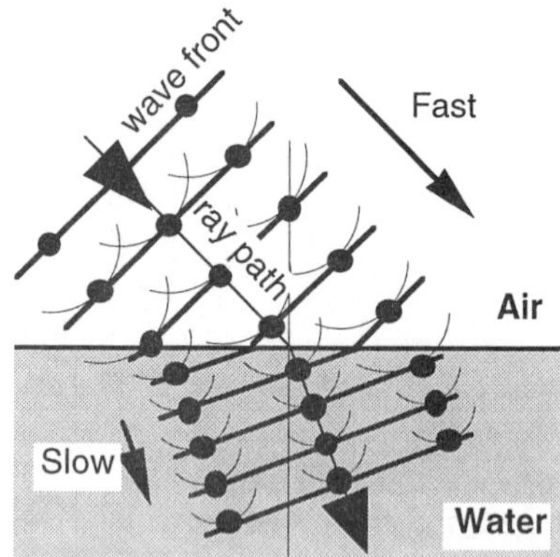

Figure 19.3
Propagation of wave fronts across an interface.

Critical Angle

When Snell's law is applied to light rays moving from a denser medium to a less dense medium (such as from water toward air), there is a **critical angle** at which light is bent so much that it follows the interface. At angles greater than this critical angle, light cannot refract out of the dense medium at all. Instead, all of light reflects (Fig 19.4).

The critical angle θ_c is found from:

$$\sin(\theta_c) = \frac{n_2}{n_1} \qquad \bullet(19.9)$$

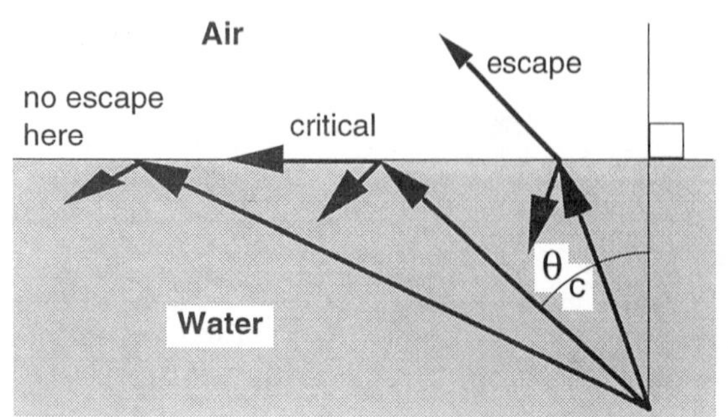

Figure 19.4
Angles greater than critical θ_c do not allow light to escape from a dense to less dense medium.

where n_1 is the refractive index for the incident ray (i.e., for the dense medium). For yellow-orange light moving from water to air, the critical angle is 48.6°. There is no critical angle for light moving into a denser medium.

LIQUID DROP OPTICS

Most liquid water optics are found by looking away from the sun. Rainbows are circles or portions of circles (Fig 19.5) that are centered on the **antisolar** point, which is the point corresponding to the shadow of your head or camera. Primary rainbows have red on the outside of the circle, and are the brightest and most easily seen. Secondary rainbows have red on the inside. Supernumerary bows are very faint, and touch the inside of the primary rainbow. To first approximation, we will assume that liquid rain drops and cloud droplets are spherical.

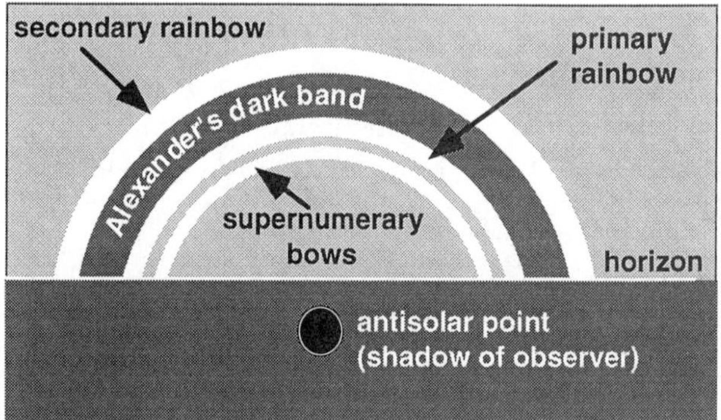

Figure 19.5
Rainbows.

Reflection from Water

Rays of light that hit a water surface at a shallow angle are reflected more than those hitting straight on. Reflectivity r vs. elevation angle Ψ above the water surface is approximately:

$$r = r_o + r_1 \cdot e^{-\Psi / a} \qquad (19.10)$$

where $r_o = 0.02$, $r_1 = 0.98$, and $a = 9.3°$. This is plotted in Fig 19.6.

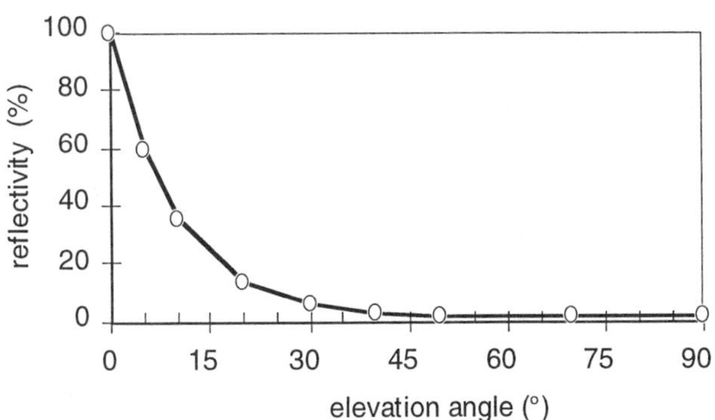

Figure 19.6
Reflectivity from a water surface as a function of ray elevation angle.

Rays of sunlight can enter a spherical drop at any distance x from the centerline (Fig 19.7). The ratio of this distance to the drop radius R is called the **impact parameter** (x/R). Because of curvature of the drop surface, rays arriving at larger impact parameters strike the drop at smaller elevation angles:

$$\psi = \arccos\left(\frac{x}{R}\right) \qquad \bullet(19.11)$$

Thus, the amount of reflected light from a drop increases with increasing impact parameter.

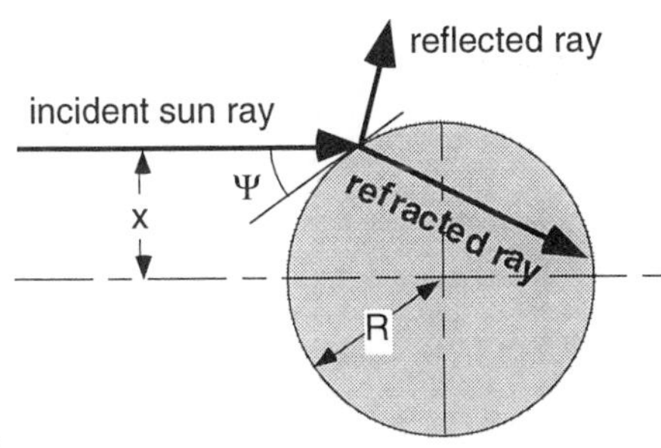

Figure 19.7
Fate of incident ray after first impact with drop.

Primary Rainbow

Rays that are not reflected from the outside of the drop can make zero or more reflections inside the drop before leaving. Those entering rays that make one reflection (in addition to the two refractions during entry and exit) cause the primary rainbow (Fig 19.8). White light is dispersed by the refractions such that reds appear on the outside and violets on the inside of the rainbow circle.

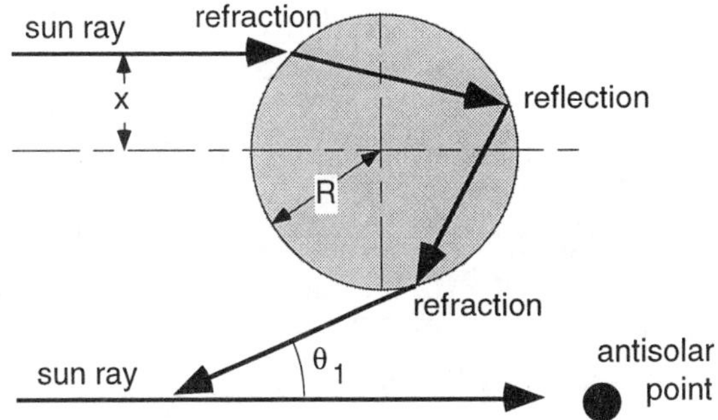

Figure 19.8
Ray geometry for primary rainbow.

Variations of impact parameter cause a range of output viewing angles θ_1 according to:

$$\theta_1 = 4 \cdot \arcsin\left(\frac{n_{air}}{n_{water}} \cdot \frac{x}{R}\right) - 2 \cdot \arcsin\left(\frac{x}{R}\right) \qquad \bullet(19.12)$$

In other words there is not a single magic angle of 42° for the primary rainbow. Instead, there is a superposition of many rays of different colors with a wide range of viewing angles.

To learn how these rays interact to form the rainbow, we can solve eq. (19.12) on a spreadsheet for a large number of evenly spaced values of the impact parameter, such as intervals of 0.02 for x/R. Output viewing angles, measured from the antisolar point, range between 0° and 42.7°, as plotted in Fig 19.9. This is why the sky is brighter inside the primary rainbow (i.e., for viewing angles of 0 to 42.7°), than just outside of it.

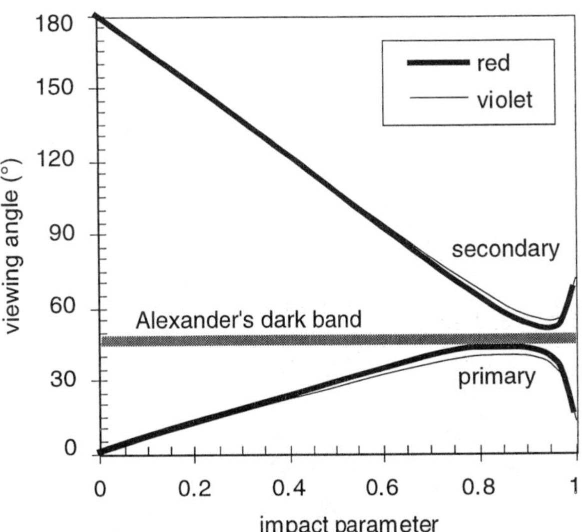

Figure 19.9
Viewing angles for rainbow rays.

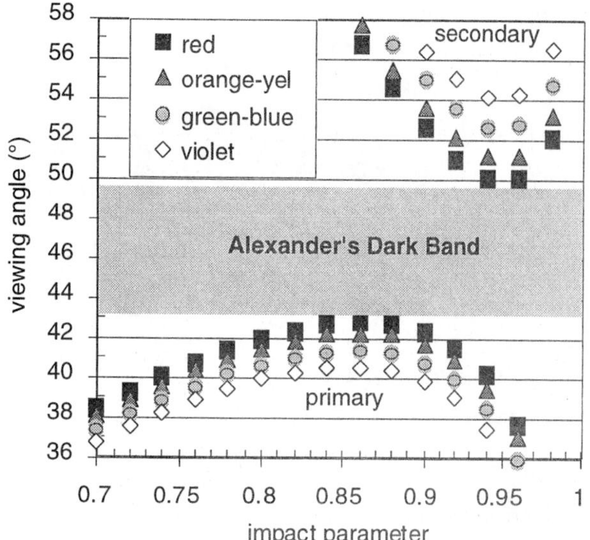

Figure 19.10
Viewing angles for rainbow rays (expanded version).

Fig 19.10 is a blow-up of Fig 19.9. Three of the data points have a viewing angle of about 42.5°, and these all correspond to rays of red light. Hence, the primary rainbow looks red at that viewing angle. At 42.0°, there are mostly orange-yellow data points, and just a couple red points. Hence, at this angle the rainbow looks orange. Similar arguments explain the other colors. Inside the primary rainbow the sky looks white, because there are roughly equal amounts of all colors returned at each angle.

Light intensity is not fully accounted for in Figs 19.9 and 19.10. Those figures assume an equal distribution of impact parameters, with one ray per parameter. However, the number of rays that can enter at any impact parameter is proportional to the circumference of a circle of radius x/R. [You can think of the cross section of the drop as being similar to a dart board. If many darts are thrown with little skill at the board, more darts will hit the outer rings of the dart board, because they have more area.]

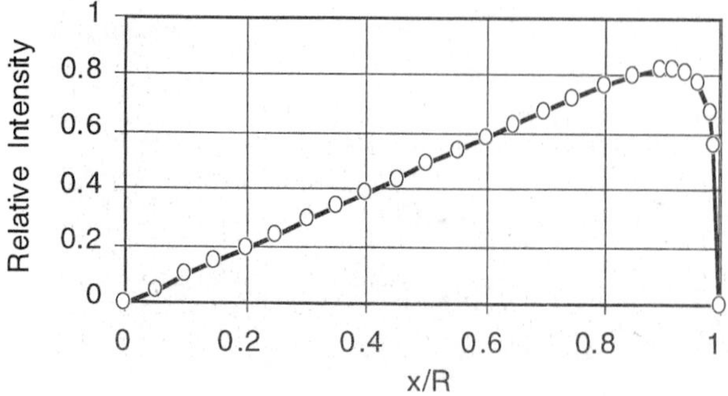

Figure 19.11
Relative intensity of light entering a rain drop at various impact-parameter values.

Circle circumference is $2\pi x$, and the drop circumference is $2\pi R$; hence, the the number of rays striking the drop is proportional to x/R. At large x/R, however, much of the ray that strikes the drop is reflected according to Fig 19.6. The combination of these two opposing factors gives the number of rays that can enter the drop to cause a rainbow (Fig 19.11).

Using Fig 19.11 with 19.10 shows that the largest intensity of light coincides with those impact parameters that give the greatest color separation. This is why rainbows can be quite vivid.

Secondary Rainbow

Entering light rays can also make two reflections before leaving the raindrop, as sketched in Fig 19.12. The extra reflection reverses the colors, putting red on the inside of the circle of a secondary rainbow.

The relationship between impact parameter and viewing angle for a secondary rainbow is:

$$\theta_1 = 180° + 2 \cdot \arcsin\left(\frac{x}{R}\right) - 6 \cdot \arcsin\left(\frac{n_{air}}{n_{water}} \cdot \frac{x}{R}\right) \qquad \bullet(19.13)$$

Again, a range of output viewing angles occurs because light enters over the full range of impact parameters. In this case, the sky is dark inside of the secondary rainbow (about 50° viewing angle), and bright outside, as sketched in Figs 19.9 and 19.10 based on spreadsheet calculations.

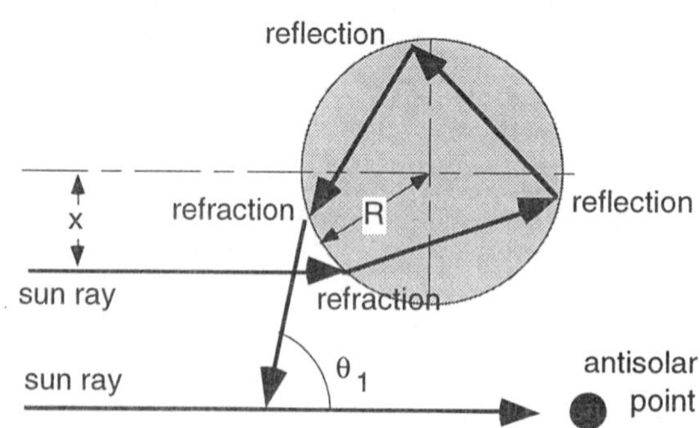

Figure 19.12
Ray geometry for secondary rainbow.

Alexander's Dark Band

Neither the primary nor secondary rainbows return light in the viewing angle range between 42.7° and 50°. Hence, the sky is noticeably darker between the primary and secondary rainbows (Figs 19.9 & 19.10). This dark region is called Alexander's dark band, after Greek philosopher Alexander of Aphrodisias who described it during the third century A.D.

Solved Example

For red light calculate the viewing angle from both primary and secondary rainbows for an impact parameter of 0.5 .

Solution

Given: $x/R = 0.5$

n_{air}=1.0002753, n_{water}=1.329 from Table 19-1.

Find: $\theta_1 = ?°$

Use eqs. (19.12) for primary rainbow:

$$\theta_1 = 4 \cdot \arcsin(0.75265 \cdot 0.5) - 2 \cdot \arcsin(0.5)$$
$$= \underline{28.42°}$$

Use eqs. (19.13) for secondary rainbow:

$$\theta_1 = 180° + 2 \cdot \arcsin(0.5) - 6 \cdot \arcsin(0.75265 \cdot 0.5)$$
$$= \underline{107.36°}$$

Check: Units OK. Physics OK.

Discussion: These rays contribute to the brightness inside the primary rainbow and outside the secondary.

FOCUS • Rainbows and the Renaissance

German monk Theodoric von Freiberg created a physical model of rainbow physics in 1304 by filling a glass sphere with water. By carefully shining light through it at different impact parameter values, he could measure the various output angles. However, the physics behind these observations were not explained until the renaissance, after Willebrod Snell van Royen had discovered the law for refraction in the early 1600's.

Twenty years after Snell's discovery, Descartes used it to demonstrate the capabilities of his scientific method by mathematically solving the rainbow problem. His calculations were very similar to our spreadsheet calculations, except that he manually performed calculations for a large number of different impact parameters. Through these repetitive "brute-force" calculations he verified the magic number of 42° for the brightest returned light, and 50° for the secondary rainbow.

About 30 years later, Newton applied his knowledge of color dispersion to explain why the rainbow has colors. Also, he was able to use his invention of calculus to elegantly derive the magic angle of 42°.

ICE CRYSTAL OPTICS

Ice-crystal optical phenomena are seen by looking more-or-less toward the sun (Fig 19.13). To observe them, shield your eyes from the direct rays of the sun to avoid blinding yourself. Also, wear sunglasses so your eyes are not dazzled by the bright sky close to the sun.

Subsuns, sun pillars, and parhelic circles are caused by simple reflection from the outside surface of ice crystals. Sundogs, halos, circumzenith arcs, and upper tangent arcs are caused by refraction through the ice crystals, with the red color closest to the sun. Subsun dogs are caused by both refraction and reflection in the ice. There are many other optical phenomena that we do not have space to describe here.

Ice crystals (types of snow) can form as high-altitude cirrus clouds, can fall from lower clouds, can form (as "**diamond dust**") within cloudless cold air of high relative humidity, or can be stirred up by the wind from fresh surface snow. They have a variety of shapes with hexagonal cross section. The two shapes most important for atmospheric optical phenomena are the hexagonal plate and column (Fig 19.14).

These plates or columns, if larger than about 30 µm, gently fall through the air with the orientation relative to horizontal as sketched in Fig 19.14. Smaller crystals tend to tumble as they fall. The most important angles of this crystal are 60° and 90°.

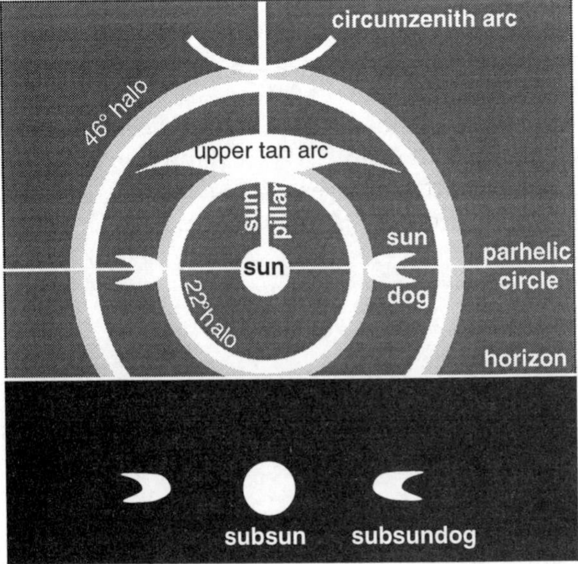

Figure 19.13
Some optical phenomena associated with ice crystals in air.

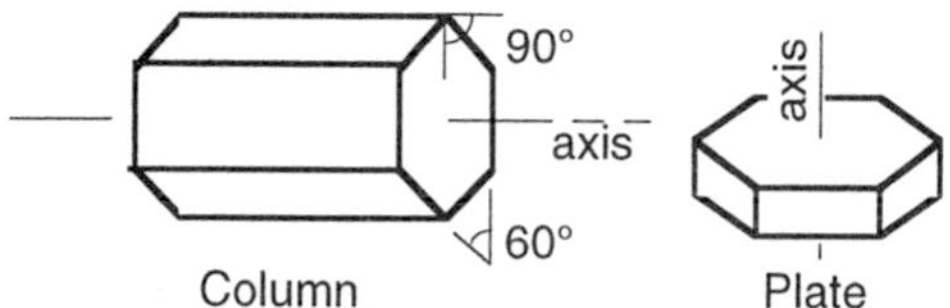

Figure 19.14
Crystal geometry relevant to atmospheric optics.

Sun Pillar

A sun pillar is a vertical column of light that appears to come out of the top of the sun. It forms by reflection of sunlight off of the outside faces of large hexagonal-column ice crystals (Fig 19.15) that are precipitating out of clouds. Wind shear can align these columns to point in the same direction, like pencils are aligned in a pencil box.

An end view of this phenomenon is sketched in Fig 19.16. The crystals are free to rotate about their column axis as they gently fall to earth. As a result, some of the ice crystals have faces that by chance reflect the sunlight to your eye, while other ice crystals reflect the sunlight elsewhere.

So few sunbeams reflect to your eye that the sun pillar appears very faint. The best time to observe sun pillars is at sunrise or sunset when the sun is hidden just below the horizon, but is still able to illuminate the ice precipitation. Sun pillars appear to be red not because of any refraction in the crystal, but because light from the rising or setting sun has lost much of its blue components due to scattering from air molecules before it reaches the ice crystals.

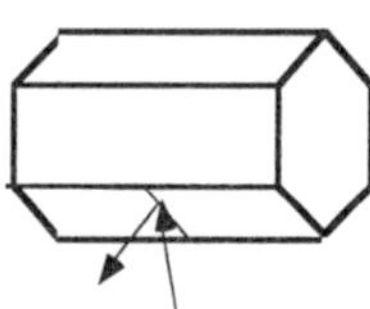

Figure 19.15
Reflection off a face of an ice column.

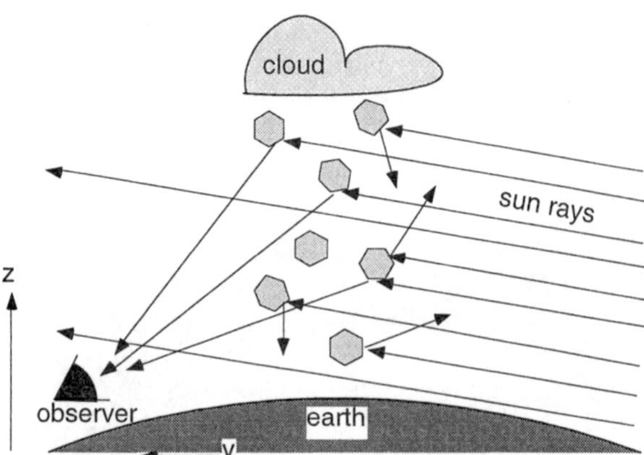

Figure 19.16
Geometry of sun pillar optics (end view) reflecting from large hexagonal-column ice crystals.

Parhelic Circle

The parhelic circle is a horizontal band of white light extending left and right through the sun. It is formed by reflection off of the vertical faces of large hexagonal plates (Fig 19.17). Like the sun pillar, these plates can have any rotation orientation about their axis, causing reflections from many different ice crystals to reach your eye from many different angles left and right from the sun (Fig 19.18).

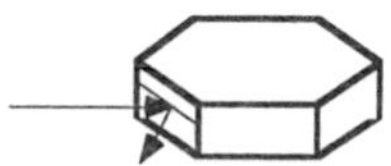

Figure 19.17
Ray geometry of parhelic circle optics.

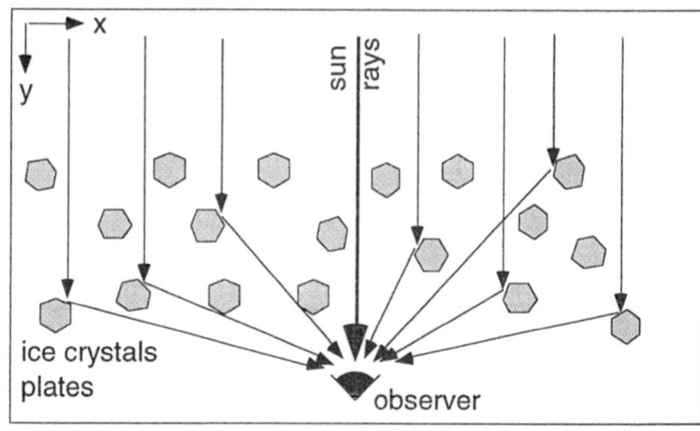

Figure 19.18
Reflection from many large hexagonal plates to form a parhelic circle (top or plan view).

Subsun

The subsun is a spot of white light seen below the horizon as viewed from a bridge or mountain. It forms by reflection of sunlight off of the top surfaces of large hexagonal plates (Fig 19.19). If all the hexagonal plates are oriented horizontally, then they act like mirrors to produce a simple reflection of the sun (Fig 19.20). As given by eq. (19.1), the subsun appears an equal angle below the horizon as the sun is above.

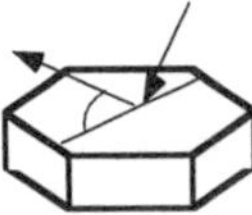

Figure 19.19
Ray geometry of subsun optics.

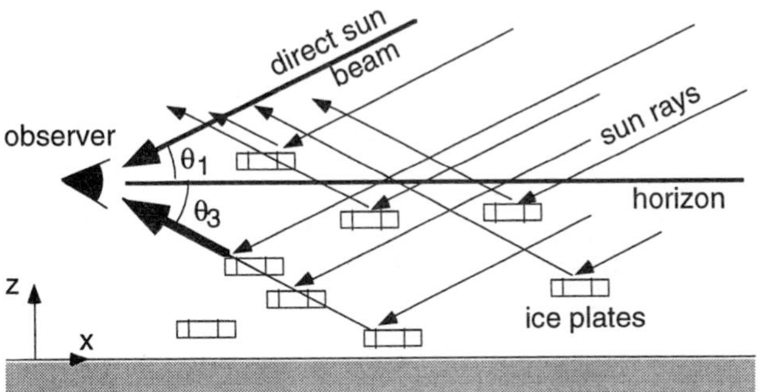

Figure 19.20
Reflection from many ice-crystal plates to form a subsun (side view).

22° Halo

Optics for 22° halos are quite interesting and not as simple as might be expected. Small ice-crystal columns are free to tumble in all directions. As a result, light rays can enter the crystal at a wide range of angles, and can be refracted over a wide range of sky.

The brightest light that we identify as the halo is caused by refraction through those ice crystals that happen to be oriented with their column axis perpendicular to the sun rays (Fig 19.21). We will study this special case. Different crystals can nevertheless have different rotation angles about their axis (the dark point in the center of the ice crystal of Fig 19.21). As a result, the angle of incidence θ_1 can vary from crystal to crystal, thereby causing different viewing (output) angle θ_2 values.

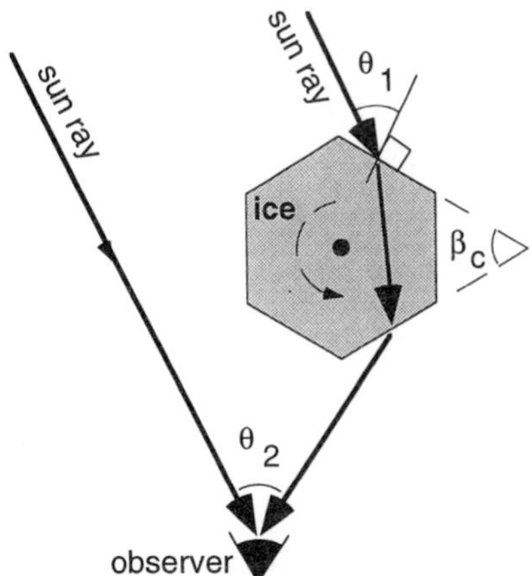

Figure 19.21
Ray geometry of 22° halo optics (top or plan view).

For the special case of β_c= 60° and light rays approaching normal to the column axis:

•(19.14)

$$\theta_2 = \theta_1 - \beta_c + \arcsin\left\{\frac{n_{ice}}{n_{air}} \cdot \sin\left[\beta_c - \arcsin\left(\frac{n_{air}}{n_{ice}} \cdot \sin\theta_1\right)\right]\right\}$$

Solved Example

What is θ_2 if $\theta_1 = 80°$ and $\beta_c = 60°$ for red light in a 22° halo?

Solution

Given: $\theta_1 = 80°$

$n_{air} = 1.0002753$, $n_{ice} = 1.307$ from Table 19-1.

Find: $\theta_2 = ?°$

Use eq. (19.14) with $\beta_c = 60°$ for hexagonal crystal:

$$\theta_2 = 80° - 60° + \text{asin}\left\{\frac{1.307 \cdot \sin}{1.0002753}\left[60° - \text{asin}\left(\frac{1.0002753}{1.307} \cdot \sin 80°\right)\right]\right\}$$

= **34.55°**

Check: Units OK. Physics OK.
Discussion: Quite different from 22°.

The 22° halo observed by a person is the result of superposition of rays of light from many ice crystals possessing many different rotation angles. Hence, it is not obvious why 22° is a magic number for θ_2. Also, if refraction is involved, then why are the red to yellow colors seen most vividly, while the blues and violets are fading to white?

To answer these questions, suppose ice crystals are randomly rotated about their axes. There would be an equal chance of finding an ice crystal having any incidence angle θ_1. We can simulate this on a computer spreadsheet by doing calculations for a large number of evenly-spaced incidence angles (for example, every 2°), and then calculating the set of θ_2 angles as an outcome. The result is plotted in Fig 19.22 for two of the colors (because n_{air}/n_{ice} varies with color).

Colors and angles of a 22° halo can be explained using Fig 19.23. For viewing (output) angles less than about 21.5°, there is no refracted light of any color; hence, the halo is dark inside (the edge closest to the sun). In the range of viewing angles averaging 22° (i.e., from 21.5° to 22.5°), there are 25 data points plotted, which cause the bright ring of light we call the halo. There are also data points at larger angles, but they are more sparse — causing the brightness of the halo to gradually fade at greater viewing angles from the sun.

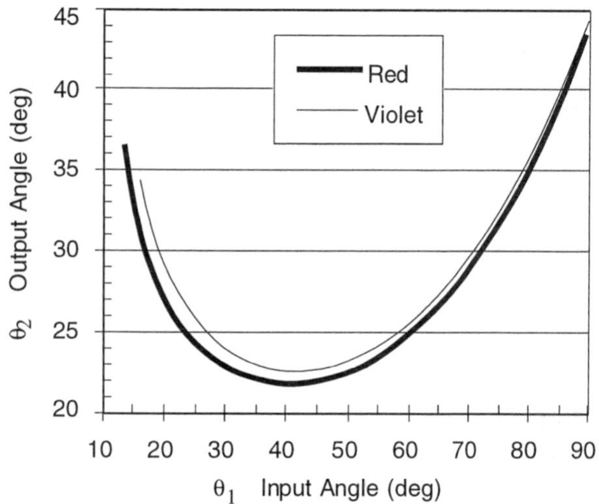

Figure 19.22
Viewing (output) angles θ_2 from hexagonal ice crystals in a 22° halo, for various incident angles θ_1 of sunlight.

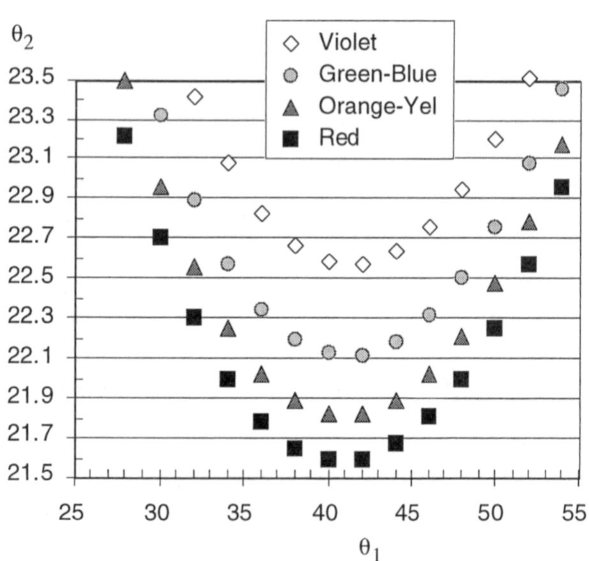

Figure 19.23
Blow-up of viewing (output) angles θ_2 from hexagonal ice crystals in a 22° halo, for various incident angles θ_1.

For a viewing angle range of θ_2 = 21.5° to 21.7°, we see four "red light" data points in Fig 19.23, but no other colors. Thus, the portion of the halo closest to the sun looks bright red. In the next range of viewing angles (21.7° - 21.9°), we see four "yellow-orange" and two "red" data points. Thus, in this range of viewing angles we see bright orange light.

In the angle range 22.1 - 22.3°, there are four blue-green data points, and two orange-yellows and two reds. These colors combine to make a bluish white color. By an angle of 23° , there are roughly equal portions of all colors, creating white light.

46° Halo

The 46° halo forms by sun rays shining through the side and end of small hexagonal columns (Fig 19.24). Equation (19.14) applies, but with β_c= 90°. The resulting viewing angles are shown in Fig 19.25. Except for the different radius, this halo has visual characteristics similar to those of the 22° halo.

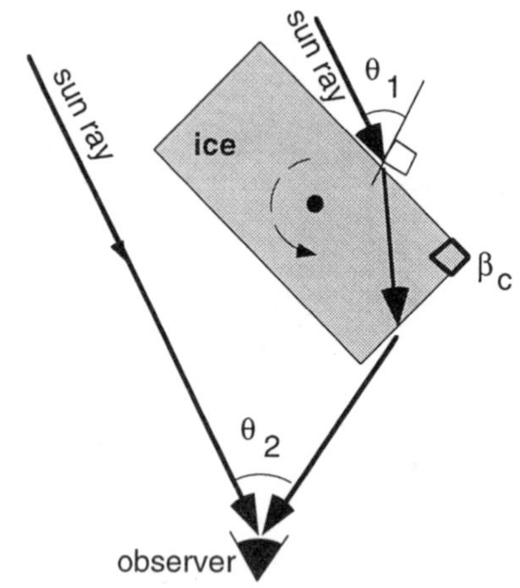

Figure 19.24
Ray geometry for 46° halo.

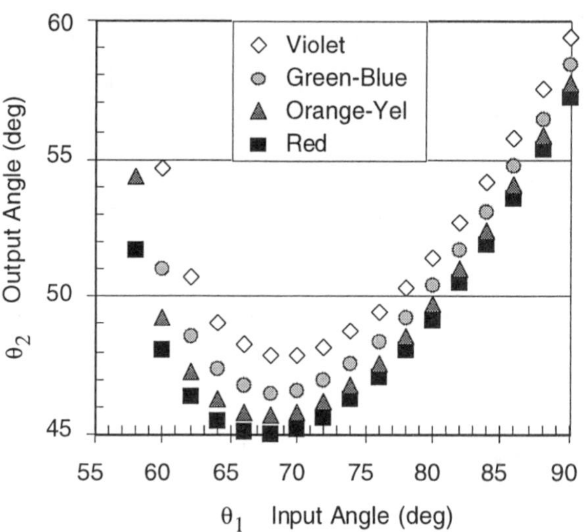

Figure 19.25
Viewing (output) angles for a 46° halo.

Solved Example

Find the viewing angle θ_2 for a 46° halo, given input angles $\theta_1 = 80°$ & $\beta_c = 90°$ for red.

Solution

Given: $\theta_1 = 80°$
Find: $\theta_2 = ?°$

Use eq. (19.14):

$$\theta_2 = 80° - 90° + \text{asin}\left\{\frac{1.307 \cdot \sin}{1.0002753}\left[90° - \text{asin}\left(\frac{1.0002753}{1.307} \cdot \sin 80°\right)\right]\right\}$$

$= \underline{\mathbf{49.18°}}$

Check: Units OK. Physics OK.
Discussion: This agrees with Fig 19.25.

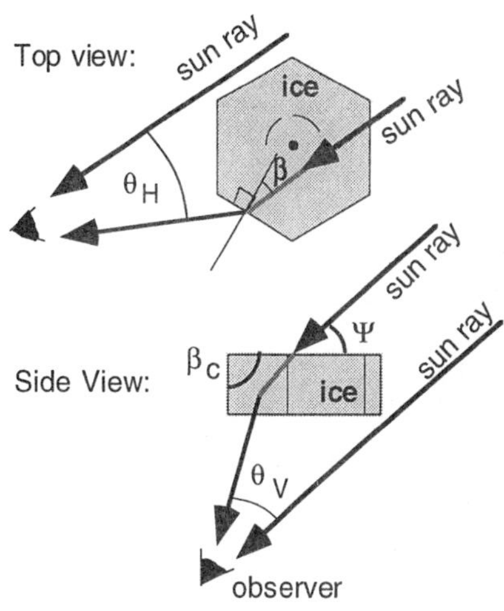

Figure 19.26
Geometry for circumzenith arc.

Science Graffito

"The skipper he stood beside the helm,
His pipe was in his mouth,
And he watched how the veering flaw did blow
The smoke now West, now South.

Then up and spake an old Sailor
Had sailed to the Spanish Main,
'I pray thee, put into yonder port,
For I fear a Hurricane.

Last night the moon had a golden ring,
And tonight no moon we see!'
The skipper he blew a whiff from his pipe,
And a scornful laugh laughed he."

– by Henry Wadsworth Longfellow, from *The Wreck of the Hesperus.*

Circumzenith Arc

The circumzenith arc is formed by refraction of sunlight through the top and side of large hexagonal plates (Fig 19.26). These plates can assume any rotation angle β about their vertically-oriented column axis. Also, at different times of day, there are different solar elevation angles Ψ. The ray equations for vertical viewing angle θ_V (elevation above the sun) and horizontal angle θ_H (azimuth from the sun) are:

$$\theta_V = S_\alpha[\arccos(\mu_{ai} \cdot \cos\psi) \ , \ \beta \ , \ \mu_{ia}] - \psi$$

$$\theta_H = S_\beta[\arccos(\mu_{ai} \cdot \cos\psi) \ , \ \beta \ , \ \mu_{ia}] - \beta \quad (19.15)$$

Subscripts "ai" are for rays going from air to ice (e.g.; $\mu_{ai} = n_{air}/n_{ice}$), while subscripts "ia" are from ice to air (e.g.; $\mu_{ia} = n_{ice}/n_{air}$).

The circumzenith arc appears as a portion of a circle about the zenith point (the point directly overhead). The angular radius of this circle as viewed from the ground is not constant, but varies with solar elevation angle as is plotted in Fig 19.27. For solar elevation angles above about 32°, no circumzenith arc is possible. For solar elevations just slightly below this maximum, the arc is so faint that it is difficult to see.

A spreadsheet can be used with a large number of evenly spaced values of β to simulate the superposition of rays from a large number of randomly-rotated ice plates. The resulting locus of output viewing angles traces the circumzenith arc in Fig 19.28.

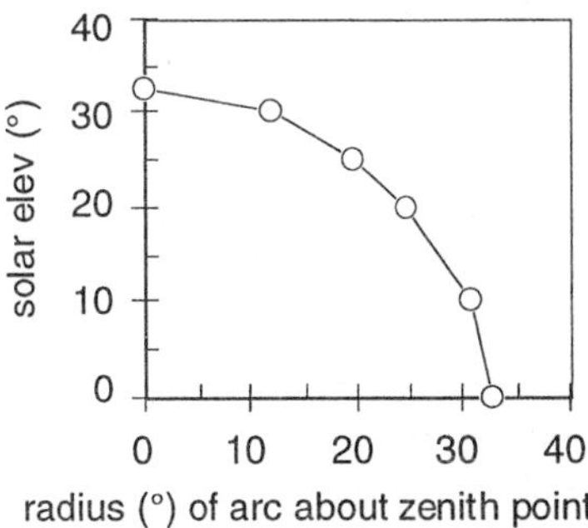

Figure 19.27
Radius of circumzenith arc about the zenith point, as viewed from the ground.

Solved Example

For a circumzenith arc, what are the elevation and azimuth viewing angles (relative to the sun) for red light, given a solar elevation of 10°, and a rotation angle of $\beta = 20°$?

Solution

Given: $\Psi = 10°$, $\beta = 20°$

$n_{air} = 1.0002753$, $n_{ice} = 1.307$ from Table 19-1.

Find: $\theta_V = ?°$, and $\theta_H = ?°$.

First, find the refraction parameters:

$\mu_{ai} = n_{air}/n_{ice} = 0.7653$,

and $\mu_{ia} = 1/\mu_{ai} = 1.3067$

Next: $\arccos[\mu_{ai}\cdot\cos(\Psi)] = \arccos(0.7653\cdot\cos 10°) = 41.09°$

Use eq. (19.15):

$\theta_V = S_\alpha[41.09°\ ,\ 20°\ ,\ 1.3067] - 10°$

$\theta_H = S_\beta[41.09°\ ,\ 20°\ ,\ 1.3067] - 20°$

Use eq. (19.6): $c_{\alpha\beta} =$

$$= \frac{(1.3067)^2}{1 + [1-(1.3067)^2]\cdot\left\{[\tan(41.09°)]^2 + [\tan(20°)]^2\right\}}$$

$= 4.634$

Use eq. (19.8):

$$S_\alpha(\alpha_1, \beta_1, \mu_{12}) = \arctan\left[\left\{4.634 \cdot [\tan(41.09)]^2\right\}^{1/2}\right]$$

$= 61.95°$

Similarly: $S_\beta = 38.08°$

Finally: $\theta_V = 61.95° - 10° = \underline{\mathbf{51.95}}°$

$\theta_H = 38.08° - 20° = \underline{\mathbf{18.08}}°$

Check: Units OK. Physics OK.

Discussion: This answer falls on the thick line plotted in Fig 19.28. Obviously this calculation is tedious for more than one ray, but it is easy on a spreadsheet.

Eq. (19.15) does not include reflection of some or all of the rays from the outside of the crystal as a function of incidence angle, nor does it include the critical angle for rays already inside the crystal. Thus, some of the points plotted in Fig 19.28 might not be possible, because the sun rays corresponding to those points either cannot enter the crystal, or cannot leave it.

The circumzenith arc is always outside of the 46° halo. Arc coloring is similar to that for the halos, with red at the bottom, closest to the sun.

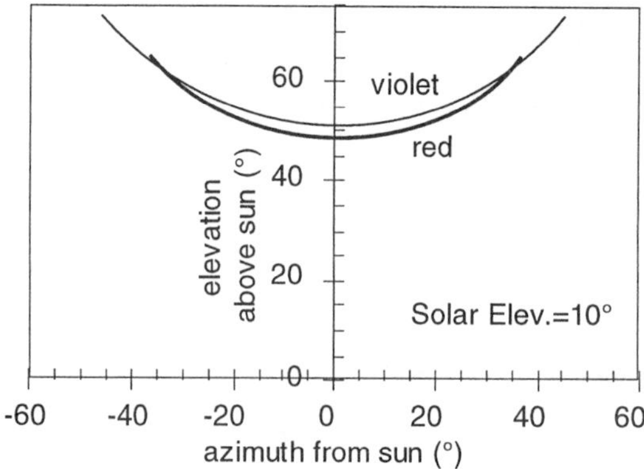

Figure 19.28
Circumzenith arc found by solving eq. (19.15).

Sun Dogs

Sun dogs are the bright spots of light to the right and left of the sun, roughly 22° from the sun. They are formed when light shines through the sides of large horizontal ice-crystal plates that gently fall through the air with their column axes vertical. The crystals are free to rotate about their column axes.

Because of the preferred orientation of these large crystals, the sun-dog optics depend on the solar elevation angle. When the sun is on the horizon, sun rays shine through the crystal as sketched in Fig 19.29a, causing bright spots on the 22° halo directly to the right and left of the sun.

For higher sun angles (Fig 19.29b), ice-crystal geometry causes the sun dogs to move horizontally further away from the sun. For example, at a solar elevation angle of $\Psi = 47°$, the sun dogs are at a horizontal viewing angle of 31° from the sun. As the solar elevation angle becomes greater than about 60°, the sun dog is so faint and so far from the 22° halo that it effectively ceases to exist.

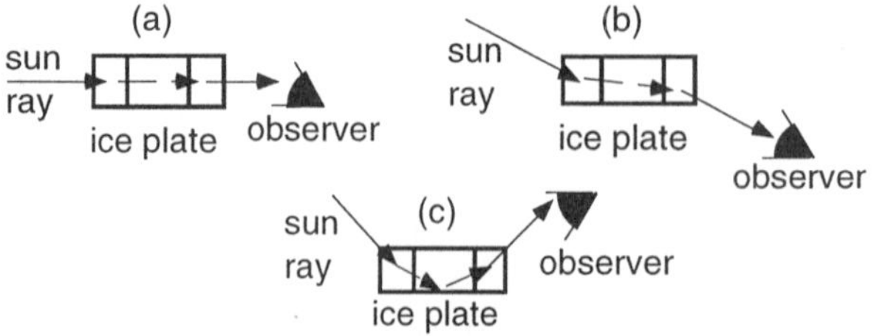

Figure 19.29
Geometry of sun dog optics (side view), for (a) low sun elevation angle, (b) higher sun angle, and for (c) subsun dogs.

Science Graffito

"It is clear and cold...And all about is snow and silence. And in the sky are three suns, and all the air is flashing with the dust of diamonds." – Jack London, on a trip down the Yukon River in search of gold in Alaska.

For a single ray input from the sun, let β be the rotation angle of the ice-crystal plate about its column axis. The ray equation for a sun dog is:

$$\theta_V = S_\alpha[S_\alpha(\psi,\beta,\mu_{ai})\ ,\ 60° - S_\beta(\psi,\beta,\mu_{ai})\ ,\ \mu_{ia}] - \psi$$

$$\theta_H = S_\beta[S_\alpha(\psi,\beta,\mu_{ai})\ ,\ 60° - S_\beta(\psi,\beta,\mu_{ai})\ ,\ \mu_{ia}] + \beta - 60° \qquad (19.16)$$

where θ_V is the viewing angle of the output ray (elevation above the sun), and θ_H is the viewing angle (azimuth to the right of the sun).

The sun dog as viewed by an observer is the superposition of rays from many ice crystals having a variety of rotation angles. For randomly rotated crystals, there is an equal chance of any β between 0° and 90°. This is simulated on a spreadsheet by solving the sun-dog equation for a large number of equally-spaced β angles. The results are plotted in Fig 19.30. Again, no account was made for rays that might be unable to enter the crystal due to reflection, or might be trapped inside due to the critical angle.

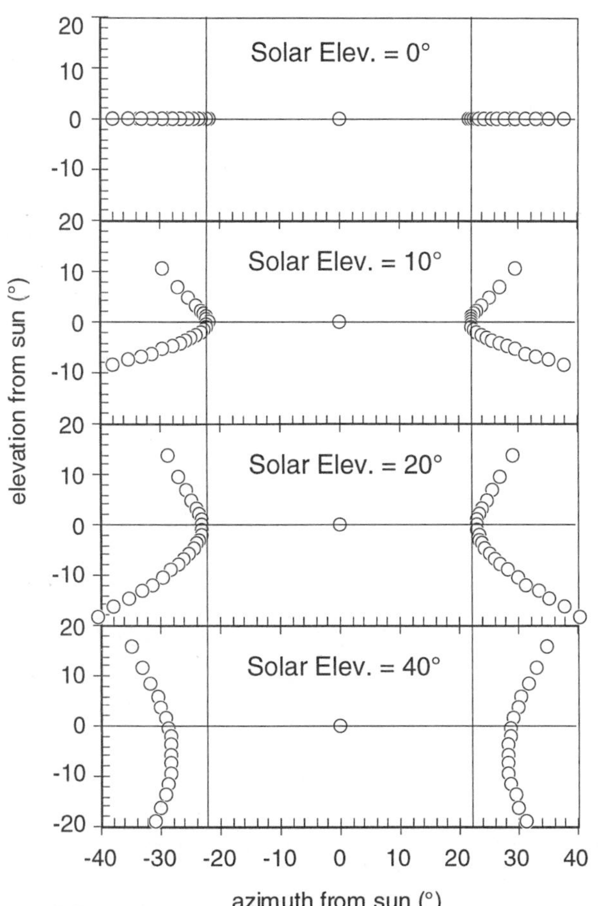

Figure 19.30
Variation of sun dog appearance with solar elevation angle. Vertical lines are 22° from sun (dot in center of graphs).

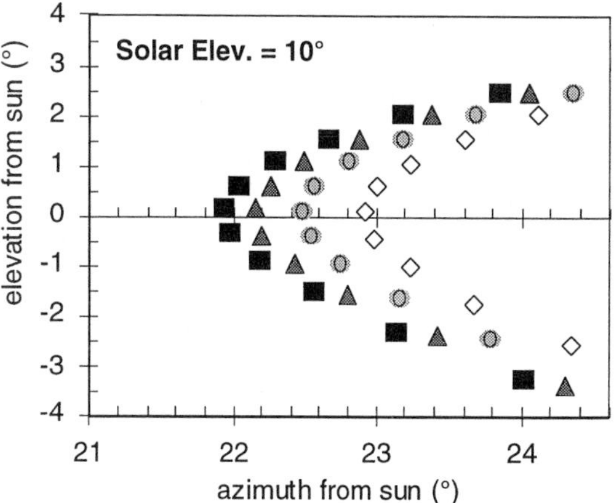

Figure 19.31
Dispersion of colors in a sun dog. (see Fig 19.25 legend)

Sun dog colors are similar to the halo, with bright red on the inside (closest to the sun), then orange and yellow, but with green, blue, and violet colors fading to white (Fig 19.31, an expanded view of Fig 19.30.)

Solved Example

Find the vertical and horizontal viewing angles of a sun-dog ray of red light for solar elevation 40° and crystal rotation $\beta = 45°$.

Solution

Given: $\Psi = 40°$, $\beta = 45°$

$n_{air} = 1.0002753$, $n_{ice} = 1.307$ from Table 19-1.

Find: $\theta_2 = ?°$

Use eq. (19.16) :

$$\theta_V = S_\alpha[S_\alpha(40°,45°,0.7653)\ ,\ 60° - S_\beta(40°,45°,0.7653)\ ,\ 1.3067] - 40°$$

$$\theta_H = S_\beta[S_\alpha(40°,45°,0.7653)\ ,\ \{60° - S_\beta(40°,45°,0.7653)\}\ ,\ 1.3067] + 45° - 60°$$

For the inside S_α and S_β:

$c_{\alpha\beta} = 0.3433$, $S_\alpha = 26.18°$, $S_\beta = 30.37°$

For the outside: $c_{\alpha\beta} = 2.8448$, because

$S_\alpha[26.18°, (60°–30.37°), 1.3067] =$

$S_\alpha[26.18°, 29.63°, 1.3067] = 39.67°$

$S_\beta[26.18°, 29.63°, 1.3067] = 43.81°$

Finally:

$\theta_v = 39.67° - 40° =$ **<u>–0.332°</u>**

$\theta_H = 43.81° - 15° =$ **<u>28.81°</u>**

Check: Units OK. Physics OK.

Discussion: This portion of the sun dog is nearly at the same elevation as the sun, but the sun dog is 28.81° to the side of the sun. This is 6.81° outside of the 22° halo.

Subsun Dogs

Ray geometry for the subsundog is identical to that for the sundog, except that the ray within the ice reflects from the bottom face (Fig 19.29c).

Upper Tangent Arcs

Large hexagonal column ice crystals generally fall with their column axis horizontal (Fig 19.14). In the absence of a wind shear, the column axes can point in any compass direction within that horizontal plane. Thus, the crystals can have any directional orientation γ with respect to the compass direction of the incoming sun ray. The crystal can also have any rotation angle α_1 about the column axis (see Fig 19.32). The upper tangent arc is also affected by the solar elevation angle, Ψ.

Because there are three input angles for this case, the solution of ray paths is quite nasty. The equations below show how one can calculate the ray output vertical viewing angle θ_V (elevation above the sun), and the horizontal viewing angle θ_H (azimuth from the sun), for any input ray. All other variables below show intermediate steps.

As before, one must examine rays entering many crystals having many angles in order to generate the locus of points that is the upper tangent arc. Even on a spreadsheet, this solution is tedious. Again, no consideration was made for rays that cannot enter the crystal due to reflection, or rays that are trapped inside because of the critical angle.

Given: Ψ, α_1, and γ:

$$\beta_1 = \arctan[\cos(\gamma)\cdot\tan(90^\circ - \psi)] \qquad (19.17)$$

$$\alpha_3 = 60^\circ - S_\alpha(\alpha_1, \beta_1, \mu_{ai})$$

$$\beta_3 = S_\beta(\alpha_1, \beta_1, \mu_{ai})$$

$$\alpha_5 = 60^\circ - S_\alpha(\alpha_3, \beta_3, \mu_{ia}) - \alpha_1 + \arctan[\sin(\gamma)\cdot\tan(90^\circ - \psi)]$$

$$\beta_5 = S_\beta(\alpha_3, \beta_3, \mu_{ia})$$

$$G = [\tan(\alpha_5)]^2 + [\tan(\beta_5)]^2$$

$$\varepsilon = \arctan\left[\frac{\tan(\beta_5)}{\tan(\alpha_5)}\right]$$

$$\phi = 90^\circ - \gamma - \varepsilon$$

$$\theta_V = 90^\circ - \psi - \arctan\left[G^{1/2}\cdot\cos(\phi)\right]$$

$$\theta_H = \arctan\left[\sin(\phi)\cdot\left(\frac{G}{1 + G\cdot[\cos(\phi)]^2}\right)^{1/2}\right]$$

To keep the spreadsheet calculations to a finite size, we used γ in the range of 55° to 90°, and α_1 in the range of 0° to 90°. The results for two solar elevations are shown in Fig 19.33. Only the left half of the upper tangent arc is shown, because it is symmetric about the vertical axis.

The locus of points of the upper tangent arc looks like the wing of a bird. For low solar elevations, the wing is up in the air. As the sun rises in the sky, the wing gently lowers. At elevations of about 45° or more the wing is wrapped closely around the 22° halo. Above about 60°, there is no solution to the upper tangent arc that is physically realistic.

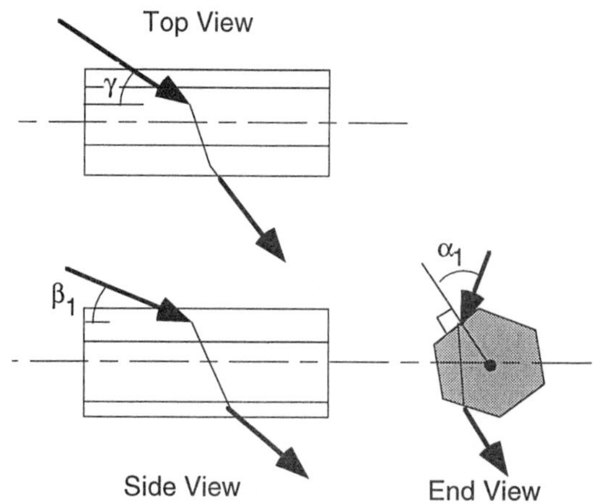

Figure 19.32
Ice crystal geometry affecting upper tangent arcs.

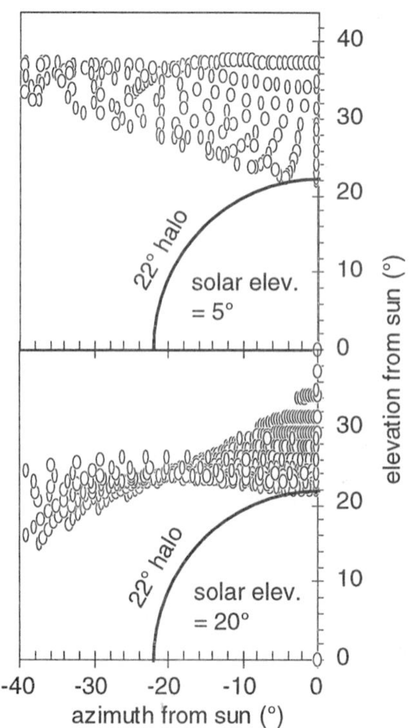

Figure 19.33
The locus of points shows the upper tangent arc (left half) for low solar elevation angles. Solid circle is the 22° halo.

Solved Example

Find the vertical and horizontal viewing angles of a ray of red light in an upper tangent arc for solar elevation $\Psi = 10°$, crystal rotation $\alpha_1 = 76°$, and crystal axis direction $\gamma = 45°$.

Solution

Given: $\Psi = 10°$, $\alpha_1 = 76°$, $\gamma = 45°$.
$n_{air} = 1.0002753$, $n_{ice} = 1.307$ from Table 19-1.
Find: $\theta_V = ?°$, $\theta_H = ?°$

Use eq. (19.17). Various intermediate results are:
$\beta_1 = 76°$
$c_{\alpha\beta} = 0.0402$ (1st calculation)
$\alpha_3 = 21.2°$, $\beta_3 = 38.8°$
$c_{\alpha\beta} = 3.9142$ (2nd calculation)
$\alpha_5 = 22.50°$, $\beta_5 = 57.84°$
$G = 2.701$, $\varepsilon = 75.4°$, $\phi = -30.4°$

Finally:
θ_v = <u>25.20°</u>
θ_H = <u>39.25°</u>

Check: Units OK. Physics OK.
Discussion: This portion of the upper tangent arc is above the sun and above the top of the 22° halo, but is off to the right. Such a position fits in well for an upper-tangent arc.

SCATTERING

Light can scatter off of air molecules, pollutant particles, dust, and cloud droplets. The type of scattering depends on the size D of the particle relative to the wavelength λ of the light. Table 19-2 summarizes the types of scattering.

Rayleigh Scattering

Air molecules have sizes of $D \cong 0.0001$ to 0.001 µm, which are much smaller than the wavelength of light (λ = 0.4 to 0.7 µm). These particles cause **Rayleigh scattering**, which makes the **sky** blue. The ratio of scattered intensity of radiation I_{scat} to the incident radiation intensity I_o is:

$$\frac{I_{scat}}{I_o} \approx 1 - \exp\left[-\frac{a \cdot (n_{air}-1)^2}{\rho \cdot \lambda^4} \cdot x\right] \qquad (19.18)$$

where a = 1.59x10^{-23} kg, ρ is air density, n_{air} is the refractive index, and x is the path length of light through the air.

Fig 19.34a shows the relative amount of scattering vs. wavelength. Because of the λ^{-4} dependence, shorter wavelengths such as blue and violet are scattered much more (about a factor of 10) than red light. Sunlight varies in intensity according to Planck's law, described in Chapter 2 and replotted here in Fig 19.34b. The product of these two curves (Fig 19.34c) shows the amount of sunlight that is scattered in the atmosphere. All curves have been normalized to have a maximum of 1.0.

Solved Example

What fraction of incident violet light is scattered by air molecules along a 20 km horizontal ray path near the earth?

Solution

Given: $x = 5\text{x}10^6$ m, $\lambda = 4\text{x}10^{-7}$ m for violet
Find: $I_{scat}/I_o = ?$
Assume: $\rho = 1$ kg·m^{-3} for simplicity.

Use eq. (19.18): $I_{scat}/I_o = 1 -$

$$\exp\left[-\frac{(1.59\times10^{-23}\text{kg})\cdot(0.0002817)^2}{(1\ \text{kg}\cdot\text{m}^{-3})\cdot(4\times10^{-7}\text{m})^4}\cdot(2\times10^4\text{m})\right]$$

$$\frac{I_{scat}}{I_o} \approx 1 - \exp[-0.9857]$$

= <u>0.627</u>

Check: Units OK. Physics OK.
Discussion: Objects far away are difficult to see because some of the light is lost. Vertical rays experience less scattering, because the air density decreases with height.

Table 19-2. Scattering of visible light.

D/λ	Particles	Size, D (µm)	Type	Phenomena	Scattering Varies With: λ	Direction	Polarize
< 1	air molecules	0.0001 to 0.001	Rayleigh	blue sky, red sunsets	X		X
≅ 1	aerosols (smog)	0.01 to 1.0	Mie	brown smog	X	X	X
> 1	cloud droplets	10 to 100	Geometric	white clouds		X	

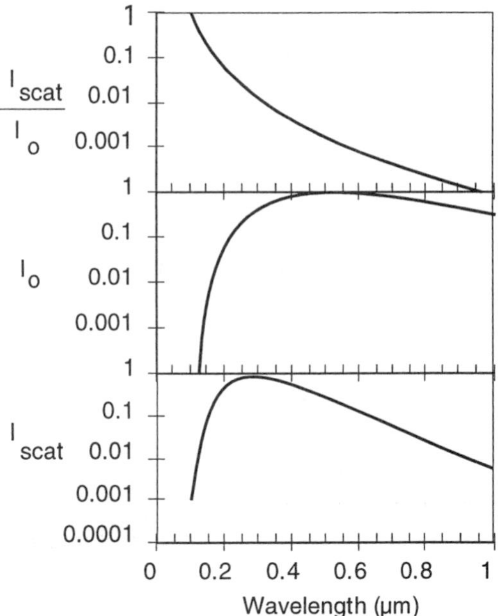

Figure 19.34
Top: Relative scattering. Middle: Planck spectrum for the sun. Bottom: product of the top two curves, indicating the amount of sunlight scattered in air.

The curve in Fig 19.34c peaks in the ultraviolet portion of the spectrum. Although this is not visible to the naked eye, the scattered ultraviolet light can affect photographic films to produce a picture that looks very hazy. Haze filters on cameras can filter out this unwanted scattered ultraviolet light.

ON DOING SCIENCE • Great Scientists Make Big Mistakes

Isaac Newton explained many atmospheric optical phenomena in his books on optics. Not all his theories were correct. One of his failures was his explanation of the blue sky.

He thought that the sky was blue for the same reason that soap bubbles or oil slicks have colors. For blue sky, he thought that there is interference between light reflecting from the backs of small water droplets and the light from the front of the drops.

Although Newton's theory was accepted for about 175 years, eventually observations were made of the polarization of sky light that were inconsistent with the theory. Lord Rayleigh proposed the presently accepted theory in 1871.

Like Newton, many great scientists are not afraid to propose radical theories. Although most radical theories prove to be wrong, the few correct theories are often so significant as to eventually create paradigm shifts in scientific thought. Unfortunately, "publish or perish" demands on modern scientists discourage such "high risk, high gain" science.

Polarization

Light propagating in the x-direction can be thought of as having oscillations in the y and z directions (Fig 19.35). This is unpolarized light. Polarizing filters eliminate the oscillations in one direction (for example, the shaded curve) while passing the other oscillations. What remains is **polarized** light, which has half the intensity of the unpolarized ray.

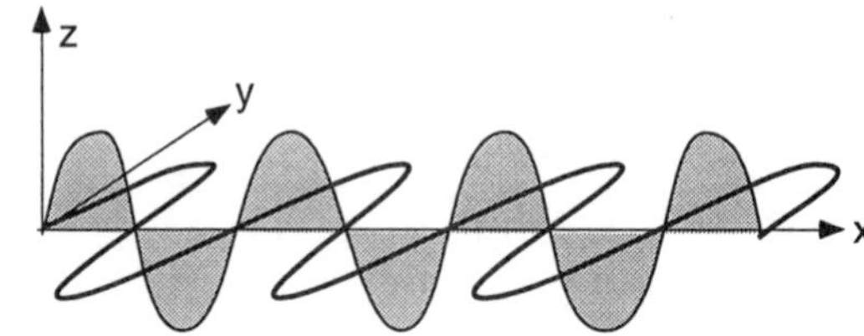

Figure 19.35
Unpolarized light consists of two cross-polarized parts.

Sunlight becomes polarized when it is scattered from air molecules. The maximum amount of polarization occurs along an arc in the sky that is 90° from the sun, as viewed from the ground (Fig 19.36).

If the sky scatters light with one polarity, and a polarizing camera filter or polarized sunglasses are rotated to eliminate the other polarity, then virtually no light reaches the camera or observer. This makes the sky look very deep blue, which provides a very striking background in photographs of clouds or other objects. If you want to maximize this effect, pick a camera angle looking toward the 90° arc from the sun.

Scattering of light by small dust particles in ice also explains why glaciers are often magnificently blue.

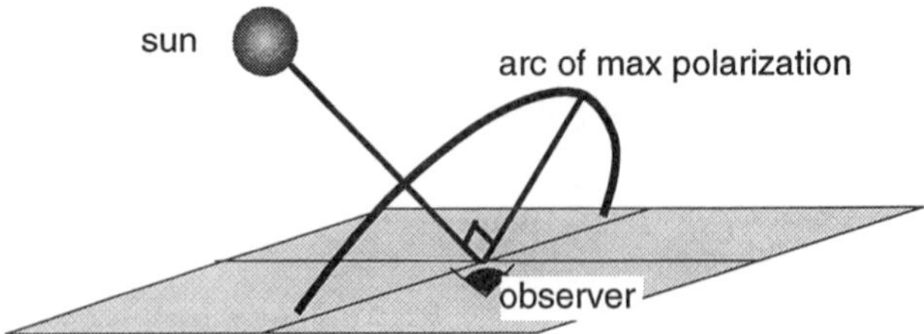

Figure 19.36
Arc of maximum polarization in the sky due to Rayleigh scattering.

Geometric Scattering

For large particles such as cloud droplets, light reflects from them according to geometric optics. Because all wavelengths are reflected equally, clouds look white when illuminated by white sunlight.

There is, however, a directional dependence. Light that is scattered in the same direction that the incident ray is pointing is called **forward scattering**. The opposite is **backward scattering**. For clouds, forward scattering is usually greater than backscattering. Thus, clouds you see in the direction of the sun look bright, while those in the direction of your shadow look darker.

Mie Scattering

Gustov Mie proposed a comprehensive theory that describes reflection, scattering, polarization, absorption, and other processes for all size particles. This theory is too complex to discuss here. The theory reduces to Rayleigh scattering for small particles, and to geometric scattering for large particles. Unfortunately, smog particles are middle size, so no simplification of Mie theory is possible.

Forward scattering is usually greater than back-scattering. Uniform smog particles can produce bluish or reddish colors. Scattering can be polarized.

DIFFRACTION

When sunlight or moonlight passes through a thin cloud of water droplets, diffraction can produce colored rings or **fringes** around the luminary. Although just one ring is sketched in Fig 19.37, there are often many concentric rings. The first ring touches the luminary.

The angle θ to each ring depends on the radius R of the droplet and the wavelength λ of light:

$$\theta = \arcsin\left(\frac{m \cdot \lambda}{R}\right) \qquad (19.19)$$

where m is a dimensionless diffraction parameter given in Table 19-3.

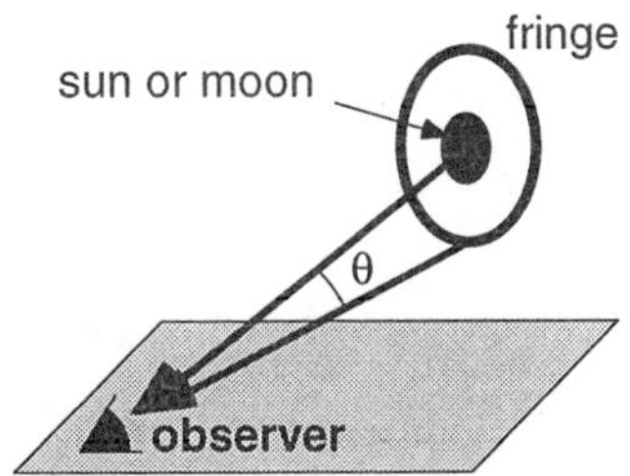

Figure 19.37
Diffraction fringe around a luminary.

Table 19-3. Diffraction fringes.

Fringe Index	m	Intensity
1	0	1.0
2	0.819	0.01745
3	1.346	0.00415
4	1.858	0.00165
5	2.362	0.00078
6	2.862	0.00043
7	3.362	0.00027
8	3.862	0.00018

For any one color, there are faint rings of light separated by darker gray background. The different colors of the spectrum have different fringe radii, causing the fringes of any one color (e.g. blue-green) to appear in the dark spaces between fringes of another color (e.g., red). Fringes further from the sun (with higher fringe index) are less bright, as given in Table 19–3 and plotted in Fig 19.38.

Some clouds such as **wave clouds** (standing **lenticular**) have extremely uniform drop sizes. This means that all the drops in the cloud produce the same fringe angles, for any one color. Hence, the diffraction from all the droplets reinforce each other to produce bright colorful fringes called **corona**. Most other clouds contain drops with a wide range of sizes, causing the colors to smear together to form a whitish disk (called an **aureole**) touching the sun or moon.

Eq. (19.19) can be solved on a spreadsheet for various cloud droplet sizes and colors. The results in Fig 19.39 show that smaller droplets produce larger-diameter fringes. For reference, typical cloud droplets have 10 μm radii, and the viewing angle subtended by the sun is 0.534°.

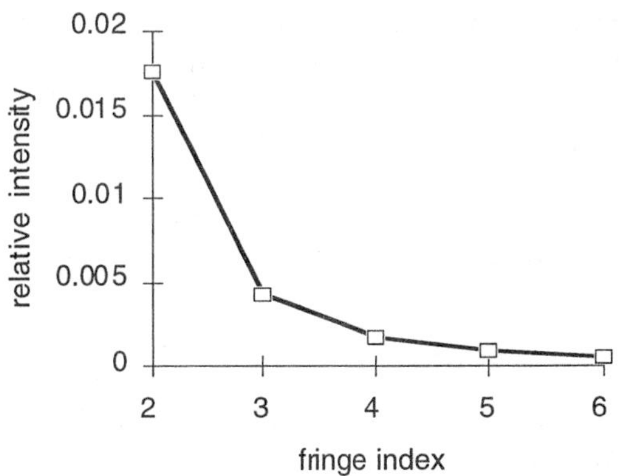

Figure 19.38
Diffraction fringe brightness. Larger indices indicate fringes that are further from the luminary.

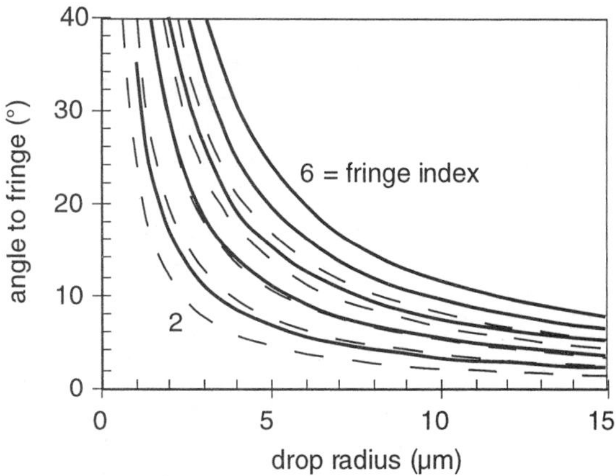

Figure 19.39
Viewing angle vs. cloud droplet size, for red (solid) and blue-green (dashed) diffraction fringes.

Droplets near the edges of scattered clouds (particularly wave clouds) can be very small as they form or evaporate, and can cause fringes of very large angular radius. As a result, the edges of clouds near the sun or moon are often colorful, a phenomenon called **iridescence**.

When looking down from above a cloud, diffraction patterns are sometimes seen around the shadow of the observer. This is called **glory**, and is associated with both diffraction and reflection from the cloud droplets.

Finally, to help discriminate between corona and halos, remember that corona are bright disks of light that touch the sun or moon, while halos have a dark region between the luminary and the halo ring. Also, corona are formed from liquid water droplets, while halos are formed from ice crystals.

Solved Example

The first visible fringe (index = 2) of red is 10° from the moon. What is the cloud drop radius?

Solution

Given: $\theta = 10°$, $\lambda = 0.7$ µm.

$n_{air} = 1.0002753$ from Table 19-1 for red light.

$m = 0.819$ from Table 19-3 for fringe #2.

Find: $R = ?$ µm

Rearrange eq. (19.19):

$$R = \frac{m \cdot \lambda}{\sin\theta} = \frac{0.819 \cdot (0.7\mu\text{m})}{\sin(10°)} = \underline{3.3\ \mu\text{m}}$$

Check: Units OK. Physics OK.
Agrees with Fig 19.39.
Discussion: Smaller than typical drop size, perhaps associated with a wave cloud.

MIRAGES

Refractive index n_{air} varies with air density ρ:

$$n_{air} - 1 \;\cong\; (n_{STP} - 1) \cdot \rho \,/\, \rho_{STP} \qquad (19.20)$$

where STP denotes standard temperature and pressure (T_{STP} = 15°C = 288 K, P_{STP} = 101.325 kPa). The standard refractive indices n_{STP} are those listed in Table 19-1 for air, and standard density is ρ_{STP} = 1.225 kg·m^{-3} (see Chapt. 1). Eq. (19.20) can be rewritten in terms of absolute temperature and pressure using the ideal gas law:

$$n_{air} - 1 = (n_{STP} - 1) \cdot \frac{T_{STP}}{T} \cdot \frac{P}{P_{STP}} \qquad (19.21)$$

A sharp change in density between two media causes a sharp kink in the ray path (Fig 19.1). A gradual change of density causes a smoothly curving ray path (Fig 19.40). The radius of curvature R_c (positive for concave up) is

$$R_c \approx \frac{\rho_{STP}}{(n_{STP} - 1) \cdot (\cos\alpha) \cdot (\Delta\rho / \Delta z)} \qquad (19.22)$$

where α is the angle of the ray above horizontal, and the gradient of density $\Delta\rho/\Delta z$ is assumed to be perpendicular to the earth's surface.

Substituting the ideal gas law and the hydrostatic relationship into eq. (19.22) yields

$$R_c \approx \frac{-(T / T_{STP}) \cdot (P_{STP} / P)}{(n_{STP} - 1) \cdot (\cos\alpha) \cdot \left\{ \frac{1}{T} \left[\frac{\Delta T}{\Delta z} + a \right] \right\}} \qquad (19.23)$$

where a = 0.0342 K·m^{-1} .

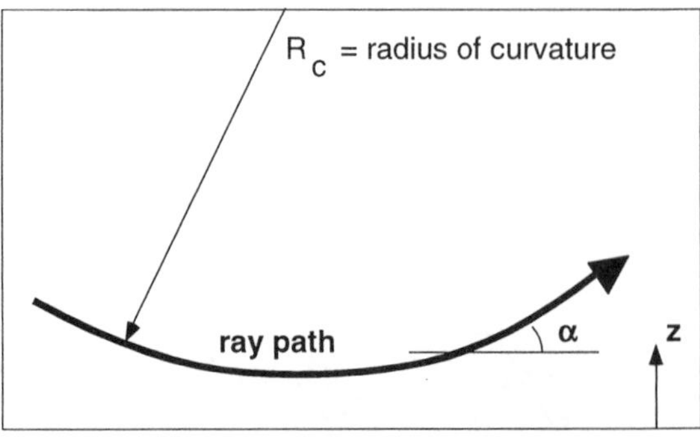

Figure 19.40
Ray curvature within a vertical gradient of density.

Because average density decreases with height in the atmosphere (see Chapt. 1), eq. (19.23) gives a negative radius of curvature in the presence of weak temperature gradients. In other words, the ray is bent downward in a standard atmosphere. This agrees with Huygens' principle, that says light rays are bent toward media of higher density.

For ray paths near horizontal, the total bending through the whole atmosphere is about 0.567°, which is why apparent sunrise occurs before geometric sunrise (see Chapt. 2). In everyday life, we rarely notice refraction associated with the standard atmosphere. However, in the presence of strong temperature gradients we can see mirages.

For rays to bend up instead of down, the term in square brackets in eq. (19.23) must be negative. This is possible when $\Delta T/\Delta z < -a$. A temperature decrease of at least 3.5°C per 10 cm height rise is necessary. Such a strong gradient is possible on hot sunny days in the air touching strongly-absorbing (black) surfaces. This condition of warm air under cold causes **inferior mirages**, where the objects appear lower than they really are. However, warm air under cold is statically unstable (Chapt. 3), and would disappear due to turbulent mixing unless continuously regenerated by solar heating of the surface. This type of mirage is common above black roads on a hot summer day.

Cold air under warm air (as in early morning, or in arctic regions) causes **superior mirages**, where objects appear higher than they are. This accentuates the downward bending of rays, causing the image of the object to loom or stretch vertically. **Fata Morgana** is an example of the latter mirage.

Solved Example

What is the radius of curvature of a horizontal indigo ray near the ground in a standard atmosphere?

Solution

Given: $z = 0$, $\alpha = 0°$, P and $T = 288$ K are STP.
$n_{air} = 1.00028$ from Table 19-1 for indigo light.
Find: $R_c = ?$ km

Assume the temperature gradient is negligible.
Rearrange eq. (19.23):

$$R_c \approx \frac{-1}{(0.00028)\cdot(\cos 0°)\cdot(0.0342\text{K}\cdot\text{m}^{-1}/288\text{K})}$$

$$= \underline{-30{,}000 \text{ km}}$$

Check: Units OK. Physics OK.
Discussion: Such a large negative radius implies very small downward curvature, which agrees with the small angle of with apparent sunrise.

SUMMARY

Atmospheric optical phenomena can be caused by reflection, refraction, scattering, and diffraction. These processes are caused by the difference in density between **hydrometeors** (liquid and solid water particles) and air.

Ice-crystal optical phenomena include the sun pillar, parhelic circle, subsun, 22° halo, 46° halo, circumzenith arc, sun dogs, subsun dogs and upper tangent arcs. They are usually found by looking toward the sun. The hexagonal column shape of many ice crystals creates 60° and 90° prismatic effects. The somewhat random orientation of these crystals causes sunlight to be returned in different directions from different crystals. The phenomena we see are the superposition from many crystals.

Rain-drop phenomena are a bit easier to describe, because the approximate spherical shape eliminates orientational dependence. Rainbows are seen by looking away from the sun.

Cloud droplets produce corona, iridescence, glory, and aureole. Air molecules cause blue sky and red sunsets.

Threads

This brings you to the end of this book. Along the way, you examined the physics of atmospheric pressure, wind, temperature, and humidity. You saw how these threads weave together to make cyclones, precipitation, thunderstorms, fronts, and hurricanes that we experience as weather. You have the quantitative wherewithal to critically evaluate weather-related issues, and to make sound decisions that affect society and the future of our planet.

If this is your last course on meteorology, be confident that you have the background to delve into the meteorology journals to learn the latest details that might help you in your engineering or scientific work. If you plan to continue your study of meteorology, the overview that you have gained here will help you put into context the advanced but tightly focused material that you will soon be learning.

Regardless of your plans, I hope that you do more than understand, appreciate, and utilize the weather. I encourage you to enjoy it.

Sincerely, *R. Stull.*

Science Graffito

"The right to fail [is] as liberating as the right to succeed." W. Zinsser, 1998: *On Writing Well*, 6 Ed.

EXERCISES

Numerical Problems

N1. Calculate the angle of reflection, given the following angles of incidence:

a. 10° b. 20° c. 30° d. 45°
e. 60° f. 70° g. 80° h. 90°

N2. What is the speed of violet light through

a. air b. water c. ice

N3. What are the angles of refraction in water, given an incident angle of 45° in air for λ (µm) of

a. 0.4 b. 0.5 c. 0.6 d. 0.7

N4. For orange-yellow light, calculate the angle of refraction for: (i) water, and (ii) ice, given the following incident angles in air:

a. 10° b. 20° c. 30° d. 45°
e. 60° f. 70° g. 80° h. 90°

N5. For an interface between water and ice, what are the angles of refraction in ice given red light with incident angles in water of:

a. 10° b. 20° c. 30° d. 45°
e. 60° f. 70° g. 80° h. 90°

N6. What is the total angle of incidence in air, given 3-D Cartesian components of incidence angles of:

a. 30°, 0° b. 30°, 30° c. 30°, 45° c. 30°, 90°
e. 45°, 0° f. 45°, 30° g. 45°, 45° h. 45°, 90°

N7. Using the incident components in air from the previous problem, find the refraction components of blue-green light in ice. Check your answers using Snell's law for the total angle.

N8. Calculate the critical angles for light moving from ice to air, for λ (µm) of

a. 0.4 b. 0.5 c. 0.6 d. 0.7

N9. What is the viewing angle between the violet bands of the primary and secondary rainbows?

N10. Find the reflectivity from spherical rain drops, given impact parameters of:

a. 0 b. 0.1 c. 0.2 d. 0.4
e. 0.6 f. 0.8 g. 0.9 h. 0.95

N11. What value(s) of the impact parameter gives a primary-rainbow viewing angle of $\theta_1 = 42.0°$ for red light?

N12.(§) Use a spreadsheet to calculate and plot primary-rainbow viewing angles for violet light for a full range of impact parameters from 0 to 1.

N13.(§) Same as previous problem but for secondary rainbow.

N14. For violet light in a 22° halo, what is the viewing angle for input angles of

a. 15° b. 20° c. 30° d. 45°
e. 60° f. 70° g. 80° h. 90°

N15.(§) Use a spreadsheet to generate curves for orange-yellow and for blue-green light, similar to those drawn in Fig 19.22 for the 22° halo.

N16.(§) Same as previous problem, but for the 46° halo.

N17. Calculate the viewing angle for a 46° halo for rays of blue-green light with input angles of

a. 59° b. 62° c. 65° d. 70°
e. 75° f. 80° g. 85° h. 90°

N18. For a solar elevation of 10°, calculate the viewing angle components for a circumzenith arc for red light with a crystal rotation of $\beta =$

a. 10° b. 12° c. 15° d. 18°
e. 20° f. 25° g. 30° h. 35°

N19.(§) Calculate and plot circumzenith arcs similar to Fig 19.28, but for solar elevation angles of

a. 0° b. 5° c. 10° d. 15°
e. 20° f. 25° g. 30° h. 35°

N20. For a solar elevation of 10°, calculate the viewing angle components for a sun dog for red light with a crystal rotation of $\beta =$

a. 15° b. 20° c. 30° d. 45°
e. 60° f. 70° g. 80° h. 85°

N21.(§) Calculate and plot sun dog arcs similar to Fig 19.30, but for solar elevation angles of

a. 0° b. 5° c. 10° d. 15°
e. 20° f. 25° g. 30° h. 35°

N22.(§) For $\gamma = 80°$, plot upper tangent arc elevation and azimuth angles for a ray of orange-yellow light, for a variety of evenly-space values of rotation angle α. Use a solar elevation of 20°.

N23. What fraction of incident red light is scattered from air molecules along a horizontal path near the earth's surface, with path length of

a. 10 m b. 100 m c. 0.5 km d. 1 km
e. 2 km f. 5 km g. 10 km h. 20 km

N24.(§) Calculate and plot the relative fraction of scattering as a function of wavelength.

N25. For cloud droplets of 5 μm radius, find the corona fringe viewing angles for the 2nd through 8th fringes, for violet light.

N26.(§) Calculate and plot the viewing angle vs. wavelength for the:

a. second corona fringe.
b. third corona fringe.
c. fourth corona fringe.

N27. If the temperature decreases 10°C over the following altitude, find the mirage radius of curvature.

a. 1 mm	b. 2 mm	c. 5 mm	d. 1 cm
e. 2 cm	f. 5 cm	g. 10 cm	h. 20 cm

Understanding & Critical Evaluation

U1. Salt water is more dense than fresh water. Sketch how light wave fronts behave as they approach the salt-water interface from the fresh water side.

U2. If neither the primary or secondary rainbows return light within Alexander's dark band, why is it not totally black?

U3. How does the intensity of light entering the rain drop affect the brightness of colors we see in the rainbow?

U4. Why might you see only one of the sun dogs?

U5. Sometimes it is possible to see multiple phenomena in the sky. List all of the optical phenomena associated with

a. large hexagonal plates
b. large hexagonal columns
c. small hexagonal columns

U6. Eq. (19.14) does not consider the finite size of a hexagon. What range of input angles would actually allow rays to exit from the face sketched in Fig 19.21?

U7. Use geometry to derive eq. (19.14) for the ice crystal shown in Fig 19.21.

U8.(§) Suppose a halo of 35° was discovered. What prism angle β_c for ice would cause this?

U9. Using geometry, derive the minimum thickness to diameter ratio of hexagonal plates that can create sun dogs, as a function of solar elevation angle.

U10. When the sun is on the horizon, what is the angle between the 22° halo and the bottom of the circumzenith arc?

U11. Contrast and compare the refraction of light in mirages, and of sound in thunder (see Chapt. 15). Discuss how the version of Snell's law in the thunderstorm chapter can be applied to light in mirages.

U12. Microwaves are refracted by changes in atmospheric humidity. Use Snell's law or Huygens' principle to describe how ducting and trapping of microwaves might occur, and how it could affect weather radar and air-traffic control radar.

U13. Use the relationship for reflection of light from water to describe the variations of brightness of a wavy sea surface during a sunny day.

U14. During the one reflection or two reflections of light inside a raindrop for primary and secondary rainbows, what happens to the portion of light that is not reflected? Who would be able to see it, and where must they look for it?

U15. For both sun dogs and rainbows, the red color comes via a path from the sun to your eyes that does not allow other colors to be superimposed. However, for both phenomena, as the wavelength gets shorter, wider and wider ranges of colors are superimposed at any viewing angle. Why, then, do rainbows have bright, distinct colors from red through violet, but sundogs show only the reds through yellows, while the large viewing angles yield white color rather than blue or violet?

U16. For hexagonal columns, what other crystal angles exist besides 60° and 90°? For these other crystal angles, at what viewing angles would you expect to see light from the crystal?

U17. In Chapt. 2, Beer's law was introduced, which related incident to transmitted light. Assume that transmitted light is 1 minus scattered light.

a. Relate the Rayleigh scattering equation to Beer's law, to find the absorption coefficient associated with air molecules.

b. If **visibility** is defined as the distance traveled by light where the intensity has decreased to 2% of the incident intensity, then find the visibility for clean air molecules.

U18. Discuss the problems and limitations of using visibility measurements to estimate the concentration of aerosol pollutants in air.

U19. If you take photographs using a polarizing filter on your camera, what is the angle between a line from the subject to your camera, and a line from the subject to his/her shadow, which would be in the proper direction to see the sky at nearly maximum polarization? By determining this angle now, you can use it quickly when you align and frame subjects for your photographs.

U20. What vertical temperature profile is needed to see the **Fata Morgana** mirage?

U21. Green, forested mountains in the distance sometimes seem purple or black to an observer. Also, the mountains sometimes seem to **loom** higher than they actually are. Discuss the different optical processes that explain these two phenomena.

U22. The sun dog simulation of Fig 19.30 does not look exactly like the sun dogs that are observed in nature. In particular, real sun dogs do not appear to be spread out in the vertical as much as indicated in that figure. Discuss the limitations of the equation and/or assumptions that went into the sun dog analysis in this chapter, and suggest ways to improve the simulation.

U23. Knowing the relationship between optical phenomena and the cloud (liquid or water) micro-physics, and the relationship between clouds and atmospheric vertical structure, cyclones and fronts, create a table that tells what kind of weather would be expected after seeing various optical phenomena.

Web-Enhanced Questions

W1. Search the web for images of the following atmospheric optical phenomena:

a. 22° halo	b. 46° halo
c. sun dogs	d. sub sun dogs
e. sub sun	f. sun pillar
g. supersun	h. upper tangent arc
i. parhelic circle	j. lower tangent arc
k. white clouds	l. circumzenith arc
m. Perry arc	n. primary rainbow
o. red sunset	p. secondary rainbow
q. blue sky	r. Alexander's dark band
s. corona	t. crepuscular rays
u. iridescence	v. anti-crepuscular rays
w. glory	x. fata morgana
y. green flash	z. mirages

W2. Search the web for images and descriptions of additional atmospheric phenomena that are not listed in the previous question.

W3. Search the web for a physical explanation of the green flash.

W4. Search the web for microphotographs of ice crystals. Find images of hexagonal plates and hexagonal columns.

W5. a. Search the web for lists of indices of refraction of light through different materials.
b. Find a more complete list of indices of refraction through ice and liquid water, for more wavelengths than were listed in Table 19–1.

W6. Search the web for highway camera or racetrack imagery, showing inferior mirages on the roadway.

W7. Search the web for computer programs to simulate optical phenomena. If you can download this software, try running it and experimenting with different conditions to produce different optical displays.

W8. Search the web for images that have exceptionally large numbers of optical phenomena present in the same photograph.

W9. Search the web for information on linear vs. circular polarization of sky light. Polarizing filters for cameras can be either circular or linear polarized. The reason for using circular polarization filters is that many automatic cameras loose the ability to auto focus or auto meter light through a linear polarizing filter.

W10. Search the web for literature, music, art, or historical references to optical phenomena, other than the ones already listed in this chapter.

Synthesis Questions

S1. Suppose atmospheric density were (a) constant with height; or (b) increasing with height. How would optical phenomena be different, if at all?

S2. How would optical phenomena be different if ice crystals were octagonal instead of hexagonal?

S3. Suppose the speed of light through liquid and solid water was faster than through air. How would optical phenomena be different, if at all?

S4. After a nuclear war, if lots of fine earth debris were thrown into the atmosphere, what optical phenomena would cockroaches (as the only remaining life form on earth) be able to enjoy?

FUNDAMENTALS OF SCIENCE

CONTENTS

A Many physical sciences including atmospheric science share the same fundamental-definitions and analysis techniques. These fundamentals include problem-solving methods, standard units, ways of expressing relationships, and formats for plotting the results. The fundamentals reviewed here are used throughout this book.

ON DOING SCIENCE • Problem Solving

The following method aids problem understanding, speeds solution, and helps to avoid errors. This method is used throughout the book in the various solved examples.

1) List the "Given" variables with their symbols, values and units.
2) List the unknown variables to "Find", with units, etc.
3) Sketch the objects, velocities, etc. if appropriate.
4) Determine which equation(s) contains the unknown variable as a function of the knowns. This equation might need to be rearranged to solve for the unknown. If the solution equation contains more unknowns, find additional equations for them.
5) Make assumptions, if necessary, for any of the unknowns for which you have no equations. Clearly state your assumptions, and justify them.
6) Solve the equations using the known or assumed values, being sure to carry along the units. Show your intermediate steps.
7) Identify the final answer by putting a box around it, underlining it, or making it bold face.
8) Check your answer. If the solved units don't match the desired units of the unknown, then either a mistake was made, or unit conversion might be needed (e.g., convert from knots to m/s). Also, certain functions such as "ln" and "exp" require arguments that are dimensionless, while trig functions like "sine" need an argument in degrees or radians. These are clues to help catch mistakes. Also, compare your answer with your sketch, to check if it is physically reasonable. Check other physical constraints (e.g., humidities cannot be negative, speeds cannot be infinite).
9) Discuss the significance of the answer.

DIMENSIONS AND UNITS

Definitions

There are seven basic dimensions in science, from which all other dimensions are derived (Table A-1). The first letters of the first three units are 'm, k, s'; hence, this system of units is sometimes called the MKS system. It has been adopted as the international system (SI) of units. Two supplementary units are listed in Table A-2.

Derived units are formed from combinations of basic units. Examples of derived units that are used frequently in meteorology are listed in Table A-3.

A prefix can be added to these units to indicate larger or smaller values, such as kilometer (km), which is 1000 meters. The most commonly used prefixes are given in Table A-4.

Table A-1. Basic dimensions.

Dimension	Unit	Abbreviation
length	meter	m
mass	kilogram	kg
time	second	s
electrical current	ampere	A
temperature	kelvin	K
amount	mole	mol
luminous intensity	candela	cd

Table A-2. Supplementary dimensions.

Dimension	Unit	Abbreviation
plane angle	radian	rad
solid angle	steradian	sr

Table A-3. Derived dimensions.

Dimension	Unit (Abbrev.)	Composition
force	newton (N)	$kg \cdot m \cdot s^{-2}$
energy	joule (J)	$kg \cdot m^2 \cdot s^{-2}$
power	watt (W)	$kg \cdot m^2 \cdot s^{-3}$
pressure	pascal (Pa)	$kg \cdot m^{-1} \cdot s^{-2}$

Science Graffito

"When you make the finding yourself — even if you're the last person on Earth to see the light — you'll never forget it." — Carl Sagan.

Table A-4. Prefixes. (*USA size designations. International designations are shown in italics, if different.)

Multiplier	Size*	Name	Abbrev
10^{18}	quintillion (*trillion*)	exa	E
10^{15}	quadrillion (*billiard*)	peta	P
10^{12}	trillion (*billion*)	tera	T
10^{9}	billion (*milliard*)	giga	G
10^{6}	million	mega	M
10^{3}	thousand	kilo	k
10^{-3}	thousandth	milli	m
10^{-6}	millionth	micro	µ
10^{-9}	billionth	nano	n
10^{-12}	trillionth	pico	p
10^{-15}	quadrillionth	femto	f
10^{-18}	quintillionth	atto	a

Unit Conversion

Form ratios of equivalent units to develop trivial methods for unit conversion. For example, a velocity of 1 knot equals 0.51 m/s. Because these two quantities equal, their ratio must be one:

$$1 = \frac{(1 \text{ knot})}{(0.51 \text{ m/s})} \tag{A.1}$$

The inverse of 1 is 1, thus, it makes no difference which quantity appears in the numerator. For example:

$$1 = \frac{(0.51 \text{ m/s})}{(1 \text{ knot})} \tag{A.2}$$

When any quantity is multiplied by 1, its value does not change. Hence, by multiplying a velocity by the ratio in eqs. (A.1) or (A.2), we can change the units without changing the physics.

Even if a conversion relationship is not known, you can sometimes figure it out if you know the values of different units that apply to a specific situation. For example, what is the conversion between pressure in "pounds per square inch" (PSI) and in "inches of mercury" (in Hg). Perhaps you might already know that the average pressure at sea level is 14.7 lb/in^2. You might also know that standard sea-level pressure is 29.92 in Hg. Thus, these two quantities are equivalent, and their ratio gives the conversion between them:

$$1 = \frac{14.7\ \text{lb/in}^2}{29.92\ \text{inHg}} = 0.49\ \frac{\text{lb/in}^2}{\text{inHg}}$$

or $1\ \text{in Hg} = 0.49\ \text{lb/in}^2$.

Ratios can also be formed to add prefixes to units. For example, 1 milligram (mg) equals 0.001 grams, by definition. Thus their ratio is one. We can use this ratio to find 5×10^7 mg in units of kg. For example:

Solved Example

Sometimes you can use units to guess the form that an equation should have. For example, the rate of metabolic heat production by humans sitting quietly is about 100 watts. Find the number of calories produced in half a day.

Solution

Given: $C = 100$ W power
Find: $B = ?$ cal daily heat production

Eqs: From tables of unit conversion in other books

1 W = 14.3353 cal/min
1 hour = 60 min
1 day = 24 hours

By forming each of these equivalences as ratios, you can convert from watts into calories, and then from minutes into hours into days:

$$C = 100\ \text{W}$$
$$= (100\ \text{W}) \cdot \left(\frac{14.3353\ \text{cal/min}}{1\ \text{W}}\right)$$
$$= 1433.53\ \text{cal/min}$$
$$= \left(1433.53 \frac{\text{cal}}{\text{min}}\right) \cdot \left(\frac{60\ \text{min}}{1\ \text{h}}\right) \cdot \left(\frac{24\ \text{h}}{\text{d}}\right)$$
$$C = 2.06 \times 10^6\ \text{cal/day}$$

By looking at the units of the last line it is obvious that if we multiply it by the time t in days, then we will be left with our desired units of calories. Thus, the final equation is:

$$B(\text{cal}) = C(\text{cal/day}) \cdot t(\text{day})$$
$$= \left(2.06 \times 10^6\ \frac{\text{cal}}{\text{day}}\right) \cdot (0.5\ \text{day})$$
$$= \underline{\mathbf{1.03 \times 10^6\ cal}}$$

Check: Units OK. Physically reasonable.
Discussion: A bit over one million calories of heat would be given off by a human sitting still for half a day. The number of calories of food we eat should be sufficient to replace those calories burned metabolically. Note that the "calories" listed on food packages are really kilocalories.

$$5 \times 10^7\ \text{mg} \cdot \left(\frac{0.001\ \text{g}}{1\ \text{mg}}\right) \cdot \left(\frac{1\ \text{kg}}{1000\ \text{g}}\right) = 50\ \text{kg}$$

The trick of forming ratios to do conversions works only when both units have the same zero point. In the example above, $0\ \text{lb/in}^2 = 0$ inHg. Similarly, 0 m/s = 0 knots. This trick fails for temperature conversions, because °F, °C, and K all have different zero points. For temperature conversions, you must use special conversion formulae, as described in the "Relationships and Graphs" section of this appendix.

Solved Example

A pilot reports a wind speed of 10 knots. Find the wind speed in m/s.

Solution:

Given: $M = 10$ kt wind speed
Find: $M = ?$ m/s
Eq: Use eq. (A.2)

$$M = 10\ \text{kt}$$
$$= (10\ \text{kt}) \cdot 1$$
$$= (10\text{kt}) \cdot \left(\frac{0.51\text{m/s}}{1\text{kt}}\right)$$
$$= \left(\frac{10 \cdot 0.51}{1}\right) \cdot \left(\frac{\text{kt} \cdot \text{m/s}}{\text{kt}}\right)$$
$$= \underline{\mathbf{5.1\ m/s}}$$

Check: Units OK. Physically reasonable.
Discussion: How do you know whether to use (A.1) or (A.2)? Given a value with knots in the numerator, we want to multiply it by a ratio that has knots in the denominator, so that the knots will cancel. Eq. (A.2) has knots in the denominator.

FUNCTIONS AND FINITE DIFFERENCE

As in other fields of science and engineering, functional relationships describe how one variable (the **dependent** variable) changes when one or more other variables (the **independent** variables) change. Suppose that P_2 is the pressure at time t_2, and P_1 is the pressure at time t_1. Pressure varies with time, or is a function of time. Such functional dependence is written generically as $P(t)$. For a single value of pressure that occurs at specific time, such as $t = 50$ s, the notation $P(50\ \text{s})$ is used.

The symbol Δ means change or difference. Differences must always be taken in the same direction relative to the independent variable. For example, if temperature T and pressure P both vary with time t, then ΔT and ΔP are defined as their values at the later time minus their values at the earlier time. For example, $\Delta T = T(t_2) - T(t_1)$, and $\Delta P = P(t_2) - P(t_1)$, where t_2 is later than t_1. The notation is sometimes simplified to be $\Delta T = T_2 - T_1$, and $\Delta P = P_2 - P_1$.

In a different example, let pressure T depend on independent variable height z. Then

$$\Delta T = T(z_2) - T(z_1) = T_2 - T_1$$

and

$$\Delta z = z_2 - z_1$$

where z_2 is higher than z_1. Furthermore, a ratio such as $\Delta T/\Delta z$ is equivalent to $(T_2 - T_1)/(z_2 - z_1)$, or $[T(z_2) - T(z_1)]/(z_2 - z_1)$, where the differences in the numerator and denominator must always be taken in the same direction. The change of something with distance is called a **gradient**. Thus, $\Delta T/\Delta z$ is a **vertical temperature gradient**.

Although calculus is a useful mathematical tool for studying the physics of the atmosphere, this book is designed for an audience who might not have had calculus. In the place of differential calculus we will use finite differences, Δ. In place of integral calculus, we will use sums or graphically examine the area under curves.

However, for those students with a calculus background, "Beyond Algebra" boxes are scattered here and there in the book to provide a taste of theoretical meteorology. These "Beyond Algebra" boxes are surrounded by a double line as shown below, and indicate a segment that may be safely skipped by students wishing to avoid calculus.

BEYOND ALGEBRA • Calculus

In theoretical meteorology, the physics of the atmosphere is described by differential equations. Outside of these "Beyond Algebra" boxes, we utilize the following approximations to avoid calculus:

$$\frac{\partial T}{\partial z} \cong \frac{\Delta T}{\Delta z} \qquad \text{for small } \Delta z$$

and

$$\frac{dT}{dt} \equiv \frac{\partial T}{\partial t} + U\frac{\partial T}{\partial x} + V\frac{\partial T}{\partial y} + W\frac{\partial T}{\partial z}$$

$$\cong \frac{\Delta T}{\Delta t} + U\frac{\Delta T}{\Delta x} + V\frac{\Delta T}{\Delta y} + W\frac{\Delta T}{\Delta z}$$

Also

$$\int T\ dz \cong \sum T(z) \cdot \Delta z\ .$$

Solved Example

Suppose the air at the ground has a temperature of 20°C, while the air at height 500 m is 15 °C. Find the vertical temperature gradient.

Solution

Given:

$z_2 = 500$ m top altitude
$z_1 = 0$ m ground altitude
$T_2 = T(z_2) = 15$ °C temperature at 500 m
$T_1 = T(z_1) = 20$ °C temperature at ground

Find:

$\Delta T / \Delta z = ?$ °C/m temperature gradient

Sketch:

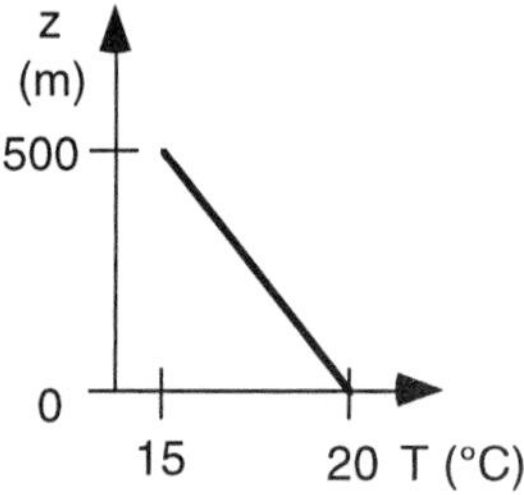

$\Delta T/\Delta z = (T_2 - T_1) / (z_2 - z_1)$
$= (15°C - 20°C) / (500\text{ m} - 0\text{ m})$
$= (-5\ °C) / (500\text{ m}) =$ **-0.01 °C/m**

Check: Units OK. Sketch OK. Sign negative.
Discussion: If this gradient is constant with height, then the temperature at the top of your head is 0.02°C colder than at your toes, assuming you are roughly 2m tall.

RELATIONSHIPS AND GRAPHS

Although $T(z)$ says that there is some functional relationship between temperature T and height z, it does not specify what that relationship is. Temperature could increase as height increases. It could decrease as height increases. It could increase with the square of height. It could vary logarithmically with height. It could be invariant with height.

The particular form of the function might be governed by some underlying physics, and have a fixed functional form that is sometimes called a physical "law" or a mathematical definition. Other functions might not be governed by underlying physics, but might vary with the particular atmospheric conditions at hand. This is true for many weather observations.

Linear, semi-log and **log-log** graphs are frequently used to discover and display relationships between dependent and independent variables.

Linear

The relationship between temperature in degrees Celsius and Fahrenheit is **linear**; namely, temperature in Fahrenheit is proportional to the **first** power of temperature in Celsius

$$T_{°F} = a \cdot T_{°C} + b \qquad \bullet(A.3)$$

where the parameters in this equation are $a = 9/5$ (°F/°C) and $b = 32$ °F. It yields a straight line on a linear graph (values along both axes of the graph increase at constant rates with distance from the origin).

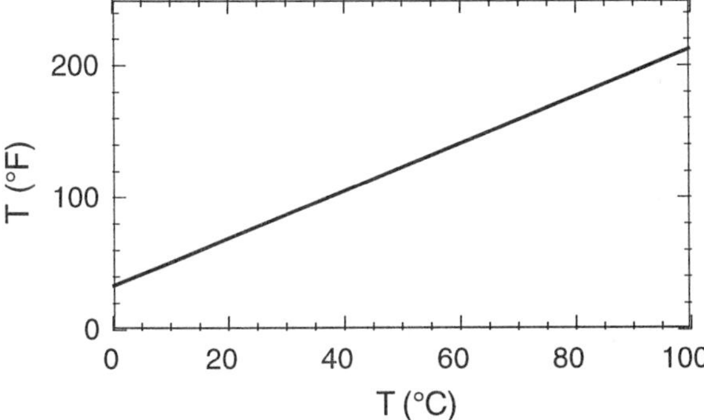

Figure A.1
Relationship between Celsius and Fahrenheit temperature, plotted on a linear graph.

> **Solved Example**
> Convert 10, 20, and 30°C to Fahrenheit.
>
> **Solution:**
> Given: $T_{°C}$ = 10°C, 20°C, 30°C
> Find: $T_{°F}$ = ? °F
>
> Sketch: (see Fig A.1)
> Use eq. (A.3):
> $T_{°F} = a \cdot T_{°C} + b$
> $= (9/5\ °F/°C)\cdot(10°C) + 32°F$
> $= 18°F + 32°F =$ **50°F** (etc. for other T)
>
> **Check**: Units OK. Sketch OK. Physics OK.
> **Discussion**: Temperatures of 10, 20, and 30°C correspond to **50**, **68**, and **86°F**, respectively. As the temperature in Celsius increases by equal amounts of 10°C, the corresponding Fahrenheit values increase in equal amounts of 18°F. Such a constant rate of increase of the dependent variable as a function of a constant rate of increase of independent variable is the mark of a linear relationship.

The slope (change in values along the vertical axis per change of values along the horizontal axis) of the line equals the factor a, which is 9/5 in this case. As shown in Fig A.1, the nonzero parameter b causes the plotted line to cross the vertical axis not at the origin $(T_{°C}, T_{°F}) = (0, 0)$, but at an intercept of T =32°F [i.e., at (0°C, 32°F].

The relationship between temperature in Celsius and absolute temperature in Kelvins is also linear:

$$T(\text{K}) = T(°\text{C}) + 273\ \text{K} \qquad \bullet(A.4)$$

If plotted on a linear graph, the slope would be 1 and the intercept is 273 K. Most equations using temperature require the use of absolute temperature.

Logarithmic

An **exponential** or **logarithmic** relationship gives a straight line when plotted on a **semi-log** graph (one axis is linear, the other is logarithmic). For example, the decrease of pressure with height is logarithmic in atmospheres where the temperature is constant with height:

$$\ln\left(\frac{P}{P_o}\right) = \frac{-a}{T} \cdot z \qquad (A.5)$$

where $a = 0.0342$ K/m, and $P_o = 101.3$ kPa for earth, on the average.

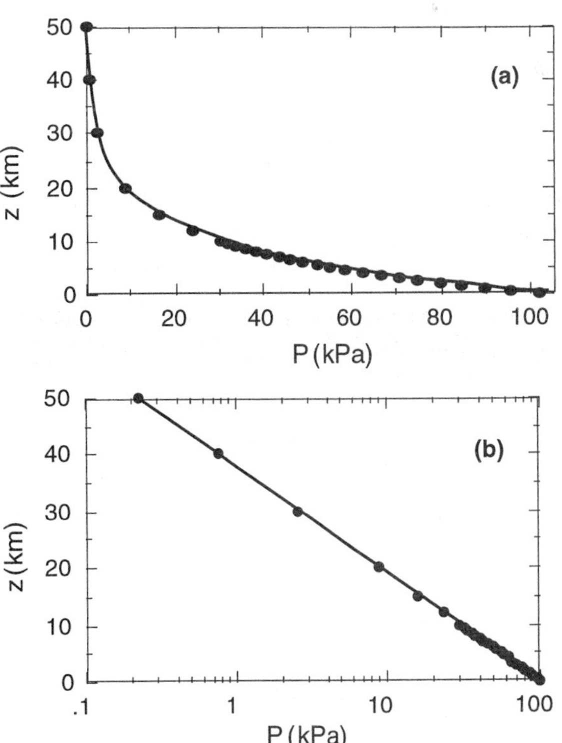

Figure A.2
Height z vs. pressure P in the atmosphere, plotted on (a) linear; and (b) semi-log graphs.

Fig A.2 plots the relationship between P and z in both linear and semi-log graphs, for $T = 280$ K. Pressure decreases rapidly with height near the ground, but decreases at a lesser rate higher in the atmosphere.

From the top graph of Fig. A.2 we see that the portion of the curve in the lowest 3 km of the atmosphere is almost a straight line, and conclude that the variation of pressure with height is nearly linear in that region. In fact, near the ground, the pressure decreases roughly 10 kPa with each 1 km increase in altitude.

The bottom graph of Fig A.2 has P plotted on a logarithmic scale along the bottom axis, but with z plotted on a linear scale on the vertical axis. All of the data points at all heights fall on a straight line on this graph. This is evidence that the logarithm of P is proportional to z.

Solved Example

Pressurized aircraft can maintain a higher pressure inside the cabin than outside. The maximum pressure difference between in and out depends on the construction of the particular aircraft. Cabin pressure is often reported in terms of altitude units instead of pressure units. Thus, a cabin pressure altitude of 2 km means that the pressure in the cabin is the same as that which would be found outside at an altitude of 2 km.

Suppose a hypothetical aircraft can maintain sea-level pressure inside the cabin up to a physical flight (outside) altitude of 5 km, beyond which the aircraft bursts open like a popped balloon. If the flight engineer reduces cabin pressure to a pressure altitude of 3 km, then how high can the aircraft climb before it explodes?

Solution: If there were a linear relationship between pressure and altitude, then the aircraft would not explode until it climbs above 5 + 3 = 8 km above the ground. However, we know that the pressure changes logarithmically rather than linearly with height.

Assume: Burst depends on pressure **difference** ΔP between cabin and outside, not on the absolute pressure or height anywhere.

For simplicity, assume $T = 280$ K every where; thus, $a/T = 0.122\ \text{km}^{-1}$ in eq. (A.5).

Given: $\Delta P = P_{cabin} - P_{outside}$ by definition.

ΔP_{max} = pressure difference when burst $= P(z{=}0) - P(z{=}5\text{ km})$

(continued next column)

Solved Example *(continuation)*

Find: $z_{outside}$ such that

$\Delta P_{max} = P(z{=}3\text{ km}) - P(z_{outside} = ?)$

Sketch:

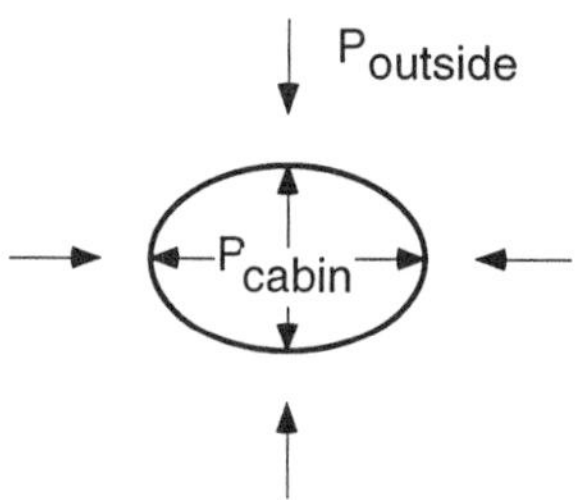

Take the exponential of both sides of eq. (A.5) to solve for pressure. The maximum pressure difference between sea-level pressure in the cabin and outside-air pressure before catastrophic failure is:

$$\Delta P_{max} = P(z{=}0) - P(z{=}5\text{ km})$$

$$\Delta P_{max} = P_o - P_{air} = P_o \cdot \left[1 - e^{-(a/T)\cdot(5\text{ km})}\right]$$

$$\Delta P_{max} = 0.4566\ P_o.$$

where P_o is the parameter from eq. (A.5), which happens to equal average sea-level pressure.

If the cabin pressure is reduced to 3 km, then the pressure difference between cabin and outside is:

$$\Delta P = P_{cabin} - P_{outside}$$

$$= P_o \cdot [\,e^{-(a/T)\cdot(3\text{km})} - e^{-(a/T)\cdot z_{outside}}]$$

Setting this pressure difference to ΔP_{max} from above allows us to solve for $z_{outside}$:

$$0.4566 = \left[e^{-(a/T)\cdot(3\text{ km})} - e^{-(a/T)\cdot z}\right]$$

Take of log of both sides and solve for z:

$$z = -(T/a)\cdot \ln\left[e^{-(a/T)\cdot(3\text{ km})} - 0.4566\right]$$

$$= [-1 / (0.122\ \text{km}^{-1})] \cdot \ln[0.2369]$$

= <u>**11.8 km**</u> = highest safe altitude

Check: Units OK. Sketch OK. Physics OK.

Discussion: This altitude is much greater than the 8 km first guess if a linear pressure variation were assumed. Thus, the safe altitude gain by an aircraft is much greater than the cabin-pressure altitude change that is allowed inside the plane. Too high a pressure in the cabin would require too much extra weight to strengthen the airframe. Too little pressure would not supply sufficient air to the passengers to breathe, inducing **hypoxia**.

Power

If the dependent variable is proportional to a **power** of the independent variable, then the data will appear as a straight line on a **log-log** graph. For example, Johannes Kepler, the 17th century astronomer, discovered that planets in the solar system have elliptical orbits around the sun, and that the time period Y of each orbit is related to the average distance R of the planet from the sun by:

$$Y = a \cdot R^{3/2} \qquad \text{(A.6)}$$

Parameter $a = 0.1996$ d·(Gm)$^{-3/2}$, approximately, where d is the abbreviation for earth days and Gm is gigameters (= 10^6 km).

Using a table of the average distance of the planets from the sun in millions of kilometers (i.e., in Gm), one can calculate orbital period using eq. (A.6). These are plotted in Fig A.3a on a linear-linear graph, and in Fig A.3b on a log-log graph. The slope of the straight line on the log-log graph (Fig A.3b) equals the power of the exponent. Namely, the range of periods between Mercury and Pluto is about 3 decades (i.e., 100 to 1,000 to 10,000 to 100,000), while the range of distances from the sun is 2 decades. Thus, the slope is 3 to 2, as indicated in eq. (A.6).

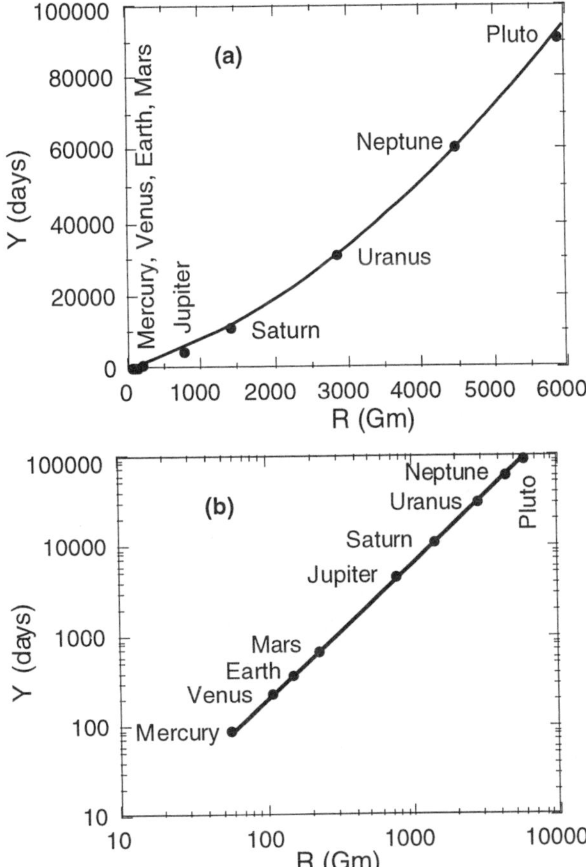

Figure A.3
Plot of planetary orbital periods in Earth days versus distance from sun in billions of meters, on (a) linear, (b) log-log graphs.

> **Solved Example**
> Verify that eq. (A.6) gives the correct orbital period for earth.
>
> **Solution**:
> Given: R = 149.6 Gm avg. distance sun to earth.
> Find: Y = ? days orbital period for earth
> Sketch: (see Fig A.3)
>
> Use eq. (A.6):
> $Y = (0.1988 \text{ d}\cdot(\text{Gm})^{-3/2}) \cdot [(150 \text{ Gm})^{1.5}]$
> = <u>**365.2 days**</u>.
>
> **Check**: Units OK. Sketch OK. Answer = 1 yr.
> **Discussion**: Time flies. (see Chapter 2).

ON DOING SCIENCE

Science is a philosophy. It is faith in a set of principles that guide the actions of scientists. It is a faith based on observation. Scientists try to explain what they observe. Theories that are not verified by observations are discarded. This philosophy applies to atmospheric science, also known as meteorology.

A good theory is one that works anywhere, anytime. Such a theory is said to be **universal**. Engineers utilize universal theories with the expectation they will continue working in the future. The structures, machines, circuits, and chemicals designed by engineers that we use in every-day life are evidence of the success of this philosophy.

But we scientists and engineers are people, and share the same virtues and foibles as others. Those of you planning to become scientists or engineers might appreciate learning some of the pitfalls so that you can avoid them, and learning some of the tools so that you can use them to good advantage.

For this reason, scattered throughout the book are boxes called "On Doing Science", summarized in Table A–5. These go beyond the mathematical preciseness and objective coldness that is the stereotype of scientists. These boxes cover issues and ideas that form the fabric of the philosophy of science. As such, they are subjective. While they give you one scientist's (my) perspectives, I encourage you to discuss and debate these issues with other scientists, colleagues, and teachers.

ON DOING SCIENCE • Have Passion

The best scientists and engineers need more than the good habits of diligence and meticulousness. They need passion for their field, and they need creativity. In this regard, they are kindred spirits to artists, composers, musicians, authors, and poets.

While an observation is something that can usually be quantified, the explanation or theory for it comes from the mind of people. For example, does light consist of particles (photons) or waves? Probably it consists of neither, but those are two theories from the creative imagination of scientists that have proved useful in explaining the observations.

The joy that a scientist feels after successfully explaining an observation, and pride that an engineer feels for making something work within the constraints of the physics and economics, are no less intense than the joy and pride felt by an artist who has just completed his or her masterpiece. Approach your work with passion, evaluate your result objectively, and enjoy your travel through life as you help society.

Table A-5. Guidelines and issues for scientists. Locations and topics of "On Doing Science" boxes.

Chapter	Issues
1	Descartes and the Scientific Method
2	Scientific Laws — The Myth
2	Seek Solutions
5	Look for Patterns
7	Cargo Cult Science
8	Consequences
9	Be Creative
10	Simple is Best
11	Toy Models
13	Truth vs. Uncertainty
14	Mathematics
14	Scientific Revolutions
17	The Citizen Scientist
18	Ethics and Data Abuse
19	Great Scientists Make Big Mistakes
A	Problem Solving
A	Have Passion
H	Parameterization Rules

SUMMARY

The MKS system of units is used by atmospheric scientists. Functional relationships between variables can sometimes be discerned when the data is plotted on linear, semi-log, or log-log graphs. An organized approach to problem solving is recommended. The philosophy of science blends the passions of people with the objective analysis of observations. All of these principles will be used throughout this book.

Threads

Problem solving methods are used in every "Solved Example" box in this book. Graphs are frequently used to illustration functional relationships.

The international system (SI) of units is also used extensively. Unit conversions are done as needed in many of the solved examples.

EXERCISES

Numerical Problems

N1. Convert the values on the left to the units at right, using a table of conversions.

a. 50 miles =? km   b. 15 knots = ? m/s
c. 30 lb/in^2 =? kPa   d. 5000 kW =? horsepower
e. 150 lb_{Mass} =? kg   f. 150 lb_{Force} =? N
g. 12 ft = ? m   h. 50 km/h = ? m/s

N2. Solve the expression on the left, and give the answer in the units at right, using a table of conversions, and the basic definitions of units:

a. (55. knots) x (36. inches) =? $m^2 \cdot s^{-1}$
b. (14. lb_F/in^2) x $(2.5\ m)^2$ = ? N
c. (120. lb_M) x (3. knots) / day =? mN
d. (15. inHg) x (2. ft^3) = ? J
e. (500. mb) x (3. knots) x $(5.\ in)^2$ =? kW
f. (9.8 $m \cdot s^{-2}$) x (6 kg) / (ft^2) = ? mb
g. (4200 $J \cdot kg^{-1} \cdot K^{-1}$) x (5°C) x (3.3 g) = ? ergs
h. $(2\ ha)^{1/2}$ / 3 weeks = ? m/s

N3. Find the gradients between each pair of neighboring heights. What is the net gradient between the top and bottom heights? How does this net gradient compare to the average of the individual gradients? Discuss.

z(m)	T(°C)
1000	10
500	15
200	17
100	17
50	15
0	10

N4. Convert the following temperatures:

a. 15°C = ? K b. 50°F = ? °C c. 70 °F =? K
d. 48°C = ? °F e. 400 K = ? °F f. 250 K=?°C

N5(§). Plot the following relationships on linear, semi-log, and log-log graphs. Use a spreadsheet (§) on a personal computer to make this easier. Assume that any variable with subscript $_o$ represents a constant.

a. $I = I_o \cdot (R_o/R)^2$
b. $U = U_o \cdot \ln(z/z_o)$
c. $E = E_o \cdot \exp(-z/z_o)$
d. $f = c_o / \lambda$
e. $q = e / e_o$
f. $c = c_o \cdot \exp[-(z/z_o)^2]$
g. $w = [2 \cdot (F/m)_o \cdot z]^{1/2}$
h. $\Delta P/\Delta P_o = (1/5) \cdot (R/R_o)^4$

N6. Find the pressure in kPa at the following heights above sea level:

a. z = -100 m (below sea level)
b. z = 20 m
c. z = 1 km
d. z = 6.5 km
e. z = 11 km
f. z = 18 km
g. z = 25 km
h. z = 30,000 ft

Understanding & Critical Evaluation

U1. Is anything dimensionally wrong with the following equations, given: $P = 100$ kPa, $z = 2$ km, $T = 30$°C, $W = 0.5$ m/s? If so, why?

a. $\log(P) = z$ b. $\sin(T) = W$
c. $\arccos(P) = T$ d. $\exp(-P/z) = 1$
e. $\cos(W) = 2$ f. $\ln(0) = z$
g. $\ln(-10) = T$ h. $\exp(0) = 1$

U2. What is the difference between $e^{2.5}$ and exp(2.5)?

U3. What is the difference between a good assumption and a bad one? What can you do to detect bad assumptions?

U4. Use tables of conversion factors to describe the following variables in terms of the basic dimensions:

a. Volts b. Coulombs c. Farads
d. Angstroms e. light years f. BTUs
g. lumens h. horsepower i. liters
j. ohms k. yards l. quarts

U5. Meteorologists have used units of dm for height, mb for pressure, hPa for pressure.

a. Discuss why these units do not follow the SI standards.

b. What are the preferred units in meteorology?

U6. Plot the following mathematical functions on linear, semi-log, and log-log graphs (you may either do it by hand on graph paper, or use spreadsheet or other plotting programs).

a. $y = x^2$ b. $y = x^5$
c. $y = x^{1/2}$ d. $y = x^{1/5}$
e. $y = x^{-2}$ f. $y = x^{-5}$
g. $y = x^{-1/2}$ h. $y = x^{-1/5}$
i. $y = \ln(x)$ j. $y = \ln(x^3)$
k. $y = \ln(x^{-3})$ l. $y = \exp(x)$
m. $y = \exp(x^3)$ n. $y = \exp(x^{-3})$
o. $y = \exp(-x^3)$ p. $y = [\ln(x)]^2$
q. $\ln(y) = \ln(x)$

Web Enhanced Questions

W1. Search the web for recommended units to use in meteorology. [Hint, on the American Meteorological Society web site: www.ametsoc.org, search for a guide for authors.]

W2. Search the web for weather maps and products from different countries, and compare the different units used for the same variables.

W3. a. Search the web for examples of how to use logarithms and exponentials.

b. Search the web for information on linear graphs, semi-log graphs, and log-log graphs.

W4. Search the web for tables of conversion factors, especially ones used in meteorology. Some sites might have interactive conversion "engines", where you enter one value and it does the conversion for you.

W5. Search the web for current official values of physical constants.

W6. Search the web for discussions about the scientific method, and about problem solving.

Synthesis Questions

S1. a. Suppose a new planet was found midway between Mars and Jupiter. Using only the graphs (Fig A.3) and not the equation, which graph would you prefer to use to estimate by interpolation the orbital period? Fig A.3a which is a curved line, or Fig A.3b which is a straight line? What is the value of the resulting period that you find?

b. Suppose that another new plant was found outside of Pluto, at a distance twice as far from the sun as is Pluto on average. Which graph would you prefer to use to extrapolate to estimate the orbital period? What is your resulting estimated period?

S2. Some units discriminate in magnitudes. For example, mm are smaller than km.

Other units discriminate in the type of physical quantity. For example, grams are different than meters.

What if units of any kind had not been invented. In other words, when scientists take measurements, they would record only the number and not the units.

a. How would the workings of science be different than now? How would science have advanced?

b. How would engineering design be different than now? How would new bridges or machines be built?

S3. Suppose there was not an organized way to approach and solve problems in science. Discuss how that would have affected the advancement of science during past history. How would it affect future science?

S4. Suppose that you discovered a new physical characteristic of nature, and that you devised a new dimension to explain it. However, also assume that the list of basic dimensions in Table A–1 is still valid, which means that your new dimension must be able to be described in terms of the basic dimensions. Describe the steps that you could take to determine the relationship between your new physical dimension and the basic units.

S5. What if science did not involve human creativity. What physical "laws" might have been described differently than they are now, or might have not been describable at all?

CONSTANTS AND CONVERSION FACTORS

B

CONTENTS

UNIVERSAL CONSTANTS

c_o = 299,792,458 m/s = speed of light in a vacuum

e = 2.7182818285 = base of natural logarithms

$G = 6.67\text{x}10^{-11}\ \text{m}^3\cdot\text{s}^{-2}\cdot\text{kg}^{-1}$ = gravitational constant

T = –273.15°C = 0 K = absolute zero

π = 3.14159265358979323846 = pi

$\sigma_{SB} = 5.67\text{x}10^{-8}\ \text{W}\cdot\text{m}^{-2}\cdot\text{K}^{-4}$ = Stefan-Boltzmann const.

$k_B = 1.3806\text{x}10^{-23}\ \text{J}\cdot\text{K}^{-1}\cdot\text{molecule}^{-1}$ = Boltzmann const

EARTH CHARACTERISTICS

1° latitude = 111 km

a = 149.457 Gm = semi-major axis of earth orbit

A = 0.3 = average global albedo

b = 149.090 Gm = semi-minor axis of earth orbit

d = 149.59787 Gm = average sun-earth distance = 1 Astronomical Unit (AU)

$d_{aphelion}$ = 151.96 Gm = furthest sun-earth distance, and occurs on 4 July

$d_{perihelion}$ = 146.96 Gm = closest sun-earth distance, and occurs on 3 January

d_r = 173 = 22 June = day of summer solstice

e = 0.0167 = eccentricity of earth orbit around sun

$g = -9.807\ \text{m}\cdot\text{s}^{-2}$ = gravitational acceleration on earth at sea level (negative = downward)

P_{moon} = 27.32 days = lunar orbital period

$m = 5.9742\text{x}10^{24}$ kg = mass of earth

P_{earth} = 365.25463 days = earth orbital period

$P_{sidereal}$ = 23.94 h = sidereal day = period for one revolution of the earth about its axis, relative to fixed stars

R_{earth} = 6356.766 km = average earth radius

R_{earth} = 6378 km = earth radius at equator

$S = 1368\ \text{W}\cdot\text{m}^{-2}$ = solar constant at top of atmos.
$= 1.125\ \text{K}\cdot\text{m}\cdot\text{s}^{-1}$ = kinematic solar constant

T_e = 255 K = effective radiation emission temperature

Φ_r = 23.45° = 0.409 radians = tilt of earth axis

$\Omega = 0.729\text{x}10^{-4}\ \text{s}^{-1}$ = rotation frequency of earth

$2\cdot\Omega = 1.458\text{x}10^{-4}\ \text{s}^{-1}$ = Coriolis factor

$2\cdot\Omega\ /\ R_{earth} = 2.287\text{x}10^{-11}\ \text{m}^{-1}\cdot\text{s}^{-1}$ = beta factor

AIR AND WATER CHARACTERISTICS

$a = 0.0337\ (\text{mm/day})\cdot(\text{W/m}^2)^{-1}$ = water depth evaporation per latent heat flux

$B = 3\text{x}10^9$ V/km = breakdown potential for air

$C_{pd} = 1004.67\ \text{J}\cdot\text{kg}^{-1}\cdot\text{K}^{-1}$ = specific heat for dry air at constant pressure at 0°C

$C_{pv} = 1850\ \text{J}\cdot\text{kg}^{-1}\cdot\text{K}^{-1}$ = specific heat for water vapor at constant pressure at 0°C

$C_{pv} = 1875\ \text{J}\cdot\text{kg}^{-1}\cdot\text{K}^{-1}$ = specific heat for water vapor at constant pressure at 15°C

$C_{liq} = 4217.6\ \text{J}\cdot\text{kg}^{-1}\cdot\text{K}^{-1}$ = specific heat of liquid water at 0°C

$C_{ice} = 2106\ \text{J}\cdot\text{kg}^{-1}\cdot\text{K}^{-1}$ = specific heat of ice at 0°C

$D = 2.11\text{x}10^{-5}\ \text{m}^2\cdot\text{s}^{-1}$ = molecular diffusivity of water vapor in air in standard conditions

e_o = 0.611 kPa = reference vapor pressure at 0°C

$k = 0.0253\ \text{W}\cdot\text{m}^{-1}\cdot\text{K}^{-1}$ = molecular conductivity of air at sea level, standard conditions

$L_d = 2.834\text{x}10^6\ \text{J}\cdot\text{kg}^{-1}$ =latent heat c

$L_f = 3.34\text{x}10^5\ \text{J}\cdot\text{kg}^{-1}$ = latent heat of

$L_v = 2.501\text{x}10^6\ \text{J}\cdot\text{kg}^{-1}$ = latent heat (

n_{air} = 1.000277 = index of refractior

n_{water} = 1.336 = index of refraction for liquid water

n_{ice} = 1.312 = index of refraction for ice

P_{STP} = 101.325 kPa = standard sea-level pressure

$\Re_d = 0.287053\ \text{kPa·K}^{-1}\text{·m}^3\text{·kg}^{-1}$

$= 287.053\ \text{J·K}^{-1}\text{·kg}^{-1}$ = gas constant for dry air

$\Re_v = 461.5\ \text{J·K}^{-1}\text{·kg}^{-1}$ = water-vapor gas constant

Ri_c = 0.25 = critical Richardson number

s_o = 343.15 m/s = sound speed in standard, calm air

T_{STP} = 15°C = standard sea-level temperature

$\varepsilon = 0.622\ g_{water}\cdot g_{air}^{-1} = \Re_d\ /\ \Re_v$

$\gamma = 0.0004\ (g_{water}\cdot g_{air}^{-1})\cdot\text{K}^{-1} = C_p\ /\ L_v$

$= 0.4\ (g_{water}\cdot kg_{air}^{-1})\cdot\text{K}^{-1}$ = psychrometric constant

$\Gamma_d = 9.75\ \text{K·km}^{-1} = g/C_p$ = dry adiabatic lapse rate

$\rho_{liq} = 1025\ \text{kg·m}^{-3}$ = density of liquid water

$\rho_{STP} = 1.225\ \text{kg·m}^{-3}$ = standard sea-level air density

$\rho_{avg} = 0.689\ \text{kg·m}^{-3}$ = air density averaged over troposphere (over z = 0 to 11 km)

CONVERSION FACTORS & COMBINED PARAMETERS

$C_{pd}\ /\ g = 102.52\ \text{m·K}^{-1}$

$C_{pd}\ /\ L_v = 0.0004\ (g_{water}\cdot g_{air}^{-1})\cdot\text{K}^{-1} = \gamma$

$= 0.4\ (g_{water}\cdot kg_{air}^{-1})\cdot\text{K}^{-1}$ = psychrometric constant

$C_{pd}\ /\ \Re_d = 3.50$ (dimensionless)

$g/C_{pd} = \Gamma_d = 9.8\ \text{K·km}^{-1}$ = dry adiabatic lapse rate

$g\ /\ \Re_d = 0.0342\ \text{K·m}^{-1}$ = 1/(hypsometric constant)

$L_v\ /\ C_{pd} = 2.5\ \text{K}\ /\ (g_{water}\cdot kg_{air}^{-1})$

$L_v\ /\ \Re_v = 5423$ K = Clausius-Clapeyron parameter

$\Re_d\ /\ C_{pd} = 0.28571$ (dimensionless) = potential-temperature constant

$\Re_d\ /\ R_v = \varepsilon = 0.622\ g_{water}\cdot g_{air}^{-1}$

$\Re_d\ /\ g = 29.29\ \text{m·K}^{-1}$ = hypsometric constant

$\rho_{air}\cdot C_{pd\ air} = 1231\ (\text{W·m}^{-2})\ /\ (\text{K·m·s}^{-1})$ at sea level

$= 12.31\ \text{mb·K}^{-1}$ at sea level

$= 1.231\ \text{kPa·K}^{-1}$ at sea level

$\rho_{air}\cdot g = 12.0\ \text{kg·m}^{-2}\text{·s}^{-2}$ at sea level

$= 0.12\ \text{mb·m}^{-1}$ at sea level

$= 0.012\ \text{kPa·m}^{-1}$ at sea level

$\rho_{air}\cdot L_v = 3013.5\ (\text{W·m}^{-2})\ /\ [(g_{water}\cdot kg_{air}^{-1})\cdot(\text{m·s}^{-1})]$ at sea level

$\rho_{liq}\cdot C_{liq} = 4.295\text{x}10^6\ (\text{W·m}^{-2})\ /\ (\text{K·m·s}^{-1})$

2π radians = 360°

[illegible]$)/\varepsilon = 0.61$ = virtual temperature constant

OTHER INFORMATION

Air quality standards: Table 17–1
Albedoes of various substances: Table 2-3
Beaufort wind scale: Focus Table 16–3
Boundary Layer Components: Fig 4.8
Climate change sensitivities: Table 18-2
Composition of the Atmosphere: Table 1–3
Computation speed: Fig 14.2
Density (air at sea level, in various units): Table 1-2
 and vs. height (standard atmosphere): Table 1–4
Diffraction fringes in corona: Table 19-3
Drop sizes: Fig 8.1
Emissivities of various substances: Table 2-2
Fujita tornado scale: Table 15–2
Gases in the atmosphere: Table 1–3
Global climate model attributes: Table 18-1
Heat index, apparent temperature: Table 3–2
Ice crystal habits (shapes): Fig 8.6
Ice nucleation thresholds: Table 8-2
Index of refraction of light: Table 19-1
Isopleth names: Table 1-5
Measurement errors: Table 14-1
Meteors in meteorology: Focus p 160
Molecular weights: Table 1-3
Newton's laws of motion: Focus p 181
Pasquill-Gifford turbulence types: Table 17-2
Pressure (air at sea level, in various units): Table 1-1
 and vs. height (standard atmosphere): Table 1–4
Rainfall intensity vs. radar echo: Fig 8.12
Roughness lengths: Table 4-1
Saffir-Simpson hurricane scale: Focus Table 16–2
Saturation humidities: Table 5-1
Scales of atmospheric motion: Table 10-1
Scattering of light in the atmosphere: Table 19-2
Solute properties: Table 8-1
Sunrise, sunset & twilight definitions: Table 2-1
Temperature (standard atmosphere): Table 1-4, and Figs 1.9, 4.2, 4.3
Thermodynamic diagrams
 Complete, for troposphere: Fig 6.2 and 6.3
 Complete, for boundary layers: Figs 6.12 & 6.13
 Dry adiabatic component: Fig 3.2 & 6.1
 Moisture components: Figs 5.4, 5.6, 6.1
TORRO tornado scale: Table 15–1
Units and dimensions: Tables A–1 to A–4
Weather instruments: p 257-258
 and observation errors: Table 14–1
Weather map examples: Figs 13.18 to 13.24
Weather station locations: Fig 12.9
Wind chill: Table 3–1, Fig 3.10
Zones of weather: Fig 11.2

NOTATION

CONTENTS

Most of these symbols follow the accepted standards of the meteorological literature. Some symbols are overloaded; namely, one symbol can mean many things. Usually the appropriate meaning is evident from context.

ROMAN SYMBOLS, ACRONYMS & CONTRACTIONS

A

A	Ampere, a unit of electrical current
	area
	albedo
	north-south amplitude of wave
	initial analysis
	amplitude
	advection of TKE
	Ball ratio ≅ 0.2
	parameter or constant (varies by context)
AAM	absolute angular momentum
ABL	atmospheric boundary layer
ACARS	aircraft communication and reporting system
AIREP	aviation report
AMDAR	aircraft meteorological data relay
Ar	argon
ASDAR	aircraft to satellite data relay
ASOS	automatic surface observing system
AT	apparent temperature (heat index)
AVN	aviation
AWOS	automatic weather observing system
A_B	albedo of black daisies
A_G	albedo of bare ground
A_W	albedo of white daisies
A_w	global albedo corresponding to an atmospheric window
a	parameter or constant (varies by context)
	absorptivity
	acceleration
	surface albedo
	length of semi-major axis of earth's orbit = 149.457 Gm
	spillage distance of front (= external Rossby radius of deformation)
$\vec{a}$	vector acceleration
a_H	mixed-layer transport coefficient
arccos	arc cosine
arcsin	arc sine
arctan	arc tangent
asin	arc sine
atm	atmospheres, a pressure unit

B

B	Beaufort wind scale
	Bowen ratio
	breakdown potential for dry air (=3x10^9 V km^{-1})
	buoyant production or consumption of TKE
	parameter or constant (varies by context)
BC	boundary condition
BLG	boundary-layer gradient (wind)
b	parameter or constant (varies by context)
	absorption cross section
	length of semi-minor axis = 149.090 Gm
	damping factor
b_1	parameter or constant (varies by context)
b_D	dimensionless coefficient (=5x10^{-4}, 1.83x10^{-3})
b_H	convective transport coefficient

C

C	circle circumference (360° or 2· radians)
	climatological condition
	circulation
	fractional area covered
	dimensionless concentration
	parameter or constant (varies by context)
$[C]$	transilient matrix

CAN Canada
CAPE convective available potential energy
C_B fraction of area covered by black daisies
Bernoulli's constant
CC cell circulation
CCN cloud condensation nuclei
C_D drag coefficient
CFL Courant-Friedrichs-Levy (condition for numerical stability)
C_G fractional area covered by bare ground
C_H bulk heat transfer coefficient
CH_4 methane
CI capping inversion
CMC Canadian Meteorological Centre
CO carbon monoxide
CO_2 carbon dioxide
C_p specific heat of air at constant pressure
C_{pd} specific heat of dry air at constant pressure $= 1004.67\ J \cdot kg^{-1} \cdot K^{-1}$
C_{pv} specific heat for water vapor at constant pressure $= 1875\ J \cdot kg^{-1} \cdot K^{-1}$
CRT video screen. Literally: Cathode Ray Tube
Curv curvature
C_s seed coverage
C_W fraction of area covered by white daisies
C_w vertical drag coefficient
C_y dimensionless crosswind-integrated concentration
°C Celsius or centigrade, a unit of temperature
c speed of light
half the distance between two foci of an ellipse
concentration of pollutant
maximum concentration
phase speed or shallow-water wave speed
parameter or constant (varies by context)
c_o speed of light in a vacuum= 299,792,458 m/s
intrinsic phase speed of a wave
c_1 parameter or constant (varies by context)
c_2 parameter or constant (varies by context)
c_3 parameter or constant (varies by context)
c_4 parameter or constant (varies by context)
$c_{\alpha\beta}$ intermediate calculation parameter for atmospheric optical ray paths
c_i speed of light through medium i
c_{ij} transilient coefficient
cd candela, a unit of luminous intensity
cm centimeter, a length unit
cos cosine
covar covariance
cP continental polar air mass
c_y crosswind-integrated concentration

D

D duration of positive heat flux
distance
size of particle
modern description
fractal dimension
diffusivity $= 2.11 \times 10^{-5}\ m^2 \cdot s^{-1}$ for water vapor in air under standard conditions
drop diameter
width of flow
death rate for daisies
parameter or constant (varies by context)
DMSP defense meteorological satellite program
DW diabatic warming rate
D_ϕ net radiative flux, or differential heating, at latitude ϕ
d diagonal length
depth
distance between earth and sun = 149.6 Gm
distance between crest and trough of a wave
Julian day
plume spread
d days, a time unit
dBZ decibels of radar reflectivity factor
d_r Julian day of summer solstice = 173
d_w precipitable water
d_y number of days in a year

E

E East
E efficiency
electric field
turbine efficiency
incoming or outgoing radiative flux
irradiance
energy
E^* total irradiance
E_1 parameter or constant (varies by context)
ECMWF European Centre for Medium-Range Weather Forecasts
E_{in} zonally and annually averaged incoming radiative flux
E_{net} zonally and annually averaged net radiative flux
E_{out} zonally and annually averaged outgoing radiative flux
EPA U.S. Environmental Protection Agency
EQ equator
Evap evaporation rate
EZ entrainment zone
E_λ^* irradiance
E_ϕ incoming or outgoing radiative flux at latitude ϕ

e base of natural logarithms = 2.7182818285
emissivity
eccentricity of earth's orbit = 0.0167
vapor pressure
current or voltage
e_o saturation vapor pressure = 0.611 kPa at 0°C
e_s saturation vapor pressure over flat pure water
e_s^* actual saturation vapor pressure
e_w equilibrium emissivity corresponding to an atmospheric window
exp base of natural logarithms = 2.71828

F

F kinematic flux
force
apparent reduced gravity
forecast
Fujita tornado damage scale
F^* net radiative flux at earth's surface (in kinematic units of K·m/s)
FA free atmosphere
FAS Federation of American Scientists
F_C tracer flux
FG flux gradient
F_E latent heat flux from earth's surface up into the air (in kinematic units of K·m/s)
F_G heat flux from conducted from the ground up to the earth's surface (in kinematic units)
F_H sensible heat flux from earth's surface up into the air (in kinematic units of K·m/s)
FNMOC Fleet Numerical Meteorological and Oceanographic Center
Fr Froude number
F_{rad} radiative flux at a surface
FS frontal strength
F_x component of net vector force toward the east
F_x kinematic heat flux component in x direction
F_y kinematic heat flux component in y direction
F_z kinematic heat flux component in z direction
°F Fahrenheit, a unit of temperature
$\vec{F}$ vector force
$\vec{F}_{net}$ net vector force
$\Im$ flux (dynamic)
$\Im^*$ net radiative flux at earth's surface (in dynamic units of W/m^2)
$\Im_{dif}$ diffusive moisture flux
$\Im_E$ latent heat flux from earth's surface up into the air (in dynamic units of W/m^2)
$\Im_G$ heat flux from conducted from the ground up to the earth's surface (in dynamic units)
$\Im_H$ sensible heat flux from earth's surface up into the air (in dynamic units of W/m^2)
$\Im_{rad}^*$ net radiative flux
$\Im_x$ dynamic heat flux component in x direction
$\Im_y$ dynamic heat flux component in y direction
$\Im_z$ dynamic heat flux component in z direction
f oscillation frequency
f_c centrifugal force
Coriolis parameter
f_G pull (force) by earth's gravity
f_o reference Coriolis parameter at the center of the beta plane
ft feet, a length unit
ft H_2O feet of water, a pressure unit
$f(X)$ fraction of cloud sizes
$F_{z\,eff\,sfc}$ effective surface heat flux

G

G gravitational constant $=6.67\text{x}10^{-11}\ m^3{\cdot}s^{-2}{\cdot}kg^{-1}$
geostrophic wind speed
intermediate calculation parameter
GB gigabyte
GCM global climate model
general circulation model
GEM Global Environmental Multiscale (Canadian numerical model)
Gm gigameter, a length unit
GPS global positioning system (via satellite)
GTS global telecommunication system
GSM global spectral model
g gravitational acceleration $= -9.8\ m{\cdot}s^{-2}$ (negative means downward)
g' reduced gravity
g grams, a unit of mass

H

H height
geopotential height
heating rate
helicity
unperturbed ocean depth
H high-pressure center
H_2 hydrogen
H_2O water
H_2O_2 hydrogen peroxide
H_2SO_4 sulfuric acid
HC hurricane category
He helium
H_e e-folding depth
HI heat index apparent temperature
HNO_3 nitric acid

H_p scale height for pressure
H_ρ scale height for density
Hz hertz, a unit of frequency
h depth of stable boundary layer
depth of cold air
wave height
h hours, a unit of time
h_o hour angle
h_2 parameter or constant (varies by context)
h_{gust} gust front depth
h_w depth of air containing waves

I

I current
$I\downarrow$ downwelling longwave radiation
$I\uparrow$ upwelling longwave radiation
I^* net longwave radiation
[I] identity matrix
I_o intensity of incident radiation
parameter or constant (varies by context)
I_1 parameter or constant (varies by context)
IC initial condition
IPV Isentropic Potential Vorticity
IR infrared
I_{scat} intensity of scattered radiation
i index
number of ions per molecule
grid index
time index
in inch, a unit of length
in Hg inches of mercury, a pressure unit

J

J Joule, a unit of energy
JNWPU Joint Numerical Weather Prediction Unit
j grid index

K

K Kelvins, a unit of absolute temperature
K eddy diffusivity
eddy viscosity
total number of height levels or layers
$K\uparrow$ upwelling solar radiation
$K\downarrow$ downwelling solar radiation
KE kinetic energy
K-H Kelvin Helmholtz

k absorption coefficient
molecular conductivity = $0.0253\ \text{W·m}^{-1}\text{·K}^{-1}$
von Kármán constant (= 0.4)
grid index
height index
parameter or constant (varies by context)
vertical data point index
k_1 parameter or constant (varies by context)
k_2 parameter or constant (varies by context)
k_B Boltzmann constant = $1.3806\text{x}10^{-23}$ $\text{J·K}^{-1}\text{·molecule}^{-1}$
kg kilograms, a mass unit
kg_f kilograms of force = $g\cdot kg_m$
kg_m kilograms of mass
km kilometers, a unit of length
kPa kiloPascals, a pressure unit
kt knot (unit of speed = 1 nautical mile per hour)

L

L length
latent heat
horizontal distribution of temperature
solar luminosity
Obukhov length
radiation out
L low-pressure center
LCL Lifting Condensation Level
LFC level of free convection
LFM limited-area fine-mesh
LHS left hand side
LOC limit of convection
L_v latent heat of condensation or vaporization = $\pm 2.5\text{x}10^6\ \text{J·kg}^{-1}$
L_f latent heat of fusion or melting = $\pm$-$3.34\text{x}10^5\ \text{J·kg}^{-1}$
L_d latent heat of deposition or sublimation = $\pm 2.83\text{x}10^6\ \text{J·kg}^{-1}$
L_x location parameter
L_ε dissipation length scale
l liter, a unit of volume
lb pound, a unit of mass. Also used as a unit of weight (pounds-force), which is the force experienced by a pound-mass under the influence of gravity
l_b buoyancy length scale
liter a unit of volume
l_m momentum length scale
ln natural logarithm (base e)
log common logarithm (base 10)

M

M wind speed
mean anomaly
mass of earth
number of tiles per side
vertical distribution of temperature
M_o wind speed in residual layer
walking speed = 2 m·s^{-1}
location parameter
M_2 parameter or constant (varies by context)
MB megabyte
M_{BL} boundary layer wind speed
M_c speed of movement of a column of air
M_{cs} cyclostrophic wind speed
ME mechanical energy
METAR meteorological observation code
MG meridional gradient of zonal momentum
MIPS millions of (computer) instructions per second
ML mixed layer
MKS metric system (SI) of units. Literally: M=meters, K=kilograms, S=seconds
MOS Model Output Statistics
MRF medium-range forecast
MSL mean sea level
M_{gust} gust front advancement speed
M_{max} maximum tangential wind
M_r total gradient wind speed
radial velocity
M_{rad} radial (inflow) component of velocity
M_s molecular weight of solute
M_t translation speed
M_{tan} tangential component of velocity
m mass
mass of satellite
Mach number
dimensionless diffraction parameter
mass of earth = 5.9742×10^{24} kg
m meters, a length unit
m_{air} mass of air
mb millibars, a pressure unit
$m_{condensing}$ mass of water vapor that condensed
min minutes, a unit of time
m H_2O meters of water, a pressure unit
mm Hg millimeters of mercury, a pressure unit
mole amount of substance
mph miles per hour
m_{planet} mass of a planet
m_s mass of solute
mT maritime tropical air mass
m_v mass of a water molecule
m_{water} mass of water
m_w molecular weight

N

N Newtons, a unit of force
North
N number of data points
number of tiles through which the perimeter passes
number of drops greater than a given radius
energy
N_o parameter or constant (varies by context)
N_1 abundance parameter
N_2 nitrogen molecule
NAAQS National Ambient Air Quality Standards
NaCl salt
N_{BV} Brünt Väisälä frequency
NCEP National Centers for Environmental Prediction
Ne neon
NGM nested-grid model
$(NH_4)_2SO_4$ ammonium sulfate
NL number of layers
NMC National Meteorological Center
NOAA National Oceanographic and Atmospheric Administration
NO_2 nitrogen dioxide
N_2O nitrous oxide
NP North Pole
NSSL National Severe Storms Laboratory
NVA negative vorticity advection
NWP numerical weather prediction
NWS National Weather Service
n number density
direction pointing toward the center of curvature
number of grid points being averaged
index of refraction
total number
n_i refractive index for medium i

O

O_2 oxygen molecule
O_3 ozone
OFM operational numerical weather forecast model

P

P pressure
period
power
perimeter
probability

Po sea-level pressure = 101.325 kPa
reference pressure = 100 kPa
mean background pressure
Pa Pascal, a pressure unit
P_B ambient pressure at bottom of column
Pb chemical abbreviation for lead
P_{BV} period of oscillation at the Brunt-Väisälä frequency
P_c reference pressure at the center of an anticyclone
P_d dynamic pressure
PE primitive equation
P_{earth} earth orbital period = 365.25463 days
PG Pasquill-Gifford
PIBAL pilot balloon
PIREPS pilot reports (of the weather in flight)
P_{moon} lunar orbital period= 27.32 days
PPI plan position indicator
PPM perfect prog method
Pr precipitation rate
probability of having a given wind speed
P_s static pressure
$P_{sidereal}$ sidereal day = 23.94 h
P_{STP} standard sea-level pressure = 101.325 kPa
P_T ambient pressure at top of column
PVA positive vorticity advection
PVU potential vorticity units (= 10^{-6}K·m^2·s^{-1}·kg^{-1})
p_s sea-level pressure
ppb parts per billion
ppm parts per million
psf pounds-force per square foot, a pressure unit
psi pounds-force per square inch, a pressure unit

Q

Q amount
source emission rate of pollutant
Q_1 total amount of pollutant emitted
Q_A cumulative heating or cooling
Q_{Ak} cumulative heating or cooling in kinematic units
Q_H heat added
q specific humidity
solar forcing ratio
q_s saturation specific humidity

R

R distance
ration of e-folding height to inversion strength in the stable boundary layer
impact parameter
radius
radius of curvature
distance of gust front from downburst center
distance from center of earth
distance from axis of rotation of earth
distance from center of a hurricane
$\Re$ gas constant in ideal gas law
R^* critical radius
RAOB radiosonde observation
RASS radio acoustic sounding system
R_o critical radius or distance
stack-top radius
mesocyclone radius
parameter or constant (varies by context)
radius of earth (see R_{earth})
distance of hurricane eye wall wind maximum from the center of the eye
R_1 parameter or constant (varies by context)
average drop radius
R_c radius of curvature
$\Re_d$ gas constant for dry air = 0.287053 kPa·K^{-1}·m^3·kg^{-1} = 287.053 J·K^{-1}·kg^{-1}
R_{earth} average earth radius = 6356.766 km
R_{earth} earth radius at equator = 6378 km
R_f flux Richardson number
Ri Richardson number
Ri_c critical Richardson number = 0.25
RH relative humidity
RHI range height indicator (a radar display)
RHS right hand side
RL Residual Layer
RMS root mean square (error)
Ro_c curvature Rossby number
RR rainfall rate
RxL radix layer
$\Re_v$ gas constant for water vapor=461.5 J·K^{-1}·kg^{-1}
r_o tornado radius
constant = 0.02
earth radius = 6356.766 km
r_1 parameter or constant (varies by context)
r_c critical radius (for escape)
r water mixing ratio, a form of humidity
radius
reflectivity
correlation coefficient
radius of downburst
parameter or constant (varies by context)
rad radians
r_s saturation mixing ratio

r_i	ice mixing ratio
r_L	liquid water mixing ratio
r_{sat}	saturation mixing ratio
r_T	total water mixing ratio
r_w	wet bulb mixing ratio
r_λ	reflectivity at one wavelength
r_∞	background mixing ratio

S

S	South
S	solar constant = 1368 ±7 W·m^{-2}
	Snell's Law for components
	supersaturation
	swirl ratio
	shear generation of TKE
	storm surge height of ocean surface above normal sea level
	radiation in
$S1$	S1 score
S^*	Eulerian net sources
	critical supersaturation
S^{**}	Lagrangian net source (of water)
S_o	internal source of heat per unit mass
	reference solar constant
SBL	Stable Boundary Layer
SI	system international (metric units)
S_I	parameter or constant (varies by context)
SL	surface layer
SO_2	sulfur dioxide
SOI	southern oscillation index
SP	South Pole
SR	equilibrium supersaturation adjacent to a drop
SSM/I	Special Sensor Microwave Imager
SST	stably-stratified turbulence
	sea surface temperature
STP	Standard Temperature and Pressure
S_x	dimensionless spread parameter
S_∞	background supersaturation far from drop
s	length of side of square
	path length
	slope
	sign coefficient
	horizontal distance
	sound velocity
	total entropy
s	seconds, a time unit
s_o	parameter or constant (varies by context)
sin	sine
sr	steradian

T

T	TORRO tornado wind speed scale
T	temperature
	net sky transmissivity
T_o	parameter or constant (varies by context)
T_o	freezing temperature in Kelvins (273.15 K)
T_1	parameter or constant (varies by context)
T_2	parameter or constant (varies by context)
T_A	temperature of atmosphere
TD	tropical depression
T_d	dew point temperature
T_e	temperature of environment
	effective radiation emission temperature = 255 K
TH	thickness
TIROS	Television and Infrared Operational (satellite) System
T_{kat}	temperature of katabatic layer
TKE	turbulence kinetic energy per unit mass
TNT	trinitrotoluene (a high explosive)
Torr	a unit of measure of pressure
TOVS	TIROS Operational Vertical Sounder
T_p	temperature of parcel
Tr	total atmospheric and oceanic heat transport needed to compensate radiation
	transport of TKE by turbulence
	horizontal transport parameter
	net atmospheric transmissivity
T_{RL}	residual layer temperature
T_r	reference temperature
TS	tropical storm
T_s	surface temperature
T_{skin}	surface skin temperature of the top few molecules on the earth's surface
T_{STP}	standard sea-level temperature = 15°C = 288 K
T_{SW}	equilibrium surface temperature with an atmospheric window
T_v	virtual temperature
TV	television
T_w	wet bulb temperature
t	time
	transmissivity
tan	tangent
tcu	towering cumulus clouds
t_d	length of day
t_L	Lagrangian time scale
t_{orbit}	orbital time period
t_{SR}	sunrise time
t_{UTC}	Coordinated Universal Time

U

U wind component toward east
tangential velocity of earth
U' velocity relative to earth
U_o mean zonal wind speed
U_{ag} ageostrophic wind component toward the east
U_g geostrophic wind component toward the east
U_{jet} jet stream velocity
UK United Kingdom
U_r gradient wind component toward the east
URL universal resource locator (internet web address)
US United States of America
USA United States of America
USSR (former) Union of Soviet Socialist Republics
UTC coordinated universal time
u_* friction velocity

V

V wind component toward north
volume
verifying analysis
voltage
V volts
V_{ag} ageostrophic wind component toward the north
V_g geostrophic wind component toward the north
V_{in} inflow velocity
VLF very low frequency
V_r gradient wind component toward the North
v molecular speed
total velocity along a streamline
$\vec{V}$ vector wind velocity
var variance
ve escape velocity
v_L velocity of current through the return-stroke path in lightning
Vol volume of water

W

W upward wind component
width
W watt, a power unit
West
W_0 stack-top exit velocity
WBF Wegener-Bergeron-Findeisen (cold-cloud precipitation formation process)
WD wave drag
W_{mid} vertical velocity across 50 kPa surface
WMO World Meteorological Organization
W_s subsidence velocity in hurricane eye
WWW World Wide Web (an internet resource)
WX weather
w terminal velocity
updraft or downdraft speed
w_o parameter or constant (varies by context)
w_* Deardorff convective velocity
w_b buoyancy velocity scale
w_d downburst velocity
w_e entrainment velocity
w_s synoptic-scale mean vertical velocity (negative for subsidence)
w_T transport velocity

X

X cloud diameter or depth
dimensionless downwind distance of receptor from source
ratio of heat flux into the ground to the net radiative flux at the earth's surface
Xe xenon
x distance toward east
distance from centerline
path length
travel distance
distance downwind
distance from front
abscissa value
x' distance east from arbitrary longitude

Y

Y time period of orbit
dimensionless crosswind distance of receptor from plume centerline
y year
y distance toward north
ordinate value
y' north-south displacement distance from center latitude
y_o center latitude
yr year

Z

Z radar reflectivity factor
dimensionless receptor height
Z_{CL} dimensionless plume centerline height
Z_s dimensionless source height
Z_T depth of troposphere (=11 km)

z — vertical distance
depth
height
z_o — aerodynamic roughness length
z_1 — initial amplitude
z_c — reference height at the center of an anticyclone
z_{CL} — plume centerline height
z_i — depth of convective mixed layer
z_s — physical stack height
z_T — depth of troposphere (=11 km)
height of topography
z_{Trop} — depth of troposphere (=11 km)

GREEK SYMBOLS

A α Alpha

α — wind direction relative to north
local azimuth angle
component angle
angle of ray above horizontal
parameter or constant (varies by context)
spread parameter
slope angle
angle of tilt of wave crests relative to vertical
elevation angle
α_1 — rotation angle about the column axis (of an ice crystal)
α_3 — intermediate calculation parameter
α_5 — intermediate calculation parameter

B β Beta

β — component angle
rotation angle
rate of change of Coriolis parameter with latitude
divergence
growth rate
constant or parameter (varies by context)
β_c — constant or parameter (varies by context)
β_1 — intermediate calculation parameter
β_3 — intermediate calculation parameter
β_5 — intermediate calculation parameter

Γ γ Gamma

Γ_d — "dry" adiabatic lapse rate = 9.8 $K \cdot km^{-1}$
Γ_{ps} — pre-storm lapse rate
Γ_s — moist lapse rate
Γ_{sa} — standard-atmosphere lapse rate = 6.5 $K \cdot km^{-1}$
γ — psychrometric constant = 0.4 $(g_{water\ vapor}/kg_{air}) \cdot K^{-1}$
γ — depth parameter
potential temperature gradient above the boundary layer
lapse rate
crystal axis orientation with respect to the compass direction of the incoming light ray

Δ δ Delta

Δ — change of
δ_{ij} — Kronecker delta
δ_s — solar declination angle

E ε Epsilon

ε — emissivity
ratio of gas constants for dry air and water vapor = 0.622 $g_{vapor}/g_{dry\ air}$
dissipation rate of TKE
intermediate calculation parameter
ε_o — permittivity of free space = 8.854×10^{-12} $A \cdot s \cdot (V \cdot m)^{-1}$

Z ζ Zeta

ζ — zenith angle
vertical vorticity
ζ_a — absolute vorticity
ζ_g — geostrophic vorticity
ζ_{IPV} — isentropic potential vorticity
ζ_p — potential vorticity
ζ_r — relative vorticity
ζ^* — dimensionless height

H η Eta

η' — vertical displacement (of a wave)

Θ θ Theta

θ — potential temperature
angle of light ray
θ_1 — incident angle
viewing angle
θ_2 — refracted angle
viewing angle
θ_3 — reflected angle
θ_a — ambient potential temperature
θ_c — critical angle

θ_e equivalent potential temperature
θ_e potential temperature of environment
θ_P initial stack gas potential temperature
θ_H horizontal angle
θ_L liquid water potential temperature
θ_p potential temperature of parcel
θ_v virtual potential temperature
θ_V vertical angle

I ι Iota

K κ Kappa

Λ λ Lambda

Λ parameter or constant (varies by context)
λ wavelength
horizontal scale
dimensionless wave parameter
climate sensitivity factor
λ_{max} wavelength of peak emission
λ_2 parameter or constant (varies by context)
λ_r longitude
Rossby radius of deformation

M μ Mu

μ_{ij} ratio of refractive indices n_i and n_j
μm micron, a length unit

N ν Nu

ν frequency
true anomaly

Ξ ξ Xi

O o Omicron

Π π Pi

3.14159265358979323846226

P ρ Rho

ρ density
ρ sea-level density = 1.225 kg·m^{-3}
ρ_o density of air at sea level
density of object
ρ_d density of dry air
ρ_f density of fluid
ρ_L density of liquid water
ρ_m number density of molecules
ρ_v absolute humidity = density of water vapor
ρ_{vs} saturation absolute humidity
ρ_{STP} standard sea-level density = 1.225 kg·m^{-3}

Σ σ Sigma

Σ summation
σ wavenumber
standard deviation
parameter or constant (varies by context)
σ^2 variance
σ_o standard deviation of raw observation
σ_1 standard deviation of drop radii
σ_g standard deviation associated with first guess
σ_L fraction of low clouds
σ_H fraction of high clouds
σ_M fraction of middle clouds
σ_{SB} Stefan-Boltzmann constant =5.67x10^{-8} W·m^{-2}·K^{-4}
σ_y lateral standard deviation of pollutant
σ_{yd} dimensionless lateral standard deviation
σ_z vertical standard deviation of pollutant
σ_z^2 vertical variance of pollutant
σ_{zc} vertical standard deviation of crosswind-integrated concentration of pollutant
σ_{zdc} dimensionless vertical standard deviation of crosswind-integrated concentration

T τ Tau

τ perihelion date
time scale
e-folding time
stress

Y υ Upsilon

Φ ϕ Phi

Φ latitude
Φ_r tilt of the earth's axis relative to the ecliptic =23.45°
ϕ latitude
angle between wind direction and sound-propagation direction
intermediate calculation parameter

X χ Chi

Ψ ψ Psi

Ψ elevation angle

Ω ω Omega

Ω earth rotation rate (=0.729x10^{-4}s^{-1})
ω circular frequency
vertical velocity

OPERATORS

d( )	total derivative
∂( )	partial derivative
Δ( )	change of or difference of
$\overline{(\ )}$	(overbar) average of
( )′	(prime) deviation from the mean
[ ]	matrix
$\|_z$	at height z
| |	absolute value
^	amplitude of wave
∫( )	integral

SUPERSCRIPTS

′	turbulent variation (deviation from mean)
	perturbation
	first guess

SUBSCRIPTS

A	any quantity
AD	advection
adv	advection
B	of parcel B
	bottom of troposphere
	black daisies
BL	boundary layer
BLG	boundary-layer gradient (wind)
C	of parcel C
°C	in degrees Celsius
CCN	cloud condensation nuclei
CF	Coriolis force
CN	centrifugal force
E	latent heat
G	ground
	bare ground
H	of heat
K	in degrees Kelvin
L	liquid
LCL	lifting condensation level
left	at the left side of a volume or box
ML	mixed layer
PG	pressure gradient
R	reference
right	at the right side of a volume or box
RL	residual layer
RxL	radix layer
skin	at the top molecules of the earth's surface
SL	surface layer
SST	sea surface temperature
STP	standard temperature and pressure
T	top of troposphere
TD	turbulent drag
TH	thickness
	thermal
turb	turbulent or turbulence
X	of parcel X
W	white daisies
a	analysis
air	of air
avg	average
cond	conduction
d	destination
	flow in the most narrow part of a channel or mountain pass
e	environment
earth	of the earth
eff	effective
eq	equilibrium
final	final
g	geostrophic
	first guess
i	destination index
init	initial
j	source index
k	data point index
liq	liquid
max	maximum
mid	at midpoint of column
	initial value
o	mean background value or reference state
	observation
	initial condition
p	parcel
rad	of radiation
s	surface
	source
	upstream flow
sfc	surface
SL or *sl*	sea level
sun	of the sun
t	at time *t*
turb	turbulent
u	*u* component of wind
v	*v* component of wind
	virtual
w	*w* component of wind
x	variable in the *x* direction
	component toward the east

y	variable in the y direction component toward the north
z	variable in the z direction
λ	at one wavelength
1	at height 1
2	at height 2
θ	potential temperature
ϕ	at latitude ϕ
∞	at a far or infinite distance away

SPECIAL SYMBOLS

§	computer spreadsheet should be used
$\mathfrak{J}$	flux (dynamic)
$\Re$	gas constant in ideal gas law
$\Re_d$	gas constant in ideal gas law for dry air
$\Re_v$	gas constant in ideal gas law for water vapor
•	key equation

ADDITIONAL READING MATERIAL

CONTENTS

D The references are organized by major subject. Some of these subjects include more than one chapter. The contents above indicate which chapters correspond to the various subjects.

THE ATMOSPHERE

Aguado, E. and Burt, J.E. 1999: *Understanding Weather and Climate*. Prentice Hall. 474 pp

Ahrens, C.D. 2000: *Meteorology Today, An Introduction to Weather, Climate, and the Environment*, 6th Ed. Brooks/Cole Thomson Learning. 602pp

Ahrens, C.D. 1998: *Essentials of Meteorology, An Invitation to the Atmosphere*. 2nd Ed. (Wadsworth) Brooks/Cole Thomson Learning. 444pp

Aristotle, 340BC: *Meteorologica* . Translated by H.D.P. Lee, Reprinted 1978. Loeb Classical Library, Harvard University Press. 433pp

Burroughs, W.J., Crowder, B. , Robertson, T. , Vallier-Talbot, E. , and Whitaker,R. 1996: *The Nature Company Guides: Weather*. Time-Life Books. 288pp

Geer, I.W. 1996: *Glossary of Weather and Climate, with Related Oceanic and Hydrologic Terms*. Amer. Meteor. Soc., 272pp

Houghton, D.D. (Ed.), 1985: *Handbook of Applied Meteorology*. Wiley-Interscience. 1461pp

Huschke, R.E. 1959: *Glossary of Meteorology*. American Meteorological Society, 45 Beacon St., Boston, MA 02108. 638pp

Ludlam, D.M. 1982: *The American Weather Book*. Houghton Mifflin Co. 296pp

Ludlam, D.M. 1991: *The Audubon Society Field Guide to North American Weather*. Knopf. 656pp

Lutgens, F.K. and Tarbuck, E.J. 1998: *The Atmosphere* (7 Ed). Prentice Hall. 434 pp

Moran, J.M. and Morgan M.D. 1997: *Meteorology. The Atmosphere and the Science of Weather* (5 Ed.). Prentice Hall. 530 pp

Moran, J.M. and Morgan M.D. 1995: *Essentials of Weather*. Prentice Hall. 351 pp

Schaefer, V.J. and Day, J.A. 1981: *A Field guide to the Atmosphere*. The Peterson Field Guide Series. Houghton Mifflin Co. 359pp

Sorbjan, Z. 1996: *Hands-on Meteorology, Stories, Theories, and Simple Experiments*. Project Atmosphere. Amer. Meteor. Soc., 306pp

U.S. Government, 1976: *U.S. Standard Atmosphere, 1976*. Superintendent of Documents, U.S. Government Printing Office, Washington DC 20402 (Stock No. 003-017-00323-0) NOAA-S/T 76-1562. 227pp

Wallace, J.M. and Hobbs, P.V. 1977: *Atmospheric Science, An Introductory Survey*. Academic Press. 467pp

Williams, J. 1992: *The Weather Book, An Easy-to-Understand Guide to the USA's Weather*. USA Today. Vintage Books, Random House. 212pp

RADIATION, THERMODYNAMICS & MOISTURE

Bohren, C.F. and Albrecht, B.A. 1998: *Atmospheric Thermodynamics*. Oxford Univ. Press. 402pp

Byers, H.R. 1959: *General Meteorology*, 3rd Ed. McGraw-Hill. 540pp

Fleagle, R.G. and Businger, J.A. 1980: *An Introduction to Atmospheric Physics*, 2 Ed. Acad. Press. 346pp

Liou, K.N. 1992: *Radiation and Cloud Processes in the Atmosphere*. Oxford University Press. 487pp

Stephens, G.L. 1994: *Remote Sensing of the Lower Atmosphere*. Oxford University Press. 544pp

Wallace, J.M. and Hobbs, P.V. 1977: *Atmospheric Science, An Introductory Survey*.Acad.Press 467pp

CLOUD & SATELLITE PICTURES

Audubon 1995: *Clouds and Storms*. National Audubon Society Pocket Guide. 192 pp

Bader, M.J., Forbes, G.S., Grant, J.R., Lilley, R.B.E., and Water, A.J. 1995: *Images in Weather Forecasting. A practical guide for interpreting satellite and radar imagery*. Cambridge Univ. Press. 499pp

Day, J.A. and V.J. Schaefer 1991: *Clouds and Weather*. Peterson First Guides. Houghton Mifflin. 128 pp

de Bont, G.W.Th.M. 1985: *De Wolken en het Weer*. (in Dutch). Uitgeverij Terra Zutphen. ISBN 90-6255-227-7. 128pp

Ludlam, D.M. 1991: *The Audubon Society Field Guide to North American Weather*. Knopf. 656pp

Mandelbrot, B.B. 1983: *The Fractal Geometry of Nature*. Freeman. 468pp

Schertzer, D. and Lovejoy, S. (Ed.) 1991: *Non-linear Variability in Geophysics, Scaling and Fractals*. Kluwer. 318pp

Scorer, R. 1972: *Clouds of the World, A Complete Colour Encyclopedia*. David & Charles Pubs. 176pp

Stephens, G.L. 1994: *Remote Sensing of the Lower Atmosphere*. Oxford University Press. 544pp

World Meteorological Organization, 1987: *International Cloud Atlas*, Vol II. World Meteor. Organization. ISBN 92-63-12407-8. 212pp

CLOUD & STORM PROCESSES

Cotton, W.R. and Anthes, R.A. 1989: *Storm and Cloud Dynamics*. Academic Press. 883pp

Doviak, R.J., and Zrnic, D.S. 1993: *Doppler Radar and Weather Observations*, 2 Ed. Acad. Press. 562pp

Emanuel, K.A. 1994: *Atmospheric Convection*. Oxford University Press. 592pp

Emanuel, K.A., and Raymond, D.J. (Eds.) 1993: *The Representation of Cumulus Convection in Numerical Models*. American Meteorological Society. 246pp

Houze, R.A. Jr. 1993: *Cloud Dynamics*. Academic Press. 573pp

Liou, K.N. 1992: *Radiation and Cloud Processes in the Atmosphere*. Oxford University Press. 487pp

Marshall, T. 1995: *Storm Talk*. Tim Marshall, 1336 Brazos Blvd., Lewisville, TX 75067, USA. 223 pp

Ray, P.S. (Ed.) 1986: *Mesoscale Meteorology and Forecasting*. American Meteorological Society, 45 Beacon St., Boston, MA 02108. 793pp

Rinehart, R.E. 1991: *Radar for Meteorologists*, 2nd Ed. Ronald E. Rinehart, P.O. Box 6124, Grand Forks, ND 58206-6124, USA. 334pp

Rogers, R.R. and Yau, M.K. 1989: *A Short Course in Cloud Physics*, 3rd Ed., Pergamon. 293pp.

Tornado Project, 1993, 1994: *Tornado Video Classics* (parts I & II videotapes & guides). The Tornado Project, P.O. Box 302, St. Johnsbury, VT 05819.

Wallace, J.M. and Hobbs, P.V. 1977: *Atmospheric Science, An Introductory Survey*. Academic Press. 467pp

Young, K.C. 1993: *Microphysical Processes in Clouds*. Oxford University Press. 448pp

DYNAMICS OF ATMOSPHERES & OCEANS

Cushman-Roisin, B. 1994: *Introduction to Geophysical Fluid Dynamics*. Prentice Hall. 320pp

Dutton, J.A. 1976: *The Ceaseless Wind, An Introduction to the Theory of Atmospheric Motion*. McGraw Hill. 579pp

Gill, A.E. 1982: *Atmosphere–Ocean Dynamics*. Academic Press. 662pp

Haltiner, G.J. and Williams, R.T. 1980: *Numerical Prediction and Dynamic Meteorology*. John Wiley & Sons. 477pp

Holton, J.R. 1992: *An Introduction to Dynamic Meteorology*, 3rd Ed. Academic Press. 511pp

Kinsman, B. 1965: *Wind Waves, Their Generation and Propagation on the Ocean Surface*. Prentice-Hall. 676pp

Kraus, E.B. and Businger, J.A. 1994: *Atmosphere-Ocean Interaction*. Oxford Univ. Press. 352pp

Lighthill, J. 1978: *Waves in Fluids*. Cambridge University Press. 504pp

Open University Course Team, 1989: *Ocean Circulation*. Pergamon Press. 238pp

Pedlosky, J. 1987: *Geophysical Fluid Dynamics*. Springer-Verlag. 710 pp

BOUNDARY LAYERS & AIR POLLUTION

Arya, S.P. 1988: *Introduction to Micrometeorology*. Academic Press. 307pp

Arya, S.P. 1999: *Air Pollution Meteorology and Dispersion*. Oxford Univ. Press. 310pp

Azad, R.S. 1993: *The Atmospheric Boundary Layer for Engineers*. Kluwer Academic Publ. 565pp

Bailey, W.G., Oke, T.R., Rouse, W.R. 1997: *The Surface Climates of Canada*. McGill-Queen's Univ. Press. 369 pp

Brutsaert, W. 1982: *Evaporation into the Atmosphere*. Kluwer Academic Publ. 299pp

Frisch, U. 1995: *Turbulence*. Cambridge Univ. Press. 296pp

Garratt, J.R. 1992: *The Atmospheric Boundary Layer*. Cambridge University Press. 316pp

Hanna, S.R., Briggs, G.A. and Hosker, R.P.,Jr. 1982: *Handbook on Atmospheric Diffusion*. Technical Information Center, Dept. of Energy. DOE/TIC-11223. Available from National Technical Information Service (Document # DE82002045), Springfield, VA 22161. (ISBN 0-87079-127-3) 102pp

Haugen, D.A. (Ed.) 1973: *Workshop on Micrometeorology*. American Meteorological Society, 45 Beacon St., Boston, MA 02108. 392pp

Haugen, D.A. (Ed.) 1975: *Lectures on Air Pollution and Environmental Impact Analyses*. American Meteorological Society, 45 Beacon St., Boston, MA 02108. 296pp

Holtslag, A.A.M., and Duynkerke, P.G. (Eds.) 1998: *Clear and Cloudy Boundary Layers*. Royal Netherlands Academy of Arts and Sciences. ISBN 90-6984-235-1. 372pp

Houghton, D.D.(Ed.) 1985: *Handbook of Applied Meteorology*. John Wiley–Interscience. 1461pp

Kaimal, J.C. and Finnegan, J.J. 1994: *Atmospheric Boundary Layer Flows*. Oxford Univ. Press. 304pp

Lenschow, D.H. (Ed.) 1986: *Probing the Atmospheric Boundary Layer*. American Meteorological Society, 45 Beacon St., Boston, MA 02108. 269pp

Lyons, T.J. and Scott, W.D. 1990: *Principles of Air Pollution Meteorology*. CRC Press. 224pp

Nieuwstadt, F.T.M. and van Dop, H. (Ed.) 1982: *Atmospheric Turbulence and Air Pollution Modelling*. Kluwer Academic Publ. 358pp

Oke, T.R. 1987: *Boundary Layer Climates*.(2nd Ed.), Routledge. 435pp

Pasquill, F. 1974: *Atmospheric Diffusion* (2nd Ed.), A study of the dispersion of windborne material from industrial and other sources. Ellis Horwood Publisher. 429pp

Panofsky, H.A. and Dutton, J.A. 1984: *Atmospheric Turbulence, Models and Methods for Engineering Applications*. John Wiley. 397pp

Scorer, R.S. 1978: *Environmental Aerodynamics*. Ellis Horwood Publisher, Halsted Press, Wiley. 488pp

Seinfeld, J.H. 1986: *Atmospheric Chemistry and Physics of Air Pollution*. Wiley-Interscience. 738pp

Sorbjan, Z. 1989: *Structure of the Atmospheric Boundary Layer*. Prentice-Hall.

Stull, R.B. 1988: *An Introduction to Boundary Layer Meteorology*. Kluwer Academic Publishers. 666pp.

Turco, R.P. 1997: *Earth Under Siege, From Air Pollution to Global Change*. Oxford Univ. Press. 527pp

Venkatram, A. and Wyngaard, J.C. 1988: *Lectures on Air Pollution Modeling*. American Meteorological Society, 45 Beacon St., Boston, MA 02108. 390pp

Vinnichenko, N.K., Pinus, N.Z., Shmeter, S.M. and Shur, G.N. 1980: *Turbulence in the Free Atmosphere*, 2nd Ed. translated from Russian by Sinclair, F.L., Consultants Bureau, Plenum Publ. 310pp

SYNOPTICS

Bader, M.J., Forbes, G.S., Grant, J.R., Lilley, R.B.E., and Water, A.J. 1995: *Images in Weather Forecasting. A practical guide for interpreting satellite and radar imagery*. Cambridge Univ. Press. 499pp

Carlson, T.N. 1991: *Mid-Latitude Weather Systems*. Routledge. 507pp

Bluestein, H.B. 1992: *Synoptic-Dynamic Meteorology in Midlatitudes. Vol 1 Principles of Kinematic and Dynamics. Vol 2 Observations and Theory of Weather Systems*. Oxford University Press. 431pp & 594pp

NUMERICAL WEATHER PREDICTION & CHAOS

Baker, G.L. and Gollub, J.P. 1996: *Chaotic Dynamics, An Introduction* (2nd Ed.), Cambridge Univ. Press. 256pp

Cvitanovic, P. 1984: *Universality in Chaos,* A Reprint Selection. Adam Hilger Ltd. 513pp

European Center for Medium-Range Weather Forecasts, 1994: *User Guide to ECMWF Products,* 2nd Ed., Meteorological Bulletin M3.2. ECMWF, Shinfield Park, Reading, Berkshire RG2 9AX, England. 59pp

Haltiner, G.J. and Williams, R.T. 1980: *Numerical Prediction and Dynamic Meteorology.* John Wiley & Sons. 477pp

Krishnamurti, T.N. and Bounoua, L. 1996: *An Introduction to Numerical Weather Prediction Techniques.* CRC Press. 293pp

Lin, C.A., Laprise, R., and Ritchie, H. (Eds.) 1997: *Numerical Methods in Atmospheric and Oceanic Modelling.* The André J. Robert Memorial Volume. Canadian Meteorological and Oceanographic Society. NRC Research Press. 633pp

Lorenz, E. 1993: *The Essence of Chaos.* Univ. of Washington Press, Seattle. 227 pp

Nebeker, F. 1995: *Calculating the Weather. Meteorology in the 20th Century.* Academic Press. 255pp

Press, W.H., Teukolsky, S.A., Vetterling, W.T. and Flannery, B.P. 1992: *Numerical Recipes in FORTRAN, The Art of Scientific Computing,* 2nd Ed. Cambridge University Press. 963pp

Thompson, J.M.T. and Steward, H.B. 1986: *Nonlinear Dynamics and Chaos.* John Wiley. 376pp

HURRICANES

Anthes, R.A. 1982: *Tropical Cyclones: Their Evolution, Structure and Effect.* Meteorological Monographs, 19, American Meteorological Society, 45 Beacon St., Boston, MA 02108. 208pp

Burpee, R.W. 1986: Mesoscale structure of hurricanes. *Mesoscale Meteorology and Forecasting,* P.S. Ray, editor. American Meteorological Society, 45 Beacon St., Boston, MA 02108. 311-330.

Emanuel, K.A. 1988: Toward a general theory of hurricanes. *American Scientist,* **76**, 370-379.

Hastenrath, S. 1991: *Climate Dynamics of the Tropics.* Kluwer Academic Publ. 488pp

Hess, J.C. and Elsner, J.B. 1994: Historical developments leading to current forecast models of annual Atlantic hurricane activity. *Bulletin of the American Meteorological Society,* **75**, 1611-1621.

Houze, R.A. Jr. 1993: *Cloud Dynamics.* Academic Press. 573pp

Pielke, R.A. 1990: *The Hurricane.* Routledge. 228pp

CLIMATE CHANGE

Brown, LR., Kane, H., and Roodman, D.M. 1994: *Vital Signs 1994, Trends that are Shaping our Future.* Worldwatch Institute, W.W. Norton & Co. 160pp

Bryson, R.A. and Murray, T.J. 1977: *Climates of Hunger, Mankind and the World's Changing Weather.* University of Wisconsin Press. 171pp

Cotton, W.R. and Pielke, R.A. 1995: *Human Impacts on Weather and Climate.* Cambridge Univ. Press, 288pp

Gonick, L., and Outwater, A. 1996: *A Cartoon Guide to the Environment.* HarperPerennial. 229pp

Gurney, R.J., Foster, R.J., and Parkinson, C.L. (Eds.) 1993: *Atlas of Satellite Observations Related to Global Change.* Cambridge Univ. Press. 470pp

Harte, J. 1988: *Consider a Spherical Cow. A Course in Environmental Problem Solving.* University Science Books. 283pp

Hartmann, D.L. 1994: *Global Physical Climatology.* Academic Press. 411pp

Parkinson, C.L. 1997: *Earth From Above. Using Color-Coded Satellite Images to Examine the Global Environment.* University Science Books. 175pp

Peixoto, J.P., and Oort, A.H., 1992: *Physics of Climate.* American Inst. of Physics. 520pp

Trenberth, K.E. (editor) 1992: *Climate System Modeling.* Cambridge University Press. 788pp

OPTICS

Fraser, A.B. 1975: *Meteorological Optics* (a booklet of captions from a 1970s photographic exhibit of *Meteorological Optics* at the National Center for Atmospheric Research, P.O. Box 3000, Boulder, CO 80307) 23pp

Greenler, R. 1980: *Rainbows, Halos, and Glories.* Cambridge University Press. 195pp

Humphreys, W.J. 1964: *Physics of the Air,* 4th Ed. Dover. 676 pp.

Liou, K.N. 1992: *Radiation and Cloud Processes in the Atmosphere.* Oxford University Press. 487pp

Lynch, D.K. 1980: *Atmospheric Phenomena, Readings from Scientific American.* W.H. Freeman and Co., 175pp

Meinel, A. and Meinel, M. 1983: *Sunsets, Twilights, and Evening Skies.* Cambridge Univ. Press. 163pp

Minnaert, M.G.J. 1993: *Light and Color in the Outdoors,* translated by Len Seymour. Springer-Verlag. 417pp

Tape, W. 1994: *Atmospheric Halos.* American Geophysical Union, 2000 Florida Ave. NW, Washington, DC 20009. 143pp

ANSWERS TO SELECTED EXERCISES

CHAPTER 1

N1a. α=90°, M=5 knots
N2a. U=10 knots, V=0 knots
N3a. P=1084 mb=108.4 kPa
N4a. (i) ΔF_{window}=3.5x10^3 N, ΔF_{door}=2x10^5 N
N5a. P=102.721 kPa
N6a. m=10,340 kg
N7a. T_v=21.8°C
N8a. (i) v_e=10,400 m·s^{-1}
N9a. T=288 K
N10a. ρ=1.02 kg·m^{-3}
N11. P=114 kPa
N13a. isobars
N14a. z=10 m
N15a. T=278 K, P=84.6 kPa, ρ=1.06 kg·m^{-3}
N16a. 0 km $\le z \le$ 11.02 km
U6a. 78% (assume T=280 K)
U7. $m_{air\ in\ plane}$=919 kg
U8a. P=84 kPa
U11a. T=-41.7°C, P=0.0125 kPa, ρ=0.00018 kg·m^{-3}

CHAPTER 2

N1a. Y=88 d
N2a. t_{orbit}=1.7 h
N5a. d=74
N6a. δ_s=-3.12°
N9a. Δt=0.44 h
N11a. Sunrise=14.37 UTC (for Seattle, WA)
N12a. F_H=0.81 K·m·s^{-1}
N13a. ν=4.3x10^{14} Hz, ϖ=2.7x10^{15} s^{-1}, σ=1.4x10^6 m^{-1}
N15a. S=1418 W·m^{-2}
N16a. Ψ=45°
N17. $E_{absorbed}$=$E_{emitted}$=274 W·m^{-2}
N18. $n \cdot b$=1.4x10^{-4} m^{-1}
N19a. $K\downarrow$=-851 W·m^{-2} (for Vancouver, Canada)
N20a. $I\uparrow$=419 W·m^{-2}
U1a. t=2 April

CHAPTER 3

N1a. Γ=6.5 K·km^{-1}
N2. T=60°C
N3. T=-17°C
N5a. θ=22°C
N6a. θ_v=23°C
N7a. θ=29°C
N8a. $\Delta F_{x\ adv}/\Delta x$=-8x10^{-5} °C·s^{-1}
N9a. $|\mathfrak{I}_{z\ cond}|$ =0.05 W·m^{-2}
N10a. F_H=0.02 K·m·s^{-1}
N11a. $\Delta F_{z\ turb}/\Delta z$=-0.00024 K·s^{-1}
N12a. $\Delta F_{z\ turb}/\Delta z$=0.0021 K·s^{-1}
N13a. $\mathfrak{I}_G$=-40 W·m^{-2} (day)
N14. B=3
N15. $\mathfrak{I}_H$=-56 W·m^{-2}, $\mathfrak{I}_E$=-34 W·m^{-2}
N16a. $T_{wind\ chill}$=14°C
U1. T=-30°C

CHAPTER 4

N5a. w_e=0.4 m·s^{-1}
N6a. Δz_i=8 km
N7a. τ=0.01 N·m^{-2}
N8a. z_0=0.0002 m, C_D=0.0014
N9a. u^*=0.37 m·s^{-1}
N11a. u^*=0.22 m·s^{-1}
N13a. C_D=0.0019
N15a. M_2=1.5 m·s^{-1}
N16. z_0=0.2 m
N17a. M_2=1.7 m·s^{-1}
N18a. M=0.5 m·s^{-1}
N20a. w^*=1.9 m·s^{-1}
N28a. S=0.16 m^2·s^{-3}
U35a. z_i=0.89 km, θ_{ML}=6.8°C

CHAPTER 5

N2a. e_s=4.4 kPa, T_d=13°C, r=9.3 g·kg^{-1}, r_s=28 g·kg^{-1}, q=9.2 g·kg^{-1}, q_s=27 g·kg^{-1}, ρ_v=11 g·m^{-3}, RH=34%, T_w=16°C
N3. RH=26%
N5a. z_{LCL}=2.5 km
N6a. r=11 g·kg^{-1}, T_d=15°C, RH=39%
N7a. r_T=5 g·kg^{-1}
N8a. r_L=0 g·kg^{-1}
N9a. T_d=19°C
N10a. T_d =16°C
N11a. RH=25%
N12a. r=13 g·kg^{-1}, T=18°C
N13a. $\Delta r_T/\Delta t$=8 g·kg^{-1}·h^{-1}
N14a. $Evap$=3.37 mm·day^{-1}
N15a. $\Delta r_T/\Delta t$=-1.8 g·kg^{-1}·h^{-1}
N16a. $\Delta r_T/\Delta t$=-0.18 g·kg^{-1}·h^{-1}
N17a. Γ_s=3.4 K·km^{-1}
N18a. T=-31°C
N19a. T=11°C
N20a. θ_L=-10°C (for N7) θ_L=19°C (for N8)
N21a. θ_L=10°C

CHAPTER 6

N4. z_i=1.8 km
N6. z_{LCL}=1.4 km
N7. $P_{cloud\ base}$=85 kPa, $P_{cloud\ top}$=65 kPa, r=4 g·kg^{-1}, r_L=3.5 g·kg^{-1}
N8a. T=30°C, T_d=1°C
N9a. RH=14%
N10a. F/m=0.4 m·s^{-2}, N_{BV}=0.02 s^{-1}

N12a. N_{BV}=0.02 s^{-1}

CHAPTER 7

N1a. active; $P_{cloud\ base}$=91 kPa, $P_{cloud\ top}$=82 kPa
N2a. T_v=21°C
N4a. M=16, N=93
N5a. yes (r_X>r_s)
N7a. 1) z_{LCL}=1.2 km
2) Δq=3.8 g·kg^{-1}
3) ΔT=9.9°C
N8a. x=13 km
N9a. t_0=2 h

CHAPTER 8

N1a. 2x10^9
N2a. n=2.5x10^9 /m^3
N5a. R*=0.24µm, S*=0.35%
N6a. R=0.045 µm
N7a. n_{CCN}=3x10^7 /m^3
N8a. x=3 mm
N9a. T=-36°C
N10a. silver iodide or metaldehyde
N11a. R=47 µm (assume T=273 K)
N12a. D=2.9x10^{-5} m^2·s^{-1}
N13a. F=-0.05 kg·kg^{-1}·m·s^{-1}
N16a. column
N18a. w=-3 mm·s^{-1}
N19a. N=1.2x10^3 /m^3
N20a. w=-2.0 m·s^{-1} (assume c=1)
N21a. d_W=3.6 cm
N22a. RR=0.15 mm·h^{-1}
N23. Total rainfall=2.1 mm

CHAPTER 9

N1a. a=0.013 m·s^{-2}
N2a. V_f=1000 m·s^{-1}
N3. $F_{y\,PG}/m$=+4.9x10^{-3} m·s^{-2} (assume ρ=1.2 kg·m^{-3})
N4a. $F_{x\,TD}/m$=-2.5x10^{-3} m·s^{-2}
$F_{y\,TD}/m$=-0.013 m·s^{-2}
(assume z_i=1000 m)
N5a. f_c=1.08x10^{-4} s^{-1}
N6a. $F_{y\,CF}/m$=-9.7x10^{-4} m·s^{-2}
N8a. $F_{x\,TD}/m$=-5.5x10^{-4} m·s^{-2}
$F_{y\,TD}/m$=-1.1x10^{-4} m·s^{-2}
(assume z_i=1000 m)
N9a. $F_{x\,CN}/m$=-2.5x10^{-4} m·s^{-2}
$F_{y\,CN}/m$=-6.3x10^{-5} m·s^{-2}
N10a. G=71 m·s^{-1}
N11a. G=39 m·s^{-1}
N12a. M_r=12 m·s^{-1}
N13. U_{BL}=3.4 m·s^{-1}, V_{BL}=5.9 m·s^{-1}
N14a. U_{BL}=-1.0 m·s^{-1}, V_{BL}=2.2 m·s^{-1}
N16a. M_{CS}=10 m·s^{-1} (assume ρ=1 kg·m^{-3})
N17a. M_{CS}=3 m·s^{-1}
N18a. W=4 cm·s^{-1}
N19a. z_i=267 m (assume T=273 K above the BL)
N21a. λ_r=2.6x10^6 m

CHAPTER 10

N1a. scale=micro γ, τ=5 s
N3a. $Power$=1.8 kW
N4a. W=5.2 m·s^{-1}
N6a. U_{eq}=2.7 m·s^{-1}
N7a. ΔP=-2.5 Pa
N8a. Δz=-11 m (assume T_v=273 K)
N9a. v_d=13 m·s^{-1}, ΔP=-34 Pa (assume ρ=1.2 kg·m^{-3})
N10a. v_{Bora}=22 m·s^{-1}
N11a. λ=555 m (assume T=283 K)
N12a. Fr=0.011
N13a. α=89°, $F_{x\,WD}/m$=-6.5x10^{-5} m·s^{-2}
N16. T=19°C, T_d=-5°C, RH=19%

CHAPTER 11

N2a. E_{in}=310 W·m^{-2}, E_{out}=260 W·m^{-2}, E_{net}=50 W·m^{-2}
N3a. $E_{\phi\,in}$=12.2 GW·m^{-1},
$E_{\phi\,out}$=10.2 GW·m^{-1}
N4a. D_ϕ=2.0 GW·m^{-1}
N6a. U_{TH}=-4.8 m·s^{-1}, V_{TH}=0
N7a. $\Delta U_g/\Delta z$=1.4 (m·s^{-1})·km^{-1},
$\Delta V_g/\Delta z$=0
N8a. U_g=0, V_g=-6.9 m·s^{-1}
N11a. U'=-232 m·s^{-1}
N12a. ζ_r=-1.0x10^{-5} s^{-1} (assume no curvature)
N13a. ζ_r=-1.0x10^{-4} s^{-1} (assume no wind shear)
N14a. ζ_a=1.2x10^{-4} s^{-1} (N12)
ζ_a=2.6x10^{-5} s^{-1} (N13)
N15a. ζ_p=1.1x10^{-8} m^{-1}·s^{-1} (N12)
ζ_p=2.4x10^{-9} m^{-1}·s^{-1} (N13)
N17a. ζ_p=5 m^{-1}·s^{-1} (assume f_c=0)
N18a. ζ_r=1.0x10^{-4} s^{-1}
N20a. β=2.3x10^{-11} m^{-1}·s^{-1}
N21a. ζ_r=-5.6x10^{-5} s^{-1}
N22a. λ_R=2.5x10^3 km
N23a. λ=5.9x10^3 km
N24a. c=-8.0 m·s^{-1} (assume U_o=0)
N28a. CC=6.6x10^{-8} s^{-1}
U1a. 7x10^{12} hp

CHAPTER 12

N1a. τ=6 days
N4a. U=9 m·s^{-1} (Interior)
N5a. T=3°C, T_d=3°C (at the top of the Olympic Mtns)
N6a. λ_R=1.0x10^5 m
N8a. Frontogenesis contribution from confluence = 0.05°C·km^{-1}·d^{-1}

CHAPTER 13

N1a. clockwise
N3a. λ=4.9x10^3 km
N4a. A=290 km
N5a. ζ_p=8.4x10^{-9} m^{-1}·s^{-1}
N6a. ζ_p=1.0x10^{-8} m^{-1}·s^{-1}, R_C=-5.9x10^3 km, R_E=5.9x10^3 km
N7a. $\Delta\zeta_r$=6x10^{-7} s^{-1}
N8a. 0
N9a. $\Delta\zeta_g/\Delta t$=8.3x10^{-11} s^{-2}
N10a. ω=-0.20 Pa·s^{-1}
N11a. ω=-1.2 Pa·s^{-1}
N14a. ΔM=9.9 m·s^{-1}
N15a. W_{mid}=0.012 m·s^{-1}
N16a. $\Delta P/\Delta t$=-0.13 kPa·h^{-1}
N17a. V_{ag}=-72 m·s^{-1}
N18 a. (i) W_{mid}=0.90 m·s^{-1}
N19a. (i) $\Delta P/\Delta t$=-9.5 kPa·h^{-1} (assume ρ=0.3 kg·m^{-3})
N20a. W_{mid}=2.7 m·s^{-1}
N21a. m=3.3x10^{12} kg
N22a. $\Delta T_v/\Delta t$=0.3 K h^{-1} (assume Δz=11 km, ρ_{air}=0.7 kg·m^{-3})

N23a. $\Delta P_s/\Delta t$=-0.08 kPa·h^{-1}
N24a. $\Delta P_s/\Delta t$=0.04 kPa·h^{-1}

CHAPTER 14

N1a. θ_{32}=14.6°C
N2a. numerically stable
N3a. P_a=101.6 kPa
N5a. T_{min}=2.9°C
N7a. Mean forecast error=105 m

CHAPTER 15

N2a. T_v=8.9°C, F/m=-0.097 m·s^{-2}, w=-20 m·s^{-1}
N3a. T=-15°C
N4a. M_{gust}=1.6 m·s^{-1}, h_{gust}=185 m
N6a. $\Delta V_{lightning}$=3x10^9 V
N7a. P=100%
N8a. t=50 μs
N9a. I=8 kA (assume v_L=2.0x10^8 m·s^{-1})
N11a. Δx_{max}=25 km
N14. H=10^{-6} m·s^{-2}
N15a. Δv=-2000 s^{-1}

CHAPTER 16

N1a. AAM=2.5x10^6 m^2·s^{-1}
N2a. M_{tan}=1.3 m·s^{-1}
N3a. M_{CS}=71 m·s^{-1}
N4a. M_{tan}=68 m·s^{-1}
N5a. ΔP_T=-2.5 kPa
N6a. $\Delta T/\Delta R$=-0.32 K·km^{-1}
N7a. s=275 J·kg^{-1}·K^{-1}
N9a. T=T_d=-34°C, r=1.2 g·kg^{-1}, s=341 J·kg^{-1}·K^{-1}
N10a. ME=12 kJ·kg^{-1}
N11a. P_{eye}=87 kPa
N12a. M_{max}=75 m·s^{-1}
N13a. (i) M_{right}=85 m·s^{-1}, M_{left}=65 m·s^{-1}
N14a. P=95 kPa (assume P_∞=100 kPa), M_{tan}=2.8 m·s^{-1}, M_{rad}=0.47 m·s^{-1} (assume z_i=1 km), W=-0.17 m·s^{-1}, T=-8.2°C
N16a. Δz=1.4 m
N17a. $Vol/(\Delta t \cdot \Delta y)$=8.4 m^2·s^{-1}, $\Delta z/\Delta x$=1.3x10^{-6}
N18a. $\Delta A/\Delta t$=2.8x10^{-5} m·s^{-1}
N19a. c=44 m·s^{-1}
N20a. h=4 m, λ=35 m
N21a. (i) B5

CHAPTER 17

N1a. σ_u^2=4.4 m^2·s^{-2}, σ_v^2=2.8 m^2·s^{-2}, σ_w^2=0.8 m^2·s^{-2}
N2a. PG turbulence type A
N3a. free convection
N4a. z_{CL}=0.92 km
N6a. σ_y^2=18x10^3 m^2, σ_z^2=8.8x10^3 m^2
N10a. 1.65 km
N11a. l_m=3 m, l_b=2.4 m
N13a. $z_{CL\,eq}$=227 m
N14a. c=12 μg·m^{-3}
N16a. X=0.072
N17a. X=0.4
N20a. c(x=1 km)=0.60 mg·m^{-3}, c(x=2 km)=0.51 mg·m^{-3}, c(x=3 km)=0.19 mg·m^{-3}, c(x=4 km)=0.12 mg·m^{-3}

CHAPTER 18

N1a. T_e=-1.6°C
N2a. T_e=168°C
N3a. T_e=-21°C
N4a. (i) T_s=50°C (ii) T_s=26°C
N5a. T_s=1.0°C

CHAPTER 19

N1a. θ_3=10°
N2a. c_{air}=2.999x10^5 km·s^{-1}
N3a. θ_2=31.7°
N4a. (i) θ_2=7.5° (ii) θ_2=7.6°
N5a. θ_2=10.2°
N6a. θ_1=30°
N7a. α_2=22.4°, β_2=0°
N8a. θ_c=49.3°
N9a. $\Delta\theta_1$=14°
N10a. r=0.02006
N11. x/R=0.807 and 0.908
N14a. θ_2=37.4°
N17a. θ_2=53.5°
N18a. θ_V=49.8°, θ_H=9.2°
N20a. θ_V=22.3°, θ_H=34.6°
N23a. I_{scat}/I_o=0.004% (assume ρ=1.225 kg·m^{-3})
N25a. θ=3.8°
N27a. R_C=105 m (assume a standard atmosphere with T=288 K; assume red light)

APPENDIX A

N1a. 80.5 km
N2a. 26 m^2·s^{-1}
N3a. $\Delta T/\Delta z$=0°C (between top and bottom)
N4a. T=288 K
N6a. P=102.5 kPa (assume T=280 K)

APPENDIX H

N2a. θ_1=13.5°C, θ_2=17.7°C, θ_3=19.9°C, θ_4=20.9°C
N3a. θ_1=13.9°C, θ_2=17.5°C, θ_3=19.8°C, θ_4=20.8°C
N4. $\Delta\theta/\Delta z$=0.005°C·m^{-1} (for z=500 to 700 m), $\Delta\theta/\Delta z$=0.010°C·m^{-1} (for z=300 to 500 m), $\Delta\theta/\Delta z$=0.025°C·m^{-1} (for z=100 to 300 m),
N5a. K(z=600 m)=0 m^2·s^{-1}, K(z=400 m)=256 m^2·s^{-1}, K(z=200 m)=96 m^2·s^{-1},
N6a. θ_1=11.4°C, θ_2=11.8°C, θ_3=12.4°C, θ_4=14.4°C

Answers calculated by Maria Furberg.

SYLLABUS

CONTENTS

If this book is used in a course with only science and engineering students, then a technically-oriented syllabus should be easy for the instructor to design. When selecting material to fit into one term, instructors should try to include the key equations that are marked with a bullet (•).

If the course is taught to a mix of students of wide range of math abilities and interests, such as a first-year survey course for non-science majors, then incorporation of the technical aspects requires care. One such syllabus is suggested below.

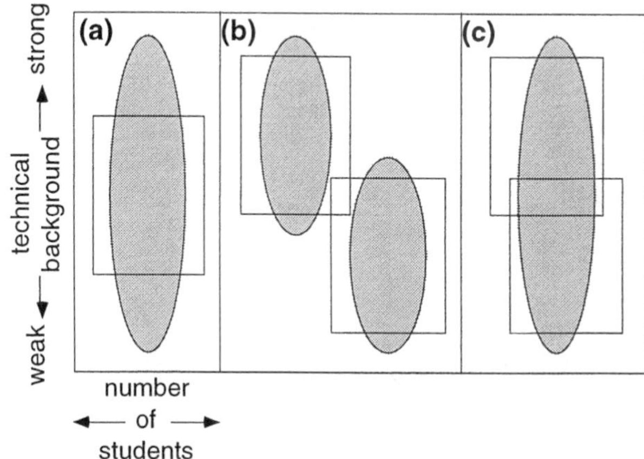

Figure F.1
The range of student skills in a class is shown by the height of the shaded oval, while the relative number of students at any skill level is indicated by the oval width. The subset of students served by the lectures is indicated by the rectangle. (a) Single course, aimed at an "average" student. (b) Two courses aimed at different student backgrounds. (c) Single course tailored to two distinct sub-groups (which is the approach described in detail here).

THE STUDENT MIX

One way to teach students with diverse backgrounds is to teach one course, and aim it at the middle or "average" student (Fig F.1a). A disadvantage is that the range of student backgrounds might be so great as to leave a large portion of the class unhappy (either bored or lost) because they are too far from the "average".

Perhaps the best way is to split the students into separate courses, each of relatively homogeneous background (Fig F.1b). A simple scenario might be to have two courses, one called "Weather for Poets 101", and another called "Atmospheric Sciences 201". However, multiple survey courses might not be practical at many universities.

A third approach is to have one course, but aim the course to two "archetypical" student levels (Fig F.1c). With the right mix of textbooks, lectures, and tests, this approach is feasible. The following proposals utilize both this book and Ahrens' *Meteorology Today* or equivalent to achieve this third approach.

TESTS & GRADING

Test design is often critical in survey courses because some of the students study for the test rather than study for the knowledge. The following suggestion is fair for the non-science majors, stimulating for the science majors, and encourages all the students to study both qualitative and quantitative aspects of meteorology.

To get 100 points on the test, students need answer <u>any</u> 2/3 of the questions (plus a few more to compensate for guessing). Each test has two parts. One part is descriptive, and the other is quantitative. The descriptive part is 2/3 of the whole test.

There is quite a lot of flexibility within this framework. For example, students with no technical skills can still achieve a perfect score if they successfully answer all descriptive questions, and guess the quantitative questions. Engineering and science majors might choose to answer all of the quantitative questions, and a bit more than half of the descriptive ones to get a perfect score. Students in the middle might choose a mix of descriptive and quantitative questions that feels right for them.

More importantly, students can answer more than 2/3 questions if they wish, because all questions are graded and the total number of correct questions is used to compute the score. Truncate the score at 100 if greater than 100 points.

Thus, students can have "test insurance" by answering most of the questions. This latter feature aids both the students and the instructor, because students quickly discover that they can maximize their score by studying both descriptive and quantitative aspects. It is the "carrot" approach rather than the "stick" approach.

The following illustrates such test formats. Describe the test characteristics by a set of numbers (D, Q, P, C), where:

- D = number of descriptive questions
- Q = number of quantitative questions
- P = points per question
- C = number of choices (if multiple choice).

In this framework, first pick the number of descriptive questions. Then, use

$$Q \approx D/2 \tag{F.1}$$

and

$$P \approx \frac{100/D}{1 + 1/(2 \cdot C)} \tag{F.2}$$

where the term in parentheses compensates for random guessing.

For example, write a test with $D = 23$ descriptive questions, $Q = 11$ quantitative questions, with $C = 5$ choices per answer. This works out to roughly $P \cong 4$ points per question, and students must answer any 25 or more correctly to get 100 points. Non-math students can achieve this by answering correctly all of the descriptive questions, and by correctly answering or guessing any two of the quantitative questions.

TEXTBOOKS & HOMEWORK

Ahrens' *Meteorology Today* (6th Ed.) would be the required text, and Stull's *Meteorology for Scientists and Engineers* would be optional. Based on the test structure defined above, most students would likely want to use the optional text to help insure higher grades.

Reading assignments are easy, because the chapters of Ahrens' and Stull's books are designed to work together (see Appendix G listing the chapter correspondences). Assign the "chapter of the week." All students are required to read Ahrens' chapter, and can optionally read the corresponding chapter in Stull's book. A short quiz or graded homework on Ahrens each week helps ensure that all students maintain regular study habits.

LECTURES & WEEKLY SCHEDULE

The lectures on different days each week are devoted to the different student sub-groups. The following example is for a course that meets 3 days per week, for 15 weeks during the semester.

Before Wednesday, for example, students are assigned to read the "chapter of the week" in Ahrens. Starting the cycle on Wednesdays is a short quiz on key terms from Ahrens to ensure that all students have done the reading. Class discussion is seeded on issues relating meteorology to society and the environment. All students are required to attend Wednesday lectures and to take the quizzes.

On Friday the lecture is descriptive. Slides of weather phenomena are presented, as are demonstrations, and discussion extending or clarifying aspects of Ahrens. On Monday the lecture is quantitative, where discussions center on numerical applications, and a selection of exercises are solved from *Meteorology for Scientists and Engineers*. Based on the test structure previously described, most students quickly learn that they can maximize their test scores by attending all lectures.

CORRESPONDENCES

CONTENTS

Below are cross-reference tables, showing how chapters of this book correspond to chapters of the descriptive companion textbooks by Ahrens (1998, 2000). Cross-reference tables are also given for the calculus-based introductory textbooks of Holton (1992), and Wallace and Hobbs (1977).

AHRENS' *Essentials**

Table G-1. Correspondence with *Essentials*.

Ahrens' Chapters	Stull's Chapters
1. The Earth's Atmosphere	1. The Atmosphere
2. Warming Earth & Atm.	2. Radiation
3. Air Temperature	3. Heat 4. Boundary Layer
4. Humidity, Condensation & Clouds	5. Moisture 7. Clouds
5. Cloud Development & Precipitation	6. Stability 8. Precipitation
6. Air Pressure & Winds	9. Dynamics
7. Atmospheric Circulations	10. Local Winds 11. Global Circulat.
8. Air Masses, Fronts, & Mid-Lat. Storms	12. Air Mass, Fronts 13. Cyclones
9. Weather Forecasting	14. Num.WX Pred.
10. Tstorms & Tornadoes	15. Thunderstorms
11. Hurricanes	16. Hurricanes
12. Air Pollution	17. Air Pollut. Disp.
13. Climate Change	18. Climate Change
14. Global Climate	(none)
15. Light, Color & Optics	19. Optics

*Ahrens, C.D. 1998: *Essentials of Meteorology. An Invitation to the Atmosphere.* (2nd Ed.) (Wadsworth) Brooks/Cole Thomson Learning. 443 pp
Stull, R.B. 2000: *Meteorology for Scientists and Engineers.* Brooks/Cole Thomson Learning. 495 pp

AHRENS' *Meteorology Today***

Table G-2. Correspondence with *Meteor. Today*

Ahrens' Chapters	Stull's Chapters
1. Earth & Atmosphere	1. The Atmosphere
2. Energy: Warming	2. Radiation
3. Seasonal and Daily Temperatures	3. Heat 4. Boundary Layer
4. Light, Color, Optics	19. Optics
5. Atmos. Moisture	5. Moisture
6. Condensation, Cloud	7. Clouds
7. Stability, Cloud Dev.	6. Stability
8. Precipitation	8. Precipitation
9. Atm. in Motion	9. Dynamics
10. Wind: Small-scale	10. Local Winds
11. Wind: Global Sys.	11. Global Circulation
12. Air Mass.& Fronts	12. Air Mass. & Fronts
13. Mid-Lat. Cyclones	13. Cyclones
14. Weather Forecasting	14. Num. WX Predict.
15. Tstorms & Tornado	15. Thunderstorms
16. Hurricanes	16. Hurricanes
17. Air Pollution	17. Air Pollution Disp.
18. Climate Change	18. Climate Change
19. Global Climate	(none)

** Ahrens, C.D. 2000: *Meteorology Today. An Introduction to Weather, Climate, and the Environment.* (6th Ed.) Brooks/Cole Thomson Learning. 602pp

HOLTON'S *Dynamic Meteorology****

Table G-3. Correspondence with *Dynamic Met.*

Holton's Chapters	Stull's Chapters
1. Introduction	1. The Atmosphere 9. Dynamics
2. Basic Conserv. Laws	3. Heat 5. Moisture 6. Stability 9. Dynamics
3. Elementary Applic.	9. Dynamics
4. Circulat. & Vorticity	11. General Circulation
5. Planetary Bound.Lay.	4. Boundary Layer
6. Synoptic Motion I: Quasi-geostrophic	13. Cyclones
7. Oscillations: Linear	10. Local Winds
8. Synoptic Motion II: Baroclinic Instability	11. General Circulation
9. Mesoscale Circulations	10. Local Winds 12. Air Mass.& Fronts 15. Thunderstorms 16. Hurricanes
10. General Circulation	11. General Circulation
11. Tropical Dynamics	(none)
12. Middle Atmos. Dyn.	(none)
13. Numerical Predict.	14. Num.WX Predict.
(none)	7. Clouds 8. Precipitation 17. Air Pollution Disp. 18. Climate Change 19. Optics

***Holton, J.R., 1992: *An Introduction to Dynamic Meteorology* (3rd Ed.). Academic Press. 511pp

WALLACE & HOBBS' *Atmospheric Science* ‡

Table G-4. Correspondence with *Atmos. Sci.*

Wallace & Hobbs' Chapters	Stull's Chapters
1. Survey of Atmos.	1. The Atmosphere
2. Thermodynamics	1. The Atmosphere 3. Heat 5. Moisture 6. Stability
3. Cyclones	12. Air Masses & Fronts 13. Cyclones
4. Microphysics	7. Clouds 8. Precipitation
5. Clouds & Storms	15. Thunderstorms 19. Optics
6. Radiative Transfer	2. Radiation
7. Global Energy Bal.	2. Radiation 4. Boundary Layer 18. Climate Change
8. Dynamics	9. Dynamics
9. General Circ.	11. General Circulation 16. Hurricanes
(none)	10. Local Winds 14. Num.WX Prediction 17. Air Pollution Disp.

‡ Wallace, J.M. and Hobbs, P.V., 1977: *Atmospheric Science, An Introductory Survey*. Academic Press. 467pp

TURBULENCE CLOSURE

CONTENTS

H The governing equations for the atmosphere (i.e., the first law of thermodynamics, ideal gas law, equations of motion, water conservation, and the continuity equation) have more unknown variables than equations, primarily due to turbulence. Hence, this set of equations is mathematically not closed, which means they cannot be solved. To be a closed system of equations, the number of unknowns must equal the number of equations.

The reason for this **closure problem** is that it is impossible to accurately forecast each swirl and eddy in the wind. To work around this problem, meteorologists **parameterize** (i.e., approximate) the net effect of all the eddies. Such an approximation is called **turbulence closure**, because it mathematically closes the governing equations so that useful weather forecasts and engineering designs can be made.

TURBULENCE CLOSURE TYPES

For common weather situations, turbulent transport in any horizontal direction nearly cancels transport in the opposite direction, but vertical transport is significant. Medium and large size turbulent eddies can transport air parcels from many different source heights to any destination height within the turbulent domain, where the smaller eddies mix the parcels together.

Different approximations of turbulent transport consider the role of small and large eddies differently. **Local closures**, which neglect the large eddies, are most common. This gives turbulent heat fluxes that flow down the local gradient of potential temperature, analogous to molecular diffusion or conduction (see Chapts. 3 & 17). One such turbulence closure is called **K-theory**.

A **nonlocal closure** alternative that accounts for the superposition of both large and small eddies is called **transilient turbulence theory** (T3). While this is more accurate, it is also more complicated. There are many other closures that have been proposed. Only K-theory and T3 are reviewed here.

ON DOING SCIENCE • Parameterization Rules

For many situations in science and engineering, the true governing equations are either not known or are so complicated as to be unwieldy. Nonetheless, an approximation might be good enough to give useful answers.

A **parameterization** is an approximation to an unknown term by one or more known terms or factors. Because these other terms do not come from first principles, one or more fudge factors are often included in the substitute term to make it have the correct behavior or order-of-magnitude. These fudge factors are called **parameters**, and are not known from first principles. They must be found **empirically** (from field or laboratory experiments).

Different scientists and engineers can invent different parameterizations for the same unknown term. While none are perfect, each would have different advantages and disadvantages. However, every parameterization must follow certain rules in order to be acceptable.

The parameterization should:
(1) be physically reasonable,
(2) have the same dimensions as the unknown term,
(3) have the same scalar or vector properties,
(4) have the same symmetries,
(5) be invariant under an arbitrary transformation of coordinate system,
(6) be invariant under an inertial or Newtonian transformation (e.g., a coordinate system moving at constant speed and direction),
(7) satisfy the same constraints and budget equations.

Even if the parameterization satisfies the above rules, it often will be successful for only a limited range of conditions. An example from this book is the transport velocity, w_T. One parameterization was developed for statically neutral boundary layers, while a different one worked better for statically unstable mixed layers. Thus, every parameterization should state the limitations on its use, including accuracy and range of validity.

In summary, parameterizations only approximate nature. As a result, they will never work perfectly. But they can be designed to work satisfactorily.

K-THEORY

One approximation to turbulent transport considers only small eddies. This approach, called **K-theory, gradient transport theory**, or **eddy-diffusion theory**, models turbulent mixing analogous to molecular diffusion Using heat flux F_H for example:

$$F_H = \overline{w'\theta'} = -K \cdot \frac{\Delta\theta}{\Delta z} \qquad \bullet\text{(H.1)}$$

This parameterization says that heat flows down the gradient of potential temperature, from warm to cold. The rate of this turbulent transfer is proportional to the parameter K, called the **eddy viscosity** or **eddy diffusivity**, with units $\text{m}^2{\cdot}\text{s}^{-1}$.

K is expected to be larger for more intense turbulence. In the surface layer, turbulence is generated by wind shear. Prandtl made a **mixing-length** suggestion that:

$$K = k^2 \cdot z^2 \cdot \left|\frac{\Delta M}{\Delta z}\right| \qquad \text{(H.2)}$$

where $k = 0.4$ is von Kármán's constant (dimensionless), z is height above ground, and $\Delta M/\Delta z$ is wind shear.

When used in the Eulerian heat budget equation, neglecting all other terms except turbulence, the result gives the heating rate of air at height z due to turbulent **flux divergence** (i.e., change of flux with height):

$$\frac{\Delta\theta(z)}{\Delta t} = K \cdot \frac{\theta(z+\Delta z) - 2\theta(z) + \theta(z-\Delta z)}{(\Delta z)^2} \qquad \text{(H.3)}$$

Although the example above was for heat flux, it can also be used for moisture or momentum flux by substituting r or M in place of θ.

K-theory works best for windy surface layers, where turbulent eddy sizes are relatively small. Fig H.1 shows that heat flux flows "down" the temperature gradient from warm to colder potential temperature, which gives a negative (downward) heat flux in the statically stable surface layer. K-theory does not apply at the ground, but only within the air where turbulence exists. For heat fluxes at the surface, use approximations given in Chapts. 3 and 4.

K-theory has difficulty for convective ABLs and should not be used there. Figs H.2a & b illustrate these difficulties, giving typical values in the atmosphere, and the resulting K values backed out using eq. (H.2). Negative and infinite K values are unphysical.

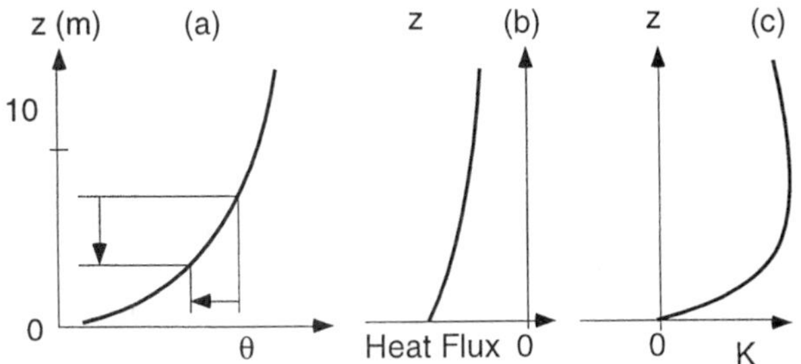

Figure H.1
Typical profiles of (a) potential temperature; (b) heat flux; and (c) eddy diffusivity in the statically stable surface layer.

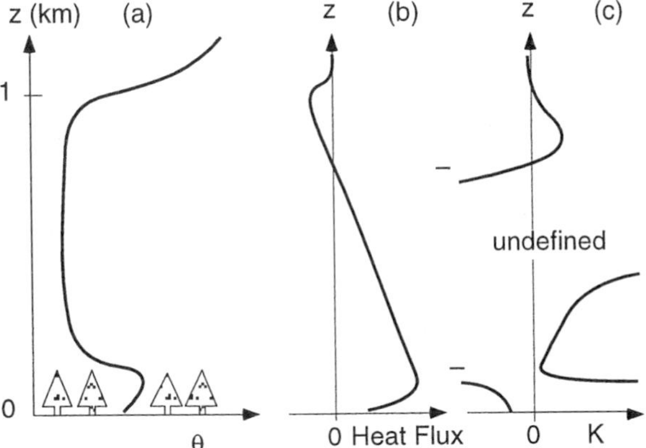

Figure H.2
The value (c) of K needed to get (b) the observed heat flux F_H from (a) the observed potential temperature θ over a forest for an unstable (convective) ABL.

Solved Example

Instruments on a tower measure $\theta = 15°C$ and $M = 5$ m/s at z = 4 m, and $\theta = 16°C$ and $M = 8$ m/s at z = 10 m. What is the vertical heat flux?

Solution:

Given:

z (m)	θ (°C)	M (m/s)
10	16	8
4	15	5

Find: $F_H = ?$ K·m/s

First, use eq. (H.2), at average z = (10+4)/2 = 7 m

$$K = k^2 \cdot z^2 \cdot \left|\frac{\Delta M}{\Delta z}\right| = (0.4 \cdot 7\text{m})^2 \left|\frac{(8-5)\text{m/s}}{(10-4)\text{m}}\right|$$

$= 3.92 \text{ m}^2/\text{s}$

Use eq. (H.1):

$$F_H = -K \cdot \frac{\Delta\theta}{\Delta z} = -(3.92 m^2 s^{-1}) \cdot \frac{(16-15)°\text{C}}{(10-4)\text{m}}$$

$=$ **-0.65 K·m/s**

Check: Units OK. Physics OK.

Discussion: The negative sign means a downward heat flux, from hot to cold. This is typical for statically stable ABL.

TRANSILIENT TURBULENCE THEORY

Instead of looking at local down-gradient transport, as was done in K-theory, one can look at all ranges of distances across which air parcels move during turbulence (Fig H.3). This is an approach called **nonlocal closure**.

Nonlocal Closure

Recalling the nonlocal explanation of static stability from Chapt. 6, we see that air parcels will rise from each relative maximum in the profile, and will sink from every relative minimum (e.g., Fig 6.10). This profile is statically unstable from the surface to the top of the upward-pointing air-parcel arrow.

For heat flux at the altitude of the dashed line in Fig H.3, K-theory would utilize the local gradient of θ at that altitude, and conclude that the heat flux should be downward and of a certain magnitude. However, if the larger-size eddies are also included, such as the parcel rising from tree-top level, we see that it is bringing warm air upward (a positive contribution to heat flux). This could partially counteract, or even overwhelm, the negative contribution to flux caused by the local small eddies.

As you can probably anticipate, a better approach would be to consider all size eddies and nonlocal air-parcel movement. One such approach is called **transilient turbulence theory (T3)**, and is described next.

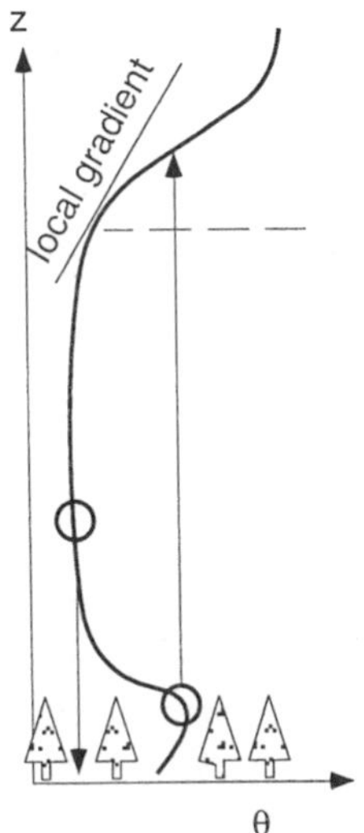

Figure H.3
Typical potential temperature θ profile in the ABL over a forest, showing nonlocal convective movement of air parcels.

Concept

Fig H.4 illustrates turbulent transport during a short time interval, Δt, where all eddy sizes can be included. Picture four layers of air labeled 1 through 4, as sketched in Fig H.4a. Each layer has some average initial potential temperature, as illustrated. Each layer can be conceptually broken into smaller air parcels (Fig H.4b). This figure illustrates the starting (**source**) locations for these parcels.

Large eddies shuffle some of these parcels, as indicated with the arrows in Fig H.4b. For example, one eddy acting on the left column of air parcels moves a parcel from layer 1 to 3, one from 3 to 2, and one from 2 to 1. In this column, none of the original air parcels have been lost, and after the eddy is finished, all heights in that column are again holding an air parcel. Other shuffles are shown in the other columns.

By the end of the time interval, the parcels are rearranged as illustrated in Fig H.4c, in various **destination** layers. The final average potential temperature in each layer is easily computed, based on the parcels within it. For example, the five air parcels in layer 1 give a layer-average potential temperature of 12°C. The final mean state of the air at the end of the time interval of turbulent mixing is shown in Fig H.4d.

In general, turbulence tends to homogenize the air. Potential temperature, wind, and humidity differences in different layers gradually become mixed toward a more uniform state by the action of turbulence. In Fig H.4, the cool layers have warmed and a warm layer has cooled toward a common potential temperature. The amount of mixing between different layers varies with time and location, as turbulence intensity changes.

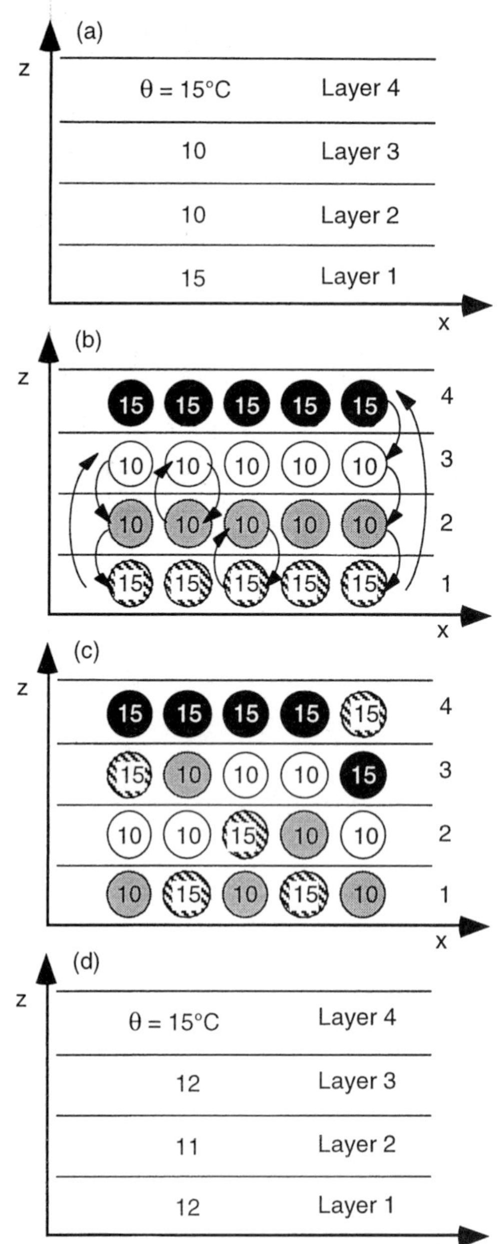

Figure H.4
Illustration of turbulent mixing.

Mathematical Framework

The amount of air moving between various sources and destinations is described by a matrix of numbers called a **transilient matrix [C]**. In that matrix, each **transilient coefficient** c_{ij} describes the fraction of air arriving at destination layer i from source layer j during time interval Δt.

For example, in Fig H.4 each air parcel represents 20% of the whole layer, because there are 5 parcels in each layer. Count the number of parcels moving from source j to destination i, multiply by 0.2, and insert the result in the proper row and column of the transilient matrix. Do this for each combination of source and destination, including cases where source and destination are in the same layer. The result is:

To Destination Layer i	From Source Layer j: **4**	**3**	**2**	**1**
4	0.8	0	0	0.2
3	0.2	0.4	0.2	0.2
2	0	0.6	0.2	0.2
1	0	0	0.6	0.4

For example, $c_{12} = 0.6$, and $c_{44} = 0.8$ here.

BEYOND ALGEBRA • Matrices

A **matrix** or **array** of numbers or variables is often used to describe a single phenomenon that has different values in different physical places, or in different directions. The array is usually written inside square brackets [...]. These numbers or variables inside the brackets are called **matrix elements.**

One simple matrix is for wind velocity:

$$\vec{V} = [U, V, W]$$

where the "V" with arrow over top represents a vector. The single physical phenomenon is motion, and the components of the matrix represent the motion in each of the three Cartesian directions.

The matrix example above is called a **row matrix**, because the elements are written in a row. **Column matrices** also exist. The example at right is a column matrix of temperatures at different heights in the atmosphere. The whole matrix represents a sounding.

$$\vec{T} = \begin{bmatrix} 25°C \\ 19°C \\ 20°C \\ 22°C \end{bmatrix}$$

Matrices can also exist with both rows and columns. The example below represents the fraction of air that mixes from any source height to any destination height in the atmosphere.

$$\vec{C} = \begin{bmatrix} 0.4 & 0.5 & 0.1 \\ 0.3 & 0.2 & 0.5 \\ 0.3 & 0.3 & 0.4 \end{bmatrix}$$

Each column represents a source height, and each row represents a destination height. Because air can move from any source to any destination, the matrix is a **square matrix** rather than a single column or row.

Matrices can be used in operations, such as addition, subtraction, or multiplication. Such operations on matrices are generically called **linear algebra**.

For addition, the matrices must be the same size and orientation. The sum is element by element. Namely, add the first element of the first matrix to the first element of the second matrix. Do the same for the second element of both matrices, and so forth. An example is: $[1, 2, -3] + [5, 7, 1] = [6, 9, -2]$ Subtraction works the same way, element by element. Sometimes, spaces are used instead of commas to separate the elements in any matrix.

Often, elements of a matrix are represented by subscripts. For the addition example, suppose the first matrix was $\mathbf{A} = [a_1 \ a_2 \ a_3]$, and the second was $\mathbf{B} = [b_1 \ b_2 \ b_3]$. Then the sum of the two matrices $\mathbf{A} + \mathbf{B} = \mathbf{C}$ can be defined by operations on the elements: $a_i + b_i = c_i$, where subscript i is the index of the matrix element. That is $i = 1, 2, 3$ to represent the first, second, or third element of the matrix. *(continued next column)*

BEYOND ALGEBRA • Matrices (part 2)

(continuation)

When multiplying two matrices, the number of columns of the first matrix must equal the number of rows of the second matrix. Therefore, the order that the matrices appear is significant (in other words, matrices are not commutative, because you can't exchange the order and expect to get the same answer). Matrix multiplication is illustrated here:

$$\begin{bmatrix} a & b \\ c & d \\ e & f \end{bmatrix} \begin{bmatrix} p & q & r \\ s & t & u \end{bmatrix} = \begin{bmatrix} (ap+bs) & (aq+bt) & (ar+bu) \\ (cp+ds) & (cq+dt) & (cr+du) \\ (ep+fs) & (eq+ft) & (er+fu) \end{bmatrix}$$

For this operation, elements in the first row of the first matrix are multiplied by respective elements in the first column of the second matrix, to give a result that appears in row 1, column 1 of the product matrix. Similarly, the first row of the first matrix times the second column of the second matrix goes into the element in row 1, column 2, of the product. The procedure appears for every combination of row in the first matrix and column in the second matrix.

Matrix multiplication can also be abbreviated using subscripts. Let the first matrix be $\mathbf{D} = [d_{ij}]$, where the first subscript (i in this case) represents the row, and the second (j in this example) the column. For example:

$$\mathbf{D} = \left[d_{ij}\right] = \begin{bmatrix} d_{11} & d_{12} \\ d_{21} & d_{22} \\ d_{31} & d_{32} \end{bmatrix}$$

The second matrix is $\mathbf{E} = [e_{jk}]$ where again the first subscript (j in this case) is the row, and the second (k in this case) is the column. Matrix multiplication can now be abbreviated as: **D E** = **F**, or:

$$\sum_{j=1}^{n} d_{ij} e_{jk} = f_{ik}$$

where n is the total number of j columns in the first matrix.

In our example, the size of matrix **D** was 3x2 (rows by columns), and the size of **E** was 2x3 (again, rows by columns), creating a result **F** that was size 3x3. Although such matrix operations are tedious, they are easy to do on a computer. Some spreadsheet programs also do matrix algebra.

Finally, it is also possible to find the log, exp, and power of matrices. It is also possible to find the inverse of matrices, which is utilized instead of doing division (dividing is the same as multiplying by the inverse).

You can learn more about matrix operations by taking a linear algebra course.

Transilient matrices have interesting characteristics. Each row sums to 1.0, as does each column, for the case where each layer contains the same mass of air. Each element in the matrix must be $0 \le c_{ij} \le 1.0$. This matrix (following the meteorological convention of height increasing upward) is flipped upside-down and left-right from the usual mathematical convention, to correspond to atmospheric layer indices starting at one at the bottom (Fig H.4).

The final potential temperature of the air in any layer i after any single time interval of turbulent mixing is found from two quantities: the initial potential temperature of the air at all layers j, and the transilient turbulence matrix c_{ij}:

$$\overline{\theta}_i = \sum_{j=1}^{NL} \left[c_{ij}\, \overline{\theta}_j \right] \qquad \bullet\text{(H.4)}$$

where NL is the number of layers. This equation gives only the turbulence contribution to temperature change. Other processes such as conduction, radiation, and advection can also change the temperature.

Nonlocal Flux

The **turbulent heat flux** across any level is the sum of the contributions from each air parcel that moves across that level. Each air parcel that crosses the line separating two layers of air contributes an amount of flux equal to its relative amount of mass (0.2 in our previous example) times its potential temperature, using a positive sign for upward-moving parcels and a negative sign for downward-movers.

For example, crossing the line between layers 1 and 2 in Fig H.4b are three air parcels moving upward a distance of Δz during time Δt, each carrying 20% of the layer having initial potential temperature 15°C. There are also three parcels of mass 0.2 moving downward a distance of Δz during Δt, with air of 10°C. Thus, the total transport across the level that separates layers 1 and 2 is:

$$F_{1,2} = [\,3 \times 0.2 \times (15°\text{C}) - 3 \times 0.2 \times (10°\text{C})\,]\cdot\Delta z/\Delta t$$

$$= +3\ °\text{C}\cdot\Delta z/\Delta t$$

Some of the air parcels from layer 1 also cross the next level, and must be counted again to find the flux between layers 2 and 3.

$$F_{2,3} = [2 \times 0.2 \times (15°\text{C}) + 1 \times 0.2 \times (10°\text{C}) - 3 \times 0.2 \times (10°\text{C})\,]\cdot\Delta z/\Delta t$$

$$= +2\ °\text{C}\cdot\Delta z/\Delta t$$

Between layers 3 and 4:

$$F_{3,4} = [1 \times 0.2 \times (15°\text{C}) - 1 \times 0.2 \times (15°\text{C})\,]\cdot\Delta z/\Delta t$$

$$= 0\ °\text{C}\cdot\Delta z/\Delta t$$

Solved Example

Given the following transilient matrix and initial potential temperature profile, find the final potential temperature profile after two time steps of Δt = 10 minutes each.

$$c_{ij}(10\,\text{min}) = \begin{bmatrix} 0.4 & 0.2 & 0.4 \\ 0.4 & 0.4 & 0.2 \\ 0.2 & 0.4 & 0.4 \end{bmatrix} \quad \text{and}$$

z(m)	θ(°C)
500	10
300	5
100	5

Solution:

Eq. (H.4) describes an ordinary matrix multiplication of $[\theta_{new}] = [C]\cdot[\theta_{old}]$. Although the matrix approach is shown here, the same answer will be obtained with eq. (H.4).

For the first time step:

$$\begin{bmatrix} \theta_{new}(500m) \\ \theta_{new}(300m) \\ \theta_{new}(100m) \end{bmatrix} = \begin{bmatrix} 0.4 & 0.2 & 0.4 \\ 0.4 & 0.4 & 0.2 \\ 0.2 & 0.4 & 0.4 \end{bmatrix} \begin{bmatrix} 10°\text{C} \\ 5°\text{C} \\ 5°\text{C} \end{bmatrix}$$

which gives the result:

$$\begin{bmatrix} \theta_{new}(500m) \\ \theta_{new}(300m) \\ \theta_{new}(100m) \end{bmatrix} = \begin{bmatrix} 7°\text{C} \\ 7°\text{C} \\ 6°\text{C} \end{bmatrix}$$

For the second time step:

$$\begin{bmatrix} 6.6°\text{C} \\ 6.8°\text{C} \\ 6.6°\text{C} \end{bmatrix} = \begin{bmatrix} 0.4 & 0.2 & 0.4 \\ 0.4 & 0.4 & 0.2 \\ 0.2 & 0.4 & 0.4 \end{bmatrix} \begin{bmatrix} 7°\text{C} \\ 7°\text{C} \\ 6°\text{C} \end{bmatrix}$$

The final profile after two steps is thus, θ = **6.6°C** at 500 m, θ = **6.8°C** at 300 m, and θ = **6.6°C** at 100 m.

Check: Units OK. Physics Reasonable.

Discussion: As expected, turbulence is making the potential temperature profile more uniform with height. Also, heat is conserved; namely, the sum of the initial temperatures (i.e., proportional to the total heat) equals the sum of the intermediate temperatures, equals the sum of the final temperatures, equals 20.

Are these fluxes consistent with the temperature changes that occurred? Zero turbulent flux enters the bottom of layer 1 by definition (assuming the ground doesn't dance a turbulent dance), but an upward (positive) flux of 3°C·$\Delta z/\Delta t$ leaves from the top, which should cause the temperature of layer 1 to decrease by 3°C. That answer agrees with the temperature drop from 15°C in Fig H.4a to 12°C in Fig H.4d.

For layer 2, an upward flux of 3 °C·$\Delta z/\Delta t$ enters the bottom while an upward flux of 2°C·$\Delta z/\Delta t$ leaves from the top, which should cause a net warming of 1 °C. Yes, the temperature in layer 2 rises from 10 to 11 °C in Fig H.4. Similarly a 2°C·$\Delta z/\Delta t$ flux enters the bottom of layer 3 while nothing leaves through the top, causing a net warming of 2°C. Finally, layer 4 has zero flux across both boundaries, and its temperature does not change.

The corresponding turbulent vertical heat-flux divergence caused by turbulence at any destination index i (corresponding to height $z = i \cdot \Delta z$) is:

$$-\left.\frac{\Delta F}{\Delta z}\right|_{z=i\cdot\Delta z} = \frac{1}{\Delta t}\cdot\sum_{j=1}^{NL}\left[\left(c_{ij}-\delta_{ij}\right)\cdot\overline{\theta}_j\right] \qquad \text{(H.5)}$$

where δ_{ij} is called the **Kronecker delta** ($\delta_{ij} = 1$ whenever $i = j$, and is zero otherwise), and NL is the total number of layers.

As turbulence continues, the process repeats itself, but with a new initial state of the air. Also the transilient matrix can change a bit each time step as the nature of turbulence changes. Thus, to make a temperature forecast, one must also forecast the corresponding transilient matrix.

The amount of turbulence, as measured by the transilient matrix or by other measures, varies depending on the wind and temperature profiles in the atmosphere. Hence, there is not a single constant transilient matrix that can be used everywhere at every time.

While turbulence is difficult to describe, we nonetheless need to know the effects of turbulence on the mean wind, temperature, and humidity. Instead of solving the transilient equations or using K-theory, atmospheric scientists have found that they can produce acceptable practical results by parameterizing the mean profiles directly. These are usually calibrated against empirical observations, and the result is called **similarity theory**. The surface layer and radix layer wind profiles of Chapt. 4 are examples of two similarity theories.

Finally, T3 can be used to calculate fluxes of moisture or momentum by substituting r or M, respectively, in place of θ.

Solved Example

Using the transilient matrix from the previous example, find the flux gradient across the each layer during the first time step.

Solution:

For the flux gradient (FG= $\Delta F/\Delta z$) part of the problem, we must solve eq. (H.5). It too can be interpreted as a matrix operation:

$$-[\mathbf{FG}] = (1/\Delta t)\cdot[\,\mathbf{C} - \mathbf{I}\,][\,\mathbf{T}\,]$$

where **I** is the **identity matrix** (defined by the Kronecker delta), and T is the temperature matrix.

$$-[\mathrm{FG}] = \frac{1}{600\ \mathrm{s}}\cdot\begin{bmatrix}0.4-1 & 0.2 & 0.4\\ 0.4 & 0.4-1 & 0.2\\ 0.2 & 0.4 & 0.4-1\end{bmatrix}\begin{bmatrix}10°\mathrm{C}\\ 5°\mathrm{C}\\ 5°\mathrm{C}\end{bmatrix}$$

$$-[\mathrm{FG}] = \frac{1}{600\ \mathrm{s}}\cdot\begin{bmatrix}-0.6 & 0.2 & 0.4\\ 0.4 & -0.6 & 0.2\\ 0.2 & 0.4 & -0.6\end{bmatrix}\begin{bmatrix}10°\mathrm{C}\\ 5°\mathrm{C}\\ 5°\mathrm{C}\end{bmatrix}$$

$$-[\mathrm{FG}] = \frac{1}{600\ \mathrm{s}}\cdot\begin{bmatrix}-3°\mathrm{C}\\ 2°\mathrm{C}\\ 1°\mathrm{C}\end{bmatrix}$$

where the column matrix on the right says that the top layer cooled by 3°C during the first time step, the middle layer warmed 2 °C, and the bottom layer warmed 1 °C.

The final set of flux gradient values are:

$-\Delta F/\Delta z =$ <u>**−0.0050 K/s**</u> at z = 500 m
$-\Delta F/\Delta z =$ <u>**0.0033 K/s**</u> at z = 300 m
$-\Delta F/\Delta z =$ <u>**0.0017 K/s**</u> at z = 100 m

which can be used in the Eulerian heat budget equation of Chapt. 3 if needed.

Check: Units OK. The sum of the warmings and coolings must always equal zero (this is a good check for math mistakes) because turbulence can only mix heat around from one place to another, it cannot create nor destroy heat.

Discussion: The answer says that the top layer should have cooled, and the bottom two layers should have warmed, which indeed they did.

Science Graffito

Seen on a bumper sticker: "**Lottery**: A tax on people who are bad at math."

SUMMARY

Turbulence is common in the atmospheric boundary layer. Turbulence is so complex that it cannot be solved exactly for each swirl and eddy. Instead, parameterizations are devised to allow approximate solutions for the net statistical effect of all turbulent eddies. Parameterizations, while not perfect, are acceptable if they satisfy certain rules.

One type of local parameterization, called K-theory, neglects the large eddies, but gives good answers for special regions such as the surface layer in the bottom 10% of the atmospheric boundary layer. It is popular because of its simplicity. Another type of parameterization is called transilient turbulence theory (T3), which is a nonlocal closure that includes all eddy sizes. It is more accurate, more complicated, and works well for free convection.

Threads

The atmospheric boundary layer (Chapt. 4) is often turbulent. Turbulent transport is one of the terms in the Eulerian heat budget (Chapt. 3). Turbulence forms in statically or dynamically unstable air (Chapt. 6). Heat, moisture, and momentum that are transferred from the ground to the air by conduction and viscosity (Chapt. 3) are then moved vertically within the air by turbulence. Air pollution (Chapt. 17) is dispersed by turbulence.

Moisture transport by turbulence affects the Eulerian water budget (Chapt. 5), which ultimately controls how much water can get into clouds (Chapt. 7) to make precipitation (Chapt. 8). Turbulence is a microscale motion (Chapt. 10), that can be composed of thermals during sunny days of free convection.

Turbulence distributes surface drag effects throughout the boundary layer, which causes boundary layer winds to converge into low-pressure centers (Chapt. 9), leading, via continuity, to updrafts, clouds, and precipitation. Turbulent mixing in high pressure centers causes the boundary layer to be identified as an air mass, and causes frontal zones to have finite thickness (Chapt. 12). The wind shear caused by turbulent drag causes the rotation that leads to tornadoes (Chapt. 15), and helps mix water vapor from the sea surface into hurricanes (Chapt. 16).

EXERCISES

Numerical Problems

N1. Verify that the following transilient matrices satisfy the requirement that each row and each column sum to 1.0 .

a

To Destination Layer i	From Source Layer j: 4	3	2	1
4	0.9	0.1	0	0
3	0.1	0.8	0.1	0
2	0	0.1	0.8	0.1
1	0	0	0.1	0.9

b

To Destination Layer i	From Source Layer j: 4	3	2	1
4	1.0	0	0	0
3	0	0.3	0.4	0.3
2	0	0.2	0.3	0.5
1	0	0.5	0.3	0.2

N2. Given the following initial sounding

z (m)	j	θ (°C)	M (m/s)
700	4	21	10
500	3	20	10
300	2	18	8
100	1	13	5

a., b. Compute the new potential temperature sounding after one time step of turbulent mixing, using the previous transilient matrices (a, b).

N3. a., b. For the previous problem, take one more time step from where you left off.

N4. For problem N2, compute the temperature flux gradient for each layer.

N5. Use K-theory with the data from exercise N2 to compute at each height the:

a. values of K b. heat fluxes

c. new θ after mixing

N6. a. Using Fig H.4, what is the average state of the layers of air after one more time interval. That is, apply the same eddies as sketched in Fig H.4b, except start with an initial state from Fig H.4d.

b. Speculate (but do not actually calculate) the final state of the layers of air after many time intervals.

Understanding & Critical Evaluation

U1. Draw a sketch similar to Fig H.4b of the air parcel movement that corresponds to the solved example in the turbulence section.

U2. What transilient matrix is necessary to cause complete mixing during one time step? Illustrate using a 3x3 matrix for simplicity.

U3. Repeat questions N2 through N4, except use T3 to calculate the mixing of wind speed rather than the mixing of potential temperature.

U4. Use K-theory with the data from exercise N2 and results from N5 to compute at each height the:

a. momentum fluxes   b. new *M* after mixing

Web-Enhanced Questions

W1. Search the web for research measurements of turbulent (heat, moisture, pollutant, or momentum) fluxes, and compare with the gradient of mean potential temperature, mixing ratio, pollutant concentration, or wind. [Hint: Search for "eddy correlation".]

W2. Search the web for eddy correlation, scintillometers, sonic anemometers, or other instruments used for measuring vertical turbulent fluxes.

W3. Search the web for eddy viscosity, eddy diffusivity, or K-theory K values in the atmosphere.

Synthesis Questions

S1. a. Suppose that turbulent layers of air mix only with their immediate neighbors above and below, rather than participating in nonlocal mixing.

a. How would the transilient matrix look?

b. How does the result relate to K-theory?

S2. Design a small, sample transilient matrix that does <u>un</u>mixing. Discuss the implications.

S3. Verify that K-theory and T3 satisfy the rules of parameterization.

S4. Positive values of *K* imply down-gradient transport of heat (i.e., heat flows from hot to cold, a process known as the Zeroth Law of Thermodynamics). What is the physical interpretation of negative values of *K*?

INDEX

C

D

E

F

G

H

I

J

K

L

M

N

O

Q

R

S

U

V

W

Index compiled by Stephanie Meyn.